1

AUTONOMIC NEUROEFFECTOR MECHANISMS

The Autonomic Nervous System

A series of books discussing all aspects of the autonomic nervous system. Edited by G. Burnstock, Department of Anatomy and Developmental Biology, University College London, UK.

Volume 1
Autonomic Neuroeffector Mechanisms
edited by *G. Burnstock and C. H. V. Hoyle*

Additional volumes in preparation

Nervous Control of the Urogenital
System
C. A. Maggi

Disorders of the Autonomic
Nervous System
D. Robertson and I. Biaggioni

Development, Regeneration and
Plasticity of the Autonomic
Nervous System
I. Hendry and C. Hill

Nervous Control of Blood Vessels
T. Bennett and S. Gardiner

Comparative Physiology and
Evolution of the Autonomic
Nervous System
S. Nilsson and S. Holmgren

Autonomic Ganglia
E. M. McLachlan

Central Nervous Control of
Autonomic Function
D. Jordan

Autonomic Nerves and the
Respiratory System
P. Barnes

Autonomic Innervation of the Skin
I. L. Gibbins and J. L. Morris

Nervous Control of the Gut and
Associated Organs
M. Costa

Autonomic-Endocrine Interactions
K. Unsicker

Nervous Control of the Heart
J. T. Shepherd and S. Vatner

Nervous Control of the Eye
N. N. Osborne and A. Bill

AUTONOMIC NEUROEFFECTOR MECHANISMS

Edited by

Geoffrey Burnstock
and
Charles H. V. Hoyle

Department of Anatomy and Developmental Biology
University College London, UK

harwood academic publishers
chur • reading • paris • philadelphia • tokyo • melbourne

Copyright © 1992 by Harwood Academic Publishers GmbH, Poststrasse 22, 7000 Chur, Switzerland. All rights reserved.

Harwood Academic Publishers

Post Office Box 90
Reading, Berkshire RG1 8JL
United Kingdom

58, rue Lhomond
75005 Paris
France

5301 Tacony Street, Drawer 330,
Philadelphia, Pennsylvania 19137
United States of America

3-14-9, Okubo
Shinjuku-ku, Tokyo 169
Japan

Private Bag 8
Camberwell, Victoria 3124
Australia

Library of Congress Cataloging-in-Publication Data

Autonomic neuroeffector mechanisms / edited by Geoffrey Burnstock and Charles H.V. Hoyle.
  p. cm. -- (Autonomic nervous system)
  Includes bibliographical references and index.
  ISBN 3-7186-5135-1
  1. Autonomic nervous system. 2. Neurotransmission. 3. Autonomic ganglia. I. Burnstock, Geoffrey. II. Hoyle, Charles H. V., 1955- . III. Series: Autonomic nervous system (Chur, Switzerland)
  [DNLM: 1. Autonomic Nervous System--chemistry. 2. Autonomic Nervous System--physiology. 3. Neural Transmission--drug effects. 4. Neural Transmission--physiology.   WL 600 A9396]
  QP368.A934   1992
  612.8'2--dc20
  DNLM/DLC
  for Library of Congress   91-7045
                                                                              CIP

Contents

Preface – Historical and Conceptual Perspective of The Autonomic Nervous System Book Series

The pioneering studies of Gaskell (1886), Bayliss and Starling (1899), and Langley and Anderson (see Langley, 1921) formed the basis of the earlier and, to a large extent, current concepts of the structure and function of the autonomic nervous system; the major division of the autonomic nervous system into sympathetic, parasympathetic and enteric subdivisions still holds. The pharmacology of autonomic neuroeffector transmission was dominated by the brilliant studies of Elliott (1905), Loewi (1921), von Euler and Gaddum (1931) and Dale (1935), and for over 50 years the idea of antagonistic parasympathetic cholinergic and sympathetic adrenergic control of most organs in visceral and cardiovascular systems formed the working basis of all studies. However, major advances have been made since the early 1960's that make it necessary to revise our thinking about the mechanisms of autonomic transmission, and that have significant implications for our understanding of diseases involving the autonomic nervous system and their treatment. These advances include:

(1) Recognition that the autonomic neuromuscular junction is not a 'synapse' in the usual sense of the term where there is a fixed junction with both pre- and post-junctional specialization, but rather that transmitter is released from mobile varicosities in extensive terminal branching fibres at variable distances from effector cells or bundles of smooth muscle cells which are in electrical contact with each other and which have a diffuse distribution of receptors (see Hillarp, 1959; Burnstock, 1986a).

(2) The discovery of non-adrenergic, non-cholinergic nerves and the later recognition of a multiplicity of neurotransmitter substances in autonomic nerves, including monoamines, purines, amino acids, a variety of different peptides and possibly nitric oxide (Burnstock *et al.*, 1964; see Burnstock, 1986b).

(3) The concept of neuromodulation, where locally released agents can alter neurotransmission either by prejunctional modulation of the amount of transmitter released or by postjunctional modulation of the time course or intensity of action

of the transmitter (Marrazzi, 1939; Brown and Gillespie, 1957; Löffelholz and Muscholl, 1969; see Vizi, 1979; Kaczmarek and Leviton, 1987).

(4) The concept of cotransmission that proposes that most, if not all, nerves release more than one transmitter (Burnstock, 1976; Hökfelt, Fuxe and Pernow, 1986; Burnstock, 1990a) and the important follow-up of this concept, termed 'chemical coding', in which the combinations of neurotransmitters contained in individual neurones are established, and whose projections and central connections are identified (see Furness and Costa, 1987).

(5) Recognition of the importance of 'sensory-motor' nerve regulation of activity in many organs, including gut, lungs, heart and ganglia, as well as in many blood vessels (Maggi and Meli, 1988; Burnstock, 1990a), although the concept of antidromic impulses in sensory nerve collaterals forming part of 'axon reflex' vasodilation of skin vessels was described many years ago (Lewis, 1927).

(6) Recognition that many intrinsic ganglia (e.g. those in heart, airways and bladder) contain integrative circuits that are capable of sustaining and modulating sophisticated local activities (see Burnstock et al., 1987). Although the ability of the enteric nervous system to sustain local reflex activity independent of the central nervous system has been recognized for many years (see Kosterlitz, 1968), it had been generally assumed that the intrinsic ganglia in peripheral organs consisted of parasympathetic neurons that provided simple nicotinic relay stations.

(7) The major subclasses of receptors to acetylcholine and noradrenaline have been recognized for many years (Dale, 1914; Alhquist, 1948), but in recent years it has become evident that there is an astonishing variety of receptor subtypes for autonomic transmitters (see Br. J. Pharmacol., (1991), 102, 560–561). Their associated transduction mechanisms are also being characterized. These advances offer the possibility for more selective drug therapy.

(8) Recognition of the plasticity of the autonomic nervous system not only in the changes that occur during development and aging, but also the changes in expression of transmitters and receptors that occur in fully mature adults under the influence of hormones and growth factors following trauma and surgery, and in a variety of disease situations (see Burnstock, 1990b).

(9) Advances in the understanding of 'vasomotor' centres in the central nervous system. For example, the traditional concept of control being exerted by discrete centres such as the vasomotor centre (Bayliss, 1923) has been supplanted by the belief that control is mediated by the action of longitudinally arranged parallel pathways involving forebrain, brainstem and spinal cord (Loewy and Spyer, 1990).

In addition to these major new concepts concerning autonomic function, the discovery by Furchgott that substances released from endothelial cells play an important role, in addition to autonomic nerves, in local control of blood flow has also made a significant impact on our analysis and understanding of cardiovascular function (Furchgott and Zawadski, 1980; Lincoln and Burnstock, 1990). The later identification of nitric oxide as the major endothelium-derived relaxing factor

(Palmer *et al.*, 1988) (confirming the independent suggestions by Ignarro and by Furchgott) and endothelin as an endothelin-derived constricting factor (Yanagisawa *et al.*, 1988) have also had a major impact in this area.

In broad terms these new concepts shift the earlier emphasis on central control mechanisms towards greater consideration of the sophisticated local peripheral control mechanisms.

Although these new concepts should have a profound influence on our considerations of the autonomic control of cardiovascular, genito-urinal, gastrointestinal and reproductive systems and other organs like the skin and eye in both normal and disease situations, few of the current textbooks take them into account. This is largely because revision of our understanding of all these different specialist areas in one volume by one author is a near impossibility. Thus, this Book Series of 14 volumes is designed to try to overcome this dilemma by dealing in depth with each major area in separate volumes and by calling upon the knowledge and expertise of leading figures in the field. Volume I deals with the basic mechanisms of *Autonomic Neuroeffector Transmission* which sets the stage for later volumes devoted to autonomic nervous control of particular organ systems, including *Bladder, Heart, Blood Vessels, Airways, Urinogenital organs, Eye* and *Skin*. Another group of volumes will deal with *Central Nervous Control of Autonomic Function, Interaction of Autonomic Mechanisms with the Endocrine System, Development Regeneration and Plasticity and Comparative Physiology and Evolution of the Autonomic Nervous System*.

Abnormal as well as normal mechanisms will be covered to a variable extent in all these volumes, depending on the topic and the particular wishes of the Volume Editor, but one (or two) volumes will be specifically devoted to *Diseases of the Autonomic Nervous System*.

A general philosophy followed in the design of this Book Series has been to encourage individual expression by Volume Editors and Chapter Contributors in the presentation of the separate topics within the general framework of the series. This was demanded by the different ways that the various fields have developed historically and the differing styles of the individuals who have made the most impact in each area. Hopefully, this deliberate lack of uniformity will add to, rather than detract from, the appeal of these books.

G. Burnstock
Series Editor

REFERENCES

Alhquist, R.P. (1948). A study of the adrenotropic receptors. *Am. J. Physiol.*, **153**, 586–600.

Bayliss, W.B. (1923). *The Vasomotor System*, London: Longman.

Bayliss, W.M. and Starling, E.H. (1899). *J. Physiol. (Lond.)*, **24**, 99–143.

Brown, G.L. and Gillespie, J.S. (1957). The output of sympathetic transmitter from the spleen of the cat. *J. Physiol. (Lond.)*, **138**, 81–102.

Burnstock, G., Campbell, G., Bennett, M. and Holman, M.E. (1964). Innervation of the guinea-pig taenia coli: are there intrinsic inhibitory nerves which are distinct from sympathetic nerves? *Int. J. Neuropharmacol.*, **3**, 163–166.

Burnstock, G. (1976). Do some nerve cells release more than one transmitter? *Neuroscience*, **1**, 239–248.

Burnstock, G. (1986a). Autonomic neuromuscular junctions: current developments and future directions. *J. Anat.*, **146**, 1–30.

Burnstock, G. (1986b). The changing face of autonomic neurotransmission. (The first von Euler Lecture in Physiology). *Acta Physiol. Scand.*, **126**, 67–91.

Burnstock, G. (1990a). Cotransmission. The Fifth Heymans Lecture – Ghent, February 17, 1990. *Arch. Int. Pharmacodyn. Ther.*, **304**, 7–33.

Burnstock, G. (1990b). Changes in expression of autonomic nerves in aging and disease. *J. Auton. Nerv. Syst.*, **30**, 525–534.

Burnstock, G., Allen, T.G.J., Hassall, C.J.S. and Pittam, B.S. (1987). Properties of intramural neurones cultured from the heart and bladder. In *Histochemistry and Cell Biology of Autonomic Neurons and Paraganglia. Exp. Brain Res. Ser. 16*, edited by C. Heym, pp. 323–328. Heidelberg: Springer Verlag.

Dale, H. (1914). The action of certain esters and ethers of choline and their reaction to muscarine. *J. Pharmacol. Exp. Ther.*, **6**, 147–190.

Dale, H. (1935). Pharmacology and nerve endings. *Proc. Roy. Soc. Med.*, **28**, 319–332.

Elliot, T.R. (1905). The action of adrenalin. *J. Physiol. (Lond.)*, **32**, 401–467.

Furchgott, R.F. and Zawadski, J.V. (1980). The obligatory role of endothelial cells in the relaxation of arterial smooth muscle by acetylcholine. *Nature*, **288**, 373–376.

Furness, J.B. and Costa, M. (1987). *The Enteric Nervous System*, Edinburgh: Churchill Livingstone.

Gaskell, W.H. (1886). On the structure, distribution and function of the nerves which innervate the visceral and vascular systems. *J. Physiol. (Lond.)*, **7**, 1–80.

Hillarp, N.-Å. (1959). The construction and functional organisation of the autonomic innervation apparatus. *Acta Physiol. Scand.* (Suppl. 157), **46**, 1–38.

Hökfelt, T., Fuxe, K. and Pernow, B. (Eds). (1986). *Coexistence of Neuronal Messengers: A New Principle in Chemical Transmission. Progress in Brain Research, Vol. 68*, Amsterdam: Elsevier.

Kaczmorek, L.K. and Leviton, I.B. (1987). *Neuromodulation. The biochemical control of neuronal excitability*, pp. 1–286. Oxford: Oxford University Press.

Kosterlitz, H.W. (1968). Intrinsic and extrinsic nervous control of motility of the stomach and the intestines. In *Handbook of Physiology, Vol. IV, Sect. 6. The Alimentary Canal*, edited by C.F. Code, pp. 2147–2172. Washington DC: American Physiological Society.

Langley, J.N. (1921). *The Autonomic Nervous System, Part 1*, Cambridge: W. Heffer.

Lewis, J. (1927). *The Blood Vessels of the Human Skin and their Responses*, London: Shaw & Sons.

Lincoln, J. and Burnstock, G. (1990). Neural-endothelial interactions in control of local blood flow. In *The Endothelium: An Introduction to Current Research*, edited by J. Warren, pp. 21–32. New York: Wiley-Liss.

Loewi, O. (1921). Über humorale Übertragbarkeit der Herznervenwirkung. XI. Mitteilung. *Pflugers Arch. Gesamte Physiol.*, **189**, 239–242.

Loewy, A.D. and Spyer, K.M. (1990). *Central Regulations of Autonomic Functions*, New York: Oxford University Press.

Löffelholz, K. and Muscholl, E. (1969). A muscarinic inhibition of the noradrenaline release evoked by postganglionic nerve stimulation. *Naunyn Schmiedeberg's Arch. Pharmacol.*, **265**, 1–15.

Maggi, C.A. and Meli, A. (1988). The sensory-efferent function of capsaicin-sensitive sensory nerves. *Gen. Pharmacol.*, **19**, 1–43.

Marrazzi, A.S. (1939). Electrical studies on the pharmacology of autonomic synapses. II. The action of a sympathomimetic drug (epinephrine) on sympathetic ganglia. *J. Pharmacol. Exp. Ther.*, **65**, 395–404.

Palmer, R.M.J., Rees, D.D., Ashton, D.S. and Moncada, S. (1988). L-arginine is the physiological precursor for the formation of nitric oxide in endothelium-dependent relaxation. *Biochem. Biophys. Res. Commun.*, **153**, 1251–1256.

Vizi, E.S. (1979). Prejunctional modulation of neurochemical transmission. *Prog. Neurobiol.*, **12**, 181–290.

von Euler, U.S. and Gaddum, J.H. (1931). An unidentified depressor substance in certain tissue extracts. *J. Physiol.*, **72**, 74–87.

Yanagisawa, M., Kurihara, H., Kimura, S., Tomobe, Y., Kobayashi, M., Mitsui, Y., Yazaki, Y., Goto, K. and Masaki, T. (1988). A novel potent vasoconstrictor peptide produced by vascular endothelial cells. *Nature*, **332**, 411–415.

Contributors

Benham, C.D.
Department of Pharmacology,
SmithKline Beecham Pharmaceuticals,
Coldharbour Road, The Pinnacles,
Harlow, Essex CM19 5AD, UK

Brock, James A.
University Department of Pharmacology,
South Parks Road, Oxford OX1 3QT, UK

Buckley, Noel J.
Department of Physical Biochemistry,
National Institute for Medical Research,
The Ridgeway, Mill Hill, London
NW7 1AA, UK

Burnstock, Geoffrey
Department of Anatomy and
Developmental Biology, University
College London, Gower Street, London
WC1E 6BT, UK

Caulfield, Malcolm
Department of Pharmacology, University
College London, Gower Street, London
WC1E 6BT, UK

Cunnane, Thomas C.
University Department of Pharmacology,
South Parks Road, Oxford OX1 3QT, UK

Dockray, G. J.
Physiological Laboratory, University of
Liverpool, Brownlow Hill, P.O. Box 147,
Liverpool L69 3BX, UK

Fillenz, Marianne
University Laboratory of Physiology,
Parks Road, Oxford OX1 3PT, UK

Gabella, Giorgio
Department of Anatomy and
Developmental Biology, University
College London, Gower Street,
London WC1E 6BT, UK

Gibbins, Ian L.
School of Medicine, Flinders
University, Bedford Park, S.A. 5042,
Australia

Hills, Judith M.
SmithKline Beecham
Pharmaceuticals, The Frythe,
Welwyn, Herts AL6 9AR, UK

Hoyle, Charles H. V.
Department of Anatomy and
Developmental Biology, University
College London, Gower Street,
London WC1E 6BT, UK

Jessen, Kristjan R.
Department of Anatomy and
Developmental Biology, University
College London, Gower Street,
London WC1E 6BT, UK

Morris, Judy L.
Department of Anatomy and
Histology, Flinders Medical Centre,
Bedford Park, S.A. 5042, Australia

1 Fine Structure of Post-Ganglionic Nerve Fibres and Autonomic Neuro-effector Junctions

Giorgio Gabella

Department of Anatomy, University College London

Most of the axons originating from ganglion neurons (the so-called post-ganglionic fibres) reach peripheral organs and provide efferent innervation. They do so by various, often variable, sometimes complex, routes. Some fibres exit the ganglion within a small nerve – a grey ramus communicans – that reaches a somatic nerve, and then travel peripherally, mainly distributing to blood vessels and skin, mixed with somatic afferent and efferent fibres. Other post-ganglionic nerves attach themselves to a large artery (paravascular nerves) and reach the periphery travelling along the arterial tree.

TOPOGRAPHICAL REPRESENTATION WITHIN AUTONOMIC GANGLIA

Large autonomic ganglia contain thousands or tens of thousands of neurons, without compartmentalization or presence of sub-groups. Are the target organs topographically represented within ganglia, or are the neurons providing the fibres for a given target grouped together at a specific location within the ganglion? Generally speaking, the degree of topographical representation of the target is modest, when it exists at all; neurons tend to cluster in the proximity of the origin of the nerve into which they project (Bowers and Zigmond, 1979). In the superior cervical ganglion of several mammalian species, neurons projecting axons into the internal carotid nerve are located in the cranial part of the ganglion, whereas those projecting into the external carotid nerve are located in the caudal part of the ganglion (Bowers and Zigmond, 1979; Kiraly *et al.*, 1979). Similarly, there is no topographical representation of the

1

periphery in the coeliac (Macrae, Furness and Costa, 1986) or the pelvic ganglion (Keast, Booth and de Groat, 1989).

POST-GANGLIONIC FIBRES

The axons issued by most ganglion neurons leave the ganglion and reach the target organ. The possibility that some ganglion neurons (in addition to SIF cells) may issue axons that do not leave the ganglion has been raised (Kiraly et al., 1989). Some ganglion neurons have local collateral axons, e.g. in frog sympathetic ganglia (Forehand and Konopka, 1989), although this is uncommon in mammalian autonomic ganglia (Purves and Lichtman, 1985).

Invariably, the postganglionic fibres have a slower conduction velocity than the corresponding preganglionic fibres. In some species, e.g. the cat, some postganglionic fibres are myelinated, but myelin sheath thickness and axon diameter are smaller than in the preganglionic fibres. More commonly, post-ganglionic fibres are unmyelinated and measure on average 0.5–0.8 μm. Their size is smaller than that of the corresponding preganglionic fibres. A post-ganglionic nerve can contain many thousands of unmyelinated axons: in the left inferior cardiac nerve of the cat up to 42,000 unmyelinated axons were counted and the diameter of the individual fibres ranges between 0.1 and 1.8 μm (average 0.61) (Emery et al., 1976).

The conduction velocity of the non-terminal portion of unmyelinated postganglionic fibres in mesenteric nerves is about 0.7 m.sec^{-1} (Bessou and Perl, 1966; Cottrell, 1984).

The length is very variable; enteric efferent fibres reach the target tissue (smooth muscle) as soon as they exit the myenteric plexus – but they travel nonetheless some distance within the muscle. Some vasomotor fibres from lumbar sympathetic ganglia first exit from the abdominal cavity, then travel the length of the lower limb. In most cases the volume of the axon is many times larger than the volume of its cell body. Considering for example the sympathetic supply to the human leg, an axon 50 cm long and 1.2 μm in diameter would have a volume of 565,000 μm^3, whereas its perikaryon, measuring about 30 μm diameter, would have a volume of 14,000 μm^3.

Although the branching patterns of post-ganglionic fibres have never been properly investigated, some axons may divide before they reach the target organ. This branching is very limited, if it takes place at all, but its occurrence may be important, for example in allowing projections of a single neuron to reach two separate targets.

DISTRIBUTION OF POST-GANGLIONIC FIBRES

The post-ganglionic fibres reach the target tissue, which is by definition the tissue in which they branch extensively and develop varicosities. Autonomic fibres are found in every organ as perivascular vasomotor fibres. Although not always clearly distinguished from these, there are autonomic fibres which are unrelated to standard effector tissues, smooth and cardiac muscle and exocrine secretory units. There are, for example, autonomic fibres close to adipocytes, mast cells, melanophores, interstitial cells, sensory and autonomic ganglion neurons, and motor endplates. In skeletal muscles, sympathetic fibres are intermingled with the dense somatic inner-vation (Santini and Ibata, 1971; Barker and Saito, 1981). The close relation between somatic and sympathetic fibres in skeletal muscle had already been demonstrated with silver impregnation by Boeke (1927), but his conclusion had been dismissed until recently. Some of these locomotor autonomic fibres arise from perivascular bundles, lie over the surface of extrafusal muscle fibres and a few intrafusal fibres, and are identified as noradrenergic (Barker and Saito, 1981). The spindles also receive some autonomic fibres identified ultrastructurally as cholinergic (Barker and Saito, 1981).

There are autonomic fibres close to secretory cells in all endocrine glands, for example in the thyroid (Melander *et al.*, 1974; Melander and Sundler, 1979) and the parathyroid (Altenahr, 1971), in the adrenal cortex (Kleitman and Holzwarth, 1985), in the ovary (Aguado & Ojeda, 1984) and in the endocrine component of the pancreas (Miller, 1981). Adrenergic fibres, which are rare or absent in white adipose tissue (Slavin and Ballard, 1978), are abundant in brown adipose tissue (Cannon *et al.*, 1986; Lever *et al.*, 1987), and the role of adrenergic nerves in brown fat metabolism is well documented (Smith and Horwitz, 1969).

NERVE BUNDLES

Within the target organ, all or most of the post-ganglionic fibres are grouped into nerve bundles. These can contain a hundred or more axons or only a few axons. The bundles are formed by terminal and preterminal portions of post-ganglionic fibres, and some of the afferent fibres that may be present in the tissue also travel in them. In some tissues, individual terminal fibres issue from nerve bundles as single fibres.

TERMINAL FIBRES

The terminal portion of the axon is, by definition, the portion that contains varicosities. (The preterminal portion is a vaguer notion in the limited light of present knowledge. The term may be used to indicate the portion of the axon that is already intramural, i.e. within the target organ, but not yet varicose, although it may be already close to, and surrounded by, terminal axons.) Even in the rare cases when the post-ganglionic fibre is myelinated at its origin from the perikaryon, the thickness of the myelin sheath decreases centrifugally, and preterminal and terminal portions of the axon are invariably unmyelinated.

The presence of varicosities is a firm criterion to identify the terminal portion of an axon, called a *terminal axon* for short. Since the varicosities are the sites where most of the synaptic vesicles of an axon are found and are therefore the sites where it is most likely that junctions are formed, they are commonly (and unobjectionably) regarded as *nerve terminals*. One varicosity in each branch is the terminal one, while all the others are nerve terminals in the sense that they are specialized points of potential transmitter release.

DENSITY OF INNERVATION

The density of innervation of a target organ depends on the number of incoming fibres, on the extent of the arborization, on the length of the terminal tree, and on the spatial frequency of varicosities. In spite of their possible importance, quantitative data on these questions are not available. While differences between tissues are evident, they are only vaguely understood. We do not have an operative definition of density of innervation. Indeed, the very notion of innervation is conceptually vague, except in the generic sense of presence of nerve fibres within a tissue or an organ. It should be mentioned that in this definition are included preterminal fibres and fibres that are in transit to another target, and all the fibre types are pooled together. It has been estimated that in the adrenergic plexus around the rabbit ear artery or the rat tail artery there are 10,000 varicosities/mm^2 (Griffith *et al.*, 1982; Sittiracha *et al.*, 1987). In arterioles of the guinea-pig submucosa 8,000–150,000 varicosities/mm^2 of outer vessel surface (Luff and McLachlan, 1989) are estimated to form neuromuscular junctions, while in the rodent vas deferens the density of adrenergic varicosities has been estimated to be as high as 2×10^6 per mm^3 of tissue (Stjärne, 1989).

BRANCHING OF NERVE BUNDLES

The nerve bundles within target organs branch repeatedly, thus decreasing in size and in number of component fibres. However, nearly as frequently they also merge with one another. In most tissues, therefore, there is not a true arborization of the bundles but rather the formation of a network that includes thick and thin links and has complete continuity. In the taenia coli of the guinea-pig, a strip of muscle up to 30 cm long, the bundles form a very elongated meshwork whose appearance remains unchanged over the entire length of the muscle. Similarly, the distribution of nerve bundles in the circular layer of the intestine or in the sphincter pupillae is not markedly different at any point around the circumference of the muscle. From the meshes of nerve bundles emerge a few individual fibres that terminate within a short distance. This happens in all smooth muscles, but it is a rare occurrence in some and a frequent occurrence in others. Typically, the intestinal and the ureteral musculature are regarded as being innervated by nerve bundles (innervation fasciculée) (Taxi, 1965) (Figure 1.1), although individual nerve fibres abutting on muscle cells can also be found. Innervation by nerve bundles is also predominant in blood vessels (Figure 1.5). In contrast, muscles such as those found in the iris, the vas deferens and the urinary bladder contain a predominance of single nerve fibres (Figure 1.2).

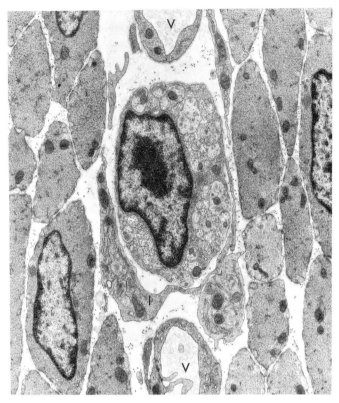

FIGURE 1.1   Transverse section of a bundle of circular musculature in the ileum of a guinea-pig, photographed in the electron microscope. A large nerve bundle, displaying the nucleus of its Schwann cell, lies at the centre. *V* capillaries, *I* interstitial cell. Magnification: 6,500 x

BRANCHING OF TERMINAL AXONS

Within a target organ, the varicose axons in the nerve bundles undergo extensive branching. Most of the branching is dichotomic and at a small angle so that the tree formed is usually very long, and is intertwined with the trees of several fibres. That the individual fibres form a true terminal arborization descends from the neuronal theory and is not in conflict with any experimental evidence. To demonstrate the full anatomical course of single axons and to prove that they are fully anatomically independent from adjacent axons remains a difficult task with the techniques available.

ORIENTATION OF NERVE BUNDLES

Within a regular layer of smooth muscle, e.g. in the wall of the gut, post-ganglionic fibres run almost invariably near-parallel to the bundles of muscle cells. The fibres

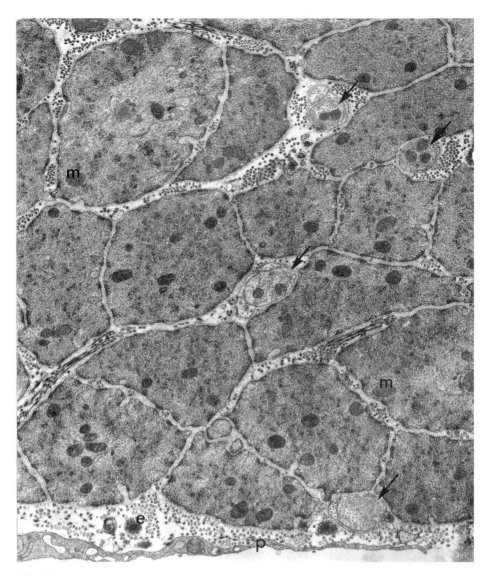

FIGURE 1.2   Transverse section of a muscle bundle in the urinary bladder of a rat, photographed in the electron microscope. Among the muscle cell profiles, four vesicle-containing axons are visible (*arrows*). *c* collagen fibrils, *e* elastic fibres, *m* mitochondria, *p* peritoneal lining. Magnification: 6,500 x

FIGURE 1.4   Three nerve endings (varicosities) in the urinary bladder of a rat. The varicosities contain small clear vesicles (*v*), mitochondria (*m*), microtubules (*t*) and endoplasmic reticulum (*r*). In A and B, the varicosity is accompanied by a slender process of Schwann cell; in C the axon is devoid of Schwann cell support. Note the closeness of the varicosity in A to a muscle cell. The varicosity in C is in intimate contact with two muscle cells. A much larger separation between varicosity and nearest muscle cell is found in B. Magnification, A: 36,000 x; B: 27,000 x; C: 18,000 x

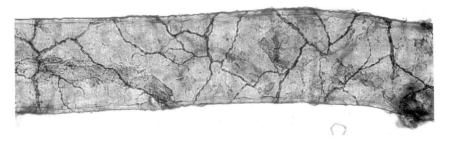

FIGURE 1.3   Superior cerebellar artery of a rat, prepared as a whole-mount and stained with antibodies against neuropeptide Y (peroxidase-antiperoxidase method). The plexus of nerve bundles has no preferential orientation with respect to the long axis of the vessel. (Photograph kindly provided by Dr M. Mione).
Magnification: 60 x

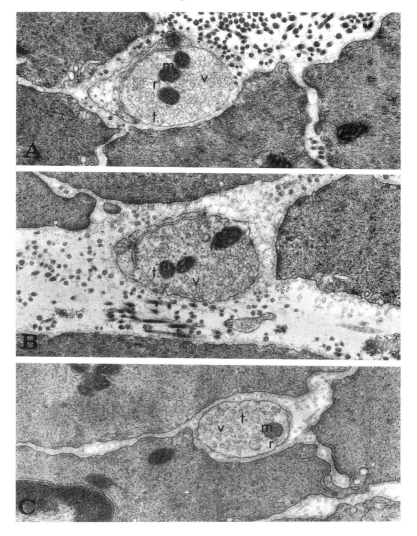

that traverse a muscle layer form a very small angle with the muscle bundles so that it may take more than 1 mm for a fibre to move across a 50 μm thick layer of muscle (the circular muscle of the guinea-pig ileum, for example).

Where the arrangement of the musculature is more complex, for example in the urinary bladder, the course of the intramuscular nerve bundles is more complex too. However, there is always a strong tendency of the nerve fibres – especially the terminal ones – to lie parallel to bundles of muscle cells.

In blood vessels, in which the nerve fibres are usually confined to the adventitia, a pattern of nerve fibres is less obvious (Figure 1.3); the fibres form meshes with only a moderate predominance of a longitudinal orientation.

Nerve bundles situated within smooth muscles are exposed to extensive longitudinal compression and stretch during muscle contraction. Upon isotonic muscle contraction, the nerve bundles acquire a slightly wavy course; the individual axons all curve at the same points, so that a distinct and regular waviness can be seen within the nerve bundles in these conditions. The effects of contraction on intramuscular nerves remain to be investigated. Since contraction can reduce the muscle length to a quarter of the resting length, there may occur significant changes in the small nerve bundles and individual axons. Considerable structural deformation can be

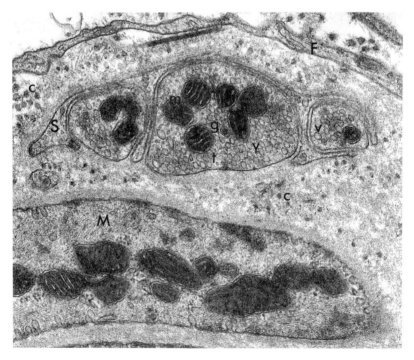

FIGURE 1.5 A small nerve bundle, made of three vesicle-containing axons, in the adventitia of an arteriole of the rat myocardium. The axons contain small agranular vesicles (v), mitochondria (m), microtubules (t). A thin process of a Schwann cell (S) forms a partial wrapping. The distance from the muscle cell (M) is about 0.2 μm. c collagen fibrils, F fibroblast process, g large granular vesicle. Magnification: 22,000 x

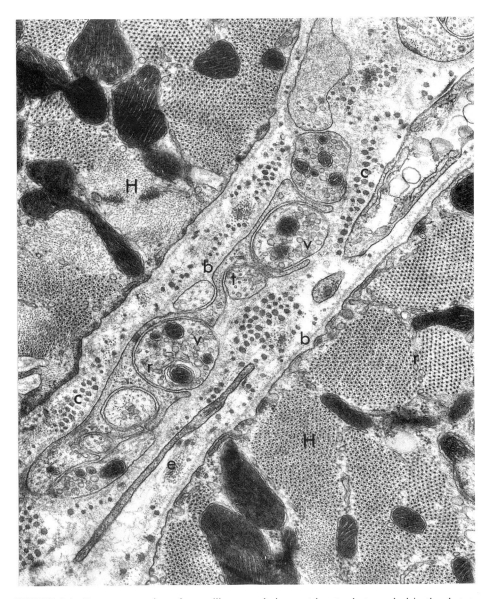

FIGURE 1.6 Transverse section of a papillary muscle in a rat heart, photographed in the electron microscope. A small nerve bundle lies between two cardiac muscle cells. (*H*). Some of the axons contain small agranular vesicles (*v*), large granular vesicles and mitochondria. A prominent basal lamina (*b*) is found around the muscle cells and around the nerve bundle. *c* collagen fibrils, *e* elastic fibres, *r* endoplasmic reticulum, *t* microtubules. Magnification: 15,000 x

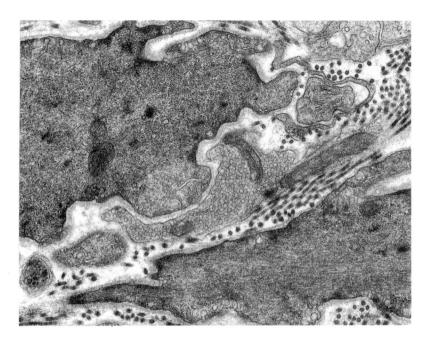

FIGURE 1.7   Muscle cells and nerve endings in the tracheal muscle of a rat. The muscle is isotonically contracted (note the corrugated appearance of the cell outlines), and the nerve endings, which are packed with vesicles, are markedly deformed. Magnification: 16,000 x

seen in nerve endings within contracted muscles (Figure 1.7). Govyrin (1976) observed that perivascular, adventitial nerve fibres of the auricular artery of the rabbit change their spatial distribution with dilatation and constriction of the vessel. In the constricted vessel, single axons and axon bundles occupy an area of the adventitia about 7 μm thick, whereas in the distended artery they occupy an area about 4 μm thick. Furthermore, the average distance of varicosities from medial muscle cells is markedly decreased in distended vessels.

STRUCTURE OF NERVE BUNDLES

Schwann cells and axons constitute the nerve bundles. Only one Schwann cell is found in any transverse section of a bundle, even in bundles containing a hundred or more axons. In large bundles, the soma of the Schwann cell resides in its deeper

FIGURE 1.9   Transverse section of the taenia coli of a newborn guinea-pig, photographed in the electron miscroscope. A small nerve bundle is surrounded by muscle cells. Axons and muscle cells lie very close to one another. Note the axons indenting the surface of a muscle cell and lying within a deep groove. Magnification 17,000 x

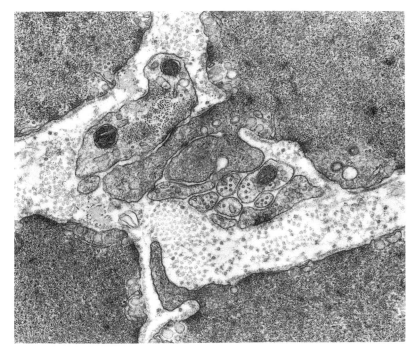

FIGURE 1.8 Muscle cells and a small nerve bundle in the taenia coli of a guinea-pig. A complex lamellar process from a muscle cell penetrates into the nerve bundle and lies close to some of its axons. Magnification: 22,000 x

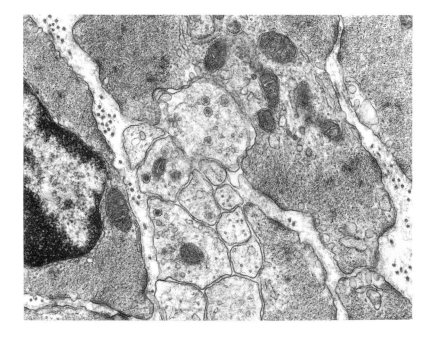

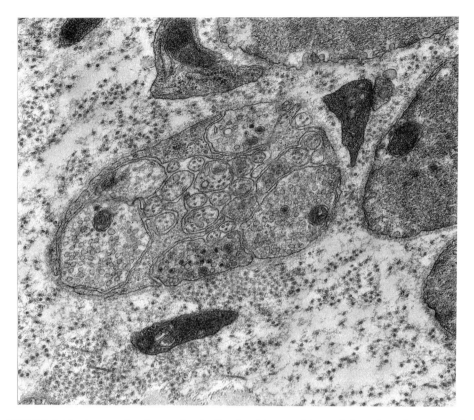

FIGURE 1.10 A nerve bundle running in the circular musculature of the colon of a guinea-pig, photographed in the electron microscope. The bundle is made of about 24 axons, some of which contain vesicles. Two axons are packed with small agranular vesicles, and two contain many small granular vesicles with different types of varicosities. All the axons lie at a considerable distance from an effector cell.
Magnification: 20,000 x

portion, and laminar cytoplasmic processes spread out to the bundle surface, which is coated by a basal lamina. There is no endoneurium nor capsular structure, but fibroblasts (and, in the intestine, also interstitial cells) are associated with the surface of a bundle. Cytoplasmic processes from muscle cells can pierce the basal lamina of the muscle cell and of the nerve bundle, and terminate among the axons (Figure 1.8). In developing visceral smooth muscles, it is common to find muscle cells partly surrounding, and closely apposed to, a nerve bundle (Figure 1.9).

Schwann cells do not provide a complete wrapping for each axon (Figures 1.2, 1.4, 1.5). So, it is common for axons to lie in membrane-to-membrane contact with one another, and for axons near the surface to be in direct contact with the basal lamina. Partial wrapping by the glial cells is common in all post-ganglionic nerves, even near their emergence for the ganglia (e.g., in cardiac sympathetic nerves, Emery *et al.*, 1976).

As the nerve bundles become smaller through branching, there is a relative decrease of the Schwann cell component and a relative increase of the axonal component. Direct contacts between axons become more common and include varicosities. The presence of direct contacts between vesicle-containing portions of two axons has prompted the suggestion that there are sites of communication between fibres, including possible axo-axonic synapses, for example between cholinergic and adrenergic axons (Ehinger *et al.*, 1970). While there is no evidence of the occurrence of axo-axonic synapses within nerve bundles, the possibility of chemical or electrical interaction between apposed neurons should not be dismissed. It is possible, however, that these axon-axonic associations are non-selective, and immunohistochemical work suggests that many fibre types, including afferent fibres, can travel together in the same bundle (Figure 1.10).

DISTRIBUTION OF VARICOSITIES

The pattern of varicosities is best seen in whole-mount preparations, stained with silver methods and methylene blue (e.g., Hillarp 1946) or histochemically stained for some neuronal marker (e.g., Falck *et al.*, 1962; Furness and Costa, 1987). The preparative procedure (freeze-drying, air-drying, dehydration rehydration etc.) has considerable influence on their appearance. Varicosities are often regularly spaced and occupy long lengths of the axon (Figure 1.11). Adrenergic varicosities are the best example of this, a varicosity occurring every 3–5 microns over hundreds of microns of axonal length. Counts of varicosities in adrenergic fibres have been made,

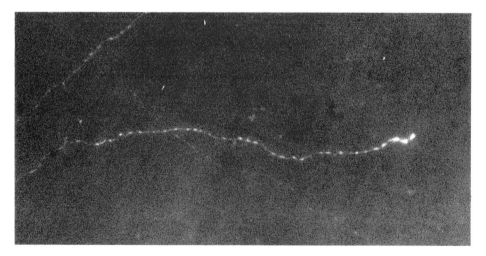

FIGURE 1.11   Whole-mount preparation of the mesentery of a guinea-pig, showing formaldehyde-induced fluorescence for catecholamine. The terminal portion of a noradrenergic axon is visible and characterized by a beaded appearance. Note that the last three varicosities at the very end of the fibre are larger and show a brighter fluorescence. Magnification: 225 x

and it has been estimated that each adrenergic fibre to the rat iris contains about 26,000 varicosities, spread over a total length of terminal axon arborization of about 100 mm (Dahlström and Haggendal, 1966). A beaded appearance is noted in all post-ganglionic fibres, and it seems unlikely that any of them should contain less than many hundreds of varicosities, but whether their numbers are as high as has been suggested for adrenergic fibre varicosities remains to be proved. Varicosities are also found in small unmyelinated afferent fibres, such as C fibres, as evidenced by immunohistochemistry for substance P and calcitonin gene-related peptide; these fibres, however, although originating from sensory ganglia, are capable of releasing neurotransmitters at the periphery and their functions may include an efferent role (see *Axonal vesicles in afferent fibres*).

Generally speaking, the distribution of varicosities is fairly regular over a short stretch of axon. However, it would be useful to have more data on this aspect of axonal morphology, in relation to the type of fibre, type of transmitter, age, and other conditions. Varicosities measure about 1–3 μm in diameter and up to 4 μm in length, and are spherical or, more commonly, fusiform (Burnstock and Costa, 1975). In adrenergic varicosities in submucosal arterioles of the guinea-pig ileum the mean volume is 1.2 μm^3 and the surface area ranges between 1.8 and 12 μm^2 (Luff *et al.*, 1987). Fluorescence microscopy provides only a vague indication of the size of stained varicosities, while electron microscopy provides accurate data only when many consecutive sections are examined. The extent of variability in size and shape of varicosities and its correlation with other features of the axon are poorly known. There are no investigations as to whether their size is related to the type of fibre, the type of transmitter, the type of organ, the animal species, or the age of the animal. A certain uniformity in the size of varicosity over a given length of axon is usually well in evidence. However, the question of varicosity uniformity over axon length remains a major area to explore. The more proximal varicosities (the ones nearest to the cell body) are often slim and histochemically weakly stained, whereas the last 3–4 varicosities, in the proximity of the fibre's end, are larger and more intensely histochemically stained (Figure 1.11).

Many varicosities are found opposite an effector such as a smooth muscle cell (see *Neuromuscular junctions* below), but some are found fully embedded in collagen, several microns away from the nearest cell (Figure 1.10), or deep within a nerve bundle, or opposite a fibroblast or a capillary endothelial cell.

Since varicosities appear arranged in a rather regular sequence, it could be argued that they can develop within a terminal axon even when an effector is not lying adjacent to each of them (uncoupled varicosities). It would be interesting to see whether 'unattached varicosities' do indeed exist, and conversely whether the closeness of an effector influences the amount of growth of a varicosity and the amount of transmitter accumulated and released.

STRUCTURE OF VARICOSITIES

Vesicles are invariably present in varicosities. They can be few in number and may not even appear in an individual section of a varicosity, or they can almost pack the varicosity. Tubules and small sacs of smooth endoplasmic reticulum, but no rough endoplasmic reticulum or ribosomes, can be found (Figure 1.4). Small but numerous mitochondria are constantly present (Figures 1.4 and 1.6). Microtubules are seen entering the varicosity from the intervaricose segment at each end, but they are only occasionally seen within the varicosity. With fixation techniques designed to enhance preservation of these organelles, however, some microtubules can be seen to traverse the varicosity without interruption from one intervaricose segment to the next (Gordon-Weeks, 1988). Similarly, intermediate filaments appear less prominent within a varicosity than in intervaricose segments. However, the features of the cytoskeleton in varicosities remain to be explored.

SIGNIFICANCE OF VARICOSITIES

Two distinct events occur in varicosities: conduction of an action potential and release of neurotransmitters. Transmitter release is intermittent, that is during nerve stimulation there is a high rate of release failure, as detected electrophysiologically. Upon nerve stimulation, a single varicosity releases between zero and ten 'packets' of transmitters, i.e. the contents of between nought and ten vesicles (Blakely and Cunnane, 1979). The intermittence of transmitter release arises from a low probability of release and not from a failure of the action potential to invade the nerve terminal (Brock and Cunnane, 1988) (see Chapter 3, this volume). Indeed, one could argue that if transmitter release intermittence was accounted for by a conduction failure at a varicosity, then nerve transmission in these fibres would be extremely inefficient and the most peripheral varicosities would hardly ever be visited by an action potential. However, Cheung (1990) has shown that in some cases transmission failure may also occur. This possibility should therefore be taken into account; however, what the rate of transmission failure may be at branching points and at varicosities of the terminal region remains as yet unknown.

Since we regard varicosities as nerve endings, the beaded structure of the terminal post-ganglionic axons allows an extremely large number of endings to be deployed by each axon. In somatic motor neurons, in which the endings are mostly located at the anatomical end-points of axonal branches onto skeletal muscle fibres, the number of terminals does not exceed 200. To obtain thousands of endings with this type of innervation, a much greater length of axons and a much greater number of branching points should be produced. By putting terminals in series along a single stretch of axon, there is economy in branching points and in axonal length. This advantage, however, is not fully utilized if the rate of release failure is as high as observed in electrophysiological studies. It is suggested that only 1 or 2% of all varicosities release noradrenaline when sympathetic nerves to the vas deferens or to an arteriole are stimulated (Hirst and Neild, 1980; Cunnane and Stjärne, 1984; Stjärne *et al.*, 1990).

AXONAL VESICLES

Axonal vesicles, usually called synaptic vesicles, occupy a large part of a varicosity (Figures 4, 5 and 6). Vesicles store neurotransmitters and other releasable substances. Many attempts have been made to identify the type of transmitter – or the content in general – of a vesicle population on the basis of their morphology. Only fifteen years ago the question of axonal vesicles was considered a relatively simple one, in so far as it was expected: i. that each transmitter (and the number of known transmitters could then be listed in a single line of text) imparted a characteristic morphology to the axonal vesicles storing it, so that the ultrastructural distribution of transmitters could be systematically worked out; ii. that each nerve ending stored and released one transmitter, or one secretory substance; iii. that the vesicle content was constant within an axon; iv. that the vesicle content was relatively uniform along the terminal portion of an axon. These expectations turned out to be somewhat unwarranted, as discussed in the following four sections.

VESICLE MORPHOLOGY

The main ultrastructural variations of axonal vesicles are size and electron-density. Vesicles measuring between 40 and 70 nm are 'small' vesicles, those over 70 nm are 'large' vesicles.

Vesicles are spherical. However, flattened vesicles (30 nm × 80–100 nm) have been found in noradrenergic nerve terminals in enteric ganglia of the small intestine of guinea-pig and man (Llewellyn-Smith *et al.*, 1984) and in the ano-coccygeus muscle (Gibbins & Haller, 1979). The influence of the preparative procedure on these appearances remains to be fully explored.

Small agranular vesicles are characteristic of identified cholinergic endings (motor endplate, efferent endings in the sphincter pupillae, preganglionic nerve endings). There are, however, small agranular vesicles that do not contain acetylcholine, for example vesicles that occur in large numbers in some non-cholinergic endings. Furthermore, virtually all nerve endings, including those packed with large granular vesicles, contain at least some agranular vesicles. When prejunctional specializations are present, they are almost invariably associated with small agranular vesicles rather than with large granular vesicles.

Small granular vesicles are characteristic of noradrenergic nerve endings. With many preparative procedures the dense cores are not preserved and these vesicles are then difficult to distinguish from those of cholinergic endings. Better preservation of dense cores of adrenergic small granular vesicles is obtained with permanganate fixation (Hökfelt, 1968; Hökfelt & Jonsson, 1968), at the price of a certain loss of general quality of tissue preservation. A more reproducible approach

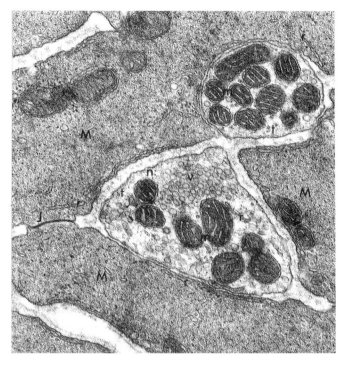

FIGURE 1.12   A nerve ending (varicosity) surrounded by three muscle cells (*M*) of the sphincter pupillae of a guinea-pig. The muscle cells are connected by gap junctions (*J*) and are rich in sarcoplasmic reticulum (*r*). The axons are completely devoid of Schwann cell wrapping and contain small clear vesicles (*v*), mitochondria (*m*), microtubules (*t*), endoplasmic reticulum (*r*), neurofilaments (*n*) and a coated vesicle (arrow). Magnification: 20,500 x

involves loading, through specific uptake, of adrenergic endings (and adrenergic vesicles) with a false transmitter such as 5- or 6-hydroxydopamine, as originally proposed by Tranzer and Thoenen (1967, 1969) (Figure 1.13).

Large granular vesicles are rare in some identified autonomic cholinergic endings, for example in the sphincter pupillae, and may contain acetylcholine (Agostoni and Whittaker, 1989). Large granular vesicles are more numerous, and sometimes even predominant, in adrenergic endings, and they contain noradrenaline. These vesicles predominate in the non-terminal portion of adrenergic axons. While the relation between small vesicles and large granular vesicles is not yet clearly explained, most neuropeptides are believed to be stored in large granular vesicles, with or without other transmitters.

The correlation between vesicle morphology and chemical content is limited (a conclusion, incidentally, that applies to all kinds of secretory vesicles). This is evident through the literature and was clearly stated in a study by Gibbins (1982). This conclusion, whether openly accepted or not, has probably inhibited more systematic study of vesicle types, serial section studies and more studies on effect

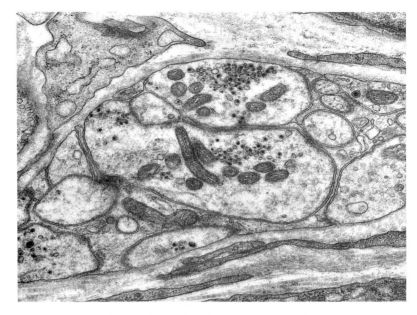

FIGURE 1.13   Adventitia of an arteriole in the submucosa of guinea-pig ileum. Preparation incubated *in vitro* with 6-hydroxy-dopamine. Axons within this nerve bundle contain small granular vesicles of the adrenergic type. The electron density of the vesicle granule is enhanced by the false transmitter. Magnification: 17,000 x

of preparative procedures on vesicle morphology. Identification of the vesicle population in a varicosity cannot be done on a single section, as stated for example by Cook and Burnstock (1976), but few have undertaken serial section studies (Gordon-Weeks, 1981).

ONE VARICOSITY – ONE TRANSMITTER

The number of substances that are released by nerves and participate in neuro-effector transmission has grown enormously and there is strong evidence that more than one secretory product can be stored in a single varicosity. By double immuno-fluorescence labelling many examples of co-localization of transmitters and neuro-peptides were demonstrated (in spite of the low resolution of fluorescence microscopy). This is in good agreement with strong pharmacological evidence of release of more than one neurotransmitter or other neuroregulatory substance from the same type of ending (see Burnstock, 1990). Well-established examples of localization of vasoactive intestinal polypeptide and acetylcholine are the cranial parasympathetic ganglion neurons (Lundberg *et al.*, 1979). While noradrenaline and neuropeptide Y are localized in the same nerve fibres in the vas deferens (Lundberg and Stjärne, 1984). Co-localization of neuropeptides and other neurotransmitters is well documented in neurons of the gut (Furness and Costa, 1987).

CONSTANCY OF TRANSMITTER CONTENT

There are many examples of transient expression of a neurotransmitter in a neuron. For example, some neurons of the myenteric plexus of the rat express a catecholaminergic phenotype only for a few days during embryonic life (Cochard *et al.*, 1978, 1979; Teitelman, Joh and Reis 1978; Teitelman *et al.*, 1981). This presumably also involves changes in vesicle populations at the nerve endings, changes which, however, are not yet documented. At least in one case there is good evidence of such a change in vesicle content. A small percentage of the neurons in paravertebral sympathetic ganglia are cholinergic and innervate eccrine sweat glands and some blood vessels. The cholinergic sympathetic neurons innervating sweat glands in adult rats are neurons that were adrenergic in the early post-natal days and gradually acquired a cholinergic phenotype, while losing the adrenergic one (Landis and Keefe, 1983). Seven days after birth small granular vesicles are present in all nerve endings associated with the sweat glands and constitute over 50% of all the vesicles; at 10 days their percentage decreases to 29%, at 14 days to 4%, and at 21 days no small granular vesicles are found in axons associated with the glands, while they were abundant in the adjacent perivascular nerves. The transition involves the expression of choline acetyltransferase, soon before the onset of the secretory response (Stevens and Landis, 1987), and the appearance of the peptides vasoactive intestinal polypeptide and calcitonin gene-related peptide within the sudomotor fibres (Landis, Siegel and Schwab, 1988). The experiments strongly suggest coexistence of transmitters and age-related changes in vesicle population in a terminal.

VESICLE DISTRIBUTION ALONG AXON LENGTH

The variability in structure and in functional properties of varicosities along the axon length remains to be investigated. Some histochemical evidence suggests accumulation of transmitters in the most peripheral varicosities (Figure 1.11). To the extent that the chemical content of a varicosity depends on transport from the cell body (membrane components, peptides) the long chain of varicosities of a terminal axon may be expected to display some heterogeneity and gradients.

In adrenergic nerve fibres, the non-terminal axons contain only vesicles of the large granular type, whereas the varicose portion of the axons contains both small and large granular vesicles (Schwarzenbrunner *et al.*, 1990). Whether there are substantial differences in this respect between varicosities of the same axon remains to be established. The possibility of a difference in transmitter content even in successive varicosities is discussed by Burnstock (1990) as one of the ways to account for variation in the relative amount of noradrenaline and ATP released from sympathetic nerve fibres of the vas deferens with different stimuli.

AXONAL VESICLE MEMBRANE

The composition of the membrane of synaptic vesicles has been studied mainly outside the autonomic nervous system, but some of the features apply to all endings. Several integral components of the vesicle membrane have been isolated. Among them the calcium-binding glycoprotein synaptophysin (or p38) (Wiedenmann and Franke, 1985; Jahn et al., 1985). This glycoprotein has been sequenced (Buckley et al., 1987; Südhof et al., 1987), has 307 amino acids, and both termini of the molecule are on the cytoplasmic (P) surface of the membrane, where the binding site for calcium is also located. Synaptophysin is present in all small synaptic vesicles (in brain), but not in large granular vesicles (Navone et al., 1986). Others have, however, detected synaptophysin in large vesicles in adrenal medullary cells and in neuroendocrine cells (Fournier and Trifaró, 1988; Lowe, Madeddu and Kelly, 1988).

Another component of synaptic vesicles is the phosphoprotein synapsin I, a peripheral protein located on the cytoplasmic (P) side of the vesicle (Huttner et al., 1983; De Camilli et al., 1983; Ueda and Greengard, 1977). Immunocytochemical studies have shown that this protein is not only nerve-specific but is localized only on small synaptic vesicles (Navone et al., 1984). Furthermore, synapsin I can be phosphorylated at multiple sites by cAMP-dependent and Ca^{2+}-dependent protein kinases (Huttner et al., 1981) and can bind to actin (Petrucci and Morrow, 1987), thus providing a possible mechanism for an interaction between synaptic vesicles and the cytoskeleton.

Synaptobrevin (Baumert et al., 1989), protein p65 (Matthew et al., 1981) and synapto-porin (Knaus et al., 1990), are some of the more recently uncovered integral membrane proteins of synaptic vesicles. Each of these has been claimed to be localized in small synaptic vesicles and not in large granular vesicles. Some authors have, however, contested the latter view (Trifaró et al., 1989). The fact that proteins such as synapsin I and synaptophysin are found in vesicles from any part of the brain and from afferent and efferent peripheral nerve endings (De Camilli et al., 1988) suggests that they are not related to a specific type of transmitter.

The important question of the relation between large and small vesicles remains unsolved. Evidence suggesting that small vesicles originate from large ones has been presented, mainly for adrenergic endings. More recently a contrasting view has been put forward, mainly from studies of brain (Navone et al., 1986).

Increased resolution in the analysis of at least four membrane components at nerve endings (plasma membrane, endoplasmic reticulum, small agranular vesicles, and large granular vesicles) is needed to clarify any traffic between them. A complete independence of these four membrane structures is unlikely, but how they are interrelated is unknown.

AXONAL VESICLE CONTENT

The release of more than one transmitter from the same type of nerve ending – and in certain circumstances from the same ending, and the presence of different 'types' of axonal vesicles, have opened the question of possible storage of two transmitters in the same vesicle. Substance P and serotonin, demonstrated by immunocyto-chemistry of consecutive sections of nerve endings in the spinal cord, are localized within the same vesicle type, the large granular vesicles, of a single ending (Pelletier *et al.*, 1981). In perivascular endings of the guinea-pig, by double immunostaining with gold particles of different size, Gulbekian and colleagues (1986) showed calcitonin gene-related peptide and substance P being present in the very same vesicles, invariably of the large granular type.

The localization of several substances within the same vesicles is therefore supported by experimental evidence. In physiological terms, the extent of co-localization in, and co-release from, the same vesicles – as opposed to co-localization in different vesicles of the same ending, or co-localization in the same axon but in different varicosities – remains to be explored.

Vesicles store other substances in addition to the transmitters. In adrenergic nerve endings, the small granular vesicles contain dopamine-β-hydroxylase and cytochrome b-561 (Fried *et al.*, 1985; Neuman *et al.*, 1984), but not chromogranin (Neuman *et al.*, 1984; Fried *et al.*, 1985). In contrast, the large granular vesicles of adrenergic endings, resembling in composition adrenal chromaffin granules (Lagercrantz, 1976; Hagn *et al.*, 1986), contain, in addition to noradrenaline, dopamine-β-hydroxylase and cytochrome b-561, also chromogranin A and the neuropeptides enkephalin and neuropeptide Y (Neuman *et al.*, 1984).

In nerve endings of other types, the content of large granular vesicles is less well understood. They contain various neuropeptides (Dahlström, 1986) and, at least in cholinergic nerve endings, they have been shown to contain chromogranin A (Volknandt *et al.*, 1987; Bööj *et al.*, 1989).

AXONAL VESICLES IN AFFERENT FIBRES

Afferent fibres issued by cranial and dorsal root ganglia include small unmyelinated fibres. In target organs these fibres arborize extensively and develop varicosities, as documented by immunohistochemical studies of afferent fibre peptides in whole mount preparations. Ultrastructurally, these endings show similar features to those of post-ganglionic nerve fibres, including varicosities and accumulations of axonal vesicles. These observations fit nicely with studies on axonal reflexes via afferent fibres. Somatic sensory fibres in the skin and the eye have for a long time been known to mediate an inflammatory reaction (known as neurogenic inflammation) in response to irritant stimuli (Jancso *et al.*, 1967; Foreman and Jordan, 1984). Lundberg and Saria (1982, 1983) showed that the occurrence of neurogenic inflam-mation in the airways, and the involvement of vagal afferent fibres has been con-firmed (Persson *et al.*, 1986; McDonald, 1988). This response is well documented in the rat and guinea-pig, and in other systems besides the airways (Lundberg *et al.*, 1984), and it involves afferent unmyelinated axons containing substance P (Lundberg *et al.*, 1983).

Insofar as they can transmit centrifugally and release transmitters, these afferent fibres, issued by neurons in spinal and cranial sensory ganglia, behave as efferent fibres (Maggi and Meli, 1988). On wonders if, conversely, efferent fibres can be stimulated peripherally and conduct centripetally and release transmitter at more proximal sites.

SYNAPTIC VESICLE DISCHARGE

The release of synaptic vesicles, as all forms of regulated secretion, is mediated by exocytosis. The details of this process are not yet fully explained, but the mechanism involves fusion of vesicle membrane and cell membrane, triggered by an increase of Ca^{2+} concentration in the cytoplasm. The ubiquitous Ca^{2+}-binding protein calmodulin is probably involved in the process. Calmodulin binds to synaptic vesicles and to secretory vesicles of neurohypophysial cells and chromaffin cells (Hikita et al., 1984; Hooper and Kelly, 1984). Another component of synaptic vesicle membrane, the 65,000 kDa protein p65, binds calmodulin with high affinity. There are very few data on exocytosis in varicosities of autonomic fibres, and the process is interpreted on the basis of data, also incomplete, obtained elsewhere.

NEUROMUSCULAR JUNCTIONS

Axons influence muscle cells and the sites where this transmission occurs are defined as neuromuscular junctions. While there is no doubt about the occurrence of transmission between axons and muscle cells, the structural features that allow identification of the junctions are difficult to define. Some of the criteria are a matter of convention or convenience.

Transmission can occur in the absence of junctions or even of closeness, and even when the two elements are situated in different organs, for example between endocrine glands and target cells. However, when transmission takes place between structures lying adjacent to each other, and the process is *local*, in the full sense of the word, as is the case at the autonomic nerve endings, then it seems correct to talk of *junctions* and of *junctional transmission*, even when prominent structural specializations are not evident. It would be natural to expect that the functional specialization of transmission is accompanied, and is supported, by structural specializations. However, it is unfortunate that structural specializations can as yet be proved only by ambiguous or controversial criteria. It seems impossible to assess fully the impact of the preparative procedures used and it is difficult to broaden the range of structural parameters to include three-dimensional extent, statistical variability, sample variability, and certain sub-structural features, such as the spatial arrangement of receptors and channels. In this article, I will assume that where there is neuromuscular transmission there are neuromuscular junctions. To temper this broad use of the term junctions, one could distinguish very specialized junctions – that vouchsafe efficiency and reliability of transmission – and less specialized junctions. The motor endplate of skeletal muscles is a very specialized junction, remarkable, in general terms, for its size, consistency of structure, and complexity of structure – at the nerve ending, in the intercellular gap and in the post-junctional membrane. Even among striated muscles there are examples of simpler junctions, with fewer axonal branches and no sub-junctional folds. In cardiac muscle one finds what we could call minimally specialized junctions. Generally speaking, neuromuscular junctions in smooth muscles are structurally inconspicuous (see, for example, Bennett, 1972; Burnstock, 1986).

What are 'structural specializations' in the varicosities and the adjacent portions of the muscle cell? The varicosity itself is a structural specialization, as described above. Other features that may have relevance to the process of transmission are the synaptic vesicles, the pre-junctional membrane densities, the extent of Schwann cell covering, the basal laminae, the intercellular gap, and the post-junctional membrane specializations.

VESICLES

Vesicles (axonal vesicles), which are few and far apart in non-terminal portions of the axons, are abundant and often densely packed, in varicosities. As to the total number of vesicles in varicosities of post-ganglionic fibres, there are approximately between 100 and 1,000 vesicles per varicosity in adrenergic fibres of the dilator pupillae of the rat (Hökfelt, 1969) and of submucosal blood vessels in the guinea-pig ileum (Luff et al., 1987); in the latter case the average is 559 vesicles. (By comparison, Birks (1974) has calculated that the average number of synaptic vesicles in the preganglionic endings in cat sympathetic ganglia is about 6000.) The variability in number and packing density is a consistent feature of varicosities in all types of post-ganglionic fibres.

PREJUNCTIONAL DENSITIES

Dense projections beneath the prejunctional membrane are observed only occasionally: they are small and of moderate electron density (by comparison with those found in synapses and somatic neuro-muscular junctions), and a few small vesicles are clustered around them. Clustering of vesicles at the prejunctional membrane is, however, more common than the occurrence of dense projections, so that one would assume the existence of a cytoskeletal apparatus for holding small vesicles near the prejunctional membrane even when dense projections are not visualized in the electron microscope. The membrane sites where dense projections are situated or around which vesicles are clustered, are regarded as release points, although the evidence for this is indirect at best.

SCHWANN CELL COVERING

The extent of the Schwann cell cytoplasm decreases from larger to smaller nerve bundles, both in absolute and relative terms. Axons only partially surrounded by a Schwann cell are found in post-ganglionic nerves, but they become more numerous in the intramural nerve bundles. Single axons are accompanied by tenuous Schwann cell processes, and some varicosities are devoid altogether of Schwann cell covering (Figure 1.4C).

BASAL LAMINA

A basal lamina surrounds every nerve bundle and covers also any exposed axons that may be present. At contacts between varicosities and muscle cells that are regarded as effective transmission points, the basal lamina of the nerve bundle and that of the muscle cell are intact (Figure 1.5). At other junctions, a single basal lamina is present, probably arising from the fusion of the lamina of the two contacting elements (Figures 1.4A, 1.4B). At other junctions the two basal laminae are absent (Figure 1.4C).

INTERCELLULAR GAP

The variability of the intercellular gap in autonomic neuro-muscular junctions is apparent in any set of micrographs (Figures 1.4 A–C). In muscles where the autonomic plexus issues single axons, close junctions between nerve endings and muscle cells are more common, although examples of the opposite are commonly encountered.

Luff *et al.* found that between 1 in 10 and 1 in 5 of the varicosities in the axons around arterioles of $50\,\mu$m diameter of the submucosa of the guinea-pig ileum, make a contact with a muscle cell within less than 100 nm (nearly 40% of the total lie between 100 and 200 nm). The smallest neuromuscular distances found in this study on blood vessels are 70 nm (Luff *et al.*, 1987), whereas in visceral muscles there are junctions where the gaps are reduced to less than 20 nm (Figure 1.4), amid junctions where the gaps are comparable to those found in blood vessels.

POST-JUNCTIONAL MEMBRANE SPECIALIZATIONS

The cell membrane of smooth muscle cells has no specializations that are characteristic of neuromuscular junctions. Caveolae and dense bands can be found beneath a varicosity, and sub-plasmalemmal cisternae of sarcoplasmic reticulum are no less common at these sites than elsewhere in the muscle cell. Membrane infoldings and post-junctional densities are absent, and were it not for the presence of the varicosity, it would be impossible to recognize the existence of a junction. When the neuromuscular gap is narrow, a small symmetrical junction of the adherens type is sometimes encountered.

RECEPTOR LOCALIZATION

The action of a transmitter depends on the presence of appropriate receptors; the action of the occupied receptors, and not the nature of the transmitter itself, determines the specific effect of a transmitter.

The search for the localization of receptors in autonomic effectors is fraught with technical difficulties. With the present techniques it remains difficult to localize receptors to identified cell types and cellular regions in organs *in situ*. Semi-quantitative comparisons of the binding in different organs or in different tissue layers can, however, be made (Kuhar, De Souza and Unnerstall, 1986). Problems of tracer degradation, tracer absorbance to unspecific sites, and tracer access (diffusion, competition with endogenous ligands), compound the difficulty of *in vivo* studies. To avoid them, some investigators have studied tissue sections labelled *in vitro* (Young and Kuhar, 1979) and nerve cell cultures (James and Burnstock, 1989).

Muscarinic receptors have been localized in several smooth muscles, as was expected from pharmacological studies. In the guinea-pig intestine the receptors detected by autoradiography have a uniform distribution throughout the muscle (Buckley and Burnstock, 1984), whereas in the trachea they are localized in the part of the muscle furthest away from the lumen (Basbaum *et al.*, 1984).

There are no data on the distribution of receptors over the muscle cell membrane and there is at present no evidence showing that receptors are, or are not, clustered at specific sites. Hirst and Neild (1981), however, report that ionophoretic application of noradrenaline on muscle cells of arterioles mimics the response to nerve stimulation, only at sites on the muscle cell membrane a few microns away from a nerve bundle; the implication being that the adrenoreceptors are localized mainly at these sites. Also unknown is the distribution of prejunctional receptors, which are known to exist and to play a role in modulating transmission at autonomic endings (Gillespie, 1980).

MUSCLE EFFECTOR

It has been suggested that the contractile unit should be a bundle of muscle cells rather than individual muscle cells, although Bennett (1972) stressed that the anatomical equivalent of a bundle is not immediately apparent and may well not be a static structure.

Although the mechanical activity of smooth muscle cells is clearly influenced by nerves, the relation between smooth muscle and nerves is more complex than that between skeletal muscle fibres and somatic nerves. In both cases a nerve fibre reaches a large number of muscle units, although the number of varicosities of an autonomic fibre is probably much larger that the number of endings of a somatic fibre. However, a mechanism in smooth muscle that has no counterpart in skeletal muscles is the presence of coupling between the muscle cells so that current can spread from innervated cells to non-innervated cells via electrotonic junctions. In many smooth muscles the sites of electrotonic coupling are gap junctions (Dewey and Barr, 1964). Their composition and structure have been extensively studied.

While some smooth muscles contain numerous gap junctions, in others the junctions are consistently rare or absent. Muscles very rich in gap junctions include some with a dense innervation and nerve endings abutting on every muscle cell, for example the sphincter pupillae (Uehara and Burnstock, 1972); muscles poor in gap junctions include some in which there is good coupling between the muscle cells, for example the vas deferens. It is still unclear how coupling is achieved in these situations (Daniel, 1986). In the myometrium a large supply of gap junctions develop under hormonal control toward the end of pregnancy (Garfield *et al.*, 1977). The possibility of smooth muscle gap junctions forming or disappearing within a short time is suggested also by *in vitro* experiments (Daniel, 1986).

The existence of a good coupling between the muscle cells insures that nerve-mediated stimulations reach almost simultaneously a significant number of muscle cells, and possibly all the cells within a selected area; furthermore, it allows a certain amount of nerve-mediated transmission failure (i.e. in a percentage of varicosities) to occur without the muscle failing to contract, since each muscle cell can be stimulated via more than one route. This mechanism reduces the burden on the individual nerve endings and goes some way in accounting for the lack of regularity in the geometry of neuromuscular junctions, at least by comparison with those of skeletal muscle. In cardiac muscle the distribution of nerve endings is more irregular still than in smooth muscles, and close junctions between nerve endings and muscle are rarer; conversely, gap junctions are extremely numerous and there is a high degree of electrotonic coupling between the muscle cells (Matter, 1973; Chen *et al.*, 1989).

MATCHING OF VARICOSITIES AND EFFECTORS

There is a very large variability in every structural feature of the junctions between autonomic endings and smooth muscle cells. The source of this variability is multifold. Some of it is accounted for by uncertainty in the effects of the preparative procedures, and some by differences between tissues and between animal species. Some of the variability may be related to the different position of varicosities along the length of the terminal axon. This is an interesting possibility, which remains to be explored.

Some of the variability may be a genuine feature of the autonomic peripheral innervation. Even within the same tissue, some varicosities are separated from the muscle cell membrane by a gap of only 20 nm, while identical varicosities from the same type of fibre lie microns away from the nearest cell. In contrast, the varicosities often appear to have a regular distribution along the length of the axon, as if the formation and the maintenance of a varicosity was not strictly dependent from the formation of an intimate contact with an effector.

Many factors are involved in the control of smooth muscle contraction: not just the nerve control via neurotransmitters, but also the myogenic activity, the hormonal sensitivity, the stretch sensitivity, and the ionic coupling. The relative consistence of performance of the effector may arise from the equilibrium of a variety of factors, rather than from a strictly determined pattern of nerve-muscle contacts.

REFERENCES

Agostoni, D.V. and Whittaker, V.P. (1989) Characterization by size, density, osmotic fragility and immunoaffinity of acetylcholine and vasoactive intestinal polypeptide-containing storage particles from myenteric neurons of the guinea-pig. *J. Neurochem.* **52**, 1474–1480.

Aguado, L.I. & Ojeda, S.R. (1984) Prepubertal ovarian function is finely regulated by direct adrenergic influences. Role of adrenergic innervation. *Endocrinology* **114**, 1845–1853.

Altenahr, E. (1971) Electron microscopical evidence for innervation of chief cells in human parathyroid glands. *Experientia* **22**, 1077–1078.

Barker, D. and Saito, M. (1981) Autonomic innervation of receptors and muscle fibres in cat skeletal muscle. *Proc. R. Soc. Lond.* B **212**, 317–332.

Basbaum, C.B., Grillo, M.A. and Widdicombe, J.H. (1984) Muscarinic receptors: evidence for a nonuniform distribution in tracheal smooth muscle and exocrine glands. *J. Neurosci.* **4**, 508–520.

Baumert, M., Maycox, P.R., Navone, F., De Camilli, P. and Jahn, R. (1989) Synaptobrevin: an integral membrane protein of 18000 dalton present in small synaptic vesicles of rat brain. *EMBO J.* **8**, 379–384.

Bennett, M.R. (1972) *Autonomic Neuromuscular Transmission*. Cambridge University Press.

Bessou, P. and Perl, E.R. (1966) A movement receptor for the small intestine. *J. Physiol. (Lond.)* **182**, 404–426.

Birks, R.I. (1974) The relationship of transmitter release and storage to fine structure in a sympathetic ganglion. *J. Neurocytol.* **3**, 133–160.

Blakeley, A.G.H. and Cunnane, T. (1979) The packeted release to transmitter from sympathetic nerves of the guinea-pig vas deferens: an electrophysiological study. *J. Physiol. (Lond.)* **296**, 85–96.

Boeke, J. (1927) Die morphologische Grundlage der sympathischen Innervation der quergestreiften Muskelfasern. *Z. Mikrosk. anat. Forsch.* **8**, 561–639.

Bööj, S., Goldstein, M., Fischer-Colbrie, R. and Dahlström, A. (1989) Calcitonin gene-related peptide and chromogranin A: presence and intra-axonal transport in lumbar motor neurons in the rat, a comparison with synaptic vesicle antigens in immunohistochemical studies. *Neuroscience* **30**, 479–501.

Bowers, C.W. and Zigmond, R.E. (1979) Localization of neurons in the rat superior cervical ganglion which project into different post-ganglionic trunks. *J. Comp. Neurol.* **185**, 381–392.

Brock, A.J. and Cunnane, T. (1988) Electrical activity at the sympathetic neuroeffector junction in the guinea-pig vas deferens. *J. Physiol. (Lond.)* **399**, 607–632.

Buckley, K., Floor, E. and Kelly, R.B. (1987) Cloning and sequence analysis of cDNA encoding p38, a major synaptic vesicle protein. *J. Cell Biol.* **105**, 2447–2456.

Buckley, N.J. and Burnstock, G. (1984) Autoradiographic localization of muscarinic receptors in guinea-pig intestine: distribution of high and low affinity binding sites. *Brain Res.* **294**, 15–22.

Burnstock, G. (1986) Autonomic neuromuscular junctions: current developments and future directions. *J. Anat. (Lond.)* **146**, 1–30.

Burnstock, G. (1990) Co-transmission. *Arch. int. Pharmacodyn, Thér.* **304**, 7–33.

Burnstock, G. and Costa, M. (1975) *Adrenergic Neurons: Their Organization, Function and Development in the Peripheral Nervous System*. London: Chapman & Hall.

Cannon, B., Nedegaard, Lundberg, J.M., Hökfelt, T., Terenius, L. and Goldstein, M. (1986) Neuropeptide tyrosine (NPY) co-stored with noradrenaline in vascular but not in parenchymal sympathetic nerves of brown adipose tissue. *Exp. Cell Res.* **164**, 546–550.

Chen, L., Goings, G.E., Upshaw-Earley, J. and Page, E. (1989) Cardiac junctions and gap junction-associated vesicles: ultrastructural comparison of in situ negative staining with conventional positive staining. *Circulation Res.* **64**, 501–514.

Cheung, D.W. (1990) Synaptic transmission in the guinea-pig vas deferens: the role of nerve action potentials. *Neuroscience* **37**, 127–134.

Cochard, P. Goldstein, M and Black, I.B. (1978) Ontogenetic appearance and disappearance of tyrosine hydroxylase and catecholamines in the rat embryo. *Proc. Nat. Acad. Sci. USA* **75**, 2986–2990.

Cochard, P. Goldstein, M and Black, I.B. (1979) Initial development of the noradrenergic phenotype in autonomic neuroblasts of the rat embryo *in vivo*. *Develop. Biol.* **71**, 100–114.

Cook, R.D. and Burnstock, G. (1976) The ultrastructure of Auerbach's plexus in the guinea-pig. I. Neuronal elements. *J. Neurocytol.* **5**, 195–206.

Cottrell, D.F. (1984) Conduction velocity and axonal diameter of alimentary C fibres. *Q.J. Exp. Physiol.* **69**, 355–364.

Cunnane, T.C. and Stjärne, L. (1984) Transmitter secretion from individual varicosities of guinea-pig and mouse vas deferens: highly intermittent and monoquantal. *Neuroscience* **13**, 1–20.

Dahlström, A. (1986) Axonal transport of transmitter organelles in adrenergic, cholinergic and peptidergic neurons. In: *Axoplasmic Transport*, ed. by Z. Iqbal, pp. 119-146. CRC Press, Boca Rato, Florida.

Dahlström, A. and Haggendal, J. (1966) Some quantitative studies on the noradrenaline content in the cell bodies and terminals of a sympathetic adrenergic neuron system. *Acta Physiol. Scand.* **67**, 271-277.

Daniel, E.E. (1986) Gap junctions in smooth muscle. In: *Cell-to-Cell Communication*. Edited by W.C. De Mello, pp. 149-185. Plenum, New York.

De Camilli, P., Cameron, R. and Greengard, P. (1983) Synapsin I (protein I), a nerve-terminal specific phosphoprotein. I. Its general distribution in synapses of the central and peripheral nervous system demonstrated by immunofluorescence in frozen and plastic sections. *J. Cell Biol.* **96**, 1331-1354.

De Camilli, P., Vittadello, M., Canevini, M.P., Zanoni, R., Jahn, R. and Gorio, A. (1988) The synaptic vesicle proteins synapsin I and synaptophysin (protein p38) are concentrated both in efferent and afferent nerve endings of the skeletal muscle. *J. Neurosci.* **8**, 1625-1631.

Dewey, M.M. and Barr, L. (1964) A study of the structure and distribution of the nexus. *J. Cell Biol.* **23**, 553-585.

Ehinger, B., Falck, B. and Sporrong, B. (1970) Possible axo-axonal synapses between peripheral adrenergic and cholinergic nerve terminals. *Z. Zellforsch.* **107**, 508-521.

Emery, D.G., Foreman, R.D. and Coggeshall, R.E. (1976) Fiber analysis of the feline inferior cardiac sympathetic nerve. *J. Comp. Neurol.* **166**, 457-468.

Falck, B., Hillarp, N.-A., Thieme, G. and Torp, A. (1962) Fluorescence of catecholamines and related components condensed with formaldehyde. *J. Histochem. Cytochem.* **10**, 348-354

Forehand, C.J. and Konopka, L.M. (1989) Frog sympathetic cells have local axon collaterals. *J. Comp. Neurol.* **289**, 294-303.

Foreman, J.C. and Jordan, C.C. (1984) Neurogenic inflammation. *Trends Pharmacol. Sci.* **5**, 116-119.

Fournier, S. and Trifaró, J.-M. (1988) A similar calmodulin-binding protein expressed in chromaffin, synaptic and neurohypophyseal secretory vesicles. *J. Neurochem.* **50**, 27-37.

Fried, G., Terenius, L., Hökfelt, T. and Goldstein, M. (1985) Evidence for differential localization of noradrenaline and neuropeptide Y in neuronal storage vesicles isolated from rat vas deferens. *J. Neurosci.* **5**, 450-458.

Furness, J.B. and Costa, M. (1987) *The Enteric Nervous System*. Edinburgh, Churchill Livingstone.

Garfield, R.E., Sims, S.M. and Daniel, E.E. (1977) Gap junctions: their presence and necessity in myometrium during parturition. *Science* **198**, 958-959.

Gibbins, I.L. & Haller, C.J. (1979) Ultrastructural identification of non-adrenergic, non-cholinergic nerves in the rat anococcygeus muscle. *Cell Tissue Research* **200**, 257-271.

Gillespie, J.S. (1980) Presynaptic receptors in the autonomic nervous system. In: *Adrenergic Activators and Inhibitors*. Edited by L. Szerkerco, pp. 353-425. Springer, Berlin.

Gordon-Weeks, P.R. (1981) Properties of nerve endings with small granular vesicles in the distal colon and rectum of the guinea-pig. *Neuroscience* **6**, 1793-1811.

Gordon-Weeks, P.R. (1988) The ultrastructure of noradrenergic and cholinergic neurons in the autonomic nervous system. In: Handbook of Chemical Neuroanatomy. Vol. 6: *The Peripheral Nervous System*, edited by A. Björklund, T. Hökfelt and C. Owman. Amsterdam, Elsevier.

Govyrin, V.A. (1976) Spatial neuromuscular relations in rabbit ear arteries. In: *Physiology of Smooth Muscle*. Edited by E. Bülbring and M.F. Shuba, pp. 279-285. Raven Press, New York.

Griffith, S.G., Crowe, R., Lincoln, J., Haven, A. and Burnstock, G. (1982) Regional differences in the density of perivascular nerves and varicosities, noradrenaline content and responses to nerve stimulation in the rabbit ear artery. *Blood Vess.* **19**, 41-52.

Gulbekian, S., Merighi, A., Wharton, J., Varndell, I.M. and Polak, J.M. (1986) Ultrastructural evidence for the coexistence of calcitonin gene-related peptide and substance P in secretory vesicles of peripheral nerves in the guinea pig. *J. Neurocytol.* **15**, 535-542.

Hagn, C., Klein, R.L., Douglas, B.H. and Winkler, H. (1986) An immunological characterization of five common antigens of chromaffin granules and of large dense-cored vesicles of sympathetic nerves. *Neurosci. Lett.* **67**, 295-300.

Hikita, T., Bader, M.F. and Trifarò, J.M. (1984) Adrenal chromaffin cell calmodulin: its subcellular distribution and binding to chromaffin membrane proteins. *J. Neurochem.* **43**: 1087-1097.

Hillarp, N.-A. (1946) Structure of the synapses and the peripheral innervation apparatus of the autonomic nervous system. *Acta Physiol. Scand.* suppl. 46.

Hirst, G.S.D. and Neild, T.O. (1980) Some properties of spontaneous excitatory junction potentials recorded from arterioles of guinea-pigs. *J. Physiol. (Lond.)* **303**: 43-60.

Hirst, G.S.D. and Neild, T.O. (1981) Localization of specialized noradrenaline receptors at neuromuscular junctions on arterioles of the guinea-pig. *J. Physiol. (Lond.)* **313**, 343–350.

Hökfelt, T (1968) *In vitro* studies on central and peripheral monoamine neurons at the ultrastructural level. *Zeit. Zellforsch.* **91**, 1–74.

Hökfelt, T. (1969) Distribution of noradrenaline storing particles in peripheral adrenergic neurons as revealed by electron microscopy. *Acta Physiol. Scand.* **76**, 427–440.

Hökfelt, T. and Jonsson, G. (1968) Studies on reaction and binding of monoamines after fixation and processing for electron microscopy with special reference to fixation with potassium permanganate. *Histochemie* **16**, 45–67.

Hooper, J.E. and Kelly, R.B. (1984) Calcium-dependent calmodulin binding to cholinergic synaptic vesicles. *J. Biol. Chem.* **259**, 141–147.

Huttner, W.B., DeGennaro, L.J. and Greengard, P. (1981) Differential phosphorylation of multiple sites in purified protein I by cyclic AMP-dependent and calcium-dependent protein kinases. *J. Biol. Chem.* **256**, 1482–1488.

Huttner, W.B., Schiebler, W., Greengard, P. and De Camilli, P. (1983) Synapsin I (protein I), a nerve terminal-specific phosphoprotein. III. Its association with synaptic vesicles studied in a highly purified synaptic vesicle preparation. *J. Cell Biol.* **96**, 1374–1388.

Jahn, R., Schiebler, W., Ouimet, C. and Greengard, P. (1985) A 38,000-dalton membrane protein (p38) present in synaptic vesicles. *Proc. Nat. Acad. Sci. USA* **82**, 4137–4141.

James, S. and Burnstock, G. (1989) Autoradiographic localization of muscarinic receptors on cultured, peptide-containing neurones from newborn rat superior cervical ganglion. *Brain Res.* **498**, 205–214.

Jancso, N., Jancso-Gabor, A. and Szolcsanyi, J. (1967) Direct evidence for neurogenic inflammation and its prevention by denervation and by pretreatment with capsaicin. *Br. J. Pharmacol.* **31**, 138–151.

Keast, J.R., Booth, A.M. and de Groat, W.C. (1989) Distribution of neurons in the major pelvic ganglion of the rat which supply the bladder, colon and penis. *Cell Tissue Res.* **256**, 105–112.

Kiraly, M., Favrod, P. and Matthews, M.R. (1989) Neuroneuronal interconnections in the rat superior cervical ganglion; possible anatomical bases for modulatory interactions revealed by intracellular horseradish peroxidase labelling. *Neuroscience* **33**, 617–642.

Kleitman, N. & Holzwarth, M.A. (1985) Catecholaminergic innervation of the rat adrenal cortex. *Cell Tissue Res.* **241**, 139–147.

Knaus, P., Marquèze-Pouey, B., Scherer, H. and Betz, H. (1990) Synaptoporin, a novel putative channel protein of synaptic vesicles. *Neuron* **5**, 453–462.

Kuhar, M.J., De Souza, E.B. and Unnerstall, J.R. (1986) Neurotransmitter mapping by autoradiography and other methods. *Annu. Rev. Neurosci.* **9**, 27–59.

Lagercrantz, H. (1976) On the composition and function of large dense cored vesicles in sympathetic nerves. *Neuroscience* **1**, 81–92.

Landis, S.C. and Keefe, D. (1983) Evidence for neurotransmitter plasticity *in vivo*: Developmental changes in properties of cholinergic sympathetic neurons. *Devel. Biol.* **98**, 349–372.

Landis, S.C., Siegel, R. and Schwab, M. (1988) Evidence for neurotransmitter plasticity *in vivo*. II. Immunocytochemical studies of rat sweat gland innervation during development. *Devel. Biol.* **126**, 129–140.

Lever, J.D., Jung, R.T., Norman, D., Symons, D., Leslie, P.J. and Nnodim, J.O. (1987) Peptidergic and catecholaminergic innervation in rat and human brown adipose tissue. *Exp. Brain Res.* **16**, 107–112.

Llewellyn-Smith, I.J., Furness, J.B., O'Brien, P.E. and Costa, M (1984) Noradrenergic nerves in human small intestine: distribution and ultrastructure. *Gastroenterology* **87**, 513–529.

Lowe, A.W., Madeddu, L. and Kelly, R.B. (1988) Endocrine secretory granules and neuronal synaptic vesicles have three integral membrane proteins in common. *J. Cell Biol.* **106**, 51–59.

Luff, S.E. and McLachlan, E.M. (1989) Frequency of neuromuscular junctions on arteries of different dimensions in the rabbit, guinea pig and rat. *Blood Vess.* **26**, 95–100.

Luff, S.E., McLachlan, E.M. and Hirst, G.D.S. (1987) An ultrastructural analysis of the sympathetic neuromuscular junctions on arterioles of the submucosa of the guinea pig ileum. *J. Comp. Neurol.* **257**, 578–594.

Lundberg, J.M. and Saria, A. (1982) Capsaicin-sensitive vagal neurons involved in control of vascular permeability in rat trachea. *Acta Physiol. Scand.* **115**, 521–523.

Lundberg, J.M. and Saria, A. (1983) Capsaicin-induced desensitization of airway mucosa to cigarette smoke, mechanical and chemical irritants. *Nature (London)* **302**, 251–253.

Lundberg, J.M. and Stjärne, L. (1984) Neuropeptide Y (NPY) depresses the secretion of ^{3}H-noradrenaline and the contractile response evoked by field stimulation in rat vas deferens *Acta Physiol. Scand.* **120**, 477–479.

Lundberg, J.M., Hökfelt, T., Schultzberg, M., Uvnas-Wallenstein, K., Kohler, C. and Said, S.I. (1979)

Occurrence of vasoactive intestinal polypeptide (VIP)-like immunoreactivity in certain cholinergic neurons of the cat: Evidence from combined immunocytochemistry and acetylcholinesterase staining. *Neuroscience* **4**, 1539–1559.

Lundberg, J.M., Saria, A., Brodin, E., Rosell, S. and Folkers, K.A. (1983) A substance P antagonist inhibits vagally induced increase in vascular permeability and bronchial smooth muscle contraction in the guinea pig. *Proc. Nat. Acad. Sci. USA* **80**, 1120–1124.

Lundberg, J.M., Brodin, E., Hua, X. and Saria, A. (1984) Vascular permeability changes and smooth muscle contraction in relation to capsaicin-sensitive substance P afferents in the guinea-pig. *Acta Physiol. Scand.* **120**, 217–227.

Macrae, I.M., Furness, J.B. and Costa, M. (1986) Distribution of subgroups of norarenaline neurons in the coeliac ganglion of the guinea-pig. *Cell Tissue Res.* **244**, 173–180.

Maggi, C.A. and Meli, A. (1988) The sensory-efferent function of capsaicin-sensitive sensory neurons. *Gen. Pharmacol.* **19**, 1–43.

Matter, A. (1973) A morphometric study on the nexus of cardiac muscle. *J. Cell Biol.* **56**, 690–696.

Matthew, W.D., Tsavaler, L. and Reichardt, L.F. (1981) Identification of a synaptic vesicle-specific membrane protein with a wide distribution in neuronal and neurosecretory tissue. *J. Cell Biol.* **91**, 257–269.

McDonald, D.M. (1988) Neurogenic inflammation in the rat trachea. I. Changes in venules, leukocytes, and epithelial cells. *J. Neurocytol.* **17**, 583–603.

Melander, A. and Sundler, F. (1979) Presence and influence of cholinergic nerves in the mouse thyroid. *Endocrinology* **105**, 7–9.

Melander, A., Ericson, L.E. and Sundler, F. (1974) Sympathetic regulation of thyroid hormone secretion. *Life Sci.* **14**, 237–246.

Miller, R.E. (1981) Pancreatic neuroendocrinology: peripheral neural mechanisms in the regulation of the islets of Langerhans. *Endocr. Rev.* **2**, 471–494.

Navone, F., Greengard, P. and De Camilli, P. (1984) Synapsin I in nerve terminals: selective association with small synaptic vesicles. *Science* **226**, 1209–1211.

Navone, F., Jahn, R., DiGioia, G., Stukenbrok, H., Greengard, P. and De Camilli, P. (1986) Protein p38: an integral membrane protein specific for small vesicles of neurons and neuroendocrine cells. *J. Cell Biol.* **103**, 2511–2527.

Neuman, B., Wiederman, C.J., Fischer-Colbrie, R., Schober, M., Sperk, G. and Winkler, H. (1984) Biochemical and functional properties of large and small dense-core vesicles in sympathetic nerves of rat amd ox vas deferens. *Neuroscience* **13**, 921–931.

Pelletier, G., Steinbusch, H.W.M. and Verhofstad, A.A.J. (1981) Immunoreactive substance P and serotonin present in the same densecore vesicles. *Nature (London)* **293**, 71–72.

Persson, C.G.A., Erjefalt, I. and Andersson, P. (1986) Leakage of macromolecules from guinea-pig tracheo-bronchial microcirculation. Effects of allergen, leukotrienes, tachykinins, and anti-asthma drugs. *Acta Physiol. Scand.* **127**, 95–105.

Petrucci, T.C. and Morrow, J.S. (1987) Synapsin I: an actin-bundling protein under phosphorylation control. *J. Cell. Biol.* **105**, 1355–1363.

Purves, D. and Lichtman, J.W. (1985) Geometrical differences among homologous neurons in mammals. *Science* **228**, 298–302.

Santini, M. and Ibata, Y. (1971) The fine structure of thin unmyelinated axons within muscle spindles. *Brain Res.* **33**, 289–302.

Schwarzenbrunner, U., Schmidle, T., Obendort, D., Scherman, D., Hook, V., Fischer-Colbrie, R. and Winckler, H. (1990) Sympathetic axons and nerve terminals: the protein composition of small and large dense-core and of a third type of vesicles. *Neuroscience* **37**, 819–827.

Sittiracha, T., McLachlan, E.M. and Bell, C. (1987) The innervation of the caudal artery of the rat. *Neuroscience* **21**, 647–659.

Slavin, B.G. and Ballard, K.W. (1978) Morphological studies on the adrenergic innervation of white adipose tissue. *Anat. Rec.* **191**, 377–390.

Smith, R.E. and Horwitz, B.A. (1969) Brown fat and thermogenesis. *Physiol. Rev.* **49**, 330–425.

Stevens, L.M. and Landis, S.C. (1987) Development and properties of the secretory response in rat sweat gland: Relationship to the induction of cholinergic function in sweat gland innervation. *Develop. Biol.* **123**, 179–192.

Stjärne, L. (1989) Basic mechanisms and local modulation of nerve impulse-induced secretion of neurotransmitters from individual sympathetic nerve varicosities. *Rev. Physiol. Biochem. Pharmacol.* **112**, 1–138.

Stjärne, L., Msghina, M. and Stjärne, E. (1990) 'Upstream' regulation of the release probability in sympathetic nerve varicosities. *Neuroscience* **36**, 571–587.

Südhof, T.C., Lottspeich, F., Greengard, P., Mehl, E. and Jahn, R. (1987) A synaptic vesicle protein with a novel cytoplasmic domain and four transmembrane regions. *Science* **238**, 1144–1146.

Taxi, J. (1965) Contribution a l'étude des connexions des neurones moteurs du système nerveux autonome. *Ann. Sci. Nat. Zool. (Paris)* **7**, 413–674.

Teitelman, G., Joh, T.H. and Reis, D.S. (1978) Transient expression of a noradrenergic phenotype in cells of the rat embryonic gut. *Brain Res.* **158**, 229–234.

Teitelman, G., Gershon, M.D., Rothman, T.P. Joh, T.H. and Reis, D.J. (1981) Proliferation and distribution of cells that transiently express a catecholaminergic phenotype during development in mice and rats. *Develop. Biol.* **86**, 348–355.

Tranzer, J.P. and Thoenen, H. (1967) Significance of 'empty vesicles' in postganglionic sympathetic nerve terminals. *Experientia* **23**, 123–124.

Tranzer, J.P. and Thoenen, H. (1969) Ultramorphologische Varänderungen der sympatischen Nervenendigungen der Katze nach Vorbehandlung mit 5- und 6-Hydroxy-Dopamin. *Naunyn-Schmiedebergs Arch. Pharmac.* **257**, 343–344.

Trifaró, J.-M., Fournier, S. and Novas, M.L. (1989) The p65 protein is a calmodulin-binding protein present in several types of secretory vesicles. *Neuroscience* **29**, 1–8.

Ueda, T. and Greengard, P. (1977) Adenosine 3′, 5′ monophsphate-regulated phosphoprotein system in neuronal membranes. I. Solubilization, purification, and some properties of an endogenous phosphoprotein. *J. Biol. Chem.* **252**, 5155–5163.

Uehara, Y. and Burnstock, G. (1972) Postsynaptic specialization of smooth muscle at close neuromuscular junctions in the guinea-pig sphincter pupillae. *J. Cell Biol.* **53**, 849–853.

Volknandt, W., Schober, M., Fischer-Colbrie, R., Zimmerman, H. and Winkler, H. (1987) Cholinergic nerve terminals in the rat diaphragm are chromogranin A immunoreactive. *Neurosci. Lett.* **81**, 241–244.

Wiedenmann, B. and Franke, W.W. (1985) Identification and localization of synaptophysin, an integral membrane glycoprotein of M 38,000 characteristic of presynaptic vesicles. *Cell* **41**, 1017–1028.

Young, W.S. Jr. and Kuhar, M.J. (1979) Autoradiographic localization of benzodiazepine receptors in brains of humans and animals. *Nature (Lond.)* **280**, 393–395.

2 Co-Transmission and Neuromodulation

J.L. Morris and I.L. Gibbins

Department of Anatomy & Histology, and Centre for Neuroscience, School of Medicine, Flinders University of South Australia

The co-existence of transmitters is a widespread phenomenon in the peripheral autonomic nervous system. In this chapter, we review the evidence for functional co-transmission from autonomic neurons to a wide variety of effector tissues, including the heart and blood vessels, the vas deferens and urinary bladder, gastro-intestinal smooth muscle, and salivary glands. We also examine the involvement of neuropeptides in ganglionic co-transmission. Commonly, the responses to autonomic nerve stimulation consist of several phases with distinctive time courses, each of which is mediated by a different co-transmitter. Furthermore, co-transmitters usually interact at many different levels, ranging from the modulation of transmitter release to the regulation of calcium concentrations in the effector cells. Thus, co-transmission represents a fundamental mechanism employed by autonomic neurons to achieve efficient and precise control of their target tissues over a broad spectrum of functional demands.

KEY WORDS co-existence; co-transmitters; smooth muscle; cardiac muscle; secretion; ganglionic transmission

INTRODUCTION

In the last ten years, we have witnessed a revolution in the way we must think about neurotransmission in the autonomic nervous system. Until the late 1960s, transmission from autonomic neurons to their peripheral effector tissues generally had been thought to be mediated either by a catecholamine or by acetylcholine (ACh). However, evidence that had been steadily accumulating since the end of the last century finally became accepted as indicating that some autonomic neurons utilize a transmitter that is neither noradrenaline (NA) nor ACh (see Burnstock, 1969; Campbell, 1970). During the 1980s, an avalanche of new information, generated by a combination of powerful new biochemical, histochemical and pharmacological techniques, has demonstrated unequivocally that there is not just one 'non-adrenergic, non-cholinergic' transmitter, but that there are perhaps dozens of potential candidates. They range from substances with other well-known functions, such as adenosine $5'$ triphosphate (ATP), to a bewildering variety of biologically active neuropeptides. Nevertheless, there are still many instances of autonomic neuroeffector transmission in which the true transmitters have not been positively identified.

As exciting as all these discoveries have been, the most astounding realization has been that individual autonomic neurons may contain and release several different neurotransmitters (see Burnstock, 1976; Hökfelt *et al.*, 1977). For example, a single sympathetic neuron may alter the activity of its effector tissues via the actions of a purine, a catecholamine, and a peptide, all of which contribute to a different aspects of the transmission process. Equally surprising have been observations that individual autonomic neurons may contain different transmitters, especially neuropeptides, with diverse and sometimes apparently opposing effects on the target cells.

For the following discussion, we have defined a neurotransmitter as any substance that is released from a nerve terminal when invaded by an action potential, and that, once released, influences the excitability or metabolic state of a target cell (which may include the neuron from which it was released). Co-transmitters, by this definition, are simply two or more transmitters that can be released from the same nerve terminal. These definitions do not make an explicit distinction between so-called 'classical' transmitters, 'false' transmitters and 'neuromodulators'. Furthermore, such a definition allows trophic actions for transmitters that do not have acute effects on their target cells (see Furness *et al.*, 1989a).

In this chapter, we will discuss many different examples of co-transmission in the autonomic nervous system. Each example has been chosen to illustrate a particular aspect of the extraordinary range of options that, in principle, could exist, and that have been discussed in detail elsewhere (e.g. Changeux, 1986; Hökfelt *et al.*, 1986; Campbell, 1987; Furness *et al.*, 1989a; Gibbins, 1989a). We will not be attempting to describe in detail the complete autonomic control of the organs and tissues concerned, as these matters will be discussed in subsequent volumes of this series. However, in cases where the identity of some of the co-transmitters is unknown, or is at least controversial, we will present some of the evidence for and against the involvement of any particular transmitter candidate.

We hope to convince the reader that co-transmission is widespread, and, indeed, may be the norm in the peripheral autonomic nervous system. Furthermore, we shall provide evidence that the details of the co-transmission process vary in sometimes subtle and complex ways from tissue to tissue, and from species to species. From this information, we will begin to build up a picture of how the autonomic nervous system is able to provide highly tuned and efficient regulation of a diverse range of effector functions.

CO-EXISTENCE OF POTENTIAL NEUROTRANSMITTERS IN PERIPHERAL AUTONOMIC NEURONS

The differential distributions of the transmitters NA and ACh within autonomic neurons has been appreciated for many years, and, indeed, underpinned the earliest attempts at a functional classification of the autonomic nervous system. Some autonomic neurons may share the biochemical characteristics of both catecholamine-synthesizing neurons and cholinergic neurons, especially during their development (Landis, 1988). However, there are no well documented examples of NA and ACh acting as co-transmitters *in vivo*. More recent biochemical studies have indicated that autonomic neurons containing NA or ACh contain additional potential neuro-transmitters, including purines, amino acids or neuropeptides. The widespread use of immunohistochemical techniques has been critical for the demonstration of the co-existence of potential co-transmitters in the same neurons. Such techniques have been particularly useful for demonstrating the co-existence of neuropeptides. We will set the scene for our detailed discussion of the pharmacology of co-transmission

TABLE 2.1
Co-existence of neuropeptides in sympathetic neurons.

Peptide combination	CA/ACh*	Target	Species	Ref.
NPY/Dyn A(1–17)/Dyn A(1–8)**NA		iris, vas deferens	guinea-pig	Gibbins & Morris, 1987
NPY/Dyn A(1–8)	NA	small blood vessels	guinea-pig	Gibbins & Morris, 1990
Dyn A(1–8)/Dyn A(1–17)	NA	piloerector muscles	guinea-pig	Gibbins, 1989b
NPY/Enk	NA	spleen	cattle	Fried et al., 1986
NPY	NA	large blood vessels	guinea-pig, mouse, rat	Morris et al., 1986c Gibbins et al., 1988
GAL	NA	blood vessels	possum, gekko	Gibbins, unpub. obs.
NPY/GAL	NA	blood vessels	cat, dog, goanna	Kummer, 1987 Gibbins, unpub. obs.
NPY/GAL	AD	blood vessels	toad	Morris et al., 1989
SOM	NA	submucous ganglia	guinea-pig	Costa & Furness, 1984
VIP	ACh	blood vessels	cat	Lindh et al., 1989
VIP/NPY/Dyn	(ACh)	blood vessels	guinea-pig	Gibbins, unpub. obs.
VIP/CGRP	ACh	sweat glands	rat	Landis et al., 1988
VIP/CGRP/SP	ACh	sweat glands	cat, dog,	Lindh et al., 1989
no known peptides	NA	salivary glands	guinea-pig, cat, rat, mouse	Gibbins, 1990 Lundberg et al., 1988b Gibbins, 1991
	NA	piloerector muscles	rat human	Schotzinger and Landis 1990 Gibbins, unpub. obs.
	NA	myenteric ganglia	guinea-pig	Costa & Furness, 1984
	NA	blood vessels	skinks	Gibbins, unpub. obs.

* Neurons contain a catecholamine (CA, either adrenaline, AD, or noradrenaline, NA) or acetylcholine (ACh).
** In guinea-pigs, all sympathetic neurons containing Dyn A(1–8) and/or Dyn A(1–17) also contain Dyn B and α-neo-endorphin.

by a brief description of recent work demonstrating the patterns of neuropeptide co-existence. These results provide the best illustration of the potential complexity and diversity of neurotransmission in the autonomic nervous system.

Most autonomic neurons contain one or more neuropeptides, usually in addition to a non-peptide transmitter. Thus, various combinations of neuropeptides occur within sympathetic neurons (Table 2.1), parasympathetic neurons (Table 2.2), pelvic autonomic neurons (Table 2.3) and enteric neurons (Furness & Costa, 1987). They also occur in many populations of preganglionic neurons (Table 2.7). The number of biochemically unrelated peptides found in various populations of autonomic neurons varies considerably, and can be surprisingly high. For example, non-noradrenergic neurons containing at least four different peptides occur in cranial and pelvic autonomic ganglia of guinea-pigs (Figures 2.1, 2.2; Morris & Gibbins, 1987; Gibbins, 1990). The most remarkable example of the multiple co-existence

TABLE 2.2
Co-existence of neuropeptides in cranial parasympathetic neurons of guinea-pigs.

Peptide combination	Target	Source ganglion	Function
SP	iris	ciliary	constrict pupil
VIP	blood vessels	sphenopalatine, otic	vasodilatation
NPY/VIP	ciliary bodies	sphenopalatine	?vasodilatation
NPY/Enk	submandibular gland	local micro-ganglia	secretion
NPY/VIP/Enk	hair follicles on side of face	sphenopalatine	?
NPY/VIP/SP	hair follicles on eyelids, lips	sphenopalatine	?
	sublingual gland	local micro-ganglia	secretion
NPY/VIP/Enk/SP	hair follicles	sphenopalatine	?
	lacrimal & zygomatic glands		secretion
	parotid gland	otic, intra-cranial	secretion

References: Gibbins and Morris, 1987, 1988; Gibbins, 1990.

TABLE 2.3
Co-existence of neuropeptides in non-noradrenergic neurons innervating some pelvic viscera of guinea-pigs.

Peptide combination	Target tissue	Probable function
NPY/VIP/±SOM/±Dyn	blood vessels	vasodilatation
	circular muscle of vas deferens	relaxation
NPY/SOM/SP	urinary bladder	contraction
NPY/SOM/VIP	urinary bladder	relaxation?

References: Morris et al., 1985, 1987; Morris and Gibbins, 1987; Gibbins et al. 1987b; Gibbins, unpub. obs.

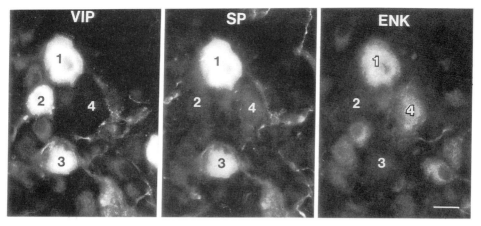

FIGURE 2.1   Co-localization of neuropeptides in nerve cell bodies located in guinea-pig sphenopalatine ganglion. Section of ganglion triple-labelled for simultaneous demonstration of immunoreactivity (IR) to vasoactive intestinal peptide (VIP), substance P (SP) and enkephalin (ENK). Some neurons contain all three peptides (cell 1). These neurons also contain NPY. Other neurons have VIP-IR alone (cell 2), VIP-IR and SP-IR (cell 3), or ENK-IR alone (cell 4). For details, see Gibbins, 1990. Scale bar = 20 μm.

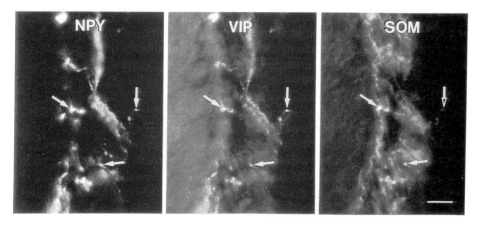

FIGURE 2.2   Co-localization of neuropeptides in vasodilator neurons projecting from paracervical ganglia to the main uterine artery of guinea-pigs. Section of uterine artery triple-labelled for stimultaneous demonstration of immunoreactivity (IR) to neuropeptide Y (NPY), vasoactive intestinal peptide (VIP) and somatostatin (SOM). Many axons have IR to all three peptides (white arrows). These axons also contain dynorphin A–IR. Some axons have IR to NPY and VIP, but not SOM (open arrows). For details, see Morris et al., 1985; 1987, Scale bar = 20 μm.

of peptides documented to date is a population of cholinergic secretomotor neurons in the small intestine of guinea-pigs: some of these cells contain neuropeptide Y (NPY), somatostatin (SOM), cholecystokinin (CCK), calcitonin gene-related peptide (CGRP), galanin (GAL) and neuromedin U in addition to ACh (Furness et al., 1985, 1989b). The list of co-existing peptides within a population of neurons becomes even greater when the various post-translational products of neuropeptide-coding genes are considered. For example, most neurons with vasoactive intestinal peptide (VIP) also contain another peptide, peptide histidine isoleucine (PHI), which is derived from the same gene (Tatemoto and Mutt, 1981; Lundberg *et al.*, 1984b).

Detailed and systematic studies in several species have demonstrated that there is a surprisingly high degree of order in the patterns of co-existence of neuropeptides within different populations of autonomic neurons. These observations have led to the development of the concept of 'chemical coding' of autonomic neurons. The central tenet of 'chemical coding' is that neuropeptides co-exist in pathway-specific combinations within individual neurons (Costa *et al.*, 1986a; Furness *et al.*, 1989a; Gibbins, 1989a). In other words, the precise combination of peptides found within a neuron is highly correlated with its peripheral projection. Thus, it is often possible to identify specific functional groups of post-ganglionic neurons on the basis of their peptide content alone. For example, in the sympathetic ganglia of guinea-pigs, pilomotor, vasoconstrictor, vasodilator and different classes of secretomotor neurons all can be distinguished from each other by their peptide content (Figure 2.3; Gibbins, 1989b, 1990a; Gibbins & Morris, 1987, 1990; Morris & Gibbins, 1989).

There are many functional consequences of the chemical coding of autonomic neurons (Costa *et al.*, 1986; Gibbins, 1989a; Furness *et al.*, 1989a). Thus, the same neuropeptide may occur in different populations of neurons that innervate the

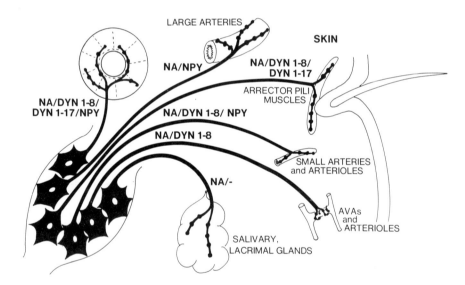

FIGURE 2.3 Chemical coding of sympathetic neurons projecting from the superior cervical ganglion to various targets in the head of guinea-pigs. Each population of neurons has a specific combination of neuropeptides. Note that all neurons containing a form of dynorphin A (DYN 1–8 or DYN 1–17) also contain dynorphin B and α-neo-endorphin. No neuropeptides have been found in neurons projecting to secretory tissue in the salivary or lacrimal glands. Neurons with similar peptide combinations also occur in other paravertebral ganglia of guinea-pigs, except that the salivary secretomotor neurons are absent. Conversely, the paravertebral ganglia have many non-noradrenergic vasodilator neurons containing prodynorphin-derived peptides, VIP and NPY. AVAs, arterio-venous anastomoses; NA, noradrenaline; NPY, neuropeptide Y. For original data see Gibbins and Morris, 1987; Gibbins, 1989b; Gibbins and Morris, 1990; Gibbins, 1990.

same target tissue and that have opposing actions on that tissue. On the other hand, a particular population of neurons may contain different neuropeptides that have opposing actions on the target tissue. Finally, neurons with apparently similar functions may contain different combinations of peptides (Tables 2.1, 2.2). The last point becomes even more apparent when homologous neurons in different species are examined: it is often almost impossible to predict correctly the precise combination of peptides that will occur in a given population of neurons that has not been examined previously (Gibbins, 1989a). The discovery of the factors that lead to the differential expression of neuropeptides in the autonomic nervous system remains a major challenge for the future.

CO-TRANSMISSION TO SMOOTH MUSCLE AND CARDIAC MUSCLE

SYMPATHETIC VASOCONSTRICTION

Most arteries and arterioles, and many veins, receive vasoconstrictor innervation from postganglionic autonomic neurons whose cell bodies are located in the prevertebral or paravertebral sympathetic ganglia. Studies *in vitro* and *in situ* have demonstrated that transmission from sympathetic vasoconstrictor neurons to vascular smooth muscle can involve up to three phases with distinct temporal and pharmacological properties. The relative contribution of each phase of neurotransmission to the sympathetic vasoconstrictor response varies with the particular blood vessel and species examined, with the parameters of electrical stimulation of the sympathetic neurons, and with the method of experimental analysis (see Bolton and Large, 1986; Burnstock and Kennedy, 1986; Hirst and Edwards, 1989).

First phase of vasoconstriction

Electrophysiological studies have demonstrated that, in many blood vessels, transmural electrical stimulation with single pulses, or with short trains of pulses, produces a rapid depolarization of smooth muscle membranes. This response peaks within 30–100 ms and lasts 200–1500 ms. With increasing strengths of stimulation, this electrical event, known as the fast excitatory junction potential (EJP), can generate an action potential and contraction of the smooth muscle (see Hirst and Edwards, 1989).

Invariably, fast EJPs are greatly reduced or abolished by treatment with drugs such as guanethidine or bretylium, which prevent transmitter release from noradrenergic nerve terminals (Bell, 1969; Cheung, 1982; Suzuki and Kou, 1983; Åstrand and Stjärne, 1989). They are also abolished by 6-hydroxydopamine (6-OHDA), which depletes noradrenergic nerve terminals of neurotransmitter and ultimately leads to their degeneration (Cheung, 1982). These observations demonstrate that the EJP is mediated by a transmitter released from noradrenergic neurons. Nevertheless, the EJP is not blocked by α-adrenoceptor antagonists (Neild and Zelcer, 1982; Table 2.4; Figure 2.4).

TABLE 2.4
Pharmacological characteristics of fast EJPs and fast contractions elicited in vascular smooth muscle by sympathetic nerve stimulation.

Blood Vessel	Electrical recording	Mechanical recording	Blocked by	Not blocked by	Reference
Rat basilar artery	EJP			Phentolamine Prazosin	Hirst et al., 1982
Dog basilar artery	EJP		Guanethidine Reduced by ATP, NA	Prazosin Enhanced by Phentolamine Yohimbine	Fujiwara et al., 1982
Rabbit ear artery	EJP			Labetolol Phentolamine Phenoxybenzamine Prazosin Yohimbine	Holman and Surprenant, 1980 Suzuki and Kou, 1983
		Fast, prazosin-resistant contraction	Guanethidine 6-OHDA α, β-Me ATP	Reserpine	Kennedy et al., 1986; Saville and Burnstock, 1988
	EJP		α, β-Me ATP Reduced by short exposure to ATP	Enhanced by long exposure to ATP	Miyahara and Suzuki, 1987; Muir and Wardle, 1988
Guinea-pig mesenteric arterioles	EJP			Phentolamine Tolazoline Prazosin Labetolol WB 4101 Indoramin	Hirst and Neild, 1980; Neild and Zelcer, 1982
Guinea-pig mesenteric artery	EJP	Fast contraction	Reduced by Clonidine Yohimbine	Prazosin Phentolamine	Kuriyama and Makita, 1983
	EJP		α, β-Me ATP ATP	Prazosin Benextramine	Nagao and Suzuki, 1988; Ishikawa, 1985; Hottenstein and Kreulen, 1987; Hirst and Jobling, 1989
Rabbit mesenteric artery	EJP		α, β-Me ATP ATP	Prazosin Enhanced by Phentolamine Phenoxybenzamine Yohimbine	Mishima et al., 1984; Ishikawa, 1985
		Phentolamine-resistant contraction	α, β-Me ATP	Reserpine Prazosin Phenoxybenzamine	von Kügelgen and Starke, 1985
Rabbit jejunal arteries	EJP	Fast	α, β-Me ATP	Prazosin	Ramme et al., 1987

TABLE 2.4 *Continued*

Blood Vessel	Electrical recording	Mechanical recording	Blocked by	Not blocked by	Reference
Rat mesenteric arteries	EJP	Fast contraction	α, β-Me ATP	Prazosin	Angus *et al.*, 1988; Yamamoto *et al.*, 1988
Dog mesenteric artery		Prazosin-resistant contraction	6-OHDA Guanethidine α, β-Me ATP	Reserpine Phentolamine Phenoxybenzamine Tolazoline DG-5128	Muramatsu *et al.*, 1984; Muramatsu, 1987
Dog mesenteric vein	EJP		Slightly reduced by yohimbine	Prazosin Phentolamine	Kou *et al.*, 1984
Cat intestinal circulation		Fast, prazosin-resistant contraction	α, β-Me ATP	Yohimbine	Taylor and Parsons, 1989
Rat renal vasculature		Contraction to low frequency stimulation	α, β-Me ATP	Prazosin Corynanthine Phentolamine	Schwartz and Malik, 1989
Guinea-pig saphenous artery	EJP		Reduced by ANAPP₃	Prazosin Yohimbine	Cheung and Fujioka, 1986; Fujioka and Cheung, 1987
Rabbit saphenous artery		Fast contraction	Guanethidine 6-OHDA α, β-Me ATP	Prazosin	Burnstock and Warland, 1987; Warland and Burnstock, 1987
Rat saphenous vein	EJP			Prazosin Yohimbine	Cheung, 1985
Guinea-pig uterine artery	EJP		Bretylium		Bell, 1969
Rat tail artery	EJP			Labetolol Phenoxybenzamine Phentolamine Prazosin	Holman and Surprenant, 1980
	EJP		α, β-Me ATP	Phentolamine Prazosin Yohimbine	Sneddon and Burnstock, 1985; Neild and Kotecha, 1986; Ono and Suzuki, 1988
	EJP	Fast contraction		Prazosin	Cheung, 1984

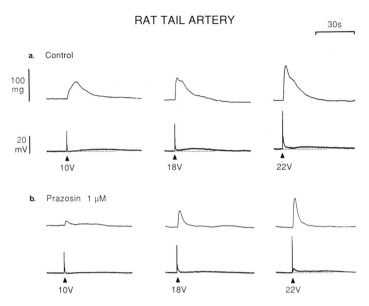

RAT TAIL ARTERY

FIGURE 2.4   Mechanical and electrical responses of a ring segment rat tail artery to transmural stimulation with single current pulses of 0.1 ms duration and increasing voltage (10, 18, 22V), applied 1 min apart. Resting membrane potential for all traces was between −55 and −60 mV. a. In control arteries the force transient (upper trace) is clearly biphasic. The intracellular electrical recordings (lower trace) show a fast active membrane response which is triggered by the EJP, followed by a slow depolarization. Note that the slower contraction occurs before the slow depolarization. b. After treatment with 1 μM prazosin, the slower phase of the force response, and the slow depolarization, are reduced selectively. Traces kindly provided by Dr. N. Kotecha and Dr. T.O. Neild.

When the mechanical responses of isolated blood vessels to sympathetic nerve stimulation have been examined carefully, a rapid constriction, resistant to α-adrenoceptor blockade, has been observed in many cases (Table 2.4). The fast component is more prominent after stimulation with single pulses or short bursts of low frequency pulses than after continuous, high frequency, electrical stimulation (Kennedy et al., 1986; Taylor and Parsons, 1989). Sometimes, this phase of constriction is not detected without blockade of presynaptic α_2-adrenoceptors (Yamamoto et al., 1988).

Second phase of vasoconstriction

In some arteries, notably those supplying thermoregulatory vascular beds, single current pulses can elicit a slow depolarization in addition to an EJP (Figure 2.4; Hirst and Edwards, 1989). However, in most other arteries, repetitive stimulation of sympathetic nerves is required to produce a slow depolarization following the EJP (Table 2.5). A slow depolarization lasting 20–40 s can occur without an EJP (see Bolton and Large, 1986). A slow contraction of the smooth muscle may be associated with the slow depolarization. When present, both the electrical and the mechanical responses are blocked by α-adrenoceptor antagonists (Table 2.5; Figure 2.5a). The

TABLE 2.5

Pharmacological characteristics of slow depolarizations and slow contractions elicited in vascular smooth muscle by sympathetic nerve stimulation.

Blood Vessel	Electrical recording	Mechanical recording	Blocked by	Not blocked by	Reference
Rabbit basilar artery		First phase of contraction	Phenoxybenz.		Lee et al., 1976, 1980
Rabbit ear main artery		Contraction	Reduced by Phentolamine Phenoxybenz.	Prazosin Labetolol	Holman and Surprenant, 1980
	Slow depol.	Contraction to high frequency stimulation	Reduced by Phentolamine Prazosin	α, β-Me ATP	Suzuki and Kou, 1983; Muir and Wardle, 1988
		Contraction	Phentolamine Reduced by Prazosin		Owen et al., 1985; Kennedy et al., 1986; Saville and Burnstock, 1988
Rabbit ear small artery		Contraction	Reduced by Phentolamine		Owen et al., 1985
Pig nasal mucosa		Constriction to single pulses	Phenoxybenz.		Lacroix et al., 1988a
		Constrictions to trains of pulses	Slightly reduced by Phenoxybenz.		Lacroix et al., 1988a
Rat aorta		Contraction	Phentolamine		Nilsson, 1984
Guinea-pig aorta		Contraction	Phentolamine Prazosin		Morris, 1991 Figure 2.5a
Dog and cat pulmonary artery		Contraction	Prazosin	Yohimbine	Constantine et al., 1980 Hyman and Kadowitz, 1989
Guinea-pig pulmonary artery	Fast and slow depol.		Phentolamine		Suzuki, 1983
	Slow depol.		Reduced by Benextramine Phentolamine		Petrou et al., 1989
Rabbit mesenteric artery		Constriction	Reduced by Phentolamine Prazosin		von Kügelgen and Starke, 1985; Muir and Wardle, 1988
Rat mesenteric arteries	Slow depol. to high frequency stimulation	Slow contraction	Prazosin	α, β-Me ATP	Angus et al., 1988 Yamamoto et al., 1988

TABLE 2.5 *Continued*

Blood Vessel	Electrical recording	Mechanical recording	Blocked by	Not blocked by	Reference
Rat superior mesenteric artery		Contraction	Phentolamine		Nilsson, 1984
Dog mesenteric artery		Contraction	Reduced by Prazosin		Muramatsu *et al.*, 1984 Muramatsu, 1987
Dog mesenteric vein	Slow depol.	Contraction	Phentolamine Yohimbine	Prazosin	Kou *et al.*, 1984; Suzuki, 1984
Guinea-pig mesenteric arteries	Slow depol.	Contraction	Greatly reduced by Prazosin Phentolamine		Hottenstein and Kreulen, 1987
Guinea-pig mesenteric vein	Fast and slow depol.	Contraction	Prazosin		Van Helden, 1988
Cat intestinal circulation		Slower contraction	Reduced by Prazosin and Yohimbine		Taylor and Parsons, 1989
Rat and rabbit renal vasculature		Contraction to high frequency stimulation	Reduced by Prazosin		Schwartz and Malik, 1989
Pig renal vasculature		Contraction to low frequency stimulation	Phenoxybenz.		Pernow and Lundberg, 1989a
Cat and pig spleen		Constriction	Reduced by Phenoxybenz. Phentolamine		Lundberg *et al.*, 1984a, 1989a
Guinea-pig uterine artery	Slow	Constriction	Bretylium		Bell, 1968, 1969
		First phase of contraction	Guanethidine Phentolamine Prazosin		Morris and Murphy, 1988
Guinea-pig saphenous artery	Slow depol.	Contraction	Prazosin	ANAPP$_3$	Fujioka and Cheung, 1987
Rabbit saphenous artery		Contraction	Reduced by Phentolamine	Prazosin Labetolol	Holman and Surprenant, 1980
		Slow contraction to sustained stimulation	Reduced by Prazosin		Burnstock and Warland, 1987
Rat saphenous vein	Slow depol.		Yohimbine	Prazosin	Cheung, 1985

TABLE 2.5 *Continued*

Blood Vessel	Electrical recording	Mechanical recording	Blocked by	Not blocked by	Reference
Rat tail artery		Contraction		Phentolamine Prazosin Labetolol	Holman and Surprenant, 1980
	Slow depol.		Phentolamine	α, β-Me ATP	Cheung, 1982; Sneddon and Burnstock, 1985
	Slow depol.	Contraction	Prazosin Phentolamine Yohimbine	α, β-Me ATP	Cheung *et al.*, 1984; Ono and Suzuki, 1988

depolarization and contraction usually are more prominent with increasing train length, or frequency, of electrical stimulation (Kennedy *et al.*, 1986; Burnstock and Warland, 1987; Schwartz and Malik, 1989). Despite the apparent association between the slow depolarization and the slow contraction, the contraction can precede the depolarization (Figure 2.4). Therefore, membrane depolarization is not necessarily a prerequisite for contraction of these vascular smooth muscles (Bolton and Large, 1986; Hirst and Edwards, 1989).

Notable exceptions to this pattern of responses occur in the main pulmonary artery and in small mesenteric veins of guinea-pigs (Suzuki, 1983; Van Helden, 1988). In these vessels, low levels of sympathetic nerve stimulation produce a slow depolarization, whereas higher frequencies of nerve stimulation produce a faster phase of depolarization lasting 2–3 s, which is essentially a slow EJP. The faster electrical event is associated with constriction.

Third phase of vasoconstriction

Slow vasoconstrictions which are resistant to α-adrenoceptor blockade have been reported in a variety of vascular beds after maintained sympathetic nerve stimulation (Table 2.6). These constrictions are slow in onset and time to reach peak amplitude (up to 2 min), and can last many minutes after cessation of stimulation. They increase in prominence with increasing train length and frequency of stimulation. The slow constrictions always are abolished by guanethidine (Figure 2.5; Lee *et al.*, 1976; Lundberg *et al.*, 1984a; Morris and Murphy, 1988), and after degeneration of noradrenergic neurons following 6-hydroxydopamine treatment (Morris *et al.*, 1986c). Very slow constrictions may occur without phases 1 or 2 of constriction being apparent (Figure 2.5c; Morris, 1991); phase 3 may be present as an event clearly separable in time from phase 2 (Figure 2.5b; Lee *et al.*, 1976; Morris and Murphy, 1988); or phase 3 may only be revealed after pharmacological blockade of the other phases (Lee *et al.*, 1980; Lundberg and Tatemoto, 1982a; Lundberg *et al.*, 1984a; Lacroix *et al.*, 1988a; Pernow and Lundberg, 1989a). It is not yet known whether there is an electrical correlate for this very slow vaso constriction, as the stimulation parameters used to elicit the responses are not routinely used in electrophysiological studies.

TABLE 2.6

Pharmacological characteristics of very slow contractions elicited in vascular smooth muscle by sympathetic nerve stimulation.

Blood vessel	Recording	Blocked by	Not blocked by	Reference
Rabbit basilar artery	Contraction to trains of pulses	Reserpine Guanethidine Bretylium	Enhanced by Phenoxybenzamine Phentolamine Tolazoline	Lee *et al.*, 1976, 1980
Rabbit uveal vasculature	Constriction to high frequency stimulation		Largely resistant to Phenoxybenazamine	Granstam and Nilsson, 1990
Rabbit central ear artery	Contractions to continuous stimulation	Guanethidine	Prazosin α, β-Me ATP	Kennedy *et al.*, 1986
Cat and pig nasal mucosa	Constriction to high frequency stimulation	Guanethidine	Largely resistant to Phenoxybenz. Reserpine	Lacroix *et al.*, 1988a Lundblad *et al.*, 1987
Cat oral mucosa and dental pulp vasculature	Constriction	Guanethidine	Partly resistant to Phentolamine	Edwall *et al.*, 1985
Cat submandibular gland vasculature	Constrictions	Guanethidine	Phenoxybenzamine	Lundberg and Tatemoto, 1982a
Guinea-pig thoracic vena cava	Contraction	Guanethidine NPY	Enhanced by Phentolamine Yohimbine Phenoxybenzamine	Morris, 1991 Figure 2.6
Cat spleen	Constrictions to long pulse trains	Guanethidine	Phenoxybenzamine	Lundberg *et al.*, 1984a
Pig spleen	Constriction after reserpine and decentralization	Guanethidine	Reduced slightly by m-ATP	Lundberg *et al.*, 1989a
Pig kidney	Vasoconstriction to high frequency stimulation		Phenoxybenzamine Reserpine and decentralization	Pernow and Lundberg, 1989a
Cat colonic vasculature	Constriction after α- and β- blockade	Guanethidine		Hellström *et al.*, 1985
Guinea-pig uterine artery	Slow phase of contraction to high frequency stimulation	Guanethidine Trypsin NPY	Phentolamine Prazosin	Morris and Murphy, 1988
Dog gracilis muscle vasculature	Vasoconstriction after reserpine and decentralization		Phenoxybenzamine	Pernow *et al.*, 1988
Rabbit tenuissimus muscle vasculature	Vasoconstriction to high frequency stimulation	Guanethidine NPY	Partly reduced by Phentolamine	Öhlén *et al.*, 1990

Not all vessels have been examined pharmacologically after intense sympathetic stimulation, so in many cases the presence or absence of the very slow constriction has not been established. Furthermore, as transmural stimulation of vessels *in vitro* can activate vasodilator neurons in addition to vasoconstrictor neurons, a very slow vasoconstriction may be masked by a simultaneous slow vasodilatation (Morris and Murphy, 1988). In these cases, slow constrictions may only be revealed if sympathetic neurons are stimulated selectively. On the other hand, in some arteries, such as the guinea-pig thoracic aorta, a very slow phase of the sympathetic constriction cannot be demonstrated even under intense stimulation conditions (Figure 2.5a).

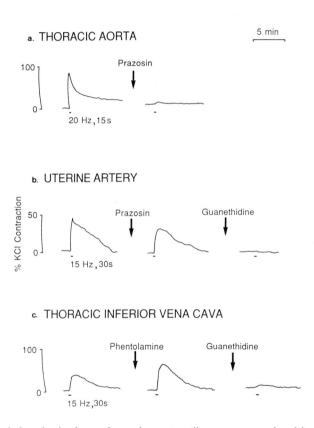

FIGURE 2.5 Variations in the form of vascular contractile responses produced by prolonged sympathetic nerve stimulation in three different blood vessels from guinea-pigs. Contractions of isolated ring segments of vessels were produced by transmural stimulation with trains of 300–450 pulses of 0.3 ms duration. Responses are expressed as a percentage of the tonic contraction produced by 0.126 M KCl. a. Contractions of descending portion of thoracic aorta are monophasic (phase 2), and are abolished by 1 μM prazosin. b. The faster (phase 2) contraction of a mid segment of the main uterine artery is blocked selectively by 1 μM prazosin. The slower (phase 3) contraction is resistant to prazosin, but is abolished by 1 μM guanethidine. c. Very slow (phase 3) contraction of thoracic portion of inferior vena cava is enhanced by 1 μM phentolamine, and is blocked by 1 μM guanethidine. Contractions are also enhanced by 1–2 μM phenoxy benzamine, 1 μM yohimbine, or 1 μM prazosin. For details, see Morris & Murphy, 1988; Morris, 1991.

Identity of neurotransmitters mediating three phases of vasoconstriction

Pharmacological and biochemical studies have indicated that three different neurotransmitters may mediate the three phases of sympathetic vasoconstriction. ATP is thought to be the mediator of the first phase; NA is generally accepted as the mediator of the second phase; and a neuropeptide, specifically NPY, is thought to be the mediator of the very slow phase. However, objections have been raised to interpretation of many studies claiming unequivocal identification of these substances as vasoconstrictor neurotransmitters. Below is a summary of the main evidence for and against the participation of ATP, NA and NPY in sympathetic neurotransmission to blood vessels.

Localization in sympathetic neurons. There is good evidence for the localization of more than one of these neurotransmitter candidates in sympathetic neurons supplying mammalian blood vessels. NA and ATP can be co-localized in vascular sympathetic nerves (Head *et al.*, 1977; Muramatsu *et al.*, 1981; Katsuragi and Su, 1980; Klein, 1982) where they occur in both large and small storage vesicles (Klein, 1982, Lagercrantz and Fried, 1982). NA and NPY have been co-localized immunohistochemically in many sympathetic vasoconstrictor neurons (Lundberg and Hökfelt, 1986; Morris, 1989a). However, NPY has been detected selectively in the large vesicle fraction of homogenized sympathetic neurons (Fried *et al.*, 1985a; De Deyn *et al.*, 1989). Although co-localization of multiple transmitters in individual storage vesicles has not been demonstrated directly, it is likely that NA and ATP occur together in small vesicles, and that NA, ATP and NPY co-exist in large vesicles, during at least some stage of the vesicle life-cycle. Some experimental evidence suggests that NA and ATP have separate vesicular pools in sympathetic neurons (Trachte *et al.*, 1989). However, it is also possible that the ratio of NA:ATP varies between vesicles in the same, or in different, sympathetic nerve terminals (Trachte *et al.*, 1989).

Quantitative studies on vesicle populations from noradrenergic neurons suggest that the total transmitter storage capacity of the large vesicles may approach 80–90% of the capacity of all of the small vesicles, even in varicosities containing only 10–15% large vesicles (Klein and Lagercrantz, 1982). Therefore, the restricted localization of NPY to the large vesicles in sympathetic nerve terminals does not necessarily result in the peptide making up a minor component of the total transmitter pool.

Release. Release of ATP, NA or NPY in response to transmural electrical stimulation, or depolarization by KCl, has been demonstrated biochemically in a number of vascular preparations *in vitro*. Due to the instability of ATP, release of ATP from vascular neurons has been demonstrated after uptake of exogenous tritiated adenosine (Su, 1975; Burnstock *et al.*, 1979; Levitt and Westfall, 1982; Katsuragi and Su, 1980; Muramatsu *et al.*, 1981). Only one recent study has measured release of endogenous purines from a blood vessel in response to transmural nerve stimulation (Westfall *et al.*, 1987a). As purines can be released from actively contracting smooth muscle cells (Su, 1975, 1983; Muramatsu *et al.*, 1981; Levitt and Westfall, 1982), from endothelial cells (Sedaa *et al.*, 1990) and from non-noradrenergic neurons (Burnstock *et al.*, 1979; Levitt and Westfall, 1982), the contribution of purine released from noradrenergic neurons to the total purine release, is likely to vary (Fredholm and Hedqvist, 1980; Levitt and Westfall, 1982; Westfall *et al.*, 1987a). Su (1975) detected bretylium-sensitive release of radiolabelled purines after transmural stimulation of the aortic adventitia. The aortic media released very little purine, which was not blocked by bretylium. In this case at least, the source of most of the ATP must be neuronal.

Use of high performance liquid chromatography and radioimmunoassay in recent years has allowed demonstration of release of both endogenous NA and endogenous NPY from sympathetic neurons (Lundberg *et al.*, 1984a, 1989a; Pernow *et al.*, 1988; Sheikh *et al.*, 1988; Pernow and Lundberg, 1989a; Lacroix *et al.*, 1989). Although release of both NA and NPY from perivascular sympathetic neurons often occurs, the two substances can be released differentially. Release of NA can be detected after low frequency (< 1 Hz) stimulation of sympathetic neurons, but NPY release often is not significant until stimulation frequencies are five to 10 times higher (Lundberg *et al.*, 1986, 1989b; Pernow *et al.*, 1988, 1989; Pernow and Lundberg, 1989a; Lacroix 1989; Lacroix *et al.*, 1989). Indeed, the ratio of NA/NPY released decreases with increasing frequencies of stimulation (Lundberg *et al*, 1986, 1989b; Pernow *et al.*, 1989). Furthermore, the pattern of delivery of impulses to the sympathetic neurons also affects the ratio of released transmitters (Pernow *et al.*, 1989). Thus, the amount of NPY released by a given number of pulses is higher with intermittent stimulation than with continuous stimulation (Allen *et al.*, 1984; Bloom *et al.*, 1987; Lacroix *et al.*, 1988a). The differential release of NA and NPY is most likely to be a consequence of the differential vesicular storage of the co-transmitters.

Depletion. The use of drugs such as reserpine has allowed a correlation to be made between the selective depletion of potential sympathetic neurotransmitters, and the selective disappearance of the various phases of vasoconstriction. Reserpine treatment depletes NA stores from sympathetic neurons supplying many tissues, including blood vessels (Lundberg *et al.*, 1985c; Morris *et al.*, 1986c). The loss of NA is accompanied by a substantial reduction of the slow depolarization and slow vasoconstriction in response to sympathetic nerve stimulation (Suzuki *et al.*, 1984; Saville and Burnstock, 1988; Lundberg *et al.*, 1989a,b). However, the EJP and fast vasoconstrictions remain after reserpine treatment (Muramatsu *et al.*, 1984; von Kügelgen and Starke, 1985; Muramatsu, 1987; Saville and Burnstock, 1988). Furthermore, purine release from sympathetic neurons supplying non-vascular smooth muscle is not reduced after reserpine treatment (Kirkpatrick and Burnstock, 1987).

Depletion of NPY from vascular sympathetic neurons also has been demonstrated after reserpine treatment (Lundberg *et al.*, 1985c; Morris *et al.*, 1986c). However, depletion of NPY generally is slower than depletion of NA, and can be prevented by resection of the preganglionic sympathetic neurons (Lundberg *et al.*, 1987). This observation has been attributed to the NPY loss being secondary to increased sympathetic nerve activity, which occurs in response to the primary depletion of NA. After reserpine treatment, with or without decentralization, very slow vasoconstrictions which are resistant to adrenoceptor blockade can be detected in response to sympathetic nerve stimulation (Lacroix *et al.*, 1988b; Pernow *et al.*, 1988; Pernow and Lundberg, 1989a). The constrictions are accompanied by release of NPY, but not of NA (Lacroix, 1989; Lundberg *et al.*, 1989b; Pernow and Lundberg, 1989a). These results are consistent with ATP, NA and NPY all contributing to different phases of sympathetic vasoconstriction.

Mimicry and pharmacological blockade. In many instances, the electrical and mechanical responses of vascular smooth muscle to sympathetic vasoconstrictor nerve stimulation can be mimicked by various combinations of ATP, or its stable analogue α,β-methylene ATP (α,β-Me ATP), NA and NPY. Furthermore, use of receptor agonists and antagonists can lead to selective blockade of the three phases of sympathetic vasoconstriction (Tables 2.4–2.6).

It is now generally believed that the EJP and fast contraction are not mediated by NA acting on α-adrenoceptors. The only studies which have reported reduction of these responses by α-adrenoceptor antagonists have used very high concentrations of the drugs, which are known to have non-specific effects.

In many blood vessels the EJP and fast vasoconstriction can be mimicked by ATP, or its analogues (Suzuki, 1985; Sneddon and Burnstock, 1985; Ishikawa, 1985; Byrne and Large, 1986; Neild and Kotecha, 1986; Nagao and Suzuki, 1988; Hirst and Jobling, 1989). Both responses are substantially reduced after treatment with the P_2 purinoceptor agonist α,β-Me ATP, or by the purinoceptor antagonist arylazido aminopropionyl ATP (ANAPP$_3$) (Cheung and Fujioka, 1986; Fujioka and Cheung, 1987). An initial application of α,β-Me ATP produces depolarization and contraction of the vascular smooth muscle, which rapidly desensitizes. Further application of α,β-Me ATP or ATP, or sympathetic nerve stimulation, no longer produces fast responses, presumably because the purinoceptors have been desensitized. This result has been widely accepted as evidence that the fast responses are mediated by ATP released from sympathetic neurons (Burnstock and Kennedy, 1986; Stjärne, 1989).

While there are many studies in which the blockade by α,β-Me ATP of nerve mediated events seems to be via an action on postsynaptic purinoceptors (see Table 2.4), in other studies α,β-Me ATP has been reported to have additional effects: α,β-Me ATP can produce depolarization and associated decrease in membrane resistance (Nagao and Suzuki, 1988; Ono and Suzuki, 1988); it can block depolarizations or contractions produced by NA (Byrne and Large, 1986; Taylor and Parsons, 1989; Dalziel *et al.*, 1990); and it can block constrictions induced by depolarizing current in denervated arteries (Neild and Kotecha, 1986). Furthermore, α,β-Me ATP can decrease the release of NPY from sympathetic neurons after reserpine treatment (Lundberg *et al.*, 1989b). These reports indicate that the pharmacological effects of α,β-Me ATP may differ between different parts of the vasculature, and therefore, the specificity of action of α,β-Me ATP should not be extrapolated from one tissue to another.

An alternative explanation for the resistance of the EJP and fast vasoconstrictions to α-adrenoceptor blockade is that NA produces these responses by acting on a population of receptors which is located selectively on the smooth muscle membrane closest to the sympathetic nerve terminals (Hirst and Neild, 1980, 1981; Luff *et al.*, 1987; Hirst and Edwards, 1989; Luff and McLachlan, 1989). These adrenoceptors, which are distinct from α- and β-adrenoceptors, have been called γ receptors. In vessels with γ receptors, notably mesenteric arterioles and cerebral arteries, application of high concentrations of NA to vascular smooth muscle in the vicinity of sympathetic nerve terminals can mimic the fast electrical and mechanical responses to sympathetic nerve stimulation (Hirst and Neild, 1980, 1981; Hirst *et al.*, 1982; Edwards *et al.*, 1989). None of these responses is affected by α-adrenoceptor blockade.

It has been suggested that, since α,β-Me ATP antagonizes responses both to ATP and to NA in cerebral arteries, the P_2-purinoceptor and the γ receptor might be identical (Byrne and Large, 1986). However, there is no experimental evidence supporting this proposal. P_2-purinoceptors and γ receptors have different distributions along innervated and non-innervated segments of rat cerebral arteries (Edwards *et al.*, 1989), and on guinea-pig mesenteric arteries and veins (Hirst and Jobling, 1989). Furthermore, α,β-Me ATP blocks the EJP produced by perivascular nerve stimulation, and the depolarization and contraction produced by exogenous ATP, but does not affect responses of the mesenteric artery mediated by NA acting on γ receptors (Hirst and Jobling, 1989).

The results summarised above indicate that ATP acting on purinoceptors, and NA acting on γ receptors, both may be involved in sympathetic neurotransmission to vascular smooth muscle. However, it seems that only one or the other of these two transmitters might mediate the fast phases of sympathetic neurotransmission in different blood vessels.

The slow depolarization and slow constriction produced by sympathetic nerve stimulation of most blood vessels are mimicked by NA and are blocked by α-adrenoceptor antagonists. Therefore, there is now general agreement that NA is the mediator of these slow events. The slow events in arterial smooth muscle are inhibited primarily by α_1-adrenoceptor antagonists (Suzuki and Kou, 1983; Kennedy *et al.*, 1986; Burnstock and Warland, 1987; Muramatsu *et al.*, 1989; Schwartz and Malik, 1989), and in some veins they are inhibited by α_2-adrenoceptor antagonists (Suzuki, 1984; Kou *et al.*, 1984; Cheung, 1985 c.f. Van Helden, 1988). However, a combination of both types of adrenoceptor antagonists may be required to abolish noradrenergic neurogenic responses in certain vessels (Ono and Suzuki, 1988; Taylor and Parsons, 1989). The recent recognition that slow contractions are not related to the EJPs, has dispelled previous doubts about the role of α-adrenoceptors in mediating some sympathetic vasoconstrictions (see Holman and Surprenant, 1980; Neild and Zelcer, 1982).

The very slow sympathetic vasoconstriction which is resistant to α-adrenoceptor blockade has been found in some, but not all, vascular beds after prolonged stimulation (Figure 2.5). Those vessels possessing very slow neurogenic vasoconstriction (Table 2.6) can be constricted by low concentrations of exogenous NPY (Lundberg and Tatemoto, 1982a; Lundberg *et al.*, 1984a; Lacroix *et al.*, 1988b; Morris and Murphy, 1988; Lundberg *et al.*, 1989a,b; Pernow and Lundberg, 1989a; Morris, 1991). Release of NPY from noradrenergic neurons also has been measured in the same vascular beds. Therefore, it has been asserted that NPY is the mediator of the slow sympathetic response persisting after α-adrenoceptor blockade (Lundberg and Hökfelt, 1986). The proposed role of a peptide in the mediation of this response is strengthened by selective blockade of very slow vasoconstrictions by the endopeptidase, trypsin (Morris and Murphy, 1988). However, the lack of receptor antagonists for NPY has hampered demonstration of a neurotransmitter role for this peptide. This problem has been partly overcome in the guinea-pig uterine artery and thoracic vena cava, where high concentrations of exogenous NPY selectively inhibit contractions produced by subsequent application of NPY, and inhibit the slow sympathetic contractions (Figure 2.6; Morris and Murphy, 1988; Morris, 1991). In rabbit tenuissimus muscle arterioles, lower concentrations of NPY (10^{-7} M) abolish sympathetic vasoconstrictions resistant to α-adrenoceptor blockade (Öhlén et al., 1990). The blockade of slow contractions by NPY is assumed to be due to desensitization of postsynaptic NPY receptors. However, as was the case for α,β-Me ATP, it must be demonstrated that NPY has no other actions on the neurotransmission process before this conclusion can be accepted unequivocally.

To date, this type of direct pharmacological evidence supporting a role for a peptide in mediation of slow sympathetic vasoconstriction has been obtained mainly in guinea-pig vessels. The sympathetic vasoconstrictor neurons supplying systemic and major distributing vessels in guinea-pigs contain NPY, but no other known

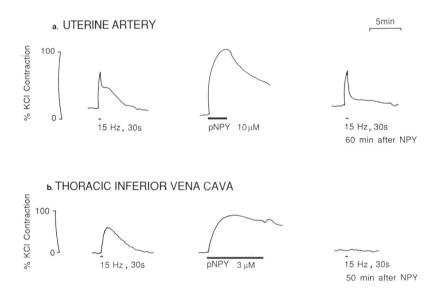

FIGURE 2.6   Blockade of very slow sympathetic vascular contractions after desensitization produced by high concentrations of NPY. a. The slow (phase 3) contraction of the guinea-pig main uterine artery in response to transmural stimulation is blocked selectively 60 min after application and washout of 10 μM porcine NPY. b. Very slow (phase 3) contraction of the guinea-pig thoracic vena cava in response to transmural stimulation is abolished 50 min after application and washout of 3 μM porcine NPY. All responses obtained in the presence of 0.6 μM propranolol and 1 μM phentolamine. For details, see Morris & Murphy, 1988; Morris, 1991.

neuropeptides (Gibbins *et al.*, 1988; Morris 1989a). However, sympathetic vasoconstrictor neurons supplying cutaneous vessels in guinea-pigs, and many vessels in other species, contain opioid peptides or galanin, with or without NPY (Kummer, 1987; Kummer *et al.*, 1986, 1988; Morris *et al.*, 1986d; Morris, 1989a; Gibbins and Morris, 1990). This raises the possibility that other neuropeptides may contribute to the very slow constrictions of some vessels, either alone, or in combination with NPY.

Physiological significance of three phases of vasoconstriction

Studies in vitro. In many cases, appropriate physiological and pharmacological studies have not been performed to determine the contributions of the three phases of vasoconstriction to the total sympathetic response. Electrophysiological studies usually have concentrated on the faster responses to single pulses or short trains of pulses, while studies employing mechanical recording of vascular smooth muscle tension usually have examined responses to long trains of pulses. However, the available information indicates that some vessels do not exhibit all three phases in response to sympathetic nerve stimulation, or to application of combinations of neurotransmitter candidates.

The contribution of the fast electrical and mechanical responses to sympathetic vasoconstriction seems to be maximal at lower levels of sympathetic stimulation (Kennedy *et al.*, 1986; Saville and Burnstock, 1988; Taylor and Parsons, 1989; Schwartz and Malik, 1989). With trains of higher frequency stimulation, a fast, α-antagonist-resistant response often is not detected. In some vessels, such as the guinea-pig pulmonary artery and thoracic aorta, these fast electrical and mechanical events are not detected with trains of low frequency stimulation (Suzuki, 1983; Petrou *et al.*, 1989; Morris and Gibbins, 1990; Morris, 1991).

Clear differences between responses to more intense sympathetic nerve stimulation can be shown in three guinea-pig blood vessels which are very different structurally and functionally: the thoracic descending aorta, a large elastic artery; the thoracic inferior vena cava, a large capacitance vessel; and the uterine artery, a medium sized muscular artery. All three vessels receive a dense innervation from sympathetic neurons containing NA and NPY. None of the vessels contracts in response to transmural stimulation with single current pulses or short pulse trains. Long trains of pulses delivered at frequencies of 1 to 5 Hz produced a monophasic contraction of the aorta and uterine artery, but caused mixed relaxation and contraction, or a contraction with a long latency, in the vena cava. The relaxation, or delay before contraction, of the vena cava was abolished by β-adrenoceptor antagonists. Stimulation frequencies above 10 Hz resulted in a large, monophasic contraction of the aorta, a biphasic contraction of the uterine artery, and a very slow contraction of the vena cava (Figure 2.5). All of the aortic contraction, and the first phase of the uterine artery contraction, could be attributed to NA acting on α-adrenoceptors (Fig. 2.5). However, α-adrenoceptors contributed very little to sympathetic contraction of the vena cava. This response was very slow, and is likely to be mediated primarily by NPY.

The pharmacological characteristics of these sympathetic responses correlate well with the distribution of postsynaptic receptors for NA and NPY, or the coupling of the receptors to the contractile apparatus. Thus, even if both NA and NPY are released from sympathetic neurons supplying all three vessels, selective localization of postsynaptic receptors will determine the contribution of the two substances to the neurotransmission process. The physiological roles of sympathetic vasomotor responses to these vessels have not been fully elucidated (Gibbins *et al.*, 1988; Morris and Gibbins, 1990; Morris, 1991). Nevertheless, it does not seem surprising that sympathetic neurons should utilize vasoconstrictor transmitters with different time courses of action to control vessels with such divergent functional demands in the circulatory system. Indeed, veins commonly are more sensitive than arteries to the direct vasoconstrictor action of NPY (Ekblad *et al.*, 1984a; Edvinsson *et al.*, 1989). Consequently, a given pattern of sympathetic impulses may normally produce constrictions of veins which are slower in onset, and longer in duration, than the constrictions of most arteries to the same level of sympathetic activity.

Studies in vivo. The preceding evidence in favour of co-transmission from sympathetic vasoconstrictor neurons largely has been obtained from vessel segments or vascular beds isolated from the rest of the circulation, with sympathetic neurons stimulated artificially. Pressor responses resistant to α-adrenoceptor blockade also occur in anaesthetized animals after electrical stimulation of sympathetic nerve pathways (Flavahan *et al.*, 1985; Hirst and Lew, 1987; Bulloch and McGrath, 1988b; Lacroix *et al.*, 1988a; Taylor and Parsons, 1989; Dalziel *et al.*, 1990). However, the synchronous stimulation of all nerve fibres in a particular pathway, often at high frequencies, is still unlikely to represent the pattern of stimulation *in vivo* (Bell, 1985; Jänig and McLachlan, 1987). Electrophysiological studies have demonstrated that the tonic firing rate of single vasoconstrictor nerve fibres is 0.3–2 Hz, whilst firing of these neurons during reflex stimulation (eg. after activation of chemoreceptors) does not exceed 15 Hz (Jänig, 1984).

Regardless of the physiological firing rates of sympathetic vasoconstrictor neurons, reflex stimulation of sympathetic pathways can reveal non-α-adrenoceptor mediated vasoconstriction *in vivo* (Hirst and Lew, 1987; Taddei *et al.*, 1989). Furthermore, there is now abundant evidence that both NA and NPY are released from nerves into the circulation after reflex sympathetic stimulation induced by haemorrhagic shock (M. Morris *et al.*, 1987; Rudehill *et al.*, 1987), thoracic surgery (Lundberg *et al.*, 1985d), the cold pressor test (M. Morris *et al.*, 1986) and moderate to intense exercise (Lundberg *et al.*, 1985a; Pernow *et al.*, 1986; M. Morris *et al.*, 1986). Although the systemic plasma levels of both NA and NPY after intense reflex sympathetic activation are an order of magnitude lower than threshold concentrations for vasoconstrictor actions *in vitro* (Pernow *et al.*, 1986), the local concentrations of these transmitters close to their release sites are high enough to have direct effects on the nearby vascular smooth muscle cells (Rudehill *et al.*, 1987).

In summary, it is now clear that sympathetic vasoconstriction should not be considered as a homogeneous event. The electrical and mechanical responses of vascular smooth muscle cells vary with the intensity of sympathetic stimulation. This is likely to be at least partly due to differences in the substances released from sympathetic nerve terminals. Co-transmission involving ATP and NA may mediate vasoconstrictor responses to tonic, or low levels of stimulation. There is strong evidence that co-transmission involving NA and a neuropeptide, particularly NPY, is responsible for vasoconstriction after intense or prolonged sympathetic stimulation. Large differences are also apparent between vasoconstrictor responses of different vascular beds, and in different species. Many of these variations can be attributed to differences in the presence, or sensitivity, of postsynaptic receptors for the three co-transmitter candidates.

SYMPATHETIC CONTRACTION OF THE VAS DEFERENS

The vas deferens of many mammalian species, particularly rodents, receives a dense innervation by sympathetic, noradrenergic neurons (Sjöstrand, 1965; Furness and Iwayama, 1972). Activation of these neurons produces strong contraction of both the longitudinal and the circular smooth muscle layers of the vas deferens. Electrical stimulation with single pulses, or with trains of pulses, often produces biphasic contractions (Swedin, 1971), the first, rapid phase of which is associated with an EJP (Burnstock and Holman, 1961; McGrath, 1978; Stjärne, 1989). The second, slower phase of the contraction seems to occur independently of membrane potential changes. The relative sizes of the two contractile phases depend on the species examined, and the portion of vas deferens examined (McGrath, 1978).

The EJP and contraction are greatly reduced, or abolished, after degeneration of noradrenergic neurons produced by 6-OHDA (Wadsworth, 1973; Fedan et al., 1981; Allcorn et al., 1986; Kirkpatrick and Burnstock, 1987). Furthermore, all excitatory responses of the vas deferens to nerve stimulation are abolished by noradrenergic neuron blocking drugs such as guanethidine and bretylium (Kuriyama, 1963; Burnstock and Holman, 1964; Ambache and Zar, 1971; Furness, 1974; Kirkpatrick and Burnstock, 1987). However, only the second phase of the contractile response is reduced after depleting neuronal stores of noradrenaline with reserpine, or after application of α-adrenoceptor antagonists (Figure 2.7). The EJP and fast contraction are largely unaffected by these drug treatments (Burnstock and Holman, 1964; Ambache and Zar, 1971; Swedin, 1971; von Euler and Hedqvist, 1975; Fedan et al., 1981; McGrath 1978; Meldrum and Burnstock 1983; Sneddon and Westfall, 1984; Stjärne and Åstrand, 1984; Allcorn et al., 1986; Suzuki and Gomi, 1987; Muir and Wardle, 1988; Cunnane and Manchanda, 1989a). These pharmacological analyses have led to general agreement that both phases of contraction of the vas deferens are mediated by transmitter released from noradrenergic nerve terminals.

The inability of exogenous NA to mimic the EJP and fast contraction suggests that these responses are not due to NA acting on novel adrenoceptors (Meldrum and Burnstock, 1983; Sneddon and Westfall, 1984; Cunnane and Manchanda, 1989b). Instead, the resistance of the EJP and fast contraction to α-adrenoceptor antagonists is likely to be due to their mediation by a substance, other than NA, that is released from noradrenergic neurons (Ambache and Zar, 1971; von Euler and Hedqvist, 1975; Stjärne, 1977, 1989; Nakanishi and Takeda, 1973). Recent studies have

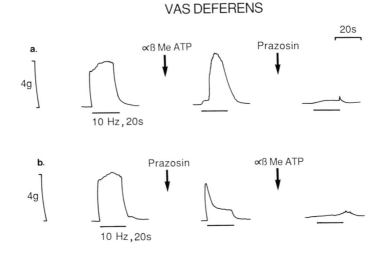

VAS DEFERENS

FIGURE 2.7   Biphasic neurogenic contractions of prostatic ends of both vasa deferentia isolated from a guinea-pig. Nerve terminals were stimulated transmurally with 200 current pulses of 0.1 ms duration. a. Application of 100 μM α,β-Me ATP caused a transient contraction (not shown), and selective blockade of the first phase of contraction. Subsequent addition of 2 μM prazosin blocked the second phase of contraction. b. Treatment of the contralateral vas deferens with 2 μM prazosin, and then 100 μM α, β-Me ATP, also abolished the neurogenic contraction. Lower concentrations of α, β-Me ATP (10–30 μM) reduce, but do not abolish, the first phase of contraction.

demonstrated coexistence of NA and NPY in sympathetic neurons supplying the vas deferens (Lundberg *et al.*, 1982e; Fried *et al.*, 1985b; Stjernquist *et al.*, 1987), and ATP, NA and NPY all are released on transmural or hypogastric nerve stimulation (Kirkpatrick and Burnstock, 1987; Lew and White, 1987; Vizi and Burnstock, 1988; Kasakov *et al.*, 1988; Katsuragi *et al.*, 1988). In the guinea-pig vas deferens, ATP appears to be released from neurons rather than from smooth muscle cells (Kasakov *et al.*, 1988). However, in the rat vas deferens, only a portion of the ^{14}C-labelled ATP released on transmural nerve stimulation seems to originate from sympathetic neurons (Vizi and Burnstock, 1988).

The EJP and fast contraction are mimicked by exogenous ATP or α,β-Me ATP (Meldrum and Burnstock, 1983; Sneddon and Westfall, 1984; MacKenzie *et al.*, 1988a,b; Cunnane and Manchanda, 1989b), but not by NPY (Stjärne *et al.*, 1986). Furthermore, the fast electrical and mechanical responses are blocked selectively by the ATP analogues ANAPP$_3$ (Fedan *et al.*, 1981; Sneddon and Westfall, 1984; Trachte *et al.*, 1989), α,β-Me ATP (Meldrum and Burnstock, 1983; Stjärne and Åstrand, 1984; Allcorn *et al.*, 1986; Bulloch and McGrath, 1988a; Muir and Wardle, 1988; MacKenzie *et al.*, 1988a,b; Cunnane and Manchanda, 1989b) or P^1, P^5-di-(adenosine-5′) pentaphosphate (AP$_5$A) (MacKenzie *et al.*, 1988b), and by the P$_2$ purinoceptor antagonist, suramin (Satchell, 1986; von Kügelgen *et al.*, 1989). So far, there is no indication that these drugs have non-specific effects in the vas deferens (Hogaboom *et al.*, 1980; Fedan *et al.*, 1981; Stjärne *et al.*, 1986; Bulloch and McGrath, 1988b; Kasakov *et al.*, 1988; von Kügelgen *et al.*, 1989). Therefore, the evidence for ATP as a sympathetic co-transmitter in this tissue must be considered to be good.

Although there is now general acceptance that ATP is the primary mediator of the fast contraction and EJP, and that NA is the primary mediator of the slow contraction, there have been suggestions that both transmitters contribute to both phases of contraction (Fedan *et al.*, 1981; Stjärne and Åstrand, 1985). This idea is based on observations of antagonists reducing both phases of contraction (Swedin, 1971; Stjärne and Åstrand, 1985). However, such observations are difficult to interpret, because NA and ATP each can potentiate the postsynaptic action of the other (Nakanishi and Takeda, 1973; Holck and Marks, 1978), and NA has presynaptic inhibitory effects on the release of both transmitters (see *Neuromodulation*). Furthermore, McGrath (1978) points out that confusion may arise in trying to relate the biphasic contraction recorded after trains of pulses, to the biphasic response to a single pulse. He argues that, regardless of the clearly separable contributions of each transmitter to the response to a single pulse, summation of biphasic events will inevitably lead to some contribution by NA to the first phase of contraction elicited by repetitive nerve stimulation.

All of the NPY released on transmural stimulation of the guinea-pig vas deferens has been assumed to originate from noradrenergic nerves (Kasakov *et al.*, 1988). Although it is likely that the dense supply of noradrenergic neurons provides a major contribution to the NPY release, some portion of the NPY could originate from the non-noradrenergic pelvic neurons which, in guinea-pigs, contain NPY and VIP (Morris *et al.*, 1985). The NPY release was demonstrated after stimulation with long trains of pulses (Kasakov *et al.*, 1988), but it is not yet known whether NPY is released after single pulses or short pulse trains. Exogenous NPY can have three actions on the vas deferens: NPY can produce a tonic contraction via a direct action on postsynaptic receptors; it can potentiate the postsynaptic actions of NA and ATP; and it can inhibit the neuronal release of NA and ATP (Lundberg *et al.*, 1982e; Lundberg and Stjärne, 1984; Stjärne *et al.*, 1986). Therefore, although there is no direct evidence for a contribution by NPY to sympathetic contraction of the vas deferens, NPY released from either sympathetic or pelvic neurons has the potential to modify the actions of the sympathetic co-transmitters NA and ATP.

PARASYMPATHETIC INNERVATION OF THE URINARY BLADDER

Although the smooth muscle of the urinary bladder is under both sympathetic and parasympathetic control, the predominant innervation of most of the smooth muscle in the body of the bladder (detrusor muscle) arises from local ganglia lying in pelvic parasympathetic pathways. When stimulated, these nerves cause a rapid phasic contraction of the muscle, which may be followed by a more sustained contractile response. Since the studies of Langley and Anderson in 1895, it has been known that the overall response to parasympathetic nerve stimulation is only partially antagonized by atropine.

Atropine-resistant responses have been reported in most species examined so far, including cats (Langley and Anderson, 1895; Edge, 1955), dogs (Henderson and

URINARY BLADDER

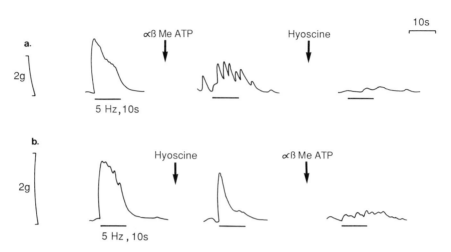

FIGURE 2.8 Neurogenic contractions of strips of detrusor smooth muscle isolated from the guinea-pig urinary bladder. Nerves were stimulated transmurally with 50 current pulses of 0.1 ms duration. a. Application of 100 μM α,β-Me ATP produces a large, phasic contraction (not shown), and blocks the initial phase of contraction. The remaining tonic contraction is abolished by 1 μM hyoscine. b. In another muscle strip, 1 μM hyoscine selectively blocks the slower phase of contraction. Subsequent application of 100 μM α,β-Me ATP abolishes the phasic contraction. Lower concentrations of α,β-Me ATP (10–30 μM) reduce, but do not abolish, the phasic contractions.

Roepke, 1934; Ursillo and Clark, 1956), rats (Hukovic *et al.*, 1965; Vanov, 1965), rabbits (Ursillo and Clark, 1956; Levin *et al.*, 1990), guinea-pigs (Figure 2.8; Chesher and Thorp, 1965), ferrets (Moss and Burnstock, 1985), pigs (Sibley, 1984; Fujii, 1988), new world monkeys (marmosets, cebus monkeys: Moss and Burnstock, 1985; Craggs *et al.*, 1986), possums (*Pseudocheirus peregrinus*: Burnstock and Campbell, 1963), lizards (*Trachydosaurus rugosus*: Burnstock and Wood, 1967), and anuran amphibians (*Bufo marinus*: Burnstock *et al.*, 1963). Normal bladders from old-world primates, including humans, generally show little of the atropine-resistant part of the response (Cowan and Daniel, 1983; Husted *et al.*, 1983; Sibley 1984; Craggs *et al.*, 1986). Whether the atropine-sensitive and atropine-resistant phases of the response are both due to the actions of ACh, or are due to the release of co-transmitters, has been debated for more than 50 years.

The two phases of the contractile response

The phasic component of the response of the detrusor muscle to parasympathetic nerve stimulation usually is not reduced by more than 50% by atropine treatment (Figure 2.8). It may be almost completely atropine-resistant at low frequencies of stimulation (e.g. 1 Hz or less) (Ambache and Zar, 1970; Dumsday, 1971; Chesher and Thorp, 1965; Hukovic *et al.*, 1965; Carpenter, 1977; Downie and Dean, 1977; Krell *et al.*, 1981; Maggi *et al.*, 1984; Levin *et al.*, 1986). The mechanical response is accompanied by EJPs which summate to produce action potentials. Neither of these membrane responses is reduced by atropine (Creed *et al.*, 1983; Callahan and Creed, 1986; Fujii, 1988; Brading and Mostwin, 1989). The atropine-resistant phasic contraction by itself probably is not sufficient to generate emptying of the bladder (Craggs and Stephenson, 1982; Levin *et al.*, 1986, 1990).

The second phase of the contractile response is more prominent at higher frequencies of stimulation (e.g. more than 5–10 Hz) and can be blocked completely by atropine (Figure 2.8; Henderson and Roepke, 1934; Hukovic *et al.*, 1965; Downie and Dean, 1977; Carpenter, 1977; Choo and Mitchelson, 1980a; Maggi *et al.*, 1984; Moss and Burnstock, 1985; Callahan and Creed, 1986; Levin *et al.*, 1986; Brading and Mostwin, 1989). In rabbits, the second contractile phase may be accompanied by a slow depolarization, which is also atropine-sensitive (Creed *et al.*, 1983; Fujii, 1988). Thus, this part of the response must be due to the action of ACh acting upon muscarinic receptors.

Acetylcholine release from post-ganglionic nerve terminals within the bladder can be demonstrated over a wide range of stimulation frequencies (Carpenter and Rand, 1965; Chesher, 1967; Choo and Mitchelson, 1980a,b; Krell *et al.*, 1981). Nevertheless, inhibition of ACh release, following blockade of choline uptake by pretreatment with hemicholinium, results primarily in the reduction of the second phase of the response (Hukovic *et al.*, 1965; Downie and Dean, 1977; Choo and Mitchelson, 1980a).

The atropine-resistance of the first phase of contraction could be explained by ACh acting on a specialized population of non-muscarinic, non-nicotinic cholinergic receptors, located within close neuromuscular junctions. However, it seems more likely that a transmitter other than ACh is responsible for the phasic contractile response of the detrusor muscle to low frequency nerve stimulation (Henderson and Roepke, 1934; Ambache and Zar, 1970; Dumsday, 1971; Carpenter, 1977; Downie and Dean, 1977). There is now increasing evidence that the fast phase of the neurogenic contraction is mediated by a purine nucleotide (Burnstock *et al.*, 1972). Exogenous ATP generally produces a phasic contraction of the detrusor muscle *in vitro* (Ambache and Zar, 1970; Burnstock *et al.*, 1972, 1978a; Dean and Downie, 1978; Husted *et al.*, 1980). ATP also produces a depolarization of the smooth muscle membrane associated with an increase in membrane conductance (Creed *et al.*, 1983; Fujii 1988). Furthermore, there is some evidence that ATP is released from nerve terminals within the bladder wall (Burnstock *et al.*, 1978a,b; but see Chaudry *et al.*, 1984 for counter evidence).

Initial attempts to test the involvement of ATP in neurotransmission were hampered by the lack of adequate pharmacological tools to manipulate purines and their receptors (see Campbell and Gibbins, 1979). In some cases, desensitization of the preparation to exogenous ATP can result in a selective inhibition of the atropine-resistant phase of the contractile response (Burnstock *et al.*, 1972; Dean and Downie, 1978; Choo and Mitchelson, 1980a; Dahlen and Hedqvist, 1980; Husted *et al.*, 1983). More recent studies have shown that desensitization of the P_2 purine receptor by α,β-Me ATP or β,γ-Me ATP is accompanied by the selective inhibition of the atropine-resistant EJP and the phasic component of the contractile response to nerve stimulation in guinea-pigs, rats, marmosets, ferrets, rabbits, and pigs (Figure 2.8;

Dahlen and Hedqvist, 1980; Kasakov and Burnstock, 1983; Hourani, 1984; Hoyle and Burnstock, 1985; Moss and Burnstock, 1985; Fujii, 1988; Brading and Mostwin, 1989; Peterson and Noronha-Blob, 1989). Combined pretreatment with the ATP analogues and atropine eliminates virtually all responses in most species. Thus, the whole transmission process may be explained by the combined actions of neurally-released ATP and ACh (Figure 2.8; Levin et al., 1986; Fujii, 1988; Brading and Mostwin, 1989).

Involvement of neuropeptides and amino acids

Despite the preceding conclusion, the neurons in the bladder do contain other potential transmitters that may contribute to co-transmission processes through mechanisms not necessarily directly related to acute post-synaptic phenomena. Neuropeptides have been reported to occur within the parasympathetic neurons innervating the detrusor muscle of the bladder of some species, although no systematic cross-species studies have been made. In toads, these neurons contain SOM and GAL (Gibbins, 1983; Morris et al., 1989); in guinea-pigs, they mostly contain SOM, NPY and substance P (Hökfelt et al., 1978; Crowe et al., 1986; Gibbins, 1989a; Gibbins et al., 1987a; James & Burnstock, 1988, 1990; Table 2.3); and in rats, they contain at least NPY (Mattiasson et al., 1985).

None of the peptides found in the intrinsic neurons has a dramatic effect on the detrusor smooth muscle itself. Substance P can cause a relatively slow contraction (Erspamer et al., 1981; Husted et al., 1981; Sjögren et al., 1982; Hills et al., 1984; Longhurst et al., 1984; Mackenzie and Burnstock, 1984; Callahan and Creed, 1986; Maggi et al., 1988 Shirakawa et al., 1989). Similarly, the contractile effects of SOM are, at best, only weak (Erspamer et al., 1981; Husted et al., 1981; Sjögren et al., 1982; Hills et al., 1984; Callahan and Creed, 1986). Nevertheless, substance P may augment the effects of non-peptide transmitters by increasing release of both ACh and the non-cholinergic transmitter (Sjögren et al., 1982; Hourani, 1984; Callahan and Creed, 1986; Shirakawa et al., 1989). Furthermore, both substance P and SOM have been reported to facilitate the excitatory effect of ATP on the detrusor smooth muscle of rabbits (Husted et al., 1981). The primary effects of NPY and GAL are likely to include the presynaptic inhibition of transmitter release (Lundberg et al., 1984d; Maggi et al., 1987).

In guinea-pig bladder, there is evidence that γ-aminobutyric acid (GABA) may be involved in the presynaptic inhibition of transmitter release (Taniyama et al., 1983; Kusunoki et al., 1984). GABA selectively inhibits the atropine-sensitive component of the contractile response to nerve stimulation and concurrently inhibits ACh release. Conversely, the GABA antagonist, bicuculline, enhances both ACh release and the atropine-sensitive part of the contractile response to nerve stimulation. Following loading of the tissue, [^{3}H]-GABA can be released in a Ca^{++}-dependent, tetrodotoxin sensitive manner (Kusunoki et al., 1984; Shirakawa et al., 1989). These experiments suggest that a population of nerve endings within the bladder at least have the capacity to take up and release GABA. Whether they also have the ability to synthesize an endogenous neurotransmitter pool of GABA remains to be seen.

How many populations of neurons mediate neurogenic contraction of bladder smooth muscle?

A crucial assumption in the discussion of co-transmission in the bladder is that all the observed effects of nerve stimulation are indeed mediated by a single population of postganglionic neurons. There are no agents which selectively and reliably inhibit the actions of cholinergic nerves in the way that 6-hydroxydopamine or guanethidine can be used to abolish the effects of catecholamine-containing neurons. It is, therefore, extremely difficult to prove that ACh and non-cholinergic transmitters are actually released from the same neurons. However, neither noradrenergic sympathetic neurons nor unmyelinated sensory neurons are likely to be involved in the non-cholinergic contraction (Vanov, 1965; Ambache and Zar, 1970; Burnstock *et al.*, 1972; Creed, 1979; Husted *et al.*, 1981; Krell *et al.*, 1981; Creed *et al.*, 1983; Maggi *et al.*, 1988). While the cholinergic and non-cholinergic components of the neurogenic contractions of bladder could be mediated by separate populations of parasympathetic neurons, immunohistochemical studies indicate that the intrinsic neurons likely to be responsible for the excitatory response in the guinea-pig and toad bladders form one homogeneous population (Gibbins, 1983, 1989a; Morris *et al.*, 1989; James and Burnstock, 1990). Therefore, it is probable that the mediators of the different phases of the bladder contraction are indeed co-transmitters released from the same nerve terminals.

NON-ADRENERGIC AUTONOMIC VASODILATATION

Powerful vasodilatations can be elicited by stimulation of non-adrenergic autonomic neurons supplying many regions of the vasculature. These responses are prominent in cephalic, pelvic, gastrointestinal and skeletal muscle vascular beds, particularly after treatment with vasoconstrictor agents. There is good evidence that ACh is the transmitter mediating autonomic vasodilatation in some blood vessels (Bülbring and Burn, 1935; Bell, 1975; Burnstock, 1980). However, reports of non-cholinergic autonomic vasodilatations have persisted for more than a century. In many cases, autonomic vasodilatations have two distinct phases: one which is blocked by atropine, and one which is resistant to atropine treatment. The data summarized below support the notion that non-adrenergic vasodilatation often involves the release of two or more transmitters from the same autonomic neurons.

Cephalic arteries

Cerebral and extra-cerebral cephalic arteries typically show both a cholinergic phase of parasympathetic dilatation and an atropine-resistant dilatation, but the size of each phase varies greatly between individual animals and between different arteries (Bevan *et al.*, 1982). In the rabbit lingual artery, the parasympathetic vasodilatation is reduced by 50% by atropine in concentrations that abolish the associated inhibitory junction potential (Brayden and Large, 1986). Thus, the non-cholinergic transmitter produces dilatation without a change in membrane potential. Parasympathetic nerve stimulation produces a different outcome in the cat infraorbital artery (Brayden,

1987). Here, although a cholinergic hyperpolarization is sometimes seen, atropine greatly enhances the size of the vasodilatation, presumably by blockade of presynaptic receptors inhibiting neurotransmitter release. Because of the enhancement by atropine of the vasodilatation, it is not possible to determine whether the hyperpolarization produced by the postsynaptic action of ACh contributes to any part of the vasodilatation.

Recent studies indicate that a fast, non-cholinergic dilatation is present in some cerebral arteries, and is blocked by haemoglobin or haemolysate (Toda, 1988), or by L-nitro-arginine (Toda et al., 1990). This is consistant with the hypothesis that nitric oxide is involved in the neurogenic vasodilatation, perhaps as a neurotransmitter (Toda et al., 1990). Indeed, nitric oxide, or a substance causing its production, has been implicated in fast, neurogenic inhibitory responses in other smooth muscle tissues (Gibson and Mirzazadeh, 1988; Gillespie and Sheng, 1988; see *Excitatory and Inhibitory Innervation of Gastrointestinal Smooth Muscle*).

In many studies demonstrating both a cholinergic and a non-cholinergic component of parasympathetic vasodilatation, the non-cholinergic dilatation is a slower event than the cholinergic component. The most complete evidence for this type of co-transmission from autonomic vasodilator neurons has been obtained for the cat submandibular salivary gland. Heidenhain (1872) first reported that vasodilatation in this organ in response to parasympathetic nerve stimulation is largely resistant to atropine, although salivary secretion is abolished by atropine (see *Co-transmission from Secretomotor Neurons*). The degree of atropine-resistance varies with the frequency of parasympathetic nerve stimulation (Lundberg 1981; Lundberg et al., 1981a,b). The transient dilatation after short bursts of low frequency stimulation (2 Hz) is abolished by atropine. However, increasing the frequency or the period of stimulation results in the appearance of a slow, sustained atropine-resistant dilatation.

The immunohistochemical discovery of the vasodilator neuropeptide, VIP, in parasympathetic neurons supplying the salivary glands (Lundberg et al., 1979, 1980; Uddman et al., 1980a), and the biochemical demonstration of VIP release into the submandibular gland circulation on stimulation of the chorda tympani nerve (Bloom and Edwards, 1980a; Uddman et al., 1980a; Lundberg et al., 1981a), indicated that VIP might mediate the non-cholinergic vasodilatations. Indeed, the size of the decrease in vascular resistance following nerve stimulation in the presence of atropine is closely related to the output of VIP from the neurons (Andersson et al., 1982). This hypothesis was further supported by mimicry of the slow dilatation by infused VIP (Bloom and Edwards 1980b; Lundberg et al., 1980), and selective blockade of the non-cholinergic neurogenic response by VIP antiserum (Lundberg et al., 1981b).

The combined use of atropine and VIP antiserum demonstrates that ACh and VIP each contributes to both phases of the dilatation in response to prolonged, or higher frequency stimulation (Lundberg et al., 1981b). At frequencies of 10 Hz or more, the proportion of VIP to ACh released from the neurons is up to five times higher than at low frequencies of stimulation (Lundberg et al., 1982b). Atropine almost abolishes the fast dilatation, and slightly reduces the sustained dilatation after stimulation at moderate frequencies (6 Hz). Conversely VIP antiserum almost abolishes the slow phase, and slightly reduces the fast phase of the parasympathetic dilatation. However at stimulation frequencies producing maximal dilatation (15 Hz), atropine prolongs the slow dilatation (Lundberg *et al.*, 1981b). As VIP release is also increased after atropine treatment (Lundberg *et al.*, 1981a; 1982b), it is likely that ACh has a presynaptic inhibitory action on transmitter release at high frequencies of stimulation. The low levels of VIP released into the venous effluent after short bursts of low frequency stimulation (Andersson *et al.*, 1982; Lundberg *et al.*, 1982b) are consistent with the conclusion reached from pharmacological studies, that ACh alone mediates parasympathetic vasodilatation in the submandibular gland after low frequency stimulation.

The evidence that parasympathetic vasodilatation is due to co-transmission rests with the morphological demonstration of VIP and acetylcholinesterase in nearly all submandibular ganglion nerve cell bodies (Lundberg *et al.*, 1979, 1980; Uddman *et al.*, 1980a; Gibbins *et al.*, 1984). Furthermore, ultrastructural examination of nerve terminals in the submandibular gland demonstrates only one type of terminal containing both small agranular vesicles, and large granular vesicles. Only the large vesicles are immunoreactive for VIP (Johansson and Lundberg, 1981). Subcellular fractionation studies confirm that VIP is associated exclusively with the large vesicle fraction, and ACh predominately with the small vesicles. It is not clear whether acetylcholine also is contained in large vesicles (Lundberg *et al.*, 1981c). As was the case for sympathetic neurons containing NA and NPY, the intracellular segregation of co-transmitters provides an explanation for differential release of transmitters from the same nerve terminals with varying stimulation regimes.

Subsequent to these studies, the VIP-containing neurons were found to contain its gene-related product, PHI (Lundberg *et al.*, 1984b). VIP and PHI are released in equal quantities into the submandibular gland circulation on parasympathetic nerve stimulation, and like VIP, the release of PHI is enhanced after atropine treatment (Lundberg *et al.*, 1984c). PHI also dilates the vasculature of the cat submandibular gland, but is less potent than VIP (Lundberg and Tatemoto, 1982b). Thus, co-transmission from parasympathetic vasodilator neurons supplying the cat submandibular gland may involve VIP and PHI. However, PHI has 50% sequence homology with VIP (Tatemoto and Mutt, 1981), and the vasodilator effects of VIP and PHI seem to be mediated through the same receptor (Edvinsson and McCulloch, 1985; Huang *et al.*, 1989). Consequently, it will be difficult to demonstrate the relative contributions of these two peptides to the neurotransmission process.

Nearly all parasympathetic nerve cell bodies in the head of cats contain VIP and PHI, (Lundberg *et al.*, 1979, 1980, 1984b; Uddman *et al.*, 1980b; Gibbins *et al.*, 1984), and nerve terminals containing VIP and PHI supply all arteries with atropine-resistant parasympathetic vasodilatations (Gibbins *et al.*, 1984; Edvinsson and McCulloch, 1985; Brayden and Bevan, 1986). Therefore, VIP and PHI also are likely mediators of the slow non-cholinergic vasodilatation in other cephalic arteries. This has been confirmed in some cephalic arteries of cats by blockade of atropine-resistant dilatations with VIP antiserum (Bevan *et al.*, 1984; Goadsby and MacDonald, 1985; Brayden and Bevan, 1986). Another study has discounted the involvement of VIP in these responses because desensitization to dilatation produced by exogenous VIP did not affect the neurogenic dilatation (Toda, 1982).

Pelvic vasculature

Vasodilatation of the penile vasculature in response to pelvic nerve stimulation also has been reported to consist of both cholinergic and non-cholinergic components (Dorr and Brody, 1967). However, there appear to be species differences in the prominence of the cholinergic and non-cholinergic components of vasodilatation (cf. Sjöstrand and Klinge, 1979; Creed *et al.*, 1988). As in cephalic arteries, there is evidence for fast and slow phases of non-cholinergic vasodilatation in the pelvic vasculature.

Biphasic dilatation in response to pelvic nerve stimulation occurs in the main penile artery of dogs and cattle, and the main uterine artery of virgin guinea-pigs (Figure 2.9; Bowman and Gillespie, 1983; Morris, 1989b c.f. Bell, 1968). However, neither phase of the dilatation of these vessels seems to be mediated by ACh (Figure 2.9; Bell, 1968, 1969; Bowman and Gillespie, 1983; Morris and Murphy, 1988). The fast phase of the vasodilatation can be mimicked by an inhibitory factor extracted from the bull retractor penis (Ambache *et al.*, 1975; Gillespie and Martin, 1980), and these dilatations can be blocked selectively by haemoglobin (Bowman and Gillespie, 1983). Recent studies have indicated that the inhibitory factor may be a substance similar to nitric oxide (Martin *et al.*, 1988). It remains possible that the fast, non-cholinergic

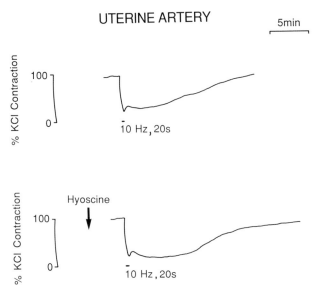

FIGURE 2.9   Biphasic neurogenic vasodilatation of the guinea-pig main uterine artery. Arterial ring segment was treated with 1 μM guanethidine, which increased the resting tension. Transmural stimulation with 200 pulses of 0.3 ms duration. Hyoscine (1 μM) does not reduce the responses, and even enhances the duration of the slow response (lower trace). The magnitude of relaxations was compared with contractions produced by 0.126 M KCl.

vasodilatation of the guinea-pig uterine artery also may be mediated by the same transmitter.

VIP is released into the penile circulation of dogs and cats on pelvic nerve stimulation, and its release is highly correlated with the non-cholinergic vasodilatation in this vascular bed (Andersson *et al.*, 1984, 1987). Furthermore, local injection of VIP into the penile circulation causes vasodilatation (Andersson *et al.*, 1984; Carati *et al.*, 1988; Juenemann *et al.*, 1987), but both ACh and VIP are necessary for the full erectile response (Carati *et al.*, 1988). VIP antiserum blocks the maintained phase of the increase in blood flow and increase in corpus cavernosum pressure resulting from cavernous nerve stimulation in dogs (Juenemann *et al.*, 1987). Again, the evidence for co-transmission involving ACh and VIP relies on the immunohistochemical demonstration of VIP in a large population of pelvic cholinergic neurons (Dail *et al.*, 1983, 1986; Dail and Hamill, 1989).

The slow, non-cholinergic dilatation of the guinea-pig uterine artery is blocked by the endopeptidase, trypsin, suggesting that a neuropeptide may be responsible for this response (Figure 2.10; Morris, 1989b). Indeed, the pelvic neurons innervating the uterine artery contain several peptides: VIP, NPY, SOM and dynorphin (Fig. 2.2). VIP is the only one of these peptides that mimics the slow phase of the neurogenic vasodilatation, and hence it is a good candidate for mediating this response. However, various analogues of VIP which act as receptor antagonists in the lung and

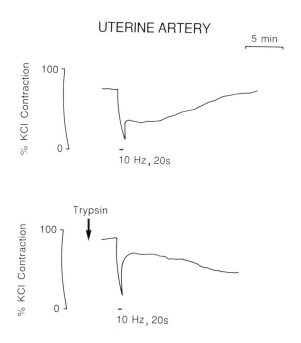

FIGURE 2.10   Blockade of slow neurogenic vasodilatation of the guinea-pig main uterine artery. A biphasic response was evident in a ring segment precontracted with prostaglandin $F_{2\alpha}$ (3 μM), after transmural stimulation with 200 pulses (top trace). Thirty minutes after addition of the endopeptidase trypsin (1.4 μg/ml), the slow phase of the neurogenic response is blocked selectively.

pancreas do not antagonize the effects of exogenous VIP on the uterine artery (Morris and Murphy, 1989). Therefore, firm evidence for VIP acting as a neurotransmitter in the uterine artery has not yet been obtained.

Despite the non-cholinergic nature of the biphasic dilatations of the uterine artery from virgin guinea-pigs, ACh does appear to mediate about 50% of the dilatation to pelvic nerve stimulation in arteries from pregnant animals (Bell, 1968). Appearance of this cholinergic component during pregnancy is correlated with a dramatic increase in the size of the vasodilatation produced by exogenous ACh, and thus seems to be due to a change in the postsynaptic cholinergic receptors (Bell, 1968). However, enhancement by atropine of the slow neurogenic vasodilatation in uterine arteries from virgin animals (Figure 2.9) indicates that ACh is also released from pelvic neurons in the non-pregnant state.

At least six co-transmitters probably co-exist in the pelvic neurons supplying the uterine artery; four peptides; ACh; and the transmitter reponsible for the fast dilatation. There are likely to be complex interactions between co-transmitters in the uterine artery, but without appropriate pharmacological tools it will be difficult to establish the precise roles of these substances. For example, NPY is contained in both the vasoconstrictor and the vasodilator neurons supplying uterine artery (Morris et al., 1985), and in addition to its potent vasoconstrictor action, it inhibits vasodilatations produced by VIP (Morris, 1990). Recently NPY also has been found together with VIP and choline acetyltransferase in perivascular neurons supplying rat cerebral arteries (Suzuki et al., 1990), suggesting that the localization of NPY in vasodilator neurons may be a widespread phenomenon. However, until the receptors mediating vascular effects of neuropeptides are indentified, and antagonists are obtained, it will be impossible to establish whether NPY acts as a co-transmitter from two different populations of neurons supplying the same blood vessel, and whether the same transmitter has different actions when released from different neurons.

EXCITATORY AND INHIBITORY INNERVATION OF GASTRO-INTESTINAL SMOOTH MUSCLE

The most complex circuitry in the peripheral autonomic nervous system lies within its enteric division. The control of gastro-intestinal motility and secretion relies upon many functionally and anatomically discrete populations of intrinsic neurons (see Furness and Costa, 1987). The smooth muscle is controlled by two main functional classes of enteric neurons: excitatory motor neurons and inhibitory motor neurons. By 1970, it was clear that a major transmitter of the excitatory neurons is ACh, but that the inhibitory neurons are neither cholinergic nor noradrenergic (Campbell, 1970). However, enteric neurons contain a wide variety of neuropeptides, as well as purines, and amino acids, which might also contribute to excitatory and inhibitory transmission to gut smooth muscle.

The assessment of the involvement of any particular substance in transmission to gut smooth muscle is made much more difficult by the presence of many different functional types of neurons within a segment of gut (Furness and Costa, 1980; 1987). Furthermore, the same substance can be a transmitter in more than one of these neuronal classes: for example, ACh may be utilized by vagal preganglionic neurons, smooth muscle motor neurons, secretomotor neurons, ascending and descending interneurons, and intrinsic sensory neurons (see Furness and Costa, 1987). Because of these problems, we will concentrate our discussion of co-transmission in the gut on the small intestine (ileum) of guinea-pigs, where the most complete body of anatomical, histochemical, pharmacological, and physiological information is available.

Recent immunohistochemical studies have demonstrated that nearly all of the intrinsic varicose axons in the circular muscle layer of the guinea-pig ileum contain immunoreactivity to either substance P or VIP: no axons contain both (Llewellyn-Smith et al., 1988). At least some of these axons must represent the terminals of cholinergic excitatory neurons and others must be the terminals of the inhibitory neurons. Comparisons of the projections of immunohistochemically-identified neurons with those of functionally-identified neurons indicate that the substance P-containing neurons must include the cholinergic motor neurons and that the VIP-containing neurons include the inhibitory ones (for further discussion, see Bornstein et al., 1986; Furness and Costa, 1987; Llewellyn-Smith et al., 1988; Smith et al., 1988; 1990).

Excitatory motor innervation

Transmural stimulation of the ileum with single pulses or with short trains of low frequency pulses, results in a fast EJP which is accompanied by contraction of the circular muscle. Both of these responses are abolished, or largely reduced, by atropine, indicating that they are mediated by the actions of ACh on muscarinic receptors. At higher frequencies of stimulation, however, another contractile response occurs which is also accompanied by an equally rapid EJP (Bywater et al., 1981; Bywater and Taylor, 1983; Bauer and Kuriyama, 1982a). These responses are not reduced by atropine or other cholinergic antagonists (Ambache et al., 1970), so they are not due the actions of neurally-released ACh. The best candidate for the mediator of the atropine-resistant response is substance P or a related tachykinin. Substance P causes a strong contraction of intestinal smooth muscle, which is accompanied by a membrane depolarization (Bauer and Kuriyama, 1982a,b). This peptide also may potentiate the effects of ACh on muscarinic receptors (Holzer and Lembeck, 1979; Fujisawa and Ito, 1982; Holzer, 1989). Desensitization or blockade of substance P receptors causes an almost complete inhibition of the atropine-resistant response (Franco et al., 1979; Gintzler and Hyde, 1983; Costa et al. 1985; Taylor and Bywater, 1986). Thus, ACh and substance P probably act as excitatory co-transmitters on the circular muscle of the guinea-pig ileum.

The cholinergic and non-cholinergic components of the excitatory response also can be elicited by generating a myenteric reflex *in vitro* following a distention stimulus to the small intestine (Tonini et al., 1981). In parallel with the effects of electrical stimulation, low levels of distention result in an excitatory response in the smooth muscle that is largely due to the actions of ACh acting upon muscarinic receptors (Figures 2.11, 2.12). A greater degree of distension causes an additional component of the response that is not affected by muscarinic antagonists: this

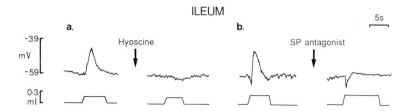

FIGURE 2.11   Excitatory electrical responses detected intracellularly in smooth muscle cells of the guinea-pig ileum, in response to distension of a segment of ileum anal to the recording site. The ileum was distended by a balloon filled with various volumes of fluid, as shown in the lower trace. a. The electrical response to an initial distension is a compound EJP. In the presence of 1 μM hyoscine, the same degree of distension no longer evokes a response. b. By increasing the distension volume, a compound EJP can be evoked in the presence of hyoscine. The excitatory response to increased distension is abolished by the substance P (SP) antagonist [D-Arg^1, D-Pro^2, D-Trp^{7,9}, Leu^{11}]-SP (20 μM). The remaining IJP is abolished by 100 μM hexamethonium. Traces were kindly provided by Dr. T.K. Smith, Dr. J.C. Bornstein and Prof. J.B. Furness. For details, see Smith et al., 1990.

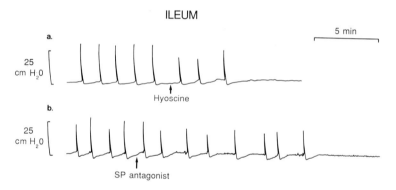

FIGURE 2.12   Peristaltic contractions of the isolated guinea-pig ileum, elicited by fluid distension. Contractions are apparent as transient increases in intraluminal pressure. a. Contractions in response to a given fluid pressure are blocked by hyoscine (1 μg/ml). b. In the presence of hyoscine, increasing the distending fluid pressure results in reappearance of the peristaltic contractions. These contractions are abolished by the SP antagonist [D-Arg^1, D-Pro^2, D-Trp^{7,9}, Leu^{11}]-SP (10 μg/ml). Traces kindly provided by Prof. M. Costa.

non-cholinergic excitatory response is associated with a concomitant increase in substance P release (Donnerer et al., 1984). The non-cholinergic response can be inhibited by blockade of substance P receptors and, therefore, is due largely to the effects of neurally-released substance P (Figures 2.11, 2.12; Bartho et al., 1982; Donnerer et al., 1984; Costa et al., 1985; Grider, 1989; Holzer, 1989; Smith and Furness, 1988; Smith et al., 1988, 1990). This is an important observation, since it clearly demonstrates that neuropeptides can be released and contribute to transmission processes under reflex stimulation conditions that must closely mimic those occurring in vivo.

Inhibitory motor innervation

Although the inhibitory neurons projecting to the circular muscle of the guinea-pig ileum contain VIP, and although VIP has an inhibitory effect on the activity of the muscle, VIP is most unlikely to be the sole transmitter released from these neurons. Stimulation of the inhibitory neurons either electrically, or via an intrinsic reflex response to gut distension, causes an inhibitory compound junction potential (IJP) as well as a phasic relaxation of the smooth muscle (Furness and Costa, 1987). In contrast, VIP generally causes a slow relaxation, and has, at best, only a slight hyperpolarizing effect on the resting membrane potential of the muscle cells (Cocks and Burnstock, 1979; Mackenzie and Burnstock, 1980; Hills *et al.*, 1983; Daniel *et al.*, 1989). Nevertheless, VIP can be released by direct or reflex activation of the inhibitory neurons in many areas of the gut of several different species (e.g. pig: Fahrenkrug *et al.*, 1978a; cat: Fahrenkrug *et al.*, 1978b; Bloom and Edwards, 1980a; guinea-pig: Grider and Makhlouf, 1987a; dog: Ito *et al.*, 1988). Furthermore, the neurogenic inhibition has been reported to be reduced by pretreatment with antibodies to VIP (opossum lower oesophageal sphincter: Goyal *et al.*, 1980; guinea-pig taenia caeci: Grider *et al.*, 1985; rat colon: Grider and Makhlouf, 1986; 1987a). Consequently, VIP itself is probably not responsible for the initial phase of the inhibitory mechanical response or for the IJP, but it may well be involved in the generation of a more maintained mechanical phase.

The transmitter mediating the IJP and the initial phase of the smooth muscle relaxation remains enigmatic, and will not be discussed in detail here (see Hoyle and Burnstock, 1989). Indeed, it has been suggested that different transmitters may be involved in different parts of the gut (Manzini *et al.*, 1986; Costa *et al.*, 1986b). A long standing contender for the transmitter is ATP, which causes relaxation of the muscle, accompanied by a membrane hyperpolarization (Burnstock, 1972; Tomita & Watanabe, 1973). However, there are many areas of the gut where ATP does not effectively mimic the responses to inhibitory nerve stimulation (Campbell and Gibbins, 1979; Bauer *et al.*, 1982; Costa and Furness, 1982; Furness and Costa, 1987). More recently, nitric oxide (NO), or an agent stimulating its production, has been suggested to be a rapidly acting inhibitory transmitter in a variety of smooth muscle preparations including the gut (Gillespie and Sheng, 1988; Gibson and Mirzazadeh, 1989; Bult *et al.*, 1990). At present, there is insufficient information to decide whether or not NO and ATP represent genuine alternative transmitters employed by different populations of inhibitory neurons in different regions of the gut. Whatever the case, however, the transmitter responsible for the fast IJP and the initial phase of the relaxation almost certainly is released from VIP-containing neurons.

Combinations of neuropeptides

In many parts of the gut, immunohistochemical studies have shown the presence of more than one neuropeptide in neurons projecting to the smooth muscle layers. One reason for this is that there may be two or more products of the same gene. Thus, neurons containing substance P also contain neurokinin A (NKA), and those containing VIP also contain PHI or one of its variants (Yanaihara *et al.*, 1983; Christophides *et al.*, 1984; Dimaline and Vowles, 1988). In at least some regions of the gut, two gene-related peptides have been shown to participate in transmission to smooth muscle. For example, both substance P and NKA are released during reflex excitation of the rat colon, and together contribute to most of the non-cholinergic component of the response (Grider, 1989). Similarly, PHI can be co-released with VIP (dog stomach: Yasui *et al.*, 1987) and both peptides may be involved in the generation of the inhibitory response to nerve stimulation (Biancini *et al.*, 1989). PHI may also contribute to the presynaptic inhibition of VIP release (Grider and Makhlouf, 1987a).

In the small intestine of rats and guinea-pigs, the VIP-containing neurons projecting to the circular muscle additionally contain NPY (Ekblad *et al.*, 1984b; Costa *et al.*, 1986a). Where it has been tested, exogenous NPY has little or no direct effect on intestinal smooth muscle, but it can inhibit both the cholinergic and non-cholinergic components of the excitatory neurogenic response (Holzer *et al.*, 1987). It is not known if this inhibitory effect occurs at the level of the cell bodies or the terminals of the excitatory neurons. If the latter is so, then NPY, co-released with VIP and the fast inhibitory transmitter, would be expected to inhibit the effects of any concurrent activity of the excitatory neurons innervating the same target cells.

In the circular muscle of the guinea-pig ileum, enkephalin occurs in subpopulations of both the substance P-containing neurons and the VIP-containing ones (Costa *et al.*, 1987; Llewellyn-Smith *et al.*, 1988). The predominant effect of opioid peptides in the intestine is to inhibit the release of other transmitters including ACh, substance P and VIP (Gintzler and Scalisi, 1982; Holzer 1984; Vizi *et al.*, 1984; Yoshimura *et al.*, 1982; Grider and Makhlouf, 1987b, c). The release of opioid peptides has been demonstrated during the propagation of peristaltic reflexes *in vitro* (Grider and Makhlouf, 1987b), but it is impossible to say if the peptides are being released from neurons projecting to the muscle, or from opioid-containing interneurons in the myenteric plexus. However, it seems probable that at least some enkephalin will be released from excitatory and inhibitory motor neurons under conditions of reflex stimulation that release tachykinins and VIP, respectively.

AUTONOMIC INNERVATION OF THE HEART

The autonomic innervation of the heart in amphibians has a prominent place in development of the theory of chemical transmission. Loewi first demonstrated that a choline ester was released from vagal cardiac neurons of the frog and toad heart to produce inhibition, and that adrenaline was released from sympathetic cardiac neurons to cause excitation of the toad heart (Loewi, 1921, 1922, 1936). It is therefore somewhat ironic that the heart of another amphibian, the toad *Bufo marinus*, provided the first evidence for 'non-classical' transmission from cardiac autonomic neurons, both sympathetic and parasympathetic (Morris *et al.*, 1981; Campbell *et al.*, 1982). Co-transmission from the vagal neurons has been demonstrated clearly in this species, although co-transmission may not be the only explanation for adrenoceptor-resistant sympathetic transmission in amphibian hearts. The evidence for co-transmission from autonomic neurons in mammalian hearts is still patchy, but there is increasing evidence that neuropeptides found in adrenergic and cholinergic neurons innervating the heart have many potent effects on cardiac function.

Vagal innervation

Stimulation of the vagus nerves causes rapid negative inotropic and chronotropic responses in spontaneously beating hearts from most vertebrates. In toad sinus venosus and guinea-pig sinoatrial node, vagal nerve stimulation causes cardiac arrest, and the membrane potential becomes positive with respect to the maximum diastolic potential (Bywater *et al.*, 1989; Campbell *et al.*, 1989). The effects of low frequency, or short bursts, of vagal stimulation are abolished by atropine, and therefore can be attributed wholly to the release of ACh from vagal neurons (Campbell *et al.*, 1982; Bywater *et al.*, 1989; Campbell *et al.*, 1989). However, the muscarinic receptors mediating vagal effects in toad and guinea-pig hearts are different from the muscarinic receptors occupied by exogenous ACh, and are likely to have a restricted location close to the vagal postganglionic nerve terminals (Bywater *et al.*, 1989; Campbell *et al.*, 1989). In toad and mudpuppy (*Necturus maculosus*) hearts, prolonged vagal stimulation at frequencies greater than 2 Hz produces a slow, atropine-resistant inhibition of both rate and force of beat (Campbell *et al.*, 1982; Axelsson and Nilsson, 1985; Campbell and Jackson, 1985; Bywater *et al.*, 1989). There is no evidence for atropine-resistant inhibition of the guinea-pig sinoatrial node after vagal stimulation with several hundred pulses at frequencies up to 30 Hz (Campbell *et al.*, 1989).

The discovery that SOM occurs in nearly all of the vagal post-ganglionic neurons of the toad, and that exogenous SOM has both negative chronotropic and inotropic effects, lead to the proposal that SOM might mediate non-cholinergic vagal actions in the toad heart. In the absence of specific SOM receptor antagonists, strong evidence for this hypothesis was obtained by producing cross-desensitization between vagal inhibition of the sinus venosus and atria, and inhibition produced by high concentrations of exogenous SOM (Campbell *et al.*, 1982; G.D.S. Hirst, personal communication). Co-transmission by ACh and SOM released from the same neurons is likely to occur during vagal stimulation at frequencies greater than 2 Hz (Campbell *et al.*, 1982). Another neuropeptide, GAL, recently was co-localized with SOM in many of the toad intracardiac neurons (Morris *et al.*, 1989), but no postsynaptic actions of GAL on the toad sinus venosus have yet been detected (G.D.S. Hirst, personal communication).

SOM cannot be detected immunohistochemically in the heart of the mudpuppy, and therefore is unlikely to mediate non-cholinergic vagal inhibitory effects in this amphibian species (Axelsson and Nilsson, 1985). However, GAL is localized in many intracardiac neurons, and has negative inotropic and chronotropic effects on sinus venosus-atrial preparations from the mudpuppy (Parsons *et al.*, 1989). It is possible, therefore, that GAL and ACh might be inhibitory co-transmitters in the mudpuppy heart.

SOM is also present in intracardiac neurons in some mammals (Day *et al.*, 1985;

Franco-Cereceda *et al.*, 1986), and exogenous SOM has a variety of inhibitory cardiac actions in mammals (Quirion *et al.*, 1979; Greco *et al.*, 1984; Diez *et al.*, 1985; Franco-Cereceda *et al.*, 1986, 1987; Hou *et al.*, 1987; Lin *et al.*, 1988). SOM also can cause release of ACh from intracardiac neurons (Wiley *et al.*, 1989). There are some inconsistencies between the effects of SOM in different mammalian species which may be related to the different molecular forms of SOM found in cardiac neurons (Day *et al.*, 1985; Franco-Cereceda *et al.*, 1986). However, SOM also can have different cardiac effects in the same species depending on the diastolic potential at the time of SOM application (Hou *et al.*, 1987). Although release of SOM from intracardiac neurons in mammals has not yet been demonstrated, SOM is well placed to regulate cardiac sympathetic activity (Franco-Cereceda *et al.*, 1986), and to suppress abnormal automaticity in both atrial and ventricular muscle (Greco *et al.*, 1984; Hou *et al.*, 1987; Lin *et al.*, 1988). If SOM is a co-transmitter with ACh in intracardiac vagal neurons in mammals, then the appropriate experiments have not been performed to reveal such subtle effects of neuronally-released SOM.

Sympathetic innervation

Sympathetic excitation of the heart is partly resistant to blockade by adrenoceptor antagonists both in the toad, *Bufo marinus* (Morris *et al.*, 1981; Bramich *et al.*, 1990), and in the frog, *Rana temporaria* (Hoyle and Burnstock, 1986). In both amphibian species, the positive inotropic and chronotropic effects of sympathetic nerve stimulation are biphasic. Both phases are abolished after chemical sympathectomy produced by 6-OHDA, or by treatment with the adrenergic neuron-blocking drug, bretylium (Morris *et al.*, 1981; Bramich *et al.*, 1990), and therefore are mediated by sympathetic adrenergic neurons. The slower excitatory effects are abolished by propranolol (Morris *et al.*, 1981; Hoyle and Burnstock, 1986), and can be attributed to adrenaline acting on β-adrenoceptors (O'Donnell and Wanstall, 1982; Bramich *et al.*, 1990). In some situations, a very slow phase of sympathetic tachycardia can be observed (Morris *et al.*, 1981; Bramich *et al.*, 1990). This response also is resistant to α- and β-adrenoceptor blockade.

In the frog heart, the fast sympathetic excitation can be mimicked by ATP or by α,β-Me ATP, and is blocked after desensitization of P_2-purinoceptors with α,β-Me ATP. The slower sympathetic response, and the positive inotropic effect of exogenous adrenaline, remain unaffected by α,β-Me ATP (Hoyle and Burnstock, 1986). These findings are consistent with co-transmission by adrenaline and ATP released from sympathetic neurons supplying the frog heart.

The fast sympathetic excitatory responses in the toad sinus venosus can be mimicked both by exogenous adrenaline and by exogenous ATP (Bramich *et al.*, 1990). However, in *B. marinus*, cardiac excitation in response to each of these three stimuli can be blocked after α,β-Me ATP treatment. Therefore, in contrast to the frog heart, α,β-Me ATP cannot distinguish between adrenaline or ATP as the mediator of the propranolol-resistant sympathetic responses in the toad heart. However, dihydroergotamine blocks the fast responses produced by sympathetic nerve stimulation and by adrenaline, but does not reduce the responses to exogenous α,β-Me ATP (Bramich *et al*, 1990). These results demonstrate that adrenaline released from sympathetic nerve terminals acts on a population of non-α, non-β-adrenoceptors to produce fast excitation of the toad heart. The presence of these receptors has been confirmed by quantitative pharmacological analysis of responses to exogenous adrenaline (Morris *et al.*, 1981), and it is likely that these receptors are located close to the sympathetic nerve terminals (Morris *et al*, 1981; Bramich *et al.*, 1990). ATP might still contribute to the sympathetic responses in the toad heart, but confirmation of this possibility awaits the use of purine antagonists which do not block the effects of adrenaline.

The mediator of the very slow tachycardia in the toad heart is not known, but high concentrations of adrenaline can produce very slow tachycardia in the presence of β-adrenoceptor antagonists (Bramich *et al.*, 1990). Alternatively, the slow response may be produced by a neuropeptide. However, no neuropeptide has been detected so far in sympathetic nerve terminals located in the toad sinus venosus, atria and ventricle. NPY is restricted to sympathetic neurons lying in the vago-sympathetic trunks in the interatrial septum (Morris *et al.*, 1986a), and is unlikely to mediate directly excitation of the sinus venosus, atria and ventricle.

In contrast to the toad heart, NPY-containing nerve fibres are widespread throughout all chambers of the heart and around coronary blood vessels in many mammals (see Morris, 1989a). Most cardiac neurons with NPY are sympathetic and project from the stellate or inferior cervical ganglia, but a few intrinsic, non-noradrenergic cardiac neurons also contain NPY (Dalsgaard *et al.*, 1986). Stimulation of the stellate ganglia in pigs results in the release of both NPY and NA from nerve endings within the heart (Rudehill *et al.*, 1986; Haas *et al.*, 1989). NPY has been suggested to be the mediator of the positive inotropic and chronotropic responses of the pig heart observed after prolonged stellate ganglion stimulation in the presence of α- and β-adrenoceptor antagonists.

In some situations NPY has cardiac effects opposite to those of NA. NPY is a potent constrictor of coronary vessels in species where the predominant effect of NA is coronary vasodilatation mediated via β-adrenoceptors (Franco-Cereceda *et al.*, 1985; Rudehill *et al.*, 1986). Furthermore, NPY can have negative inotropic and chronotropic effects (Allen *et al.*, 1983; Franco-Cereceda *et al.*, 1985; Balasubramanian *et al.*, 1988), actions that have been attributed to secondary effects of reduced cardiac perfusion after NPY-induced coronary vasoconstriction (Rudehill *et al.*, 1986). However, experiments on rat and rabbit atrial and ventricular strips *in vitro* have demonstrated direct inhibitory effects of NPY on the myocardium (Balasubramanian *et al.*, 1988). The functional significance of these effects of NPY have not been determined, but it is possible that NPY may be involved in arrhythmic actions of sympathetic nerve stimulation (Gillis *et al.*, 1974; Rudehill *et al.*, 1986).

NPY also has presynaptic inhibitory actions on transmitter release from both parasympathetic and sympathetic cardiac neurons in mammals (Lundberg *et al.*, 1984d; Franco-Cereceda *et al.*, 1985). NPY mimics the non-noradrenergic inhibition of the cardiac vagus which occurs in reponse to reflex or electrical stimulation of the cardiac sympathetic neurons, and may be the mediator of this effect (Potter, 1985, 1987a,b; Warner and Levy, 1989). The NPY receptors mediating inhibition of the vagus seem to be located on the postganglionic vagal nerve terminals

(Lundberg *et al.*, 1984d; Potter, 1987b). Although NA does not seem to participate in sympathetic inhibition of the cardiac vagus in a manner similar to NPY (Potter, 1985; Koyanagawa *et al.*, 1989), α-adrenoceptor antagonists can enhance the sympathetic response (Warner and Levy, 1989 c.f. Koyanagawa *et al.*, 1989), indicating that NA and NPY still may act as co-transmitters at the level of the vagal nerve terminals.

In addition to NPY, opioid peptides are found in cardiac sympathetic neurons supplying the heart in some mammals (Hughes *et al.*, 1977; Spampinato and Goldstein, 1983; Lang *et al.*, 1983; Xiang *et al.*, 1984; Weihe *et al.*, 1985). Opioid peptides can have direct positive inotropic effects on cardiac myocytes (Laurent *et al.*, 1985), and it has been suggested that opioid peptides acting via κ receptors may contribute to arrhythmias during myocardial ischaemia (Sitsapesan and Parratt, 1989). Furthermore, enkephalins have inhibitory actions of the cardiac vagus (Weitzell *et al.*, 1984; Wong-Dusting and Rand, 1985; Musha *et al.*, 1989), and the sympathetic nerve-mediated inhibition of the cardiac vagus can be blocked by the opioid receptor antagonists naloxone or naltrexone (Koyanagawa *et al.*, 1989). This provides good evidence for release of opioid peptides from cardiac sympathetic neurons, and their action on opioid receptors located on the vagal nerve terminals. It is not yet clear whether opioid peptides mediate all of the sympathetic inhibition of the vagus, or whether both NPY and enkephalins are involved.

CO-TRANSMISSION FROM SECRETOMOTOR NEURONS

The autonomic control of secretion from exocrine glands has provided some of the most intriguing and illuminating examples of the subtle interactions that can take place between co-transmitters. Two tissues in particular have been studied extensively from this point of view: the parasympathetic innervation of the submandidular salivary gland of cats; and the parasympathetic innervation of the parotid salivary gland of rats. We will discuss each of these examples in turn.

CAT SUBMANDIBULAR SALIVARY GLAND

Stimulation of the parasympathetic nerves supplying the salivary glands results in a copious secretion of saliva, accompanied by a prominent increase in blood flow to the glands. In the submandibular gland of cats, the secretory part of the response is abolished by muscarinic antagonists, whilst the accompanying vasodilatation is at best only partly atropine-sensitive, and is mediated largely by VIP and PHI released from the same local cholinergic neurons (see *Non-adrenergic Autonomic Vasodilatation*).

VIP by itself does not seem to stimulate salivary secretion in this gland (Lundberg *et al.*, 1982a). Nevertheless, pretreatment with specific antibodies to VIP will reduce not only the atropine-resistant vasodilator responses, but also the atropine-sensitive secretory responses (Lundberg *et al.*, 1980; 1981b). This apparent contradiction can be explained by the observation that VIP is able to potentiate the secretomotor effects of ACh (Lundberg *et al.*, 1980; 1982a). The potentiation probably occurs via two different mechanisms. First, VIP increases the affinity of muscarinic receptors for ACh by up to 100 000 times, perhaps by switching the receptors from a low affinity to a high affinity conformation (Lundberg *et al.*, 1982c). Second, VIP itself stimulates cAMP formation in the salivary gland (Enyedi and Fredholm, 1984). This cAMP is likely to augment the rise in intracellular calcium concentration generated by muscarinic receptor activation, which in turn will tend to enhance saliva secretion (see below).

This example illustrates very clearly a case of co-transmission in which the co-transmitters have a range of direct and indirect effects on different target tissues of the same neurons. VIP (and probably PHI) is almost certainly involved in the generation of the vasodilator responses of the submandibular gland blood vessels to parasympathetic nerve stimulation, whereas the major function of these peptides in the secretory response is to potentiate the effects of ACh. The facilitatory effects of the peptides on secretion are almost entirely hidden from conventional pharmacological analysis, since they can only be seen when muscarinic receptors are activated at the same time.

RAT PAROTID SALIVARY GLAND

Unlike the case of the cat submandibular gland we have just described, atropine-resistant secretory responses to parasympathetic nerve stimulation can be obtained in the parotid gland of rats. Compared with the cat submandibular gland, postganglionic parasympathetic fibres containing VIP are more rare in the rat parotid gland (Wharton *et al.*, 1979; Uddman *et al.*, 1980a). However, many of the parasympathetic secretomotor neurons originating in the otic ganglion contain substance P (Sharkey and Templeton, 1984; Ekström *et al.*, 1988a, 1989a) and probably NKA (Ekström *et al.*, 1987). Most of these neurons also contain NPY in addition to the other peptides (Le Blanc *et al.*, 1987; Le Blanc and Landis, 1988; Gibbins, 1990 and unpub. obs). Stimulation of the auriculo-temporal nerve excites postganglionic fibres to the parotid gland, resulting in the copious secretion of watery saliva. Most of this secretion is abolished after blockade of muscarinic receptors with atropine. However, at higher frequencies (10–40 Hz), secretion can be evoked in the presence of atropine (Ekström *et al.*, 1983a, 1984, 1985, 1988b). Initially, the volume of the saliva formed in the presence of atropine is up to 35% of that secreted prior to atropine treatment, but the output soon drops to about 5–10% of control (Ekström *et al.*, 1983a, 1984, 1985). The concentration of protein, especially amylase, in the saliva formed and secreted in the presence of cholinergic receptor blockade is substantially higher than that formed under normal circumstances (Ekström *et al.*, 1984, 1985, 1988b).

The atropine-resistant saliva secretion seems to be due largely to the combined effects of neurally-released substance P and VIP. The non-cholinergic secretory response is largely abolished by substance P antagonists or by substance P receptor desensitization (Gallacher, 1983). It is also greatly reduced by depletion of substance P and VIP from the postganglionic neurons following prolonged high-intensity electrical stimulation (Ekström *et al.*, 1985). Substance P is a potent secretagogue in this tissue, stimulating secretion of both fluid and amylase from the gland (Chang and Leeman, 1970; Liang and Cascieri, 1979; Brown and Hanley, 1981; Gallacher, 1983). However, the concentration of amylase after substance P application tends to be low compared with that seen after nerve stimulation in the presence of atropine (Ekström and Olgart, 1986), suggesting that substance P is unlikely to be the only mediator of the non-cholinergic response. On the other hand, although NKA is less potent than substance P as a secretogogue, it can stimulate the secretion of saliva with a high concentration of amylase (Ekström et al., 1987). Similarly, VIP has little or no direct effect on fluid secretion but it does stimulate amylase secretion (Gallacher, 1983; Ekström et al., 1983a; Inoue et al., 1985; Ekström and Olgart, 1986). PHI has similar effects to VIP but is less potent (Inoue et al., 1985).

Whilst neither substance P nor VIP alone closely mimics the non-cholinergic secretomotor responses, their actions together can explain most of the non-cholinergic neurogenic effects. This is primarily because VIP potentiates the secretory effects of substance P and NKA, so that amylase-rich saliva is produced (Gallacher, 1983; Ekström et al., 1987). The VIP-induced potentiation of the substance P-mediated responses arises from the interaction of two separate intracellular second messenger pathways activated independently by each peptide. The substance P receptors are coupled to the same intracellular second messenger pathway and membrane cation channels as muscarinic receptors, and both receptor classes probably regulate a common calcium pool (Putney, 1977, 1986; Gallacher and Petersen, 1980; Hanley et al., 1980; Merritt and Rink, 1987; Chuang, 1989; Hulme et al., 1990). These pathways have been studied intensively in the rat parotid gland and, as a consequence, a large amount of information is available on the probable modes of actions of transmitters regulating saliva secretion. We will present only a brief overview of this complex and active area of research.

Substance P receptors are coupled via a GTP-binding protein (G-protein) to phospholipase C (Putney et al., 1986; Taylor et al., 1986; Chuang, 1989; Birnbaumer, 1990). When activated by receptor occupation, this enzyme increases phosphatidylinositol (PI) turnover (Hanley et al., 1980; Merritt and Rink, 1987), leading to increased intracellular concentrations of calcium due to calcium release from the endoplasmic reticulum and the entry of extracellular calcium via channels in the cell membrane (Hanley et al., 1980; Putney, 1986; Merritt and Rink, 1987). The increased intracellular calcium concentration triggers the opening of calcium-dependent monovalent cation channels (mostly potassium channels) in the cell membrane, leading to a change in membrane potential (Gallacher and Petersen, 1980; Brown and Hanley, 1981; Gallacher, 1983; Putney, 1986; Merritt and Rink, 1987). In addition, PI breakdown generates diacylglycerol which can activate protein kinase C, which seems to be involved in the stimulation of amylase release (Putney, 1986; Chuang, 1989). The net effect of these responses is the calcium-dependent secretion of saliva containing a moderate concentration of amylase (Brown and Hanley, 1981; Ekström and Olgart, 1986; Putney, 1986; Dreux et al., 1987).

VIP has no effect on PI turnover or membrane potential in this tissue (Hanley et al., 1980; Gallacher, 1983), but it stimulates the accumulation of cAMP within the cell due to the activation of adenylate cyclase (Inoue et al., 1985; Westlind-

Danielsson *et al.*, 1990). Raised cAMP levels themselves are likely to activate protein kinase C (McKinney and Rubin, 1988) which will itself trigger an increase in the concentration of intracellular calcium. It may also phosphorylate proteins involved in the secretion of amylase (Putney, 1986). This pathway is separate from that stimulated by substance P receptor activation and is independent of extracellular calcium levels (Gallacher, 1983; Putney *et al.*, 1986; McKinney and Rubin, 1988; McKinney *et al.*, 1989). The two intracellular pathways leading to increases in intracellular calcium concentration converge at some point beyond the steps described here (Inoue *et al.*, 1985; Putney, 1986; Putney *et al.*, 1986; Dreux *et al.*, 1987; McKinney *et al.*, 1989) and ultimately result in the enhanced secretion of amylase-rich saliva in the presence of atropine.

From the preceding discussion, we can conclude that the major function of the peptides is to potentiate the secretory effects of muscarinic receptor activation and to alter the composition of the saliva. The effects are achieved primarily via interactions of the second messenger systems activated by ACh and substance P on one hand (PI pathway) and by VIP on the other (cAMP pathway): substance P will increase the availability of intracellular calcium, whilst VIP will enhance the output of amylase. Almost certainly, both VIP and substance P have other metabolic effects on the parotid cells, such as the stimulation of ornithine decarboxylase activity (Ekström *et al.*, 1989b). The intracellular pathways involved here are not known.

There are two further factors to consider in determining the relative roles of the peptides and ACh in parasympathetic transmission. First, analysis of the changes in intracellular calcium concentrations in parotid gland cells during activation of muscarinic and substance P receptors shows that the stimulatory effects of each receptor class have different time courses. Substance P induces a transient increase in intracellular calcium, whereas ACh produces a more sustained response. Consequently, muscarinic stimulation elicits a larger production of saliva over time than does a similar concentration of substance P (Merritt and Rink, 1987). This is at least partly because the substance P receptors show rapid desensitization (Gallacher, 1983; Merritt and Rink, 1987).

The second probable reason for a small non-cholinergic response to nerve stimulation is that the peptides themselves are only likely to be released in sufficient quantities to generate a response in their own right when the nerves are stimulated at higher frequencies (e.g. greater than 8–10 Hz). At lower stimulation frequencies (e.g. 2 Hz), most of the secretory response is atropine-sensitive. Nevertheless, even at such low frequencies, there still may be sufficient release of both substance P and VIP to enhance secretion induced primarily by ACh (Ekström and Olgart, 1986). Although no information is available for rats, experiments in rabbits and dogs indicate that parasympathetic neurons projecting to the salivary glands fire in the range of 1–8 Hz during normal feeding (Emmelin and Homberg, 1967; Gjörstrup, 1980). However, in response to noxious or unpleasant stimuli, they can fire at average frequencies up to 30 Hz (Emmelin and Homberg, 1967) and may have peak firing frequencies as high as 120 Hz (Kawamura *et al.*, 1982). Such frequencies would release peptides in sufficient quantities to have a significant influence on the secretory response.

In addition to substance P and VIP, the cholinergic neurons innervating the rat parotid gland also contain NPY (Leblanc and Landis, 1988; Lundberg *et al.*, 1988b). At present, there is no information about the effects of NPY in this tissue. However, experiments on other tissues suggest two likely actions. Commonly, NPY has presynaptic inhibitory actions on transmitter release from many different functional classes of neurons (e.g. Dahlhof *et al.*, 1985; Hellström, 1987; Grundemar *et al.*, 1988). A major post-synaptic effect of NPY on many cell types, including secretory cells, is to inhibit the activation of adenylate cyclase by agents such as VIP (Fredholm *et al.*, 1985; Olasmaa *et al.*, 1987; Cox & Cuthbert, 1988; Reynolds and Yokota, 1988). Thus, we predict that the net action of NPY released from these neurons will be to down-regulate their own actions.

In the rat parotid gland, α-adrenergic receptors are also linked to the same intracellular pathway as muscarinic and substance P receptors (Putney, 1986; Chuang, 1989; McKinney *et al.*, 1989). Thus, they also stimulate PI turnover and raise intracellular calcium levels. Conversely, β-adrenergic receptors are linked to the same or a similar pathway as VIP receptors; that is, they stimulate the formation of cAMP, leading to amylase secretion, and can potentiate the effects of activation of any of the receptor types linked to the PI pathway (Hanley *et al.*, 1980; Dreux *et al.*, 1987; McKinney and Rubin, 1988; McKinney *et al.*, 1989). Consequently, stimulation of the sympathetic pathway to the parotid gland simultaneously with the parasympathetic pathway greatly enhances the production of amylase-rich saliva (Gallacher, 1983; Anderson *et al.*, 1984; Garrett, 1987; Ekström *et al.*, 1988b). The sympathetic neurons are likely to be able to alter the composition of saliva, especially amylase content, at frequencies of stimulation around 0.5–2 Hz, much lower than those required to achieve substantial release of neuropeptides from the parasympathetic neurons (Gjörstrup, 1979; 1980; Emmelin *et al.*, 1980; Anderson *et al.*, 1984; Garrett, 1987).

From the data summarized above, the innervation of the parotid gland of the rat provides an excellent example of co-transmission, in which the various putative transmitters, ACh, substance P and VIP, all interact to provide multifaceted regulation of the quantity and composition of the saliva secreted upon parasympathetic stimulation. At another level of interaction, the intracellular events controlling secretion are influenced further by the degree of concurrent sympathetic activity.

OTHER GLANDS, OTHER SPECIES

Secretomotor neurons commonly contain VIP (see Lundberg *et al.*, 1988b). In many situations, VIP co-exists with other neuropeptides, such as substance P, NPY and enkephalin in guinea-pig parotid glands (Table 2.2; Gibbins, 1990), or with substance P and CGRP in sympathetic sudomotor neurons of cats (Table 2.1; Lindh *et al.*, 1989). Almost nothing is known of the roles of these multiple peptides in secretomotor co-transmission.

Despite the widespread occurence of VIP in cholinergic secretomotor neurons, this association is not universal. For example, the submandibular glands of guinea-pigs are innervated by local parasympathetic neurons, presumed to be cholinergic, that contain enkephalin and low levels of NPY, but no detectable VIP (Table 2.2; Gibbins, 1990). Furthermore, in the submucous plexus of guinea-pig small intestine, cholinergic secretomotor neurons that contain a large number of neuropeptides (NPY, SOM, CGRP, CCK, GAL and neuromedin U) do not have VIP. On the other hand, within the same plexus there is another population of secretomotor neurons that do contain VIP (and dynorphin and GAL), but, in this case, the neurons are non-cholinergic (Furness *et al.*, 1984, 1989b).

GANGLIONIC CO-TRANSMISSION

All autonomic pre-ganglionic neurons have been assumed to be cholinergic, since ganglionic transmission is mediated primarily by the activation of nicotinic receptors generating fast excitatory post-synaptic potentials (f-EPSPs) and action potentials. Nevertheless, immunohistochemical studies have demonstrated the presence of a wide variety of other potential transmitters in autonomic preganglionic neurons. These substances include several different neuropeptides as well as the amino acids, GABA and glutamate (Table 2.7). In some cases, at least two different peptides co-exist in the same population of preganglionic neurons (e.g. substance P and enkephalin in avian ciliary ganglion: Erichsen *et al.*, 1982; a somatostatin-like peptide and substance P in toad vagal pathways: Gibbins *et al.*, 1987b). Peptide-containing preganglionic neurons usually form distinct subsets of the total preganglionic pool innervating a particular ganglion, and they commonly seem to be associated with well defined populations of postganglionic neurons (eg. Gibbins *et al.*, 1987b; Horn and Stofer, 1988, 1989; Heym *et al.*, 1984; Morris and Gibbins, 1987; Morris *et al.*, 1987). Observations such as these indicate that there is a high degree of specificity in the connectivity between peptide-containing preganglionic neurons and their postganglionic targets.

   All of the compounds co-existing with ACh in preganglionic neurons can alter neuronal excitability in some way, and so any of them also could contribute to ganglionic transmission. However, despite many studies investigating the actions of chemicals other than ACh on postganglionic neurons, there is relatively little direct functional evidence for co-transmission is autonomic ganglia. The best documented example to date comes from the sympathetic chain ganglia of frogs. We will discuss this case first.

AMPHIBIAN SYMPATHETIC GANGLIA

The postganglionic neurons in the sympathetic chains of bullfrogs and toads can be classified into two main groups (Figure 2.13; Nishi *et al.*, 1967; Honma, 1970; Dodd and Horn, 1983a). 'C cells' are relatively small and contain NPY in addition to adrenaline (the predominant catecholamine in anuran sympathetic ganglia). In toads, at least, the same neurons also contain GAL (Morris *et al.*, 1989). Retrograde tracing and double-labelling immunohistochemical studies have shown that these cells project to blood vessels. 'B cells' are larger than C cells and generally lack NPY and

TABLE 2.7
Examples of potential co-transmitters in preganglionic neurons.

Substance	Pathway/Ganglion	Reference
ANF (BNP)	rat cervical sympathetic	Papka et al., 1985; Debinski et al., 1988
	rat sympathetic & pelvic	Saper et al., 1989
CRF	cat sympathetic	Krukoff, 1986
	rat sympathetic	Merchenthaler et al., 1983
CGRP	frog sympathetic	Kuramoto & Fujita, 1986
	frog cardiac vagal	Peng & Chen, 1988
	rat cervical sympathetic	Yamamoto et al., 1989
	rat pelvic	Senba & Tohyama, 1988
Enk	guinea-pig prevertebral sympathetic	Schultzberg et al., 1979; Dalsgaard et al., 1982; Konishi et al., 1981
	guinea-pig paravertebral sympathetic	Lindh et al., 1986
	cat pelvic	Glazer & Basbaum, 1980; Kawatani et al., 1983
	cat sympathetic	Heym et al., 1984 Krukoff et al., 1985
	rat cervical sympathetic	Häppölä et al., 1987
	rat sympathetic	Kondo et al., 1985
	rat pelvic	Papka et al., 1985
	human sympathetic	Hervonen et al., 1981
GABA	rat cervical sympathetic	Eugène, 1987; Kåsa et al., 1988
glutamate	rat sympathetic	Morrison et al., 1989
LHRH	frog sympathetic	Jan et al., 1980
NT	cat sympathetic	Heym et al., 1984; Krukoff et al., 1985
	cat cervical sympathetic	Caverson et al., 1989
	cat adrenal	Lundberg et al., 1982d
	guinea-pig sympathetic	Reinecke et al., 1983
	guinea-pig prevertebral sympathetic	Stapelfeldt & Szurszewski, 1989a
SOM	cat ciliary	Kondo et al., 1982
	cat sympathetic	Krukoff et al., 1985
	rat pelvic	Dalsgaard et al., 1981
SP	frog cardiac vagal	Bowers et al., 1986
	cat sympathetic	Krukoff et al., 1985
	guinea-pig pelvic	Dalsgaard et al., 1983a; Morris & Gibbins, 1987
SP/Enk	avian ciliary	Erichsen et al., 1982
SP/SOM	toad vagal inhibitory	Gibbins et al., 1987a
VIP	human sympathetic	Järvi et al., 1989
	cat sympathetic	Krukoff, 1986

Abbreviations: ANF, atrial natriuretic factor; BNP, brain natriuretic peptide; CRF, corticotropin-releasing factor; CGRP, calcitonin gene-related peptide; Enk, enkephalin; GABA, gamma-amino butyric acid; LHRH, luteinizing hormone releasing hormone (luliberin); NT, neurotensin; SOM, somatostatin; SP, substance P; VIP, vasoactive intestinal peptide.
Two peptides separated by a slash (/) have been shown to co-exist within a single population of preganglionic neurons. Some of the other substances may co-exist with each other in particular groups of preganglionic neurons, but no evidence is available. It is generally assumed that all of these neurons are also cholinergic (see text).

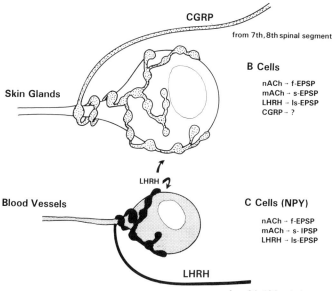

FIGURE 2.13   Summary of the specific projections between peptide-containing preganglionic nerve terminals and distinct classes of postganglionic neurons in frog sympathetic ganglia. In toads, the C cells also contain galanin in addition to NPY (Morris *et al.*, 1989). The agents contributing to the various phases of ganglionic transmission in each cell type are also summarized here. Note that muscarinic ACh receptors mediate a slow-EPSP in B cells, but a slow-IPSP in C cells. Also note that, although LHRH mediates late, slow EPSPs in both B cells and C cells, preganglionic nerve terminals with LHRH only occur around C cells. See text for details.

GAL (Morris *et al.*, 1989): these cells project mainly to the skin, where they probably provide secretomotor innervation to cutaneous glands (Horn *et al.*, 1987, 1988). In frogs, the preganglionic neurons innervating the B cells contain CGRP (Kuramoto and Fujita, 1986; Horn and Stofer, 1989), whilst those projecting to C cells contain luteinizing hormone-releasing hormone (LHRH, GnRH, luliberin; Jan *et al.*, 1980; Jan and Jan, 1982; Kuramoto and Fujita, 1986; Horn and Stofer, 1988) or a closely related peptide (see Eiden *et al.*, 1982; Jones *et al.*, 1984; Jones 1987). As well as having different neuropeptide content, the preganglionic neurons projecting to B cells or to C cells can be distinguished by their conduction velocities and their spinal origins (Nishi *et al.*, 1967; Jan and Jan, 1982; Dodd and Horn, 1983a; Horn and Stofer, 1989). In the case of the C cells, nearly all of the preganglionic neurons contain LHRH (Jan and Jan, 1982). Since all neurons receive cholinergic preganglionic input, at least some of the LHRH-containing preganglionic neurons must also be cholinergic.

Upon preganglionic stimulation, both B cells and C cells show fast EPSPs, mediated by nicotinic receptors. Both cell types also develop slower synaptic potentials mediated by muscarinic ACh receptors. However, in B cells, stimulation of these receptors generates a slow *excitatory* post-synaptic potential (s-EPSP), whilst, in C cells, muscarinic receptor activation causes a slow *inhibitory* post-synaptic potential (s-IPSP) (Jan and Jan, 1982; Dodd and Horn, 1983a,b; Kuffler and Sejnowski, 1983). In addition to these cholinergic potentials, both B cells and C cells have a so-called 'late, slow' excitatory post-synaptic potential (ls-EPSP) that is not mediated by ACh. When the preganglionic neurons are stimulated at 5–10 Hz, the ls-EPSP has a latency of several seconds and with a time to peak response of about one minute, and a total duration of 5–10 minutes (Jan and Jan, 1982; Kuffler and Sejnowski, 1983). After prolonged stimulation in this frequency range, the calcium-dependent release of LHRH can be detected by radio-immunoassay (Jan and Jan, 1982). Furthermore, the ls-EPSP is abolished after pretreatment of the ganglia with specific LHRH antagonists (Jan and Jan, 1982; Jones *et al.*, 1984). Thus, the ls-EPSP is likely to be mediated by LHRH or a closely related peptide.

The effects of LHRH and the ls-EPSP on the excitability of cells are somewhat complex, as LHRH itself has at least two different interactions with cholinergic transmission in the ganglia. In general, the ls-EPSP enhances the ability of C cells to fire repetitively at frequencies up to 13 or 14 Hz. Moreover, interactions of the ls-EPSP with the muscarinic s-IPSP are able to produce patterned bursts of activity in the postganglionic neurons that are not necessarily seen in the preganglionic neurons (Dodd and Horn, 1983b). This enhanced excitability of the postganglionic neurons is due to suppression by LHRH of a voltage-dependent, outward potassium current (the M-current, I_M: Adams and Brown, 1980; Akasu *et al.*, 1983; Jones *et al.*, 1984; Jones, 1987), which, when operating, tends to act as brake against the repetitive discharge of action potentials (Brown *et al.*, 1981). In B cells, LHRH and ACh, acting on muscarinic receptors, both inhibit the M-current, and similarly increase the excitability of the cells (Adams and Brown, 1980; Brown *et al.*, 1981). In apparent contrast to its excitatory effect on the postganglionic neurons, LHRH decreases the size of the nicotininc f-EPSP, apparently by reducing the quantal content of the preganglionic neurons (Hasuo and Akasu, 1988).

Although preganglionic neurons with LHRH form morphological synapses only with C cells, ls-EPSPs, mediated by an LHRH-like peptide, occur both in B cells and C cells (Jan and Jan, 1982; Horn and Stofer, 1988, 1989). These results strongly suggest that the neurally released LHRH must diffuse a considerable distance (perhaps up to $100\mu m$) from synapses on C cells to high-affinity receptors on B cells (Jan *et al.*, 1983). If this is indeed the case, then the activity of preganigionic neurons with close synaptic relationships with one group of postganglionic neurons can influence the activity of another group of postganglionic neurons with which they lack direct synaptic connections.

GUINEA-PIG INFERIOR MESENTERIC GANGLION

Despite the localization of many neuropeptides in preganglionic neurons of mammals (see Table 2.7), there are few examples unequivocally showing their involvement in ganglionic transmission. Much work has been done on non-cholinergic transmission in the inferior mesenteric ganglion (IMG) of guinea-pigs (see Dun, 1983; Simmons, 1985). This ganglion receives cholinergic excitatory synaptic inputs from enteric neurons as well as from preganglionic neurons with cell bodies in the spinal cord. In addition to fast EPSPs mediated by nicotinic cholinergic receptors, a prominent, non-cholinergic, slow EPSP is observed in many cells (Crowcroft & Szurszewski, 1971; Neild, 1978). At least some of this response is due to the release of an excitatory agent, probably substance P or a related tachykinin, from axon collaterals of capsaicin-sensitive, unmyelinated sensory neurons projecting to the gut (Tsunoo et al., 1982; Dalsgaard et al., 1983a; Dun and Kiraly, 1983; Matthews and Cuello, 1984; Saria et al., 1987; Amann et al., 1988).

However, distension of the colon, or electrical stimulation of the colonic nerves connecting the colon and the IMG, can elicit non-cholinergic s-EPSPs that are resistant to capsaicin treatment. These responses are unlikely to be mediated by processes of primary sensory neurons. Rather, they are the result of stimulating distension-sensitive enteric neurons having cell bodies within the wall of the colon (Crowcroft and Szurszewski, 1971; Peters and Kreulen, 1984; Kreulen and Peters, 1986; Hankins and Dray, 1988; see Furness and Costa, 1987, for a detailed discussion of this reflex circuit). The neurons projecting from the gut to the prevertebral ganglia of guinea-pigs contain several different neuropeptides, including VIP, CCK, GRP (gastrin-releasing peptide; mammalian bombesin), and dynorphin (Dalsgaard et al., 1983a,b; Schultzberg, 1983; Masuko & Chiba, 1988; Webber & Heym, 1988). All these peptides co-exist in the same neurons, which are also probably cholinergic (Costa et al., 1986a; Macrae et al., 1986; Furness and Costa, 1987). Of these peptides, both VIP and CCK cause slow depolarizations when applied exogenously to IMG neurons (Mo and Dun, 1984; Schumann and Kreulen, 1986; Love and Szurszewski, 1987), and so could be potential mediators of the non-cholinergic s-EPSP generated by colon distension. Indeed, in many cells, this response is reduced by desensitization of VIP receptors or by exposure of the ganglia to VIP antiserum, suggesting a direct involvement of VIP in the generation of non-cholinergic s-EPSPs (Love & Szurszewski, 1987). Nevertheless, the observation that, in some cells, the s-EPSPs were not affected by VIP antiserum indicates that other substances also may contribute to neurotransmission in this ganglion.

Preganglionic neurons supplying the guinea-pig IMG have been reported to contain at least two different neuropeptides: enkephalin (Schultzberg et al., 1979; Dalsgaard et al., 1982, 1983a,b; Masuko and Chiba, 1988; Webber and Heym, 1988) and neurotensin (Reinecke et al., 1983). It is not yet known if these peptides exist in the same or different populations of preganglionic neurons. However, it is presumed that they all are cholinergic. The predominant effect of each peptide apparently is the modulation of other on-going synaptic activity within the ganglion. For example, neurotensin causes a slow depolarization of some IMG neurons (Stapelfeldt and Szurszewski, 1989a) and potentiates the s-EPSP seen in these neurons following distension of the colon or electrical stimulation of the colonic nerves (Stapelfeldt and Szurszewski, 1989b). This effect can be mimicked by stimulation of the ventral roots of the splanchnic nerves running to the IMG. Furthermore, the potentiation of the s-EPSP seen after ventral root stimulation is abolished after densensitization of the ganglia with high doses of neurotensin. These results strongly imply that a neurotensin-like peptide can be released from at least some preganglionic neurons and can increase the effectiveness of synaptic transmission from peptide-containing enteric neurons to IMG cells. In other words, there is heterosynaptic facilitation of transmission in this ganglion that is mediated by a neurotensin-like peptide.

The effect of enkephalin on transmission through the IMG seems to be opposite from that of neurotensin. When applied exogenously, enkephalin has little direct effect on the excitability of IMG cells themselves. However, it reduces the size of the cholinergic f-EPSPs and non-cholinergic s-EPSPs, primarily by a presynaptic inhibitory mechanism that decreases the quantal content of the EPSPs and increases the probability of transmission failure (Konishi et al., 1981; Jiang et al., 1982; Shu et al., 1987). Evidence that enkephalin (or a related opioid peptide) is released from preganglionic neurons supplying the IMG comes from observations that s-EPSPs and f-EPSPs are both enhanced following pretreatment of the ganglion with the opioid receptor antagonist, naloxone (Konishi et al., 1981; Jiang et al., 1982). At the concentrations used in these experiments, naloxone had no direct effect on the excitability of the post-ganglionic neurons. Following stimulation of preganglionic neurons running in one pathway to the IMG, there may be a naloxone-sensitive inhibition of cholinergic transmission from preganglionic neurons running in another pathway to the ganglion (Konishi et al., 1981). This result indicates that opioid peptides almost certainly mediate heterosynaptic inhibition of transmission in the IMG. A similar role for enkephalin has been reported in transmission through parasympathetic ganglia innervating the urinary bladder of cats (De Groat and Kawatani, 1989).

DO NON-CHOLINERGIC CO-TRANSMITTERS CONTRIBUTE TO GANGLIONIC TRANSMISSION UNDER PHYSIOLOGICAL CONDITIONS?

In the examples of ganglionic transmission discussed in detail above, preganglionic neurons have been stimulated electrically with trains of pulses. It is most unlikely that these stimulation parameters closely match those occurring in vivo. Therefore, we must ask if there is any evidence for non-cholinergic ganglionic transmission in response to reflex activation of preganglionic neurons, either in vivo or in an appropriate experimental system.

One potential problem that arises here is that the non-cholinergic transmitter may have effects on the postganglionic neurons that are not seen acutely under any regime of stimulation. For example, the vagal preganglionic neurons supplying parasympathetic neurons in the interatrial septum of frogs contain a substance P-like peptide. However, exogenously applied substance P itself apparently has no acute effect on the excitability of the postganglionic neurons. This simply may be because the exogenous peptide does not have access to postsynaptic receptors that are normally activated by the endogenously-released substance P-like peptide. On the other hand, there is no evidence to date for a non-cholinergic component of transmission to these ganglion cells (Bowers et al., 1986). Thus, in this system, the peptides in preganglionic neurons may alter neuronal excitability only indirectly, perhaps by altering levels of intracellular messengers involved in functions other than the mediation of acute responses to preganglionic stimulation.

Nevertheless, there are two well documented examples in which non-cholinergic ganglionic transmission has been demonstrated following reflex activation of the input neurons. First, in the IMG of guinea-pigs, slow EPSPs can be observed after distension of the colon. Although the original stimulus here (colonic distension) may only approximate the situation *in vivo*, it is likely that the response of the neurons approaches their normal behaviour. This response is at least partially resistant to capsaicin, which suggests that the input neurons are mechanoreceptive cells with the cell bodies lying in the enteric plexuses (Kreulen & Peters, 1986; Furness & Costa, 1987).

The second example involves transmission from preganglionic neurons to postganglionic sympathetic neurons innervating skeletal muscle blood vessels in cats. At least some of these postganglionic neurons, which probably cause vasoconstriction, are innervated by unmyelinated preganglionic neurons. When these preganglionic fibres are stimulated electrically (e.g. 50 pulses at 25 Hz), the postganglionic neurons show a long-lasting excitation, which is resistant to blockade of both nicotinic and muscarinic ACh receptors (Alkadhi and McIsaac, 1973; Jänig *et al.*, 1982). Similar responses also can be obtained reflexly, following stimulation of arterial chemoreceptors with 8% O_2 (Jänig *et al.*, 1983). The transmitter responsible for the non-cholinergic transmission is not known, but the prolonged time course of its action suggests that a peptide is most likely to be implicated. These results also demonstrate that, regardless of the similarity or otherwise of electrical stimulation regimes to natural patterns of stimulation occurring *in vivo*, reflex activation of preganglionic neurons can indeed result in the release of functional non-cholinergic transmitters.

NEUROMODULATION

MODULATION BY NEUROTRANSMITTERS

Many neurotransmitters have presynaptic or postsynaptic modulatory actions in addition to the direct mediation of postsynaptic events. For example, it is well known that NA and ACh act on receptors located on autonomic nerve terminals to modify release of transmitters, including themselves (see reviews by Vizi, 1979; Starke, 1987). With the emergence of more details of the co-transmission process has come the realization that several transmitters released from the same neuron can act at multiple sites to modify each others' actions. Some of these modulatory actions have been mentioned earlier in this article.

A well-studied example of the modulatory interactions between co-transmitters released from autonomic neurons involves NA, ATP and NPY, which are released from postganglionic sympathetic neurons (see above). NA, NPY, and adenosine derived from breakdown of ATP, all have presynaptic inhibitory actions on sympathetic transmitter release. Changes in the release of one or more of the three co-transmitters have been detected in various tissues with sympathetic innervation, after application of agonists (and sometimes antagonists) for α_2-adrenoceptors, NPY receptors, or P_1 purinoceptors (Table 2.8). The relative importance of these

TABLE 2.8

Changes in transmitter release from sympathetic neurons produced by the co-transmitters NA, ATP and NPY

| | Change in transmitter release or overflow | | |
	NA	ATP	NPY
Agonist or antagonist			
NA, clonidine	Decreased NA or ^{3}H-NA release in many tissues (see Starke, 1987)	Decreased EJP and fast contraction of rat vas deferens (Stjärne et al., 1989; Major et al., 1989)	Decreased NPY release in pithed guinea-pig, and heart (Dahlöf et al., 1986; Haas et al., 1989)
Yohimbine	Increased ^{3}H-NA efflux in many tissues (see Starke, 1987; Haas et al., 1989; Pernow and Lundberg, 1989b)	Increased rapid contraction of rabbit ileocolic artery (Bulloch and Starke, 1990)	Increased NPY efflux in pig kidney and guinea-pig heart (Pernow and Lundberg, 1989b; Haas et al., 1989)
ATP, adenosine, cyclohexyladenosine	Decreased ^{3}H-NA efflux (see Fredholm and Hedqvist, 1980; Paton, 1981; Su, 1983; Haas et al., 1989)	Decreased fast contraction of rat vas deferens (Major et al., 1989)	Decreased NPY efflux in guinea-pig heart (Haas et al., 1989)
NPY, PYY	Decreased NA or ^{3}H-NA efflux in many tissues (Lundberg et al., 1985b; Lundberg and Stjärne, 1984; Dahlöf et al., 1985b; Franco-Cereceda et al., 1985; Pernow et al., 1987; Westfall et al., 1987b; Wong-Dusting and Rand, 1988; Ellis and Burnstock, 1990; Donoso et al., 1988; Pernow and Lundberg, 1989b; Haas et al., 1989)	Decreased EJP in rat vas deferens (Stjärne et al., 1989) Decreased ATP efflux from guinea-pig vas deferens (Ellis and Burnstock, 1990)	Decreased NPY efflux in pig kidney (Pernow and Lundberg, 1989b)

three modes of presynaptic inhibition has not been determined. It is likely that the various presynaptic receptors will be differentially activated as the ratio of released co-transmitters changes with changes in frequency and duration of sympathetic nerve stimulation. There have been some doubts raised about the physiological significance of modulation of neurotransmission by presynaptic α-adrenoceptors (Angus and Korner, 1980; Kalsner, 1980). However, more recently it has been suggested that autoreceptors are activated by transmitter released from adjacent varicosities rather than the same varicosity, and that they are important for regulating the intermittency of transmitter release from individual release sites (Stjärne, 1986).

In addition to these presynaptic modulatory actions of sympathetic co-transmitters, NA and NPY have been reported to interact at postsynaptic sites. However, unlike the situation in parts of the central nervous system (Agnati *et al.*, 1983; Goldstein *et al.*, 1986), and renal collecting tubules of the rat (Dillingham and Anderson, 1989), NA and NPY do not seem to interact at the level of the postsynaptic receptors in the peripheral vasculature (Lundberg *et al.*, 1988a). In many vascular beds, low concentrations of NPY do not produce direct vasoconstriction, but in these vessels NPY generally enhances the rate or magnitude of vasoconstrictions produced by a variety of agonists including NA, and by sympathetic nerve stimulation (Ekblad *et al.*, 1984a; Edvinsson *et al.*, 1984; Dahlöf *et al.*, 1985a, b; Wahlestedt *et al.*, 1985; Han and Abel, 1987; Neild, 1987; Westfall *et al.*, 1987b; Andriantsitohaina and Stoclet, 1988; Revington and McCloskey, 1988; Wong-Dusting and Rand, 1988; Abel and Han, 1989; Saville *et al.*, 1990; Vu *et al.*, 1989). In contrast, it has been reported that NPY can inhibit NA contractions of dog basilar arteries (Suzuki *et al.*, 1988).

Several different modes of action of NPY have been reported in different studies to explain the potentiation of vasoconstrictions. NPY-induced enhancement of vasoconstrictions have been partly attributed to depolarization of the smooth muscle in rabbit cerebral arteries (Abel and Han, 1989) and rat mesenteric arterioles (Andriantsitohaina and Stoclet, 1988), but NPY enhances contractions of the rat tail artery without depolarizing the muscle (Neild and Kotecha, 1990). Furthermore, some studies have concluded that NPY enhances calcium influx, and only potentiates the component of vasoconstrictor response which is due to influx of extracellular calcium (Andriantsitohaina and Stoclet, 1988, 1990; Oshita *et al.*, 1989). Other workers have concluded that intracellular calcium stores are utilized by NPY to potentiate vasoconstrictions produced by other agonists (Wahlestedt *et al.*, 1985; Vu *et al.*, 1989). Possibly, NPY activates different cellular mechanisms in different parts of the vasculature.

In addition to modifying the release and actions of transmitters released from sympathetic noradrenergic neurons, NPY also has been reported to modulate the actions of non-noradrenergic autonomic neurons by inhibiting transmitter release (Stjernqvist *et al.*, 1983; Potter, 1985; Grundemar *et al.*, 1988; Matran *et al.*, 1989), and to modulate membrane events in cell bodies of sensory neurons (Ewald *et al.*, 1988, 1989; Matran *et al.*, 1989).

It has long been known that opioid receptor agonists also can modulate transmission from autonomic neurons. Indeed, these effects were well known before discovery of the endogenous opioid peptides (see Starke, 1977). Although peripheral opioid receptors may well be activated by circulating opioid peptides, the recent discovery of opioid peptides in many preganglionic neurons (Dalsgaard *et al.*, 1982; Table 2.7), in postganglionic autonomic neurons (Figures 2.1–2.3), particularly in the cardiovascular system (Lang *et al.*, 1983; Kummer *et al.*, 1986, 1988; Fried *et al.*, 1986; Bastiaensen *et al.*, 1988; Gibbins and Morris, 1990), and in enteric neurons (see

Furness *et al.*, 1988; Kromer, 1988; Steele and Costa, 1990), indicates that opioids released from autonomic neurons are well placed to modulate neurotransmission.

In general, opioid peptides have inhibitory effects on transmission in autonomic ganglia. Some of these actions have been discussed earlier in this chapter. Opioid peptides also inhibit transmitter release from postganglionic autonomic nerve terminals, effects which seem to be mediated mostly via delta opiate receptors (Henderson *et al.*, 1972; Illes *et al.*, 1980, 1985, 1986; von Kügelgen *et al.*, 1985; Chernaeva and Charakchieva, 1988; Gan and Duckles, 1988; Budai and Duckles, 1988; Musha *et al.*, 1989). Opioid peptides may also act via naloxone-sensitive receptors to mediate direct inhibitory actions in the cardiovascular system, or to inhibit the excitatory effects of NA (Moore and Dowling, 1982; Ruth *et al.*, 1984; Wahl, 1885). However, enhancement by enkephalin of spontaneous activity of the rat portal vein, is not blocked by naloxone (Yamamoto *et al.*, 1984).

These modulatory effects of NPY and opioid peptides have led to suggestions that, in many organs, the primary role of peptides released from autonomic neurons is to act as 'modulators' rather than 'neurotransmitters'. Perhaps this definition has been applied readily to neuropeptides because they are relatively large molecules and tend to be slowly acting, and because acute physiological roles have sometimes been difficult to delineate. Furthermore, the recent findings of growth promoting or inhibiting actions of neuropeptides (Nilsson *et al.*, 1985; Brenneman and Foster, 1987; Brenneman *et al.*, 1987; Bulloch, 1987; Mitsuhashi and Payan, 1987; Rebuffat *et al.*, 1988) has encouraged the classification of peptides as long-term 'trophic' substances. Whilst not dismissing the importance of modulatory actions of neuronally-released peptides, we believe that the classification of individual neuropeptides as modulators or trophic factors has discouraged attempts to demonstrate additional acute postsynaptic roles for these substances. Indeed, using a broad definition of a neurotransmitter (see Introduction), it seems to be the rule rather than the exception that transmitters, including neuropeptides, have several actions, both direct and modulatory, which may have very different time courses.

NEUROMODULATION BY SUBSTANCES PRODUCED IN NON-NEURONAL TISSUE

Transmission from autonomic neurons also can be modulated by substances which are synthesized in nearby tissues, or circulate as hormones. Some modulators, such as adrenaline (AD) or 5-hydroxytryptamine (5-HT), may be taken up and stored in autonomic nerve terminals in which they are not synthesized (Majewski *et al.*, 1981; Verbeuren *et al.*, 1983; Kawasaki and Takasaki, 1984; Levitt and Duckles, 1986; Abrahamson and Nedergaard, 1989; Jackowski *et al.*, 1989; Kawasaki *et al.*, 1989). Furthermore, AD or 5-HT can be released from sympathetic nerve terminals along with NA (Kawasaki and Takasaki, 1984; Abrahamson and Nedergaard, 1989; Kawasaki *et al.*, 1989; Misu *et al.*, 1989), and can have presynaptic (AD: Misu *et al.*, 1989) or postsynaptic actions (5-HT: Kawasaki and Takasaki, 1984). It has even been shown that sympathetic nerve terminals can take up the exogenous receptor antagonist, propranolol, and release it together with NA on stimulation of the neurons (Daniell *et al.*, 1979). If these exogenously-derived substances can be shown to be released under any physiological conditions, we must seriously consider that they should be called co-transmitters.

Presynaptic or postsynaptic effects on autonomic neurotransmission are common for prostaglandins, particularly PGE_2 (Hedqvist, 1970, 1977; Deckers *et al.*, 1989), for 5-HT (Saum and de Groat, 1973; Wallis and North, 1978; North *et al.*, 1980; Molderings *et al.*, 1989; Nishimura and Akasu, 1989), histamine (McGrath and Shepherd, 1976; Rand *et al.*, 1982; Gross *et al.*, 1984; Ichinose *et al.*, 1989), angiotensin II (Bell, 1972; Zimmerman *et al.*, 1972; Malik and Nasjletti, 1976; Böke and Malik, 1983; Ziogas and Story, 1987; Pernow and Lundberg, 1989b; Ellis and Burnstock, 1989; Rónai, 1990), atrial natriuretic peptide (Thoren *et al.*, 1986; Nakamura and Inagami, 1986; Kuchel *et al.*, 1987; Petterson *et al.*, 1989; Patel, 1989) and endothelin (Wiklund *et al.*, 1989; Tabuchi *et al.*, 1989a, 1989b; Reid *et al.*, 1989). In some cases there is evidence for a modulatory action of endogenous stores of these substances (angiotensin II: Böke and Malik, 1983; Ziogas and Story, 1987; PGE_2: Deckers *et al.*, 1989; histamine: Gross *et al.*, 1984). However, the physiological roles of many of these modulators have not yet been determined.

CONCLUDING REMARKS

We would like to conclude with a few general remarks, based on the information from the preceding examples of autonomic co-transmission. Some common threads emerge out of the diversity. First, potential co-transmitters are widespread in autonomic neurons of all functional and anatomical classes, and many examples are accruing in which co-transmission can be demonstrated. Second, the different transmitters present in a neuron often have distinct and divergent time-courses of action upon their effector tissues. In some cases, they may even have opposing actions on the target cells. Third, the effects of some co-transmitters may be very subtle indeed, perhaps involving interactions of different intracellular second messenger systems, that by themselves have little if any direct effect on the activity of the target cells. Fourth, the precise contribution of a given set of potential transmitters to the co-transmission process may vary considerably from one neuroeffector system to another, even if the neurons seem to be neurochemically identical. The simple and immediate consequence of these conclusions is that it is difficult to predict the nature of the co-transmission process purely from knowledge of the neurochemistry of the neurons concerned.

TIME COURSES OF ACTIONS OF CO-TRANSMITTERS

When the time courses of action of various co-transmitters are compared across several different tissues, it seems clear that there may be up to three distinct phases in the process: a rapid phase, with a time course measured in tenths of seconds; a slower phase measured in seconds or tens of seconds; and a very slow or sustained phase measured in minutes, and even in tens of minutes. All three phases can be seen in some blood vessels or in ganglionic transmission. In other tissues, only two of the phases may be present, for example the biphasic responses of the vas deferens or urinary bladder smooth muscle to excitatory nerve stimulation. In yet other cases, such as the excitatory innervation of the circular muscle of guinea-pig small intestine, the actions of agents contributing to the co-transmission process have similar time courses, and their relative contributions can only be identified after pharmacological manipulations.

Fast responses and changes in membrane potential

Generally, only the first, fast phase of the response is accompanied by large changes in membrane potential (EJPs, or IJPs, or f-EPSPs). Nearly everywhere they have been described in autonomic ganglia, f-EPSPs are due to the activation of nicotinic ACh receptors. One possible exception is a f-EPSP, apparently mediated by GABA, in superior cervical ganglion cells of rats (Eugène, 1987). In many smooth muscles where EJPs are unaffected by conventional adrenergic and cholinergic receptor blockade (e.g. vas deferens, urinary bladder, some blood vessels), these responses are abolished almost uniformally following desensitization of P_2 purinergic receptors with a long-acting agent such as α,β-Me ATP. The simplest interpretation of these experiments is that the EJPs are mediated by a purine nucleotide, presumably ATP or a very closely related compound. However, pharmacological studies indicate that α,β-Me ATP can also have other actions in some tissues. Therefore, although there is good evidence that ATP mediates EJPs and fast contractions of smooth muscle in some organs, there may also be other mediators of fast events. Clarification of this issue awaits the discovery and widespread use of more selective receptor antagonists for purines, and for other transmitter candidates.

In some cases, the fast response seems to be caused by the activation of a specialised population of subsynaptic receptors rather than by a unique 'co-transmitter'. Good examples of this are provided by the sympathetic innervation of the submucosal arterioles of the guinea-pig small intestine (Hirst and Neild, 1980, 1981), and of the toad sinus venosus (Bramich et al., 1990).

The transmitter responsible for the IJP seen in many smooth muscle tissues remains enigmatic (see Hoyle and Burnstock, 1989). The rapid changes in membrane potential usually are accompanied by a rapid initial phase of smooth muscle relaxation. However, the absence of suitable pharmacological tools to manipulate the inhibitory transmission process has meant that little real progress in this area has been forthcoming.

The intermediate response

Postsynaptic responses of intermediate time course are most commonly, but not universally, mediated by a catecholamine or ACh. In smooth muscle tissues, this phase of the response generally causes the largest changes in mechanical activity. For example, in the urinary bladder, or in many blood vessels, conventional pharmacological blockade results in almost total loss of the predominant mechanical responses to low or moderate levels of nerve stimulation. At least in some cases, the cell surface receptors for the transmitters are linked directly to second messenger pathways regulating levels of intracellular calcium. These responses usually are accompanied by only comparatively small alterations in membrane potential which are probably secondary to the intracellular events.

In tissues where no part of the response to nerve stimulation is affected by known receptor antagonists, it may be impossible to identify a component with an intermediate time course that can be separated from the superimposed effects of a fast component and a slow component. Indeed, where both the fast and the slow components make a significant contribution to the response of the effector tissue, a co-transmitter specifically mediating a part of the response with an intermediate time course may be absent.

Slow responses and neuropeptides

The scarcity (or even absence) of receptor antagonists for many neuropeptides has hampered definitive demonstration of their roles in autonomic neurotransmission. However, there is good evidence in some tissues that the very slow phase of the postsynaptic response to autonomic nerve stimulation is mediated by one or more neuropeptides. This is true not only in smooth and cardiac muscles, but also in secretory tissues and in postganglionic neurons. On the other hand, some responses mediated by peptides can be relatively rapid, as in the case of the contribution of tachykinins to excitatory transmission in the guinea-pig ileum.

In many cases, the involvement of neuropeptides as co-transmitters is proportionally greater at higher frequencies, or more prolonged periods, of direct nerve stimulation. It has been argued that the experimental conditions used to generate these responses are not physiological, in that rates of neuronal firing commonly used *in vitro* are not seen *in vivo*. However, although physiological firing rates of autonomic neurons often are thought to lie in the range of 1 to 8 Hz, neurons in some pathways have been recorded to fire at rates up to 15–16 Hz, following reflex activation. Conversely, in some systems, under appropriate experimental regimes, neuropeptides may contribute to autonomic transmission at stimulation frequencies as low as 2 Hz. This combination of conditions occurs in salivary glands, for example.

The release of peptides can be measured after the reflex activation of the appropriate autonomic pathways, whether in isolated preparations, such as the gut, or in intact organisms. For example, substance P is released from cholinergic motor neurons in response to distension of the small intestine of guinea-pigs. Equally convincing is the release of NPY into the circulation following intense sympathetic activity induced by sustained exercise in humans. Results like these clearly show that the firing rates of the neurons *in vivo* are sufficient to cause the release of neuropeptides in addition to non-peptide transmitters such as ACh or NA.

WHY CO-TRANSMISSION?

We are now in a position to attempt to answer the question posed at the beginning of this article: why do autonomic neurons use so many different co-transmitters? Three clues may help us here. First, as summarized above, co-transmitters often have actions with different time courses. Second, many transmitters have broadly similar actions, such as presynaptic inhibition of transmitter release, or activation of a particular intracellular messenger pathway (adenylate cyclase, for example). Third, co-transmitters can interact at several different levels in the neurotransmission process, such as the regulation of transmitter release, the modulation of membrane potential, or the alteration of intracellular calcium concentration.

When these clues are combined with the observations that autonomic neurons contain pathway-specific combinations of co-transmitters, a simple hypothesis emerges: the particular transmitter combination found in any population of neurons is adapted to match the specific functional requirements for efficient neuro-effector transmission in that pathway. Efficiency of neurotransmission may be achieved in several ways: by priming the excitability of a cell via a rapid change in membrane potential; by reducing net transmitter release through a presynaptic or heterosynaptic mechanism; or by the synergistic action of two or more transmitters at the level of receptors, second messenger activation, regulation of intracellular calcium stores, or even the control of gene expression.

During the long evolutionary history of the vertebrates (over at least 650 million years), various combinations of transmitters are likely to have developed in different neuronal populations on a somewhat opportunistic basis. Nevertheless, the chemical structures of the transmitters themselves (including most neuropeptides), as well as their corresponding receptors, have been highly conserved (see Gibbins, 1989a). The fine tuning of autonomic transmission in each effector tissue seems to have involved the recruitment of a suite of transmitters with diverse but overlapping ranges of actions. The challenge before us now is to unravel the full extent of these actions and their interactions, and to thereby reveal the underlying basis for the precise control of visceral function by the autonomic nervous system.

ACKNOWLEDGEMENTS

Work from our laboratories is supported by the National Health and Medical Research Council of Australia. We are grateful to many colleagues, in particular Profs. Graeme Campbell, Marcello Costa and John Furness, for many fruitful discussions over the years. We thank Drs. N. Kotecha. T.O. Neild, T.K. Smith, J.C. Bornstein, J.B. Furness and M. Costa for supplying figures, and Dr. G.D.S. Hirst for providing us with unpublished results. Sue Matthew and Pat Vilimas provided invaluable assistance with preparation of the manuscript.

REFERENCES

Abel, P.W. and Han, C. (1989) Effects of neuropeptide Y on contraction, relaxation, and membrane potential of rabbit cerebral arteries. *Journal of Cardiovascular Pharmacology*, **13**, 52–63.
Abrahamsen, J. and Nedergaard, O.A. (1989) Adrenaline released as a cotransmitter does not enhance stimulation-evoked ^{3}H-noradrenaline release from rabbit isolated aorta. *Journal of Autonomic Pharmacology*, **9**, 337–346.
Adams, P.R. and Brown, D.A. (1980) Luteinizing hormone-releasing factor and muscarinic agonists act on the same voltage-sensitive K + current in bullfrog sympathetic neurones. *British Journal of Pharmacology*, **68**, 353–355.

Agnati, L.F., Fuxe, K., Bonfenati, F., Battistini, N., Harfstrand, A., Tatemoto, K., Hökfelt, T. and Mutt, V. (1983) Neuropeptide Y *in vitro* selectively increases the number of α_2-adrenergic binding sites in membranes of the medulla oblongata of the rat. *Acta Physiologica Scandinavica*, **118**, 293–296.

Akasu, T., Nishimura, T. and Koketsu, K. (1983) Substance P inhibits the action potentials in bullfrog sympathetic ganglion cells. *Neuroscience Letters*, **41**, 161–166.

Alkadhi, K.A. and McIsaac, R.J. (1973) Non-nicotinic transmission during ganglionic block with chlorisondamine and nicotine. *European Journal of Pharmacology*, **124**, 78–85.

Allcorn, R.J., Cunnane, T.C. and Kirkpatrick, K. (1986) Actions of α,β-methylene ATP and 6-hydroxydopamine on sympathetic neurotransmission in the vas deferens of the guinea-pig, rat and mouse: support for co-transmission. *British Journal of Pharmacology*, **89**, 647–659.

Allen, J.M., Bircham, P.M.M., Edwards, A.V. and Bloom, S.R. (1983) Neuropeptide Y (NPY) reduces myocardial perfusion and inhibits the force of heart contraction of the isolated perfused rabbit heart. *Regulatory Peptides*, **6**, 247–254.

Allen, J.M., Bircham, P.M.M., Bloom, S.R. and Edwards, A.V. (1984) Release of neuropeptide Y in response to splanchnic nerve stimulation in the conscious calf. *Journal of Physiology*, **357**, 401–408.

Amann, R., Hankins, M.W. and Dray, A. (1988) Actions of neuropeptide K and calcitonin gene-related peptide on inferior mesenteric ganglion cells – tachykinin interactions with non-cholinergic potentials evoked by ureteric nerve stimulation. *Neuroscience Letters*, **85**, 125–130.

Ambache, N. and Zar, M.A. (1970) Non-cholinergic transmission by post-ganglionic motor neurons in the mammalian bladder. *Journal of Physiology*, **210**, 761–783.

Ambache, N. and Zar, M.A. (1971) Evidence against adrenergic motor transmission in the guinea-pig vas deferens. *Journal of Physiology*, **216**, 359–389.

Ambache, N., Verney, J. and Zar, M.A. (1970) Evidence for the release of two atropine-resistant spasmogens from Auerbach's plexus. *Journal of Physiology*, **207**, 761–782.

Ambache, N., Killick, S.W. and Zar, M.A. (1975) Extraction from ox retractor penis of an inhibitory substance which mimics its atropine resistant neurogenic relaxation. *British Journal of Pharmacology*, **54**, 409–410.

Anderson, L.C., Garrett, J.R., Johnson, D.A., Kauffman, D.L., Keller, P.J. and Thulin, A. (1984) Influence of circulatory catecholamines on protein secretion into rat parotid saliva during parasympathetic stimulation. *Journal of Physiology*, **352**, 163–171.

Andersson, P.O., Bloom, S.R., Edwards, A.V. and Järhult, J. (1982) Effects of stimulation of the chord tympani in bursts on submaxillary responses in the cat. *Journal of Physiology*, **322**, 469–483.

Andersson, P.O., Bloom, S.R. and Mellander, S. (1984) Haemodynamics of pelvic nerve induced penile erection in the dog: possible mediation by vasoactive intestinal polypeptide. *Journal of Physiology*, **350**, 209–224.

Andersson, P.O., Bjornberg, J., Bloom, S.R. and Mellander, S. (1987) Vasoactive intestinal polypeptide in relation to penile erection in the cat evoked by pelvic and by hypogastric nerve stimulation. *Journal of Urology*, **138**, 419–422.

Andriantsitohaina, R. and Stoclet, J.C. (1988) Potentiation by neuropeptide Y of vasoconstriction in rat resistance arteries. *British Journal of Pharmacology*, **95**, 419–428.

Andriantsitohaina, R. and Stoclet, J.C. (1990) Enhancement by neuropeptide Y (NPY) of the dihydropyridine-sensitive component of the response to α_1-adrenoceptor stimulation in rat isolated mesenteric arterioles. *British Journal of Pharmacology*, **99**, 389–395.

Angus, J.A. and Korner, P.I. (1980) Evidence against presynaptic α-adrenoceptor modulation of cardiac sympathetic transmission. *Nature*, **286**, 28–291.

Angus, J.A., Broughton, A. and Mulvany, M.J. (1988) Role of α-adrenoceptors in constrictor responses of rat, guinea-pig and rabbit small arteries to neural activation. *Journal of Physiology*, **403**, 495–510.

Åstrand, P. and Stjärne, L. (1989) ATP as a sympathetic co-transmitter in rat vasomotor nerves – further evidence that individual release sites respond to nerve impulses by intermittent release of single quanta. *Acta Physiologica Scandinavica*, **136**, 355–365.

Axelsson, M. and Nilsson, S. (1985) Control of the heart in the mudpuppy Necturus maculous. *Experimental Biology*, **44**, 229–239.

Balasubramaniam, A., Grupp I., Matlib, M.A., Benza, R., Jackson, R.L., Fischer, J.E. and Grupp, G. (1988) Comparison of the effects of neuropeptide Y (NPY) and 4-norleucine-NPY on isolated perfused rat hearts; effects of NPY on atrial and ventricular strips of rat heart and on rabbit heart mitochondria. *Regulatory Peptides*, **21**, 289–299.

Bartho, L., Holzer, P., Donnerer, J. and Lembeck, F. (1982) Evidence for the involvement of substance P in the atropine-resistant peristalsis of the guinea-pig ileum. *Neuroscience Letters*, **32**, 69–74.

Bastiaensen, E., Miserez, B. and De Potter, W. (1988) Subcellular fractionation of bovine ganglion stellatum: co-storage of noradrenaline, Met-enkephalin and neuropeptide Y in large 'dense-cored' vesicles. *Brain Research*, **442**, 124–130.

Bauer, V. and Kuriyama, H. (1982a) Evidence for non-cholinergic, non-adrenergic transmission in the guinea-pig ileum. *Journal of Physiology*, **330**, 95–110.

Bauer, V. and Kuriyama, H. (1982b) The nature of non-cholinergic transmission in longitudinal and circular muscles of the guinea-pig ileum. *Journal of Physiology*, **332**, 375–391.

Bauer, V., Matusák, O. and Kuriyama, H. (1982) Non-cholinergic, non-adrenergic responses to nerve stimulation of different regions of guinea-pig small intestine. *Naunyn-Schmiedeberg's Archives of Pharmacology*, **319**, 108–114.

Bell, C. (1968) Dual vasoconstrictor and vasodilator innervation of the uterine arterial supply in the guinea-pig. *Circulation Research*, **23**, 279–289.

Bell, C. (1969) Transmission from vasoconstrictor and vasodilator nerves to single smooth muscle cells of the guinea-pig uterine artery. *Journal of Physiology*, **205**, 695–708.

Bell, C. (1972) Mechanism of enhancement by angiotensin II of sympathetic adrenergic transmission in the guinea-pig. *Circulation Research*, **31**, 348–355.

Bell, C. (1975) Vasodilator nerves in regional circulatory control. *Clinical and Experimental Pharmacology and Physiology*, **Suppl 2**, 49–53.

Bell, C. (1985) Comparison of the antagonistic effects of phentolamine on vasoconstrictor responses to exogenous and neurally released noradrenaline *in vivo*. *British Journal of Pharmacology*, **85**, 249–253.

Bevan, J.A., Buga, G.M., Jope, C.A., Jope, R.S. and Moritoki, H. (1982) Further evidence for a muscarinic component to the neural vasodilator innervation of cerebral and cranial extracerebral arteries of the cat. *Circulation Research*, **51**, 421–429.

Bevan, J.A., Moscowitz, M., Said, S.I. and Buga, G. (1984) Evidence that vasoactive intestinal polypeptide is a dilator transmitter to some cerebral and extracerebral cranial arteries. *Peptides*, **5**, 385–388.

Biancini, P., Beinfeld, M.C., Hillemeier, C. and Behar, J. (1989) Role of peptide histidine isoleucine in relaxation of cat lower esophageal sphincter. *Gastroenterology*, **97**, 1083–1089.

Birnbaumer, L. (1990) G proteins in signal transduction. *Annual Review of Pharmacology and Toxicology*, **30**, 675–706.

Bloom, S.R. and Edwards, A.V. (1980a) Effects of autonomic stimulation on the release of vasoactive intestinal peptide from the gastrointestinal tract of the calf. *Journal of Physiology*, **299**, 437–452.

Bloom, S.R. and Edwards, A.V. (1980b) Vasoactive intestinal peptide in relation to atropine vasodilation in th submaxillary gland of the cat. *Journal of Physiology*, **300**, 41–53.

Bloom, S.R., Edwards, A.V. and Garrett, J.R. (1987) Effects of stimulating the sympathetic innervation in bursts on submandibular vascular and secretory function in cats. *Journal of Physiology*, **393**, 91–106.

Böke, T. and Malik, K.U. (1983) Enhancement by locally generated angiotensin II of release of the adrenergic transmitter in the isolated rat kidney. *Journal of Pharmacology and Experimental Therapeutics*, **226**, 900–907.

Bolton, T.B. and Large, W.A (1986) Are junction potentials essential? Dual mechanism of smooth muscle cell activation by transmitter released from autonomic nerves. *Quarterly Journal of Experimental Pharmacology*, **71**, 1–28.

Bornstein, J.C., Costa, M., Furness, J.B. and Lang, R.J. (1986) Electrophysiological analysis of projections of enteric inhibitory motoneurones in the guinea-pig small intestine. *Journal of Physiology*, **370**, 61–74.

Bowers, C.W., Jan, L.Y. and Jan, Y.N. (1986) A substance P-like peptide in bullfrog autonomic nerve terminals: anatomy, biochemistry and physiology. *Neuroscience*, **19**, 343–356.

Bowman, A. and Gillespie, J.S. (1983) Neurogenic vasodilatation in isolated bovine and canine penile arteries. *Journal of Physiology*, **341**, 603–616.

Brading, A.F. and Mostwin, J.L. (1989) Electrical and mechanical responses of guinea-pig bladder muscle to nerve stimulation. *British Journal of Pharmacology*, **98**, 1083–1090.

Bramich, N.J., Edwards, F.R. and Hirst, G.D.S. (1990) Sympathetic nerve stimulation and applied transmitters on the sinus venosus of the toad. *Journal of Physiology*, **429**, 349–375.

Brayden, J.E. (1987) Atropine potentiates neurogenic vasodilatation of the feline infraorbital artery: possible mechanisms. *Neuroscience Letters*, **78**, 343–348.

Brayden, J.E. and Bevan, J.A. (1986) Evidence that vasoactive intestinal polypeptide (VIP) mediates neurogenic vasodilation of feline cerebral arteries. *Stroke*, **17**, 1189–1192.

Brayden, J.E. and Large, W.A. (1986) Electrophysiological analysis of neurogenic vasodilatation in the isolated lingual artery of the rabbit. *British Journal of Pharmacology*, **89**, 163–171.

Brenneman, D.E. and Foster, G.A. (1987) Structural specificity of peptides influencing neuronal survival during development. *Peptides*, **8**, 687–694.

Brenneman, D.E., Neale, E.A., Foster, G.A., D'Autremont, S.W. and Westbrook, G.L. (1987) Non-neuronal cells mediate neurotrophic action of vasoactive intestinal peptide. *Journal of Cell Biology*, **104**, 1603–1610.

Brown, C.L. and Hanley, M.R. (1981) The effect of substance P and related peptides on amylase release from rat parotid gland slices. *British Journal of Pharmacology*, **73**, 517–523.

Brown, D.A., Constanti, A. and Adams, P.R. (1981) Slow cholinergic and peptidergic transmission in sympathetic ganglia. *Federation Proceedings*, **40**, 2625–2630.

Budai, D. and Duckles, S.P. (1988) Influence of stimulation train length on the opioid-induced inhibition of norepinephrine release in the rabbit ear artery. *Journal of Pharmacology and Experimental Therapeutics*, **247**, 839–843.

Bülbring, E. and Burn, J.H. (1935) The sympathetic dilator fibres in the muscles of the cat and dog. *Journal of Physiology*, **83**, 483–501.

Bulloch, A.G.M. (1987) Somatostatin enhances neurite outgrowth and electrical coupling of regenerating neurons in Helisoma. *Brain Research*, **412**, 6–17.

Bulloch, J.M. and McGrath, J.C. (1988a) Blockade of vasopressor and vas deferens responses by α,β-methylene ATP in the pithed rat. *British Journal of Pharmacology*, **94**, 103–108.

Bulloch, J.M. and McGrath, J.C. (1988b) Selective blockade by nifedipine of 'purinergic' rather than adrenergic nerve-mediated vasopressor responses in the pithed rat. *British Journal of Pharmacology*, **95**, 695–700.

Bulloch, J.M. and Starke, K. (1990) Presynaptic α_2-autoinhibition in a vascular neuroeffector junction where ATP and noradrenaline act as co-transmitters. *British Journal of Pharmacology*, **99**, 279–284.

Bult, H., Boeckxstaens, G.E., Pelckmans, P.A., Jordaens, F.H., Van Maercke, Y.M. and Herman, A.G. (1990) Nitric oxide as an inhibitory non-adrenergic, non-cholinergic neurotransmitter. *Nature*, **345**, 346–347.

Burnstock, G. (1969) Evolution of the autonomic innervation of visceral and cardiovascular systems in vertebrates. *Pharmacological Reviews*, **21**, 247–324.

Burnstock, G. (1972) Purinergic nerves. *Pharmacological Reviews*, **24**, 509–581.

Burnstock, G. (1976) Do some nerve cells release more than one transmitter? *Neuroscience*, **1**, 239–248.

Burnstock, G. (1980) Cholinergic and purinergic regulation of blood vessels. In *Handbook of Physiology, Section 2: The Cardiovascular System, Volume II. Vascular Smooth Muscle.*, edited by Bohr, D.F., Somlyo, A.P. and Sparks, H.V., pp. 567–612. Bethesda: American Physiological Society.

Burnstock, G. and Campbell, G. (1963) Comparative physiology of the vertebrate autonomic nervous system. II. Innervation of the urinary bladder of the ringtail possum (Pseudocheiras peregrinus). *Journal of Experimental Biology*, **40**, 421–436.

Burnstock, G. and Holman, M.E. (1961) The transmission of excitation from autonomic nerve to smooth muscle. *Journal of Physiology*, **155**, 115–133.

Burnstock, G. and Holman, M.E. (1964) An electrophysiological investigation of the actions of some autonomic blocking drugs on transmission in the guinea pig vas deferens. *British Journal of Pharmacology*, **23**, 600–612.

Burnstock, G. and Kennedy, C. (1986) A dual function for adenosine 5'-triphosphate in the regulation of vascular tone. Excitatory cotransmitter with noradrenaline from perivascular nerves and locally released inhibitory intravascular agent. *Circulation Research*, **58**, 319–330.

Burnstock, G. and Warland, J.J. (1987) A pharmacological study of the rabbit saphenous artery *in vitro*: a vessel with a large purinergic contractile response to sympathetic nerve stimulation. *British Journal of Pharmacology*, **90**, 111–120.

Burnstock, G. and Wood, M. (1967) Innervation of the urinary bladder of the sleepy lizard (Trachydosaurus rugosus). II. Physiology and pharmacology. *Comparative Biochemistry and Physiology*, **20**, 675–690.

Burnstock, G., O'Shea, J. and Wood, M. (1963) Comparative physiology of the vertebrate autonomic nervous system. I. Innervation of the urinary bladder of the toad (Bufo marinus). *Journal of Experimental Biology*, **40**, 403–419.

Burnstock, G., Dumsday, B. and Smythe, A. (1972) Atropine-resistant excitation of the urinary bladder: the possibility of transmission via nerves releasing a purine nucleotide. *British Journal of Pharmacology*, **44**, 451–461.

Burnstock, G., Cocks, T., Crowe, R. and Kasakov, L. (1978a) Purinergic innervation of the guinea pig urinary bladder. *British Journal of Pharmacology*, **63**, 125–138.

Burnstock, G., Cocks, T., Kasakov, L. and Wong, H.K. (1978b) Direct evidence for ATP release from non-adrenergic, non-cholinergic ('purinergic') nerves in the guinea-pig taenia coli and bladder. *European Journal of Pharmacology*, **49**, 145–149.

Burnstock, G., Crowe, R. and Wong, H.K. (1979) Comparative pharmacological and histochemical evidence for purinergic inhibitory innervation of the portal vein of the rabbit, but not guinea-pig. *British Journal of Pharmacology*, **65**, 377–388.

Byrne, N.G. and Large, W.A. (1986) The effect of α,β-methylene ATP on the depolarization evoked by noradrenaline (γ-adrenoceptor response) and ATP in the immature rat basilar artery. *British Journal of Pharmacology*, **88**, 6–8.

Bywater, R.A.R. and Taylor, G.S. (1983) Non-cholinergic fast and slow post-stimulus depolarization in the guinea-pig ileum. *Journal of Physiology,* **340**, 45–56.

Bywater, R.A.R., Holman, M.E. and Taylor, G.S. (1981) Atropine-resistant depolarization in the guinea-pig small intestine. *Journal of Physiology*, **316**, 369–378.

Bywater, R.A.R., Campbell, G., Edwards, F.R., Hirst, G.D.S. and O'Shea, J.E. (1989) The effects of vagal stimulation and applied acetylcholine on the sinus venosus of the toad. *Journal of Physiology*, **415**, 35–56.

Callahan, S.M. and Creed, K.E. (1986) Non-cholinergic neurotransmission and the effect of peptides on the urinary bladder of guinea-pigs and rabbits. *Journal of Physiology*, **374**, 103–115.

Campbell, G. (1970) Autonomic nervous supply to effector tissues. In *Smooth Muscle*, edited by Bülbring, E., Brading, A., Jones, A. and Tomita, T., pp. 451–495. London: Edward Arnold.

Campbell, G. (1987) Cotransmission. *Annual Review of Pharmacology and Toxicology,* **27**, 51–70.

Campbell, G. and Gibbins, I.L. (1979) Nonadrenergic, noncholinergic transmission in the autonomic nervous system. In *Trends in Autonomic Pharmacology, volume 1*, edited by Kalsner, S., pp. 103–144. Baltimore-Munich: Urban & Schwarzenberg.

Campbell, G. and Jackson, F. (1985) Independent co-release of acetylcholine and somatostatin from cardiac vagal neurones in toad. *Neuroscience Letters*, **60**, 47–50.

Campbell, G., Gibbins, I.L., Morris, J.L., Furness, J.B., Costa, M., Oliver, J.O., Beardsley, A.M. and Murphy, R. (1982) Somatostatin is contained in and released from cholinergic nerves in the heart of the toad Bufo marinus. *Neuroscience*, **7**, 2013–2023.

Campbell, G., Edwards, F.R., Hirst, G.D.S. and O'Shea, J.E. (1989) Effects of vagal stimulation and applied acetylcholine on pacemaker potentials in the guinea-pig heart. *Journal of Physiology*, **415**, 57–68.

Carati, C.J., Creed, K.E. and Keogh, E.J. (1988) Vascular changes during penile erection in the dog. *Journal of Physiology*, **400**, 75–88.

Carpenter, F.G. (1977) Atropine resistance and muscarinic receptors in the rat urinary bladder. *British Journal of Pharmacology*, **59**, 43–49.

Carpenter, F.G. and Rand, S.A. (1965) Relation of acetylcholine release to responses of the rat urinary system. *Journal of Physiology*, **180**, 371–382.

Caverson, M.M., Bachoo, M., Ciriello, J. and Polosa, C. (1989) Effect of preganglionic stimulation or chronic decentralization on neurotensin-like immunoreactivity in sympathetic ganglia of the cat. *Brain Research*, **482**, 365–370.

Changeux, J.P. (1986) Co-existence of neuronal messengers and molecular selection. *Progress in Brain Research*, **68**, 373–403.

Chang, M.M. and Leeman, S.E. (1970) Isolation of a sialogogue peptide from bovine hypothalamic tissue and its characterization as substance P. *Journal of Biological Chemistry*, 245, 4784–4790.

Chaudhry, A., Downie, J.W. and White, T.D. (1984) Tetrodotoxin-resistant release of ATP from superfused rabbit detrusor muscle during electrical field stimulation in the presence of luciferin-luciferase. *Canadian Journal of Pharmacology and Physiology*, **62**, 153–156.

Chernaeva, L. and Charakchieva, S. (1988) Leucine-enkephalin- and neuropeptide Y-modulation of [^{3}H]noradrenaline release in the oviduct of mature and juvenile rabbits. *General Pharmacology*, **19**, 137–142.

Chesher, G.B. (1967) Acetylcholine in extract and perfusates of urinary bladder. *Journal of Pharmacy and Pharmacology*, **19**, 445–455.

Chesher, G.B. and Thorp, R.H. (1965) The atropine resistance of the response to intrinsic nerve stimulation of the guinea-pig bladder. *British Journal of Pharmacology and Chemotherapy*, **25**, 288–294.

Cheung, D.W. (1982) Two components in the cellular response of rat tail arteries to nerve stimulation. *Journal of Physiology*, **328**, 461–468.

Cheung, D.W. (1984) Neural regulation of electrical and mechanical activities in the rat tail artery. *Pflügers Archive*, **400**, 335–337.

Cheung, D.W. (1985) An electrophysiological study of α-adrenoceptor mediated excitation-contraction coupling in the smooth muscle cells of the rat saphenous vein. *British Journal of Pharmacology*, **84**, 265–271.

Cheung, D.W. and Fujioka, M. (1986) Inhibition of the excitatory junction potential in the guinea-pig saphenous artery by ANAPP$_3$. *British Journal of Pharmacology*, **89**, 3–5.

Choo, L.K. and Mitchelson, F. (1980a) The effect of troxypyrrolidinium, a choline uptake inhibitor, on the excitatory innervation of the rat urinary bladder. *European Journal of Pharmacology*, **61**, 293–301.

Choo, L.K. and Mitchelson, F. (1980b) The effect of indomethacin and adenosine 5'-triphosphate on the excitatory innervation of the rat urinary bladder. *Canadian Journal of Physiology and Pharmacology*, **58**, 1042–1048.

Christofides, N.D., Yiangou, Y., Tatemoto, K. and Bloom, S.R. (1984) Characterization of peptide histidine isoleucine-like immunoreactivity in the rat, human, guinea-pig and cat gastrointestinal tracts – evidence for species differences. *Digestion*, **30**, 165–170.

Chuang, D.M. (1989) Neurotransmitter receptors and phosphoinositide turnover. *Annual Review of Pharmacology and Toxicology*, **129**, 71–110.

Cocks, T. and Burnstock, G. (1979) Effects of neuronal polypeptides on intestinal smooth muscle: a comparison with non-adrenergic, non-cholinergic nerve stimulation and ATP. *European Journal of Pharmacology*, **54**, 251–259.

Constantine, J.W., Gunnell, D. and Weeks, R.A. (1980) α_1- and α_2-vascular adrenoceptors in the dog. *European Journal of Pharmacology*, **66**, 281–286.

Costa, M. and Furness, J.B. (1982) Nervous control of intestinal motility. In *Mediators and drugs in gastrointestinal motility. I. (Handbook of Experimental Pharmacology, vol. 59)*, edited by Bertaccini, G., pp. 279–382. Berlin: Springer-Verlag.

Costa, M. and Furness, J.B. (1984) Somatostatin is present in a subpopulation of noradrenergic nerve fibres supplying the intestine. *Neuroscience*, **13**, 911–920.

Costa, M., Furness, J.B., Pullin, C.O. and Bornstein, J.C. (1985) Substance P enteric neurons mediate non-cholinergic transmission to the circular muscle of the guinea-pig intestine. *Naunyn-Schmiedeberg's Archives of Pharmacology*, **328**, 446–453.

Costa, M., Furness, J.B. and Gibbins, I.L. (1986a) Chemical coding of enteric neurons. *Progress in Brain Research*, **68**, 217–239.

Costa, M., Furness, J.B. and Humphreys, C.M.S. (1986b) Apamin distinguishes two types of relaxation mediated by enteric nerves in the guinea-pig gastrointestinal tract. *Naunyn-Schmiedeberg's Archives of Pharmacology*, **332**, 79–88.

Costa, M., Furness, J.B. and Llewellyn-Smith, I.J. (1987) Histochemistry of the enteric nervous system. In *Physiology of the Gastrointestinal Tract, 2nd edition*, edited by Johnson, L.R., pp. 1–40. New York: Raven Press.

Cowan, W.D. and Daniel, E.E. (1983) Human female bladder and its non-cholinergic contractile function. *Canadian Journal of Physiology and Pharmacology*, **61**, 1236–1246.

Cox, H.M. and Cuthbert, A.W. (1988) Neuropeptide Y antagonizes secretogogue evoked chloride transport in rat jejunal epithelium. *Pflügers Archive*, **413**, 38–42.

Craggs, M.D. and Stephenson, J.D. (1982) The effects of parasympathetic blocking agents on bladder electromyograms and function in conscious and anaesthetized cats. *Neurophysiology*, **21**, 695–703.

Craggs, M.D., Rushton, D.N. and Stephenson, J.D. (1986) A putative non-cholinergic mechanism in urinary bladders of new but not old world monkeys. *Journal of Urology*, **136**, 1348–1350.

Creed, K.E. (1979) The role of the hypogastric nerve in bladder and urethral activity of the dog. *British Journal of Pharmacology*, **65**, 367–375.

Creed, K.E., Whikawa, S. and Ito, Y. (1983) Electrical and mechanical activity recorded from rabbit urinary bladder in response to nerve stimulation. *Journal of Physiology*, **338**, 149–164.

Creed, K.E., Carati, C.J. and Keogh, E.J. (1988) Autonomic control and vascular changes during penile erection in monkeys. *British Journal of Urology*, **61**, 510–515.

Crowcroft, P.J. and Szurszewski, J.H. (1971) A study of the inferior mesenteric and pelvic plexus ganglia of guinea-pig with intracellular electrodes. *Journal of Physiology*, **219**, 421–441.

Crowe, R., Haven, A.J. and Burnstock, G. (1986) Intramural neurones of the guinea-pig urinary bladder: histochemical localization of putative neurotransmitters in cultures and newborn animals. *Journal of the Autonomic Nervous System*, **15**, 319–339.

Cunnane, T.C. and Manchanda, R. (1989a) Simultaneous intracellular and focal extracellular recording of junction potentials and currents, and the time course of quantal transmitter action in rodent vas deferens. *Neuroscience*, **30**, 563–575.

Cunnane, T.C. and Manchanda, R. (1989b) Effects of reserpine pretreatment on neuroeffector transmission in the vas deferens. *Clinical and Experimental Pharmacology and Physiology*, **16**, 451–455.

Dahlen, S.E. and Hedqvist, P. (1980) ATP, β-γ-methylene-ATP, and adenosine inhibit non-cholinergic, non-adrenergic transmission in rat urinary bladder. *Acta Physiologica Scandinavica*, **109**, 137–142.

Dahlöf, C., Dahlöf, P. and Lundberg, J.M. (1985a) Neuropeptide Y (NPY)-enhancement of blood pressure increase upon α-adrenoceptor activation and direct pressor effects in pithed rats. *European Journal of Pharmacology*, **109**, 289–292.

Dahlöf, C., Dahlöf, P., Tatemoto, K. and Lundberg, J.M. (1985b) Neuropeptide Y (NPY) reduces field stimulation-evoked release of noradrenaline and enhances force of contraction in the rat portal vein. *Naunyn-Schmiedeberg's Archives of Pharmacology*, **328**, 327–330.

Dahlöf, C., Dahlöf, P. and Lundberg, J.M. (1986) α_2-adrenoceptor-mediated inhibition of nerve stimulation-evoked release of neuropeptide Y (NPY)-like immunoreactivity in the pithed guinea-pig. *European Journal of Pharmacology*, **131**, 279–283.

Dail, W.G., and Hamill, R.W. (1989) Parasympathetic nerves in penile erectile tissue of the rat contain choline acetyltransferase. *Brain Research*, **487**, 165–170.

Dail, W.G., Moll, M.A. and Weber, K. (1983) Localization of vasoactive intestinal polypeptide in penile erectile tissue and in the major pelvic ganglion of the rat. *Neuroscience*, **10**, 1379–1386.

Dail, W.G. Minorsky, N., Moll, M.A and Manzanares, K. (1986) The hypogastric nerve pathway to penile erectile tissue: histochemical evidence supporting a vasodilator role. *Journal of the Autonomic Nervous System*, **15**, 341–349.

Dalsgaard, C.J., Hökfelt, T., Johansson, O. and Elde, R. (1981) Somatostatin immunoreactive cell bodies in the dorsal horn and the parasympathetic intermediolateral nucleus of the rat spinal cord. *Neuroscience Letters*, **27**, 335–340.

Dalsgaard, C.J., Hökfelt, T., Elfvin, L.G. and Terenius, L. (1982) Enkephalin containing sympathetic neurons projecting to the inferior mesenteric ganglion: evidence from combined retrograde tracing and immunohistochemistry. *Neuroscience*, **7**, 2039–2050.

Dalsgaard, C.J., Hökfelt, T., Schultzberg, M., Lundberg, J.M., Terenius, L., Dockray, G.J. and Goldstein, M. (1983a) Origins of peptide-containing fibers in the inferior mesenteric ganglion of the guinea-pig: immunohistochemical studies with antisera to substance P, enkephalin, vasoactive intestinal polypeptide, cholecystokinin and bombesin. *Neuroscience*, **9**, 191–211.

Dalsgaard, C.J., Vincent, S.R., Hökfelt, T., Christensson, I. and Terenius, L. (1983b) Separate origins for the dynorphin and enkephalin immunoreactive fibres in the inferior mesenteric ganglion of the guinea-pig. *Journal of Comparative Neurology*, **221**, 482–489.

Dalsgaard, C.J., Fransco-Cerecceda, A., Saria, A., Lundberg, J.M., Theordorsson-Norheim, E. and Hökfelt, T. (1986) Distribution and origin of substance P- and neuropeptide Y-immunoreactive nerves in the guinea-pig heart. *Cell and Tissue Research*, **243**, 477–485.

Dalziel, H.H., Gray, G.A., Drummond, R.M., Furman, B.L. and Sneddon, P. (1990) Investigation of the selectivity of α,β-methylene ATP in inhibiting vascular responses of the rat *in vivo* and *in vitro*. *British Journal of Pharmacology*, **99**, 820–824.

Daniel, E.E., Jager, L.P. and Jury, J. (1989) Vasoactive intestinal polypeptide and non-adrenergic, non-cholinergic inhibition in lower oesophageal sphincter of opossum. *British Journal of Pharmacology*, **96**, 746–752.

Daniell, H.B., Walle, T., Gaffney, T.E. and Webb, J.G. (1979) Stimulation-induced release of propanolol and norepinephrine from adrenergic neurons. *Journal of Pharmacology and Experimental Therapeutics*, **208**, 354–359.

Day, S., Gu, J., Polak, J. and Bloom, S. (1985) Somatostatin in the human heart and comparison with guinea-pig and rat heart. *British Heart Journal*, **53**, 153–157.

Dean, D.M. and Downie, J.W. (1978) Contribution of adrenergic and 'purinergic' neurotransmission to contraction in rabbit detrusor. *Journal of Pharmacology and Experimental Therapeutics*, **207**, 431–445.

Deckers, I.A., Rampart, M., Bult, H. and Herman, A.G. (1989) Evidence for the involvement of prostaglandins in modulation of acetylcholine release from canine bronchial tissue. *European Journal of Pharmacology*, **167**, 415–418.

De Deyn, P.P., Pickut, B.A., Verzwijvelen, A. and D'Hooge, R. (1989) Subcellular distribution and axonal transport of noradrenaline, dopamine-β-hydroxylase and neuropeptide Y in dog splenic nerve. *Neurochemistry International*, **15**, 39–47.

De Groat, W.C. and Kawatani, M. (1989) Enkephalinergic inhibition in parasympathetic ganglia of the urinary bladder of the cat. *Journal of Physiology*, **413**, 13–29.

Diez, J., Tamargo, J. and Valenzuela, C. (1985) Negative inotropic effect of somatostatin in guinea-pig atrial fibres. *British Journal of Pharmacology*, **86**, 547–555.

Dillingham, M.A. and Anderson, R.J. (1989) Mechanism of neuropeptide Y inhibition of vasopressin action in rat cortical collecting tubule. *American Journal of Physiology*, **256**, F408–F413.

Dimaline, R. and Vowles, L. (1988) Alternative processing pathways for preprovasoactive intestinal peptide in the enteric nervous system of the rat. *Regulatory Peptides*, **20**, 199–210.

Dodd, J. and Horn, J.P. (1983a) A reclassification of B and C neurones in the ninth and tenth paravertebral sympathetic ganglia of the bullfrog. *Journal of Physiology*, **334**, 255–270.

Dodd, J. and Horn, J.P. (1983b) Muscarinic inhibition of sympathetic C neurones in the bullfrog. *Journal of Physiology*, **334**, 271–292.

Donnerer, J., Barthó, L., Holzer, P. and Lembeck, F. (1984) Intestinal peristalsis associated with release of immunoreactive substance P. *Neuroscience*, **11**, 913–918.

Donoso, V., Silva, M., St-Pierre, S. and Huidobro-Toro, J.P. (1988) Neuropeptide Y (NPY), an endogenous presynaptic modulator of adrenergic neurotransmission in the rat vas deferens: structural and functional studies. *Peptides*, **9**, 545–553.

Dorr, L.D. and Brody, M.J. (1967) Hemodynamic mechanisms of erection in the canine penis. *American Journal of Physiology*, **213**, 1526–1531.

Downie, J.W. and Dean, D.M. (1977) The contribution of cholinergic post-ganglionic neurotransmission to contraction of rabbit detrusor. *Journal of Pharmacology and Experimental Therapeutics*, **203**, 417–425.

Dreux, C,. Imhoff, V. and Rossignol, B. (1987) [^{3}H] protein secretion in rat parotid gland: substance P – β-adrenergic synergism. *American Journal of Physiology*, **253**, G774–G782.

Dumsday, B.H. (1971) Atropine-resistance of the urinary bladder. *Journal of Pharmacy and Pharmacology*, **23**, 222–225.

Dun, N.J. (1983) Peptide hormones and transmission in sympathetic ganglia. In *Autonomic Ganglia*, edited by Elfvin, L.G., pp. 345–366. Chichester: John Wiley & Sons.

Dun, N.J. and Kiraly, M. (1983) Capsaicin causes release of a substance P-like peptide in guinea-pig inferior mesenteric ganglia. *Journal of Physiology*, **340**, 107–120.

Edge, N.D. (1955) A contribution to the innervation of the urinary bladder of the cat. *Journal of Physiology*, **127**, 54–68.

Edvinsson, L. and McCulloch, J. (1985) Distribution and vasomotor effects of peptide HI (PHI) in feline cerebral blood vessels *in vitro* and *in situ*. *Regulatory Peptides*, **10**, 345–356.

Edvinsson, L., Ekblad, E., Håkanson, R. and Wahlstedt, C. (1984) Neuropeptide Y potentiates the effects of various vasoconstrictor agents on rabbit blood vessels. *British Journal of Pharmacology*, **83**, 519–525.

Edvinsson, L., Gulbenkian, S., Wharton, J., Jansen, I. and Polak, J.M. (1989) Peptide-containing nerves in the rat femoral artery and vein. An immunocytochemical and vasomotor study. *Blood Vessels*, **26**, 254–271.

Edwall, B., Gazelius, B., Fazekus, A., Theodorsson-Norheim, E. and Lundberg, J.M. (1985) Neuropeptide Y (NPY) and sympathetic control of blood flow in oral mucosa and dental pulp in the cat. *Acta Physiologica Scandinavica*, **125**, 253–264.

Edwards, F.R., Hards, D., Hirst, G.D.S. and Silverberg, G.D. (1989) Noradrenaline (γ) and ATP responses of innervated and non-innervated rat cerebral arteries. *British Journal of Pharmacology*, **96**, 785–788.

Eiden, L.E., Loumaye, E., Sherwood, N. and Eskay, R.L. (1982) Two chemically and immunologically distinct forms of Lutenizing hormone-releasing hormone are differentially expressed in frog neural tissues. *Peptides*, **3**, 323–327.

Ekblad, E., Edvinsson L., Wahlstedt, C., Uddman, R., Håkanson, R. and Sundler, F. (1984a) Neuropeptide Y co-exists and co-operates with noradrenaline in perivascular nerve fibres. *Regulatory Peptides*, **8**, 225–235.

Ekbald, E., Håkanson, R. and Sundler, F. (1984b) VIP and PHI coexist with NPY-like peptide in intramural neurons of the small intestine. *Regulatory Peptides*, **10**, 47–56.

Ekström, J. and Olgart, L. (1986) Complementary action of substance P and vasoactive intestinal peptide on the rat parotid secretion. *Acta Physiologica Scandinavica*, **126**, 25–31.

Ekström, J., Månsson, B. and Tobin, G. (1983) Vasoactive intestinal polypeptide evoked secretion of fluid and protein from rat salivary glands and the development of supersensitivity. *Acta Physiologica Scandinavica*, **119**, 169–175.

Ekström, J., Månsson, B. and Tobin, G. (1984) Effects of atropine or high frequency burst excitation on the composition of parasympathetic rat parotid saliva. *Acta Physiologica Scandinavica*, **122**, 409–414.

Ekström, J., Brodin, E., Ekman, R., Håkanson, R., Månsson, R. and Tobin, G. (1985) Depletion of neuropeptides in rat parotid glands and declining atropine-resistant salivary secretion upon continuous parasympathetic nerve stimulation. *Regulatory Peptides*, **11**, 353–359.

Ekström, J., Månsson, B. and Tobin, G. (1987) Substance K and salivary secretion in the rat. *Pharmacology and Toxicology*, **60**, 104–107.

Ekström, J., Ekman, R., Håkanson, R., Sjögren, S. and Sundler, F. (1988a) Calcitonin gene-related peptide in rat salivary glands: neuronal localization, depletion upon nerve stimulation and effects on salivation in relation to substance P. *Neuroscience*, **26**, 933–949.

Ekström, J., Garrett, J.R., Månsson, B. and Tobin, G. (1988b) The effects of atropine and chronic sympathectomy on maximal parasympathetic stimulation of parotid saliva in rats. *Journal of Physiology*, **403**, 105–116.

Ekström, J., Ekman, R., Håkanson, R., Luts, A., Sundler, F. and Tobin, G. (1989a) Effects of capsaicin pretreatment on neuropeptides and salivary secretion of rat parotid glands. *British Journal of Pharmacology*, **97**, 1031–1038.

Ekström, J., Månsson, B., Nilsson, B.O., Rosengren, E. and Tobin, G. (1989b) Receptors involved in the nervous system regulations of polyamine metabolism in rat salivary glands. *Acta Physiologica Scandinavica*, **135**, 255–261.

Ellis, J.L. and Burnstock, G. (1989) Angiotensin neuromodulation of adrenergic and purinergic co-transmission in the guinea-pig vas deferens. *British Journal of Pharmacology*, **97**, 1157–1164.

Ellis, J.L. and Burnstock, G. (1990) Neuropeptide Y neuromodulation of sympathetic co-transmission in the guinea-pig vas deferens. *British Journal of Pharmacology*, **100**, 457–462.

Emmelin, N. and Homberg, J. (1967) Impulse frequency in secretory nerves of salivary glands. *Journal of Physiology*, **191**, 205–214.

Emmelin, N., Grampp, W. and Thesleff, P. (1980) Sympathetically evoked secretory potentials in the parotid gland of the cat. *Journal of Physiology*, **302**, 183–195.

Enyedi, P. and Fredholm, B.B. (1984) Calcium-dependent enhancement by carbachol of the VIP-induced cyclic AMP accumulation in cat submandibular gland. *Acta Physiologica Scandinavica*, **120**, 523–528.

Erichsen, J.T., Karten, H.J., Eldred, W.D. and Brecha, N.C. (1982) Localization of substance P-like and enkephalin-like immunoreactivity within preganglionic terminals of avian ciliary ganglion: light and electron microscopy. *Journal of Neuroscience*, **2**, 994–1003.

Erspamer, V., Ronzoni, G. and Erspamer, G.F. (1981) Effects of active peptides on the isolated muscle of the human urinary bladder. *Investigative Urology*, **18**, 302–304.

Eugéne, D. (1987) Fast non-cholinergic depolarizing post-synaptic potentials in neurons of rat superior cervical ganglion. *Neuroscience Letters*, **78**, 51–56.

Ewald. D.A., Sternweis, P.C. and Miller, R.J. (1988) Guanine nucleotide binding (G_o) induced coupling of NPY receptors to calcium channels in sensory neurons. *Proceedings of the National Academy of Science, USA*, **88**, 3633–3637.

Ewald, D.A., Pang, I.H., Sternweis, P.C. and Miller, R.J. (1989) Differential G protein-mediated coupling of neurotransmitter receptors to Ca^{2+} channels in rat dorsal root ganglion neurons *in vitro*. *Neuron*, **2**, 1185–1193.

Fahrenkrug, J., Galbo, H. and Holst, J.J. (1978a) Influence of the autonomic nervous system on the release of vasoactive intestinal polypeptide from the porcine gastrointestinal tract. *Journal of Physiology*, **280**, 405–422.

Fahrenkrug, J., Haglund, J., Jodal, M., Lundgren, O., Olbe, L. and Schaffalitzky de Muckadell, O.B. (1978b) Nervous release of vasoactive intestinal polypeptide in the gastrointestinal tract of cats – possible physiological implications. *Journal of Physiology*, **284**, 291–305.

Fedan, J.S., Hogaboom, G.K., O'Donnell, J.P., Colby, J. and Westfall, D.P. (1981) Contribution by purines to the neurogenic responses of the vas deferens of the guinea-pig. *European Journal of Pharmacology*, **69**, 41–53.

Flavahan, N.A., Grant, T.L., Greig, J. and McGrath, J.C. (1985) Analysis of the α-adrenoceptor-mediated, and other, components in the sympathetic vasopressor responses of the pithed rat. *British Journal of Pharmacology*, **86**, 265–274.

Franco, R., Costa, M. and Furness, J.B. (1979) Evidence for the release of endogenous substance P from intestinal nerves. *Naunyn-Schmiedeberg's Archives of Pharmacology*, **306**, 195–201.

Franco-Cereceda, A., Lundberg, J.M. and Dahlöf, C. (1985) Neuropeptide Y and sympathetic control of heart contractility and coronary vascular tone. *Acta Physiologica Scandinavica*, **124**, 361–369.

Franco-Cereceda, A., Lundberg, J.M. and Hökfelt, T. (1986) Somatostatin: an inhibitory parasympathetic transmitter in the human heart? *European Journal of Pharmacology*, **132**, 101–102.

Fredholm, B.B. and Hedqvist, P. (1980) Modulation of neurotransmission by purine nucleotides and nucleosides. *Biochemical Pharmacology*, **29**, 1635–1643.

Fredholm, B.B., Jansen, I. and Edvinsson, L. (1985) Neuropeptide Y is a potent inhibitor of cylic AMP accumulation in feline cerebral blood vessels. *Acta Physiologica Scandinavica*, **124**, 467–469.

Fried, G., Lundberg, J.M. and Theodorsson-Norheim, E. (1985a) Subcellular storage and axonal transport of neuropeptide Y (NPY) in relation to catecholamines in the cat. *Acta Physiologica Scandinavica*, **125**, 145–154.

Fried, G., Terenius, L., Hökfelt, T. and Goldstein, M. (1985b) Evidence for differential localization of noradrenaline and neuropeptide Y in neuronal storage vesicle isolated from rat vas deferens. *Journal of Neuroscience*, **5**, 450–458.

Fried, G., Terenius, L., Brodin, E., Efendic, S., Dockray, G., Fahrenkrug, J., Goldstein, M. and Hökfelt, T. (1986) Neuropeptide Y, enkephalin and noradrenaline coexist in sympathetic neurons innervating the bovine spleen. Biochemical and immunohistochemical evidence. *Cell and Tissue Research*, **243**, 495–508.

Fujii, K. (1988) Evidence for adenosine triphosphate as an excitatory transmitter in guinea-pig, rabbit and pig urinary bladder. *Journal of Physiology*, **404**, 39–52.

Fujioka, M. and Cheung, D.W. (1987) Autoregulation of neuromuscular transmission in the guinea-pig saphenous artery. *European Journal of Pharmacology,* **139**, 147–153.

Fujisawa, K. and Ito, Y. (1982) The effects of substance P on smooth muscle cells and on neuro-effector transmission in the guinea-pig ileum. *British Journal of Pharmacology*, **76**, 279–290.

Fujiwara, S., Itoh, T. and Suzuki, H. (1982) Membrane properties and excitatory neuromuscular transmission in the smooth muscle of dog cerebral arteries. *British Journal of Pharmacology*, **77**, 197–208.

Furness, J.B. (1974) Transmission to the longitudinal muscle of the guinea-pig vas deferens: the effect of pretreatment with guanethidine. *British Journal of Pharmacology*, **50**, 63–68.

Furness, J.B. and Costa, M. (1980) Types of nerves in the enteric nervous system. *Neuroscience*, **5**, 1–20.

Furness, J.B. and Costa, M. (1987) *The Enteric Nervous System.*, Edinburgh: Churchill Livingston.

Furness, J.B. and Iwayama, T. (1972) The arrangement and identification of axons innervating the vas deferens of the guinea-pig. *Journal of Anatomy*, **113**, 179–196.

Furness, J.B., Costa, M. and Keast, J.R. (1984) Choline acetyltransferase and peptide immunoreactivity of submucous neurons in the small intestine of the guinea-pig. *Cell and Tissue Research*, **237**, 328–336.

Furness, J.B., Costa, M., Gibbins, I.L., Llewellyn-Smith, I.J. and Oliver, J.R. (1985) Neurochemically similar myenteric and submucous neurons directly traced to the mucosa of the small intestine. *Cell and Tissue Research*, **241**, 155–163.

Furness, J.B., Llwellyn-Smith, I.J., Bornstein, J.C. and Costa, M. (1988) Chemical neuroanatomy and the analysis of neuronal circuitry in the enteric nervous system. In *Chemical Neuroanatomy. Vol 6. The Peripheral Nervous System.*, edited by Björklund, A., Hökfelt, T. and Owman, C., pp. 151–218. Amsterdam: Elsevier.

Furness, J.B., Morris, J.L., Gibbins, I.L. and Costa, M. (1989a) Chemical coding of neurons and plurichemical transmission. *Annual Reviews of Pharmacology and Toxicology*, **29**, 289–306.

Furness, J.B., Pompolo, S., Murphy, R. and Giraud, A. (1989b) Projections of neurons with neuromedin U-like immunoreactivity in the small intestine of the guinea-pig. *Cell and Tissue Research*, **257**, 415–422.

Gallacher, D.V. (1983) Substance P is a functional neurotransmitter in the rat parotid gland. *Journal of Physiology*, **342**, 483–498.

Gallacher, D.V. and Petersen, O.H. (1980) Substance P increases membrane conductance in parotid acinar cells. *Nature*, **283**, 393–395.

Gan, E.A. and Duckles, S.P. (1988) Comparison of the pre-junctional effect of opioids in two blood vessels of the rabbit. *European Journal of Pharmacology*, **158**, 21–28.

Garrett, J.R. (1987) The proper role of nerves in salivary secretion: a review. *Journal of Dental Research*, **66**, 387–397.

Gibbins, I.L. (1983) Peptide-containing nerves in the urinary bladder of the toad, Bufo marinus. *Cell and Tissue Research*, **229**, 137–144.

Gibbins, I.L. (1989a) Co-existence and co-function. In *The Comparative Physiology of Regulatory Peptides*, edited by Holmgren, S., pp. 308–343. London: Chapman & Hall.

Gibbins, I.L. (1989b) Dynorphin-containing pilomotor neurons in the superior cervical ganglion of guinea-pigs. *Neuroscience Letters*, **107**, 45–50.

Gibbins, I.L. (1990) Target-related patterns of co-existence of neuropeptide Y, vasoactive intestinal peptide, enkephalin and substance P in cranial parasympathetic neurons innervating the facial skin and exocrine glands of guinea-pigs. *Neuroscience*, **38**, 541–560.

Gibbins, I.L. (1991) Vasomotor, pilomotor and secretomotor neurons distinguished by size and neuropeptide content in superior cervical ganglia of mice. *Journal of the Autonomic Nervous System*, (in press).

Gibbins, I.L. and Morris, J.L. (1987) Co-existence of neuropeptides in sympathetic, cranial autonomic and sensory neurons innervating the iris of the guinea-pig. *Journal of the Autonomic Nervous System*, **21**, 67–82.

Gibbins, I.L. and Morris, J.L. (1988) Co-existence of immunoreactivity to neuropeptide Y and vasoactive intestinal peptide in non-noradrenergic axons innervating guinea-pig cerebral arteries after sympathectomy. *Brain Research*, **444**, 402–406.

Gibbins, I.L. and Morris, J.L. (1990) Sympathetic noradrenergic neurons containing dynorphin but not neuropeptide Y innervate small cutaneous blood vessels of guinea-pigs. *Journal of Autonomic Nervous System*, **29**, 137–149.

Gibbins, I.L., Brayden, J.E. and Bevan, J.A. (1984) Perivascular nerves with immunoreactivity to vasoactive intestinal polypeptide in cephalic arteries of the cat: distribution, possible origins and functional implications. *Neuroscience*, **13**, 1327–1346.

Gibbins, I.L., Furness, J.B. and Costa, M. (1987a) Pathway-specific patterns of co-existence of substance P, calcitonin gene-related peptide, cholecystokinin and dynorphin in neurons of the dorsal root ganglia of the guinea-pig. *Cell and Tissue Research*, **248**, 417–437.

Gibbins, I.L., Campbell, G., Morris, J.L., Nilsson, S. and Murphy, R. (1987b) Pathway-specific connections between peptide-containing preganglionic and postganglionic neurons in the vagus nerve of the toad (Bufo marinus). *Journal of the Autonomic Nervous System*, **20**, 43–55.

Gibbins, I.L., Morris, J.L., Furness, J.B. and Costa, M. (1988) Innervation of systemic blood vessels. In *Non-adrenergic Innervation of Blood Vessels. Vol. II. Regional Innervation.*, edited by Burnstock, G. and Griffith, S., pp. 1–36. Boca Raton: CRC Press.

Gibson, A. and Mirzazadeh, S. (1989) N-methylhydroxylamine inhibits and M&B 22948 potentiates relaxations of the mouse anococcygeus to non-adrenergic, non-cholinergic field stimulation and to nitrovasodilator drugs. *British Journal of Pharmacology*, **96**, 637–644.

Gillespie, J.S. and Martin, W. (1980) A smooth muscle inhibitory material from the bovine retractor penis and rat anococcygeus muscles. *Journal of Physiology*, **309**, 55–64.

Gillespie, J.S. and Sheng, H. (1988) Influence of haemoglobin and erythrocytes on the effects of EDRF, a smooth muscle inhibitory factor, and nitric oxide on vascular and non-vascular smooth muscle. *British Journal of Pharmacology*, **95**, 1151–1156.

Gillis, R.A., Pearle, D.L. and Hoekman, T. (1974) Failure of β-adrenergic blockade to prevent arrhythmias induced by sympathetic nerve stimulation. *Science*, **185**, 70–72.

Gintzler, A.R. and Scalisi, J.A. (1982) Effects of opioids on non-cholinergic excitatory responses of the guinea-pig isolated ileum: inhibition of release of enteric substance P. *British Journal of Pharmacology*, **75**, 199–206.

Gintzler, A.R. and Hyde, D. (1983) A specific substance P antagonist attenuates non-cholinergic electrically induced contradictions of the guinea-pig isolated ileum. *Neuroscience Letters*, **40**, 75–80.

Gjörstrup, P. (1979) Amylase secretion in rabbit parotid gland when stimulating the parasympathetic nerves during parasympathetic activity. *Journal of Physiology*, **296**, 443–451.

Gjörstrup, P. (1980) Parotid secretion of fluid and amylase in rabbits during feeding. *Journal of Physiology*, **309**, 101–116.

Glazer, E.J. and Basbaum, A.I. (1980) Leucine-enkephalin localization in and axoplasmic transport by sacral parasympathetic preganglionic neurons. *Science*, **208**, 1479–1481.

Goadsby, P.J. and MacDonald, G.J. (1985) Extracranial vasodilation mediated by vasoactive intestinal polypeptide (VIP). *Brain Research*, **329**, 285–288.

Goldstein, M., Kusano, N., Adler, C. and Meller, E. (1986) Characterization of central neuropeptide Y receptor binding sites and possible interactions with α_2-adrenoceptors. *Progress in Brain Research*, **68**, 331–335.

Goyal, R.K., Rattan, S. and Said, S.I. (1980) VIP as a possible neurotransmitter of non-cholinerigic, non-adrenergic inhibitory neurones. *Nature*, **288**, 378–380.

Granstam, E. and Nilsson, F.E. (1990) Non-adrenergic sympathetic vasoconstriction in the eye and some other facial tissue in the rabbit. *European Journal of Pharmacology*, **175**, 175–186.

Greco, A., Ghirlanda, G., Barone, C., Bertoli, A., Caputo, S., Uccioli, L. and Manna, R. (1984) Somatostatin in paroxysmal supraventricular and junctional tachycardia. *British Medical Journal*, **288**, 28.

Grider, J.R. (1989) Tachykinins as transmitters of ascending contractile component of the peristaltic reflex. *American Journal of Physiology*, **257**, G709–G714.

Grider, J.R. and Makhlouf, G.M. (1986) Colonic peristaltic reflex: identification of vasoactive intestinal peptide as mediator of descending relaxations. *American Journal of Physiology*, **251**, G40–G45.

Grider, J.R. and Makhlouf, G.M. (1987a) Prejunctional inhibition of vasoactive intestinal peptide release. *American Journal of Physiology*, **253**, G7–G12.

Grider, J.R. and Makhlouf, G.M. (1987b) Role of opioid neurons in the regulation of intestinal peristalsis. *American Journal of Physiology*, 253, G226–G231.

Grider, J.R. and Makhlouf, G.M. (1987c) Suppression of inhibitory neural input to colonic circular muscle by opioid peptides. *Journal of Pharmacology and Experimental Therapeutics*, 243, 205–210.

Grider, J.R., Sable, M.B., Bitar, K.N., Said, S.I. and Makhlouf, G.M. (1985) Vasoactive intestinal peptide. Relaxant neurotransmitter in taenia coli of the guinea-pig. *Gastroenterology*, 89, 36–42.

Gross, S.S., Guo, Z.G., Levi, R., Bailey, W.H. and Chenouda, A.A. (1984) Release of histamine by sympathetic nerve stimulation in the guinea-pig heart and modulation of adrenergic responses. A physiological role for cardiac histamine? *Circulation Research*, 54, 516–526.

Grundemar, L., Widmark, E., Waldeck, B. and Håkanson, R. (1988) Neuropeptide Y: prejunctional inhibition of vagally induced contractions in the guinea-pig trachea. *Regulatory Peptides*, 23, 309–313.

Haas, M., Cheng, B., Richardt, G., Lang, R.E. and Schömig, A. (1989) Characterization and presynaptic modulation of stimulation-evoked exocytotic co-release of noradrenaline and neuropeptide Y in guinea-pig heart. *Naunyn-Schmiedeberg's Archives of Pharmacology*, 339, 71–78.

Han, C. and Abel, P.W. (1987) Neuropeptide Y potentiates contraction and inhibits relaxation of rabbit coronary arteries. *Journal of Cardiovascular Pharmacology*, 9, 675–681.

Hankins, M.W. and Dray, A. (1988) Non-cholinergic synaptic potentials mediated by lumbar colonic nerve in the guinea-pig inferior mesenteric ganglion *in vitro*. *Neuroscience*, 26, 1073–1081.

Hanley, M.R., Lee, C.M., Jones, L.M. and Michell, R.H. (1980) Similar effects of substance P and related peptides on salivation and on phosphatidylinositol turnover in rat salivary glands. *Molecular Pharmacology*, 18, 78–83.

Häppölä, O., Soinila, S., Päivärinta, H. and Panula, P. (1987) [Met^5]enkephalin-Arg^6-Phe^7- and [Met^5]enkephalin-Arg^6-Gly^7-Leu^8-immunoreactive nerve fibres and neurons in the superior cervical ganglion of the rat. *Neuroscience*, 21, 283–295.

Hasuo, H. and Akasu, T. (1988) Presynaptic inhibition of cholinergic transmission by peptidergic neurons in bullfrog sympathetic ganglia. *Pflügers Archive*, 413, 206–208.

Head, R.J., Stitzel, R.E., De La Lande, I.S. and Johnson, S.M. (1977) Effect of chronic denervation on the activities of monoamine oxidase and catechol-O-methyltransferase and on the contents of noradrenaline and adenosine triphosphate in the rabbit ear artery. *Blood Vessels*, 14, 229–239.

Hedqvist, P. (1970) Studies on the effect of prostaglandins E1 and E2 on the sympathetic neuromusclular transmission in some animal tissues. *Acta Physiologica Scandinavica*, Suppl. 345, 1–40.

Hedqvist, P. (1977) Basic mechanisms of prostaglandins on autonomic neurotransmission. *Annual Review of Pharmacology and Toxicology*, 17, 259–279.

Heidenhain, R. (1872) Uber die Wirkung einiger Gifte auf die Nerven der Glandula Submaxillaris. *Pflügers Archives gesamte Physiologie*, 5, 309–318.

Hellström, P.M. (1987) Mechanisms involved in colonic vasoconstriction and inhibition of motility induced by neuropeptide Y. *Acta Physiologica Scandinavica*, 129, 549–556.

Henderson, G., Hughes, J. and Kosterlitz, H.W. (1972) A new example of a morphine-sensitive neuro-effector junction: adrenergic transmission in the mouse vas deferens. *British Journal of Pharmacology*, 46, 764–766.

Henderson, V.E. and Roepke, M.H. (1934) The role of acetylcholine in bladder contractile mechanisms and in parasympathetic ganglia. *Journal of Pharmacology and Experimental Therapeutics*, 51, 97–111.

Hervonen, A., Linnoila, I., Pickel, V.M., Helen, P., Pelto-Huikko, M., Alho, H. and Miller, R.J. (1981) Localization of [Met^5]- and [Leu^5]-enkephalin-like immunoreactivity in nerve terminals in human paravertebral sympathetic ganglia. *Neuroscience*, 6, 323–330.

Heym, C., Reinecke, M., Weihe, E. and Forssmann, W.G. (1984) Dopamine-β-hydroxylase-, neurotensin-, substance P-, vasoactive intestinal polypeptide-, and enkephalin- immuno-histochemistry of paravertebral and prevertebral ganglia in the cat. *Cell and Tissue Research*, 235, 411–418.

Hills, J.M., Collis, C.S. and Burnstock, G. (1983) The effects of vasoactive intestinal polypeptide on the electrical activity of guinea-pig intestinal smooth muscle. *European Journal of Pharmacology*, 88, 371–376.

Hills, J.M., Meldrum, L.A., Klarskov, P. and Burnstock, G. (1984) A novel non-adrenergic, non-cholinergic nerve-mediated relaxation of the pig bladder neck: an examination of possible transmitter candidates. *European Journal of Pharmacology*, 99, 287–294.

Hirst, G.D.S. and Edwards, F.R. (1989) Sympathetic neuroeffector transmission in arteries and arterioles. *Physiological Reviews*, 69, 546–604.

Hirst, G.D.S. and Jobling, P. (1989) The distribution of γ-adrenoceptors and P$_2$ purinoceptors in mesenteric arteries and veins of the guinea-pig. *British Journal of Pharmacology*, **96**, 993–999.

Hirst, G.D.S. and Lew, M.J. (1987) Lack of involvement of α-adrenoceptors in sympathetic neural vasoconstriction in the hind quarter of the rabbit. *British Journal of Pharmacology*, **90**, 51–60.

Hirst, G.D.S. and Neild, T.O. (1980) Evidence for two populations of excitatory receptors for noradrenaline on arteriolar smooth muscle. *Nature*, **283**, 767–768.

Hirst, G.D.S. and Neild, T.O. (1981) Localization of specialized noradrenaline receptors at neuromuscular junctions on arterioles of the guinea-pig. *Journal of Physiology*, **313**, 343–350.

Hirst, G.D.S., Neild, T.O. and Silverberg, G.D. (1982) Noradrenaline receptors on the rat basilar artery. *Journal of Physiology*, **328**, 351–360.

Hökfelt, T., Elfvin, L.G., Elde, R., Schultzberg, M., Goldstein, M., and Luft, R. (1977) Occurrence of somatostatin-like immunoreactivity in some peripheral sympathetic noradrenergic neurones. *Proceedings of the National Academy of Sciences, U.S.A.*, **74**, 3587–3591.

Hökfelt, T., Schultzberg, M., Elde, R., Nilsson, G., Terenius, L., Said, S. and Goldstein, M. (1978) Peptide neurons in the peripheral tissues including the urinary tract: immunohistochemical studies. *Acta Pharmacologica et Toxicologica*, **43**, **Suppl II**, 79–89.

Hökfelt, T., Fuxe, K. and Pernow, B. (1986) *Co-existence of neuronal messengers: a new principle in chemical transmission*, Amsterdam: Elsevier.

Hogaboom, G.K.O., Donnell, J.P. and Fedan J.S. (1980) Purinergic receptors: photoaffinity analog of adenosine triphospate is a specific adenosine triphosphate antagonist. *Science*, **208**, 1273.

Holck, M.I. and Marks, B.H. (1978) Purine nucleotide and nucleoside interactions on normal and subsensitive α adrenoceptor responsiveness in guinea-pig vas deferens. *Journal of Pharmacology and Experimental Therapeutics*, **205**, 104–117.

Holman, M.E. and Surprenant, A. (1980) An electrophysiological analysis of the effects of noradrenaline and α-receptor antagonists on neuromuscular transmission in mammalian arteries. *British Journal of Pharmacology*, **71**, 651–661.

Holzer, P. (1984) Characterization of the stimulus-induced release of immunoreactive substance P from the myenteric plexus of the guinea-pig small intestine. *Brain Research*, **297**, 127–136.

Holzer, P. (1989) Ascending enteric reflex: multiple neurotransmitter systems and interactions. *American Journal of Physiology*, **256**, G540–G545.

Holzer, P. and Lembeck, F. (1979) Effects of neuropeptides on the efficiency of the peristaltic reflex perfused intact. *Naunyn-Schmiedeberg's Archives of Pharmacology*, **307**, 257–264.

Holzer, P., Lippe, I.T., Barthó, L. and Saria, A. (1987) Neuropeptide Y inhibits excitatory neurons supplying the circular muscle of the guinea pig small intestine. *Gastroenterology*, **92**, 1944–1950.

Honma, S. (1970) Functional differentiation in sB and sC neurons in toad sympathetic ganglia. *Japanese Journal of Physiology*, **20**, 281–295.

Horn, J.P. and Stofer, W.D. (1988) Double labeling of the paravertebral sympathetic C system in the bullfrog with antisera to LHRH and NPY. *Journal of the Autonomic Nervous System*, **23**, 17–24.

Horn, J.P. and Stofer, W.D. (1989) Preganglionic and sensory origins of calcitonin gene-related peptide-like and substance P-like immunoreactivities in bullfrog sympathetic ganglia. *Journal of Neuroscience*, **9**, 2543–2561.

Horn, J.P., Stofer, W.D. and Fatherazi, S. (1989) Neuropeptide Y-like immunoreactivity in bullfrog sympathetic ganglia is restricted to C cells. *Journal of Neuroscience*, **7**, 1717–1727.

Horn, J.P., Fatherazi, S. and Stofer, W.D. (1988) Differential projections of B and C sympathetic axons in peripheral nerves of the bullfrog. *Journal of Comparative Neurology*, **278**, 570–580.

Hottenstein, O.D. and Kreulen, D.L. (1987) Comparison of the frequency dependence of venous and arterial responses to sympathetic nerve stimulation in guinea-pigs. *Journal of Physiology*, **383**, 153–167.

Hourani, S.M.O. (1984) Desentization of the guinea-pig urinary bladder by the enantiomers of adenylyl 5′-(β,γ-methylene)-diphosphate and by substance P. *British Journal of Pharmacology*, **82**, 161–164.

Hou, Z.Y., Lin, C.I., Chiu, T.H., Chiang, B.N., Cheng, K.K. and Ho, L.T. (1987) Somatostatin effects in isolated human atrial fibres. *Journal of Molecular and Cellular Cardiology*, **19**, 177–185.

Hoyle, C.H.V. and Burnstock, G. (1985) Atropine-resistant excitatory junction potentials in rabbit bladder are blocked by α,β-methylene ATP. *European Journal of Pharmacology*, **114**, 239–240.

Hoyle, C.H.V. and Burnstock, G. (1986) Evidence that ATP is a neurotransmitter in the frog heart. *European Journal of Pharmacology*, **124**, 285–289.

Hoyle, C.H.V. and Burnstock, G. (1989) Neuromuscular transmission in the gastrointestinal tract. In *The Handbook of Physiology. Section 6: The Gastrointestinal System. Volume I. Motility and Circulation*, edited by Wood, J.D., pp. 435–464. Baltimore: American Physiological Society.

Huang, M., Itoh, H., Lederis, K. and Rorstad, O. (1989) Evidence that vascular actions of PHI are mediated by a VIP-preferring receptor. *Peptides*, **10**, 993–1001.

Huang, M. and Rorstad, O.P. (1984) Cerebral vascular adenylate cyclase: evidence for coupling to receptors for vasoactive intestinal peptide and parathyroid hormone. *Journal of Neurochemistry*, **43**, 849–856.

Hughes, J., Kosterlitz, H.W. and Smith, T.W. (1977) The distribution of methionine-enkephalin and leucine-enkephalin in the brain and peripheral tissues. *British Journal of Pharmacology*, **61**, 639–647.

Hukovic, S., Rand, M.J. and Vanov, S. (1965) Observations on an isolated preparation of rat urinary bladder. *British Journal of Pharmacology and Chemotherapy*, **24**, 178–188.

Hulme, E.C., Birdsall, N.J.M. and Buckley, N.J. (1990) Muscarinic receptor subtypes. *Annual Review of Pharmacology and Toxicology*, **30**, 633–673.

Husted, S., Sjögren, C. and Andersson, K.E. (1980) Mechanisms of the responses to non-cholinergic, non-adrenergic nerve stimulation and to ATP in isolated rabbit urinary bladder: evidence for ADP evoked prostaglandin release. *Acta Pharmacologica et Toxicologica*, **47**, 84–92.

Husted, S., Sjögren, C. and Andersson, K.E. (1981) Substance P and somatostatin and excitatory neurotransmission in rabbit urinary bladder. *Archive Internationale de Pharmacodynamie*, **252**, 72–85.

Husted, S., Sjögren, C. and Andersson K.E. (1983) Direct effects of adenosine and adenine nucleotides on isolated human urinary bladder and their influence on electrically induced contradictions. *Journal of Urology*, **130**, 392–398.

Hyman, A.L. and Kadowitz, P.J. (1989) Analysis of responses to sympathetic nerve stimulation in the feline pulmonary vascular bed. *Journal of Applied Physiology*, **67**, 371–376.

Ichinose, M., Stretton, C.D., Schwartz, J-C, and Barnes, P.J. (1989) Histamine H_3-receptors inhibit cholinergic neurotransmission in guinea-pig airways. *British Journal of Pharmacology*, **97**, 13–15.

Illes, P., Zieglgänsberger, W. and Herz, A. (1980) Calcium reverses the inhibitory action of morphine neuroeffector transmission in the mouse vas deferens. *Brain Research*, **191**, 511–522.

Illes., P., Pfeiffer, N., von Kügelgen, I. and Starke, K. (1985) Presynaptic opioid receptor subtypes in the rabbit ear artery. *Journal of Pharmacology and Experimental Therapeutics*, **232**, 526–533.

Illes, P., Ramme, D. and Starke, K. (1986) Presynaptic opioid δ-receptors in the rabbit mesenteric artery. *Journal of Physiology*, **379**, 217–228.

Inoue, Y., Kaku, K., Kaneko, T., Yanaihara, N. and Kanno, T. (1985) Vasoactive intestinal peptide binding to specific receptors on rat parotid acinar cells induces amylase secretion accompanied by intracellular accumulation of cyclic adenosine 3′-5′ monophosphate. *Endocrinology*, **116**, 686–692.

Ishikawa, S. (1985) Actions of ATP and α,β-methylene ATP on neuromuscular transmission and smooth muscle membrane of the rabbit and guinea-pig mesenteric arteries. *British Journal of Pharmacology*, **86**, 777–787.

Ito, S., Ohga, A. and Ohta, T. (1988) Gastric relaxation and vasoactive intestinal peptide output in response to reflex vagal stimulation in the dog. *Journal of Physiology*, **404**, 683–693.

Jackowski, A., Crockard, A. and Burnstock, G. (1989) 5-hydroxytryptamine demonstrated immunohistochemically in rat cerebrovascular nerves largely represents 5-hydroxytryptamine uptake into sympathetic nerve fibres. *Neuroscience*, **29**, 453–462.

James, S. and Burnstock, G. (1988) Neuropeptide Y-like immunoreactivity in intramural ganglia of the newborn guinea-pig bladder. *Regulatory Peptides*, **23**, 237–245.

James, S. and Burnstock, G. (1990) Colocalization of peptides and a catecholamine- synthesizing enzyme in intraluminal neurons of the newborn guinea-pig urinary bladder in culture. *Regulatory Peptides*, **28**, 177–188.

Jan, L.Y. and Jan, Y.N. (1982) Peptidergic transmission in sympathetic ganglia of the frog. *Journal of Physiology*, **327**, 219–246.

Jan, L.Y., Jan, Y.N. and Brownfield, M.S. (1980) Peptidergic transmitters in synaptic boutons of sympathetic ganglia. *Nature*, **288**, 380–382.

Jan, Y.N., Bowers, C.W., Branton, D., Evans, L. and Jan, L.Y. (1983) Peptides in neuronal function: studies using frog autonomic ganglia. *Cold Spring Harbor Symposium of Quantitative Biology*, **43**, 363–374.

Jiang, Z.G., Simmons, M.A. and Dun, N.J. (1982) Enkephalinergic modulation of non-cholinergic transmission in mammalian prevertebral ganglia. *Brain Research*, **235**, 185–191.

Jänig, W. (1984) Patterns of activity in postganglionic vasoconstrictor neurones *in vivo*. In *Progress in Microcirculation Research II*, edited by Courtice, F.C., Garlick, D.G. and Perry, M.A., pp. 336–345. Sydney: Committee in Postgraduate Medical Education, University of N.S.W.

Jänig, W. and McLachlan, E.M. (1987) Organization of lumbar spinal outflow to distal colon and pelvic organs. *Physiological Reviews*, **67**, 1332–1404.

Jänig, W., Krauspe, R. and Wiedersatz, G. (1982) Transmission of impulses from pre- to postganglionic vasoconstrictor and sudomotor neurons. *Journal of the Autonomic Nervous System*, **6**, 95–106.

Jänig, W., Krauspe, R. and Wiedersatz, G. (1983) Reflex activation of postganglionic vasoconstrictor neurones supplying skeletal muscle by stimulation of arterial chemoreceptors via non-nicotinic synaptic mechanisms in sympathetic ganglia. *Pflügers Archive*, **396**, 95–100.

Johansson, O. and Lundberg, J.M. (1981) Ultrastructural localization of VIP-like immunoreactivity in large dense cored vesicles of 'cholinergic type' nerve terminals in cat exocrine glands. *Neuroscience*, **5**, 847–862.

Jones, J.W. (1987) Chicken II luteinizing hormone-releasing hormone inhibits the M current of bullfrog sympathetic neurons. *Neuroscience Letters*, **80**, 180–184.

Jones, S.W., Adams, P.R., Brownstein, M.J. and Rivier, J.E. (1984) Teleost luteinizing hormone-releasing hormone: action on bullfrog sympathetic ganglia is consistent with role as neurotransmitter. *Journal of Neuroscience*, **4**, 420–429.

Järvi, R., Helen, P., Hervonen, A. and Pelto-Huikko, M. (1989) Vasoactive intestinal peptide (VIP) – like immunoreactivity in the human sympathetic ganglia. *Histochemistry*, **90**, 347–351.

Juenemann, K.P., Lue, T.F., Luo,. J.A., Jadallah, S.A., Nunes, L.L. and Tanagho, E.A. (1987) The role of vasoactive intestinal polypeptide as a neurotransmitter in canine penile erection: a combined *in vivo* and immunohistochemical study. *Journal of Urology*, **138**, 871–877.

Kalsner, S. (1980) Limitation of presynaptic adrenoceptor theory: the characteristics of the effects of noradrenaline and phenoxybenzamine on stimulation-induced efflux of [^{3}H]-noradrenaline in vas deferens. *Journal of Pharmacology and Experimental Therapeutics*, **212**, 232–239.

Kasakov, L. and Burnstock, G. (1983) The use of the slowly degradable analog, α,β,-methylene ATP, to produce desensitization of the P_2-purinoceptor: effect on non-adrenergic, non-cholinergic responses of the guinea-pig urinary bladder. *European Journal of Pharmacology*, **86**, 291–294.

Kasakov, L., Ellis, J., Kirkpatrick, K., Milner, P. and Burnstock, G. (1988) Direct evidence for concomitant release of noradrenaline, adenosine 5′-triphosphate and neuropeptide Y from sympathetic nerves supplying the guinea-pig vas deferens. *Journal of the Autonomic Nervous System*, **22**, 75–82.

Katsuragi, T. and Su, C. (1980) Purine release from vascular adrenergic nerves by high potassium and a calcium ionophore, A-23187. *Journal of Pharmacology and Experimental Therapeutics*, **215**, 685–690.

Katsuragi, T., Tokunaga, T., Miyamoto, K., Kuratomi, L. and Furukawa, T. (1988) Norepinephrine and adenosine triphosphate release in different ratio from guinea-pig vas deferens by high potassium chloride, ouabain and monensin. *Journal of Pharmacology and Experimental Therapeutics*, **247**, 302–308.

Kawamura, Y., Matsuo, A. and Yamamoto, T. (1982) Analysis of reflex responses in preganglionic parasympathetic fibres innervating submandibular glands of rabbits. *Journal of Physiology*, **322**, 241–255.

Kawasaki, H. and Takasaki, K. (1984) Vasoconstrictor response induced by 5-hydroxytryptamine released from vascular adrenergic nerves by periarterial nerve stimulation. *Journal of Pharmacology and Experimental Therapeutics*, **229**, 816–822.

Kawasaki, H., Urabe, M. and Takasaki, K. (1989) Presynaptic α_2-adrenoceptor modulation of 5-hydroxytryptamine and noradrenaline release from vascular adrenergic nerves. *European Journal of Pharmacology*, **164**, 35–43.

Kawatani, M., Lowe, I.P., Booth, A.M., Backes, M.G., Erdman, S.L. and De Groat, W.C. (1983) The presence of leucine-enkephalin in the sacral preganglionic pathway to the urinary bladder of the cat. *Neuroscience Letters*, **39**, 143–148.

Kennedy, C., Saville, V.L. and Burnstock, G. (1986) The contributions of noradrenaline and ATP to the responses of the rabbit central ear artery to sympathetic nerve stimulation depend on the parameters of stimulation. *European Journal of Pharmacology*, **122**, 291–300.

Kirkpatrick, K. and Burnstock, G. (1987) Sympathetic nerve-mediated release of ATP from the guinea-pig vas deferens is uneffected by reserpine. *European Journal of Pharmacology*, **138**, 207–214.

Klein, R.L. (1982) Chemical composition of the large noradrenergic vesicles. In *Neurotransmitter Vesicles*, edited by Klein, R.L., Lagercrantz, H. and Zimmermann, H., pp. 133–174. London: Academic Press.

Klein, R.L. and Lagercrantz, H. (1982) Insights into the functional role of the noradrenergic vesicles. In *Neurotransmitter Vesicles*, edited by Klein, R.L., Lagercrantz, H. and Zimmerman, H., pp. 219–239. London: Academic press.

Kondo, H., Kataya, Y. and Yui, R. (1982) On the occurrence and physiological effect of somatostatin in the ciliary ganglion of cats. *Brain Research*, **247**, 141–144.

Kondo, H., Kuramoto, H., Wainer, B.H. and Yanaihara, N. (1985) Evidence for the co-existence of acetyl choline and enkephalin in the sympathetic preganglionic neurons of rats. *Brain Research*, **335**, 309–314.

Konishi, S., Tsundo, A. and Otsuka, M. (1981) Enkephalin as a transmitter for presynaptic inhibition in sympathetic ganglia. *Nature*, **294**, 80–82.

Kou, K., Ibengwe, J. and Suzuki, H. (1984) Effects of α-adrenoceptor antagonists on electrical and mechanical responses of the isolated dog mesenteric vein to perivascular nerve stimulation and exogenous noradrenaline. *Naunyn-Schmiedebergs Archives of Pharmacology*, **326**, 7–13.

Koyanagawa, H., Musha, T., Kanda, A., Kimura, T. and Satoh, S. (1989) Inhibition of vagal transmission by cardiac sympathetic nerve stimulation in the dog: possible involvement of opioid receptor. *Journal of Pharmacology and Experimental Therapeutics*, **250**, 1092–1096.

Krell, R.D., McCoy, J.L. and Ridley, P.T. (1981) Pharmacological characterization of the excitatory innervation to the guinea-pig urinary bladder *in vitro*: evidence for both cholinergic and non-adrenergic, non-cholinergic neurotransmission. *British Journal of Pharmacology*, **74**, 15–22.

Kreulen, D.L. and Peters, S. (1986) Non-cholinergic transmission in a sympathetic ganglion of the guinea-pig elicited by colon distension. *Journal of Physiology*, **374**, 315–334.

Kromer, W. (1988) Endogenous and exogenous opioids in the control of gastrointestinal motility and secretion. *Pharmacological Reviews*, **40**, 121–162.

Krukoff, T.L. (1986) Segmental distribution of corticotropin-releasing factor-like and vasoactive intestinal peptide-like immunoreactivities in presumptive sympathetic preganglionic neurons of the cat. *Brain Research*, **382**, 153–157.

Krukoff, T.L., Ciriello, J. and Calaresu, F.R. (1985) Segmental distribution of peptide-like immunoreactivity in cell bodies of the thoracic lumbar sympathetic nuclei of the cat. *Journal of Comparative Neurology*, **40**, 90–102.

Kása, P., Joó, F., Dobó, E., Wenthold, R.J., Ottersen, O.P., Storm-Mathisen, J. and Wolff, J.R. (1988) Heterogenous distributions of GABA-immunoreactive nerve fibres and axon terminals in the superior cervical ganglion of adult rat. *Neuroscience*, **26**, 635–644.

Kuchel, O., Debinski, W., Racz, K., Bun, N.T., Garcia, R., Cusson, J., Lardchelle, P., Cantin, M. and Genest, J. (1987) An emerging relationship between the peripheral sympathetic nervous system activity and the atrial natriuretic factor. *Life Sciences*, **40**, 1545–1551.

Kuffler, S.W. and Sejnowski, T.J. (1983) Peptidergic and muscarinic excitation at amphibian sympathetic synapses. *Journal of Physiology*, **341**, 257–278.

Kummer, W. (1987) Galanin- and neuropeptide Y-like immunoreactivities coexist in paravertebral sympathetic neurons of the cat. *Neuroscience Letters*, **78**, 127–131.

Kummer, W., Heym, C., Colombo, M. and Lang, R. (1986) Immunohistochemical evidence for extrinsic and intrinsic opioid systems in the guinea pig superior cervical ganglion. *Anatomy and Embryology*, **174**, 401–405.

Kummer, W., Reinecke, M., Heym, C. and Forssmann, W.G. (1988) Distribution of opioid functionally related to the cardiovascular system. In *Opioid Peptides and Blood Pressure Control.*, edited by Stumpe, K.O., Kraft, K. and Faden, A.I., pp. 5–12. Berlin, Heidelberg, New York, London, Paris, Tokyo: Springer-Verlag.

Kuramoto, H. and Fujita, T. (1986) An immunohistochemical study of calcitonin gene-related peptide (CGRP) containing nerve fibres in the sympathetic ganglia of bullfrogs. *Biomedical Research*, **7**, 349–357.

Kuriyama, H. (1963) Electrophysiological observations on the motor innervation of the smooth muscle cells in the guinea-pig vas deferens. *Journal of Physiology*, **169**, 213–228.

Kuriyama, H. and Makita, Y. (1983) Modulation of noradrenergic transmission in the guinea-pig mesenteric artery: an electrophysiological study. *Journal of Physiology*, **335**, 609–627.

Kusunoki, M., Taniyama, K. and Tanaka, C. (1984) Neuronal GABA release and GABA inhibition of ACh release in guinea-pig urinary bladder. *American Journal of Physiology*, **246**, R502–R509.

Lacroix, J.S. (1989) Adrenergic and non-adrenergic mechanisms in sympathetic vascular control of the nasal mucosa. *Acta Physiologica Scandinavica*, **136 (Suppl. 581)**, 1–63.

Lacroix, J.S., Stjärne, P., Ånggård, A. and Lundberg, J.M. (1988a) Sympathetic vascular control of the pig nasal mucosa (I): increased resistance and capacitance vessel responses upon stimulation with irregular bursts compared to continuous impulses. *Acta Physiologica Scandinavica*, **132**, 83–90.

Lacroix, J.S., Stjärne, P., Ånggård, A. and Lundberg, J.M. (1988b) Sympathetic vascular control of the pig nasal mucosa (II): reserpine-resistant, non-adrenergic nervous response in relation to neuropeptide Y and ATP. *Acta Physiologica Scandinavica*, **133**, 183–197.

Lacroix, J.S., Stjärne, P., Ånggård, A. and Lundberg, J.M. (1989) Sympathetic vascular control of the pig nasal mucosa (III): co-release of noradrenaline and neuropeptide Y. *Acta Physiologica Scandinavica*, **135**, 17–28.

Lagercrantz, H. and Fried, G. (1982) Chemical composition of the small noradrenergic vesicles. In *Neurotransmitter Vesicles*, edited by Klein, R.L., Lagercrantz, H. and Zimmerman, H., pp. 175–188. London: Academic Press.

Landis, S.C., Siegel, R.E. and Schwab, M. (1988) Evidence for neurotransmitter plasticity *in vivo*. II. Immunocytochemical studies of rat sweat gland innervation during development. *Developmental Biology*, **126**, 129–140.

Landis, S.C. (1988) Neurotransmitter plasticity in sympathetic neurons and its regulation by environmental factors *in vitro* and *in vivo*. In *Handbook of Chemical Neuroanatomy, Volume 6. The Peripheral Nervous System*, edited by Björklund, A., Hökfelt, T., and Owman, C., pp. 65–115. Amsterdam: Elsevier.

Langley, J.N. and Anderson, H.K. (1985) The innervation of the pelvic and adjoining viscera. II. The bladder. *Journal of Physiology*, **19**, 71–84.

Lang, R.E., Hermann, K., Dietz, R., Gaida, W., Ganten, D., Kraft, K. and Unger, T. (1983) Evidence for the presence of enkephalins in the heart. *Life Sciences*, **32**, 399–406.

Laurent, S., Marsh, J.D. and Smith, T.W. (1985) Enkephalins have a direct positive inotropic effect on cultured cardiac myocytes. *Proceedings of the National Academy of Science, USA*, **82**, 5930–5934.

Leblanc, G.G. and Landis, S.C. (1988) Target specificity of neuropeptide Y-immunoreactive parasympathetic neurons. *Journal of Neuroscience*, **8**, 146–155.

Leblanc, G.G., Trimmer, B.A. and Landis, S.C. (1987) Neuropeptide Y-like immunoreactivity in rat cranial parasympathetic neurons: coexistence with vasoactive intestinal peptide and choline acetyltransferase. *Proceedings National Academy of Science, USA*, **84**, 3511–3515.

Lee, T.J.F., Su, C. and Bevan, J.A. (1976) Neurogenic sympathetic vasoconstriction of the rabbit basilar artery. *Circulation Research*, **39**, 120–126.

Lee, T.J.F., Chiueh, C.C. and Adams, M. (1980) Synaptic transmission of vasoconstrictor nerves in rabbit basilar artery. *European Journal of Pharmacology*, **61**, 55–70.

Levin, R.M., Ruggieri, M.R. and Wein, A.J. (1986) Functional effects of the purinergic innervation of the rabbit bladder. *Journal of Pharmacology and Experimental Therapeutics*, **236**, 452–457.

Levin, R.M., Longhurst, P.A., Kato, K., McGuire, E.J., Elbadawi, A. and Wein, A.J. (1990) Comparative physiology and pharmacology of the cat and rabbit urinary bladder. *Journal of Urology*, **143**, 848–852.

Levitt, B. and Duckles, S.P. (1986) Evidence against serotonin as a vasoconstrictor neurotransmitter in the rabbit basilar artery. *Journal of Pharmacology and Experimental Therapeutics*, **238**, 880–885.

Levitt, B. and Westfall, D.P. (1982) Factors influencing the release of purines and norepinephrine in the rabbit portal vein. *Blood Vessels*, **19**, 30–40.

Lew, M.J. and White, T.D. (1987) Release of endogenous ATP during sympathetic nerve stimulation. *British Journal of Pharmacology*, **92**, 349–355.

Liang, T. and Cascieri, M.A. (1979) Substance P stimulation of amylase release by isolated parotid cells and inhibition of substance P induction of salivation by vasoactive intestinal polypeptide. *Molecular and Cellular Endocrinology*, **15**, 151–162.

Lin, C.I., Su, M.T., Luk, H.N. and Wu, H.L. (1988) Suppressive effects of somatostatin in dog Purkinge fibres. *British Journal of Pharmacology*, **93**, 192–198.

Lindh, B., Staines, W., Hökfelt, T., Terenius, L. and Salvaterra, P.M. (1986) Immunohistochemical demonstration of choline acetyltransferase-immunoreactive preganglionic nerve fibres in guinea pig autonomic ganglia. *Proceedings of the National Academy of Science, USA*, **83**, 5316–5320.

Lindh, B., Lundberg, J.M. and Hökfelt, T. (1989) NPY-, galanin-, VIP/PHI-, CGRP- and substance P- immunoreactive neuronal subpopulations in cat autonomic and sensory ganglia and their projections. *Cell and Tissue Research*, **256**, 259–273.

Llewellyn-Smith, I.J., Furness J.B., Gibbins, I.L. and Costa, M. (1988) Quantitative ultrastructural analysis of enkephalin-, substance P-, and VIP-inmmunoreactive nerve fibers in the circular muscle of the guinea pig small intestine. *Journal of Comparative Neurology*, **272**, 139–148.

Loewi, O. (1921) Uber humorale Ubertragbarkeit der Herznervenwirkung. I. Mitteilung. *Pflügers Archives gesamte physiologie*, **189**, 239–242.

Loewi, O. (1922) Uber humorale Ubertragbarkeit der Herznervenwirkung. II. Mitteilung. *Pflügers Archives gesamte Physiologie*, **193**, 201–213.

Loewi, O. (1936) Quantitative und Qualitative Untersuchungen uber den Sympathicusstoff. *Pflügers Archives gesamte Physiologie*, **237**, 504–514.

Longhurst, P.A., Belis, J.A., O'Donnell, J.P., Galie, J.R. and Westfall, D.P. (1984) A study of the atropine-resistant component of the neurogenic response of the rabbit urinary bladder. *European Journal of Pharmacology*, **99**, 295–302.

Love, J.A. and Szurszewski, J.H. (1987) The electrophysiological effects of vasoactive intestinal polypeptide in the guinea-pig inferior mesenteric ganglion. *Journal of Physiology*, **394**, 67–84.

Luff, S.E. and McLachlan, E.M. (1989) Frequency of neuromuscular junctions on arteries of different dimensions in the rabbit, guinea-pig and rat. *Blood Vessels*, **26**, 95–106.

Luff, S.E., McLachlan, E.M. and Hirst, G.D.S. (1987) An ultrastructural analysis of the sympathetic neuromuscular junctions on arterioles of the submuscosa of the guinea-pig ileum. *Journal of Comparative Neurology*, **257**, 578–594.

Lundberg, J.M. (1981) Evidence of the co-existence of vasoactive intestinal polypeptide (VIP) and acetylcholine in neurons of cat exocrine glands. Morphological, biochemical and functional studies. *Acta Physiologica Scandinavica*, **suppl. 496**, 1–57.

Lundberg, J.M. and Hökfelt, T. (1986) Multiple co-existence of peptides and classical transmitters in peripheral autonomic and sensory neurons – functional and pharmacological implications. *Progress in Brain Research*, **68**, 241–262.

Lundberg, J.M. and Stjärne, L. (1984) Neuropeptide Y (NPY) depresses the secretion of ^{3}H-noradrenaline and the contractile responses evoked by field stimulation, in rat vas deferens. *Acta Physiologica Scandinavica*, **120**, 477–479.

Lundberg, J.M. and Tatemoto, K. (1982a) Pancreatic polypeptide family (APP, BPP, NPY and PYY) in relation to sympathetic vasoconstriction resistant to α-adrenoceptor blockade. *Acta Physiologica Scandinavica*, **116**, 393–402.

Lundberg, J.M. and Tatemoto, K. (1982b) Vascular effects of the peptides PYY and PHI: comparison with APP and VIP. *European Journal of Pharmacology*, **83**, 143–146.

Lundberg, J.M. Hökfelt, T., Schultzberg, M., Uvnas-Wallensten, K., Kohler, L. and Said, S. (1979) Occurrence of VIP-like immunoreactivity in cholinergic neurons of the cat: evidence from combined immunohistochemistry and acetylcholinesterase staining. *Neuroscience*, **4**, 1539–1559.

Lundberg, J.M., Ånggård, A., Fahrenkrug, J., Hökfelt, T. and Mutt, V. (1980) Vasoactive intestinal polypeptide in cholinergic neurons of exocrine glands: functional significance of coexisting transmitters for vasodilation and secretion. *Proceedings of the National Academy of Science, USA*, **77**, 1651–1655.

Lundberg, J.M., Ånggård, A., and Fahrenkrug, J. (1981a) Complementary role of vasoactive intestinal polypeptide (VIP) and acetylcholine for cat submandibular gland blood flow and secretion. I. VIP release. *Acta Physiologica Scandinavica*, **113**, 317–327.

Lundberg, J.M., Ånggård, A., and Fahrenkrug, J. (1981b) Complementary role of vasoactive intestinal polypeptide (VIP) and acetylcholine for cat submandibular gland blood flow and secretion. II. Effects of cholinergic antagonists and VIP antiserum. *Acta Physiologica Scandinavica*, **113**, 329–336.

Lundberg, J.M., Fried, G., Fahrenkrug, J., Hamstedt, B., Hökfelt, T., Lagercrantz, H., Lundgren, G. and Ånggård, A. (1981c) Subcellular fractionation of cat submandibular gland: comparative studies on the distribution of acetylcholine and vasoactive intestinal polypeptide (VIP). *Neuroscience*, **6**, 1001–1010.

Lundberg, J.M., Ånggård, A., and Fahrenkrug, J. (1982a) Complementary role of vasoactive intestinal polypeptide (VIP) and acetylcholine for cat submandibular gland blood flow and secretion. III. Effects of local infusions. *Acta Physiologica Scandinavica*, **114**, 329–337.

Lundberg, J.M., Ånggård, A., Fahrenkrug, J., Lundgren, G. and Holmstedt, B. (1982b) Corelease of VIP and acetylcholine in relation to blood flow and salivary secretion in cat submandibular salivary gland. *Acta Physiologica Scandinavica*, **115**, 525–528.

Lundberg, J.M., Hedlund, B. and Bartfai, T. (1982c) Vasoactive intestinal polypeptide enhances muscarinic ligand binding in cat submandibular salivary gland. *Nature*, **295**, 147–149.

Lundberg, J.M., Rökaeus, Å., Hökfelt, T., Rosell, S., Brown, M. and Goldstein, M. (1982d) Neurotensin-like immunoreactivity in the preganglionic sympathetic nerves and in the adrenal medulla of the cat. *Acta Physiologica Scandinavica*, **114**, 153–155.

Lundberg, J.M., Terenius, L., Hökfelt, T., Martling, C.R., Tatemoto, K., Mutt, V., Polak, J., Bloom, S. and Goldstein, M. (1982e) Neuropeptide Y (NPY)-like immunoreactivity in peripheral noradrenergic neurons and effects of NPY on sympathetic function. *Acta Physiologica Scandinavica*, **116**, 477–480.

Lundberg, J.M., Ånggård, A., Theodorsson-Norheim, E. and Pernow, J. (1984a) Guanethidine-sensitive release of neuropeptide Y-like immunoreactivity in the cat spleen by sympathetic nerve stimulation. *Neuroscience Letters*, **52**, 175–180.

Lundberg, J.M., Fahrenkrug, J., Hökfelt, T., Martling, C.R., Larsson, O., Tatemoto, K. and Ånggård, A. (1984b) Co-existence of peptide HI (PHI) and VIP in nerves regulatory blood flow and bronchial smooth muscle tone in various mammals including man. *Peptides*, **5**, 593–606 .

Lundberg, J.M., Fahrenkrug, J., Larsson, O. and Ånggård, A. (1984c) Corelease of vasoactive intestinal polypeptide and peptide histidine isoleucine in relation to atropine-resistant vasodilation in cat submandibular salivary gland. *Neuroscience Letters*, **52**, 37–42.

Lundberg, J.M., Hua, X.Y. and Franco-Cereceda, A. (1984d) Effects of neuropeptide Y (NPY) on mechanical activity and neurotransmission in the heart, vas deferens and urinary bladder of the guinea-pig. *Acta Physiologica Scandinavica*, **121**, 325–332.

Lundberg, J.M., Martinsson, A., Hemsen, A., Theodorsson-Norheim, E., Svedenhag, J., Ekblom, B. and Hjemdahl, P. (1985a) Corelease of neuropeptide Y and catecholamines during physical exercise in man. *Biochemical Biophysical Research Communications*, **133**, 30–36.

Lundberg, J.M., Pernow, J., Tatemoto, K. and Dahlöf, C. (1985b) Pre- and post-junctional effects of NPY on sympathetic control of rat femoral artery. *Acta Physiologica Scandinavica*, **123**, 511–513.

Lundberg, J.M., Saria, A., Franco-Cereceda, A., Hökfelt, T., Terenius, L. and Goldstein, M. (1985c) Differential effects of reserpine and 6-hydroxydopamine on neuropeptide Y (NPY) and noradrenaline in peripheral neurons. *Naunyn-Schmiedeberg's Archives of Pharmacology*, **328**, 331–341.

Lundberg, J.M., Torsell, L., Sollevi, A., Pernow, J., Theodorsson-Norheim, E. Änggård, A. and Hamberger, B. (1985d) Neuropeptide Y and sympathetic vascular control in man. *Regulatory Peptides*, **13**, 41–52.

Lundberg, J.M., Rudehill, A., Sollevi, A., Theodorsson-Norheim, E. and Hamberger, B. (1986) Frequency- and reserpine-dependent chemical coding of sympathetic transmission: differential release of noradrenaline and neuropeptide Y from pig spleen. *Neuroscience Letters*, **63**, 96–100.

Lundberg, J.M., Pernow, J., Fried, G. and Änggård, A. (1987) Neuropeptide Y and noradrenaline mechanisms in relation to reserpine induced impairment of sympathetic neurotransmission in the cat spleen. *Acta Physiologica Scandinavica*, **131**, 1–10.

Lundberg, J.M., Hemsen, A., Rudehill, A., Härfstrand, A., Larsson, O., Sollevi, A., Saria, A., Hökfelt, T., Fuxe, K. and Fredholm, B.B. (1988a) Neuropeptide Y- and α-adrenergic receptors in pig spleen: localization, binding characteristics, cyclic AMP effects and functional responses in control and denervated animals. *Neuroscience*, **24**, 659–672.

Lundberg, J.M., Martling, C.R. and Hökfelt, T. (1988b) Airways, oral cavity and salivary glands: classical transmitters and peptides in sensory and autonomic motor neurons. In *Handbook of Chemical Neuroanatomy, Vol. 6: The Peripheral Nervous System*, edited by Björklund, A., Hökfelt, T. and Owman, C., pp. 391–444. Amsterdam: Elsevier.

Lundberg, J.M., Rudehill, A. and Sollevi, A. (1989a) Pharmacological characterization of neuropeptide Y and noradrenaline mechanisms in sympathetic control of pig spleen. *European Journal of Pharmacology*, **163**, 103–113.

Lundberg, J.M., Rudehill, A. Sollevi, A. and Hamberger, B. (1989b) Evidence for co-transmitter role of neuropeptide Y in the pig spleen. *British Journal of Pharmacology*, **96**, 675–687.

Lundblad, L., Änggård, A., Saria, A. and Lundberg, J.M. (1987) Neuropeptide Y and non-adrenergic sympathetic vascular control of the cat nasal mucosa. *Journal of the Autonomic Nervous System*, **20**, 189–197.

Mackenzie, I. and Burnstock, G. (1980) Evidence against vasoactive intestinal polypeptide being the non-adrenergic, non-cholinergic inhibitory transmitter released from the nerves supplying the smooth muscle of the guinea-pig taenia coli. *European Journal of Pharmacology*, **67**, 255–264.

Mackenzie, I. and Bursntock, G. (1984) Neuropeptide action on the guinea-pig bladder – a comparison with the effects of field stimulation and ATP. *European Journal of Pharmacology*, **105**, 85–94.

Mackenzie, I., Kirkpatrick, K.A. and Burnstock, G. (1988a) Comparative study of the actions of AP_5A and α,β-methylene ATP on nonadrenergic, noncholinergic neurogenic excitation in the guinea-pig vas deferens. *British Journal of Pharmacology*, **94**, 699–706.

Mackenzie, I., Manzini, S. and Burnstock, G. (1988b) Regulation of voltage-dependent excitatory responses to α,β-methylene ATP, ATP and non-adrenergic nerve stimulation by dihydropyridines in the guinea-pig vas deferens. *Neuroscience*, **27**, 317–332.

Macrae, I.M., Furness, J.B. and Costa, M. (1986) Distribution of subgroups of noradrenaline neurons in the coeliac ganglion of the guinea-pig. *Cell and Tissue Research*, **244**, 173–180.

Maggi, C.A., Santicioli, P. and Meli, A. (1984) Postnatal development of myogenic contractile activity and excitatory innervation of rat urinary bladder. *American Journal of Physiology*, **247**, R972–R978.

Maggi, C.A., Santicioli, P., Patacchini, R., Turini, D., Barbanti, G., Benefort, P., Giuciani, S. and Meli, A. (1987) Galanin: a potent modulator of excitatory neurotransmission in the human urinary bladder. *European Journal of Pharmacology*, **143**, 135–137.

Maggi, C.A., Santicioli, P., Patacchini, R., Geppetti, P., Giuliani, S., Astolfi, M., Baldi, E., Parlani, M., Theodorsson, E., Fusco, B. and Meli, A. (1988) Regional differences in the motor response to capsaicin in the guinea-pig urinary bladder: relative role of pre- and post-junctional factors related to neuropeptide-containing sensory nerves. *Neuroscience*, **27**, 675–688.

Majewski, H., Rand, M.J. and Tang, L.H. (1981) Activation of prejunctional β-adrenoceptors in rat atria by adrenaline applied exogenously or released as a co-transmitter. *British Journal of Pharmacology*, **73**, 669–679.

Major, T.C., Weishaar, R.E. and Taylor, D.G. (1989) Two phases of contractile response in rat isolated vas deferens and their regulation by adenosine and α-receptors. *European Journal of Pharmacology*, **167**, 323–331.

Malik, K.U. and Nasjletti, A. (1976) Facilitation of adrenergic transmission by locally generated angiotensin II in rat mesenteric arteries. *Circulation Research*, **38**, 26–30.

Manzini, S., Maggi, C.A. and Meli, A. (1986) Pharmacological evidence that at least two different non-adrenergic, non-cholinergic inhibitory systems are present in the rat small intestine. *European Journal of Pharmacology*, **123**, 229–236.

Martin, W., Smith, J.A., Lewis, M.J. and Henderson, A.H. (1988) Evidence that inhibitory factor extracted from bovine retractor penis is nitrite, whose acid-activated derivative is stabilized nitric oxide. *British Journal of Pharmacology*, **93**, 579–586.

Masuko, S. and Chiba, T. (1988) Projection pathways, co-existence of peptides and synaptic organization of nerve fibres in the inferior mesenteric ganglion of the guinea-pig. *Cell and Tissue Research*, **253**, 507–516.

Matran, R., Martling, C.R. and Lundberg, J.M. (1989) Inhibition of cholinergic and non-adrenergic, non-cholinergic bronchoconstriction in the guinea-pig mediated by neuropeptide Y and α_2-adrenoceptors and opiate receptors. *European Journal of Pharmacology*, **163**, 15–23.

Matthews, M.R. and Cuello, A.C. (1984) The origin and possible significance of substance P immunoreactive networks in the prevertebral ganglia and related structures in the guinea-pig. *Philosophical Translations of the Royal Society of London, Series B.*, **306**, 247–276.

Mattiasson, A., Ekblad, E., Sundler, F. and Uvelius, B. (1985) Origin and distribution of neuropeptide Y-, vasoactive intestinal polypeptide-, and substance P-containing nerve fibres in the urinary bladder of the rat. *Cell and Tissue Research*, **239**, 141–146.

McGrath, J.C. (1978) Adrenergic and 'non-adrenergic' components in the contractile response of the vas deferens to a single indirect stimulus. *Journal of Physiology*, **283**, 23–39.

McGrath, M.A. and Shephard, J.T. (1976) Inhibition of adrenergic neurotransmission in canine vascular smooth muscle by histamine. *Circulation Research*, **39**, 566–573.

McKinney, J.S. and Rubin, R.P. (1988) Enhancement of cyclic AMP modulated salivary amylase secretion by protein kinase C activators. *Biochemical Pharmacology*, **37**, 4433–4438.

McKinney, J.S., Desole, M.S. and Rubin, R.P. (1989) Convergence of cAMP and phosphoinositide pathways during rat parotid secretion. *American Journal of Physiology*, **257**, C651–C657.

Meldrum, L.A. and Burnstock, G. (1983) Evidence that ATP acts as a co-transmitter with noradrenaline in sympathetic nerves supplying the guinea-pig vas deferens. *European Journal of Pharmacology*, **92**, 161–163.

Merchenthaler, I., Hynes, M.A., Vigh, S., Shally, A.V. and Petrusz, P. (1983) Immunocytochemical localization of corticotrophin releasing factor (CRF) in the rat spinal cord. *Brain Research*, **275**, 373–377.

Merritt, J.E. and Rink, T.J. (1987) The effects of substance P and carbachol on inositol tris- and tetrakis phosphate formation and cytosolic free calcium in rat parotid acinar cells. *Journal of Biological Chemistry*, **262**, 14912–14916.

Mishima, S., Miyahara, H. and Suzuki, H. (1984) Transmitter release modulated by α-adrenoceptor antagonists in the rabbit mesenteric artery: a comparison between noradrenaline outflow and electrical activity. *British Journal of Pharmacology*, **83**, 537–547.

Misu, Y., Kuwahara, M., Amano, H. and Kubo, T. (1989) Evidence for tonic activation of prejunctional β-adrenoceptors in guinea-pig pulmonary arteries by adrenaline derived from the adrenal medulla. *British Journal of Pharmacology*, **98**, 45–50.

Mitsuhashi, M. and Payan, D.G. (1987) The mitogenic effects of vascular neuropeptides on cultured smooth muscle cell lines. *Life Sciences*, **40**, 853–861.

Miyahara, H. and Suzuki, H. (1987) Pre- and post-junctional effects of adenosine triphosphate on noradrenergic transmission in the rabbit ear artery. *Journal of Physiology*, **389**, 423–440.

Molderings, G.J., Göthert, M., Fink, K., Roth, E. and Schlicker, E. (1989) Inhibition of noradrenaline release in the pig coronary artery via a novel serotonin receptor. *European Journal of Pharmacology*, **164**, 213–222.

Mo, N. and Dun, N.J. (1984) Vasoactive intestinal polypeptide facilitates muscarinic transmission in mammalian sympathetic ganglia. *Neuroscience Letters*, **52**, 19–23.

Moore, R.H. and Dowling, D.A. (1982) Effects of enkephalins on perfusion pressure in isolated hindlimb preparations. *Life Sciences*, **31**, 1559–1566.

Morris, J.L. (1989a) The cardiovascular system. In *The Comparative Physiology of Regulatory Peptides*, edited by Holmgren, S., pp. 272–307. London: Chapman and Hall.

Morris, J.L. (1989b) Neuropeptides and autonomic vasodilatation of the guinea-pig uterine artery. *Proceedings of the International Union of Physiological Sciences*, **17**, 62.

Morris, J.L. (1990) Neuropeptide Y inhibits relaxations of the guinea-pig uterine artery produced by vasoactive intestinal peptide. *Peptides*, **11**, 381–386.

Morris, J.L. (1991) Roles of neuropeptide Y and noradrenaline in sympathetic neurotransmission to the thoracic vena cava and aorta of guinea-pigs. *Regulatory Peptides*, **32**, 297–310.

Morris, J.L. and Gibbins, I.L. (1987) Neuronal colocalization of neuropeptides, catecholamines and catecholamine-synthesizing enzymes in the guinea-pig paracervical ganglia. *Journal of Neuroscience*, **7**, 3117–3130.

Morris, J.L. and Gibbins, I.L. (1989) Co-localization and plasticity of transmitters in peripheral autonomic and sensory neurons. *International Journal of Developmental Neuroscience*, **7**, 521–531.

Morris, J.L. and Gibbins, I.L. (1990) Structure-function relationships of the autonomic nervous system in the thoracic circulations. *Proceedings of the Australian Physiological and Pharmacological Society*, **21**, 29–39.

Morris, J.L. and Murphy, R. (1988) Evidence that neuropeptide Y released from noradrenergic axons causes prolonged contraction of the guinea-pig uterine artery. *Journal of the Autonomic Nervous System*, **24**, 241–249.

Morris, J.L. and Murphy, R. (1989) Analogues of vasoactive intestinal peptide (VIP) contract the guinea-pig uterine artery but do not antagonize VIP-induced relaxations. *European Journal of Pharmacology*, **162**, 375–379.

Morris, J.L., Gibbins, I.L. and Clevers, J. (1981) Resistance of adrenergic neurotransmission in the toad heart to adrenoceptor blockade. *Naunyn-Schmiedeberg's Archives of Pharmacology*, **317**, 331–338.

Morris, J.L., Gibbins, I.L., Furness, J.B., Costa, M. and Murphy, R. (1985) Co-localization of neuropeptide Y, vasoactive intestinal polypeptide and dynorphin in non-noradrenergic axons of the guinea-pig uterine artery. *Neuroscience Letters*, **62**, 31–37.

Morris, J.L., Gibbins, I.L., Campbell, G., Murphy, R., Furness, J.B. and Costa, M. (1986a) Innervation of the large arteries and heart of the toad (Bufo marinus) by adrenergic and peptide-containing neurons. *Cell and Tissue Research*, **243**, 171–184.

Morris, J.L., Gibbins, I.L. and Murphy, R. (1986b) Neuropeptide Y-like immunoreactivity is absent from most perivascular noradrenerngic axons in a marsupial, the brush-tailed possum. *Neuroscience Letters*, **71**, 264–270.

Morris, J.L., Murphy, R., Furness, J.B. and Costa, M. (1986c) Partial depletion of neuropeptide Y from noradrenergic perivascular and cardiac axons by 6-hydroxydopamine and reserpine. *Regulatory Peptides*, **13**, 147–162.

Morris, J.L., Gibbins, I.L. and Furness, J.B. (1987) Increased levels of dopamine-β-hydroxylase-like immunoreactivity in non-noradrenergic axons supplying the guinea-pig uterine artery after 6-hydroxydopamine treatment. *Journal of the Autonomic Nervous System*, **21**, 15–27.

Morris, J.L., Gibbins, I.L. and Osborne, P.B. (1989) Galanin-like immunoreactivity in sympathetic and parasympathetic neurons of the toad Bufo marinus. *Neuroscience Letters*, **102**, 142–148.

Morris, M.J., Elliott, J.M., Cain, M.D., Kapoor, V., West, M.J. and Chalmers, J.P. (1986) Plasma neuropeptide Y levels rise in patients undergoing exercise tests for the investigation of chest pain. *Clinical and Experimental Pharmacology and Physiology*, **13**, 437–440.

Morris, M.J., Kapoor, V. and Chalmers, J.P. (1987) Plasma neuropeptide Y concentration is increased after hemorrhage in conscious rats: relative contributions of sympathetic nerves and the adrenal medulla. *Journal of Cardiovascular Pharmacology*, **9**, 541–546.

Morrison, S.F., Callaway, J., Milner, T.A. and Reis, D.J. (1989) Glutamate in the spinal sympathetic intermediolateral nucleus: localization by light and electron microscopy. *Brain Research*, **503**, 5–15.

Moss, H.E. and Burnstock, G. (1985) A comparative study of electrical field stimulation of the guinea-pig, ferret and marmoset urinary bladder. *European Journal of Pharmacology*, **114**, 311–316.

Muir, T.C. and Wardle, K.A. (1988) The electrical and mechanical basis of co-transmission in some vascular and non-vascular smooth muscles. *Journal of Autonomic Pharmacology*, **8**, 203–218.

Muramatsu, I. (1987) The effect of reserpine on sympathetic, purinergic neurotransmission in the isolated mesenteric artery of the dog: a pharmacological study. *British Journal of Pharmacology*, **91**, 467–474.

Muramatsu, I., Fujiwara, M., Miura, A. and Sakakibara, Y. (1981) Possible involvement of adenine nucleotides in sympathetic neuroeffector mechanisms of dog basilar artery. *Journal of Pharmacology and Experimental Therapeutics*, **216**, 401–409.

Muramatsu, I., Kigoshi, S. and Oshita, M. (1984) Nonadrenergic nature of prazosin-resistant sympathetic contraction in the dog mesenteric artery. *Journal of Pharmacology and Experimental Therapeutics*, **229**, 532–538.

Muramatsu, I., Ohmura, T. and Oshita, M. (1989) Comparison between sympathetic adrenergic and purinergic transmission in the dog mesenteric artery. *Journal of Physiology*, **411**, 227–243.

Musha, T., Satoh, E., Koyanagawa, H., Kimura, T. and Satoh, S. (1989) Effects of opioid agonists on sympathetic and parasympathetic transmission to the dog heart. *Journal of Pharmacology and Experimental Therapeutics*, **250**, 1087–1091.

Nagao, T. and Suzuki, H. (1988) Effects of α, β-methylene ATP on electrical responses produced by ATP and nerve stimulation in smooth muscle cells of the guinea-pig mesenteric artery. *General Pharmacology*, **19**, 799–805.

Nakamura, N. and Inagami, T. (1986) Atrial natriuretic factor inhibits norepinephrine release evoked by sympathetic nerve stimulation in isolated perfused rat mesenteric arteries. *European Journal of Pharmacology*, **123**, 459–461.

Nakanishi, H. and Takeda, H. (1973) The possible role of adenosine triphosphate in chemical transmission between the hypogastric nerve terminal and seminal vesicle in the guinea-pig. *Japanese Journal of Physiology*, **23**, 479–490.

Neild, T.O. (1978) Slowly-developing depolarization of neurones in the guinea-pig inferior mesenteric ganglion following repetitive stimulation of the preganglionic nerves. *Brain Research*, **140**, 231–239.

Neild, T.O. (1987) Actions of neuropeptide Y on innervated and denervated rat tail arteries. *Journal of Physiology*, **386**, 19–31.

Neild, T.O. and Kotecha, N. (1986) Effects of α, β-methylene ATP on membrane potential, neuromuscular transmission and smooth muscle contraction in the rat tail artery. *General Pharmacology*, **17**, 461–464.

Neild, T.O. and Kotecha, N. (1990) Actions of neuropeptide Y on arterioles of the guinea-pig small intestine are not mediated by smooth muscle depolarization. *Journal of the Autonomic Nervous System*, **30**, 29–36.

Neild, T.O. and Zelcer, E. (1982) Noradrenergic neuromuscular transmission with special reference to arterial smooth muscle. *Progress in Neurobiology*, **19**, 141–158.

Nilsson, H. (1984) Different nerve responses in consecutive sections of the arterial system. *Acta Physiologica Scandinavica*, **121**, 353–361.

Nilsson, J., von Euler, A.M. and Dalsgaard, C.J. (1985) Stimulation of connective tissue cell growth by substance P and substance K. *Nature*, **315**, 61–63.

Nishimura, T. and Akasu, T. (1989) 5-hydroxytryptamine produces presynaptic facilitation of cholinergic transmission in rabbit parasympathetic ganglia. *Journal of the Autonomic Nervous System*, **26**, 251–260.

Nishi, S., Soeda, H. and Koketsu, K. (1967) Release of acetylcholine from sympathetic pre-ganglionic nerve terminals. *Journal of Neurophysiology*, **30**, 114–134.

North, R.A., Henderson, G., Katayama, Y. and Johnson, S.M. (1980) Electrophysiological evidence for presynaptic inhibition of acetylcholine release by 5-hydroxytryptamine in the enteric nervous system. *Neuroscience*, **5**, 581–586.

O'Donnell, S.R. and Wanstall, J.C. (1982) Pharmacological experiments demonstrate that toad (Bufo marinus) atrial β-adrenoceptors are not identical with mammalian β_2- or β_1-adrenoceptors. *Life Sciences*, **31**, 701–708.

Öhlén, A., Persson, M.G., Lindbom, L., Gustafsson, L.E. and Hedqvist, P. (1990) Nerve-induced nonadrenergic vasoconstriction and vasodilation in skeletal muscle. *American Journal of Physiology*, **258**, H1334–H1338.

Olasmaa, M., Påhlman, S. and Terenius, L. (1987) β-adrenoceptor, vasoactive intestinal polypeptide (VIP) and neuropeptide tyrosine (NPY) receptors functionally coupled to adenylate cyclase in the human neuroblastoma. *Neuroscience Letters*, **83**, 161–166.

Ono, H. and Suzuki, H. (1988) Effects of α-adrenoreceptor antagonists and α, β-methylene ATP on electrical responses produced by neurotransmitters in smooth muscle cells of the rat tail artery. *Biomedical Research*, **9**, 457–466.

Oshita, M., Kigoshi, S. and Muramatsu, I. (1989) Selective potentiation of extracellular Ca^{2+}-dependent contraction by neuropeptide Y in rabbit mesenteric arteries. *General Pharmacology*, **20**, 363–367.

Owen, M.P., Quinn, C. and Bevan, J.A. (1985) Phentolamine-resistant neurogenic constriction occurs in small arteries at higher frequencies. *American Journal of Physiology*, **249**, H404–H414.

Papka, R.E., Taurig, H.M. and Wekstein, M. (1985) Localization of peptides in nerve terminals in the paracervical ganglia of the rat by light and electron microscopic immunohistochemistry: enkephalin and atrial natriuretic factor. *Neuroscience Letters*, **61**, 285–290.

Parsons, R.L., Neel, D.S., Konopka, L.M. and McKeon, T.W. (1989) The presence and possible role of a galanin-like peptide in the mudpuppy heart. *Neuroscience*, **29**, 749–759.

Patel, K.P. (1989) Atrial natriuretic factor attenuates sympathetic neuroeffector responses in hindlimb vasculature of rabbits. *Canadian Journal of Physiology and Pharmacology*, **67**, 1101–1105.

Paton, D.M. (1981) Presynaptic neuromodulation mediated by purinergic receptors. In *Purinergic Receptors*, edited by Burnstock, G., pp. 199–219. London: Chapman & Hall.

Peng, H.B. and Chen, Q. (1988) Localization of calcitonin gene-related peptide (CGRP) at a neuronal nicotinic synapse. *Neuroscience Letters*, **95** 75–80.

Pernow, J. and Lundberg, J.M. (1989a) Release and vasoconstrictor effects of neuropetide Y in relation to non-adrenergic sympathetic control of renal blood flow in the pig. *Acta Physiologica Scandinavica*, **136**, 507–517.

Pernow, J. and Lundberg, J.M. (1989b) Modulation of noradrenaline and neuropeptide Y (NPY) release in pig kidney *in vivo*: involvement of α_2, NPY and angiotensin II receptors. *Naunyn-Schmiedeberg's Archives of Pharmacology*, **340**, 379–385.

Pernow, J., Lundberg, J.M., Kaijser, L., Hjemdahl, P., Theodorsson-Norheim, E., Martinson, A. and Pernow, B. (1986) Plasma neuropeptide Y-like immunoreactivity and catecholamines during various degrees of sympathetic activation in man. *Clinical Physiology*, **6**, 561–578.

Pernow, J., Svenberg, T. and Lundberg, J.M. (1987) Actions of calcium antagonists on pre- and postjunctional effects of neuropeptide Y on human peripheral blood vessels *in vitro*. *European Journal of Pharmacology*, **136**, 207–218.

Pernow, J., Kahan, T. and Lundberg, J.M. (1988) Neuropeptide Y and reserpine-resistant vasoconstriction evoked by sympathetic nerve stimulation in the dog skeletal muscle. *British Journal of Pharmacology*, **94**, 952–960.

Pernow, J., Schweiler, J., Kahan, T., Hjemdahl, P., Oberle, J., Wallin, B.G. and Lundberg, J.M. (1989) Influence of sympathetic discharge pattern on noradrenaline and neuropeptide Y release. *American Journal of Physiology*, **257**, H866–H872.

Peterson, J. and Noronha-Blob, L. (1989) Effects of selective cholinergic antagonists and α, β-methylene ATP on guinea-pig urinary bladder contractions *in vivo* following pelvic nerve stimulation. *Journal of Autonomic Pharmacology*, **9**, 303–313.

Peters, S. and Kreulen, D.L. (1984) A slow EPSP in mammalian inferior mesenteric ganglion persists after *in vivo* capsaicin. *Brain Research*, **303**, 186–189.

Petrou, S., Jobling, P., Clarke, A.L. and Osborne, P.B. (1989) Structure and function of pulmonary arterial innervation in the rat and guinea-pig. *Proceedings of the Australian Physiology and Pharmacology Society*, **20**, 38P.

Pettersson, A., Hedner, J. and Hedner, T. (1989) Relationship between renal sympathetic activity and diuretic effects of atrial natriuretic peptide (ANP) in the rat. *Acta Physiologica Scandinavica*, **135**, 323–333.

Potter, E.K. (1985) Prolonged non-adrenergic inhibition of cardiac vagal action following sympathetic stimulation: neuromodulation by neuropeptide Y? *Neuroscience Letters*, **54**, 117–122.

Potter, E.K. (1987a) Guanethidine blocks neuropeptide Y-like inhibitory action of sympathetic nerves on cardiac vagus. *Journal of the Autonomic Nervous System*, **21**, 87–90.

Potter, E.K. (1987b) Presynaptic inhibition of cardiac vagal postganglionic nerves by neuropeptide Y. Neuroscience Letters, **83**, 101–106.

Putney, J.W. (1977) Muscarinic, α-adrenergic and peptide receptors regulate the same calcium influx sites in the parotid gland. *Journal of Physiology*, **268**, 139–149.

Putney, J.W. (1986) Identification of cellular activation mechanisms associated with salivary secretion. *Annual Review of Physiology*, **48**, 75–84.

Quirion, R., Regoli, D., Rioux, F. and St-Pierre, S. (1979) An analysis of the negative inotropic action of somatostatin. *British Journal of Pharmacology*, **66**, 251–257.

Ramme, D., Regenold, J.T., Starke, K., Busse, R. and Illes, P. (1987) Identification of the neuroeffector transmitter in jejunal branches of the rabbit mesenteric artery. *Naunyn-Schmiedeberg's Archives of Pharmacology*, **336**, 267–272.

Rand, M.J., Storey, D.F. and Wong-Dusting, H.K. (1982) Effect of histamine on the resting and stimulation induced release of (^{3}H)-noradrenaline in guinea-pig isolated atria. *British Journal of Pharmacology*, **75**, 57–64.

Rebuffat, P., Malendowicz, L.K., Belloni, A.S., Mazzochi, G. and Nussdorfer, G.G. (1988) Long-term stimulatory effect of neuropeptide Y on the growth and steroidogenic capacity of rat adrenal zona glomerulosa. *Neuropeptides*, **11**, 133–136.

Reid, J.J., Wong-Dusting, H.K. and Rand, M.J. (1989) The effect of endothelin on noradrenergic transmission in rat and guinea-pig atria. *European Journal of Pharmacology*, **168**, 93–96.

Reinecke, M., Forssmann, W.G., Thiekötter, G. and Triepel, J. (1983) Localization of neurotensin – immunoreactivity in the spinal cord and peripheral nervous system of the guinea-pig. *Neuroscience Letters*, **37**, 37–42.

Revington, M. and McCloskey, D.I. (1988) Neuropeptide Y and control of vascular resistance in skeletal muscle. *Regulatory Peptides*, **23**, 331–342.

Reynolds, E.E. and Yokota, S. (1988) Neuropeptide Y receptor-effector coupling mechanisms in cultured vascular smooth muscle cells. *Biochemical and Biophysical Research Communications*, **151**, 919–925.

Ronai, A.Z. (1990) Inhibition of neurotransmission by angiotensin I and II in rabbit isolated ear artery. *European Journal of Pharmacology*, **179**, 281–286.

Rudehill, A., Sollevi, A., Franco-Cereceda, A. and Lundberg, J.M. (1986) Neuropeptide Y (NPY) and the pig heart: release and coronary vasoconstrictor effects. *Peptides*, **7**, 821–826.

Rudehill, A., Olcen, M., Sollevi, A., Hamberger, B. and Lundberg, J.M. (1987) Release of neuropeptide Y upon haemorrhagic hypovolaemia in relation to vasoconstrictor effects in the pig. *Acta Physiologica Scandinavica*, **131**, 517–523.

Ruth, J.A., Doerr, A.L. and Eiden, L.E. (1984) [Leu⁵]enkephalin inhibits norepinephrine–induced contraction of rat aorta. *European Journal of Pharmacology*, **105**, 189–191.

Saper, C.B., Hurley, K.M., Moga, M.M., Houmes, H.R., Adams, S.A., Leahy, K.M. and Needleman, P. (1989) Brain natriuretic peptides: differential localization of a new family of neuropeptides. *Neuroscience Letters*, **96**, 29–34.

Saria, A., Ma, R.C., Dun, N.J., Theodorsson–Norheim, E. and Lundberg, J.M. (1987) Neurokinin A in capsaicin-sensitive neurons of the guinea-pig inferior mesenteric ganglia: an additional putative mediator for the non-cholinergic excitatory postsynaptic potential. *Neuroscience*, **21**, 951–958.

Satchell, D. (1986) Antagonism of the ATP component of sympathetic co-transmission in the rat vas deferens by AMP. *European Journal of Pharmacology*, **132**, 305–308.

Saum, W.R. and de Groat, W.C. (1973) The actions of 5-hydroxytryptamine on the urinary bladder and on vesical autonomic ganglia in the cat. *Journal of Pharmacology and Experimental Therapeutics*, **185**, 70–83.

Saville, V.L. and Burnstock, G. (1988) Use of reserpine and 6-hydroxydopamine supports evidence for purinergic cotransmission in the rabbit ear artery. *European Journal of Pharmacology*, **155**, 271–277.

Saville, V.L., Maynard, K.I. and Burnstock, G. (1990) Neuropeptide Y potentiates purinergic as well as adrenergic responses of the rabbit ear artery. *European Journal of Pharmacology*, **176**, 117–125.

Schotzinger, R.J. and Landis, S.C. (1990) Post-natal development of autonomic and sensory innervation of thoracic hairy skin in the rat. A histochemical, immunocytochemical and radioenzymatic study. *Cell and Tissue Research*, **260**, 575–587.

Schultzberg, M. (1983) Bombesin-like immunoreactivity in sympathetic ganglia. *Neuroscience*, **8**, 363–374.

Schultzberg, M., Hökfelt, T., Terenius, L., Elfvin, L.G., Lundberg, J.M., Brandt, J., Elde, R.P. and Goldstein, M. (1979) Enkephalin immunoreactive nerve fibres and cell bodies in sympathetic ganglia of the guinea-pig and rat. *Neuroscience*, **4**, 249–270.

Schumann, M.A. and Kreulen, D.L. (1986) Action of cholecystokinin octapeptide and CCK-related peptides on neurones in inferior mesenteric ganglion of guinea-pig. *Journal of Pharmacology and Experimental Therapeutics*, **239**, 618–625.

Schwartz, D.D. and Malik, K.U. (1989) Renal periarterial nerve stimulation-induced vasoconstriction at low frequencies is primarily due to release of a purinergic transmitter in the rat. *Journal of Pharmacology and Experimental Therapeutics*, **250**, 764–771.

Sedaa, K.O., Bjur, R.A., Shinozuka, K. and Westfall, D.P. (1990) Nerve and drug-induced release of adenine nucleosides and nucleotides from rabbit aorta. *Journal of Pharmacology and Experimental Therapeutics*, **252**, 1060–1067.

Senba, E. and Tohyama, M. (1988) Calcitonin gene-related peptide containing autonomic efferent pathways to the pelvic ganglia of the rat. *Brain Research*, **449**, 386–390.

Sharkey, K.A. and Templeton, D. (1984) Substance P in the rat parotid gland: evidence for a dual origin from the otic and trigeminal ganglia. *Brain Research*, **304**, 392–396.

Sheikh, S.P., Holst, J.J., Skak-Nielsen, Y., Knigge, U., Warberg, J., Theodorsson-Norheim, E., Hökfelt, T., Lundberg, J.M. and Schwartz, T.W. (1988) Release of NPY in pig pancreas: dual parasympathetic and sympathetic regulation. *American Journal of Physiology*, **255**, G46–G54.

Shirakawa, J., Nakanishi, T., Taniyama, K., Kamidono, S. and Tanaka, C. (1989) Regulation of the substance P-induced contraction via the release of acetylcholine and γ-aminobutyric acid in the guinea-pig urinary bladder. *British Journal of Pharmacology*, **98**, 437–444.

Shu, H.D., Love, J.A. and Szurszewski, J.H. (1987) Effect of enkephalins on colonic mechanoreceptor input to inferior mesenteric ganglion. *American Journal of Physiology*, **252**, G128–G135.

Sibley, C.N.A. (1984) A comparison of spontaneous and nerve-mediated activity in bladder muscle from man, pig and rabbit. *Journal of Physiology*, **354**, 431–443.

Simmons, M.A. (1985) The complexity and diversity of synaptic transmission in the prevertebral sympathetic ganglia. *Progress in Neurobiology*, **24**, 43–93.

Sitsapesan, R. and Parratt, J.R. (1989) The effects of drugs interacting with opioid receptors on the early ventricular arrhythmias arising from myocardial ischemia. *British Journal of Pharmacology*, **97**, 795–800.

Sjögren, C., Andersson, K.E. and Husted, S. (1982) Contractile effects of some polypeptides on the isolated urinary bladder of guinea-pig, rabbit and rat. *Acta Pharmacologica Toxicologica*, **50**, 175–184.

Sjöstrand, N.O. (1965) The adrenergic innervation of the vas deferens and the accessory male genital glands. *Acta Physiologica Scandinavica*, **Suppl. 257**, 1–82.

Sjöstrand, N.O. and Klinge, E. (1979) Principal mechanisms controlling penile retraction and protrusion in rabbits. *Acta Physiologica Scandinavica*, **106**, 199–214.

Smith, T.K. and Furness, J.B. (1988) Relex changes in circular muscle activity elicited by stroking the mucosa: an electrophysiological analysis in the isolated guinea-pig ileum. *Journal of the Autonomic Nervous System*, **25**, 205–218.

Smith, T.K., Furness, J.B., Costa, M. and Bornstein, J.C. (1988) An electrophysiological study of the projections of motor neurones that mediate non-cholinergic excitation in the circular muscle of the guinea-pig small intestine. *Journal of the Autonomic Nervous System*, **22**, 115–128.

Smith, T.K., Bornstein, J.C. and Furness, J.B. (1990) Distension evoked ascending and descending reflexes in the circular muscle of guinea-pig ileum: an intracellular study. *Journal of the Autonomic Nervous System*, **29**, 203–218.

Sneddon, P. and Burnstock, G. (1985) ATP as a co-transmitter in rat tail artery. *European Journal of Pharmacology*, **106**, 149–152.

Sneddon, P. and Westfall, D.P. (1984) Pharmacological evidence that adenosine triphosphate and noradrenaline are co-transmitters in the guinea-pig vas deferens. *Journal of Physiology*, **347**, 561–580.

Spampinato, S. and Goldstein, A. (1983) Immunoreactive dynorphin in rat tissues and plasma. *Neuropeptides*, **3**, 193–212.

Stapelfeldt, W.H. and Szurszewski, J.H. (1989a) Neurotensin facilitates release of substance P in the guinea-pig inferior mesenteric ganglion. *Journal of Physiology*, **411**, 325–345.

Stapelfeldt, W.H. and Szurszewski, J.H. (1989b) Central neurotensin nerves modulate colo-colonic reflex activity in the guinea-pig inferior mesenteric ganglion. *Journal of Physiology*, **411**, 347–365.

Starke, K. (1977) Regulation of noradrenaline release by presynaptic receptor systems. *Reviews of Physiology, Biochemistry and Pharmacology*, **77**, 1–124.

Starke, K. (1987) Presynaptic α-adrenoceptors. *Reviews of Physiology, Biochemistry and Pharmacology*, **107**, 73–146.

Steele, P.A. and Costa, M. (1990) Opioid-like immunoreactive neurones in secretomotor pathways of the guinea-pig ileum. *Neuroscience*, **38**, 771–786.

Stjernquist, M., Emson, P., Owman, C., Sjöberg, N.O., Sundler, F. and Tatemoto, K. (1983) Neuropeptide Y in the female reproductive tract of the rat. Distribution of nerve fibres and motor effects. *Neuroscience Letters*, **39**, 279–284.

Stjernquist, M., Owman, C., Sjöberg, N.O. and Sundler, F. (1987) Coexistence and cooperation between neuropeptide Y and norepinephrine in nerve fibers of guinea-pig vas deferens and seminal vesicle. *Biological Reproduction*, **36**, 149–155.

Stjärne, L. (1977) Do potassium ions released from nerves modulate the sensitivity to transmitter in 'close' neuro-effector junctions of the vas deferens? *Neuroscience*, **2**, 373–387.

Stjärne, L. (1986) New paradigm: Sympathetic neurotransmission by lateral interaction between secretory units? *News in Physiological Sciences*, **1**, 103–106.

Stjärne, L. (1989) Basic mechanisms and local modulation of nerve impulse-induced secretion of neurotransmitters from individual sympathetic nerve varicosities. *Review of Physiology, Biochemistry and Pharmacology*, **112**, 1–137.

Stjärne, L. and Åstrand, P. (1984) Discrete events measure single quanta of adenosine 5'-triphosphate secreted from sympathetic nerves of guinea-pig and mouse vas deferens. *Neuroscience*, **13**, 21–28.

Stjärne, L. and Åstrand, P. (1985) Relative pre- and postjunctional roles of noradrenaline and adenosine 5'-triphosphate as neurotransmitters of the sympathetic nerves of guinea-pig and mouse vas deferens. *Neuroscience*, **14**, 929–946.

Stjärne, L., Lundberg, J.M. and Åstrand, P. (1986) Neuropeptide Y – A co-transmitter with noradrenaline and adenosine 5'-triphosphate in the sympathetic nerves of the mouse vas deferens? A biochemical, physiological and electropharmacological study. Neuroscience, 18, 151–166.

Stjärne, L., Stjärne, E. and Mshgina, M. (1989) Does clonidine- or neuropeptide Y-mediated inhibition of ATP secretion from sympathetic nerves operate primarily by increasing a potassium conductance? Acta Physiologica Scandinavica, 136, 137–138.

Su, C. (1975) Neurogenic release of purine compounds in blood vessels. Journal of Pharmacology and Experimental Therapeutics, 195, 159–166.

Su, C. (1983) Purinergic neurotransmission and neuromodulation. Annual Review of Pharmacology and Toxicology, 23, 397–411.

Suzuki, H. (1983) An electrophysiological study of excitatory neuromuscular transmission in the guinea-pig main pulmonary artery. Journal of Physiology, 336, 47–59.

Suzuki, H. (1984) Adrenergic transmission in the dog mesenteric vein and its modulation by α-adrenoceptor antagonists. British Journal of Pharmacology, 81, 479–489.

Suzuki, H. (1985) Electrical responses of smooth muscle cells of the rabbit ear artery to adenosine triphosphate. Journal of Physiology, 359, 401–415.

Suzuki, H. and Kou, K. (1983) Electrical components contributing to the nerve-mediated contractions in the smooth muscle of the rabbit ear artery. Japanese Journal of Pharmacology, 33, 743–756.

Suzuki, H., Mishima, S. and Miyahara, H. (1984) Effects of reserpine treatment on electrical responses induced by perivascular nerve stimulation in the rabbit ear artery. Biomedical Research, 5, 259–265.

Suzuki, N. and Gomi, Y (1987) Neurogenic contractile responses of the circular smooth muscle of the guinea-pig vas deferens. Japanese Journal of Physiology, 45, 211–221.

Suzuki, N., Hardebo, J.E., Kåhrström, J. and Owman, C. (1990) Neuropeptide Y co-exists with vasoactive intestinal polypeptide and acetylcholine in parasympathetic cerebrovascular nerves originating in the sphenopalatine, otic and internal carotid ganglia of the rat. Neuroscience, 36, 507–520.

Suzuki, Y., Shibuya, M., Ikegaki, I., Satoh, S.I., Takayasu, M. and Asano, T. (1988) Effects of neuropeptide Y on canine cerebral circulation. European Journal of Pharmacology, 146, 271–277.

Swedin, G. (1971) Studies on neurotransmission mechanisms in the rat and guinea-pig vas deferens. Acta Physiologica Scandinavica, Suppl. 369, 1–34.

Tabuchi, Y., Nakamura, M., Rakugi, H., Nagano, M. and Ogihara, T. (1989a) Endothelin enhances adrenergic vasoconstriction in perfused rat mesenteric arteries. Biochemical and Biophysical Research Communications, 159, 1304–1308.

Tabuchi, Y., Nakamaru, M., Rakugi, H., Nagano, M., Mikami, H. and Ogihara, T. (1989b) Endothelin inhibits presynaptic adrenergic neurotransmission in rat mesenteric artery. Biochemical and Biophysical Research Communications, 161, 803–808.

Taddei, S., Salvetti, A. and Pedrinelli, R. (1989) Persistence of sympathetic mediated forearm vasoconstriction after α-blockade in hypertensive patients. Circulation, 80, 485–490.

Taniyama, K., Kusunoki, M. and Tanaka, C. (1983) γ-aminobutyric acid inhibits motility of the isolated guinea-pig urinary bladder. European Journal of Pharmacology, 89, 163–166.

Tatemoto, K. and Mutt, V. (1981) Isolation and characterization of the intestinal peptide porcine PHI (PHI-27), a new member of the glucagon-secretin family. Proceedings of the National Academy of Science, USA, 78, 6603–6607.

Taylor, C.M., Merritt, J.E., Putney, J.M. and Rubin, R.P. (1986) A guanine nucleotide-dependent regulatory protein couples substance P receptors to phospholipase C in rat parotid gland. Biochemical and Biophysical Research Communications, 136, 362–368.

Taylor, E.M. and Parsons, M.E. (1989) Adrenergic and purinergic neurotransmission in arterial resistance vessels of the cat intestinal circulation. European Journal of Pharmacology, 164, 23–33.

Taylor, G.S. and Bywater, R.A.R. (1986) Antagonism of non-cholinergic excitatory junction potentials on the guinea-pig ileum by a substance P antagonist. Neuroscience Letters, 63, 23–26.

Thoren, P., Mark, A.L., Morgan, D.A., O'Neil, T.P., Needleman, P. and Brody, M. (1986) Activation of vagal depressor reflexes by atriopeptins inhibits renal sympathetic nerve activity. American Journal of Physiology, 251, H1252–H1257.

Toda, N. (1982) Relaxant responses to transmural stimulation and nicotine of dog and monkey cerebral arteries. American Journal of Physiology, 243, H145–H153.

Toda, N. (1988) Hemolysate inhibits cerebral artery relaxation. Journal of Cerebral Blood Flow and Metabolism, 8, 46–53.

Toda, N., Ayajiki, K. and Okamura, T. (1990) Cerebral non-adrenergic, non-cholinergic vasodilator nerve function as affected by acetylcholine and nitric oxide. Blood Vessels, 27, 60–61.

Tomita, T. and Watanabe, H. (1973) A comparison of the effects of adenosine triphosphate with noradrenaline and with the inhibitory potential of the guinea-pig taenia coli. Journal of Physiology, 231, 167–177.

Tonini, M., Frigo, G., Lecchini, L., D'Angelo, L. and Crema, A. (1981) Hyoscine-resistant peristalsis in guinea-pig ileum. *European Journal of Pharmacology*, **71**, 375–381.

Trachte, G.J., Binder, S.B. and Peach, M.J. (1989) Indirect evidence for separate vesicular neuronal origins of norepinephrine and ATP in the rabbit vas deferens. *European Journal of Pharmacology*, **164**, 425–433.

Tsunoo, A., Konishi, S. and Otsuka, M. (1982) Substance P as an excitatory transmitter of primary afferent neurones in guinea-pig sympathetic ganglia. *Neuroscience*, **7**, 2025–2037.

Uddman, R., Fahrenkrug, J., Malm, L., Alumets, J., Håkanson, R. and Sundler, F. (1980a) Neuronal VIP in salivary glands: distribution and release. *Acta Physiologica Scandinavica*, **110**, 31–38.

Uddman, R., Malm, L. and Sundler, F. (1980b) The origin of vasoactive intestinal polypeptide (VIP) nerves in feline nasal mucosa. *Acta Otolaryngologica*, **89**, 152–156.

Ursillo, R.C. and Clark, B.B. (1956) The action of atropine on the urinary bladder of the dog and on the isolated nerve-bladder strip preparation of the rabbit. *Journal of Pharmacology and Experimental Therapeutics*, **118**, 338–347.

Van Helden, D.F. (1988) Electrophysiology of neuromuscular transmission in guinea-pig mesenteric veins. *Journal of Physiology*, **401**, 469–488.

Vanov, S. (1965) Responses of rat urinary bladder '*in situ*' to drugs and to nerve stimulation. *British Journal of Pharmacology and Chemotherapy*, **24**, 591–600.

Verbeuren, T.J., Jordans, F.H. and Herman, A.G. (1983) Accumulation and releases of [^{3}H]-5-hydroxytryptamine in saphenous veins and cerebral arteries of the dog. *Journal of Pharmacology and Experimental Therapeutics*, **226**, 579–588.

Vizi, E.S. (1979) Presynaptic modulation of neurochemical transmission. *Progress in Neurobiology*, **12**, 181–290.

Vizi, E.S. and Burnstock, G. (1988) Origin of ATP release in the rat vas deferens: concomitant measurement of [^{3}H]noradrenaline and [^{14}C]ATP. *European Journal of Pharmacology*, **158**, 69–77.

Vizi, E.S., Ono, K., Aram-Vizi, V., Duncalf, D. and Foldes, F.F. (1984) Presynaptic inhibitory-effect of met-enkephalin on [^{14}C] acetylcholine release from the myenteric plexus and its interaction with muscarinic negative feedback inhibition. *Journal of Pharmacology and Experimental Therapeutics*, **230**, 493–499.

von Euler, U.S. and Hedqvist, P. (1975) Evidence for an α- and β_2-receptor mediated inhibition of the twitch response in the guinea-pig vas deferens by noradrenaline. *Acta Physiologica Scandinavica*, **93**, 572–573.

von Kügelgen, I.V. and Starke, K. (1985) Noradrenaline and adenosine triphosphate co-transmitters of neurogenic vasoconstriction in rabbit mesenteric artery. *Journal of Physiology*, **367**, 435–455.

von Kügelgen, I., Illes, P., Wolf, D. and Starke, K. (1985) Presynaptic inhibitory opioid δ- and κ-receptors in a branch of the rabbit ileocolic artery. *European Journal of Pharmacology*, **118**, 97–105.

von Kügelgen, I., Bultmann, R. and Starke, K. (1989) Effects of suramin and α, β-methylene ATP indicate noradrenaline-ATP co-transmission in the response of the mouse vas deferens to single and low frequency pulses. *Naunyn-Schmiedeberg's Archives of Pharmacology*, **340**, 760–763.

Vu, H.Q., Budai, D. and Duckles, S.P. (1989) Neuropeptide Y preferentially potentiates responses to adrenergic nerve stimulation by increasing rate of contraction. *Journal of Pharmacology and Experimental Therapeutics*, **251**, 852–857.

Wadsworth, R.M. (1973) Abolition of neurally evoked motor responses of the vas deferens by 6-hydroxydopamine. *European Journal of Pharmacology*, **21**, 383–387.

Wahlestedt, C., Edvinsson, L., Ekblad, E. and Håkanson, R. (1985) Neuropeptide Y potentiates noradrenaline-evoked vasoconstriction: mode of action. *Journal of Pharmacology and Experimental Therapeutics*, **234**, 735–741.

Wahl, M. (1985) Effects of enkephalins, morphine, and naloxone on pial arteries during perivascular microapplication. *Journal of Cerebral Blood Flow and Metabolism*, **5**, 451–457.

Wallis, D.I. and North, R.A. (1978) The action of 5-hydroxytryptamine on single neurones of the rabbit superior cervical ganglion. *Neuropharmacology*, **17**, 1023–1028.

Warland, J.I. and Burnstock, G. (1987) Effects of reserpine and 6-hydroxydopamine on the adrenergic and purinergic components of sympathetic nerve responses of the rabbit saphenous artery. *British Journal of Pharmacology*, **92**, 871–880.

Warner, M.R. and Levy, M.N. (1989) Neuropeptide Y as a putative modulator of the vagal effects on heart rate. *Circulation Research*, **64**, 882–889.

Webber, R.H. and Heym, C. (1988) Immunohistochemistry of biogenic polypeptides in nerve cells and fibres of the guinea-pig inferior mesenteric ganglion after perturbations. *Histochemistry*, **88**, 287–297.

Weihe, E., McKnight, A.T., Corbett, A.D. and Kosterlitz, H.W. (1985) Proenkephalin- and prodynorphin-derived opioid peptides in guinea-pig heart. *Neuropeptides*, **5**, 453–456.

Weitzell, R., Illes, P. and Starke, K. (1984) Inhibition via opioid μ and δ-receptors of vagal transmission in rabbit isolated heart. *Naunyn-Schmiedeberg's Archives of Pharmacology*, **328**, 186–190.

Westfall, D.P., Sedaa, K. and Bjur, R.A. (1987a) Release of endogenous ATP from rat caudal artery. *Blood Vessels*, **24**, 125–127.

Westfall, T.C., Carpentier, S., Chen, X., Beinfeld, M.C., Naes, L. and Meldrum, M.J. (1987b) Prejunctional and postjunctional effects of neuropeptide Y at the noradrenergic neuroeffector junction of the perfused mesenteric arterial bed of the rat. *Journal of Cardiovascular Pharmacology*, **10**, 716–722.

Westlind-Danielsson, A., Müller, R.M. and Bartfai, T. (1990) Atropine treatment induced cholinergic supersensitivity at receptor and second messenger levels in the rat salivary gland. *Acta Physiologica Scandinavica*, **138**, 431–441.

Wharton, J., Polak, J.M., Bryant, M.G., Van Noorden, S., Bloom, S.R. and Pearse, A.G.E. (1979) Vasoactive intestinal polypeptide (VIP)-like immunoreactivity in salivary glands. *Life Sciences*, **25**, 273–280.

Wiklund, N.P., Öhlén, A. and Cederqvist, B. (1989) Adrenergic neuromodulation by endothelin in guinea-pig pulmonary artery. *Neuroscience Letters*, **101**, 269–273.

Wiley, J.W., Uccioli, L., Owyang, C. and Yamada, T. (1989) Somatostatin stimulates acetylcholine release in the canine heart. *American Journal of Physiology*, **257**, H483–H487.

Wong-Dusting, H.K. and Rand, M.J. (1985) Effect of [D-Ala^2, Met^5]-enkephalinamide and [D-Ala^2, D-Leu^5] enkephalin on cholinergic and noradrenergic neurotransmission is isolated atria. *European Journal of Pharmacology*, **111**, 65–72.

Wong-Dusting, H.K. and Rand, M.J. (1988) Pre- and postjunctional effects of neuropeptide Y on the rabbit isolated ear artery. *Clinical and Experimental Pharmacology and Physiology*, **15**, 411–418.

Xiang, J.Z., Archelos, J. and Lang, R.E. (1984) Enkephalins in the heart. *Clinical and Experimental Hypertension, Part A*, **6**, 1883–1888.

Yamamoto, K., Senba, E., Matsunage, T. and Tohyama, M. (1989). Calcitonin gene-related peptide containing sympathetic preganglionic and sensory neurons projecting to the superior cervical ganglion of the rat. *Brain Research*, **487**, 158–163.

Yamamoto, R., Cline, W.H. and Takasaki, K. (1988) Reassessment of the blocking activity of prazosin at low and high concentrations on sympathetic neurotransmission in the isolated mesenteric vasculature of rats. *Journal of Autonomic Pharmacology*, **8**, 303–309.

Yamamoto, Y., Hotta, K. and Matsuda, T. (1984) Effect of methionine-enkephalin on the spontaneous electrical and mechanical activity of the smooth muscle of the rat portal vein. *Life Sciences*, **34**, 993–999.

Yanaihara, N., Nokihara, K., Yanaihara, C., Iwanga, T. and Fujita, T. (1983) Immunocytochemical demonstration of PHI and its co-existence with VIP in intestinal nerves of the rat and pig. *Archivum Histologicum Japonicum*, **46**, 575–582.

Yasui, A., Naruse, S., Yanaihara, C., Ozaki, T., Hoshino, M., Mochizuki, T., Daniel, E.E. and Yanaihara, N. (1987) Co-release of PHI and VIP by vagal stimulation in the dog. *American Journal of Physiology*, **253**, G13–G19.

Yoshimura, K., Huidobro-Toro, J.P., Lee, N.M., Loh, H.H. and Way, E.L. (1982) κ opioid properties of dynorphin and its peptide fragments on guinea-pig ileum. *Journal of Pharmacology and Experimental Therapeutics*, **222**, 71–79.

Zimmerman, B.G., Gomer, S.K. and Liao, J.C. (1972) Action of angiotensins on vascular adrenergic nerve endings: facilitation of norepinephrine release. *Federation Proceedings*, **31**, 1344–1350.

Ziogas, J. and Story, D.F. (1987) Effect of locally generated angiotensin II on noradrenergic neuroeffector function in the rat isolated caudal artery. *Journal of Hypertension*, **5**, S47–S52.

3 Electrophysiology of Neuroeffector Transmission in Smooth Muscle

James A. Brock & Thomas C. Cunnane

University Department of Pharmacology, South Parks Road, Oxford, OX1 3QT, U.K.

In this chapter, the electrophysiology of transmission at the autonomic neuroeffector junction is reviewed. A brief historical account of the development of the concept of chemical transmission is given. Emphasis is placed on the role that electrophysiological studies have played in elucidating the mechanisms underlying transmission at many autonomic neuroeffector junctions. The structure and innervation of various tissues are described briefly in order that the basic biophysical properties of smooth muscle relating to neuroeffector transmission can be more easily understood. The key features of transmitter release mechanisms at the level of the individual varicosity in postganglionic sympathetic nerve terminals are discussed in detail. Evoked transmitter release occurs intermittently and only a single quantum is secreted when the release mechanism of a varicosity is activated by the nerve impulse. Transmission at the autonomic neuroeffector junction is compared with transmission at the skeletal neuromuscular junction. Action potential propagation in the terminal nerve network of sympathetic fibres is also described. The characteristic features of excitatory and inhibitory junction potentials in many autonomically innervated tissues are described in some detail. Brief notes on the nature of the neurotransmitter(s) generating junction potentials in each tissue are included. Nerve endings in the autonomic nervous system are endowed with a variety of prejunctional receptors which can regulate transmitter release. Perhaps, the most widely studied prejunctional receptor is the α_2-adrenoceptor. The electrophysiological evidence for α-autoinhibition is discussed and the mechanisms involved in α_2-adrenoceptor mediated inhibition of transmitter release considered. Thus attention is focused on some of the new developments in this exciting and rapidly changing field to enable readers to gain new insights into the mechanisms involved in the storage and release of neurotransmitters in the autonomic nervous system.

INTRODUCTION

The aims of this chapter are to review the electrophysiology of autonomic neuroeffector transmission and to show how electrophysiological techniques have led to a better understanding of transmitter release mechanisms at the autonomic neuroeffector junction.

We begin with a brief historical account of the development of the concept of chemical transmission. Subsequent sections concentrate primarily on electrophysiological studies at various autonomic neuroeffector junctions in smooth muscle. In order to understand the origin of the complex electrophysiological signals recorded it is essential to have a knowledge of the structure and innervation of smooth muscle. Therefore, the structure and innervation of smooth muscle will be briefly described but the reader is referred to a comprehensive account in Chapter 1. The basic biophysical properties of smooth muscle relevant to an understanding of neuroeffector transmission are described and comparisons made with the skeletal neuromuscular junction and synapses.

When smooth muscle cells are penetrated with microelectrodes and nerves stimulated, changes in membrane potential are recorded. At the autonomic neuroeffector junction these transient changes in membrane potential have been termed junction potentials. The characteristic features of excitatory and inhibitory junction potentials in many autonomically innervated tissue are described and tabulated as a source of reference for interested readers. Brief notes on the nature of the neurotransmitter(s) generating junction potentials in each tissue are included but more extensive discussions of putative transmitter candidates can be found elsewhere in this volume. Two important observations struck us when surveying the literature, namely that considerable doubt remains as to the chemical identity of the transmitter(s) generating junction potentials and that surprisingly little is known about transmitter release mechanisms at many autonomic neuroeffector junctions. Perhaps the exception is the sympathetic neuroeffector junction where much progress has been made in recent years. Much of our knowledge of transmitter release mechanisms comes from studies at the skeletal neuromuscular junction. Therefore, the key features of transmitter release mechanisms at the skeletal neuromuscular junction are summarized to provide a framework in which to discuss, in some detail, transmitter release mechanisms at the level of the individual varicosity in postganglionic sympathetic nerve terminals.

Finally, nerve endings in the autonomic nervous system are endowed with a variety of prejunctional receptors which can regulate transmitter release. The most widely studied prejunctional receptor is the α_2-adrenoceptor located on or near sympathetic nerve terminals. The last part of the chapter discusses the electrophysiological evidence for α-autoinhibition and considers the mechanisms involved in α_2-adrenoceptor-mediated inhibition of transmitter release.

We hope that this chapter will focus attention on some of the new developments in this exciting and rapidly changing field and serve to stimulate others to investigate the many puzzles remaining. We do not see this chapter as a comprehensive review of all the important papers in the field but as a selective account of the development of ideas on chemical transmission in the autonomic nervous system. We apologize now to our many colleagues for failing to cite directly so many important papers but reference is made indirectly through the use of several excellent reviews which summarise the key aspects of smooth muscle electrophysiology not covered here. In reading this chapter we hope that the reader will obtain new insights into the mechanisms involved in the storage and release of neurotransmitters in the autonomic nervous system and in particular the ways in which nerves regulate the activity of smooth muscle.

HISTORICAL PERSPECTIVES

The process of neurotransmission involves the passage of information from one nerve cell to another or from a nerve cell to a muscular, glandular or other effector cell. In most cases this information is conveyed across a narrow but discrete gap between the cells by a chemical substance released from the transmitting neuron ('chemical transmission'), and sometimes by electric current flow between the two cells through their fused membranes ('electrical transmission'). For historical reasons and for consistency, the use of the term 'synapse' is reserved for nerve-nerve contacts as originally defined by Sherrington (1897), and the more general term 'junction' is employed for other kinds of neuroeffector contacts.

Historically, much of the evidence for the idea that nerves release chemicals (i.e. neurotransmitters) orignates from studies at the autonomic neuroeffector junction. Although earlier references to the idea of chemical neurotransmission do exist (see Euler, 1981), the first suggestion that a specific chemical substance mediates transmission between the nerve and the effector tissue was provided by Elliott (1904). He noted the similarity between the effects of exogenously applied adrenaline and sympathetic nerve stimulation and therefore suggested that 'adrenalin (adrenaline) might then be the chemical stimulant liberated on each occasion when the impulse arrives at the periphery'. Barger and Dale (1910) observed that the effects of adrenaline and sympathetic stimulation were not strictly parallel and therefore questioned Elliott's conclusion. In the study of Barger and Dale the effects of a number of other sympathomimetic amines were investigated and it was noted that the action of 'particularly the amino- and ethylamino-bases of the catechol group (which included noradrenaline: amino-ethanol-catechol), corresponds more closely with that of the sympathetic nerve-impulses than does that of adrenine (adrenaline)'.

Dixon (1906) suggested that stimulation of the cranial and sacral involuntary nerves (i.e. parasympathetic) resulted in the release of a muscarine-like substance and Dale (1914) noted that the correspondence between the action of acetylcholine and stimulation of the cranio-sacral involuntary nerves was 'striking, and little, if at all, less perfect than that between the actions of the true sympathetic nerves and of adrenine (adrenaline) and its allies'.

The first unequivocal demonstration of chemical neurotransmission was provided by Loewi (1921). In his classical experiment, the saline perfusing one isolated beating frog heart was used to perfuse a second heart. Stimulation of the vagus nerve supplying the first heart caused stopping or slowing of its heartbeat as expected. The critical observation, however, was that the perfusion fluid from the first heart, after a period of stimulation, inhibited the second heart. Thus stimulation of the vagus nerve resulted in the liberation of a substance ('vagusstoff') into the perfusion media, presumably from the nerve terminals in the heart, which then inhibited the second unstimulated heart. In subsequent studies Loewi and his colleagues demonstrated that the actions of vagusstoff were mimicked by acetylcholine. In particular, Loewi and Navratil (1926) reported that the effects of stimulating the vagus, and of vagusstoff and acetylcholine, on the heart were potentiated by the anticholinesterase eserine, whereas eserine had no effect on the action of muscarine. Loewi (1921) also noted an 'accelerans' factor ('acceleransstoff') released following stimulation of the sympathetic nerves supplying the frog heart. This substance was later identified as adrenaline (Loewi, 1936).

Dale (1933) in order to promote 'clear thinking', proposed that nerves releasing an adrenaline-like substance be termed *adrenergic*, whereas those which release an acetylcholine-like substance be termed *cholinergic*. Bacq (1934; see Bacq, 1975) was the first to suggest that noradrenaline may be the sympathetic neurotransmitter. However, this view was not widely accepted until the demonstration by Euler (1946a,b) that noradrenaline was the predominate catecholamine stored in mammalian sympathetic nerve terminals. This finding was followed by the demonstration that noradrenaline was the major catecholamine collected in the venous blood of the cat spleen following splenic nerve stimulation (Peart, 1949). These early studies clearly demonstrate that transmission at the autonomic neuroeffector junction is mediated by chemicals released from nerve endings by nerve impulses.

At the skeletal neuromuscular junction and synapses the idea of chemical transmission was not so readily accepted. The transmission of excitation from nerve to muscle or from nerve to nerve was thought to occur without delay. Therefore, the idea that the nerve action potential, when it arrives in the nerve terminal, has first to evoke the release of a chemical did not fit with this preconception. For this and other reasons it was believed that transmission at both skeletal neuromuscular junctions and synapses was electrical (see Eccles, 1964).

Evidence suggesting that acetylcholine acts as a neurotransmitter at autonomic ganglia was first provided by Feldberg and Vertiainen (1934) who showed that acetylcholine was released from the superior cervical ganglion, which had been pretreated with eserine to prevent the degradation of acetylcholine, and that applied acetylcholine produced a postsynaptic response similar to that of nerve stimulation. In a similar series of experiments on the mammalian skeletal neuromuscular junction, Dale, Felberg & Vogt (1936) and Brown, Dale & Felberg (1936) suggested that transmission at this site was also due to the release of acetylcholine. However, it was only following the demonstration, using electrophysiological techniques, that acetylcholinesterase inhibitors prolonged the postjunctional response to nerve stimulation (i.e. the end plate pontential (EPP)) in various vertebrate skeletal muscles, that the transmitter role of acetylcholine at these junctions became firmly established (Eccles, Katz & Kuffler, 1942; Eccles & MacFarlane, 1949).

The application of electrophysiological techniques to record both postjunctional and postsynaptic potentials has provided unequivocal evidence for chemical transmission in the peripheral and central nervous systems (see Kuffler, Nichols & Martin, 1984).

AUTONOMIC NEUROEFFECTOR JUNCTION

ANATOMICAL CONSIDERATIONS

The majority of electrophysiological experiments described in this chapter have been carried out on isolated smooth muscle preparations and in particular the rodent vas deferens and blood vessels. It is useful to describe the key anatomical features of smooth muscle which have a bearing on the interpretation of electrophysiological records, namely (1) the structure of individual smooth muscle cells and bundles, (2) the relationship between neighbouring cells and (3) the structure of the neuroeffector junction.

Anatomy of smooth muscle cells and bundles

In many hollow visceral organs there is an outer longitudinal and an inner circular smooth muscle layer. Some hollow organs have an additional inner, thin, longitudinal muscle layer (e.g. mouse vas deferens. Furness & Iwayama, 1972; Sjöstrand, 1981). In blood vessels the smooth muscle cells are usually confined to the media and are arranged in a spiral fashion with respect to the longitudinal axis, with the angle of turn increasing with decreasing diameter of the vessel (Burnstock, 1970).

Smooth muscle of the alimentary canal and urogenital tract is penetrated by laminae of connective tissue which subdivide the muscle into groups of cells; i.e. muscle bundles (Burnstock, 1970). These bundles branch and anastomose with each other. The muscle bundles rather than individual smooth muscle cells have been suggested to be the functional units of excitation (see Bennett, 1972). The demonstration of physical connections between the individual muscle bundles indicates that the electrophysiologically defined units of excitation have no discrete morphological correlate. In large arteries muscle cells are usually organised into concentric sheets which can be only one cell thick, the sheets being separated by elastic laminae (Gabella, 1981). The walls of the terminal arterioles are arranged into a single sheet of smooth muscle only one cell diameter thick.

Individual smooth muscle cells vary considerably in length depending on the tissue (range 30–450 nm; see Burnstock 1970) and have diameters of between 2 and 6 μm. Each smooth muscle cell is surrounded by a basal lamina about 20 nm thick, which is separated from the cell membrane by an electron-lucent space. The smooth muscle cells within the bundles and sheets are separated from each other by an extracellular space which is about 100 nm wide. This space is filled with connective tissue, blood vessels, nerves, Schwann cells and macrophages (Burnstock, 1970; Bennett, 1972).

Intercellular relationships

One of the most important factors which determines the electrical properties of smooth muscle is the extent of electrical coupling between individual smooth muscle cells (Bozler, 1948; Tomita, 1970), the whole muscle behaving functionally as a syncytium. Evidence for electrical coupling between cells comes from the demonstration that electrotonic potentials generated by extracellular or intracellular current injection can be recorded at a distance much greater than one cell length away from the source (Tomita, 1967; Abe & Tomita, 1968, Hirst & Neild, 1978). Although smooth muscle cells are separated from each other for most of their length, ultra-structural studies have revealed cell-to-cell contacts, termed 'gap junctions' or 'nexuses' (Gabella, 1981). The gap junctions are characterized by the close apposition, but not fusion, of the adjacent plasma membranes, the overall thickness of the junction being approximately double the thickness of the plasma membrane. Owing to the presence of a 2–3 nm gap between the two membranes the junction appears as a seven-layered structure. Gap junctions are thought to provide the structural basis for electrical coupling between cells (see Gabella, 1981; Daniel, 1987). Freeze-fracture studies have revealed the presence of discrete intramembrane particles at the junctions, which may well represent intercellular ion channels spanning the two membranes (e.g. Fry, Devine & Burnstock, 1977).

Innervation

The majority of smooth muscles are innervated by postganglionic nerve fibres of the autonomic nervous system. The cell bodies of postganglionic sympathetic neurons lie either in the paravertebral ganglia (i.e. the sympathetic chain) or in the prevertebral ganglia (e.g. the hypogastric and abdominal ganglia), whereas those of the postganglionic parasympathetic nerves usually lie within the walls of the tissue which they innervate or in remote ganglia (e.g. the cranial autonomic ganglia). Postganglionic nerve fibres are always unmyelinated and in their terminal regions are highly branched, the axons becoming punctuated by small, irregularly shaped swellings known as varicosities. The varicosities, which are 0.5–2 μm in diameter and about 1 μm in length, are packed with vesicles and mitochondria, and usually occur at 3.5–5 μm intervals. The axon spanning the intervaricose regions is usually about 0.1–0.5 μm in diameter. Vesicles are believed to be the sites of transmitter storage and therefore the varicosities are thought to be the sites of transmitter release in the terminal nerve network.

The structural arrangement and density of innervation of different smooth muscles varies considerably (Burnstock, 1970, see Chapter 1). However, for the purpose of this discussion two quite different patterns of innervation will be described, namely that of the rodent vas deferens and that of vascular smooth muscle (arteries and aterioles).

Innervation of the rodent vas deferens

The innervation of the rodent vas deferens is unusual in that it is supplied by short postganglionic sympathetic neurons whose cell bodies lie within the hypogastric ganglia situated close to the prostatic end of the vas deferens (Sjöstrand, 1965; Ferry, 1967). The preganglionic sympathetic fibres innervating these ganglia run in the hypogastric nerve trunk. The vas deferens nerves, which contain mainly non-myelinated postganglionic sympathetic nerve fibres (0.2–1.75 μm in diameter) embedded in Schwann cells together with a small number of fine myelinated fibres (1–2 μm in diameter), divide into numerous branches as they enter the vas deferens at its prostatic end (Merrillees, Burnstock & Holman, 1963). The vas deferens nerves run within the connective tissue sheath of the organ, branches passing into the muscle coats where they split into smaller bundles of 2 to 8 axons. Fibres both within the muscle coats and in the surface bundles become varicose (Bennett, 1972). In the varicose regions the Schwann cell covering becomes incomplete, leaving some of the nerve fibres naked at points apposed to the smooth muscle cells. Finally, single fibres leave the nerve bundles in the muscle and lose their Schwann cell sheaths, the varicosities of these naked fibres making close neuroeffector contacts (10–20 nm) with the muscle cells (Merrillees, 1968); i.e. close contact varicosities. The naked fibres often end within a shallow depression in the surface of a smooth muscle cell.

Merrillees (1968) made a detailed analysis of the innervation of single muscle cells in the longitudinal muscle layer of the guinea-pig vas deferens. The majority of the varicosities occur at a distance of 50 nm or greater from the smooth muscle cell membranes and only about 20% of the cells in the guinea-pig vas deferens receive close-contact varicosities (i.e. ≤ 50 nm). Furthermore, some of the muscle cells appear to receive no direct innervation. In the rat and mouse vas deferens close-contact varicosities occur more frequently than in guinea-pig vas deferens, giving rise to the suggestion that the majority of muscle cells in these tissues are innervated by at least one close-contact varicosity (see Burnstock, 1970). It should be pointed out that in all these anatomical studies random sections were examined to arrive at these conclusions.

Innervation of arteries and arterioles

In contrast to the rodent vas deferens, most mammalian arteries and arterioles are innervated by long postganglionic sympathetic nerves whose cell bodies lie in the paravertebral ganglia. The terminal axons of these sympathetic nerves, which are normally confined to the adventitia, ramify over the smooth muscle surface of arteries and arterioles. Rarely, if ever, do the terminal axons penetrate more than 2 or 3 cells deep into the smooth muscle layer.

Previous studies showed that only a small proportion of varicosities made close neuroeffector junctions with individual smooth muscle cells in vascular tissues when random sections were examined. The separation between varicosities and smooth muscle membranes varied between 0.1 and 10 μm, observations which led to the idea that transmitter may be released at variable distances from the smooth muscle cells (Burnstock, 1970).

An important change in thinking on the pattern of innervation of blood vessels has come from recent studies in which serial sections through selected regions of nerve and smooth muscle were examined, and individual varicosities and the adjacent smooth muscle layers reconstructed. In the guinea-pig submucosal arteriole more than 80% of varicosities, at points where they are naked of Schwann cell, form a close contact (< 100 nm) with an arteriolar smooth muscle cell (Luff, McLachlan & Hirst, 1987). The neuromuscular junctions are characterized by aggregation of transmitter storage vesicles towards the prejunctional membrane and fusion of the basal laminae of the nerve axon and smooth muscle cell. At some junctions the prejunctional membranes have electron dense patches which bear a resemblance to the presynaptic specializations observed at some synapses (Luff, 1988). The proportion of close contacts found in random sections of arteriole was not much greater than that found in random sections of arteries (Luff, McLachlan & Hirst, 1987). Recent studies of serial sections in a range of different arteries provide data which indicate that all muscular arteries in rats and guinea-pigs, and most muscular arteries in rabbits are innervated by specialized sympathetic neuroeffector junctions (Luff & McLachlan, 1989). Indeed these authors suggest that the frequency of occurrence of neuromuscular junctions, rather than the absolute number of varicosities per unit area, should be used to define the innervation density of arteries. This is an important area of research and more serial section investigations need to be carried out to determine the relationship between nerve and muscle in other smooth muscles.

POSTJUNCTIONAL MECHANISMS

Basic properties

In the following section the characteristic features of electrical activity recorded from various smooth muscles in response to stimulation of excitatory or inhibitory autonomic nerves are described. Because individual smooth muscle cells are electrically coupled to their neighbours, the electrical activity recorded with an intracellular microelectrode is generated not only in the impaled cell but also in coupled cells. In a number of smooth muscles stimulation of excitatory nerves with a single stimulus evokes a transient depolarization, termed the excitatory junction potential (EJP). In other smooth muscles, depolarizations are only recorded in response to trains of stimuli. Stimulation of inhibitory nerves often elicits a hyperpolarization. If the hyperpolarization is evoked by a single stimulus it is termed an inhibitory junction potential (IJP). The amplitude of both EJPs and IJPs is graded with stimulus intensity and depends on the number of nerve fibres recruited. Thus as the stimulus intensity is increased the junction potentials increase in amplitude until all the nerve fibres are excited. This electrophysiological data provides good evidence that smooth muscle is multiply innervated.

Although in some cases it is possible to stimulate the nerve trunk innervating a smooth muscle preparation, in most the intramural nerves are activated by field or transmural electrical stimulation. In these cases not only are the nerve fibres subjected to the applied voltage but so are the muscle fibres and other cells within the preparation. In these cases it is important to establish that the recorded electrical activity results solely from nerve excitation. Normally, pulse durations of less than 1 ms duration are used to stimulate nerves selectively. In order to confirm that the electrical activity recorded is neurogenic it is essential to demonstrate that the activity is abolished by agents which block nerve impulse propagation. Since the action potential in smooth muscle is generated by the activation of voltage – dependent Ca^{2+} channels, it is possible to use the Na^+ channel blockers tetrodotoxin (TTX) or saxitoxin to block selectively nerve-mediated responses. In the case of sympathetic fibres, sensitivity to the adrenergic neuron blocking agents (e.g. guanethidine or bretylium) provides additional support for the neurogenic origin of the recorded electrical activity. Similarly botulinum toxin can be used to block transmitter release from cholinergic nerve fibres.

At a number of synapses and neuroeffector junctions, specific receptors for the neurotransmitter are concentrated in the regions of the postjunctional membrane lying close to the prejunctional nerve terminals (e.g. Dennis, Harris & Kuffler, 1971; Kuffler & Yoshikami, 1975; Takeuchi, 1976; Cull-Candy, 1976; Bowman, 1985). However, at the neuroeffector junction in smooth muscle, there is little evidence of postjunctional specialization (Burnstock, 1970; Gabella, 1981). Thus it is often assumed that exogenously applied transmitter can mimic the effects of the endogenously released transmitter. However, there may be pharmacologically and/or functionally distinct populations of receptors at the junction between nerve and muscle in the autonomic nervous system. This has recently been shown to be the case in the toad and the guinea-pig heart (Campbell *et al.* 1989; Bywater *et al.* 1989). In these tissues bath-applied acetylcholine and electrical stimulation of the vagus caused bradycardia and cessation of the heart beat, but the effects of these procedures on pacemaker action potentials were quite different. The effects of both exogenous acetylcholine and vagal nerve stimulation were blocked by hyoscine, indicating that they were mediated through muscarinic receptors. For this reason it has been concluded that the muscarinic receptors activated by neurally released acetylcholine are 'functionally distinct' from those activated by exogenously applied acetylcholine. Thus, it is clearly important to ensure that the effects of exogenously applied transmitters truly mimic the effects of nerve stimulation and in many cases this has not been convincingly demonstrated.

Smooth muscle contracts when the intracellular Ca^{2+} concentration is raised by one of several mechanisms. Calcium can enter the cell through voltage – dependent or receptor – operated Ca^{2+} channels. In addition Ca^{2+} can be released from internal stores (e.g. the sarcoplasmic reticulum) by second messengers. In smooth muscles which generate spontaneous muscle action potentials, the depolarizations elicited by excitatory nerve stimulation result in an increased rate of muscle action potential discharge. Since the action potential in smooth muscle involves the influx of Ca^{2+}, this will increase contractile activity. In electrically quiescent smooth muscle, depolarization may result in the membrane potential reaching threshold for the opening of voltage – dependent Ca^{2+} channels, although this need not necessarily be associated with the generation of muscle action potentials. Some smooth muscle do not readily generate muscle action potentials although calcium entry through voltage – dependent Ca^{2+} channels may contribute to contraction (see Bolton & Large, 1986). However, in some of these tissues, after treatment with the K^+ channel blocking agent tetraethylammonium (TEA) muscle action potentials are generated (e.g. rat anococcygeus, Creed, Gillespie & Muir, 1975; rabbit ear artery, Droogmans, Raemaekers & Casteels, 1977; rabbit main pulmonary artery, Haeusler & Thorens, 1980; rabbit saphenous artery, Holman & Surprenant, 1979). In these tissues it is probable that outward currents normally prevent the net inward currents which are necessary for the generation of muscle action potentials. In various smooth muscles contractions evoked by stimulation of excitatory nerves are, at least in part, independent of changes in membrane potential. In these tissues the cytoplasmic Ca^{2+} may be raised either by the activation of receptor operated Ca^{2+} channels or by the release of Ca^{2+} from internal stores triggered by the generation of second messengers (pharmaco-mechanical coupling) (Van Breemen & Saida, 1989). Hyperpolarization in response to inhibitory nerve stimulation results in a decreased rate of action potential firing in smooth muscles which are spontaneously active. In these tissues this is probably the principal mechanism generating relaxation. However in smooth muscles which do not readily discharge action potentials relaxation need not necessarily be associated with hyperpolarization (e.g. bovine trachea, Cameron et al. 1983; rat anococcgyeous, Creed, Gillespie & Muir, 1975; guinea-pig uterine artery, Bell, 1969).

Electrical basis of neurotransmitter action

The initial postjunctional action of many neurotransmitters is to produce a transient change in the conductance of the postjunctional membrane (Katz, 1966; Takeuchi, 1976; Cull-Candy, 1984; Kuffler, Nichols & Martin, 1984). Transmitter action through receptors may cause the membrane conductance to either increase or decrease, resulting in a transient current across the membrane. When expressed across the membrane resistance and capacitance, this current produces a change in transmembrane potential (Fatt & Katz, 1951; Takeuchi & Takeuchi, 1959; Eccles, 1964). The potential change may be measured either as a change in surface potential with an extracellular electrode (Eccles & Magladery, 1937; Eccles, Katz & Kuffler, 1941, 1942; Brooks & Eccles, 1947) or as a change in the transmembrane potential of individual cells with an intracellular microelectrode (Fatt & Katz 1951; Burnstock & Holman, 1961).

In many situations it is desirable to know the relationship between the duration of the conductance change (or membrane current) and the membrane potential change. The rate at which the potential changes develops and spreads along a cell surface is determined to a large extent by the passive electrical properties of the cell membrane; i.e. its resistance and capacitance (Katz, 1948). For instance, both the velocity of action potential propagation, and the time course and spatial spread of membrane potential changes will depend upon the passive electrical properties of the tissue. In order to establish which factors are important in determining the temporal properties of membrane currents and potentials, it is therefore necessary to have a knowledge of membrane electrical properties.

The passive electrical properties of cells, the diameters of which are small compared with their length (e.g. nerve axons and skeletal muscle fibres), have been successfully investigated and described in terms of 'cable analysis'; i.e. by treating the cells as hollow 'core-conductors' which behave electrically as uniform co-axial cables (Hodgkin & Rushton, 1946; Fatt & Katz, 1951). A coaxial cable is a structure which consists essentially of a cylinder of an electrical conductor covered by a concentric shell of higher resistance and immersed in a conducting medium. In its cellular analogue, the shell of the cable is represented by the surface membrane, and the conducting core by the cytoplasm. The impedance of the cable is represented by the resistance and capacitance of unit lengths of the membrane in parallel (inductive effects in biological tissues being negligible). In general, cable analysis involves the measurement and estimation of the electrical parameters of the cell membrane, e.g. the time and length constant, resistance and capacitance. Based on these parameters, the time courses of currents and voltages in biological membranes can be predicted theoretically and compared with experimentally determined values (e.g. Katz, 1948; Fatt & Katz, 1951; Takeuchi & Takeuchi, 1959).

Membrane potential and current at the skeletal neuromuscular junction and synapses

At the amphibian skeletal neuromuscular junction, the membrane potential changes resulting from the action of acetylcholine released spontaneously and following nerve stimulation are the miniature end-plate potential (MEPP) and the evoked end-plate potential (EPP) respectively (Eccles, Katz & Kuffler, 1941; Fatt & Katz, 1951, 1952; Del Castillo & Katz, 1956). The MEPP is caused by the spontaneous release of a multimolecular packet of acetylcholine. The EPP is the response of the muscle end-plate to the almost synchronous release of about one hundred packets of transmitter evoked by an action potential in the motor nerve terminal. The MEPP and the EPP differ in several respects, particularly with regard to the small amplitude of the MEPP compared with the EPP under normal conditions (Fatt & Katz, 1952; Del Castillo & Katz, 1954a,b), but they have very similar time courses (Fatt & Katz, 1959). At synapses transmitter-activated depolarizations are termed the excitatory post-synaptic potentials (EPSPs) and hyperpolarizations the inhibitory postsynaptic potentials (IPSPs).

Early inferences that the active phase of transmitter action at the skeletal muscle fibre motor end-plate is brief compared with the duration of the EPP were made by Eccles, Katz & Kuffler (1941), Kuffler (1942), and Katz (1948), before the advent of intracellular recording. Their 'graphical' analysis of the extracellularly recorded EPP was based on differentiating the EPP (Eccles, Katz & Kuffler, 1941) to arrive at an estimate of the time course of the underlying current. The analysis assumed that the rate of change of membrane potential (dV/dt) was proportional to the intensity of transmitter action (I) and decayed exponentially with the time constant of the membrane (t_m).

Obtaining dV/dt from the experimental records of the EPP and taking t_m to be the same as the time constant of decay of the EPP, I was estimated from an equation of the form:

$$dV/dt = I - V/t_m$$

This method of analysis indicated that the duration of the conductance change produced by acetylcholine at the muscle fibre membrane is brief, lasting only 4–5 ms. Although the procedure neglected the effect of the conductance change in the junctional region on the time constant of the muscle fibre membrane, its predictions were shown later by a variety of techniques to be essentially correct.

Katz (1948) verified the above conclusions in his theoretical analysis of the skeletal muscle fibre membrane. Using an identical 4 ms pulse of underlying membrane current, Katz was able to simulate the EPP observed in two different amphibian muscles by simply using the different values of t_m for the two muscles (inferred from the time constants of decay of the respective EPPs). These results indicated that the decay of the EPP was determined by the passive electrical properties of the muscle fibre membranes. Intracellular recording of the EPP at the frog skeletal neuromuscular junction (Fatt & Katz, 1951) offered a direct means of testing these assertions. Fatt & Katz (1951) observed that the EPP decayed exponentially with distance along the muscle fibre membrane, indicating the fibre had cable-like properties. In order to estimate the time course of displacement of membrane charge induced by transmitter action, the spatial spread from the end-plate of the EPP at various time intervals was plotted. It was found that membrane charge reached a maximum about 1.5–3 ms after the commencement of the EPP and then decayed exponentially with a time constant of about 25 ms. This value of 25 ms agreed with independent estimates of t_m which the authors obtained using the 'square-pulse' analysis of Hodgkin & Rushton (1946). Therefore, the time course of the falling phase of the EPP could be attributed to the passive decay of charge across the muscle fibre membrane.

The brief time course of transmitter action underlying the MEPP and the EPP was demonstrated directly using the techniques of focal extracellular recording and voltage – clamp. The potential difference measured by a focal extracellular electrode is not influenced by the resistance and capacitance of the muscle fibre membrane, which determine the decay of the intracellularly recorded membrane potential change. Therefore, the time course of the change of potential at sites of transmitter action in the extracellular field closely follows the time course of membrane current (see Brooks & Eccles, 1947; Fatt & Katz, 1951, 1952; Del Castillo & Katz, 1956) and provides a reasonable measure of the duration of transmitter action.

Del Castillo & Katz (1956) used simultaneous intracellular and focal extracellular recording of the MEPP, the EPP and the underlying membrane currents to demonstrate unequivocally that the time course of the current underlying the membrane potential change was brief compared with the duration of the membrane potential change. Similar conclusions have been reached in studies at other neuromuscular junctions using a combination of extracellular and intracellular recording techniques (e.g. insect, Usherwood, 1972; crustacean, Dudel & Kuffler, 1961a).

The first direct demonstration that the time course of transmitter action underlying the EPP was brief came from investigations using the voltage – clamp technique at the frog skeletal neuromuscular junction. Takeuchi & Takeuchi (1959) demonstrated that the duration of the membrane current produced by acetylcholine following nerve stimulation (end-plate current or EPC) was brief compared with the EPP. Treating the muscle fibre as a coaxial cable, they were able to simulate accurately the time course of the EPP from the observed brief EPC. These results were corroborated by voltage – clamp studies at the neuromuscular junctions of the frog (Kordas, 1972a,b; Magleby & Stevens, 1972), toad (Gage & McBurney, 1975), insect (Anwyl & Usherwood, 1974), crustacean (Takeuchi, 1976) and mammal (Head, 1983) for both stimulation – evoked and spontaneous potentials.

Observations based on (1) the analysis of fluctuations of voltage or current produced by acetylcholine ('noise analysis') (Katz & Miledi 1972, 1973, 1976; Anderson

& Stevens 1973) and (2) the patch-clamp method of observing single channel activity directly (Neher & Sakmann, 1976), established that the mean open time of the nicotinic receptor-operated ion channel is similar to the time constant of decay of the EPC and MEPC. A close correspondence between macroscopic and unitary transmitter-activated currents have also been reported at other neuromuscular junctions (Cull-Candy, 1984).

A similar development of ideas arose from analyses of the electrophysiology of the synapse. For instance Eccles and his colleagues (Curtis & Eccles, 1959; Eccles, 1964) initially estimated the duration of transmitter action underlying the EPSP in cat spinal motoneurons by a method identical to that used by Katz and his colleagues for the EPP, namely differentiation of the membrane potential change. Although it has been proposed that a primarily brief current underlies the EPSP, it has also been suggested that a slowly decaying current might be responsible for the slow EPSPs often seen in this preparation (Curtis & Eccles, 1959; Eccles, 1964). However Jack *et al.* (1971) showed that no 'residual' transmitter action was required if the effects of the dendritic tree on the recorded electrical activity were taken into account. According to this interpretation the duration of transmitter action at each release site is brief but identical currents give rise to EPSPs of considerably different durations. The presence of a non-uniform dendritic tree, which receives synaptic contact from many afferent neurons, makes it difficult to obtain an adequate voltage – clamp of the motononeuron. Despite this limitation, Araki & Terzuolo (1962) nevertheless succeeded in obtaining a reasonable voltage-clamp of spinal motoneurons and were able to show that the respective transmitter-activated membrane currents underlying the EPSP and the IPSP were brief compared with the membrane potential changes.

Membrane potential and current at the autonomic neuroeffector junction

The demonstration that current can flow between smooth muscle cells and the 'distributed' innervation makes electrophysiological signals obtained from smooth muscle difficult to interpret. To date, the biophysical analysis of the relationship between transmitter induced current and potential change has been carried out primarily in sympathetically innervated tissues. In the absence of stimulation, spontaneous excitatory junction potentials (SEJPs) can be recorded in several sympathetically innervated smooth muscles. SEJPs are thought to be caused by the release of a multimolecular packet of neurotransmitter from a varicose nerve terminal. The duration of the depolarization evoked by sympathetic nerve stimulation, the EJP, is normally more prolonged than the SEJP in the rodent vas deferens and various vascular smooth muscles (Fig. 3.1). Furthermore SEJPs may be as large as the evoked EJP (Burnstock & Holman, 1962a, 1966).

The syncytial nature of smooth muscle determines, to a large extent, the passive electrical properties. Two methods of passing current across smooth muscle membranes have been used to determine their passive electrical properties. One method is to pass current through intracellular microelectrodes (Hashimoto, Holman & Tille, 1966; Bennett, 1967; Holman, Taylor & Tomita, 1977; Tomita, 1967). The other is to use external electrodes whose dimensions are large compared to indivi-

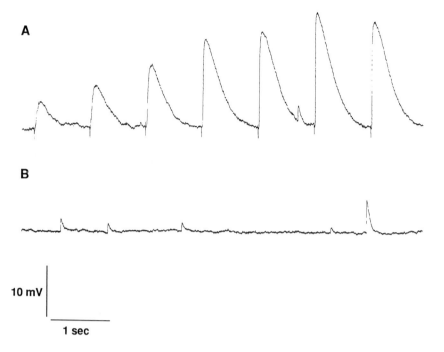

FIGURE 3.1   Intracellular recordings of spontaneous and evoked excitatory junction potentials (SEJPs & EJPs) from the guinea-pig vas deferens. (A) EJPs evoked by a train of 7 stimuli at 1 Hz. (B) SEJPs. There are two points to note; first, the brief duration of the SEJP in comparison with the EJP; second, the growth in amplitude of the EJP during a short train of stimuli (facilitation).

dual smooth muscle cells, so that current is passed through many cells simultaneously. The 'partition stimulation' method (Tomita, 1966a, 1967; Abe & Tomita, 1968), in which a piece of tissue is drawn between two polarizing plates, has been used most frequently. In this case the source of current, which has the form of a plane at right angles to the longitudinal axis of the tissue, drives the two transverse axes of the tissue towards isopotential. Hence current can only spread along the longitudinal axis of the tissue. It has been found using these techniques that individual smooth muscle cells do not possess cable-like properties when polarized intracellularly. However, when large areas of tissue are polarized by external electrodes, smooth muscle acquires cable-like properties (Tomita, 1966a,b, 1970; Bennett, 1972). Based on the responses to intracellular and extracellular polarization, various proposals have been made regarding the time course of the transmitter induced membrane current and potential changes at the sympathetic neuroeffector junction.

Although the time course of the MEPP at the skeletal neuromuscular junction is primarily determined by the passive electrical properties of the postjunctional membrane, that of the SEJP in the rodent vas deferens has been suggested by various workers to reflect the time course of transmitter action. It can be assumed that during an SEJP the electrical properties of smooth muscle are similar to those revealed by intracellular current injection, because only a single point source of current is involved. A number of workers (Hashimoto, Holman & Tille, 1966; Tomita, 1967; Bennett, 1967; Holman, Taylor & Tomita, 1977) have studied the membrane potential response in the guinea-pig and mouse vas deferens to intracellular current injection. In these tissues the 'time constant' of the electrotonic potential (range 2–7 ms) is approximately 10 times faster than the time constant of decay of the SEJP (range 20–35 ms). Hence the time constant following local injection of current into the syncytium is too brief to account for the time constant of decay of the SEJP. Thus it has been concluded that the passive electrical properties of the vas deferens do not determine the time course of decay of the SEJP and that transmitter action might continue throughout its entire duration.

During an evoked EJP it is often assumed that the whole smooth muscle is uniformly depolarized. Therefore, the time course of the potential change may be determined by the passive electrical properties of the smooth muscle membrane as estimated from the cable – like properties of the tissue. Tomita (1967) analysed the cable properties of the guinea-pig vas deferens using the partition stimulation method. In this tissue the time constant of the membrane (t_m) was estimated to be about 100 ms and the length constant (l_m) 1.5–2.5 mm. The time constant of the declining phase of the EJP (250–350 ms) is significantly greater than t_m. This estimate of t_m is too brief to account for the falling phase of the EJP. Therefore it was concluded (Tomita, 1967, 1970), as for the SEJP, that the decay phase of the EJP might reflect the time course of transmitter action rather than t_m.

In a re-investigation of the passive electrical properties of the guinea-pig vas deferens, Bywater and Taylor (1980) demonstrated that estimates of t_m depend critically on the respective lengths of tissue placed in the recording and stimulating compartments of the 'partition bath'. In particular, the calculated value of t_m was shown to be a considerable underestimate of its true value if less than three length constants of the tissue were place in the stimulating compartment. In addition to placing sufficient lengths of tissue in the stimulating compartment, Bywater & Taylor (1980) also took the following precautions to eliminate sources of error in the estimation of t_m. First, polarizing currents of small field strength were applied in order to minimize the likelihood of generating active conductance changes whilst recording electrotonic potentials. Second, electrotonic potentials were corrected for voltage drops in the extracellular field by averaging and subtracting the voltage transients recorded extracellularly from the electrotonic potentials recorded intracellularly with the same micro-electrode. Using these precautions t_m and l_m were estimated to be about 270 ms and 0.86 mm respectively.

This new estimate of t_m is similar to the time constant of decay of the EJP in the guinea-pig vas deferens. Therefore, the decay of the EJP appears to be determined by t_m. These results would suggest that a brief duration of transmitter action underlies the EJP. Bywater and Taylor (1980) estimated the duration of transmitter action by differentiation of the EJP with respect to time and employing the estimated value of t_m. The calculated curve resembled an SEJP in time course, suggesting that the current underlying both the SEJP and the EJP is similar in duration to the SEJP.

Blakeley & Cunnane (1978, 1979a,b) studied the rising phases of individual EJPs in the guinea-pig and mouse vas deferens primarily to establish the characteristics of the transmitter release process (see *Transmitter Release Studies*). They showed that there were discontinuities in the rising phases of EJPs which could be hightlighted by electronic differentiation into transient peaks of depolarization termed 'discrete events'. The possibility was discussed that discrete events might reflect the time course of transmitter-activated membrane current. In the case of the EJP, it was suggested that the duration of the conductance change might be brief and of a duration similar to that of the SEJP. However in the case of the SEJP it was not possible to define the time course of the underlying current.

It is interesting that attempts at demonstrating cable-like properties of the mouse and rat vas deferens with external polarization have not so far been successful (Holman, 1973; Holman, Taylor & Tomita, 1977; Goto *et al.* 1977). Hence speculation about the duration of transmitter action in these tissues rests on indirect evidence. Furness (1970) studied the effects of varying the external K^+ concentration on the EJP in the mouse vas deferens and came to the conclusion that the decay of the EJP depended mainly on the passive electrical properties of the smooth muscle membrane. On reducing the external K^+ concentration, which increases membrane resistance and therefore the membrane time constant, Furness (1970) noted a marked increase in the decay time of the EJP without any significant increase in the rise time. Furthermore, the time course of the SEJP did not change

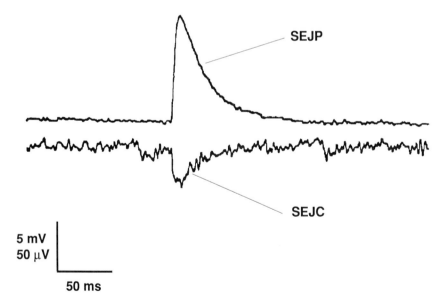

FIGURE 3.2 Simultaneously recording of a spontaneous excitatory junction potential (SEJP) and a spontaneous excitatory junction current (SEJC) in the mouse vas deferens. Note the time courses of the SEJP and the SEJC are similar indicating that the time course of the SEJP is largely determined by the duration of the underlying conductance change. The calibration bar is 5 mV for the SEJP and 50 μV for the SEJC. [From Cunnane & Manchanda, 1989a.]

with this procedure. Thus the decay phase of the EJP, but not the SEJP, was susceptible to changes in the electrical properties of the smooth muscle membrane.

It is difficult to obtain an adequate voltage-clamp of smooth muscle because of its syncytial properties (Bolton, Tomita & Vassort, 1981). Therefore direct measurements of the membrane currents underlying junction potentials in the vas deferens have not been made. However, a method has been developed recently to record extracellularly transmitter-activated currents at the sympathetic neuromuscular junction (Brock & Cunnane, 1987, 1988a). Using this method it is clear, in both the guinea-pig and mouse vas deferens, that the time course of the transmitter induced current is brief compared with the EJP (Brock & Cunnane, 1988a; Åstrand, Brock & Cunnane, 1988). Furthermore, spontaneous and evoked excitatory junction currents (SEJCs & EJCs) and SEJPs recorded in these tissues have closely similar time courses (Fig. 3.2 & 3.3). Thus, it is clear that the SEJP reflects

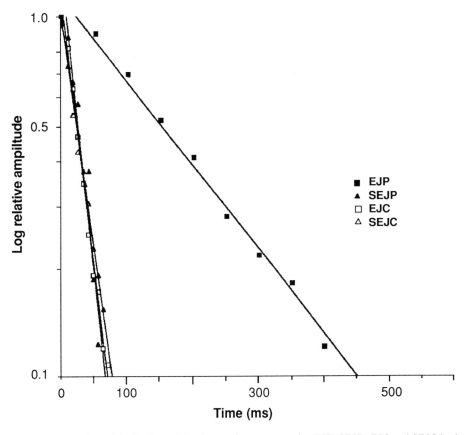

FIGURE 3.3 Semilogarithmic plots of the decay of a representative EJP, SEJP, EJC and SEJC in the guinea-pig vas deferens. The plot shows that the time courses of the extracellularly recorded SEJC and evoked EJC are similar to that of the intracellularly recorded SEJP. Thus, the time course of the current underlying the EJP is brief compared to its duration. [From Brock & Cunnane, 1988a.]

the time course of transmitter action in the vas deferens, whilst the decaying phase of the EJP is largely determined by t_m (Fig. 3.3) (Cunnane & Manchanda, 1988, 1989a).

Investigations of the electrical properties of various vascular smooth muscles have indicated that the time course of decay of EJPs is determined by t_m (Holman & Surprenant, 1979; Kajiwara, Kitamura & Kuriyama, 1981; Hirst & Neild, 1978, 1980a; Cassell, McLachlan & Sittiracha, 1988). Hirst & Neild (1978, 1980a) showed that the time constant of decay of both the EJP (Hirst & Neild, 1978) and the SEJP (Hirst & Neild, 1980a) were similar to t_m in electrically-short segments of guinea-pig submucosal arteriole. The SEJPs and EJPs in this preparation have similar time courses because both produce a uniform depolarization of the whole smooth muscle. Hirst & Neild (1980a) estimated that the duration of the conductance change underlying both the EJP and the SEJP was brief. This was subsequently confirmed by direct measurement of the SEJCs and EJCs in voltage-clamped preparations (Finkel, Hirst & Van Helden, 1984).

Extracellular recording techniques have been used to measure transmitter-induced membrane currents in various arteries of the rat (Åstrand, Brock & Cunnane, 1988; Åstrand & Stjärne, 1989a,b). In these tissues, like the rodent vas deferens, the SEJCs, EJCs and SEJPs have similar time courses. The EJCs in these tissues are also brief compared with EJPs suggesting that the time course of decay of the EJP is largely determined by the passive electrical properties of the smooth muscle. In contrast Kuriyama & Suzuki (1981) reported that the estimated values of t_m in guinea-pig jejunal arteries was consistently shorter than the time constant of decay of the EJP. Should these measured values of t_m be a true reflection of the membrane properties of this tissue then these results would indicate that transmitter action continues during the decaying phase of the EJP.

In summary, at the sympathetic neuroeffector junction in the rodent vas deferens and in various blood vessels the SEJP reflects the time course of transmitter action, whilst the decaying phase of the EJP is largely determined by the passive electrical properties of the smooth muscle. The explanation for this difference resides in the fact that the smooth muscle cells are electrically coupled. When a single packet of transmitter is released spontaneously the current generated locally spreads rapidly into neighbouring cells. Thus the charge at the point source will be rapidly dissipated and the time course of the SEJP will be similar to that of the underlying current. Following nerve stimulation, transmitter is released at a number of different points throughout the muscle. Current spreading from each point source results in depolarization of the whole muscle, which in turn prevents further current spread. Therefore, the time course of the EJP will be determined largely by the relatively slow process of charge dissipation across the muscle membrane; i.e. the time course is dependent on t_m. In support of this view Purves (1976) showed theoretically that in a three-dimensional syncytium at isopotential, a membrane conductance change similar in time course to the SEJP could give rise to a membrane potential change which lagged behind the current and which resembled, in its initial time course, the EJP in the guinea-pig vas deferens.

In other smooth muscles the time courses of junction potentials are usually of longer duration than those recorded in the rodent vas deferens and arteries. Previously it has been suggested that these differences are determined primarily by the morphological arrangement of the neuromuscular junction (Bennett, 1972). However, the mechanisms involved in signal transduction and the passive electrical properties of the smooth muscle membrane are also likely to be important. Signal transduction mechanisms are probably important in determining the latency of the junction potential (i.e. the time interval between the stimulus and the start of the junction potential). When the transmitter directly gates a conductance the latency of the junction potential is likely to be brief. On the other hand if signal transduction involves the generation of second messengers the latency of the junction potential can probably be accounted for largely by the time required for their generation; i.e. diffusion alone cannot account for the observed long latencies. The time courses of the current underlying the junction potential may be determined by the time course of inactivation of second messengers rather than the duration of transmitter action *per se*. In tissues where the duration of junction potentials are prolonged it is likely that their time courses reflect underlying changes in conductance.

BASIC ELECTROPHYSIOLOGICAL STUDIES

POSTJUNCTIONAL ELECTRICAL ACTIVITY AT THE AUTONOMIC NEUROEFFECTOR JUNCTION

In the following section the characteristic features of the postjunctional electrical activity recorded with an intracellular micro-electrode from various smooth muscles in response to excitatory and/or inhibitory nerve stimulation is described together with a brief account of the proposed mechanisms of action of some of the putative transmitters.

ELECTRICAL RESPONSES TO EXCITATORY NERVE STIMULATION

1. Sympathetic

Nerve-mediated depolarizations recorded in sympathetically innervated tissues can be divided, on the basis of their sensitivity to α-adrenoceptor antagonists, into two principal classes. In most arteries, rodent vasa deferentia and a number of other tissues electrical stimulation elicits an EJP which is not blocked by α- or β-adrenoceptor antagonists (Table 3.1). However, there are a number of veins and various other tissues in which the depolarization elicited by nerve stimulation is abolished by α-adrenoceptor antagonists (Table 3.2). In a small number of blood vessels (e.g. rat tail artery, Cheung, 1982a,b; rabbit ear artery, Suzuki & Kou, 1983) the nerve-mediated depolarizations can be divided into both α-adrenoceptor insensitive and α-adrenoceptor sensitive components.

α-Adrenoceptor antagonist resistant responses. The α-adrenoceptor insensitive EJPs are typified by those recorded in the guinea-pig vas deferens. In this tissue EJPs elicited by single stimuli have a short latency (10–20 ms), a rise time of ~ 100 ms and a

TABLE 3.1

Examples and characteristics of α-adrenoceptor resistant EJPs recorded from various sympathetically innervated smooth muscles.

Species/Tissue	Latency ms	Rise Time ms	Duration ms	Decay ms	SEJPs	Facilitation	Comments	Ref
Guinea pig								
Vas deferens	20->50	15-20(1/2 R.T.)	–	~150(1/2)	+	+	Extrinsic Stimulation	Burnstock & Holman (1961)
Seminal vesicle (C.M.)	–	–	~600	–	+	+		Ohkawa (1982)
Mesenteric artery (jejunal)	10-40	–	–	–	+	+		Kuriyama & Makita (1983)
Mesenteric artery (colonic)	–	~45	~320	~210(T.C.)	+		Extrinsic stimulation	Hottenstein & Kreulen (1987)
Submucosal arteriole	10-15	110(T.P.)	~350 (H.W.)	–	+	+		Hirst (1977)
Ear artery	–	–	–	–	+	+		Kajiwara, Kitamura & Kuriyama (1981)
Saphenous artery	–	–	–	–				Cheung & Fujioka (1986)
Basilar artery	–	–	–	–		+	No Response to single stimuli	Karashima & Kuriyama (1981)
Renal artery	–	–	–	–		+	No response to single stimuli	Makita (1983)
Rat								
Vas deferens	10-20	10-20	~150	–	+	+		Holman (1970)
Mesenteric (Ileal)	–	~100	<1000	–				Hill, Hirst & Van Helden (1983)
Basilar artery	–	~100	>1000	–				Hirst, Neild & Silverberg (1982)
Tail artery	–	26-95	66.2-290 (H.W.)	–	+	+		Cheung (1982a,b)
Saphenous vein*	–	–	>2000	–		+		Cheung (1985)
Mouse								
Vas deferens	10-20	10-20	100-250	–	+	+		Holman (1970)
Rabbit								
Prostate (capsule)	~120	–	–	–				Seki & Suzuki (1989)
Mesenteric artery	–	–	200-300	–		+		Mishima, Miyahara & Suzuki (1984)
Mesenteric artery (jejunal)	–	–	–	–				Kuriyama & Makita (1984)
Saphenous artery	~15	~100	–	~200 (T.C.)	+	+		Holman & Surprenant (1979)
Ear artery	–	–	~1000	–		+		Suzuki & Kou (1983)
Dog								
Mesenteric artery (jejunal)	–	–	–	–	+	+		Kou, Kuriyama & Suzuki (1982)
Basilar artery	–	–	–	–		+		Fujiwara, Itoh & Suzuki (1982)
Middle cerebral artery	–	–	–	–		+		Fujiwara, Itoh & Suzuki (1982)
Mesenteric vein*	–	–	>2000	–		+		Suzuki (1984b)
Bovine								
Retractor penis	–	–	–	–				Byrne & Muir (1984)

C.M. = Circular muscle
1/2 R.T. = 1/2 rise time
H.W. = Half width
T.P. = Time to peak
T.C. = Time constant
t1/2 = 1/2 decay time
* EJP of long duration

TABLE 3.2

Examples and characteristics of α-adrenoceptor antagonist sensitive EJPs recorded from various sympathetically innervated smooth muscles.

Species/Tissue	Latency ms	Time to peak sec	Duration min/sec	Stimulus Single	Stimulus Repetitive	α1/α2	Comments	Ref
Guinea-pig								
Mesenteric vein (Ileal)	–	–	–		+	α1/α(?)		Suzuki (1981); Van Helden (1988a)
Mesenteric vein (colonic)	–	–	–		+	α(?)	Extrinsic stimulation	Hottenstein & Kreulen (1987)
Mesenteric artery (colonic)	–	–	–		+	α1	Extrinsic stimulation	Hottenstein & Kreulen (1987)
Pulmonary artery	–	10–15	2–3 mins	+		α1		Suzuki (1983)
Mouse								
Anococcygeus	~500	~1.3	1–2 secs	+§	+	α1		Large (1982)
Rat								
Saphenous vein	–	–	–	+*	+	α2		Cheung (1985)
Tail artery	–	4–7	~30 secs	+		α1		Cheung (1982a,b)
Anococcygeus	100–250	0.5–1.25	1–2 secs	+		α1		Creed, Gillespie & Muir (1975); Byrne & Large (1985)
Rabbit								
Portal vein	–	–	–		+	α(?)	Sucrose gap	Holman et al. (1968)
Carotid artery	–	–	–		+	α(?)		Mekata (1984)
Ear artery	–	6–10	10–15 secs	+		α1		Suzuki & Kou (1983)
Mesenteric artery	–	–	–					Kuriyama & Makita (1984)
Dog								
Mesenteric vein	3–5 Sec	25	–	+		α2		Suzuki (1984b)
Retractor penis	26	0.12	0.38 Sec (T.C.)	+		α(?)		Kinekawa, Komori & Ohashi (1984)
Internal anal sphincter	–	–	–		+	α1		Kubota & Szurszewski (1984)
Cat								
Gastric submucosal venule	–	–	–		+	α1(?)		Morgan (1983)

* = High stimulation strengths only
§ = Some tissues only
T.C. = Time constant of decay

monoexponential decay phase with a time constant of ~300 ms (Burnstock & Holman, 1961). The latency of the EJP can be accounted for mainly by the time required for the nerve impulse to propagate from its point of initiation, since the junctional delay is only 1–3 ms (Brock & Cunnane, 1988a). EJPs in the guinea-pig vas deferens are abolished by the adrenergic neuron blocking agents guanethidine and bretylium and by the voltage-dependent Na^+ channel blocker tetrodotoxin (TTX) and are absent in tissues that have been treated with 6-hydroxydopamine to destroy sympathetic nerve terminals (Burnstock & Holman, 1964; Furness, 1974; Allcorn, Cunnane & Kirkpatrick. 1986). These observations confirm that α-adrenoceptor resistant EJPs are generated by a neurotransmitter released from sympathetic nerves. In most tissues EJPs evoked by single supramaximal stimuli show relatively little variation in latency or shape and are usually of insufficient magnitude to activate voltage – dependent Ca^{2+} channels. During short trains of low frequency stimuli the amplitude of successive EJPs increases until a constant level is reached (Fig. 3.1). The successive growth in EJP amplitude during a train of stimuli is termed facilitation. The rate of facilitation increases with the frequency of nerve stimulation. As the interval between successive stimuli is decreased the membrane potential fails to return to the resting value between stimuli and individual EJPs summate, that is, the second EJP in a train arises from the falling phase of the first. As a result of facilitation and summation, the net depolarization can become sufficiently large to reach threshold for opening voltage – dependent Ca^{2+} channels.

In the absence of stimulation, SEJPs can be recorded in several tissues in which the EJPs are resistant to α-adrenoceptor antagonists (see Table 1).

What chemical generates the α-adrenoceptor antagonist insensitive excitatory junction potential? Although most postganglionic sympathetic nerves are noradrenergic (in the sense that they contain noradrenaline), it is at present unclear whether released noradrenaline mediates the α-adrenoceptor insensitive EJPs. Much recent evidence suggests that the EJP may be generated by a second transmitter released from sympathetic nerve terminals along with noradrenaline, possibly the purine nucleotide adenosine 5′-triphosphate (ATP) (Westfall, Stitzel & Rowe, 1978; Burnstock, 1986). It has long been known that catecholamines and ATP are co-stored in chromaffin granules in the adrenal medulla (see Smith & Winkler, 1972) and there is now a considerable body of evidence which demonstrates that ATP and noradrenaline are co-stored in sympathetic nerve terminals (see Klein & Lagercrantz, 1981). Following electrical stimulation, the release of endogenous ATP from sympathetic nerves has been demonstrated (Lew & White, 1987; Kirkpatrick & Burnstock, 1987; but see White, 1988). In some sympathetically innervated tissues α-adrenoceptor antagonist insensitive EJPs are reduced in amplitude by a photoaffinity antagonist at the P_{2x}-purinoceptor, $ANAPP_3$ (Sneddon, Westfall & Fedan, 1982) or abolished after P_{2x}-purinoceptors have been desensitized by α,β-methylene ATP (Sneddon & Burnstock, 1984a,b; Burnstock & Kennedy, 1985). It is likely that both these agents act postjunctionally since neither of these agents inhibit noradrenaline release from sympathetic nerves (Fedan *et al.* 1981; Allcorn, Cunnane & Kirkpatrick, 1986; Miyahara & Suzuki, 1987). It has also been reported that another ATP analogue P^1, P^5-di-(adenosine-5') pentaphosphate desensitizes the P_{2x}-purinoceptor in the guinea-pig vas deferens and abolishes both SEJPs and EJPs (MacKenzie, Kirkpatrick & Burnstock, 1988). Thus, in some sympathetically innervated tissues, ATP or a related purine nucleotide may function as a neurotransmitter.

An alternative explanation for the resistance of EJPs to α-adrenoceptor antagonists has been offered, namely, that noradrenaline generates EJPs by acting at a unique junctional receptor, the 'γ' receptor (Hirst & Neild, 1980b,c, 1981) with a pharmacological profile quite different from that of other adrenoceptors. What follows is a consideration, from an electrophysiological point of view, of the neurotransmitter status of noradrenaline and ATP in some sympathetically innervated tissues, in particular the rodent vas deferens.

The γ-receptor hypothesis. Evidence supporting the γ-receptor hypothesis was provided by Hirst and Neild (1980b). These authors used intracellular microelectrodes to record the responses of individual smooth muscle cells of the guinea-pig submucosal arterioles to nerve stimulation and to the ionophoretic application of noradrenaline. They reported that at some points on the surface of the arteriole, noradrenaline produced a depolarization which was 'superficially similar' to the electrically evoked EJP. However, at most sites noradrenaline elicited a local contraction without altering the membrane potential. The electrical responses to noradrenaline and the EJP were resistant to treatment with the α-adrenoceptor antagonist phentolamine, whereas the membrane potential independent contraction elicited by exogenous noradrenaline was readily blocked. Hirst and Neild (1981) reported that the sites at which noradrenaline produced membrane depolarization were restricted to regions of the arteriole within $10\,\mu m$ of the sympathetic nerve terminals which innervate them. They suggested that two types of adrenoceptor exist in this preparation, a junctional receptor which differs from conventional α-adrenoceptors and an extrajunctional α-adrenoceptor. By analogy with guinea-pig mesenteric arterioles, Hirst and Neild (1980b) proposed that the α-adrenoceptor resistant EJPs recorded in the rodent vas deferens and other vascular preparations may be mediated through the specialized junctional or 'γ-receptor' (Hirst & Neild, 1980c). However, it should be noted that focally applied noradrenaline has never been shown to produce depolarization in the guinea-pig vas deferens (Sneddon & Westfall, 1984; Cunnane & Manchanda, 1988), although paradoxically bath application of noradrenaline to the whole tissue produces depolarization (Sneddon & Westfall, 1984).

In support of the γ-receptor hypothesis Hirst, Neild & Silverberg (1982) reported that in the rat basilar artery, a tissue which is devoid of postjunctional α-adrenoceptors, bath applied noradrenaline at high concentrations produces a depolarization which is resistant to both α- and β-adrenoceptor antagonists. However, in this tissue it was not possible to mimic the electrically evoked EJP by ionophoresis (Byrne, Hirst & Large, 1985). Some studies have questioned the specificity of α,β-methylene ATP at the P_{2x} purinoceptor in smooth muscle. For example, in the rat basilar artery the depolarization induced by both exogenously applied ATP and noradrenaline was blocked by α,β-methylene ATP (Byrne & Large, 1986).

However, the specificity of ANAPP$_3$ for the P$_{2x}$-purinoceptor has not been questioned (see White, 1988).

The purinergic hypothesis. The principal evidence against the γ-receptor hypothesis in the rodent vas deferens comes from the demonstration that EJPs can be recorded in tissues pretreated with reserpine (Burnstock & Holman, 1962b; Burnstock, Holman & Kuriyama, 1964; Sneddon & Westfall, 1984; Stjärne & Åstrand, 1984), which specifically depletes noradrenaline from sympathetic nerves by blocking its uptake into storage vesicles (Smith & Winkler, 1972). Indeed there is some evidence to suggest that reserpine pretreatment may leave the ATP content of vesicles unaffected. Kostron *et al.* (1977) showed that whereas reserpine inhibited the carrier-mediated uptake of catecholamines into isolated bovine chromaffin granules, the carrier-mediated uptake of ATP continued and could be specifically blocked by another agent, aractyloside. This finding indicates that the uptake of noradrenaline into chromaffin granules (and perhaps into sympathetic nerve vesicles) may not be an obligatory process for the uptake of ATP (although noradrenaline and ATP are normally incorporated together if both are available for uptake; see Winkler, Fischer-Colbrie & Weber, 1981). Therefore if ATP mediates the EJP, it follows that reserpine may not significantly alter the ATP content of the individual storage vesicle. In the majority of studies the doses of reserpine used to pre-treat the animals were adequate to almost totally deplete ($\geq$ 98 per cent) the noradrenaline content of sympathetic nerve terminals. The persistence of EJPs (and contractions) in reserpinized vas deferens strongly suggests that the EJP is mediated by a transmitter other than noradrenaline (Sneddon & Westfall, 1984). However, interpretation of some of the effects of reserpinization is the subject of keen debate (Hirst, 1987) and have been reconciled within the framework of the γ-receptor hypothesis as follows. First, even though chronic reserpine treatment induces as much as 99.7% loss of the noradrenaline content of the sympathetic nerves in the guinea-pig vas deferens (Wakade & Krusz, 1972; Kirkpatrick & Burnstock, 1987), it can be supposed that the small fraction of noradrenaline remaining after reserpinization is sufficient to sustain neurotransmission, or that the tissues studied electrophysiologically were not adequately depleted of noradrenaline (Hirst, 1987; Hirst & Edwards, 1989). Second, EJPs may be elicited by the release of recently synthesized noradrenaline following nerve stimulation (Duval, Hicks & Langer, 1986). This process has been suggested to operate *in vitro* in reserpine pre-treated tissues to a degree sufficient to maintain neurotransmission (Hirst, 1987). Third, some studies have indicated a reduction in the size of fully-facilitated EJPs after reserpinization (Burnstock & Holman, 1962b; Stjärne & Åstrand, 1984). The low amplitude of the EJPs has been suggested to be due to a decrease in the size of the sympathetic transmitter quantum because a smaller amount of noradrenaline is present in each transmitter storage vesicle (Burnstock & Holman, 1962b). Thus it may not be necessary to invoke non-noradrenergic mechanisms to explain the resistance of the EJP evoked by sympathetic nerve stimulation to reserpinization.

These issues have been re-investigated in the guinea-pig vas deferens by performing a correlated study of the effects of reserpine pre-treatment on EJP amplitude, tissue noradrenaline content and the size of the transmitter quantum released from the sympathetic nerves. The precaution was taken to use long trains of supramaximal stimuli (up to 100 pulses) to ensure that the amplitudes of fully-facilitated EJPs were measured in control and reserpinized tissues. SEJPs and SEJCs were recorded using intracellular and focal extracellular recording techniques to estimate the size of the transmitter quantum. The noradrenaline content of tissues used in these electrophysiological studies was assessed using both HPLC and histochemical methods. Furthermore, the effects on EJPs of a noradrenaline synthesis inhibitor, α-methyl-p-tyrosine, were also investigated to establish whether the release of recently-synthesized noradrenaline could account for the presence of EJPs in reserpinized vasa deferentia *in vitro* (Cunnane & Manchanda, 1989b).

The reported reduction in the size of the first and subsequent EJPs in a train after reserpinization deserves special mention (Burnstock & Holman 1962b; Stjärne & Åstrand 1984). In tissues which were almost totally depleted of their noradrenaline content by reserpinization, it was possible to record EJPs in response to nerve stimulation. The depletion of noradrenaline in these tissues was confirmed either by measuring the noradrenaline content of the tissues with HPLC or by histochemical techniques. It was shown that the amplitudes of fully facilitated EJPs in the reserpinized tissues were similar to the largest EJPs observed in control tissues (Sneddon & Westfall, 1984). The small size of EJPs reported in previous studies can probably be attributed to the use of relatively short (up to 30) trains of stimuli. In the study of Cunnane & Manchanda (1989b) 40–50 pulses were often required for EJPs to reach fully facilitated levels in reserpinized tissues. The size of the sympathetic transmitter quantum, as measured by the SEJP or SEJC, is unaffected by reserpine pre-treatment (Cunnane & Manchanda, 1989b). Facilitation simply takes longer to develop.

The presence and large amplitude of fully facilitated EJPs was maintained after superfusion of reserpinized tissues with the noradrenaline synthesis inhibitor α-methyl-p-tyrosine for > 60 min. Therefore, it is difficult to see how EJPs, at least in the guinea-pig vas deferens, could arise from the postjunctional action of newly synthesized noradrenaline acting on 'γ'-receptors (Hirst, 1987).

There is still a need for more studies with reserpine. Recently, it has been demonstrated in reserpinized guinea-pig vas deferens that EJPs evoked by short trains of stimuli facilitated to amplitudes up to 200% of control values (Brock *et al.* 1990). The pattern of facilitation was similar to that observed in untreated control tissues after antagonism of prejunctional α_2-adrenoceptors with yohimbine and presumably reflects the loss of α-autoinhibition (Brock *et al.* 1990). Interestingly, Kirkpatrick & Burnstock (1987) demonstrated that reserpine pre-treatment, whilst causing a 99% depletion of the noradrenaline content of the guinea-pig vas deferens, did not inhibit the release of ATP from the sympathetic innervation following nerve stimulation and may actually have increased it. This finding is consistent with the specific inhibition by reserpine of the incorporation of noradrenaline into storage vesicles in sympathetic nerve terminals. The reasons for the variable effects of reserpine reported in the literature may well reflect differences in the reserpinization protocols employed.

EJPs which were resistant to the combined action of reserpine and α-methyl-*p*-tyrosine were nevertheless abolished by α,β-methylene ATP. Taken in conjunction with several other separate lines of evidence (Sneddon & Westfall 1984; Allcorn, Cunnane & Kirkpatrick, 1986), it seems plausible to attribute the generation of EJPs to the release of ATP or a related purine nucleotide from sympathetic nerve terminals (Sneddon & Westfall 1984; Burnstock 1986). This idea is supported by the observation that focally applied ATP, but not noradrenaline, can evoke depolarizations in the guinea-pig vas deferens which closely resemble EJPs (Sneddon & Westfall, 1984; Cunnane & Manchanda, 1988). No evidence was found to suggest that the noradrenaline released together with ATP in normal tissues, contributes to the time course of junction potentials (Cunnane & Manchanda, 1988, 1990).

The effects of reserpine, α,β-methylene ATP and ANAPP$_3$ in other sympathetically innervated organs also points to the existence of purinergic responses. In some arteries α-adrenoceptor insensitive EJPs persist after reserpine pre-treatment (rat tail artery, Cheung, 1982a,b; rabbit ear artery, Suzuki 1984a) but, not after exposure to α,β-methylene ATP (e.g. rat ear artery, Sneddon & Burnstock, 1984b; rabbit and rat mesenteric artery, Ishikawa, 1985) or ANAPP$_3$ (guinea-pig saphenous artery, Cheung & Fujioka, 1986). In the rabbit ear artery ionophoretic application of ATP produces a depolarization which resembles the EJP (Suzuki, 1985). Similarly, mechanical responses are only partly abolished by reserpine pre-treatment, and the residual contraction is abolished by α,β-methylene ATP in the dog mesenteric artery (Muramatsu, 1986, 1987) and the rabbit mesenteric artery (Kügelgen & Starke, 1985).

Other pharmacological evidence. The results of pharmacological studies of contraction in the rodent vas deferens are also compatible with the co-transmission hypothesis. The early phase of contraction elicited by nerve stimulation is not blocked by α- and β-adrenoceptor antagonists (Swedin, 1971; Ambache & Zar, 1971; Ambache *et al.* 1972; Euler & Hedqvist, 1974, 1975). α-Adrenoceptor antagonists were however effective in abolishing the delayed phase of contraction following nerve stimulation (Swedin, 1971; McGrath, 1978) and the responses to exogenously applied noradrenaline (Boyd, Chang & Rand, 1960; Ohlin & Strömblad, 1963; Birmingham & Wilson, 1963; Ambache & Zar, 1971; Wakade & Krusz, 1972).

Further evidence supporting a transmitter role for ATP in the guinea-pig vas deferens has been provided by Cunnane and Manchanda (1988). In this study the effects of cooling on the intracellularly recorded EJPs and depolarizations evoked by focally applied ATP and α,β-methylene ATP were investigated. Cooling prolonged both the time course of the EJP and the response to exogenous ATP without affecting the already prolonged response to α,β-methylene ATP. The prolonged response to ATP and to nerve stimulation produced by cooling may result from an inhibition of a transmitter inactivation system which is temperature sensitive, perhaps an ecto-ATPase (see Gordon, 1986). The lack of effect of cooling on the time course of the response to focally applied α,β-methylene ATP might reflect the fact that this agent is a stable analogue of ATP, and therefore resistant to degradation by ATPases. One also has also to consider the possibility that the membrane properties of the smooth muscle cells are altered by cooling (Cassell, McLachlan & Sittiracha, 1988). However, the time courses of both the SEJP and the SEJC are prolonged by cooling, suggesting that there is an alteration in the duration of transmitter action.

The evidence to date supports the view that the α-adrenoceptor antagonist resistant EJPs recorded in various blood vessels and the in rodent vas deferens result from the release of ATP or a related purine nucleotide from the sympathetic nerve terminals. In cells isolated from the rabbit ear artery ionophoretic application of ATP evokes a depolarization with a latency of about 70 ms which results from a selective increase in cation conductance with a reversal potential close to 0 mV (Benham et al. 1987; Benham & Tsien, 1987). The ATP activated current increases linearly for membrane potentials between 0 and -60 mV and at potentials below -80 mV the current rises more steeply (i.e. the current displays inward rectification). Analysis of the cation selectivity of the ATP-activated channels indicates that they are more permeable to Ca^{2+} than to Na^+ (Benham & Tsien, 1987; Benham, 1989). However under physiological conditions where the Na^+ concentration in the extracellular fluid is much greater than that of Ca^{2+}, only about 10% of the inward current is likely to be carried by Ca^{2+}. Benham (1989) has suggested that Ca^{2+} entering through the ATP-gated channels may directly activate contractile proteins. However, in most tissues investigated α-adrenoceptor resistant EJPs only result in contraction following opening of voltage–dependent Ca^{2+} channels (Blakeley et al. 1981; Bolton & Large, 1986), although in the rabbit ear artery it has been suggested that Ca^{2+} entry during the EJP may elicit contraction (Suzuki & Kou, 1983). The activation of ATP-operated channels did not involve second messenger generation, indicating that these channels were directly gated by ATP (Benham & Tsien, 1987). ATP and α,β-methylene ATP were equipotent in activating this channel and both produced subsequent desensitization. Adenosine diphosphate (ADP) was less effective and adenosine was without effect. These findings suggest that a P_2-purinoceptor is involved in the response. A similar ATP-operated increase in conductance has been described in cells isolated from the rat vas deferens (Friel, 1988). This conductance also displayed inward rectification, but was only poorly activated by α,β-methylene ATP and desensitization did not occur during prolonged exposure to this agent. The response to ATP in these cells was however readily desensitized.

The current – voltage relationship for the current underlying the α-adrenoceptor antagonist resistant EJP in the guinea-pig submucosal arteriole has been determined (Finkel, Hirst & Van Helden, 1984). In this tissue over the range of membrane potentials -20 to -90 mV peak EJC amplitude increased linearly, with a reversal potential close to 0 mV. These results suggest that the EJC is generated by an increase in membrane conductance. The range of membrane potentials used in this study was too small to determine whether the voltage dependence of the conductance increase underlying both the EJC and the ATP generated current recorded in single cells, was similar. To clarify this point the current-voltage relationship of

EJCs over a larger range of membrane potentials will have to be determined. Another point which needs to be addressed concerns the latency of the current evoked by ionophoretic application of ATP to single cells (~ 70 ms), which is much longer than the junctional delay (1–3 ms) at the sympathetic neuroeffector junction (Brock & Cunnane, 1988a).

As pointed out by Hirst and Edwards (1989) whatever the outcome of this debate 'both hypotheses (i.e. the purinergic and the γ-receptor hypothesis) require that transmission occurs at points where there is little separation between the membranes of the nerve terminal and the muscle. The γ-hypothesis demands that there is a high concentration of noradrenaline near specialized junctional receptors. The ATP hypothesis potentially relies on the efficient use of only a small number of ATP molecules.' Currently the best estimate for the ATP content of a vesicle is very low, the noradrenaline/ATP ratio being between 20–60 (see Klein & Lagercrantz, 1981). Since the noradrenaline content of a small adrenergic vesicle is of the order of 1000–3000 molecules this means that the ATP content may be as low as 20–60 molecules (Stjärne & Åstrand, 1984). Finkle, Hirst & Van Helden (1984) have estimated that the conductance change underlying the SEJP, which presumably reflects the release of the transmitter content of a single vesicle, is ~ 2 nS. Since single ATP-gated channels in rabbit ear artery have a conductance of approximately 5 pS (Benham & Tsien, 1987), 400 channels would have to be opened following the release of a single vesicle. Therefore it is difficult to envisage how a small number of ATP molecules could produce the observed increase in conductance unless they interact with channels which exhibit a very large conductance. New studies are required to determine the exact ATP content of a vesicle, since early estimates may well be unreliable because ATP is a very labile molecule.

α-Adrenoceptor antagonist sensitive responses. The characteristic features of α-adrenoceptor sensitive depolarizations elicited by nerve stimulation vary considerably from tissue to tissue and depend to a large extent on the stimulation parameters used. In the rat anococcygeus single pulse field stimulation evokes an EJP which has a latency of several hundred milliseconds, a mean rise time of ~ 650 ms and a total duration of 1–2 secs (Creed, 1975). Following short trains (2–5) of stimuli at 5 or 10 Hz this EJP may be preceded by a 'fast' depolarization which has a latency of less than 100 ms and a total duration of under 1 sec (Creed, Gillespie & Muir, 1975; Cunnane, Muir & Wardle, 1987). In the experiments described by Creed, Gillespie & Muir (1975) and Cunnane, Muir & Wardle (1987) both the EJP and 'fast' depolarizations were sensitive to α-adrenoceptor antagonists. Neither the EJP nor the 'fast' depolarization were affected by α,β methylene ATP (Cunnane, Muir & Wardle, 1987). However, Byrne and Large (1984) have reported a 'fast' depolarization in the rat anococcygeus which was evoked with a very brief latency. In this case treatment with the α-adrenoceptor antagonist phentolamine abolished the EJP but potentiated the 'fast' depolarization. The most likely explanation is that these α-adrenoceptor antagonist insensitive 'fast' depolarizations do not correspond with those reported by other workers. Since ionophoretic application of ATP elicited a

depolarization with a similarly brief latency, Byrne and Large (1984) suggested that these 'fast' depolarizations may be evoked by the same mechanism as the α-adrenoceptor insensitive EJPs recorded in other sympathetically innervated tissues. In this case the potentiating effects of phentolamine on the 'fast' depolarization may be explained by blockade of presynaptic α-adrenoceptors whose activation normally inhibits transmitter release (see *Prejunctional Receptors*). In addition to the EJP and the 'fast depolarization recorded in the rat anococcygeus, a more prolonged phase of depolarization which lasts several tens of seconds has been recorded in tissues maintained at room temperature (Bolton & Large, 1986). This 'slow' depolarization was blocked by the α-adrenoceptor antagonist prazosin. Similar slow depolarizations have been recorded in veins (e.g. rabbit portal vein, Holman *et al.* 1968; guinea-pig mesenteric vein, Suzuki, 1981) and in the few arteries which have α-sensitive depolarizations (e.g. rat tail artery, Cheung 1982a,b; rabbit ear artery, Suzuki & Kou, 1983). These depolarizations are normally only elicited by trains of stimuli and have a rise times of 10–30 seconds and durations up to 3 minutes. In guinea-pig mesenteric veins short trains of high-frequency stimuli (10–20 pulses, 20 Hz) elicit an initial 'fast' α-adrenoceptor antagonist sensitive depolarization similar in time course to the EJP elicited by single stimuli in the rat anococcygeus (Van Helden, 1988a). A similar two phase α-adrenoceptor sensitive depolarization has been reported in the guinea-pig pulmonary artery (Suzuki, 1983).

Contractions mediated through activation of α_1-adrenoceptors are not normally associated with muscle action potential discharge. Furthermore, activation of α_1-adrenoceptors can lead to contraction through mechanisms which are independent of changes in membrane potential (see Minneman, 1988). However, it is likely that if the muscle membrane is sufficiently depolarized voltage – dependent Ca^{2+} channels will be opened. In tissues which possesses postjunctional α_2-adrenoceptors, contractions evoked by activation of these receptors appear to be mediated primarily by the opening of voltage – dependent Ca^{2+} channels (Minneman, 1988). Consistent with this suggestion nerve evoked contraction of the rat saphenous vein, in which the depolarization is mediated through α_2-adrenoceptors, is dependent on Ca^{2+} entry through voltage – dependent channels (Cheung, 1985). Recent studies, however, suggest that contractions elicited by activation of postjunctional α_2-adrenoceptors may also involve the mobilization of Ca^{2+} from intracellular stores (see Dun *et al.* 1991).

Depolarizations of the rat anococcygeus elicited by bath applied noradrenaline can be divided into two α-adrenoceptor antagonist sensitive components; an initial 'fast' phase which peaks within 1–2 seconds followed by a 'slow' phase which is maintained during the period of application of noradrenaline (Byrne & Large, 1985). The 'fast' phase of depolarization and the EJP elicited by nerve stimulation are both abolished by replacing the external sodium chloride with sodium benzene-sulphonate, whereas the slow depolarization is unaffected by this treatment. In the rat anococcygeus therefore, two mechanisms of depolarization appear to be associated with α-adrenoceptor activation. Several lines of evidence suggest that the 'fast' depolarization elicited by noradrenaline in the rat anococcygeus results from an increase in Cl^- conductance. First, both the estimated reversal potential

(-27 mV) for the depolarization elicited by ionophoretic application of noradrenaline and the membrane potential at the maximum amplitude (-29 mV) of the 'fast' depolarization elicited by bath application of noradrenaline (Byrne & Large, 1985) lie close to the Cl^- equilibrium potential (-24 mV) determined in smooth muscle cells of the guinea-pig vas deferens (Aickin & Brading, 1982). Second, in cells isolated from the rat anococcygeus ionophoretic application of noradrenaline evokes an inward current by producing a decrease in membrane conductance (Byrne & Large, 1987). The reversal potential for this current and the Cl^- equilibrium potential are similar suggesting that the inward current results from an increase in Cl^- conductance. A similar increase in Cl^- conductance has been reported in cells isolated from the rabbit portal vein (Byrne & Large, 1988; Amedee & Large, 1989) and the rat portal vein (Pacaud et al. 1989). However in these cells noradrenaline also appears to increase cation conductance. In guinea-pig mesenteric vein ionophoretic application of noradrenaline evokes a 'fast' and a 'slow' depolarization which mimics the response to nerve stimulation (Van Helden, 1988a). The conductance changes underlying both 'fast' and 'slow' depolarizations have been investigated using electrically short segments of guinea-pig mesenteric vein which were effectively at isopotential and could be voltage clamped (Van Helden, 1988b). The reversal potential for the current underlying the 'fast' depolarization was about -22 mV, suggesting an increase in Cl^- conductance. During the slow depolarization there was an increase in membrane resistance. Since hyperpolarization of the membrane to values close to the K^+ equilibrium potential much reduced the size of the 'slow' depolarization, it would appear that the 'slow' depolarization results from a decrease in K^+ conductance. Suzuki (1981) suggested that depolarizations of the guinea-pig mesenteric vein induced by the bath application of noradrenaline resulted from a decrease in K^+ conductance. Furthermore, there is an increase in membrane resistance during the nerve evoked, α-adrenoceptor antagonist sensitive, 'slow' depolarization in the rat tail artery (Cassell, McLachlan & Sittiracha, 1988), again suggesting that there is a decrease in K^+ conductance.

The long latency of the α-sensitive EJP and the depolarizations elicited by focal application of noradrenaline to whole tissues (Byrne & Large, 1984; Van Helden, 1988a) and isolated cells (Byrne & Large 1987, 1988; Pacaud et al. 1989) cannot be accounted for by the time required for noradrenaline to diffuse from its sites of release. It has been suggested that the noradrenaline induced increase in Cl^- conductance is triggered by the release of Ca^{2+} from internal stores (Byrne & Large, 1987). In support of this suggestion caffeine, which releases Ca^{2+} from internal stores, increased the Cl^- conductance of cells isolated from the rat anococcygeus (Byrne & Large, 1987), the rabbit portal vein (Byrne & Large, 1988) and the rat portal vein (Pacaud et al. 1989). The inward current elicited by noradrenaline in single cells is sometimes preceded by an outward current which is not present in K^+ free conditions and is probably due to Ca^{2+}-activated K^+ channels (Byrne & Large, 1987). This finding suggests that a rise in cytoplasmic Ca^{2+} precedes the Cl^- conductance increase. Consistent with this suggestion, α-adrenoceptor activation in a number of different tissues results in the release of Ca^{2+} from intracellular stores through the hydrolysis of inositol phospholipids to generate the intracellular messenger myo-ionositol-1,4,5-trisphosphate (Minneman, 1988). The mechanism underlying the α-adrenoceptor mediated decrease in K^+ remains to be established.

2. Cholinergic

Cholinergic EJPs have been recorded primarily in intestinal smooth muscles (Table 3.3). In general, these EJPs are blocked by relatively low concentrations of atropine or hyosine (10^{-6} M or lower) and are mediated by acetylcholine acting through muscarinic receptors. EJPs evoked by single stimuli have a latency between 50–200 ms, a time to peak of 200–400 ms and a total duration of 500 to 800 ms. The amplitude of successive EJPs recorded during repetitive stimulation may either increase (colon, Gillespie, 1962a; chick oesophagus, Ohashi & Ohga, 1967) or decrease (rabbit rectococcygeus, Blakeley, Cunnane & Muir, 1979; guinea-pig ileum longitudinal muscle, Bauer & Kuriyama, 1982a; guinea-pig ileum circular muscle, Bywater & Taylor, 1986a, 1989). It has been suggested that this depression may in part be due to the prejunctional effects of endogenous opioids released during the period of nerve stimulation in the guinea-pig ileum (Puig *et al.* 1977; Fosbraey & Johnson, 1980; Davison & Najafi-Farashah, 1981). However, in guinea-pig ileum circular muscle the observed depression was not sensitive to naloxone suggesting that opioid receptors are not involved (Bywater & Taylor, 1989). In this preparation the depression of EJP amplitude appears to be mediated through mechanisms acting at the level of ganglia rather than at the level of the neuroeffector junction.

In general the application of acetylcholine to smooth muscle produces a depolarization resulting from an increase in membrane conductance and an inward current with a reversal potential at about $-10\,mV$ (Bolton 1979). Studies using the whole cell clamp technique suggest that the effects of muscarinic receptor activation on smooth muscle cells are complex. Acetylcholine evokes an inward current by decreasing membrane conductance in cells isolated from the longitudinal muscle of the rabbit jejunum (Benham, Bolton & Lang, 1985). The results of this study indicate that the inward current is generated by a non-selective increase in cation permeability. Inoue, Kitamura & Kuriyama (1987) have demonstrated that acetylcholine activates a cation channel which is permeable to both Na^+ and K^+ in cells isolated from the guinea-pig ileum. In both of these studies the current-voltage relationship of the acetylcholine-induced current was U-shaped with a peak current at approximately $-40\,mV$. In gastric smooth muscle cells of the toad *Bufo marinus*, activation of the muscarinic receptor caused a depolarization resulting from the suppression of a voltage – gated K^+ conductance (Sims, Singer & Walsh, 1985). A similar decrease in a specific K^+ current, termed the M-current, has been reported in neuronal tissue (Brown & Adams, 1980). Much of the evidence to date supports the view that the acetylcholine induced depolarization in mammalian smooth muscle is due primarily to the activation of a non-specific cation conductance similar to the one described above. However, the data do not exclude the possibility that a decrease in K^+ conductance may contribute to the net depolarization (see Bolton, 1989).

The question remains whether the effects of exogenously applied acetylcholine can faithfully mimic the effects of neurally released acetylcholine. Membrane

TABLE 3.3

Examples and characteristics of cholinergic EJPs recorded from various smooth muscles (adapted from Hoyle & Burnstock, 1989).

Species/Tissue	Latency ms	Rise Time ms	Duration ms	Decay ms	Facilitation	Comments	Ref
Guinea-pig							
Ileum L.M.	100–250	–	–	–			Bauer & Kuriyama (1982a)
Ileum C.M.	–	–	–	–			Bywater & Taylor (1986a)
Jejunum L.M.	–	–	–	–	+		Kuriyama, Osa & Toida (1967)
Jejunum L.M.	20–65	–	–	–			Hidaka & Kuriyama (1969)
Taenia ceaci	100–200	200–400	500–800	–	+		Bennett (1966)
Caecum C.M.	6 (minimum)	–	–	–			Ito & Kuriyama (1973)
Rectum L.M.	80–320	120–350	450–720	–			Ito & Kuriyama (1971)
Colon C.M./L.M.	100–400	150–200	–	100–150(t1/2)			Furness (1969)
Stomach C.M.	~160	–	~1000	–			Komori & Suzuki (1986)
Gastric sling L.M.	~140	–	~600	–		Sucrose gap	Beck & Osa (1971)
Gastric corpus C.M./L.M.	150–200	~200	400–500	–			Beani, Bianchi & Crema (1971)
Rabbit							
Jejunum C.M./L.M.	–	–	–	–	+		Kitamura (1978)
Duodenum C.M./L.M.	–	–	–	–			Cheung & Daniel (1980)
Rectum L.M.	–	–	–	–			Suzuki et al. (1979)
Colon C.M./L.M.	–	–	–	–			Furness (1969)
Colon	400	–	~600	–	+	Extrinsic stimulation	Gillespie (1962a)
Caecum C.M.	~175	–	–	–	+	Sucrose gap	Small (1971)
Prostate L.M.	~150	–	–	–			Seki & Suzuki (1989)
Rectococcygeous	~250	~300	–	533(t1/2)		Extrinsic stimulation	Blakeley, Cunnane & Muir (1979)
Bovine							
Trachea	100–250	800		1500(T.C.)		Sucrose gap	Cameron & Kirkpatrick (1977)
Cat							
Trachea	–	–	3000–5000	–		Sucrose gap	Ito & Takeda (1982)
Ferret							
Trachea	200–450	600–900(T.P)	–	200–500(t1/2)			Coburn (1984)
Chick							
Gizzard	90–350	–	430–1300	–	+		Bennett (1969)
Oesophagus	90–160	150–250	700–950	–	+	Sucrose gap	Ohashi & Ohga (1967)
Pigeon							
Gizzard	70–250	–	430–900	–	+		Bennett (1969)

L.M. = Longitudinal muscle
C.M. = Circular muscle
T.P. = Time to peak
t1/2 = 1/2 decay time
T.C. = Time constant

polarization using the partition stimulation method (Abe & Tomita, 1968) has been used to determine the relationship between membrane potential and EJP amplitude in various intestinal smooth muscles (e.g. guinea-pig stomach circular muscle, Komori & Suzuki, 1986; rabbit jejunum, Kitamura, 1978; rabbit rectum, Suzuki *et al.* 1979). Over the limited range of membrane potentials attainable using this technique, the relationship between membrane potential and EJP amplitude is apparently linear and the reversal potential, determined by extrapolation, lies somewhere between -18 and -5 mV. Care should be taken in interpreting these results since the EJPs are presumably occurring concurrently with IJPs in many of these tissues and it is not possible to determine the effects of membrane polarization on transmitter release. However, it is clear that EJPs increase in size as the membrane is hyperpolarized to values close to the K^+ equilibrium potential, indicating that EJPs do not result from a decrease in K^+ conductance. The apparently linear relationship between membrane potential and EJP amplitude questions the functional role of the non-specific cation channel described above.

3. Non-adrenergic, non-cholinergic

In addition to the α-adrenoceptor resistant EJPs recorded in sympathetically innervated tissues, EJPs have been recorded in various mammalian intestinal smooth muscles, the chick rectum and the mammalian urinary bladder which are not blocked by a combination of cholinoceptor, or α- and β-adrenoceptor antagonists.

Mammalian intestine. The response of the guinea-pig ileum circular muscle to a single stimulus in the presence of atropine is complex comprising an initial hyperpolarization (IJP) followed by a post-stimulus depolarization (PSD) (Bywater, Holman & Taylor, 1981). Treatment of the muscle with apamin abolished the IJP leaving a transient EJP with a variable latency (between 350–900 ms). During trains of low frequency stimuli (0.25 Hz) the amplitude of successive EJPs facilitate to reach the threshold for the initiation of an action potential in the smooth muscle. After short trains of stimuli (3 pulses, 10 Hz) the PSD is clearly divided into two TTX-sensitive components; an early fast phase (fPSD) which peaks in ~1 second and corresponds in time course with the EJP recorded in response to a single stimulus in the presence of apamin, and a late slow phase (sPSD) which peaks in ~ 4 seconds (Bywater & Taylor, 1983). Atropine insensitive EJPs have also been recorded in the longitudinal muscle of the guinea-pig ileum (Bauer & Kuriyama, 1982a,b) but in this case the EJPs were generated with a much briefer latency (25–100 ms) and, during trains of low frequency stimuli, facilitation did not occur. It should be noted that Bywater and Taylor (1986b) failed to demonstrate atropine insensitive EJPs in the longitudinal muscle of the guinea-pig ileum.

A number of lines of evidence suggest that both the fPSD recorded in the circular muscle and the atropine insensitive EJPs recorded in the longitudinal muscle may be evoked by the release of substance P or a related neurokinin. First, the application of substance P depolarizes smooth muscle cells in both the longitudinal and circular layers of the guinea-pig ileum, and in the continued presence of substance P both the fPSD in the circular muscle (Bywater & Taylor, 1983) and the atropine insensitive EJP in the longitudinal muscle (Bauer & Kuriyama, 1982b) are abolished. This effect can be attributed to desensitization of substance P receptors. Second, the fPSD is inhibited by the substance P analogue antagonist [D-Arg^1,D-Pro^2,D-Trp^{7,9},Leu^{11}]- substance P (Taylor & Bywater, 1986). Much additional evidence suggests that substance-P is an excitatory transmitter in the guinea-pig ileum (see Costa et al. 1985; Hoyle & Burnstock, 1989). For example, substance P immunoreactive fibres originating in the myenteric ganglia innervate both the circular and longitudinal muscle layers (Schultzberg et al. 1980; Costa et al. 1980). Immunoreactive fibres are found throughout the circular muscle layer, whereas the longitudinal muscle layer receives a diffuse plexus of fibres which do not penetrate the muscle. The release of substance-P from intestinal neurons has been demonstrated in response to electrical stimulation, nicotinic receptor agonists, high K$^+$, and liquid distension (Baron, Jaffe & Gintzler, 1983; Holzer, 1984; Donnerer et al. 1984). Contractions of the longitudinal and circular muscle elicited by electrical stimulation in the presence of muscarinic antagonists are prevented by desensitization of substance P receptors (i.e. with exogenously applied substance P) and/or by substance P antagonists (Franco, Costa & Furness, 1979; Gintzler & Hyde, 1983; Bjökroth, 1983; Costa et al. 1985). Furthermore, the contractions to substance P and the non-cholinergic contractions elicited by nerve stimulation in a number of other tissues (e.g. guinea-pig taenia caeci, Leander et al. 1981; opossum oesophageal muscularis mucosa, Domoto et al. 1983) are blocked by similar treatments, suggesting that in these tissues substance P may be the non-cholinergic excitatory transmitter.

Little is known about the ionic mechanisms underlying the generation of these non-cholinergic EJPs. In the circular muscle layer of the guinea-pig ileum, applications of substance P for periods of less than 5 minutes caused a depolarization which appeared to result from a decrease in membrane resistance (Niel, Bywater & Taylor, 1983a). When substance P was applied for periods greater than 10 minutes, the membrane potential repolarized to a value more positive than initial resting potential, and at this time the membrane resistance was increased. The authors suggested that the resistance change resulted from a decrease in K$^+$ conductance. Substance P is believed to produce a decrease in K$^+$ conductance in certain neuronal tissues (Katayama & North, 1978; Adams, Brown & Jones, 1983; Stanfield, Nakajima & Tamaguchi, 1985) and a similar mechanism of action may account for the contractile response to substance P in mammalian smooth muscle (Fugisawa & Ito, 1982; Holzer, 1982; Holzer & Petsche, 1983). Hyperpolarization of the membrane to values close to the K$^+$ equilibrium potential increased the amplitude of fPSDs recorded from the circular muscle layer (Bywater & Taylor, 1983). Furthermore hyperpolarizing electrotonic potentials recorded during the fPSD were decreased in amplitude. These findings are consistent with the fPSD being due to an increase in membrane conductance rather than to a decrease in K$^+$ conductance. Recently, Nakazawa et al. (1990) have reported that in smooth muscle cells isolated from guinea-pig ileum substance P, elicited an inward current resulting from an increased membrane conductance. In this case the inward current was carried mainly by Na$^+$.

Chick rectum. EJPs elicited by field stimulation in both the longitudinal and circular muscle of the chick rectum have a latency of 15–25 ms, a time to peak of 70–180 ms and a total duration of ~1 second (Ohashi *et al.* 1977; Takewaki & Ohashi, 1977). In the longitudinal muscle EJPs facilitate during short trains of low frequency stimuli (0.4–0.8 ms), whereas in the circular muscle the second EJP in a train of stimuli is substantially smaller than the first, and thereafter no change in EJP amplitude is observed (Obashi *et al.* 1977). These EJPs were resistant to treatment with atropine, hyosine, phentolamine and guanethidine. Based on the observation that α,β-methylene ATP abolished contractions evoked by both exogenous ATP and electrical stimulation of the excitatory nerves, it has been suggested that ATP is the excitatory transmitter in this tissue (Meldrum & Burnstock, 1985). This view has been challenged by Komori, Kwon & Ohashi (1988) who investigated the effects of α,β-methylene ATP in the chick rectum. They found that during exposure to α,β-methylene ATP there was a transient depolarization which decayed to values more positive than the resting potential. Furthermore, there was a maintained increase in membrane conductance during the period of exposure to α,β-methylene ATP. Thus, although EJPs in this tissue were markedly depressed by α,β-methylene ATP, it can be argued that this resulted from an alteration in the electrical properties of the muscle membrane rather than desensitization of P_{2x}-purinoceptors. Komori, Fukutome & Ohashi (1986) have suggested that chicken neurotensin may be the non-adrenergic, non-cholinergic excitatory transmitter in the chick rectum.

Mammalian urinary bladder. In the rabbit urinary bladder (Creed, Ishikawa & Ito, 1983) the membrane response to field stimulation is divided into two components, comprising an initial EJP followed by a late depolarization. The late depolarization was blocked by atropine and potentiated by neostigmine and resulted from the activation of muscarinic receptors. EJPs were resistant to atropine, phentolamine, guanethidine, methysergide and mepyramine indicating that they were not mediated through activation of muscarinic, adrenergic, tryptaminergic or histaminergic receptors respectively. Similar atropine resistant EJPs have been recorded in the guinea-pig and pig urinary bladder (Fujii, 1988, Brading & Mostwin, 1989). In both the guinea-pig and pig urinary bladder an atropine sensitive late depolarization was not recorded. EJPs in the rabbit urinary bladder are characterized by a brief latency and a total duration of some 500 ms (Creed, Ishikawa & Ito, 1983). During exposure to α,β-methylene ATP, EJPs and the depolarizations elicited by bath applied ATP in guinea-pig (Fujii, 1988), rabbit (Hoyle & Burnstock, 1985; Fujii, 1988) and pig (Fujii, 1988) urinary bladder are abolished, suggesting that ATP is the neurotransmitter. In all three tissues, α,β-methylene ATP appears to desensitize selectively P_{2x}-purinoceptors since the effects of this agent on both membrane potential and conductance returned to control values during the period of exposure (Fujii, 1988). Furthermore, α,β-methylene ATP did not alter the response of the guinea-pig urinary bladder to exogenous acetylcholine. Electrical stimulation has been shown to release endogenous ATP from intramural nerves in guinea-pig urinary bladder (Burnstock *et al.* 1978). In support of the idea that ATP is the transmitter responsible for the

EJP, Inoue & Brading (1990) have described an ATP-gated inward current in cells isolated from guinea-pig urinary bladder. This current has some similarities in both its pharmacological sensitivity and ionic selectivity to that described in rabbit ear artery (Benham & Tsien, 1987) and rat vas deferens (Friel, 1988). It remains to be established whether or not ATP and acetylcholine are co-secreted from the terminals of postganglionic parasympathetic nerve fibres.

ELECTRICAL RESPONSES TO INHIBITORY NERVE STIMULATION

1. Adrenergic

Repetitive stimulation of the perivascular nerves innervating many intestinal smooth muscles including the guinea-pig taenia caeci (Bennett, Burnstock & Holman, 1966) and the rabbit colon (Gillespie, 1962b) at frequencies > 5 Hz results in a small hyperpolarization and a relaxation of the tissue. In the guinea-pig taenia caeci both the hyperpolarization and the relaxation were blocked by guanethidine and bretylium, suggesting that sympathetic nerves were activated. The high frequency of stimulation required to evoke these inhibitory responses perhaps means they are unlikely to be important physiologically. Adrenergic hyperpolarizations and relaxations have also been described in cat gastric submucosal venules (Morgan, 1983) and rabbit facial vein (Prehn & Bevan, 1983). In these cases hyperpolarizations were recorded at frequencies > 2 Hz and were mediated through the activation of β-adrenoceptors.

2. Non-adrenergic, non-cholinergic

In various smooth muscles, non-adrenergic, non-cholinergic IJPs have been recorded (see Table 3.4). The characteristics of some of these IJPs will now be discussed in some detail.

Intestine. In many parts of the gastrointestinal tract a single transmural electrical stimulus, delivered in the presence of atropine, evokes a large transient membrane hyperpolarization and an associated relaxation of the smooth muscle (Table 4). In general IJPs have latencies of 100–150 ms, rise times of 200–500 ms and total durations of 1000–2000 ms. IJPs are not blocked by either guanethidine or bretylium (Burnstock, 1972), indicating that they are non-noradrenergic. IJPs are however abolished by TTX showing that they have a neurogenic origin. The following lines of evidence indicate that IJPs result solely from an increase in K^+ conductance. 1. Where determined, the relationship between membrane potential and IJP amplitude is linear with a reversal potential close to the K^+ equilibrium potential (e.g. guinea-pig ileum circular muscle, Bywater, Holman & Taylor, 1981; rabbit rectum longitudinal muscle, Suzuki et al, 1979; rabbit jejunum longitudinal muscle, Kitamura, 1978). 2. IJPs in the guinea-pig ileum (Bywater, Holman & Taylor, 1981; Bauer & Kuriyama, 1982b), colon (Shuba & Valdimirova, 1980; Maas, 1981) and stomach (Komori & Suzuki, 1986) are abolished by the K^+ channel blocking agent apamin. Although apamin is believed to block selectively Ca^{2+} activated K^+ channels, to

TABLE 3.4

Examples and characteristics of non-adrenergic, non-cholinergic IJPs recorded from various smooth muscle preparations (adapted from Hoyle & Burnstock, 1989).

Species/Tissue	Latency ms	Rice Time ms	Duration ms	Decay (t½) ms	Comments	Ref
Guinea-pig						
Ileum L.M.	–	130–360	–	–		Bauer & Kuriyama (1982a)
Ileum C.M.	~100	–	–	–		Bywater, Holman & Taylor (1981)
Jejunum L.M.	60–150	–	–	220–480		Hidaka & Kuriyama (1969)
Jejunum L.M.	50–80	120–280	–	–		Kuriyama, Osa & Toida (1967)
Taenia caeci	80–120	200–300	–	250–500(T.C.)		Bennett, Burnstock & Holman (1966)
Taenia caeci	120–160	–	–	260–390		Bülbring & Tomita (1967)
Caecum C.M.	38 (minimum)	–	750–1800	–		Ito & Kuriyama (1973)
Rectum L.M.	240–620	360–860	800–1200	250–380		Ito & Kuriyama (1971)
Colon C.M.	45–150	150–250	~1000	–		Furness (1969)
Stomach C.M.	~160	–	~600	–		Komori & Suzuki (1986)
Gastric sling L.M.	~140	–	–	–	Sucrose gap	Beck & Osa (1971)
Gastric corpus C.M./L.M.	150–200	200–300	1000–1800	–		Beani, Bianchi & Crema (1971)
Rabbit						
Jejunum C.M./L.M.	–	–	–	–		Kitamura (1978)
Duodenum C.M.	–	–	–	–		Cheung & Daniel (1980)
Rectum C.M./L.M.	80–150	150–250	800–1300	220–350		Suzuki et al. (1979)
Colon C.M./L.M.	~150	~300	–	~320	Sucrose gap	Furness (1969)
Caecum C.M.	~325	–	3000–4000	–		Small (1972)
Prostate capsule	~300	–	–	–		Seki & Suzuki (1989)
Prostate L.M.	–	–	–	–		Seki & Suzuki (1989)
Rectococcygeous	~200	~850(T.P.)	–	~1350	Extrinsic stimulation	Blakeley, Cunnane & Muir (1979)
Anococcygeus	–	~725	–	–		Creed & Gillespie (1977)
Rat						
Anococcygeus	–	–	–	–		Byrne & Large (1984)
Dog						
Middle cerebral artery	–	–	~10,000	–		Suzuki & Fujiwara (1982)
Internal anal sphincter	–	–	–	–		Kubota & Szurszewski (1984)
Ileocolonic sphincter	–	–	–	–		MacKenzie & Szurszewski (1984)
Bovine						
Retractor penis	–	–	–	–	Repetitive stimulation	Byrne & Muir (1984)
Trachea	–	–	–	–		Cameron et al (1983)
Chick						
Rectum	75–200	200–350	400–1500	–		Bennett (1969)
Gizzard						
Pigeon						
Gizzard	70–250	150–250	500–1500	–		Bennett (1969)
Opossum						
Oesophagus	~100	~500	–	–		Kannan, Jager & Daniel (1985)

L.M. = Longitudinal muscle; C.M. = Circular muscle; T.P. = Time to peak; t1/2 = 1/2 decay time; T.C. = Time constant.

date apamin sensitive Ca^{2+} activated K^+ channels have not been described in intestinal smooth muscle (Bolton, 1989). In the circular muscle of the guinea-pig ileum and distal colon, treatment with apamin reveals a slow hyperpolarization (Niel, Bywater & Taylor, 1983b; Smith & Bywater, 1983). Apamin resistant IJPs have also been recorded in the opossum oesophagus (Daniel *et al.* 1983) and in the circular muscle of the chick rectum (Komori & Ohashi, 1988). Taken together, these findings indicate that at least two different mechanisms generate membrane hyperpolarizations.

Determined efforts have been made to establish the identity of the inhibitory transmitter(s) in the intestine. The two substances for which most evidence has been advanced are ATP and vasoactive intestinal peptide (VIP). The evidence for and against these and other putative transmitters has been extensively reviewed elsewhere (Burnstock 1979, 1981, 1986; Furness & Costa, 1982, 1987; Daniel 1985; Gillespie, 1982; Holye & Burnstock 1989; Jager & Hertog, 1985; Small & Weston 1970) and will not be dealt with here. The evidence to date does not support the view that there is a single mechanism of non-adrenergic, non-cholinergic inhibition in intestinal smooth muscle. This has been clearly demonstrated by Costa, Furness & Humphreys (1986) who investigated the effectiveness of apamin in blocking relaxations to VIP, α,β-methylene ATP and non-adrenergic, non-cholinergic inhibitory nerve stimulation in various guinea-pig intestinal smooth muscles. In all the tissues examined apamin inhibited relaxations elicited by α,β-methylene ATP but had little or no effect on responses to VIP. Nerve-mediated relaxations were substantially reduced by apamin in antrum circular muscle, ileum longitudinal and circular muscle, taenia caeci and distal colon circular muscle, but this agent was without effect in the fundus, distal colon circular muscle or in the longitudinal muscle of the proximal colon. In guinea-pig intestinal smooth muscle, nerve stimulation evokes both an apamin sensitive IJP and an apamin insensitive slow hyperpolarization (see above). Costa, Furness & Humphreys (1986) therefore suggested that these two events may correspond to the apamin sensitive and the apamin insensitive relaxations.

Other smooth muscles. Non-adrenergic, non-cholinergic IJPs have been recorded in smooth muscle of the rabbit anococcygeus (Creed & Gillespie, 1977) and rectococcygeus (Blakeley, Cunnane & Muir, 1979) and bovine retractor penis (Byrne & Muir, 1984). In these tissues IJPs have a longer duration than those recorded in intestinal smooth muscle. IJPs evoked by single stimuli have a latency of ~200 ms, a time to peak of ~700 ms and a total duration around 2 secs. In all three tissues, IJPs were accompanied by relaxations and were unaffected by guanethidine. When the relationship between membrane potential and IJP amplitude in the rabbit anococcygeus was studied the results indicated that the reversal potential was close to the K^+ equilibrium potential (Creed & Gillespie, 1977), suggesting that the inhibitory transmitter produces an increase in K^+ conductance. Consistent with this suggestion, the amplitude of electrotonic potentials recorded during the IJP were reduced indicating a decrease in membrane resistance. In the rat anococcygeus, nerve mediated relaxations were not normally associated with an IJP (Creed,

Gillespie & Muir, 1975). However, in this tissue electrotonic potentials were reduced in amplitude during the period of inhibitory nerve stimulation, suggesting that the membrane resistance was decreased (Creed & Gillespie, 1977). Byrne & Large (1984) have reported that the α-adrenoceptor antagonist insensitive, transient depolarizations recorded in the rat anococcygeus were sometimes followed by a hyperpolarization which had a time to peak of ~1500 ms. It may well be that in these tissues the membrane hyperpolarizations have nothing to do with the inhibitory mechanical response. Recent advances have been made in our understanding of inhibitory transmission in the rat anococcygeus and bovine retractor penis muscle. There is good reason to believe that l-arginine is released from inhibitory nerves and is transformed into nitric oxide or a related substance which diffuses into the smooth muscle cell to activate guanylate cyclase directly. This is a novel concept of neuro-effector transmission in that we are used to the idea that a neurotransmitter once released acts through a receptor located on the surface of the effector cell. Gillespie, Liu & Martin (1990) have recently reviewed this exciting area of research.

Non-adrenergic, non-cholinergic IJPs have also been recorded in canine middle cerebral artery following field stimulation of the intramural nerve fibres (Suzuki & Fujiwara, 1982). In this tissue a single stimulus evokes an EJP (α-adrenoceptor antagonist insensitive) followed by a 'slow' hyperpolarization of about 10 sec duration. The hyperpolarization was not blocked by phentolamine, guanethidine or atropine. The amplitudes of electrotonic potentials were reduced during the period of hyperpolarization suggesting that the membrane conductance was increased. Since a decrease in the external K^+ concentration increased IJP amplitude, the IJP can be attributed to an increase in K^+ conductance. Further supporting evidence is provided by the fact that the hyperpolarization is blocked by the K^+ channel blocking agent tetraethylammonium. However, the IJP was resistant to treatment with apamin. As far as we are aware, no putative transmitter has been proposed.

In precontracted bovine trachea short trains of repetitive stimuli (10 Hz for 1 sec) evoke a cholinergic EJP and contraction, followed by relaxation when stimulation is stopped (Cameron et al. 1983). Following repetitive stimulation (10 Hz for 15–30 secs) the relaxation was coincident with a hyperpolarization of a few mV. This hyperpolarization was not blocked by propranolol, suggesting that it was not mediated through activation of β-adrenoceptors. Although both ATP and VIP caused hyperpolarization and relaxation, the effects of nerve stimulation were better mimicked by VIP. Consistent with this suggestion, bovine trachea has a high content of VIP and electrical stimulation induced a TTX sensitive increase in VIP release. Hyperpolarizations in response to non-adrenergic, non-cholinergic inhibitory nerve stimulation in cat or guinea-pig trachea have not been detected (Ito & Takeda, 1982).

TRANSMITTER RELEASE STUDIES

Surprisingly little is known about the mechanisms involved in the storage and release of neurotransmitters generating junction potentials in smooth muscle. In this section we will discuss how junction potentials can be used to gain insights into transmitter release mechanisms. It is useful first to review briefly transmitter release mechanisms at the well characterized skeletal neuromuscular junction to provide a framework for the subsequent discussion of release mechanisms at the autonomic neuroeffector junction.

PREJUNCTIONAL MECHANISMS – SKELETAL NEUROMUSCULAR JUNCTION

Our present understanding of the fundamental mechanisms involved in neuro-transmitter release processes originates with the elegant electrophysiological studies of Katz and his colleagues in the early 1950's. Intracellular recording techniques were used to measure the postjunctional response to acetylcholine released from somatic motor nerves at the amphibian skeletal neuromuscular junction. These preparations were used because they have two unique advantages. First individual muscle fibres are usually innervated by a single nerve terminal from a branch of a lower motor neuron. Second muscle fibres are electrically isolated from one another so that all of the electrical activity recorded can be attributed to transmitter released from a single nerve terminal. Fatt and Katz (1952) demonstrated that when a muscle fibre was penetrated with a micro-electrode at the end-plate region, i.e. the specialized region of the skeletal muscle fibre in close apposition to the nerve terminal, small random depolarizations were recorded. These potentials were potentiated by the anticholinesterase prostigmine, reduced by curare and absent in muscles that had previously been denervated. On the basis of these observations they suggested that the potentials, termed miniature end-plate potentials (MEPPs), resulted from the release of acetylcholine from motor nerve terminals. Studies have demonstrated that the MEPP results from the release of a multimolecular packet of acetylcholine rather than from a single molecule (Fatt & Katz, 1952; see Kuffler, Nichols & Martin, 1984).

Quantal transmission

When the motor nerve innervating the muscle was stimulated a large EPP was recorded in the impaled cell which had a similar time course to the MEPP. The relationship between the MEPP and the EPP was determined by reducing the Ca^{2+} content of the bathing solution, which resulted in a marked reduction in EPP amplitude with no corresponding effect on the amplitude of MEPPs. Individual EPPs fluctuated in amplitude in a random stepwise manner; the step unit as the EPP approached the size of the MEPP having a similar amplitude to the MEPP. In addition, a proportion of the stimuli failed to elicit EPPs. On the basis of these observations Fatt and Katz (1952) formulated the quantal hypothesis; the idea that the EPP is made up of a hundred or so all-or-none units (quanta) which are identical in size to the MEPP.

A statistical analysis of the transmitter release process was carried out by Del Castillo and Katz (1954a). The observed fluctuations in EPP amplitude, when the probability of transmitter release was lowered by increasing the Mg^{2+} and reducing the Ca^{2+} concentration, fitted the Poisson distribution. The mean quantal content was determined either by dividing the mean amplitude of the EPP by the mean MEPP amplitude or by using the method of 'failures' (see Katz, 1966). Both methods produced similar results which supported the quantal hypothesis. The authors pointed out that the observed distribution of EPP amplitudes would be expected to change from a Poisson to a binomial form as the probability of transmitter release is increased; the Poisson distribution being the limit of a binomial distribution.

It was shown that non-linear summation of the EPP can occur and this can lead to serious errors in the estimation of the quantal content of the EPP. Martin (1955) showed that when the mean quantal content was raised by adjusting the Mg^{2+}/Ca^{2+} concentrations, the observed amplitude fluctuations of EPPs could still be accounted for by the Poisson distribution. In this analysis non-linear summation of the quantal units was taken into account. Using this method the normal quantal content of the EPP at the frog skeletal neuromuscular junction was estimated to be several hundred quanta.

Evidence for quantal release of transmitter at the mammalian skeletal neuromuscular junction was first provided by Boyd and Martin (1956). They showed that the EPP in the cat tenuissimus muscle was made up of discrete units (i.e. quanta) of the same size as the MEPP. They estimated the normal quantal content of the EPP, assuming a Poisson distribution and taking into account non-linear summation, to be about 200–300 units. Liley (1956a) reported that transmitter release at the skeletal neuromuscular junction of the rat diaphragm was also quantal.

These ideas fitted well with the observation that MEPPs at the amphibian skeletal neuromuscular junction occur randomly (Fatt & Katz, 1952). It has been suggested, based on the sensitivity of the EPP to the external Ca^{2+} concentration (Katz & Miledi, 1965a) and the observation that depolarization of the nerve terminal accelerated the rate of MEPP discharge (Del Castillo & Katz, 1954b; Liley, 1956b), that the depolarization of the nerve terminal induced by the action potential causes a transient increase in the cytosolic Ca^{2+} concentration due to the opening of voltage – dependent Ca^{2+} channels, thereby increasing the probability of transmitter release. This theory does not require that the nerve impulse starts up a new secretory process but instead markedly raises the probability of observing events that occur all the time.

The random nature of MEPP discharge at the amphibian neuromuscular junction has been questioned, since the discharge of a single MEPP transiently increases the probability of observing subsequent MEPPs (Bennett & Pettigrew, 1975; Cohen, Kita & Van der Kloot, 1974a,b). At other neuromuscular junctions and synapses where the patterns of spontaneous events have been investigated, the occurrence of an event transiently increases the probability that subsequent events will be observed (Usherwood, 1972; Cohen, Kita & Van der Kloot, 1974c; Rees, 1974; Bornstein, 1978). Furthermore, the sensitivity of the MEPP and EPP to changes in the external Ca^{2+} and Mg^+ concentration, osmotic pressure and drugs differs (see Hubbard 1973). These findings suggest that the evoked release of quanta may not be simply related to the release process underlying the spontaneous discharge.

Studies at other neuromuscular junctions and at chemical synapses have confirmed the quantal nature of the transmitter release process (see Korn & Faber, 1987). However, the distribution of amplitudes of evoked postjunctional and evoked postsynaptic potentials under normal conditions are rarely fitted by Poisson statistics (see McLachlan, 1978). At some neuroeffector junctions and synapses release is often better fitted by a simple binomial model (Blackman & Purves, 1969; Korn *et al.* 1982; Wernig, 1975) where there is a population of n units with a uniform probability p of releasing a single quantum. This type of analysis is of particular interest in that both n and p may have physical correlates. It has been suggested that n may represent the number of quanta available for release or the number of active release sites and p may be a measure of the cytosolic concentration of Ca^{2+} (see McLachlan, 1978). At the frog neuromuscular junction Wernig (1975) suggested that under normal conditions p approaches 1, n being approximately 100–200. The actual value of n would be expected to depend on the size of the nerve terminal investigated (Kuno, Turkanis & Weakly, 1971). This estimate of n is of the same order of magnitude as the number of 'active zones' described morphologically at these junctions (see Wernig, 1975), which are thought to be the regions of the nerve terminal specialized for transmitter release. On the basis of these observations it has been suggested that under normal conditions each active zone has only one active release site. At the mammalian neuromuscular junction p, determined using the simple binomial model, also approaches 1 (Bennett & Florin, 1974; Bennett, Florin & Hall, 1975).

Probability of transmitter release

The assumption that all potential transmitter release sites have the same probability of discharging a quantum may not be valid. An alternative view is that provided by the compound binomial model in which the probability of release varies at different release sites on the same nerve terminal (see McLachlan, 1978; Redman, 1990). At the amphibian neuromuscular junction Bennett and Lavidis (1979) have demonstrated a considerable non-uniformity in the probability of release at different sites along the nerve terminal when the external Ca^{2+} concentration was reduced. Focal extracellular electrodes were employed in this study to record activity along a 30 μm length of nerve terminal, which may be up to 1000 μm in total length. The area of the nerve terminal 'seen' by the extracellular electrode would be expected to contain only a small proportion of the total number of release sites. However, on occasion it was found that the mean quantal content of release at some of these sites was comparable to that of the whole junction and the conclusion was reached that the majority of released quanta come from only a small population of the available sites which have a relatively high probability of release. Raising the external Ca^{2+} concentration and delivering short trains of electrical stimuli increased the probability of release at each of the release sites in a manner which was independent of the initial probability. Therefore, under normal conditions non-uniform release would be expected. Bennett and Lavidis (1979) suggested that the number of voltage – dependent Ca^{2+} channels associated with a particular release site may determine its probability of secretion.

In the above analyses the assumption was made that following stimulation single quanta are released in an all-or-none manner, and that each release site behaves as an independent unit. There is little evidence to suggest that these assumptions are not valid.

Morphological correlates of the quantum

In the mid 1950's several workers reported the presence of membrane bound vesicles in close association with the prejunctional or presynaptic membrane in number of different nerve terminals (Palade, 1954; Palay, 1954; De Robertis & Bennett, 1955; Robertson, 1956). It was suggested that these vesicles might be involved in transmitter storage and transport (De Robertis & Bennett, 1955). Del Castillo and Katz (1955) suggested that at the neuromuscular junction 'it is possible to imagine a mechanism by which each particle (i.e. vesicle) loses its charge of acetylcholine ions in an all-or-none manner when it collides with, or penetrates the membrane of the nerve terminal'. The same suggestion was made by Robertson (1956). This idea gave rise to the vesicle hypothesis, ie. that the transmitter content of a 'synaptic vesicle' represents the quantum.

Synaptic vesicles prepared from several different tissues by homogenization and centrifugation have been shown to contain neurotransmitters (De Robertis, 1967). In addition, prolonged electrical nerve stimulation can deplete the vesicle numbers in nerve terminals of several different tissues (see Zimmermann, 1979). At the frog neuromuscular junction the best morphological evidence for the vesicular release of transmitter comes from the studies of Heuser *et al.* (1979) and Heuser & Reese (1981). They developed a technique in which the morphological changes associated with transmitter release could be investigated with a high degree of temporal resolution by rapidly freezing the tissue shortly after stimulation. In these studies the number of vesicle openings at the active zones was correlated with the quantal content of the EPP following treatment with 4-aminopyridine (4-AP) to increase the number of quanta released. There was a surprisingly good agreement between the number of vesicles fusing with the surface of the nerve terminal and the estimated quantal content of the evoked potential in the presence of a range of 4-AP concentrations, suggesting that the vesicle is the morphological correlate of the transmitter quantum.

Non-vesicular quantal transmission

Several workers (Tauc, 1982; Dunant & Israël, 1985) have questioned whether the available evidence supports the vesicle hypothesis. Studies with tritiated choline in the Torpedo electric organ demonstrate that recently synthesized acetylcholine is preferentially released upon nerve stimulation (Dunant & Israel, 1985). Furthermore, prolonged stimulation, that depressed the electrical response to 10% of the initial value, caused no change in the number of vesicles or in the specific activity of labelled vesicular acetylcholine, indicating that newly synthesized acetylcholine was not taken up into vesicles during the period of stimulation, as would be expected if vesicles were recycled. However, the cytoplasmic pool of acetylcholine was depleted to about 50% of its initial value, and the specific radioactivity of the remaining transmitter was high, indicating a high rate of synthesis. These findings suggest that the released transmiter comes from the cytoplasm and not from the vesicle. In this case the vesicle is not the morphological correlate of the quantum but instead a membrane-bound macromolecular structure (or 'vesigate') which binds several hundred molecules of acetylcholine, release occurring following a conformational change in the macromolecule which results in transport of the bound acetylcholine to the exterior (Tauc, 1982). In these latter experiments it was assumed that the population of vesicles was homogeneous, but Zimmermann & Whittaker (1977) have demonstrated that the vesicle population of the nerve terminals of the Torpedo electric organ changes upon electrical stimulation. A second type of vesicle, smaller in diameter, was formed presumably from vesicles which have already undergone one or more cycles of endocytosis/exocytosis. When the nerves were stimulated in the presence of tritiated acetate the smaller vesicles showed a marked accumulation of labelled acetylcholine, while the amount of labelled acetylcholine in the larger vesicles was reduced. These findings have recently been supported by Stadler and Kiene (1987), who also reported a third population of vesicles. It is fair to say that although the electrophysiological evidence for quantal transmission is strong at the skeletal neuromuscular junction, the biochemical evidence for the vesicular origin of the released acetylcholine is not. However, in recent years advances have been made in this area with the introduction of vesamicol.

Studies with vesamicol (AH5183)

Perhaps the best evidence for the vesicle hypothesis comes from studies using the drug AH5183 (2-(4)phenylpiperidino) cyclohexanol) now termed vesamicol. On the basis of the unusual characteristics of the frequency-dependent neuromuscular block Marshall (1970) suggested that vesamicol acted by inhibiting the uptake of acetylcholine into vesicles. Anderson, King & Parsons (1983) confirmed this hypothesis showing that vesamicol acted stereoselectively to produce a reversible, non-competitive inhibition of the uptake of acetylcholine into isolated synaptic vesicles from the Torpedo electric organ, the (−) isomer being the active agent. Evidence from studies in other tissues indicates that vesamicol does not inhibit acetylcholine synthesis (Michaelson, Licht & Burstein, 1987).

Suszkiw & Manalis (1987) reported that vesamicol produces a marked decrease in EPP amplitude and mobilization of transmitter during short trains of stimuli at the frog neuromuscular junction. These authors suggest that a subpopulation of 'active' vesicles are involved in exocytosis, the remaining vesicles providing a reserve transmitter pool which can be used to maintain the cytoplasmic levels of transmitter during periods of prolonged activity. Searl, Prior & Marshall (1990) found at the snake neuromuscular junction that stimulation in control experiments resulted in the production of two populations of MEPCs, consisting of both pre-stimulation amplitude MEPCs and a second population of small amplitude MEPCs whose numbers were stimulation dependent. In the presence of (-) vesamicol stimulation produced a concentration dependent reduction in the amplitude of the stimulation dependent population of MEPCs, leading to their abolition at concentrations greater than 5μM. At all concentrations used, pre-stimulation amplitude MEPCs were still present following stimulation, although their frequency of occurrence was reduced. Similar results have been reported at the rat neuromuscular junction (Marshall, Prior & Searl, 1990). However, in untreated tissues stimulation had no effect on MEPC amplitudes, a bi-modal distribution to MEPC amplitudes, consisting of both pre-stimulation amplitude MEPCs and a second population of small amplitude MEPCs, only being revealed in the presence of vesamicol. The two populations of MEPCs in both the rat and the snake have been suggested to represent a population of pre-formed previously unreleased quanta and therefore vesamicol insensitive MEPCs, and a population of recycled quanta whose filling is impaired by vesamicol. The reasons for the differences between the snake and rat neuromuscular junction remain to be established. These studies provide no evidence for a reserve pool of storage vesicles.

Interestingly, in synaptosomes made from nerve terminals of the Torpedo electric organ, vesamicol selectively inhibits the release of tritium in tissues treated with tritiated acetate indicating that newly synthesized acetylcholine requires uptake into the vesicles prior to release (Michaelson, Licht & Burstein, 1987). Vesamicol also inhibits the spontaneous non-quantal release of transmitter at the mammalian neuromuscular junction (Edwards et al. 1985), which accounts for over 95% of the transmitter overflow in the absence of stimulation. It has therefore been suggested that non-vesicular release of transmitter results from the inclusion of the vesicular acetylcholine carrier into the nerve terminal membrane following exocytosis (Edwards et al. 1985). It will be interesting to apply this or similar agents to parasympathetic nerve terminals and autonomic ganglia to gain insights into the fundamental mechanisms involved in the storage and release of acetylcholine in the autonomic nervous system.

Is the transmitter contents of one vesicle the quantum?

The vesicle hypothesis is now widely accepted, and the presence of vesicles is used to define the regions of the nerve terminal specialized for transmitter release (see Kuffler, Nichols & Martin, 1984). However, the fundamental idea that the transmitter contents of a single vesicle represents the quantum has been questioned (see Tremblay, Laurie & Colonnier, 1983). When recordings are made from the end-plate region of high resistance skeletal muscle cells a second population of subminiature potentials has been identified which are between 1/7th and 1/15th the size of the MEPP (Kriebel & Gross, 1974). These potentials have been termed subminiature end-plate potentials (s-MEPPs). In certain preparations it has been reported that the MEPP amplitude distribution displays multiple peaks which are regularly spaced suggesting that they are made up of sub-units the same size as the s-MEPP (Kriebel & Gross, 1974; Kriebel, Llados & Matteson, 1976; Wernig & Stirner, 1977). These observations led to the suggestion that the s-MEPP represents the release of a single vesicle and that the MEPP (i.e. quantum) results from the synchronous release of approximately 10 vesicles (the multivesicular hypothesis). This hypothesis has been vigorously challenged by Magleby and Miller (1981), who concluded that the intervals between the amplitude classes in the MEPP amplitude distribution were not uniform. However, in a recent study of the amplitude distributions of miniature end-plate currents (Erxleben & Kriebel, 1988), the intervals between the amplitude classes both within and between different preparations were found to be highly uniform. Katz and Miledi (1979) after investigating the potentiating effects of diaminopyridine and tetraethylammonium on the EPP questioned whether, under conditions of enhanced release, it would be possible to mobilize a sufficient number of vesicles if each quantal unit resulted from the synchronous discharge of several vesicles. An alternative view is that released transmitter is not vesicular (Erxleben & Kriebel, 1988; see above). Korn (1984; see also Korn & Faber, 1987) in an elegant series of experiments on transmitter release mechanisms in inhibitory nerves innervating the teleost Mauthner cell, concluded that the evidence was consistent with the idea that only a single vesicle was released following activation of the transmitter release mechanism in a bouton. There may well be physical constraints on the release mechanism whereby only a single vesicle can undergo exocytosis from a particular release site. Under normal conditions the value of n derived using binomial statistics at several neuromuscular junctions and synapses is similar to the number of putative release sites described morphologically (see Korn & Faber, 1987). Korn (1984) has therefore suggested that the 'one vesicle hypothesis' may be a general feature of release sites in both the peripheral and central nervous systems.

Thus, it is at the skeletal neuromuscular junction that neurotransmission has been most thoroughly investigated and the most complete evidence obtained for the processes involved in neurotransmitter release. In the absence of stimulation, individual packets of transmitter are secreted. When the motor nerve is stimulated, about one hundred transmitter quanta are released almost synchronously to generate the end-plate potential. The focal depolarization spreads to the surrounding sarcolemma, activates voltage – dependent Na^+ channels and a muscle action potential is initiated, which spreads along the muscle fibre membrane to elicit contraction. There is a high safety factor for the transmission of excitation from somatic motor nerve to skeletal muscle. We will now describe as far as possible, the corresponding events at the autonomic neuroeffector junction and consider the technical problems which have to be overcome to achieve a similar degree of resolution of the transmitter release process.

PREJUNCTIONAL MECHANISMS – AUTONOMIC NEUROEFFECTOR JUNCTION

There are difficulties in analysing junction potentials to study transmitter release at the autonomic neuroeffector junction. For this reason a combination of biochemical and electrophysiological methods have been used to elucidate the fundamental mechanisms involved in transmitter release. Much of our understanding of transmitter release mechanisms at the autonomic neuroeffector junction depend to a large extent on studies at the sympathetic neuroeffector juntion and these will now be described in some detail.

Use of EJPs to study transmitter release

The first direct electrophysiological evidence for the packeted release of transmitter at the autonomic neuroeffector junction comes from the studies of Burnstock & Holman (1961, 1962a,b) who recorded SEJPs with intracellular microelectrodes in the guinea-pig vas deferens. However, it was not possible to determine the quantal content of the EJP in smooth muscle for three main reasons. First, the difference in time course of the SEJP and EJP means that they are not directly comparable. Second, the amplitude distribution of the SEJPs is not unimodal but skewed towards the noise level of the recording system, making it impossible to determine the mean quantal size. Third, statistical tests are only valid if the activity of a single junction can be identified or if the innervation is limited to a specific area of the muscle fibre (see Burnstock & Holman, 1962a). As previously stated, smooth muscle cells receive a multiple innervation and, owing to electrical coupling between adjacent cells, the electrical activity recorded in any one cell reflects the activity of transmitter released both locally and at sites remote from the recording electrode. These properties of smooth muscle in part explain why the SEJP amplitude distribution is skewed; transmitter action close to the recording electrode producing a larger SEJP than that occurring at distant release sites owing to attenuation of signals spreading electrotonically. However, variable numbers of transmitter molecules in the packets, the width of the junctional cleft and variations in the input impedances of different smooth muscle cells in the tissue may all contribute to the form of the SEJP amplitude distribution (Holman, 1970).

Therefore, using conventional electrophysiological techniques the relationship between spontaneous and evoked transmitter release could not be established with any degree of certainty at the sympathetic neuroeffector junction. The situation is worse at many other autonomic neuroeffector junctions principally because spontaneous junction potentials are rarely observed. It is fair to say that much of our understanding of the mechanisms involved in the storage and release of transmitters at the sympathetic neuroeffector junction came initially from biochemical studies. It is useful at this point to review this evidence and show how subsequent electrophysiological studies have helped to answer some of the questions raised.

Biochemical studies of transmitter release

Using biochemical methods it has been shown that most of the noradrenaline stored in postganglionic sympathetic neurons is present in storage granules, and that the granular store is heterogeneous; i.e. noradrenaline is found in both a 'light' and a 'heavy' particulate fraction (see Smith & Winkler, 1972). Electron microscopy has revealed the presence of at least two types of vesicle in the varicosities of sympathetic nerves, the small (25–60 nm) and the large (70–160 nm) dense cored vesicles. The 'light' and 'heavy' particulate fractions isolated from sympathetic nerves probably correspond to the small and the large dense cored vesicles respectively. The proportions of small and large dense cored vesicles present in varicosities varies from species to species. For example in the rat vas deferens the large dense cored vesicles account for only about 5% of the vesicle population whereas in the ox spleen they account for up to 50% (Klein & Lagercrantz, 1981).

There is good pharmacological evidence to suggest that only noradrenaline stored in vesicles is released by nerve impulses. Reserpine depletes the vesicular noradrenaline store by preventing the uptake and/or binding of the amine by the storage vesicles. In the presence of drugs which prevent the metabolism of cytosolic noradrenaline (monoamine oxidase inhibitors), reserpinized tissues are able to sequester tritiated noradrenaline via the neuronal uptake mechanism (i.e. uptake$_1$). The available evidence suggests that most of the sequestered noradrenaline is free in the cytosol and is not taken up by the vesicles (Iversen, Glowinsky & Axelrod, 1965; Potter, 1967; Häggendal & Malmfors, 1969). In these tissues the evoked release of tritiated noradrenaline is markedly inhibited compared with control tissues (Potter, 1967), suggesting that only noradrenaline contained in storage vesicles can be released by nerve impulses.

Further evidence to support the vesicular origin of the released noradrenaline was provided by De Potter *et al.* (1969) and Smith *et al.* (1970) who demonstrated that two proteins, dopamine β-hydroxylase and chromogranin A, present in the noradrenaline storage granules isolated from dog and calf splenic nerve axons, were released in a Ca^{2+} dependent manner following nerve stimulation. Furthermore, the amount of protein released was correlated with the amount of noradrenaline released. By analogy with the adrenal medulla where all the soluble constituents of the chromaffin granule are released during stimulation (see Smith & Winkler, 1972), this evidence is taken to support the view that noradrenaline is released from sympathetic nerves by an exocytotic mechanism. Indeed, in the adrenal medulla adrenaline, ATP and its metabolites are secreted in a molar ratio similar to that found in the chromaffin granule (Douglas, Poisner & Rubin, 1965; Douglas & Poisner, 1966; Banks, 1966) there being approximately 4 molecules of adrenaline for each molecule of ATP (Hillarp, 1958; Winkler & Westhead, 1980). ATP is also co-localized with noradrenaline in both the 'light' and 'heavy' particulate fractions derived from sympathetic nerves (Lagercrantz, 1976; Fried, 1980). In these particulate fractions the noradrenaline/ATP ratio is higher than in the chromaffin granule, being between 20–60 in both the 'heavy' and 'light' particulate fractions prepared from nerve terminals (see Klein & Lagercrantz, 1981). However, a comparison between the stoichiometry of noradrenaline and ATP release from sympathetic nerves has not been possible owing to the difficulty in confirming the prejunctional origin of the released ATP (see Fried, 1980).

Evidence suggesting that both the small and large dense cored vesicles are involved in transmitter release comes from the study of Smith *et al.* (1970). They found that the dopamine β-hydroxylase activity/noradrenaline ratio in tissue perfusates from stimulated dog and calf spleen was lower than in the large dense cored vesicles derived from splenic nerve axons. Fried (1980) demonstrated that the 'light' particulate fraction, by comparison with the 'heavy' particulate fraction, contains a very low dopamine β-hydroxylase content in relation to its noradrenaline content. Therefore, it has been suggested that dopamine β-hydroxylase is secreted mainly from large dense cored vesicles, while noradrenaline is secreted from both the large and the small dense cored vesicles (see Thureson-Klein, 1983). An alternative view is that most of the noradrenaline is secreted by a mechanism other than exocytosis, possibly by a cation exchange process between the noradrenaline storage vesicles attached to the plasma membrane and the exterior of the nerve (see Uvnäs & Åborg, 1980). However, most of the evidence to date does not appear to support this view.

How much transmitter is released from a varicosity by an action potential?

The biochemical evidence supports the hypothesis that vesicles are the sites of transmitter storage and that only the vesicular pool of transmitter is released by nerve stimulation. The question then arises how much transmitter is released from each varicosity per stimulus? The noradrenaline content of a varicosity has been calculated by a number of different authors. Dahlström, Häggendal & Hökfelt (1966) determined the total tissue noradrenaline content of both the rat iris and rat vas deferens and estimated the number of noradrenergic varicosities in these tissues using the histochemical technique of Hillarp and his colleagues (see Falck *et al.* 1962). Assuming that all the noradrenaline was present in varicosities (see Malmfors, 1965), these authors calculated that the noradrenaline content of a single varicosity was 4.2×10^{-15} g and 5.6×10^{-15} g in the rat iris and rat vas deferens respectively. Similar values have been obtained for sympathetic varicosities in the rabbit pulmonary artery (10^{-14} g/varicosity, Bevan, Chesher & Su, 1969) and the main uterine artery of the guinea-pig (3.4×10^{-14} g/varicosity, Bell & Vogt, 1971). The amount of endogenous noradrenaline secreted per stimulus in relation to the total tissue content (fractional release) has been determined in a number of sympathetically innervated tissues, values varying between 2×10^{-5} (Folkow, Häggendal & Lisander, 1967) and 2×10^{-4} (Bell & Vogt, 1971) for stimulation at 5–6 Hz. In these studies neuronal uptake (uptake$_1$) was inhibited and when extraneuronal uptake (uptake$_2$) was also inhibited there was no further increase in noradrenaline overflow (Bell & Vogt, 1971). Taking the range of values reported and assuming that the fractional release is the same at the level of the individual varicosity, then the amount of noradrenaline released from a varicosity per stimulus ranges between 1×10^{-19} g (Folkow, Häggendal & Lisander, 1967) and 7×10^{-18} g (Bell & Vogt, 1971). As the weight of a noradrenaline molecule is 2.6×10^{-22} g, the number of molecules released from a varicosity per stimulus would range between 400 and 25000. Assuming that the noradrenaline content of a single vesicle was approximately 15000 molecules (i.e. the same as a chromaffin granule), Folkow, Häggendal & Lisander (1967) used their estimate of 400 molecules of noradrenaline being released from a varicosity per stimulus (6 Hz stimulation) to suggest that on average only about 1–3% of the transmitter content of a vesicle was released from a varicosity per stimulus. Similarly, Bevan, Chesher & Su (1969) estimated that each varicosity when stimulated at 10 Hz secreted 2000 molecules of noradrenaline per stimulus, which would be equal to only 13% of the contents of a vesicle. In contrast, Bell & Vogt (1971) proposed that their value of 25000 molecules of noradrenaline being released from a varicosity per stimulus (5 Hz stimulation) was consistent with the contents of 1 or more vesicles being secreted from each varicosity per stimulus.

A major source of error in the above calculations appears to be the estimate of the number of noradrenaline molecules present in a vesicle. The present estimate for the noradrenaline content of a 'large' and 'small' dense-cored vesicle in a post-ganglionic sympathetic nerve terminal is 8000–15000 molecules and 500–1000 molecules respectively (see Klein & Lagercrantz, 1981). Since the released noradrenaline is believed to originate from both small and large dense-cored vesicles (see above), the estimates for number of vesicles secreted from a varicosity per stimulus based on the value of 15000 molecules of noradrenaline in a vesicle would appear to be inaccurate. Another source of error is the estimate of the number of varicosities in the tissues investigated. Luff, McLachlan & Hirst (1987) have suggested that the histochemical techniques (e.g. Falck-Hillarp fluorescence microscopy) used to determine the number of varicosities in these studies may underestimate their numbers by a factor of two or three.

Another way to approach this question is to consider the relationship between the number of vesicles in a varicosity and the fractional release per stimulus. Luff, McLachlan & Hirst (1987) made a detailed analysis of the varicosities on the post-ganglionic sympathetic nerves innervating the guinea-pig submucosal arterioles. In an analysis of 11 varicosities which made close contacts with smooth muscle cells these authors found between 200 and 1500 vesicles per varicosity (mean = 559), the number being related to the volume of the varicosity. Hökfelt (1969) reported a similar number of vesicles in the noradrenergic varicosities innervating the dilator muscle of the rat iris. However, Hökfelt (1969) also reported that in some large varicosities (no number given) the number of vesicles was much higher but since they could not be followed in serial section counts of numbers were not made. If it is assumed that the average number of vesicles per varicosity is 1000 and that all these vesicles are involved in the storage and release of noradrenaline, then if each stimulus releases 1 vesicle per varicosity the fractional release would be 10^{-3}. Since the fractional release of noradrenaline per stimulus is between 2×10^{-4} and 2×10^{-5} these observations would suggest that the equivalent of 1 vesicle is secreted from a varicosity every 5 to 50 stimuli. This estimate is similar to that given by Folkow, Häggendal & Lisander (1967) and Bevan, Chesher & Su (1969) but is derived by using the number of vesicles in a varicosity and makes no assumptions about the amount the noradrenaline in either a varicosity or a vesicle.

The conclusion to be drawn from the above studies is that the amount of noradrenaline released from the average varicosity per stimulus is much less than the amine content of a single vesicle. Folkow, Häggendal & Lisander (1967) suggested two possible mechanisms to account for this observation, namely: 1) each varicosity secretes a fraction of the transmitter content of a vesicle per stimulus or; 2) varicosities secrete the whole transmitter content of a vesicle intermittently. At the time the former hypothesis was favoured since it could not easily be envisaged how an intermittent transmitter release process could uniformly activate a smooth muscle but recent studies suggest that the latter hypothesis was correct.

Recent electrophysiological studies

New electrophysiological approaches have been developed to study depolarization-secretion coupling in sympathetic nerve terminals. These approaches have shown that action potential evoked transmitter release is intermittent and that normally only a single quantum is secreted when the release mechanism of a varicosity is activated.

Intermittent transmitter release mechanism

Blakeley & Cunnane (1978, 1979a,b) used intracellular recording techniques to record EJPs in the guinea-pig and mouse vas deferens. When the rising phases of EJPs were differentiated, transient peaks in the rate of depolarization were detected which the authors termed 'discrete events'. Discrete events occurred at one or a few fixed latencies after the stimulus, but the important observation was made that events at any one latency occurred intermittently (Fig. 3.4). The authors concluded that discrete events occurring at a single latency measure the release of transmitter from one or a few release sites located close to the recording electrode. It was assumed that the anatomical basis of the single release site was the varicosity. Differentiated SEJPs had a similar time course to the evoked discrete events recorded in the same impalement. Furthermore, it was possible to find differentiated SEJPs which closely matched in both amplitude and time course selected discrete events recorded in the same cell (Fig. 3.5). These findings suggest that both the SEJP and the discrete event have a similar anatomical basis; i.e. the transmitter content of a single vesicle. The amplitude distribution of discrete events occurring at any one latency displayed several preferred values, and in some cells the larger preferred values were multiples of the smallest preferred value. Therefore, it was proposed that the number of packets of transmitter released from a single varicosity per stimulus was small, varying between 0 and 10 (see also Blakeley, Mathie & Petersen, 1984). However, Blakeley and Cunnane (1979a) stated that 'as we have at present only the conduction velocity to assign the discrete event to any particular release site our data will overestimate the frequency with which transmitter release occurs when a single release site is activated'. These observations provide direct evidence for intermittent release of packets of transmitter from individual varicosities during nerve stimulation. Burnstock and Holman (1962a) were the first to describe fluctuations in the rising phases of EJPs and suggested that 'not all the nerve endings were activated each time the nerve was stimulated'.

A more rigorous analysis of individual discrete events in the guinea-pig vas deferens was undertaken by Cunnane and Stjärne (1982, 1984a). They demonstrated that when transmitter release was reduced by decreasing the Ca^{2+} content of the bathing solution, some cells had a very low probability of discrete event occurrence. In one so called 'simple' cell only 3 discrete events were recorded in response to about 500 stimuli at 1 Hz (Fig. 3.6A). A comparison between the differentiated SEJPs and electrically evoked discrete events in this cell revealed that the largest SEJP was closely matched by the largest discrete event (Fig. 3.6B). In 'complex' cells, in which the probability of discrete event occurrence was much higher, some

Guinea-pig   **Mouse**

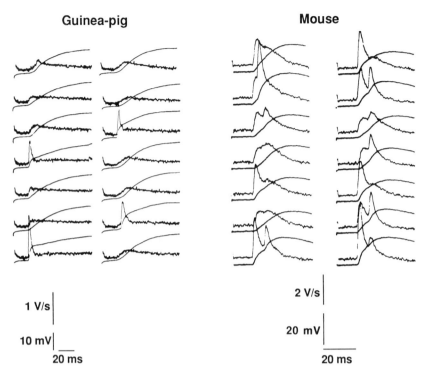

FIGURE 3.4   Intracellular recordings of the membrane potential and its first time derivative ($\partial V/\partial t$, noisy records) of single smooth muscle cells in the guinea-pig and mouse vas deferens. EJPs were evoked by trains of stimuli at 0.91 Hz. The transient peaks in the rate of depolarization, termed discrete events, occur intermittently showing that action potential evoked transmitter release is intermittent. Calibration bars, EJPs, mVs; discrete events, V/s. [Adapted from Blakeley & Cunnane, 1979a]

of the events within a particular latency band formed distinct amplitude classes. Each amplitude class was composed of two or three discrete events which were not only matched in amplitude but also in time course. In these cells it was possible to match the discrete events in a particular amplitude class with differentiated SEJPs recorded in the same cell. Indeed the authors report that it 'seemed possible in principle, to find a matching spontaneous discrete event for each amplitude class of evoked discrete event'. The simplest interpretation of these findings is that discrete events in a particular amplitude class reflect the secretion of transmitter from the same or closely related release site(s). Furthermore, as discrete events are similar to the differentiated SEJPs occurring in the same cell, it would seem that transmitter release is monoquantal. An interesting observation made by the authors was that during a long train of stimuli at 0.5–2 Hz (up to 1200 stimuli) only 2 or 3 matching discrete events within any particular amplitude class were recorded, and these often occurred within 1–5 pulses of one another. This finding

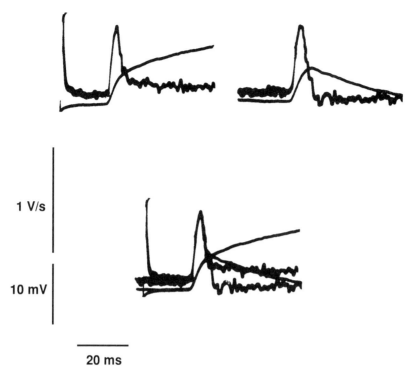

1 V/s

10 mV

20 ms

FIGURE 3.5 Intracellular records of the membrane potential of a smooth muscle cell in the guinea-pig vas deferens comparing a selected evoked discrete event and a SEJP. The EJP was selected from a train evoked at 0.91 Hz. Top left record, EJP and discrete event, top right SEJP and first time derivative ($\partial V/\partial t$). Bottom record, evoked discrete event and differentiated SEJP superimposed photographically showing them to be similar in both amplitude and time course. Calibration bars, EJPs, mVs; $\partial V/\partial t$, V/s. [From Blakeley & Cunnane, 1979.]

suggests that transmitter secretion from a particular release site is highly intermittent, and that when release occurs there may be a transient increase in the probability that transmitter release will occur from the same or closely related release site(s). It is important to note here that an inhibition was expected since it is generally believed from pharmacological experiments that released transmitter inhibits subsequent transmitter release locally through activation of prejunctional receptors (see *Local Regulation*). Cunnane and Stjärne (1984a) reported that the characteristic features of discrete events recorded in the mouse vas deferens were similar to those in the guinea-pig vas deferens, except that normally they had a higher probability of occurrence. This difference may well be due to the different densities of innervation of the mouse and guinea-pig vas deferens (Burnstock, 1970).

Hirst and Neild (1980a) also reported that transmitter release from the postganglionic sympathetic nerves innervating the guinea-pig submucosal arterioles was intermittent. These authors used a short segment of the arteriole cut from the branching arteriolar arcade. In these segments, with two electrodes impaled in cells at

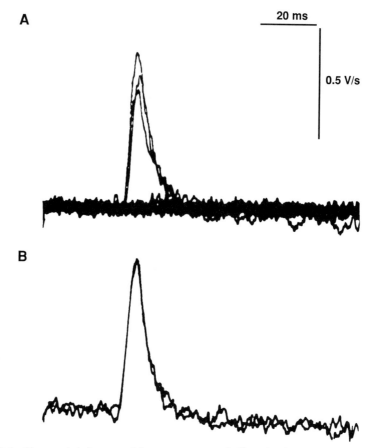

FIGURE 3.6   Characteristic features of discrete events recorded in a 'simple cell'. (A) Discrete events were evoked by 4 trains of 125 stimuli at 1 Hz in a single smooth muscle cell of the guinea-pig vas deferens. Only 3 discrete events were evoked, all similar in latency, amplitude and time course. (B) An evoked discrete event from record A, photographically superimposed on a differentiated SEJP recorded in the same cell showing them to be virtually identical in amplitude and time course. [From Cunnane & Stjärne, 1984a.]

either end of the preparation, SEJPs and EJPs were recorded with virtually identical amplitude; i.e. there was little spatial attenuation of the signals. Furthermore, the SEJPs and EJPs had similar time courses. This similarity between the time course of the SEJPs and the EJPs is because the cut ends of the arteriolar segment seal to form a high resistance between the cytoplasm and the extracellular fluid. Therefore, current can only escape locally in this preparation, and the time course of both the SEJP and EJP will be determined by the rate of charge loss across the arteriolar membrane (see *Membrane Potential & Current at the Autonomic Neuroeffector Junction*). Thus using this preparation it is possible to determine directly the relationship between the SEJP and the EJP. It is interesting to note that the amplitude distribution of SEJPs is unimodal (and not skewed as in most smooth muscle preparations) and the amplitudes correspond to the smallest EJPs recorded. It is likely that the uniform distribution of SEJP amplitudes reflects quantal transmitter release from a population of varicosities which make similar 'close contacts' with smooth muscle (see Hirst & Edward, 1989). Assuming that the EJPs are composed of quantal units the same size as SEJPs, analysis of the distribution of EJP amplitudes suggests that the majority of EJPs have a quantal content of only 2 or 3. In these preparations the number of varicosities was estimated, using fluorescence microscopy, at between 112 and 224. The study of Luff, McLachlan & Hirst (1987) suggests that fluorescent microscopy underestimates the number of varicosities in this preparation by two or threefold. These results therefore suggest that the probability of release of a single quantum from individual varicosities in these arterioles is less than 0.01.

Therefore, both biochemical and electrophysiological studies of transmitter release in a number of vascular and non-vascular smooth muscles suggest that transmitter release at the level of the individual varicosity is intermittent (see Cunnane, 1984). The electrophysiological studies also confirm that evoked transmitter release is quantal, and that the quantum probably represents the whole transmitter contents of a vesicle. These conclusions may not be valid for noradrenaline, since it now seems likely that the EJP and SEJP may be generated by the release of ATP or a related purine nucleotide (see *Electrical Responses to Excitatory Nerve Stimulation*). The evidence to date suggests that both noradrenaline and ATP are co-stored and co-released, since ATP is present in the same vesicles as noradrenaline (see Smith & Winkler, 1972). Therefore, it follows that the SEJP or the discrete event will also measure indirectly the quantal release of noradrenaline.

Causes of intermittence

There are two possible causes of intermittence: (1) intermittent failure of the action potential to propagate into the nerve terminal network or (2) a low probability of transmitter release in the invaded varicosity or a combination of both of these processes. To address these questions at the sympathetic neuroeffector junction a new approach, using focally applied extracellular suction electrodes, has been developed (Fig. 3.7A) (Brock & Cunnane, 1987, 1988a; Brock, 1988).

Characteristic features of transmitter release-focal extracellular recording. The principles involved in the use of a suction electrode can be briefly summarized; a resistance is created at the junction between the edge of the electrode and the smooth muscle surface. A potential change proportional to the postjunctional current is created over this 'seal' resistance (Stühmer, Roberts & Almers, 1983). The polarity of the potential is determined by the direction of current flow over the 'seal' resistance. A negative-going potential reflects postjunctional current originating inside the suction electrode, while transmitter action outside the electrode generates a positive-going signal. The use of suction electrodes therefore allows the electrical activity generated by a small population of varicosities and smooth muscle cells located beneath the suction electrode to be investigated.

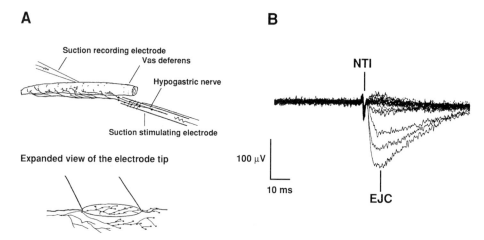

FIGURE 3.7 Recording of electrical activity from the surface of the guinea-pig vas deferens. (A) Schematic representation of extracellular recording from the surface of the vas deferens using a suction electrode (tip diameter N 50μm). (B) Simultaneous measurement of the nerve terminal impulse (NTI) and evoked excitatory junction currents (EJCs) evoked by a train of 25 pulses at 1 Hz. [FIGURE 7B from Brock & Cunnane, 1987.]

Spontaneous electrical activity These methods were first applied to the guinea-pig vas deferens. The extracellular potentials recorded from the surface of the vas deferens in the absence of stimulation (SEJCs) are produced by the spontaneous release of transmitter, since they have similar characteristics to intracellularly recorded SEJPs (see Brock & Cunnane, 1988a). The overall time courses of the SEJC and the SEJP are similar and there is no obvious relationship between the amplitude and the time constant of decay. These findings support the view that the time course of the SEJP is determined by the time course of transmitter action (see *Membrane Potential & Current at the Autonomic Neuroeffector Junction*). The amplitude distribution of SEJCs and SEJPs are similarly skewed, suggesting that the spread of the electrotonic potential may not be the only factor determining the shape of the SEJP amplitude distribution. The size of SEJCs recorded with a suction electrode should reflect the magnitude of the current generated at the site of transmitter action. However, other factors which may be important in determining the amplitude of SEJCs include (1) the different relationships between individual varicosities and the smooth muscle cells which they innervate, (2) a variation in the size of the transmitter quantum released from different sites, (3) a variation in the density of the postjunctional receptors and/or (4) different relationships between the sites of transmitter release and the recording electrode.

Evoked electrical activity The pharmacological profile of electrically evoked EJCs, recorded extracellularly in the guinea-pig vas deferens, was largely the same as that of the SEJCs (Brock & Cunnane, 1988a). EJCs were, however, Ca^{2+}-dependent and readily abolished by treatment with the inorganic Ca^{2+} entry blocker Co^{2+} and TTX. In this respect EJCs are similar to the intracellularly recorded EJPs. The configuration of the EJCs recorded from the guinea-pig vas deferens depends on the intensity of stimulation and the site of recording. When only the release of transmitter from varicosities located beneath the suction electrode is recorded, EJCs are similar in both amplitude and time course to the SEJCs recorded in the same attachment. Activity outside the electrode produces a positive-going signal which in some attachments distorts the negative-going EJC. If an SEJC represents the release of a single quantum, then these findings suggest that the EJC results from the evoked quantal release of transmitter from the sympathetic nerve terminals. Negative-going EJCs at a fixed latency occur intermittently (Fig. 3.7B) supporting the previous observation that transmitter release from individual varicosities is intermittent (Blakeley & Cunnane, 1979a; Cunnane & Stjärne, 1984a).

The evidence obtained from focal extracellular recording techniques supports the view that only a single quantum is secreted by a varicosity (Brock & Cunnane, 1988a). The key evidence can be summarized thus: First, the amplitude distributions of SEJCs and EJCs recorded in the same attachment during low frequency stimulation are similar; multimodal distributions of the amplitudes of evoked events are not observed (Fig. 3.8). Second, during trains of stimuli EJCs are recorded with closely similar amplitudes and time courses; indeed, pairs of matching EJCs are often recorded within a few stimuli of each other suggesting, as with the discrete event, that previous release facilitates subsequent release from the same site. Similar observations have been made elsewhere and may be a characteristic feature of transmitter release from single or closely associated release sites (Robitaille & Tremblay, 1987). Third, individual EJCs can be matched, both in amplitude and time course, with selected SEJCs in the same attachment. Taking the simplest interpretation of the quantal hypothesis, that the SEJC represents the postjunctional response to the release of a single quantum, then individual varicosities normally secrete only one quantum, intermittently, when the release mechanism is activated by the nerve impulse during trains of low frequency stimuli. It seems unlikely in a low probability system that evoked EJCs of identical form, which occur within a few stimuli of each other, represent multiquantal releases from the same varicosity; in this case the matching SEJC would also have to be the result of a multiquantal release. It is at present not possible to rule out the possibility that individual release sites on a varicosity may occasionally release more than one quantum or to determine whether there are several release sites on a varicosity. Clearly there will be occasions when summation of individual EJCs will occur, notably at higher frequencies when the probability of EJC occurrence is raised (Brock & Cunnane 1988a). Another possible interpretation is that only a small number of receptors mediate the quantal EJC and SEJC, the amount of transmitter released from a single vesicle being much greater than that required to activate all the receptors (see Jack, Redman & Wong, 1981; Redman 1990; Edwards, Konnerth & Sakmann, 1990). In this case it will not be possible to resolve, using present techniques, the amount of transmitter released by a single varicosity. However, biochemical studies of noradrenaline release suggest that only a single packet of transmitter is released intermittently.

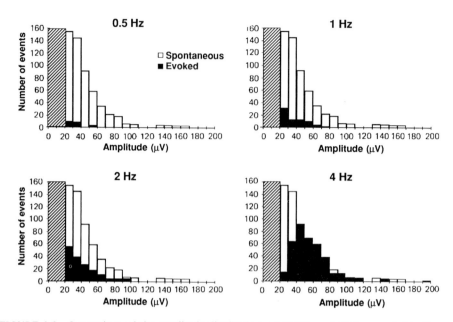

FIGURE 3.8   Comparison of the amplitude distributions of SEJCs and EJCs recorded in the same attachment from the surface of the guinea-pig vas deferens. EJCs were evoked by trains of 500 stimuli at 0.5, 1, 2 & 4 Hz. The amplitude distribution of the SEJCs is superimposed on that of the EJCs at each frequency for ease of comparison. The hatched area represents twice background noise level and is the level where the presence or absence of individual events could not be stated with any degree of confidence. At frequencies up to 2 Hz the amplitude distributions are similar. At 4 Hz the evoked amplitude distribution shifts to the right (i.e. individual EJCs become larger) presumably because of summation of coincident EJCs. [From Brock & Cunnane, 1988a.]

Negative-going EJCs are always preceded by a non-intermittent nerve impulse. This impulse is believed to be the nerve terminal impulse since the occurrence of an EJC is strictly dependent on its arrival (Fig. 3.7B & Fig. 3.9). In addition, the sensitivity of the impulse to the adrenergic neuron blocking agents, bretylium and guanethidine, supports the view that it is the impulse in a sympathetic nerve terminal (Brock & Cunnane, 1988b). The nerve impulses recorded were evoked in an all-or-none manner about the stimulus threshold and retained their shape during repetitive stimulation, suggesting that they were single unit recordings. Furthermore there was no variation in the size of the invading nerve impulse when EJCs occurred, and when they did not, which might have accounted for the intermittence of transmitter release (see Fig. 3.12). Thus intermittent transmitter release cannot be accounted for by failure of the nerve impulse to propagate in the terminals; action potentials invade sympathetic nerve terminals faithfully in an unvarying manner. It has recently been reported by Cheung (1990) that impulse propagation occurs intermittently in some sympathetic nerve terminals in the guinea-pig vas deferens. Occasional failures of impulse propagation have been reported previously (Cunnane & Stjärne, 1984b). However, it is now clear from our studies that intermittent arrival of the nerve impulse at the site of recording is due to frequency-dependent failure of

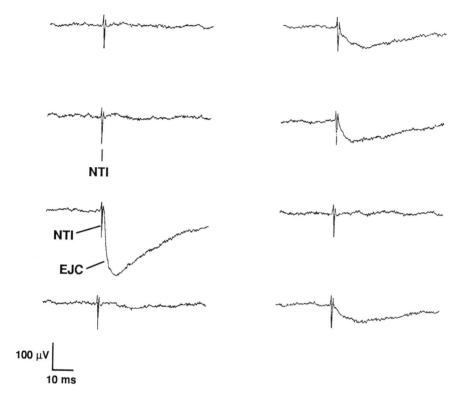

FIGURE 3.9 Simultaneous measurement of the nerve terminal impulse (NTI) and evoked excitatory junction currents (EJCs) in the guinea-pig vas deferens. A series of eight traces showing the relationship of the NTI to transmitter release. The NTIs and EJCs were evoked by stimulation of the hypogastric nerve with suprathreshold stimuli at 1 Hz. Although the action potential arrived in the nerve terminal on every occasion the transmitter release process was not activated. Intermittence of transmitter release is not therefore caused by failure of the nerve terminal action potential to invade the secretory varicosities but results from a low probability of release in the invaded varicosity. [From Brock & Cunnane, 1988a]

transmission in the hypogastric ganglion (Brock & Cunnane, 1990). When sympathetic nerve fibres are stimulated postganglionically nerve impulse propagation failure in sympathetic nerve terminals does not occur (Brock & Cunnane, 1987, 1988).

Similar focal extracellular studies of both impulse propagation and transmitter release have been carried out in the mouse vas deferens (Åstrand, Brock & Cunnane, 1988) and various arteries (Åstrand, Brock & Cunnane, 1988; Åstrand & Stjärne, 1989a,b). Essentially similar results were obtained, transmitter release was intermittent and intermittence was not due to failure of impulse propagation in sympathetic nerve terminals.

Active or passive invasion of varicose nerve terminals One question which has received much attention is whether the entire length of a nerve terminal is excitable.

In order to answer this complex question impulse propagation in the nerve terminals of several different neuromuscular preparations has been studied using focal extracellular electrodes. The problem with this approach is that extracellular records can be difficult to interpret and a potential minefield of misinterpretation. For this reason it is useful to review briefly what signals an extracellular electrode would be expected to record and why, for a uniformly propagating action potential in or near a nerve terminal.

The size of the potential recorded with a focal extracellular electrode is proportional to the net current flow across the membrane; the polarity of the potential being positive during the outward movement of current and negative during the inward movement of current. The expected changes in the membrane current can be predicted from the cable properties of the axon (Murray, 1983). Assuming that the extracellular fluid is at isopotential, then the axial current (i_a) between two points along the axon will, by Ohm's law, be given by the difference in the membrane potential between the two points divided by the sum of the external and axoplasmic resistance for that length of axon. If this distance is made very small and it is assumed that the external resistance is negligible then:

$$i_a = 1/r_a . \partial V_m / \partial_x$$

where r_a is the axoplasmic resistance per unit length of axon, V_m is the membrane potential and x is the distance along the axon. During the uniform propagation of a nerve action potential the axial current at any point is related to the rate of change of the membrane potential ($\partial V_m / \partial t$, where t = time), which is dependent on the conduction velocity (q):

$$i_a = \partial V_m / \partial t . 1/r_a . 1/q.$$

Since all the current must either flow axially or across the membrane and the axial current depends on the rate of change of the membrane potential, the membrane current (i_m) recorded will be proportional to the rate of change of the axial current $\partial i_a / \partial t$ (i.e. the second derivative of the membrane potential $\partial^2 V_m / \partial t^2$).

The membrane current recorded during the passage of an action potential consists of a capacitative current which is proportional to the rate of change of the membrane potential and the magnitude of the membrane capacity (C_m), an 'excitable' current (i_e) and a 'passive' (resistive) current (V_m / r_m) across the membrane resistance r_m:

$$i_m = C_m . \partial V_m / \partial t + V_m / r_m + i_e.$$

The triphasic nerve impulse (positive-negative-positive) provides unambiguous evidence of action potential invasion. In this case the extracellularly recorded nerve impulse is the second derivative of the intracellularly recorded action potential ($\partial^2 V_m / \partial t^2$) (Fig. 10). The initial positive-going potential being caused by current spreading distally within the axon from the 'active' region of the nerve fibre to cause an outward movement of current at the site of recording. The peak of the initial positive-going phase of the nerve impulse occurs when the 'active' region arrives at the site of recording; the now net inward movement of current generates a negative-going potential. As the 'active' region propagates beyond the site of recording, current spreading proximally from the 'active' region generates a net outward movement of current at this site and the second positive-going phase of the nerve impulse.

Diphasic and monophasic waveforms are more difficult to interpret. The diphasic impulse (positive-negative) may represent electrotonic invasion of the nerve terminal from a point of block (Dudel, 1963). In this case the membrane current:

$$i_m = C_m.\partial V_m/\partial t + V_m/r_m$$

Thus the membrane current is proportional to the first derivative of the intracellularly recorded action potential ($\partial V_m/\partial t$) (Fig. 3.10), the positive-going phase of the nerve impulse represents the rising phase of the action potential in the proximal unblocked region, whereas the negative-going phase represents repolarization. However, Brooks and Eccles (1947; see also Katz & Miledi, 1965b) pointed out that close to a nerve ending (i.e. at the closed end of a linear conductor) an extracellular electrode would record a diphasic waveform even though the action potential propagates to the end. This situation arises because at the nerve ending all the axial current must cross the membrane. Therefore, the membrane current is equal to the axial current and proportional to the first derivative of the intracellularly recorded action potential ($\partial V_m/\partial t$).

A monophasic nerve impulse usually represents the electrotonic invasion of the nerve terminal from a point of block. This situation will occur if the capacitative current in the nerve terminal is small relative to the passive ionic current. In this case the membrane current will be equal to V_m/r_m (Dudel, 1963).

Katz and Miledi (1965b) studied impulse propagation at the frog neuromuscular junction and concluded that the entire 100–200 μm length of the nerve terminal was excitable. They recorded a triphasic waveform throughout most of the length of the nerve terminal. However, in the distal regions close to the nerve ending the waveform became diphasic because of the 'cable termination' effect discussed above, and not to electrotonic invasion. Hubbard and Schmidt (1963) also suggested that the nerve terminals innervating the rat diaphragm are excitable. In this preparation on most occasions diphasic waveforms were recorded. However, as the nerve terminals in this preparation are much shorter than those at the frog neuromuscular junction, the shape of these signals may also be due to the cable termination effect. In both studies the authors investigated the effects of electrically stimulating nerve terminals. In both cases antidromic nerve impulses were elicited, providing further support that nerve terminals are excitable throughout their entire length.

At the crayfish neuromuscular junction Dudel (1963) recorded a triphasic waveform in proximal regions of the nerve terminal which became diphasic and then monophasic as the impulse approached the terminal (synaptic region). Dudel (1963) suggested that these observations reflected electrotonic invasion of the terminal from a point of block close to the nerve terminal, and that the most distal regions of the nerve terminal were not excitable. In support of this claim Dudel and Kuffler (1961b) were not able to elicit antidromic nerve impulses by electrically stimulating the terminals. However, as pointed out above, the interpretation of extracellular records is equivocal. Zucker (1974a) re-investigated the

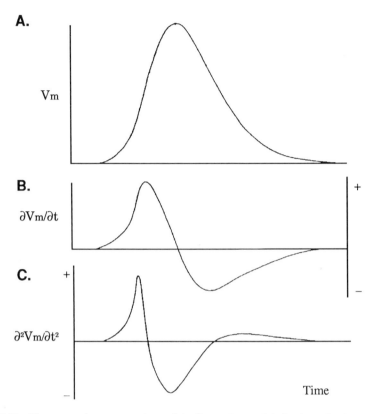

FIGURE 3.10   Diagrammatic representations of the first and second derivatives of voltage against time for an action potential. (A) A simplified action potential. (B) The rate of change of the membrane potential, $\partial V_m/\partial t$. (C) The second derivative, $\partial^2 V_m/\partial t^2$.

characteristic features of impulse conduction at the crayfish neuromuscular junction and concluded that these nerve terminals are excitable. In this later study antidromic nerve impulses were evoked by stimulating the terminals (Zucker, 1974b).

Brigant and Mallart (1982) investigated impulse propagation in the mouse triangularis sterni nerve-muscle preparation. Three different shapes of nerve impulse were recorded: 1) at the transition between the myelinated and non-myelinated parts of the axon (i.e. preterminal) the waveform consisted of a small positive-going deflection followed by a two-component negative-going deflection; 2) in the terminal branches a two-component positive-going monophasic waveform was recorded; 3) in a small region between the preterminal axon and the terminal branches a triphasic waveform was recorded. Application of K^+ channel blockers reduced both the late component of the monophasic waveform recorded at the terminal and the second negative-going component of the preterminal response. Therefore, it was suggested that these currents are generated when, following depolarization of the nerve terminal, an outward K^+ current at the terminal generates an inward movement of K^+ in the preterminal regions. Local application of TTX reduced the initial negative-going component of the nerve impulse recorded in the preterminal region, suggesting that it was caused by the inward movement of Na^+, but did not alter the shape of the waveform recorded at the terminal. Thus Brigant and Mallart (1982) concluded that Na^+ channels are restricted to the preterminal region and that the nerve terminals are not excitable. However, Konishi and Sears (1984) pointed out that the almost simultaneous activation of the penultimate and the ultimate heminode in this preparation may depolarize the nerve terminal so rapidly that the membrane potential approaches the Na^+ equilibrium potential before activation of the Na^+ current. Therefore, in this case no inward current would be recorded even though the nerve terminals bear excitable Na^+ channels.

In a recent review of the available data on whether or not nerve terminals at neuromuscular junctions are excitable, Smith (1988) states that a 'prudent conclusion from these studies is that motor nerve terminals are excitable. Action potentials invade the entire ending'.

In general, three shapes of nerve impulses are recorded from fibres running on or near the surface of the guinea-pig vas deferens using focal extracellular suction electrodes: first the classical triphasic, positive-negative-positive signal reflecting the extracellular equivalent of the propagating nerve impulse; second, a diphasic, positive-negative-going nerve impulse; third, an impulse which was largely positive-going with only a small negative-going component (Brock & Cunnane, 1988a). The shape of the nerve terminal impulse may well reflect recordings made from different regions of the terminals. However, for technical reasons it has not been possible to determine from which region of the nerve terminal these signals were recorded. It is possible that diphasic signals can be attributed to block of impulse propagation at a point proximal to the site of the recording. However, local perfusion of the electrode with TTX (see below) always abolishes the negative-going phase of the diphasic signals leaving a positive-going signal, suggesting that they result from actively propagating nerve impulses. These findings indicate that the varicose nerve terminals of postganglionic sympathetic nerves are invaded actively.

Active invasion is necessary for transmitter release to occur   Following internal perfusion of the extracellular recording electrode with TTX, active invasion of the varicosities is blocked but electrotonic invasion from the point of block still occurs (Fig. 11). Interestingly local application of TTX totally inhibits the occurrence of EJCs, showing that active invasion by a TTX-sensitive nerve impulse is a fundamental requirement for evoked transmitter release to occur (Brock & Cunnane, 1988a). Katz & Miledi (1967) demonstrated using focal extracellular electrodes that the local production of EPPs at the frog neuromuscular junction ceased distal to a site at which TTX was focally applied, while proximal to this point EPPs were not affected. They also concluded that active invasion of somatic motor nerve terminals is necessary for transmitter release to occur.

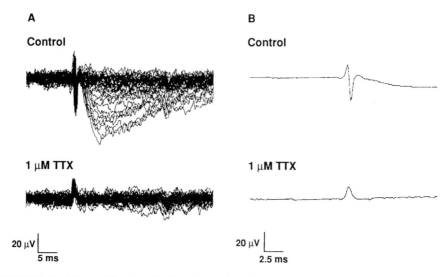

FIGURE 3.11 Effects of local application of tetrodotoxin (TTX) inside the recording suction electrode on impulse conduction in sympathetic nerve terminals in the guinea-pig vas deferens. (A) Nerve impulses and EJCs evoked by a train of 50 pulses at 1 Hz before and following perfusion with 1μM TTX for 10 minutes. (B) Averages of 80 nerve impulses recorded in the same attachment before and during the addition of TTX. Although the nerve impulse still electrotonically invaded the terminals the residual depolarization was insufficient to trigger the transmitter release mechanism. [From Brock & Cunnane, 1988a.]

Probability of transmitter release　In general the frequency of occurrence of SEJCs and the number of EJCs recorded depends on the tip diameter of the recording electrode. One can make a rough calculation about the number of varicosities located under a suction electrode which usually has a tip diameter of about 50 μm. Assuming that varicosities have similar release characteristics and that they are no more than 5 μm apart, then at least 10 may contribute to the evoked activity recorded by a 50 μm tip diameter electrode. However, if the axon branches estimates of the number of varicosities cannot readily be made without corresponding microanatomical studies of the region beneath the electrode. In the guinea-pig vas deferens the probability of recording an EJC during trains of stimuli varied considerably from attachment to attachment. This observation suggests that there may be 'hot spots' with a particularly high probability of release, similar to those found at the skeletal neuromuscular junction (Del Castillo & Katz, 1956; Bennett & Lavidis, 1979). Increasing the frequency of stimulation raises the probability of EJC occurrence at all recording sites (i.e. frequency-dependent facilitation) (Fig. 12). In order to make definitive statements about the release probability of varicosities located under the electrode, detailed microanatomical studies of the site of recording are required. These studies are at present underway in our laboratory.

Frequency-dependent facilitation　No changes in the shape or size of the nerve terminal impulse can be measured, at frequencies of stimulation up to 4 Hz, which might account for facilitation (Fig. 12). It is worth noting that extracellular record-ing will tend to emphasize the high-frequency components of the nerve impulse and it is quite possible that important slower conductance changes may be over-looked. It is clear that a large increase in action potential amplitude (Brown & Holmes, 1956) cannot account for the observed facilitation of transmitter release, at least in these nerves. At present it seems that the increased probability of release associated with facilitation is due to some mechanism acting at the level of the individual varicosity. In all experiments, facilitation was observed as a change in the number of EJCs evoked per train of stimuli (Fig. 12). Facilitation does not apparently involve the same release site secreting a variable number of quanta, since increased numbers of matching EJCs were not observed. The impression to date is that facilitation involves either the recruitment of previously silent varicosities or more release sites on the same varicosity becoming active, each secreting a single quantum (or perhaps a combination of these mechanisms).

Zucker (1974a) has reported that facilitation of transmitter release from the nerves innervating the crayfish claw opener muscle occurs without any apparent alteration in the shape of the extracellularly recorded nerve terminal impulse. In this study the effects of local depolarization of the nerve terminals in the presence of TTX were investigated. Facilitation of transmitter release was observed when paired or repetitive depolarizing pulses of a constant magnitude were given. A similar facilitation of transmitter release occurs in response to pairs of depolarizing

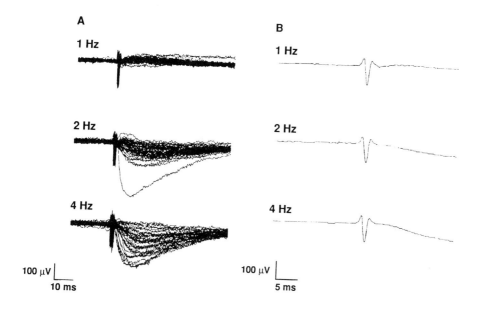

FIGURE 3.12 Frequency dependent facilitation of transmitter release in the guinea-pig vas deferens. (A) Extracellular recording of nerve terminal impulses and EJCs evoked by stimulation of the hypogastric nerve with trains of 100 suprathreshold stimuli at 0.5, 1 & 2 Hz. (B) Averages of 20 nerve terminal impulses and associated EJCs in the same attachment, showing that facilitation is not associated with a detectable alteration in the configuration of the nerve terminal impulse. [From Brock & Cunnane, 1988a.]

pulses in the presence of TTX at the frog neuromuscular junction (Katz & Miledi, 1967). These results suggest that facilitation of transmitter release results from an alteration in the process of depolarization-secretion coupling. Katz & Miledi (1965a) proposed that during a period of repetitive stimulation a residual change in the Ca^{2+} concentration in the nerve terminal may lead to facilitation of transmitter release (the residual calcium hypothesis). However, to date the exact process of facilitation remains to be fully elucidated (see Magleby, 1987).

A frequency-dependent increase in the latency of the nerve terminal impulse in some postganglionic sympathetic nerve fibres has been observed (see also Cunnane & Stjärne, 1984b). Aston-Jones, Segal & Bloom (1980) reported that antidromic nerve impulses recorded in rat locus coeruleus neurons exhibited frequency-dependent alteration in conduction velocity during trains of stimuli; initially a small decrease followed by a very pronounced increase in latency was observed. The authors suggested that this latency shift may be a general property of small diameter nerve fibres and result from reduced ion concentration gradients during trains of nerve action potentials. In particular, a loss of cytosolic K^+ may cause a depolarization of the axon, and thereby inactivate a proportion of the Na^+ channels and cause an increased K^+ conductance. Such alterations would increase the threshold for the initiation of the nerve action potential and thereby slow conduction. Similar frequency dependent increases in nerve impulse conduction have been reported in other fibres (see Ferreyra-Moyano & Cinelli, 1986).

Thus, evidence obtained with focal extracellular electrodes has extended knowledge gained from intracellular studies on transmitter release and demonstrated, unequivocally, that at the level of the individual varicosity action potential evoked transmitter release occurs intermittently. Furthermore, it has been shown that intermittence is not cause by a failure of the action potential to invade the secretory terminals. Active invasion of the nerve terminals by a sodium-dependent action potential is obligatory for evoked transmitter release to occur. Electrotonic invasion from a point of nerve impulse block does not activate the transmitter release mechanism. The action potential appears to arrive in the nerve terminal in an unvarying manner at frequencies of stimulation up to 4 Hz. The factor which determines the amount of transmitter secreted is therefore the rate of arrival of action potentials in the nerve terminals and this leads to an increase in the number of active varicosities. Thus, it is possible to achieve graded responses of smooth muscles by varying the centrally determined pattern of nerve impulse traffic arriving in the periphery.

Much work needs to be carried out to determine whether similar transmitter release mechanisms operate at other autonomic neuroeffector junctions. In particular, it would be prudent to study impulse propagation and transmitter release in nerve terminals before drawing conclusions about the mechanisms of drug action and, in particular, when trying to investigate the mechanisms involved in the modulation of transmitter release through prejunctional receptors.

Prejunctional receptors

Autonomic nerve terminals are endowed with a wide variety of prejunctional (presynaptic) receptors which when activated by appropriate agonists can either increase or decrease electrically evoked transmitter release (Starke, 1977, 1987; Gillespie, 1980). Perhaps the most widely studied of these is the prejunctional α_2-adrenoceptor which plays a key role in the regulation of transmitter release from sympathetic nerve terminals (α-autoinhibition). For this reason the discussion will be limited to the evidence for α-autoinhibition and in particular how activation of the prejunctional α_2-adrenoceptor leads to inhibition of transmitter release.

α-Autoinhibition

Various authors have shown that α-adrenoceptor antagonists increase noradrenaline overflow during sympathetic nerve stimulation (see Starke, 1977, 1987; Gillespie, 1980). Since this effect cannot be explained simply by an alteration in the uptake and/or metabolism of the released noradrenaline, it has been concluded that there is an increase in the amount of noradrenaline released (see Farnebo & Hamberger, 1971). In support of this claim De Potter *et al.* (1971) and Johnson *et al.* (1971) demonstrated that the α-adrenoceptor antagonists phenoxybenzamine and phentolamine also increased the release of the vesicular protein dopamine β-hydroxylase, for which there is no known mechanism of inactivation.

Since α-adrenoceptor antagonists increase noradrenaline release from tissues in which the effector response is mediated through β-adrenoceptors, i.e. cardiac tissues (Starke, 1971), the action of these drugs cannot be attributed to blockade of the postjunctional receptors. Starke (1971) proposed 'that adrenergic nerve terminals are endowed with structures related to the α-adrenoceptors on the effector tissues. On reaction with α-stimulants, e.g. with liberated noradrenaline, these neuronal α-adrenoceptors mediate. . . .inhibition of transmitter release; in the presence of α-blocking agents this restriction is attenuated'. Kirpekar and Puig (1971), Langer *et al.* (1971) and Farnebo and Hamberger (1971) suggested similar hypotheses. The demonstration that exogenous noradrenaline inhibits the release of neuronal noradrenaline from cardiac tissue during nerve stimulation (Starke, 1972; McCulloch, Rand & Story, 1973) and that this action is antagonized by α-adrenoceptor antagonists supports this idea (Starke, 1972). Indeed, in various sympathetically innervated tissues α-adrenoceptor agonists decrease electrically evoked release of noradrenaline (see Starke, 1977; Gillespie, 1980). The idea that α-adrenoceptors, located on the sympathetic nerve terminals, 'autoregulate' noradrenaline release has received much support (see Starke, 1977, 1987; Westfall, 1977; Langer, 1974, 1981; Gillespie, 1980) and to date remains the most widely accepted explanation for the effects of α-adrenoceptor antagonists on transmitter release. It should be noted that transmitter release from noradrenergic neurons in the central nervous system is also modulated through α-adrenoceptors (see Langer, 1981). In addition, α-adrenoceptors are present, for example, on cholinergic nerve terminals in the gut and activation of these adrenoceptors inhibits acetylcholine release (Paton & Vizi, 1969).

Based on the differential sensitivity of pre- and postjunctional α-adrenoceptors to phenoxybenzamine, Langer (1974) proposed a classification of the adrenoceptor into α_1 (postjunctional) and α_2 (prejunctional) sub-types. In support of this sub-classification, the pre- and postjunctional receptors in a number of different tissues have been shown to differ in their sensitivity to a range of agonists and antagonists (see Langer, 1981). However, α-adrenoceptors with similar pharmacological characteristics to α_2-adrenoceptors have been described at sites other than noradrenergic nerve terminals, e.g. vascular smooth muscle (Timmermans & Van Zwieten, 1981). Therefore, the α_1/α_2 sub-classification of the α-adrenoceptors based on their anatomical location is no longer valid.

Studies of the effects of α-adrenoceptor antagonists on EJPs in rodent vas deferens and various blood vessels support the hypothesis that transmitter release is regulated through prejunctional α_2-adrenoceptors (Starke, 1987) (Fig. 3.13A). EJPs in these tissues probably measure the release of ATP and not noradrenaline (see *Electrical Responses to Excitatory Nerve Stimulation*). Indeed, studies with α-adrenoceptor antagonists provide indirect evidence that noradrenaline and ATP are co-released from sympathetic nerve terminals. The studies of Forsyth & Pollock (1988), however, suggest that activation of α_2-adrenoceptors can differentially modify the release of noradrenaline and ATP. They investigated the effects of the α_2-adrenoceptor agonist, clonidine, on the tritiated noradrenaline overflow and the contraction elicited by electrical stimulation of the mouse vas deferens. They found that clonidine inhibited the first phase of contraction but potentiated the second, suggesting that the purinergic component (first phase) was inhibited and the noradrenergic component (second phase) potentiated. Surprisingly, in this study clonidine actually potentiated the release of ^{3}H-noradrenaline from these tissues (see also Stjärne, 1975).

It is often assumed that potentiation of transmitter release by α-adrenoceptor antagonists is due to removal of α-autoinhibition. The possibility remains that the effects of α-adrenoceptor antagonists are not due solely to blockade of α-adrenoceptors located on or near the nerve terminals (Kalsner & Quillan, 1984). For this reason the effects of the α-adrenoceptor antagonist yohimbine on transmitter release in vasa deferentia from guinea-pigs previously treated with reserpine to deplete noradrenaline stores were investigated (Brock et al. 1990). In control tissues yohimbine increased the amplitude of all but the first 1–3 EJPs evoked during trains of stimuli. However, yohimbine had no effect on EJP amplitude in reserpinized tissues (Fig. 3.13B) (see also Illes & Starke, 1983). These results show that yohimbine increases transmitter release by interrupting α-autoinhibition and not by a non-selective action (e.g. K^+ channel block in nerve terminals), and strongly supports the view that ATP release is modulated by co-released noradrenaline. It is interesting that in reserpinized and in yohimbine treated control tissues that the development of full facilitation of EJPs requires more stimuli, suggesting that activation of prejunctional α-adrenoceptors receptors normally modifies the mechanisms underlying facilitation.

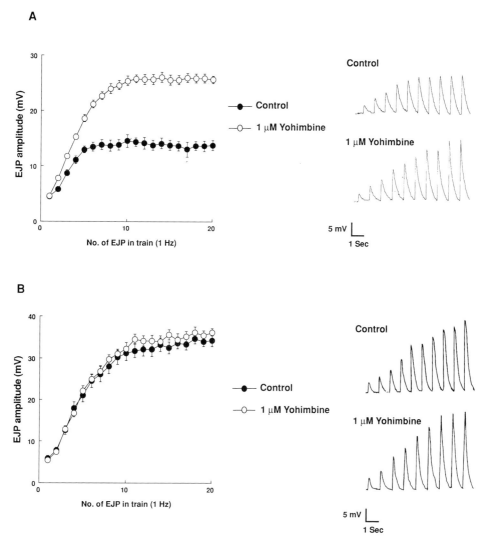

FIGURE 3.13   Effects of yohimbine (1 μM) on EJP amplitude in control and reserpinized guinea-pig vas deferentia. (A) In control tissues EJPs facilitated during the first 5–8 stimuli. (B) In reserpinized tissues, facilitation continues for 10 to 15 stimuli. Yohimbine markedly increased the amplitude of the fully facilitated EJP in control tissues but had no effect on EJPs in reserpinized tissues. Interestingly, yohimbine prolonged the time course of facilitation in control tissues suggesting that activation of prejunctional α-adrenoceptor normally limits the magnitude of facilitation. The data presented is the mean amplitude $\pm$ S.E.M. of 8 control cells and 8 cells in the presence of yohimbine in the same tissue. Trains of 10 EJPs at 1 Hz after each procedure are shown inset in the panels [From Brock *et al*. 1990; Brock & Cunnane, 1990].

Local regulation It is commonly believed that transmitter released from one varicosity acts locally to inhibit subsequent transmitter release from the same varicosity. However, the observation that transmitter release at the level of the individual varicosity is highly intermittent (see *Intermittent Transmitter Release Process*) means that transmitter is unlikely to accumulate locally in the vicinity of any one varicosity. Furthermore, it is difficult to envisage how transmitter can feedback locally, on an impulse-by-impulse basis, if release from a particular varicosity is unlikely to occur again within the next 100 or so stimuli. Studies of the occurrence of discrete events in the guinea-pig vas deferens in the absence of prejunctional α-adrenoceptor antagonists failed to demonstrate that endogenous transmitter regulates its own release locally (Blakeley, Cunnane & Petersen, 1982). Similarly studies using focal extracellular recording techniques, have not shown that the occurrence of an EJC in the guinea-pig vas deferens reduces the probability of observing subsequent EJCs (Brock & Cunnane, 1990). Thus transmitter released by nerve impulses does not apparently inhibit subsequent release from the same or closely related group of varicosities (see Brock & Cunnane, 1990, Brock *et al.* 1990). Rather it would appear that noradrenaline acts at sites remote from its own site of release (lateral inhibition).

The question is raised, therefore, whether endogenous noradrenaline is able to inhibit transmitter release *locally* when it is released from nerve terminals by the indirectly acting sympathomimetic amine, tyramine. Local application of tyramine by internal perfusion of the suction electrode powerfully inhibits the occurrence of electrically evoked EJCs in the guinea-pig vas deferens, an effect reversed by the competitive α-adrenoceptor antagonists yohimbine and phentolamine (Brock & Cunnane, 1989). In reserpinized tissues, tyramine had little or no inhibitory effect but clonidine still powerfully inhibited electrically evoked transmitter release showing that the prejunctional α-adrenoceptors were functionally intact. The simplest explanation for these findings is that noradrenaline released from sympathetic nerve terminals by tyramine stimulates prejunctional α-adrenoceptors and inhibits the electrically evoked release of transmitter. These results demonstrate that α_2-adrenoceptors are located close to the sites of transmitter release but the interesting observation remains that local inhibition by noradrenaline released by action potentials cannot be demonstrated.

How does activation of α_2-adrenoceptors inhibit transmitter release? The mechanisms whereby α-adrenoceptor agonists and antagonists modify the release of noradrenaline from postganglionic sympathetic nerve terminals has been extensively investigated (see Illes, 1986; Starke, 1987). Three different mechanisms for α-adrenoceptor mediated inhibition of transmitter release have been put forward, namely: 1) hyperpolarization of nerve terminals caused by an increase in K^+ conductance; 2) inhibition of voltage – dependent Ca^{2+} channels and thereby a decreased depolarization – induced Ca^{2+} entry; 3) direct interference with some step between Ca^{2+} entry and transmitter release. Each of these mechanisms will now be considered.

Electropharmacological studies in the central and peripheral nervous system have provided evidence both for an increased K^+ conductance and for a decreased Ca^{2+} conductance following activation of α-adrenoceptors in cell bodies. In neurons of the rat locus coeruleus (Williams, Henderson & North, 1985), the guinea-pig myenteric plexus (Surprenant & North, 1985) and the guinea-pig submucous plexus (North & Surprenant, 1985; Surprenant & North, 1985) the primary action of α-agonists was to cause an increase in K^+ conductance. An inhibition of voltage – dependent Ca^{2+} entry has been demonstrated following the application of nor-adrenaline to neurons of the rat superior cervical ganglion (Horn & McAfee, 1980) and the chick embryo dorsal root ganglion (Dunlap & Fischbach, 1981). Williams and North (1985) have questioned whether the inhibition of this voltage – dependent Ca^{2+} current was mediated through either α_1- or α_2-adrenoceptors. This conclusion was drawn because in rat locus coeruleus neurons noradrenaline and adrenaline but not the selective α_2-adrenoceptor agonist, clonidine, suppressed an inward Ca^{2+} current, and this suppression was not reversed by the selective α_1-adrenoceptor antagonist, prazosin, the selective α_2-adrenoceptor antagonist, yohimbine, or the β-adrenoceptor antagonist, propranolol. In contrast, the increase in K^+ conductance induced by both noradrenaline and clonidine in the rat locus coeruleus neurons (Williams, Henderson & North, 1985) and in the guinea-pig submucous plexus neurons (North & Surprenant, 1985) was sensitive to a range of α_2-adrenoceptor antagonists. Thus α_2-adrenoceptor mediated inhibition of trans-mitter release from the nerve terminals of locus coeruleus neurons and other neurons may operate through an increase in K^+ conductance. This conclusion assumes that α_2-adrenoceptors on nerve terminals are coupled to K^+ channels, an assumption which may not be valid.

An increase in K^+ conductance may inhibit transmitter release in two ways; (1) by shortening the duration of the action potential and thereby decreasing depo-larization-induced Ca^{2+} entry or (2) by blocking action potential propagation in the nerve terminal network. Nakamura et al. (1981) have demonstrated that clonidine decreases the excitability of the nerve terminals of locus coeruleus neurons. In these experiments antidromic action potentials were recorded, intracellularly, following electrical stimulation of the terminal fields in the frontal cortex. Local infusion of clonidine at the site of stimulation increased the threshold for stimula-tion, an effect reversed by the α-adrenoceptor antagonist, phentolamine. However, it was always possible to overcome fully the effects of clonidine on the excitability of nerve terminals by increasing the stimulus strength, suggesting that clonidine does not induce failure of action potential propagation. Ryan et al. (1985) concluded that activation of prejunctional α_2-adrenoceptors may increase K^+ conductance and thereby alter the magnitude of the action potential induced depolarization.

Morita & North (1981a) reported that clonidine prevents action potential invasion into the soma of guinea-pig myenteric plexus neurons following focal stimulation of their cell processes (nerve terminals). However, in this study they did not rigou-rously investigate the site of action of this agent. In another series of experiments Morita and North (1981b) demonstrated that morphine and enkephalin, two agents which inhibit acetylcholine release from cholinergic nerve terminals by activating prejunctional opiate receptors, hyperpolarized the soma of guinea-pig myenteric plexus neurons when they were applied to the cell processes. This finding presumably reflects the electrotonic spread of hyperpolarization from the cell processes to the cell body. Both morphine and enkephalin inhibited the release of acetylcholine from the nerve terminals of myenteric ganglia neurons and prevented action potential propagation into the soma following focal stimulation of cell processes. Failure of action potential propagation occurred even when the drugs were applied focally at a distance from the stimulating electrode. Therefore, Morita and North (1981b) suggested that inhibition of action potential propagation, due to hyperpolarization of the axonal membrane, was the principal mechanism regulating transmitter release.

Further evidence supporting an effect on impulse conduction following activation of prejunctional α-adrenoceptors was obtained by comparing the action of α-adrenoceptor antagonists on noradrenaline release evoked by electrical stimulation with that evoked by high K^+ concentrations (Stjärne, 1981; Alberts, Bartfai & Stjärne. 1981). Alberts, Bartfai & Stjärne (1981) observed that in the guinea-pig vas deferens phentolamine increased tritiated noradrenaline release evoked both by nerve stimulation and by high K^+ concentrations. However, phentolamine caused a much greater increase in ^{3}H-noradrenaline overflow following nerve stimulation. Therefore, these authors concluded that activation of the prejunctional α-adrenoceptor acts mainly to restrict action potential invasion into nerve terminals but that activation also modifies depolarization-secretion coupling. This finding is equivocal, since it assumes that potassium depolarization of a varicosity is similar to that caused by nerve stimulation, which it certainly is not, potassium depolarization, in comparison with nerve stimulation, being slow in onset and prolonged.

In sympathetically innervated tissues drugs usually have to be applied to the entire nerve terminal arborization. One advantage of focal extracellular recording techniques is that drugs can be applied locally to a few varicosities enclosed within the recording electrode. Using this method it is clear that clonidine reduces and yohimbine increases the number of evoked EJCs recorded during trains of stimuli in the guinea-pig vas deferens without any apparent alteration in the nerve terminal impulse (Brock & Cunnane, 1990; Brock et al. 1990). Furthermore, no alteration in the configuration of the nerve terminal was noted when these drugs were applied to the whole tissue in concentrations which modify transmitter release. These results suggest that activation of prejunctional α-adrenoceptors interferes directly with the depolarization-secretion coupling mechanisms and not action potential propagation in the secretory terminals.

Studies on the mechanism of α_2-adrenoceptor mediated inhibition of transmitter release from postganglionic sympathetic nerves favour an effect on Ca^{2+} entry or sequestration in the nerve terminal as the fundamental step for several reasons. First, there is an inverse relationship between the inhibition of transmitter release produced by α-adrenoceptor agonists and the external Ca^{2+} concentration (Alberts, Bartfai & Stjärne, 1981). Indeed most, if not all, alterations in the experimental conditions which would increase Ca^{2+} entry during action potential propagation (e.g. treatment with K^+ channel blockers) appear to overcome the effects of α-adrenoceptor activation (see Starke, 1987). Second, increasing the stimulation frequency, a procedure which would be expected to increase the cytosolic Ca^{2+} levels, overcomes the effect of α_2-adrenoceptor activation (Alberts, Bartfai & Stjärne, 1981). These findings suggest that activation of α_2-adrenoceptors interferes with the availability of Ca^{2+} for stimulus-secretion coupling. A reduction in Ca^{2+} availability may be caused by a decrease in the duration of the action potential due to an increase in the K^+ conductance (see above). However, in the mouse vas deferens the inhibition of the electrically evoked EJP by clonidine was little affected by varying the external K^+ concentration (Illes & Dörge, 1985). Indeed, the action of clonidine was enhanced when the external K^+ concentration was increased.

Evidence supporting the view that activation of the α_2-adrenoceptor inhibits a voltage – dependent Ca^{2+} current has been provided by Göthert (1977). When rabbit hearts were perfused with a Ca^{2+} free solution containing a high K^+ concentration, the introduction of Ca^{2+} elicited tritiated noradrenaline release. This Ca^{2+} dependent transmitter release was blocked by the α-adrenoceptor agonist oxymetazoline, and this effect was antagonized by phentolamine. As no outward K^+ current would be expected at these high K^+ concentrations, the α-adrenoceptor agonist cannot be acting by increasing the outward movement of K^+. This finding is consistent with a decrease in a voltage – dependent Ca^{2+} conductance.

Much of the evidence supporting the role of Ca^{2+} in the mechanism underlying α_2-adrenoceptor activation would be consistent with the modulation of a step between Ca^{2+} entry and the release of transmitter. Schoffelmeer & Mulder (1983) have suggested that activation of the α_2-adrenoceptor may modify the sequestration of axoplasmic Ca^{2+} but this idea has not been thoroughly investigated.

Modulation of both adenylate cyclase and protein kinase C activity may be important in the regulation of transmitter release from sympathetic nerves (see Majewski, Ishac & Musgrave, 1988). Activation of α_2-adrenoceptors inhibits cyclic AMP formation in a number of different tissues, this inhibitory effect being mediated through a GTP binding protein (G_i) (Jakobs, 1985; Dolphin, 1987). Cyclic AMP acts primarily by activating a cyclic AMP – dependent kinase, which in turn phosphorylates a range of different proteins including enzymes and ion channels. It is known that treatment with cyclic-AMP analogues and phosphodiesterase inhibitors increases electrically evoked transmitter release in a number of different sympathetically innervated tissues (see Majewski, Ishac & Musgrave, 1988). In addition, activation of prejunctional β-adrenoceptors, which facilitates transmitter release from postganglionic sympathetic nerves (see Gillespie, 1980), stimulates adenylate cyclase in various tissues (see Levitzki, 1986). In support of β-adrenoceptor-mediated facilitation of transmitter release being generated by the activation of adenylate cyclase, the action of β-adrenoceptor agonists on the rabbit pulmonary artery was potentiated by phosphodiesterase inhibitors (Johnston & Majewski, 1986).

Evidence that inhibition of adenylate cyclase plays a role in α_2-adrenoceptor-mediated inhibition of transmitter release is more circumspect. If modulation of adenylate cyclase activity is involved, then treatment with cyclic-AMP analogues should modify the activity of α-adrenoceptor agonists and antagonists. However, in mouse atria the application of 8-bromo-cyclic AMP did not affect the inhibition of transmitter release induced by clonidine or the facilitatory effect of phentolamine (Johnston, Majewski & Musgrave, 1987). Similarly, the addition of phosphodiesterase inhibitors or pertussis toxin (i.e. to inactivate G_i), did not modify the effects of these drugs (Johnston, Majewski & Musgrave, 1987; Musgrave, Marley & Majewski, 1987). Conversely, Schoffelmeer, Hogenbloom & Mulder (1985) found that 8-bromo-cyclic AMP decreased the effects of both α-adrenoceptor agonists and phentolamine on electrically evoked noradrenaline release in rat brain slices, suggesting that the α-adrenoceptor may be linked to adenylate cyclase.

Activation of protein kinase C by phorbol esters (e.g. phorbol 12-myristate 13-acetate, PMA) has been shown to increase action potential induced noradrenaline release in a number of sympathetically innervated tissues (see Majewski, Ishac & Musgrave, 1988). Similarly, the protein kinase inhibitors polymyxin B and H-7 inhibit electrically evoked noradrenaline release in various tissues (see Majewski, Ishac & Musgrave, 1988). If α_2-adrenoceptor-mediated inhibition of transmitter release involves the activation of protein kinase C, phorbol esters would be expected to modify the action of the α-adrenoceptor agonists and antagonists. Treatment of mouse atria with PMA did not alter the effect of clonidine on electrically evoked noradrenaline release (Musgrave & Majewski, 1987, 1989). Similarly, the effect of both clonidine and yohimbine on noradrenaline release from electrically stimulated rabbit hippocampal slices was not altered by treatment with PMA (Allgaier, Von Kügelgen & Hertting. 1986). The protein kinase C inhibitor polymyxin B at concentrations which alone did not inhibit noradrenaline release, reduced the facilitatory effect of the α-adrenoceptor antagonist idazoxan on noradrenaline release from mouse atria (Musgrave & Majewski, 1989). However, polymyxin B did not alter the inhibitory effects of clonidine. Since polymyxin B also reduced the facilitatory effect of of the β-adrenoceptor agonist isoprenaline, high-frequency electrical stimulation and the K^+ channel blocking agent tetraethylammonium (Musgrave & Majewski, 1989; Majewski et al. 1990), it appears that protein kinase C may play a critical role in maintaining elevated levels of transmitter release under all these conditions. For this reason protein kinase C inhibitors may be inappropriate tools with which to investigate the possible role of protein kinase C in the mechanism of action of drugs that facilitate transmitter release (see Majewski et al. 1990). These findings suggest that activation of α_2-adrenoceptors does not lead to the modulation of protein kinase C activity.

In conclusion, the mechanism(s) involved in the inhibition of transmitter release produced by the activation of α_2-adrenoceptors remain poorly understood.

CONCLUSIONS

In this chapter, we have shown how electrophysiological techniques can be used to study neuroeffector transmission in the autonomic nervous system and in particular to investigate neurotransmitter release mechanisms. Electrophysiological methods allow transmitter release mechanisms to be studied on an impulse-by-impulse basis with a high degree of temporal resolution. The key points arising from such studies are that action potential evoked transmitter release from individual varicosities occurs intermittently and that only a single quantum is released when the nerve impulse activates the release mechanism in a varicosity. Focal extracellular recording techniques have shown that intermittence cannot be attributed to failure of impulse conduction in the terminal nerve network but rather to a low probability of release in the invaded varicosity. The reasons for the low probability of transmitter release are not known. Little conclusive information is available about the molecular events underlying transmitter release or indeed how release is modified by drugs. One future approach to this problem will be to study transmitter release mechanisms using a combination of electrophysiological and biochemical techniques.

Perhaps the most important point to emerge from this review is that a clear change in thinking has emerged about the way in which the activity of smooth muscle is controlled by nerves, and the mechanisms involved in the local regulation of transmitter release through prejunctional receptors. It may be necessary to reinterpret much of the basic pharmacological data obtained from bath application of drugs in view of recent morphological and pharmacological evidence which suggest that smooth muscle may be innervated, to a much greater degree than previously believed, by close neuroeffector junctions transmitting information through specialized junctional receptors. It may well be that exogenously applied substances 'see' a different population of receptors on the smooth muscle. The picture is further complicated in that the pattern of nerve impulse traffic to different smooth muscles and glands in the periphery under different physiological circumstances, is not known, although we know that complex, frequency dependent responses, can be elicited through the release of several transmitters from the same nerve terminal.

There are however, disadvantages in the electrophysiological approach. Considerable doubt remains about the chemical identity of the neurotransmitter generating junction potentials at many neuroeffector junctions in the autonomic nervous system. A great deal of information is available about the characteristic features of junction potentials in smooth muscle but this comes largely from studies using *in vitro* preparations. Necessarily, nerves are synchronously activated when smooth muscles are stimulated through extrinsic nerves or by field stimulation, resulting in large junction potentials. However, the challenge in the immediate future will be to determine the physiological role of junction potentials *in vivo*. There is relatively little *in vivo* data available on the electrophysiology of transmitter release mechanisms in smooth muscle although this is an area which has recently begun to be addressed (Neild & Keef, 1985; Meehan *et al.* 1990). It will be important in the next decade to compare the fundamental knowledge obtained *in vitro* studies with that obtained in suitable *in vivo* preparations.

REFERENCES

Abe, Y. & Tomita, T. (1968). Cable properties of smooth muscle. *Journal of Physiology* **196**, 87–100.
Adams, P.R., Brown, D.A. & Jones, S.W. (1983) Substance P inhibits the M-current in bullfrog sympathetic neurons. *British Journal of Pharmacology* **79**, 330–333.

Aickin, C.C. & Brading, A.F. (1982) Measurement of intracellular chloride in guinea-pig vas deferens by ion analysis, chloride efflux and micro-electrodes. *Journal of Physiology* 326, 139–154.

Alberts, P., Bartfai, T. & Stjärne, L. (1981). Site(s) and ionic basis of α-autoinhibition and facilitation of [3H] noradrenaline secretion in the guinea-pig vas deferens. *Journal of Physiology*, 312, 297–334.

Allcorn, R.J., Cunnane, T.C. & Kirkpatrick, K. (1986). Actions of α,β-methylene ATP and 6-hydroxy-dopamine on sympathetic neurotransmission in the vas deferens of the guinea-pig, rat and mouse: support for co-transmission. *British Journal of Pharmacology*, 89, 647–659.

Allgaier, C., Von Kügelgen, O. & Hertting, G. (1986). Enhancement of noradrenaline release by 12-o-tetradecanoyl phorbal-13-acetate, an activator of protein kinase C. *European Journal of Pharmacology*, 129, 389–392.

Ambache, N. & Zar, M.A. (1971). Evidence against adrenergic motor transmission in the guinea-pig vas deferens. *Journal of Physiology*, 216, 359–389.

Ambache, N., Dunk, L.P., Verney, J. & Zar M.A. (1972). Inhibition of post-ganglionic motor transmission in the vas deferens by indirectly acting sympathomimetic drugs. *Journal of Physiology*, 227, 433–456.

Amedee, T & Large, W.A. (1989) Microelectrode study on the ionic mechanisms which contribute to the noradrenaline-induced depolarization in isolated cells of the rabbit portal vein. *British Journal of Pharmacology* 97, 1331–1337.

Anderson, C.R. & Stevens, C.F. (1973). Voltage-clamp analysis of acetylcholine produced end-plate current fluctuations at frog neuromuscular junction. *Journal of Physiology* 235, 655–693.

Anderson, D.C., King, S.C. & Parsons, S.M. (1983) Pharmacological characterisation of the acetyl-choline active transport system in purified Torpedo electric organ synaptic vesicles. *Molecular Pharmacology* 24, 48–54.

Anwyl, R. & Usherwood, P.N.R. (1974). Voltage clamp studies of glutamate synapse. *Nature* 252, 591–593.

Araki, T. & Terzuolo, C.A. (1962). Membrane currents in spinal motoneurones associated with the action potential and synaptic activity. *Journal of Neurophysiology* 25, 772–789.

Aston-Jones, G. Segal, M. & Bloom, F.E. (1980). Brain aminergic axons exhibit marked variability in conduction velocity. *Brain Research*, 195, 215–222.

Åstrand, P. & Stjärne, L. (1989a). On the secretory activity of single varicosities in the sympathetic nerves innervating the rat tail artery. *Journal of Physiology* 409, 207–220.

Åstrand, P. & Stjärne, L. (1989b) ATP as a sympathetic co-transmitter in rat vasomotor nerves – further evidence that individual release sites respond to nerve impulses by intermittent release of single quanta. *Acta Physiologica Scandinavica.* 136, 355–365.

Åstrand, P., Brock, J.A. & Cunnane, T.C. (1988) Time course of transmitter action at the sympathetic neuroeffector junction in vascular and non vascular smooth muscle. *Journal of Physiology.* 401, 657–670.

Bacq, Z.M. (1934). La pharmacologie due système nerveux autonome, et particulièrement du sympathique, d'après la théorie neurohumorale. *Annales de Physiol*, 10, 467.

Bacq, Z.M. (1975). *Chemical Transmission of Nerve Impulses: a Historical Survey*. Pergamon Press: Oxford, New York, Toronto, Sydney, Braunschweig.

Banks, P. (1966). The release of adenosine triphosphate catebolites during the secretion of catecholamines by bovine adrenal medulla. *Biochemical Journal*, 101, 536–541.

Barger, G. & Dale, H.H. (1910–11). Chemical structure and sympathomimetic action of amines. *Journal of Physiology*, 41, 19–59.

Baron, S.A., Jaffe, B.M. & Gintzler, A.R. (1983) Release of substance P from the enteric nervous system: direct quantification and characterisation. *Journal of Pharmacology and Experimental Therapeutics* 227, 365–370.

Bauer, V. & Kuriyama, H. (1982a) Evidence for non-cholinergic non-adrenergic transmission in the guinea-pig ileum. *Journal of Physiology* 330, 95–100.

Bauer, V. & Kuriyama, H. (1982b) The nature of non-cholinergic non-adrenergic transmission in longitudinal and circular muscle of the guinea-pig ileum. *Journal of Physiology* 332b, 375–391.

Beani, L.C., Bianchi, C. & Crema, A. (1971) Vagal non-cholinergic inhibition of guinea-pig stomach. *Journal of Physiology* 217, 259–279.

Beck, C.S. & Osa, T. (1971) Membrane activity in guinea-pig gastric sling muscle: a nerve-dependent phenomenon. *American Journal of Physiology* 220, 1397–1403.

Bell, C. (1969) Transmission from vasoconstrictor and vasodilator nerves to single smooth muscle cells of the guinea-pig uterine artery. *Journal of Physiology* 205, 695–708.

Bell, C. & Vogt, M. (1971). Release of endogenous noradrenaline from an isolated muscular artery. *Journal of Physiology*, 215, 509–520.

Benham, C.D. (1989) ATP activated channels gate calcium entry in single smooth muscle cells dissociated from rabbit ear artery. *Journal of Physiology* **419**, 689–701.

Benham, C.D. & Tsien, R.W. (1987). A novel receptor-operated Ca^{2+} permeable channel activated by ATP in smooth muscle. *Nature*, **328**, 275–278.

Benham, C.D., Bolton, T.B. & Lang, R.J. (1985) Acetylcholine activates an inward current in single mammalian smooth muscle cells. *Nature* **316**, 345–347.

Benham, C.D., Bolton, T.B., Byrne, N.G. & Large, W.A. (1987). Action of externally applied adenosine triphosphate on single smooth muscle cells dispersed from rabbit ear artery. *Journal of Physiology* **387**, 473–488.

Bennett, M.R. (1966) Transmission from intramural excitatory nerves to the smooth muscle cells of the guinea-pig taenia coli. *Journal of Physiology* **182**, 132–147.

Bennett, M.R. (1967). The effect of intracellular current pulses in smooth muscle cells of the guinea-pig vas deferens at rest and during transmission. *Journal of General Physiology* **50**, 2459–2470.

Bennett, M.R. (1972). *Autonomic Neuromuscular Transmission.* Cambridge University Press: Cambridge.

Bennett, M.R. & Florin, T. (1974). A statistical analysis of the release of acetylcholine at newly formed synapses in striated muscle. *Journal of Physiology*, **238**, 93–107.

Bennett, M.R. & Lavidis, N.A. (1979). The effect of calcium ions on the secretion of quanta evoked by an impulse at nerve terminal release sites. *Journal of General Physiology*, **74**, 429–456.

Bennett, M.R. & Pettigrew, A.G. (1975). The formation of synapses in amphibian striated muscle during development. *Journal of Physiology*, **252**, 203–239.

Bennett, M.R., Burnstock, G. & Holman, M.E. (1966) Transmission from perivascular inhibitory nerves to the guinea-pig taenia coli. *Journal of Physiology* **182**, 527–540.

Bennett, M.R., Florin, T. & Hall, R. (1975). The effect of calcium ions on the binomial statistic parameters which control acetylcholine release at synapses in striated muscle. *Journal of Physiology*, **247**, 429–446.

Bennett, T. (1969) Nerve mediated excitation and inhibition of the smooth muscle cells of the avian gizzard. *Journal of Physiology* **204**, 669–686.

Bevan, J.A., Chesher, G.B. & Su, C. (1969). Release of adrenergic transmitter from terminal plexus in artery. *Agents and Actions*, **1**, 20–26.

Birmingham, A.T. & Wilson, A.B. (1963). Preganglionic and postganglionic stimulation of the guinea-pig vas deferens preparation. *British Journal of Pharmacology and Chemotherapy*, **21**, 569–580.

Björkroth, U. (1983) Inhibition of smooth muscle contractions induced by capsaicin and electrical stimulation by a substance P antagonist. *Acta Physiologica Scandinavica* Suppl. **515**, 11–16.

Blackman, J.G. & Purves, R.D. (1969). Intracellular recordings from ganglia of the thoracic sympathetic chain in guinea-pig. *Journal of Physiology*, **203**, 173–198.

Blakeley, A.G.H. & Cunnane, T.C. (1978). Is the vesicle the quantum of sympathetic transmission? *Journal of Physiology* **280**, 30–31P.

Blakeley, A.G.H. & Cunnane, T.C. (1979a). The packeted release of transmitter from the sympathetic nerves of the guinea-pig vas deferens: an electrophysiological study. *Journal of Physiology*, **296**, 85–96.

Blakeley, A.G.H. & Cunnane, T.C. (1979b). Packeted transmitter release in the mouse vas deferens; an electrophysiological study. *Journal of Physiology* **295**, 44–45P.

Blakeley, A.G.H., Cunnane, T.C. & Muir, T.C. (1979) The electrical responses of the rabbit rectococcygeus following extrinsic parasympathetic nerve stimulation. *Journal of Physiology* **293**, 539–550.

Blakeley, A.G.H., Brown, D.A., Cunnane, T.C., French, A.M., McGrath, J.C. & Scott, N.C. (1981). Effects of nifedipine on electrical and mechanical responses of the rat and guinea-pig vas deferens. *Nature*, **294**, 759–760.

Blakeley, A.G.H., Cunnane, T.C. & Petersen, S.A. (1982). Local regulation of transmitter release from rodent sympathetic nerve terminals? *Journal of Physiology* **325**, 93–109.

Blakeley, A.G.H., Mathie, A. & Petersen, S.A. (1984). Facilitation at single release sites of a sympathetic neuroeffector junction in the mouse. *Journal of Physiology* **349**, 57–71.

Bolton, T.B. (1979) Mechanism of action of transmitters and other substances on smooth muscle. *Physiological Reviews* **59**, 606–717.

Bolton, T.B. (1989) Electrophysiology of the intestinal musculature. In *Handbook of Physiology* Section 6: The gastrointestinal system Vol 1. Motility and Circulation Part 1. Oxford University Press, New York. 217–250.

Bolton, T.B. & Large, W.A. (1986). Review Article: Are junction potentials essential? Dual mechanism of smooth muscle cell activation by transmitter released from autonomic nerves. *Quarterly Journal of Experimental Physiology* **71**, 1–28.

Bolton, T.B., Tomita, T. & Vassort, G. (1981). Voltage clamp and the measurement of ionic conductances in smooth muscle. In: *Smooth Muscle: An Assessment of Current Knowledge* pp. 47–63. Eds.: Bülbring, E., Brading, A.F., Jones, A.W. & Tomita, T. Edward Arnold, London.

Bornstein, J.C. (1978). Spontaneous multiquantal release at synapses in the guinea-pig hypogastric ganglia: evidence that release can occur in bursts. *Journal of Physiology*, **282**, 375–398.

Bowman, W.C. (1985). The neuromuscular junction: recent developments. *European Journal of Anaesthesiology* **2**, 59–93.

Boyd, H., Chang, V. & Rand, M.J. (1960) The anticholinesterase activity of some antiadrenaline drugs. *British Journal of Pharmacology and Chemotherapy.* **15**, 525–531.

Boyd, I.A. & Martin, A.R. (1956). The end-plate potential in mammalian muscle. *Journal of Physiology*, **132**, 74–91.

Bozler, E. (1948). Conduction, automaticity and tonus of visceral muscles. *Experientia* **4**, 213–218.

Brading, A.F. & Mostwin, J.L. (1989) Electrical and mechanical responses of guinea-pig bladder muscle to nerve stimulation. *British Journal of Pharmacology*, **98**, 1083–1090.

Brigant, J.L. & Mallart, A. (1982). Presynaptic currents in mouse motor endings. *Journal of Physiology*, **333**, 619–636.

Brock, J.A. (1988). *Analysis of factors controlling transmitter release from sympathetic nerves.* D.Phil Thesis. University of Oxford.

Brock, J.A. & Cunnane, T.C. (1987). Relationship between the nerve action potential and transmitter release from sympathetic postganglionic nerve terminals. *Nature* **326**, 605–607.

Brock, J.A. & Cunnane, T.C. (1988a). Electrical activity at the sympathetic neuroeffector junction in the guinea-pig vas deferens. *Journal of Physiology* **399**, 607–632.

Brock, J.A. & Cunnane, T.C. (1988b). Studies on the mode of action of bretylium and guanethidine in post-ganglionic sympathetic nerve fibres. Naunyn-Schmiedeberg's Archives of Pharmacology **338**, 504–509.

Brock, J.A. & Cunnane, T.C. (1989). Local application of clonidine, yohimbine and tyramine to sympathetic nerve terminals: role of prejunctional a-adrenoceptors. *British Journal of Pharmacology.* **98**, 868.P

Brock, J.A. & Cunnane T.C. (1990). Transmitter release from sympathetic nerve terminals on an impulse-by-impulse basis and presynaptic receptors. *Annals of the New York Academy of Sciences* **604**, 176–186.

Brock, J.A., Cunnane, T.C., Starke, K. & Wardell, C.F. (1990). a_2-adrenoceptor-mediated autoinhibition of sympathetic transmitter release in guinea-pig vas deferens studied by intracellular and focal extracellular recording of junction potentials and currents Naunyn-Schmiedeberg's Archives of Pharmacology **342**, 45–52.

Brooks, C.Mc.C. & Eccles, J.C. (1947). Electrical investigation of the monosynaptic pathway through the spinal cord. *Journal of Neurophysiology*, **10**, 251–274.

Brown, D.A., & Adams, P.R. (1980) Muscarinic suppression of a novel voltage-sensitive K^+ current in a vertebrate neurone. *Nature* **283**, 673–676.

Brown, G.L. & Holmes, O. (1956). The effects of activity on mammalian nerve fibres of low conduction velocity. *Proceedings of the Royal Society* B, **145**, 1–14.

Brown, G.L., Dale, H.H. & Feldberg, W. (1936). Reactions of the normal mammalian muscle to acetylcholine and to eserine. *Journal of Physiology* **87**, 394–424.

Bülbring, E. & Tomita, T. (1967) Properties of the inhibitory potential of smooth muscle as observed in the response to field stimulation of the guinea-pig taenia coli. *Journal of Physiology* **189**, 299–315.

Burnstock, G. (1970). Structure of smooth muscle and its innervation. In, *Smooth Muscle*. ed. Bülbring, E., Brading, A.F., Jones, A.W. & Tomita, T. Edward Arnold: London. pp. 1–70.

Burnstock G (1972) Purinergic nerves *Pharmacological Reviews* **24**, 509–581.

Burnstock G (1979) Past and current evidence for the purinergic nerve hypothesis. In. *Physiological and Regulatory Functions of Adenosine and Adenine Nucleotides*. ed. Baer, H.P. & Drummond, G.I. Raven Press, New York. 3–32.

Burnstock G (1981) Neurotransmitters and trophic Factors in the autonomic nervous system. *Journal of Physiology* **313**, 1–35.

Burnstock, G. (1986). The changing face of autonomic neurotransmission. Acta *Physiologica Scandinavica*, **126**, 67–91.

Burnstock, G., Cocks, T., Crowe, R. & Kasakov, L. (1978) Purinergic innervation of the guinea-pig bladder. *British Journal of Pharmacology* **63**, 125–138.

Burnstock, G. & Holman, M.E. (1961). The transmission of excitation from autonomic nerve to smooth muscle. *Journal of Physiology*, **155**, 115–133.

Burnstock, G. & Holman, M.E. (1962a). Spontaneous potentials at sympathetic nerve endings in smooth muscle. *Journal of Physiology*, **160**, 446-460.

Burnstock, G. & Holman, M.E. (1962b). Effect of denervation and of reserpine treatment on transmission at sympathetic nerve endings. *Journal of Physiology*, **160**, 461-496.

Burnstock, G. & Holman, M.E. (1964). An electrophysiological investigation of the actions of some autonomic blocking drugs on transmission in the guinea-pig vas deferens. *British Journal of Pharmacology* **23**, 600-612.

Burnstock, G. & Holman, M.E. (1966). Junction potentials at adrenergic synapses. *Pharmacological Reviews* **18**, 481-493.

Burnstock, G. & Kennedy, C. (1985). Is there a basis for distinguishing two types of P_2-purinoceptor? *General Pharmacology*, **16**, 433-440.

Burnstock, G., Holman, M.E. & Kuriyama, H. (1964). Facilitation of transmission from autonomic nerve to smooth muscle of guinea-pig vas deferens. *Journal of Physiology* **172**, 31-49.

Byrne, N.G. & Large, W.A. (1984) Comparison of the biphasic excitatory junction potential with membrane responses evoked by bath applied adenosine triphosphate and noradrenaline in the rat anococcygeus muscle. *British Journal of Pharmacology* **83**, 751-758.

Byrne, N.G. & Large, W.A. (1985) Evidence for two mechanisms of depolarization associated with a_1-adrenoceptor activation in the rat anococcygeus muscle. *British Journal of Pharmacology* **86**, 711-712.

Byrne, N.G. & Large, W.A. (1986). the effect of α,β-methylene ATP on the depolarization evoked by noradrenaline (γ-adrenoceptor response) and ATP in the immature rat basilar artery. *British Journal of Pharmacology*, **88**, 6-8.

Byrne, N.G. & Large, W.A. (1987) Action of noradrenaline on single smooth muscle cells freshly dispersed from the rat anococcygeus muscle. *Journal of Physiology* **389**, 513-525.

Byrne, N.G. & Large, W.A. (1988) Membrane ionic mechanisms activated by noradrenaline in cells isolated from the rabbit portal vein. *Journal of Physiology* **404**, 557-573.

Byrne, N.G. & Muir, T.C. (1984) Electrical and mechanical responses of the bovine retractor penis to nerve stimulation and to drugs. *Journal of Autonomic Pharmacology* **4**, 261-271.

Byrne, N.G., Hirst, G.D.S. & Large, W.A. (1985). Electrophysiological analysis of the nature of adrenoceptors in the rat basilar artery during development. *British Journal of Pharmacology*, **86**, 217-227.

Bywater, R.A.R & Taylor, G.S. (1980). The passive membrane properties and excitatory junction potentials of the guinea-pig vas deferens. *Journal of Physiology*, **300**, 303-316.

Bywater, R.A.R & Taylor, G.S. (1983) Non-cholinergic fast and slow poststimulus depolarizations in the guinea-pig ileum. *Journal of Physiology* **340**, 47-56.

Bywater, R.A.R & Taylor, G.S. (1986a) Cholinergic neurotransmission to the circular muscle of the guinea-pig ileum. Proceedings of IUPS 16:

Bywater, R.A.R & Taylor, G.S. (1986b) Non-cholinergic excitatory and inhibitory junction potentials in the circular smooth muscle of the guinea-pig ileum. *Journal of Physiology* **374**, 153-164.

Bywater, R.A.R & Taylor, G.S. (1989) Excitatory neurotransmission of the circular muscle in the guinea-pig ileum. *Proceedings of The Australian Physiological and Pharmacological Society* **20**, 128P.

Bywater, R.A.R., Holman, M.E. & Taylor, G.S. (1981) Atropine-resistant depolarization in the guinea-pig small intestine. *Journal of Physiology* **316**, 369-378.

Bywater, R.A.R., Campbell, G., Edwards, F.R., Hirst, G.D.S. & O'Shea J.E. (1989) The effect of vagal stimulation and applied acetylcholine on the sinus venosus of the toad. *Journal of Physiology* **415**, 35-56.

Cameron, A.R. & Kirkpatrick, C.T. (1977) A study of excitatory neuromuscular transmission in the bovine trachea. *Journal of Physiology* **270**, 733-745.

Cameron, A.R. Johnston, C.F. Kirkpatrick, C.T. & Kirkpatrick, M.C. (1983) The quest for the inhibitory neurotransmitter in bovine tracheal smooth muscle. *Quarterly Journal of Experimental Physiology* **68**, 413-426.

Campbell, G.D., Edwards, F.R., Hirst, G.D.S. & O'Shea, J.E. (1989) Effects of vagal stimulation and applied acetylcholine on pacemaker potentials in the guinea-pig heart. *Journal of Physiology* **415**, 57-68.

Cassell, J.F., McLachlan, E. & Sittiracha, T. (1988). The effect of temperature on neuromuscular transmission in the main caudal artery of the rat. *Journal of Physiology* **397**, 31-49.

Cheung, D.W. (1982a). Spontaneous and evoked excitatory junction potentials in rat tail arteries. *Journal of Physiology* **328**, 449-459.

Cheung, D.W. (1982b). Two components in the cellular response of rat tail arteries to nerve stimulation. *Journal of Physiology* **328**, 461-468.

Cheung, D.W. (1985) An electrophysiological study of α-adrenoceptor mediated excitation-contraction coupling in the smooth muscle cells of the rat saphenous vein. *British Journal of Pharmacology* **84**, 265-271

Cheung, D.W. (1990) Synaptic transmission in the guinea-pig vas deferens: the role of nerve action potentials. *Neuroscience* **37**, 127-134.

Cheung, D.W. & Daniel, E.E. (1980) Comparative study of the smooth muscle layers of the rabbit duodenum. *Journal of Physiology* **309**, 13-27.

Cheung, D.W. & Fujioka, M. (1986). Inhibition of the junction potential in the guinea-pig saphenous vein by ANAPP3. *British Journal of Pharmacology* **89**, 3-5.

Coburn, R.F. (1984) Neural coordination of excitation of ferret trachealis muscle. *American Journal of Physiology* **246**, C459-466.

Cohen, I., Kita, H. & Van der Kloot, W. (1974a). The intervals between miniature end-plate potentials in the frog are unlikely to be independently or exponentially distributed. *Journal of Physiology*, **236**, 327-339.

Cohen, I., Kita, H. & Van der Kloot, W. (1974b). The stochastic properties of spontaneous quantal release of transmitter at the frog neuromuscular junction. *Journal of Physiology*, **236**, 341-361.

Cohen, I., Kita, H. & Van der Kloot, W. (1974c). Stochastic properties of spontaneous transmitter release at the crayfish neuromuscular junction. *Journal of Physiology*, **236**, 363-371.

Costa, M., Cuello, A.C., Furness, J.B. & Franco, R. (1980) Distribution of enteric neurons showing immunoreactivity for substance P in the guinea-pig ileum. *Neuroscience* **5**, 323-331.

Costa, M., Furness, J.B., Pullin, C.O. & Bornstein, J. (1985) Substance P enteric neurones mediate non-cholinergic transmission to the circular muscle of the guinea-pig intestine. *Naunyn-Schmiedeberg's Archives of Pharmacology* **328**, 446-453.

Costa, M. Furness, J.B. & Humphreys, C.M.S. (1986) Apamin distinguishes two types of relaxation mediated by enteric nerves in the guinea-pig gastrointestinal tract. *Naunyn-Schmiedeberg's Archives of Pharmacology*.

Creed, K.E. (1975) Membrane properties of the smooth muscle cells of the rat anococcygeus. *Journal of Physiology* **245**, 49-62.

Creed, K.E. & Gillespie, J.S. (1977) Some electrical properties of the rabbit anococcygeus muscle and a comparison of the effects of inhibitory nerve stimulation in the rat and rabbit. *Journal of Physiology* **273**, 137-153

Creed, K.E., Gillespie, J.S. & Muir, T.C. (1975) The electrical basis of excitation and inhibition in the rat anococcygeus muscle. *Journal of Physiology* **245**, 33-47.

Creed, K.E., Ishikawa, S. & Ito, Y. (1983) Electrical and mechanical activity recorded from rabbit urinary bladder in response to nerve stimulation. *Journal of Physiology* **338**, 149-164.

Cull-Candy, S.G. (1976). L-glutamate receptors in locust muscle fibres. In: *Motor Innervation of Muscle*, pp. 263-288. Ed.: Thesleff, S. Academic Press, London.

Cull-Candy, S.G. (1984). New ways of looking at synaptic channels: noise analysis and patch-clamp recording. In: *Recent Advances in Physiology*, Vol 10 pp. 1-28. Ed.: Baker, P.F. Churchill Livingstone, London.

Cunnane, T.C. (1984). The mechanism of neurotransmitter release from sympathetic nerves. *Trends in Neuroscience* **7**, 248-253.

Cunnane, T.C. & Manchanda, R. (1988). Electrophysiological analysis of the inactivation of sympathetic transmitter in the guinea-pig vas deferens. *Journal of Physiology*, **404**, 349-364.

Cunnane, T.C. & Manchanda, R. (1989a). Simultaneous intracellular and focal extracellular recording of junction potentials and currents, and the time course of quantal transmitter action in rodent vas deferens. *Neuroscience*, **30**, 563-575.

Cunnane, T.C. & Manchanda, R. (1989b). Effects of reserpine pre-treatment on neuroeffector transmission in the vas deferens. *Clinical and Experimental Pharmacology and Physiology*, **16**, 451-455.

Cunnane, T.C. & Manchanda, R. (1990). On the factors which determine the time courses of junction potentials in the guinea-pig vas deferens. *Neuroscience* **37**, 507-516.

Cunnane, T.C. & Stjärne, L. (1982). Secretion of transmitter from individual varicosities of guinea-pig and mouse vas deferens: all-or-none and extremely intermittent. *Neuroscience*, **7**, 2565-2576.

Cunnane, T.C. & Stjärne, L. (1984a). Transmitter secretion from individual varicosities of guinea-pig and mouse vas deferens: highly intermittent and monoquantal. *Neuroscience*, **13**, 1-20.

Cunnane, T.C. & Stjärne, L. (1984b). Frequency dependent intermittency and ionic basis of impulse conduction in postganglionic sympathetic fibres of guinea-pig vas deferens. *Neuroscience*, **11**, 211-229.

Cunnane, T.C., Muir, T.C. & Wardle, K.A. (1987). Is co-transmission involved in the excitatory responses of the rat anococcygeus muscle? *British Journal of Pharmacology* **92**, 39-46.

Curtis, D.R. & Eccles, J.C. (1959). The time courses of excitatory and inhibitory synaptic actions. *Journal of Physiology* 145, 529–546.

Dahlström, A., Häggendal, J. & Hökfelt, T. (1966). The noradrenaline content of the varicosities of sympathetic adrenergic nerve terminals in the rat. *Acta Physiologica Scandinavica*, 67, 289–294.

Dale, H.H. (1914) The action of certain esters and ethers of choline and relation to muscarine. *Journal of Pharmacology and Experimental therapeutics*. 6, 147–190.

Dale, H.H. (1933). Nomenclature of fibres in the autonomic system and their effects. *Journal of Physiology*, 80, 10P–11P.

Dale, H.H., Feldberg, W. & Vogt, M. (1936). Release of acetylcholine at voluntary motor nerve endings. *Journal of Physiology*, 86, 353–380.

Daniel, E.E. (1985) Nonadrenergic, noncholinergic (NANC) neuronal inhibitory interactions with smooth muscle. In *Calcium and Contractility: Smooth Muscle* ed. Grover, A.K. & Daniel, E.E. Humana Press, Clifton. 385–426.

Daniel, E.E. (1987). Gap junctions in smooth muscle. In: *Cell-to-Cell Communication*, pp 149–186. Ed.: De Mello, W.C. Plenum, New York.

Daniel, E.E., Helmy-Elkholy, A., Jager, L.P., Kannan, M.S. (1983) Neither a purine nor VIP is the mediator of inhibitory nerves of opossum oesophageal smooth muscle. *Journal of Physiology* 336, 243–260.

Davison, J.S., & Najafi-Farashah, A. (1981) The effect of repetitive electrical stimulation on the response of the guinea-pig ileal longitudinal muscle to single pulse stimulation. *Journal of Physiology* 310, 58P–59P.

De Potter, W.P., De Schaepdryver, A.F., Moerman, E.J. & Smith, A..D (1969). Evidence for the release of vesicle-proteins together with noradrenaline upon stimulation of the splenic nerve. *Journal of Physiology* 204, 102–104P.

De Potter, W.P., Chubb, I.W., Put, A. & Schaepdryver, A.F. (1971). Facilitation of the release of noradrenaline and dopamine β-hydroxylase at low stimulation frequencies by α-blocking agents. *Archives Internationales Pharmacodynamie et de Thérapie Arch*, 193, 191–197.

De Robertis, E. (1967). Ultrastructure and cytochemistry of the synaptic region. *Science*, 156, 907–914.

De Robertis, E.D.P. & Bennett, H.S. (1955). Some features of the sub-microscopic morphology of synapses in frog and earthworm. *Journal of Biophysical and Biochemical Cytology*, 1, 47–58.

Del Castillo, J. & Katz, B. (1954a). Quantal components of the end-plate potential. *Journal of Physiology*, 124, 560–573.

Del Castillo, J. & Katz, B. (1954b). Changes in end-plate activity produced by pre-synaptic polarization. *Journal of Physiology*, 124, 586–604.

Del Castillo, J. & Katz, B. (1955). Local activity at a depolarized nerve-muscle junction. *Journal of Physiology*, 128, 396–411.

Del Castillo, J. & Katz, B. (1956). Localization of active sports within the neuromuscular junction of the frog. *Journal of Physiology*, 132, 630–646.

Dennis, M.J., Harris, A.J. & Kuffler, S.W. (1971). Synaptic transmission and its duplication by focally applied acetylcholine in parasympathetic neurons in the heart of the frog. *Proceedings of the Royal Society Series B* 177, 509–539.

Dixon, W.E. (1906). Vagus inhibition. *British Medical Journal*, 2, 1807.

Dolphin, A.C. (1987) Nucleotide binding proteins in signal transduction and disease. *Trends in Pharmacological Science* 10, 53–57.

Domoto, T., Jary, J., Berezin, I., Fox, J.E. & Daniel, E.E. (1983) Does substance P comediate with acetylcholine in the nerves of opossum oesophageal muscularis mucosa? *American Journal of Physiology* 245, G19–28.

Donnerer, J., Bartho, P., Holzer, P. & Lembeck, F. (1984) Intestinal peristalsis associated with the release of immunoreactive substance P. *Neuroscience* 11, 913–918.

Douglas, W.W. & Poisner, A.M. (1966). Evidence that the secreting adrenal chromaffin cell releases catecholamines directly from ATP -rich granules. *Journal of Physiology*, 183, 236–248.

Douglas, W.W., Poisner, A.M. & Rubin, R.P. (1965). Efflux of adenine nucleotides from perfused adrenal glands exposed to nicotine and other chromaffin stimulants. *Journal of Physiology*, 179, 130–137.

Droogmans, G., Raemaekers, L. & Casteels, R. (1977) Electro- and pharmacomechanical coupling in the smooth muscle cells of the rabbit ear artery. *Journal of General Physiology* 70, 129–148.

Dudel, J. (1963). Presynaptic inhibition of the excitatory nerve terminal in the neuromuscular junction of the crayfish. *Pflügers Archiv*, 277, 537–557.

Dudel, J. & Kuffler, S.W. (1961a). The quantal nature of transmission and spontaneous miniature potentials at the crayfish neuromuscular junction. *Journal of Physiology*, 155, 514–529.

Dudel, J. & Kuffler, S.W. (1961b). Presynaptic inhibition at the crayfish neuromuscular junction. *Journal of Physiology*, **155**, 543–562.

Dun, W.R., Daly, C.J., McGrath, J.C. & Wilson V.G. (1991). The effect of nifedipine on α_2-adrenoceptor-mediated contractions in several blood vessels from the rabbit. *British Journal of Pharmacology* **103**, 1493–1499.

Dunant, Y. & Israël, M. (1985). The release of acetylcholine. *Scientific American*, **252**, 58–66.

Dunlap, K. & Fischbach, G.D. (1981). Neurotransmitters decrease the calcium conductance activated by depolarization of embryonic chick sensory neurones. *Journal of Physiology*, **317**, 519–535.

Duval, N., Hicks, P.E. & Langer, S.Z. (1986). Reserpine-resistant responses to nerve stimulation in the cat nictitating membrane involve the release of newly synthesized noradrenaline. *European Journal of Pharmacology* **122**, 93–1]101.

Eccles, J.C. (1964). *The Physiology of Synapses*. Springer-Verlag: Berlin, Göttingen, Heidelberg.

Eccles, J.C. & MacFarlane, W.V. (1949). Action of anticholinesterases on end-plate potential of frog muscle. *Journal of Neurophysiology*, **12**, 59–80.

Eccles, J.C. & Magladery, J.W. (1937). The excitation and response of smooth muscle. *Journal of Physiology* **90**, 31–67.

Eccles, J.C. Katz, B. & Kuffler, S.W. (1941). Nature of the 'end-plate potential' in curarized muscle. *Journal of Neurophysiology* **4**, 362–387.

Eccles, J.C., Katz, B. & Kuffler, S.W. (1942). Effect of eserine on neuromuscular transmission. *Journal of Neurophysiology*, **5**, 211–230.

Edwards, F.A., Konnerth, A. & Sakmann, B. (1990) Quantal analysis of inhibitory synaptic transmission in the dentate gyrus of rat hippocampal slices: a patch-clamp study. *Journal of Physiology* **430**, 213–249.

Edwards, C., Dolezal, V., Tucek, S., Zemkova, H. & Vyskocil, F. (1985). Is an acetylcholine transport system responsible for nonquantal release of acetylcholine at the rodent myoneural junction? *Proceedings of the National Academy of Science*, **82**, 3514–3518.

Elliott, T.R. (1904). On the action of adrenalin. *Journal of Physiology*, **31**, xx–xi (Proceedings of the Physiological Society).

Erxleben, C. & Kriebel, M.E. (1988). Subunit composition of the spontaneous miniature end-plate currents at the mouse neuromuscular junction. *Journal of Physiology*, **400**, 659–676.

Euler, U.S. von (1946a). The presence of a substance with sympathin E properties in spleen extracts. *Acta Physiologica Scandinavica*, **11**, 168–186.

Euler, U.S. Von (1946b). A specific sympathomimetic ergone in adrenergic nerve fibres (sympathin) and its relations to adrenaline and nor-adrenaline. *Acta Physiologica Scandinavica* **12**, 73–97.

Euler, U.S. von (1981). Historical perspective: growth and impact of the concept of chemical neurotransmission. In. *Chemical Neurotransmission: 75 Years*. ed. Stjärne, L. Hedqvist, P. Lagercrantz, H. & Wennmalm, A. Academic Press: London, New York Toronto, Sydney, San Francisco. pp. 2–12.

Euler, U.S. Von & Hedqvist, P. (1974). Phenoxybenzamine blockade of the enhancing effect of substance P on the twitch induced by transmural nerve stimulation in the guinea pig vas deferens. *Acta Physiologica Scandinavica* **92**, 283–285.

Euler, U.S. Von & Hedqvist, P. (1975). Evidence for an α- and β_2-receptor mediated inhibition of the twitch response in the guinea pig vas deferens by noradrenaline. *Acta Physiologica Scandinavica* **93**, 572–573.

Falck, B., Hillarp, N.-Å., Thieme, G. & Torp, A. (1962). Fluorescence of catecholamines and related compounds condensed with formaldehyde. *Histochemistry and Cytochemistry*, **10**, 348–354.

Farnebo, L.-O. & Hamberger, B. (1971). Drug-induced changes in the release of [^{3}H]-noradrenaline from field stimulated rat iris. *British Journal of Pharmacology*, **43**, 97–106.

Fatt, P. & Katz, B. (1951). An analysis of the end-plate potential recorded with an intra-cellular electrode. *Journal of Physiology* **115**, 320–370.

Fatt, P. & Katz, B. (1952). Spontaneous subthreshold activity at motor nerve endings. *Journal of Physiology*, **117**, 109–128.

Fedan, J.S., Hogaboom, G.K., O'Donnell, J.P., Colby, J. & Westfall, D.P. (1981). Contribution by purines to the neurogenic response of the vas deferens of the guinea-pig. *European Journal of Pharmacology*, **69**, 41–53.

Feldberg, W. & Vertiainen, A. (1934). Further observations on the physiology and pharmacology of a sympathetic ganglion. *Journal of Physiology*, **83**, 103–128.

Ferreyra-Moyano, H. & Cinelli, A.R. (1986). Axonal projections and conduction properties of olfactory peduncle neurons in the armadillo (Chaetophractus vellerosus). *Experimental Brain Research*, **64**, 527–534.

Ferry, C.B. (1967). The innervation of the vas deferens of the guinea-pig. *Journal of Physiology*, **192**, 463-478.

Finkel, A.S., Hirst, G.D.S. & Van Helden, D.F. (1984). Some properties of excitatory junction currents recorded from submucous arterioles of guinea-pig ileum. *Journal of Physiology*, **351**, 87-98.

Folkow, B., Häggendal, J. & Lisander, B. (1967). Extent of release and elimination of noradrenaline at peripheral adrenergic nerve terminals. *Acta Physiologica Scandinavica*, Suppl 307.

Forsyth, F.M. & Pollock, D. (1988). Clonidine and morphine increase [3H]-noradrenaline overflow in mouse vas deferens. *British Journal of Pharmacology*, **93**, 35-44.

Fosbraey, P. & Johnson, E.S. (1980) Release-modulating acetylcholine receptors on cholinergic neurones of the guinea-pig ileum. *British Journal of Pharmacology* **68**, 289-300.

Franco, R., Costa, M. & Furness, J. (1979) Evidence for the release of endogenous substance P from intestinal nerves. *Naunyn-Schmiedeberg's Archives of Pharmacology* **306**, 185-201.

Fried, G. (1980). Small noradrenergic storage vesicles isolated from rat vas deferens – biochemical and morphological characterisation. *Acta Physiologica Scandinavica*, Suppl. **493**, 1-23.

Friel, D.D. (1988) An ATP-sensitive conductance in single smooth muscle cells from the rat vas deferens. *Journal of Physiology* **401**, 361-380.

Fry, G.N., Devine, C.E. & Burnstock, G. (1977). Freeze-fracture studies of nexuses between smooth muscle cells. *Journal of Cell Biology* **72**, 26-34.

Fugisawa, K. & Ito, Y. (1982) The effect of substance P on smooth muscle cells and on neuroeffector transmission in the guinea-pig ileum. *British Journal of Pharmacology* **80**, 409-420.

Fujii, K. (1988) Evidence for adenosine triphosphate as an excitatory transmitter in guinea-pig, rabbit and pig urinary bladder. *Journal of Physiology* **404**, 39-52.

Fujiwara, S., Itoh, T. & Suzuki, H. (1982) Membrane properties and excitatory neuromuscular transmission in the smooth muscle of dog cerebral arteries. *British Journal of Pharmacology* **77**, 197-208.

Furness, J.B. (1969) An electrophysiological study of the innervation of the smooth muscle of the colon. *Journal of Physiology* **205**, 549-562.

Furness, J.B. (1970). The effect of external potassium ion concentration on autonomic neuromuscular transmission. *Pflügers Archiv* **317**, 310-326.

Furness, J.B. (1974). Transmission to the longitudinal muscle of the guinea-pig vas deferens: the effect of pretreatment with guanethidine. *British Journal of Pharmacology*, **50**, 63-68.

Furness, J.B. & Costa, M. (1982) Indentification of gastrointestinal neurotransmitters. In. Mediators and Drugs in Gastrointestinal Motility. *Handbook of Experimental Pharmacology* 59/I ed. Bertaccini, G. Springer-Verlag. Berlin.

Furness, J.B. & Costa, M. (1987) *The Enteric Nervous System*. Churchhill Livingstone, Edinburgh.

Furness, J.B. & Iwayama, T. (1972). The arrangement and identification of axons innervating the vas deferens of the guinea-pig. *Journal of Anatomy* **113**, 179-196.

Gabella, G. (1981). Structure of smooth muscles. In: *Smooth Muscle: An Assessment of Current Knowledge* pp. 1-46. Eds.: Bülbring, E., Brading, A.F., Jones, A.W. & Tomita, T. Edward Arnold, London.

Gage, P.W. & McBurney, R.N. (1975). Effects of membrane potential, temperature and neostigmine on the conductance change caused by a quantum of acetylcholine at the toad neuromuscular junction. *Journal of Physiology* **244**, 385-407.

Gillespie, J.S. (1962a) The electrical and mechanical responses of intestinal smooth muscle cells to stimulation of their extrinsic parasympathetic nerves. *Journal of Physiology* **162**, 76-92.

Gillespie, J.S. (1962b) Spontaneous mechanical and electrical activity of stretched and unstretched intestinal muscle cells and their response to sympathetic nerve stimulation. *Journal of Physiology* **162**, 54-75.

Gillespie, J.S. (1980). Presynaptic receptors in the autonomic nervous system. In. *Handbook of Experimental Pharmacology* Vol 54: Adrenergic Activators and Inactivators. ed. L. Sezekeres. Part I, Springer-Verlag: Berlin, 352-425.

Gillespie, J.S. (1982) *Non-adrenergic non-cholinergic inhibitory control of gastrointestinal motility. I. Motility of the Digestive Tract.* ed. Weinbeck, M. Raven Press, New York. 51-66.

Gillespie, J.S., Liu, X. & Martin, W. (1990). The neurotransmitter of the non-adrenergic, non-cholinergic inhibitory nerves to smooth muscle of the genital system. In: *Nitric oxide from L-arginine. Bioregulatory system.* ed. Moncada, S. Elsevier: Science Publishers: B.V. Amsterdam.

Gintzler, A.R. & Hyde, D. (1983) A specific substance P antagonist attenuates non-cholinergic electrically induced contractures of the guinea-pig isolated ileum. *Neuroscience Letters* **40**, 75-79.

Gordon, J.L. (1986). Extracellular ATP: effects, sources and fate. *Biochemical Journal*, **233**, 309-319.

Göthert, M. (1977). Effects of presynaptic modulators on Ca2 + − induced noradrenaline release from cardiac sympathetic nerves. *Naunyn Schmiedeberg's Archives Pharmacology*, **300**, 267-272.

Goto, K., Millechia, L.E., Westfall, D.F. & Fleming, W.W. (1977). A comparison of the electrical properties and morphological characteristics of the smooth muscle of the rat and guinea-pig vas deferens. *Pflügers Archiv* **368**, 253–261.

Häggendal, J. & Malmfors, T. (1969). The effect of nerve stimulation on adrenergic nerves after reserpine pretreatment. *Acta Physiologica Scandinavica*, **75**, 33–38.

Hashimoto, Y., Holman, M.E. & Tille, J. (1966). Electrical properties of the smooth muscle membrane of the guinea-pig vas deferens. *Journal of Physiology* **186**, 27–41.

Head, S.D. (1983). Temperature and end-plate currents in rat diaphragm. *Journal of Physiology* **334**, 441–459.

Haeusler, G. & Thorens, S. (1980) Effects of tetraethylammonium chloride on contractile, membrane and cable properties of rabbit artery muscles. *Journal of Physiology* **303**, 203–224.

Heuser, J.E. & Reese, T.S. (1981). Structural changes after transmitter release at the frog neuromuscular junction. *The Journal of Cell Biology*, **88**, 564–580.

Heuser, J.E., Reese, T.S., Dennis, M.J., Jan, Y., Yan, L & Evans, L. (1979). Synaptic vesicle exocytosis captured by quick freezing and correlated with quantal transmitter release. *The Journal of Cell Biology*, **81**, 275–300.

Hidaka, T. & Kuriyama, H. (1969) Responses of the smooth muscle membrane of guinea-pig jejunum elicited by field stimulation. *Journal of General Physiology* **53**, 471–486.

Hill, C.E., Hirst, G.D.S. & Van Helden, D.F. (1983) Development of sympathetic innervation to proximal and distal arteries of the rat mesentery. *Journal of Physiology* **338**, 129–147.

Hillarp, N.Å. (1958). Adenosinephosphates and inorganic phosphate in the adrenaline and noradrenaline containing granules of the adrenal medulla. *Acta Physiologica Scandinavica*, **42**, 321–332.

Hirst, G.D.S. (1977) Neuromuscular transmission in arterioles of guinea-pig submucosal arterioles. *Journal of Physiology* **273**, 263–275.

Hirst, G.D.S. (1987) Neural control of blood vessels – one or two transmitters? In: *Vascular Neuroeffector Mechanisms*, pp. 149–160. Eds.: Bevan, J.A., Majewski, H., Maxwell, R.A. & Story, D.F. IRL Press: Oxford, Washington D.C.

Hirst, G.D.S. & Edwards, F.W. (1989) Sympathetic neuroeffector transmission in arteries and arterioles. *Physiological Reviews* **69**(2), 546–604.

Hirst, G.D.S. & Neild, T.O. (1978). An analysis of excitatory junction potentials recorded from arterioles. *Journal of Physiology*, **280**, 87–104.

Hirst, G.D.S. & Neild, T.O. (1980a). Some properties of spontaneous excitatory junction potentials recorded from arterioles of guinea-pigs. *Journal of Physiology*, **303**, 43–60.

Hirst, G.D.S. & Neild, T.O. (1980b). Evidence for two populations of excitatory receptors for noradrenaline on arteriolar smooth muscle. *Nature*, **283**, 767–768.

Hirst, G.D.S. & Neild, T.O. (1980c). Noradrenergic transmission – Reply to McGrath. *Nature*, **288**, 302.

Hirst, G.D.S. & Neild, T.O. (1981). Localization of specialized noradrenaline receptors at neuromuscular junctions on arterioles of the guinea-pig. *Journal of Physiology*, **313**, 343–350.

Hirst, G.D.S., Neild, T.O. & Silverberg, G.D. (1982). Noradrenaline receptors on the rat basilar artery. *Journal of Physiology*, **328**, 351–360.

Hodgkin, A.L. & Rushton, W.A.H. (1946). The electrical constants of a crustacean nerve fibre. *Proceedings of the Royal Society Series B* **133**, 444–479.

Hökfelt, T. (1969). Distribution of noradrenaline storage particles in peripheral adrenergic neurons as revealed by electron microscopy. *Acta Physiologica Scandinavica*, **76**, 427–440.

Holman, M.E. (1970). Junction potentials in smooth muscle. In. *Smooth Muscle*. ed. Bülbring, E., Brading, A.F., Jones, A.W. & Tomita, T. Edward Arnold: London. pp. 244–288.

Holman, M.E. (1973). Overview. *Philosophical Transactions of the Royal Society* London B. **265**, 25–34.

Holman, M.E. & Neild, T.O. (1979). Membrane properties. *British Medical Bulletin* **35** (3), 235–241.

Holman, M.E. & Suprenant, A. (1979). Some properties of the excitatory junction potentials recorded from saphenous arteries of rabbits. *Journal of Physiology* **287**, 337–351.

Holman, M.E., Kasby, C.B., Suthers, M.B. & Wilson, J.A.F. (1968) Some properties of smooth muscle of the rabbit portal vein. *Journal of Physiology* **196**, 111–132.

Holman, M.E., Taylor, G.S. & Tomita, T. (1977). Some properties of the smooth muscle of the mouse vas deferens. *Journal of Physiology* **266**, 751–764.

Holzer, P. (1982) An enquiry into the mechanism by which substance P facilitates the phasic longitudinal contractions of the rabbit ileum. *Journal of Physiology* **325**, 377–392.

Holzer, P. (1984) Characterisation of the stimulus-induced release of immunoreactive substance P from the myenteric plexus of the guinea-pig small intestine. *Brain Research* **297**, 127–136.

Holzer, P. & Petsche, U. (1983) On the mechanism of contraction and desensitization by substance P in the intestinal smooth muscle of the guinea pig. *Journal of Physiology* **342**, 549–568.

Horn, J.P. & McAfee, D.A. (1980). Alpha-adrenergic inhibition of calcium-dependent potentials in rat sympathetic neurones. *Journal of Physiology*, **301**, 191–204.

Hottenstein, O.D. & Kreulen, D.L. (1987) Frequency dependence of guinea-pig inferior mesenteric artery and vein responses to repetitive colonic nerve stimulation. *Journal of Physiology* **384**, 153–167.

Hoyle, C.H.V. & Burnstock, G. (1985) Atropine-resistant excitatory junction potentials in rabbit bladder are blocked by α, β-methylene ATP. *European Journal of Pharmacology* **114**, 239–240.

Hoyle, C.H.V. & Burnstock, G. (1989) Neuromuscular transmission in the gastrointestinal tract. In. *Handbook of Physiology* Section 6: The gastrointestinal system Vol 1. Motility and Circulation Part 1. Oxford University Press, New York. 435–464.

Hubbard, J.I. (1973). Microphysiology of vertebrate neuromuscular transmission. *Physiological Reviews*, **53**, 674–723.

Hubbard, J.I. & Schmidt, R.F. (1963). An electrophysiological investigation of mammalian motor terminals. *Journal of Physiology*, **166**, 145–167.

Illes, P. (1986). Mechanisms of receptor-mediated modulation of transmitter release in noradrenergic, cholinergic and sensory neurones. *Neuroscience*, **17**, 909–928.

Illes, P. & Dörge, L. (1985). Mechanism of $α_2$-adrenergic inhibition of neuroeffector transmission in the mouse vas deferens. *Naunyn-Schmiedeberg's Archives of Pharmacology*, **328**, 241–247.

Illes, P. & Starke, K. (1983). An electrophysiological study of presynaptic α-adrenoceptor in the vas deferens of the mouse. *British Journal of Pharmacology*, **78**, 365–373.

Inoue, R. & Brading, A.F. (1990). The properties of the ATP-induced depolarisation and current in single cells isolated from the guinea-pig urinary bladder. *British Journal of Pharmacology*, **100**, 619–625.

Inoue, R., Kitamura, K. & Kuriyama, H. (1987) Acetylcholine activates single sodium channels in smooth muscle cells. *Pflügers Archiv* **410**, 69–74.

Ishikawa, S. (1985). Actions of ATP and α, β-methylene ATP on neuromuscular transmission and smooth muscle membrane of the rabbit and guinea-pig mesenteric arteries. *British Journal of Pharmacology* **86**, 777–787.

Ito, Y. & Kuriyama, H. (1971) The properties of the rectal smooth muscle membrane of the guinea-pig in relation to the nervous influences. *Japanese Journal of Physiology* **21**, 277–294.

Ito, Y. & Kuriyama, H. (1973) Membrane properties and inhibitory innervation of the circular muscles of guinea-pig caecum. *Journal of Physiology* **231**, 455–470.

Ito, Y. & Takeda, K. (1982) Non-adrenergic inhibitory nerves and putative transmitters in the smooth muscle cat trachea. *Journal of Physiology* **330**, 497–511.

Iversen, L.L., Glowinsky, J. & Axelrod, J. (1965). The uptake and storage of 3H-norepinephrine in the reserpine-pretreated rat heart. *Journal of Pharmacology and Experimental Therapeutics*, **150**, 173–183.

Jack, J.J.B., Redman, S.J. & Wong, K. (1981). The components of synaptic potentials evoked in cat spinal motoneurones by impulses in single Ia afferents. *Journal of Physiology* **321**, 65–96.

Jack, J.J.B., Miller, S., Porter, R. & Redman, S.J. (1971). The time course of the minimal excitatory post-synaptic potentials evoked in spinal motoneurones by group Ia afferent fibres. *Journal of Physiology* **215**, 353–380.

Jager, L.P. & Hertog, A. (1985) Criteria for the involvement of adenosine and adenine nucleotides in nonadrenergic, noncholinergic transmission. In. *Methods in Pharmacology* Vol 6. ed. Paton, D.M. Plenum Publishing Corporation, New York.

Jakobs, K.H. (1985) Coupling mechanisms of $α_2$-adrenoceptors. *Journal of Cardiovascular Research.* Suppl. 6 S109–112.

Johnson, D.G., Thoa, N.B., Wienshilboum, R., Axelrod, J. & Kopin, I.J. (1971). Enhanced release of dopamine β-hydroxylase from sympathetic nerves by calcium and phenoxybenzamine and its reversal by prostaglandins. *Proceedings of the National Academy of Science*, **68**, 2227–2230.

Johnston, H. & Majewski, H. (1986). Prejunctional β-adrenoceptors in rabbit pulmonary artery and mouse atria: effect of α-adrenoceptor blockade and phosphodiesterase inhibition. *British Journal of Pharmacology*, **87**, 553–562.

Johnston, H., Majewski, H. & Musgrave, I.F. (1987). Involvement of cyclic nucleotides in prejunctional modulation of noradrenaline release in mice atria. *British Journal of Pharmacology*, **91**, 773–781.

Kajiwara, M., Kitamura, K. & Kuriyama, H. (1981). Neuromuscular transmission and smooth muscle membrane properties in the guinea-pig ear artery. *Journal of Physiology* **315**, 283–302.

Kalsner, S. & Quillan, M. 1984. A hypothesis to explain the presynaptic effects of adrenoceptor antagonists. *British Journal of Pharmacology*, **82**, 515–522.

Kannan, M.S., Jager, L.P. & Daniel, E.E. (1985) Electrical properties of smooth muscle cell membrane of opossum esophagus. *American Journal of Physiology* **248**, G342–G346.

Karashima, H. & Kuriyama, H. (1981) Electrical properties of smooth muscle cell membrane and neuromuscular transmission in the guinea-pig basilar artery. *British Journal of Pharmacology* **74**, 497–504.

Katayama, Y. & North, R.A. (1978) Does substance P mediate slow synaptic excitation within the myenteric plexus? *Nature* **274**, 387–388.

Katz, B. (1948). The electrical properties of the muscle fibre membrane. *Proceedings of the Royal Society Series B* **135**, 506–534.

Katz, B. (1966). Nerve, muscle and synapse. McGraw-Hill, New York.

Katz, B. & Miledi, R. (1965a). The effect of calcium on acetylcholine release from motor nerve endings. *Proceedings of the Royal Society B*, **161**, 496–503.

Katz, B. & Miledi, R. (1965b). Propagation of electrical activity in motor nerve terminals. *Proceedings of the Royal Society B*, **161**, 453–482.

Katz, B. & Miledi, R. (1967). Tetrodotoxin and neuromuscular transmission. *Proceedings of the Royal Society B*, **167**, 8–22.

Katz, B. & Miledi, R. (1972). The statistical nature of the acetylcholine potential and its molecular components. *Journal of Physiology* **224**, 665–699.

Katz, B. & Miledi, R. (1973). The characteristics of 'end-plate noise' produced by different depolarizing drugs. *Journal of Physiology* **230**, 707–717.

Katz, B. & Miledi, R. (1976). The analysis of end-plate noise – a new approach to the study of acetylcholine/receptor interaction. In: *Motor Innervation of Muscle*, pp 31–50. Ed.: Thesleff, S. Academic Press, London.

Katz, B. & Miledi, R. (1979). Estimates of the quantal content during 'chemical potentiation' of transmitter release. *Proceedings of the Royal Society B*, **205**, 369–378.

Kinekawa, F., Komori, S. & Ohashi, H. (1984) Cholinergic inhibition of adrenergic transmission in the dog retractor penis muscle. *Japanese Journal of Pharmacology* **34**, 343–352.

Kirkpatrick, K. & Burnstock, G. (1987). Sympathetic nerve mediated release of ATP from the guinea-pig vas deferens is unaffected by reserpine. *European Journal of Physiology*, **138**, 207–214.

Kirpekar, S.M. & Puig, M. (1971). Effect of flow stop on noradrenaline release from normal spleens and spleens treated with cocaine, phentolamine or phenoxybenzamine. *British Journal of Pharmacology*, **43**, 359–369.

Kitamura, K. (1978) Comparative aspects of membrane properties and innervation of longitudinal and circular muscle layers of rabbit jejunum. *Japanese Journal of Physiology* **28**, 583–601.

Klein, R.L. & Lagercrantz, H. (1981). Noradrenergic vesicles: composition and function. In. *Chemical Neurotransmission 75 Years*. ed. Stjärne, L., Hedqvist, P., Lagercrantz, H. & Wennmalm, Å. Academic Press: London, New York, Toronto, Sydney, San Francisco. pp. 69–83.

Komori, S. & Ohashi, M. (1988) Some membrane properties of the circular muscle of the chicken rectum and its non-adrenergic non-cholinergic innervation. *Journal of Physiology* **401**, 417–435.

Komori, S. & Suzuki, H, (1986) Distribution and properties of excitatory and inhibitory junction potentials in circular muscle of the guinea-pig stomach. *Journal of Physiology* **370**, 339–355.

Komori, S., Fukutome, T. & Ohashi, H. (1986) Isolation of a peptide material showing strong rectal muscle-contracting activity from chicken rectum and its identification as chicken neurotensin. *Japanese Journal of Pharmacology* **40**, 577–589.

Komori, S., Kwon, S.C. & Ohashi, H. (1988) Effects of prolonged exposure to α,β-methylene ATP on non-adrenergic, non-cholinergic excitatory transmission in the rectum of the chicken. *British Journal of Pharmacology* **94**, 9–18.

Konishi, T. & Sears, T.A. (1984). Electrical activity of mouse motor nerve terminals. *Proceedings of the Royal Society B*, **222**, 115–120.

Kordas, M. (1972a). An attempt at an analysis of the factors determining the time course of the end-plate current. I. The effects of prostigmine and of the ratio of $Mg2+$ to $Ca2+$. *Journal of Physiology* **224**, 317–332.

Kordas, M. (1972b). An attempt at an analysis of the factors determining the time course of the end-plate current. II. Temperature. *Journal of Physiology* **224**, 333–348.

Korn, H. (1984). What central inhibitory pathways tell us about mechanisms of transmitter release. *Experimental Brain Research*, Suppl. **9**, 201–224.

Korn, H. & Faber, D.S. (1987). Regulation and significance of the probabilistic release mechanisms at central synapses. In. *Synaptic Function* ed. G.M. Edelman., W.E. Gall. & W.M. Cowan. John Wiley & Sons, Inc: New York, Chichester, Brisbane, Toronto, Singapore. pp 57–108.

Korn, H., Mallet, A., Triller, A. & Faber, D.S. (1982). Transmission at a central synapse. II. Quantal description of release with a physical correlate for binomial n. *Journal of Neurophysiology*, **48**, 679–707.

Kostron, H., Winkler, H., Peer, L.J. & König, P. (1977). Uptake of adenosine triphosphate by isolated adrenal chromaffin granules: a carrier-mediated transport. *Neuroscience* **2**, 159-166.

Kou, K., Kuriyama, H. & Suzuki, H. (1982) Effects of 3,4-dihydro-8-(2-hydroxy-3-isopropylamino-propoxy)3-nitroxy-2H-1-benzopyran(K-351) on smooth muscle cells and neuromuscular transmission in the canine mesenteric artery. *British Journal of Pharmacology* **77**, 679-689.

Kriebel, M.E. & Gross, C.E. (1974). Multimodal distribution of frog miniature endplate potentials in adult, denervated and tadpole leg. *Journal of General Physiology*, **64**, 85-103.

Kriebel, M.E., Llados, F. & Matteson, D.R. (1976). Spontaneous subminiature end-plate potentials in mouse diaphragm muscle: evidence for synchronous release. *Journal of Physiology*, **265**, 553-581.

Kubota, M. & Szurszewski, J.H. (1984) Innervation of the canine internal anal sphincter. *Gastroenterology* **86**, 1146.

Kuffler, S.W. (1942). Electrical potential changes at an isolated nerve-muscle junction. *Journal of Neurophysiology* **5**, 18-26.

Kuffler, S.W. & Yoshikami, D. (1975). The number of transmitter molecules in a quantum: an estimate from ionophoretic application of acetylcholine at the neuromuscular synapse. *Journal of Physiology* **251**, 465-482.

Kuffler, S.W., Nichols, J.G. & Martin, A.R. (1984). *From Neuron to Brain*. 2nd Ed. Sinauer Associates Inc: Sunderland, Massachusetts.

Kügelgen, I.V. & Starke, K. (1985). Noradrenaline and adenosine triphosphate as co-transmitters of neurogenic vasoconstriction in rabbit mesentery artery. *Journal of Physiology* **367**, 435-455.

Kuno, M., Turkanis, S.A. & Weakly, J.N. (1971). Correlation between nerve terminal size and transmitter release at the neuromuscular junction of the frog. *Journal of Physiology*, **213**, 545-556.

Kuriyama. H. & Makita, Y. (1983) Modulation of noradrenergic transmission in the guinea-pig mesenteric artery: an electrophysiological study. *Journal of Physiology* **335**, 609-627.

Kuriyama, H. & Makita, Y. (1984) The presynaptic regulation of noradrenaline release differs in mesenteric arteries of the rabbit and guinea-pig. *Journal of Physiology* **351**, 379-396.

Kuriyama, H., Osa T, & Toida, N. (1967) Nervous factors influencing the membrane activity of intestinal smooth muscle. *Journal of Physiology* **191**, 257-270.

Kuriyama, H. & Suzuki, H. (1981). Adrenergic transmissions in the guinea-pig mesenteric artery and their cholinergic modulations. *Journal of Physiology* **317**, 383-396.

Lagercrantz, H. (1976). On the composition and function of large dense cored vesicles in sympathetic nerves. *Neuroscience*, **1**, 81-92.

Langer, S.Z. (1974). Presynaptic regulation of catecholamine release. *Biochemical Pharmacology*, **23**, 1793-1800.

Langer, S.Z. (1981). Presynaptic regulation of the release of catecholamines. *Pharmacological Reviews*, **32**, 337-362.

Langer, S.Z., Adler, E., Energo, A. & Stefano, F.J.E. (1971). The role of the α-receptors in regulating noradrenaline overflow by nerve stimulation. *Proceedings of the 25th International Congress of Physiological Sciences*, **335**.

Large, W.A. (1982) Membrane potential responses of the mouse anococcygeus muscle to ionophoretically applied noradrenaline. *Journal of Physiology* **326**, 385-400.

Leander, S., Håkanson, R., Rosell. S., Folkers, k., Sundler, F. & Tornqvist, K. (1981) A specific substance P antagonist blocks smooth muscle contractions induced by non-cholinergic, non-adrenergic nerve stimulation. *Nature* **294**, 467-469.

Lew, M. & White, T.D. (1987). Release of endogenous ATP during sympathetic nerve stimulation. *British Journal of Pharmacology*, **92**, 349-355.

Levitzki, A. (1986). β-Adrenergic receptors and their mode of coupling to adenylate cyclase. *Physiological Reviews*, **66**, 819-854.

Liley, A.W. (1956a). The quantal components of the mammalian end-plate potential. *Journal of Physiology*, **133**, 571-587.

Liley, A.W. (1956b). The effects of presynaptic polarization on the spontaneous activity at the mammalian neuromuscular junction. *Journal of Physiology*, **134**, 427-443.

Loewi, O. (1921). Über humorale übertragbarkeit der herznervenwirkung. *Pfügers Archiv für die gesmte Physiologie des Meschen und der Tiere*, **189**, 239-242.

Loewi, O. (1936). Über humorale übertragbarkeit der herznervenwirkung. XIV. Quantitative und qualitative untersuchungen über den sympathicusstoff. *Pfügers Archiv für die gesmte Physiologie des Meschen und der Tiere*, **237**. 504-514.

Loewi, O. & Navratil, E. (1926). Über humorale übertragbarkeit der herznervenwirkung. X. Über das schicksal des vasusstoffes. *Pfügers Archiv für die gesmte Physiologie des Meschen und der Tiere*, **214**, 678-688.

Luff, S.E. (1988) Neurovascular junctions in arterial vessels. In: *Vascular Neuroeffector Mechanisms* ed. Bevan J.A., Majewski, H., Maxwell R.A. & Story, IRL Press: Oxford Washington pp. 31-37.

Luff, S.E. & McLachlan, E.M. (1989). Frequency of neuromuscular junctions on arteries of different dimensions in the rabbit, guinea-pig and rat. *Blood Vessels.* **26**, 95-106.

Luff, S.E., McLachlan, E.M., & Hirst, G.D.S. (1987). An ultrastructural analysis of the sympathetic neuromuscular junctions on arterioles of the submucosa of the guinea pig ileum. *Journal of Comparative neurology* **257**, 578-594.

Maas, A.J.J. (1981) The effect of apamin on response evoked by field stimulation in guinea-pig taenia caeci. *European Journal of Pharmacology* **73**, 1-9.

MacKenzie, I. & Szurszewski, J.H. (1984) Effect of long periods of stimulation on the inhibitory nervous input to the smooth muscle of the ilecolonic sphincter *in vivo*. *Gastroenterology* **86**, 1169.

Mackenzie, I., Kirkpatrick, K.A. & Burnstock, G. (1988). Comparative study of the actions of AP5P and α,β methylene ATP on nonadrenergic neurogenic contractions of the guinea-pig vas deferens. *British Journal of Pharmacology.* **94**, 699-706.

Magleby, K.L. (1987). Short-term changes in synaptic efficacy. In. *Synaptic function* ed. G.M. Edelman., W.E. Gall. & W.M. Cowan, John Wiley & Sons, Inc: New York, Chichester, Brisbane, Toronto, Singapore. pp. 21-56.

Magleby, K.L. & Miller, D.C. (1981). Is the quantum of transmitter release composed of subunits? A critical analysis in the mouse and frog. *Journal of Physiology*, **311**, 267-287.

Magleby, K.L. & Stevens, C.F. (1972). A quantitative description of end-plate currents. *Journal of Physiology* **223**, 173-197.

Majewski, H., Ishac, E.J.N. & Musgrave, I.F. (1988). Intraneuronal mechanisms involved in the modulation of noradrenaline release through prejunctional receptors. In. *Vascular Neuroeffector Mechanisms*. ed. Bevan, J.A., Majewski, H., Maxwell, R.A. & Story, D.F. IRL Press: Oxford, Washington. pp 243-250.

Majewski, H., Costa, M., Foucart, S., Murphy, T.V. & Musgrave, I.F. (1990) Second messengers are involved in facilatory but not inhibitory receptor actions at sympathetic nerve endings. *Annals of the New York Academy of Sciences* **604**, 266-275.

Makita Y (1983) Effects of adrenoceptor agonists and antagonists on smooth muscle cells and neuromuscular transmission in the guinea-pig renal artery and vein. *British Journal of Pharmacology* **80**, 671-679.

Malmfors, T. (1965) Studies on adrenergic nerves; the use of rat and mouse iris for direct observations on their physiology and pharmacology at cellular and subcellular levels. *Acta Physiologica Scandinavica*, Suppl. 248.

Marshall, I.G. (1970). A comparison between the blocking actions of 2-(4-phenylpiperidino) cyclohexanol (AH5183) and its N-methyl quaternary analogue (AH5954). *British Journal of Pharmacology*, **40**, 68-77.

Marshall, I.G., Prior, C. & Searl T. (1990) The effect of (-)-vesamicol on miniature endplate current amplitudes in the rat hemidiaphram. Journal of Physiology **424**, 52P.

Martin, A.R. (1955). A further study of the statistical composition of the end-plate potential. *Journal of Physiology*, **130**, 114-122.

McCulloch, M.W., Rand, M.J. & Story, D.F. (1973). Inhibition of 3H-noradrenaline release from sympathetic nerves of guinea-pig atria by a presynaptic α-adrenoceptor mechanism. *British Journal of Pharmacology*, **46**, 523P.

McGrath, J.C. (1978). Adrenergic and 'non-adrenergic' components in the contractile response of the vas deferens to a single indirect stimulus. *Journal of Physiology* **283**, 23-39.

McLachlan, E.M. (1978). The statistics of transmitter release at chemical synapses. In. *International Review of Physiology*, Neurophysiology III. Vol 17 ed. Porter, R. University Press: Baltimore. pp. 49-117.

Meehan, A.G., Cunnane, T.C., Johnson, P.C., Hirst, G.D.S., Liu, B. & Kreulen, D.L. (1990) In vivo electrophysiological studies on sympathetic transmission in mesenteric arterioles of the guinea-pig and rat. *Blood Vessels.* **27**, 44-45.

Mekata, F. (1984) Different electrical responses of outer and inner muscle of rabbit carotid artery to noradrenaline and nerves. *Journal of Physiology* **346**, 589-598.

Meldrum, L.A. & Burnstock, G. (1985) Investigations into the identity of the non-adrenergic, noncholinergic excitatory transmitter in the smooth muscle of chicken rectum. *Comparative Biochemistry and Physiology* **81C**, 307-309.

Merrillees, N.C.R. (1968). The nervous environment of individual smooth muscle cells of the guinea-pig vas deferens. *Journal of Cell Biology*, **37**, 794-817.

Merrillees, N.C.R., Burnstock, G. & Holman, M.E. (1963). Correlation of fine structure and physiology of the innervation of smooth muscle in the guinea-pig vas deferens. *Journal of Cell Biology*, **19**, 529-550.

Michaelson, D.M., Licht, R. & Burstein, M. (1987). The effects of the vesicular acetylcholine uptake blocker AH5183 on evoked and spontaneous release of acetylcholine from the cholinergic nerve terminals. In. *Cellular and molecular basis of cholinergic function.* ed. A.J. Dowdall & J.N. Hawthorn., Ellis Horwood Ltd: Chichester, England. pp. 269-276.

Minneman, K.P. (1988) α_1-adrenoceptors, inositol phosphates, and sources of cell Ca2+. *Pharmacological Reviews* **40**, 87-119.

Mishima, S., Miyahara, H. & Suzuki, H. (1984) Transmitter release modulated by α-adrenoceptor antagonists in the rabbit mesenteric artery: a comparison between noradrenaline outflow and electrical activity. *British Journal of Pharmacology* **83**, 537-547.

Miyahara, H. & Suzuki, H. (1987) Pre- and post-junctional effects of adenosine triphosphate on noradrenergic transmission in the rabbit ear artery. *Journal of Physiology* **389**, 423-440.

Morgan, K.G. (1983) Comparison of membrane electrical activity of cat gastric submucosal arterioles and venules. *Journal of Physiology* **345**, 135-147.

Morita, K. & North, R.A. (1981a). Clonidine activates membrane potassium conductance in myenteric neurones. *British Journal of Pharmacology*, **74**, 419-428.

Morita, K. & North, R.A. (1981b). Opiates and enkephalin reduce the excitability of neuronal processes. *Neuroscience*, **6**, 1943-1951.

Muramatsu, I. (1986). Evidence for sympathetic, purinergic transmission in the mesenteric artery of the dog. *British Journal of Pharmacology* **87**, 478-480.

Muramatsu, I. (1987). The effect of reserpine on sympathetic, purinergic neurotransmission in the isolated mesenteric artery of the dog: a pharmacological study. *British Journal of Pharmacology* **91**, 467-474.

Murray, R.W. (1983). *Test Your Understanding of Neurophysiology.* Cambridge University Press: Cambridge, New York, New Rochelle Melbourne, Sydney.

Musgrave, I., Marley, P. & Majewski, H. (1987). Pertussis toxin does not attenuate α_2-adrenoceptor mediated inhibition of noradrenaline release in mice atria. *Naunyn-Schmiedeberg's Archives of Pharmacology*, **336**, 280-286.

Musgrave, I.F. & Majewski, H. (1987). Lack of effect of protein kinase C activation on the inhibition of noradrenaline release by clonidine. *Blood Vessels*, **24**, 218.

Musgrave, I.F. & Majewski, H. (1989) Effect of phorbol ester and polymyxin B on modulation of noradrenaline release in mouse atria. *Naunyn-Schmiedeberg's Archives of Pharmacology*, **339**, 48-53.

Nakamura, S., Tepper, J.M., Young, S.J. & Groves, P.M. (1981). Neurophysiologic consequences of presynaptic receptor activation: changes in noradrenergic terminal excitability. *Brain Research*, **226**, 155-170.

Nakazawa, K. Inoue, K., Fujimori, K. & Takanaka (1990) Difference between substance P and acetylcholine-induced currents in mammalian smooth muscle cells. *European Journal od Pharmacology* **179**, 453-456.

Neher, E. & Sakmann, B. (1976). Single-channel currents recorded from membrane of denervated frog muscle fibres. *Nature* **260**, 799-802.

Niel, J., Bywater, R.A.R. & Taylor, G.S. (1983a) Effect of substance P on non-cholinergic fast and slow post-stimulus depolarization in the guinea-pig ileum. *Journal of the Autonomic Nervous System* **9**, 573-584.

Niel, J., Bywater, R.A.R. & Taylor G.S. (1983b) Apamin-resistant post-stimulus hyperpolarization in the circular muscle of the guinea-pig ileum. *Journal of the Autonomic Nervous System* **9**, 565-569.

Neild, T.O. & Keef, K. (1985) Measurement of the membrane potential of arterial smooth muscle in anesthetized animals and its relationship to changes in artery diameter. *Microvascular Research*, **30**, 19-28.

North, R.A. & Surprenant, A. (1985). Inhibitory synaptic potentials resulting from α_2-adrenoceptor activation in guinea-pig submucous plexus neurones. *Journal of Physiology*, **358**, 17-34.

Ohashi, H., Niato, T., Takewaki, T. & Okada, T. (1977) Non-cholinergic, excitatory junction potentials in smooth muscle of chicken rectum. *Japanese Journal of Pharmacology* **27**, 379-387.

Ohashi, H. & Ohga, A. (1967) Transmission of excitation from the parasympathetic nerve to the smooth muscle. *Nature* **216**, 291-292.

Ohkawa, H. (1982) Excitatory junction potentials recorded from the circular smooth muscles of the guinea-pig seminal vesicle. *Tohoku Journal of Experimental Medicine* **136**, 89-102.

Ohlin, P. & Strömblad, C.R. (1963). Observations on the isolated vas deferens. *British Journal of Pharmacology* **20**, 299-306.

Pacaud, P., Loirand, G., Mironneau, C. & Mironneau, J. (1989) Noradrenaline activates a calcium-activated chloride and increases the voltage-dependent calcium current in cultured single cells of the rat portal vein. *British Journal of Pharmacology* **97**, 139–146.

Palade, G.E. (1954). Electron microscope observations of interneuronal and neuromuscular synapses. *Anatomical Record*, **118**, 335–336.

Palay, S.L. (1954). Electron microscope study of the cytoplasm of neurons. *Anatomical Record*, **118**, 336.

Paton, W.D.M. & Vizi, E.S. (1969). The inhibitory action of noradrenaline and adrenaline on acetylcholine output by guinea-pig longitudinal muscle strip. *British Journal of Pharmacology*, **35**, 10–28.

Peart, W.S. (1949). The nature of splenic sympathin. *Journal of Physiology*, **108**, 491–501.

Potter, L.T. (1967). Role of intraneuronal vesicles in the synthesis, storage and release of noradrenaline. *Circulation Research*, 21 Suppl. 3, 13–24.

Prehn, J.L. & Bevan, J.A. (1983) Facial vein of the rabbit. Intracellularly recorded hyperpolarization of smooth muscle cells induced by β-adrenergic receptor stimulation. *Circulation Research* **52**, 465–470.

Puig, M.M., Gascon, P., Craviso, G.L. & Musacchio, J.M. (1977) Endogenous opiate receptor ligand: electrically induced release in the guinea-pig ileum. *Science* **195**, 419–420.

Purves, R.D. (1976). Current flow and potential in a three-dimensional syncytium. *Journal of Theoretical Biology* **60**, 147–162.

Redman, S. (1990) Quantal analysis of synaptic potentials in neurons of the central nervous system. *Physiological Reviews.* **70**, **165**–198.

Rees, D. (1974). The spontaneous release of transmitter from insect nerve terminals as predicted by the negative binomial theorem. *Journal of Physiology*, **236**, 129–142.

Robertson, J.D. (1956). The ultrastructure of a reptilian myoneural junction. *Journal of Biophysical and Biochemical Cytology*, **2**, 381.

Robitaille, R & Tremblay, J.P. (1987). Non-uniform release at the frog neuromuscular junction: evidence of morphological and physiological plasticity. *Brain Research Reviews*, **12**, 95–116.

Ryan, L.J., Tepper, J.M., Sawyer, S.F., Young, S.J. & Groves, P.M. (1985). Autoreceptor activation in central monoamine neurons: modulation of neurotransmitter release is not mediated by inter-mittent axonal conduction. *Neuroscience*, **15**, 925–931.

Schoffelmeer, A.N.M. & Mulder, A.H. (1983). [3H] Noradrenaline release from rat neocortex in the absence of extracellular $Ca2+$ and its presynaptic alpha$_2$-adrenergic modulation: a study of the possible role of cyclic AMP. *Naunyn-Schmiedeberg's Archives of Pharmacology*, **323**, 188–192.

Schoffelmeer, A.N.M., Hogenbloom, F. & Mulder, A.H. (1985). Evidence for a presynaptic adenylate cyclase system facilitating [3H] norepinephrine release from rat brain neocortex slices and synaptosomes. *Journal of Neuroscience* **5**, 2685–2689.

Schultzberg, M., Höfelt, T., Nilsson, G., Terenius L., Refeld, J., Brown, M., Elde, R., Goldstein, M. & Said, S.I. (1980) Distribution of peptide and catecholamine neurons in the gastrointestinal tract of the rat and guinea-pig: immunohistochemical studies with antisera to substance P, VIP, enkephalins, somatostatin, gastrin, neurotensin and dopamine-β-hydroxylase. *Neuroscience* **5**, 689–744.

Searl, T., Prior, C. & Marshall, I.G. (1990) Effect of 2-(4-phenylpiperidino) cyclohexanol, an inhibitor of vesicular acetylcholine uptake, on two populations of miniature endplate currents at the snake neuromuscular junction. *Neuroscience* **35**, 145–156.

Seki, N. & Suzuki, H. (1989) Electrical and mechanical activity of rabbit prostate smooth muscles in response to nerve stimulation. *Journal of Physiology* **417**, 651–663.

Sherrington, C.S. (1897). The central nervous system Vol 3. In: *Text-Book of Physiology*, 7th Ed. ed. M. Foster. Macmillan: London.

Shuba, M.F. & Vladimirova, I.A. (1980) Effect of apamin on the electrical responses of smooth muscle to adenosine 5'-triphosphate and to non-adrenergic, non-cholinergic nerve stimulation. *Neuroscience* **5**, 853–859.

Sims, S.M., Singer, J.J. & Walsh, J.V. (1985) Cholinergic agonists suppress a potassium current in freshly dissociated smooth muscle cells of the toad. *Journal of Physiology* **367**, 503–529.

Sjöstrand, N.O. (1965). The adrenergic innervation of the vas deferens and the accessory male genital organs. *Acta Physiologica Scandinavica*, Suppl. **257**, 1–82.

Sjöstrand, N.O. (1981) Smooth muscles of the vas deferens and other organs in the male reproductive tract. In: *Smooth Muscle: An Assessment of Current Knowledge* pp. 367–376. Eds.: Bülbring, E., Brading, A.F., Jones, A.W. & Tomita, T. Edward Arnold, London.

Small, R.C. (1971) Transmission from cholinergic neurones to circular smooth muscle obtained from the rabbit caecum. *British Journal of Pharmacology* **42**, 656P–657P.

Small, R.C. (1972) Transmission from intramural inhibitory neurones to circular smooth muscle of the rabbit caecum and the effects of catecholamines. *British Journal of Pharmacology* **45**, 149P–150P.

Small, R.C. & Weston, A.H. (1979) Intramural inhibition in rabbits and guinea-pig intestine. In. *Physiological and Regulatory Functions of Adenosine and Adenine Nucleotides*. ed. Baer, M.P. & Drummond, G.I. Raven Press, New York. 45–60.

Smith, I.K. & Bywater, R.A.R. (1983) Apamin resistant hyperpolarization and noncholinergic excitation in the guinea-pig distal colon. *Neuroscience Letters* Suppl. **11**, 575.

Smith, A.D. & Winkler, H. (1972) Fundamental mechanisms of the release of catecholamines. In. *Handbook of Experimental Pharmacology Volume 33: Catecholamines*. ed. Blaschko, H. & Muscholl, E. Springer-Verlag: Berlin, Heidelberg, New York, pp. 538–617.

Smith, A.D., De Potter, W.P., Moerman, E.J. & De Schaepdryver, A.F. (1970). Release of dopamine β-hydroxylase and chromogranin A upon stimulation of the splenic nerve. *Tissue and Cell*, **2**, 547–568.

Smith, D.O. (1988). Determinants of nerve-terminal excitability. In. *Neurology and Neurobiology Volume 35: Long-term potentiation: from biophysics to behaviour*. ed. Landfield, P.W. & Deadwyler, S.A. Alan Liss, Inc: New York. pp. 411–438.

Sneddon, P. & Burnstock, G. (1984a). Inhibition of excitatory junction potentials in guinea-pig vas deferens by α, β-methylene-ATP: further evidence for ATP and noradrenaline as co-transmitters. *European Journal of Pharmacology* **100**, 85–90.

Sneddon, P. & Burnstock, G. (1984b). ATP as a co-transmitter in rat tail artery. *European Journal of Pharmacology*, **106**, 149–152.

Sneddon, P. & Westfall, D.P. (1984). Pharmacological evidence that adenosine triphosphate and noradrenaline are co-transmitters in the guinea-pig vas deferens. *Journal of Physiology*, **347**, 561–580.

Sneddon, P., Westfall, D.P. & Fedan, J.S. (1982). Cotransmitters in the motor nerves of the guinea-pig vas deferens: electrophysiological evidence. *Science*, **218**, 693–695.

Stadler, H. & Kiene, M.-L. (1987). The life cycle of cholinergic synaptic vesicles. In. *Cellular and Molecular Basis of Cholinergic Function*. ed. Dowdall, M.J. & Hawthorne, J.N. Ellis Horwood: Chichester. pp. 297–302.

Stanfield, P.R., Nakajima, Y. & Tamaguchi, K. (1985). Substance P raises neuronal excitability by inward rectification. *Nature* **315**, 498–501.

Starke, K. (1971) Influence of α-receptor stimulant on noradrenaline release. *Naturwissenshaften.* **58**, 420.

Starke, K. (1972). Influence of extracellular noradrenaline on the stimulation-evoked secretion of noradrenaline from sympathetic nerves: evidence for an α-receptor-mediated feed-back inhibition of noradrenaline release. *Naunyn-Schmiedeberg's Archives of Pharmacology*, **275**, 11–23.

Starke, K. (1977). Regulation of noradrenaline release by presynaptic receptors. *Reviews of Physiology, Biochemistry and Pharmacology*, **77**, 1–124.

Starke, K. (1987). Presynaptic α-adrenoceptors. *Reviews of Physiology, Biochemistry and Pharmacology*, **107**, 73–146.

Stjärne, L. (1975). Clonidine enhances the secretion of sympathetic neurotransmitter from isolated guinea-pig tissues. *Acta Physiologica Scandinavica*, **93**, 142–144.

Stjärne, L. (1981). On sites and mechanisms of presynaptic control of noradrenaline secretion. In. *Chemical Neurotransmission: 75 Years*. ed. Stjärne, L. Hedqvist, P. Lagercrantz, H. & Wennmalm, A. Academic Press: London, New York Toronto, Sydney, San Francisco. pp. 257–272.

Stjärne, L. & Åstrand, P. (1984). Discrete events measure single quanta of adenosine 5'-triphosphate secreted from sympathetic nerves of guinea-pig and mouse vas deferens. *Neuroscience* **13**, 21–28.

Stühmer, W., Roberts, W.M. & Almers, W. (1983). The loose patch clamp. In. *Single Channel Recording*. ed. Sakmann, B. & Neher, E. Plenum Press: New York. pp. 123–132.

Surprenant, A. & North, R.A. (1985). μ-opioid receptors and α_2-adrenoceptors coexist on myenteric but not on submucous neurones. *Neuroscience*, **16**, 425–430.

Suszkiw, J.B. & Manalis, R.S. (1987). Acetylcholine mobilization: effects of vesicular ACh uptake blocker, AH5183, on ACh release in rat brain synaptosomes, Torpedo electroplax and frog neuromuscular junction. In. *Cellular and molecular basis of cholinergic function*. ed. A.J. Dowdall & J.N. Hawthorn., Ellis Horwood Ltd: Chichester, England. pp. 323–332.

Suzuki, H. (1981) Effects of endogenous and exogenous noradrenaline on the smooth muscle of guinea-pig mesenteric vein. *Journal of Physiology* **321**, 495–512.

Suzuki, H. (1983) An electrophysiological study of excitatory neuromuscular transmission in the guinea-pig main pulmonary artery. *Journal of Physiology* **336**, 47–59.

Suzuki, H. (1984a) Effects of reserpine on electrical responses evoked by perivascular nerve stimulation in the rabbit ear artery. *Biomedical Research* **5**, 259–266.

Suzuki, H. (1984b) Adrenergic transmission in the dog mesenteric vein and its modulation by α-adrenoceptor antagonists. *British Journal of Pharmacology* **81**, 479–489.

Suzuki, H. (1985). Electrical responses of smooth muscle cells of the rabbit ear artery to adenosine triphosphate. *Journal of Physiology* **359**, 401–415.

Suzuki, H. & Fujiwara, S. (1982) Neurogenic electrical responses of single smooth muscle cells of the dog middle cerebral artery. *Circulation Research* **51**, 751–759.

Suzuki, H. & Kou, K. (1983). Electrical components contributing to the nerve mediated contractions in the smooth muscles of rabbit ear artery. *Japanese Journal of Physiology*, **33**, 743–756.

Suzuki, H., Kamata, S., Kitano, S. & Kuriyama, H. (1979) Differences in electrical properties of longitudinal and circular muscle cells of the rabbit rectum. *General Pharmacology* **10**, 511–519.

Swedin, A. (1971). Biphasic mechanical response of the isolated vas deferens to nerve stimulation. *Acta Physiologica Scandinavica* **81**, 574–576.

Takeuchi, A. & Takeuchi, N. (1959). Active phase of frog's end-plate potential. *Journal of Neurophysiology* **22**, 395–411.

Takeuchi, A. (1976). Excitatory and inhibitory transmitter actions at the crayfish neuromuscular junction. In: *Motor Innervation of Muscle*, pp. 231–261. Ed.: Thesleff, S. Academic Press, London.

Takewaki, T, & Ohashi, H. (1977) Non-cholinergic excitatory transmission to intestinal smooth muscle cells. *Nature* **268**, 749–750.

Tauc, L. (1982). Nonvesicular release of neurotransmitter. *Physiological Reviews*: **62**, 857–893.

Taylor, G.S. & Bywater, R.A.R. (1986) Antagonism of non-cholinergic excitatory junction potentials in the guinea-pig ileum by a substance P analogue antagonist. *Neuroscience Letters* **63**, 23–26.

Thureson-Klein, Å. (1983). Exocytosis from large and small dense cored vesicles in noradrenergic nerve terminal. *Neuroscience*, **10**, 245–252.

Timmermans, P.B.M.W.M. & Van Zwieten, P.A. (1981). The postsynaptic α_2-adrenoceptor. *Journal of Autonomic Pharmacology*, **1**, 171–183.

Tomita, T. (1966a). Electrical responses of smooth muscle to external stimulation in hypertonic solution. *Journal of Physiology* **183**, 450–468.

Tomita, T. (1966b). Membrane capacity and resistance of mammalian smooth muscle. *Journal of Theoretical Biology* **12**, 216–227.

Tomita, T. (1967). Current spread in the smooth muscle of the guinea-pig vas deferens. *Journal of Physiology* **189**, 163–176.

Tomita, T. (1970). Electrical properties of mammalian smooth muscle. In *Smooth Muscle*, ed Bülbring, E., Brading, A., Jones, A. & Tomita T. pp. 197–243. London: Edward Arnold Publ. Ltd.

Tremblay, J.P., Laurie, R.F. & Colonnier, M. (1983). Is the MEPP due to the release of one vesicle or to the simultaneous release of several vesicles at one active zone. *Brain Research Reviews*, **6**, 299–314.

Usherwood, P.R.N. (1972). Transmitter release from insect excitatory motor nerve terminals. *Journal of Physiology*, **227**, 527–551.

Uvnäs, B. & Åborg, C.-H. (1980). Possible role of nerve impulse induced sodium influx in a proposed multivesicular fractional release of adrenaline and noradrenaline from the chromaffin granule. *Acta Physiologica Scandinavica*, **109**, 363–368.

Van Breemen, C. & Saida, K. (1989) Cellular mechanisms regulating $[Ca^{2+}]_i$ smooth muscle. *Annual Review of Physiology* **51**, 315–329.

Van Helden, D.F. (1988a) Electrophysiology of neuromuscular transmission in guinea-pig mesenteric veins. *Journal of Physiology* **401**, 469–488.

Van Helden, D.F. (1988b) An α-adrenoceptor-mediated chloride conductance in mesenteric veins of the guinea-pig. *Journal of Physiology* **401**, 489–501.

Wakade, A.R. & Krusz, J. (1972). Effect of reserpine, phenoxybenzamine and cocaine on neuromuscular transmission in the vas deferens and seminal vesicle of the guinea-pig. *Journal of Pharmacology and Experimental Therapeutics*, **181**, 310–317.

Wernig, A. (1975). Estimates of statistical release parameters from crayfish and frog neuromuscular junctions. *Journal of Physiology*, **244**, 207–221.

Wernig, A. & Stirner, H. (1977). Quantum amplitude distribution point to functional unity of the synaptic active zone. *Nature*, **269**, 820–822.

Westfall, D.P., Stitzel, R.E. & Rowe, J.N. (1978). The postjunctional effects and neural release of purine compounds in the guinea-pig vas deferens. *European Journal of Pharmacology* **50**, 27–38.

Westfall, T.C. (1977). Local regulation of adrenergic neurotransmission. *Physiological Reviews*, **57**, 659–728.

White, T.D. (1988). The role of adenine compounds in autonomic neurotransmission. *Pharmacology & Therapeutics* **38**, 129–168.

Williams J.T. & North, R.A. (1985). Catecholamine inhibition of calcium action potentials in rat locus coeruleus neurons. *Neuroscience*, **14**, 103–109.

Williams, J.T., Henderson, G. & North, R.A. (1985). Characterization of α_2-adrenoceptors which increase potassium conductance in rat locus coeruleus neurones. *Neuroscience*, **14**, 95–101.

Winkler, H. & Westhead, E. (1980). The molecular organization of the chromaffin granules. *Neuroscience*, **5**, 1803–1823.

Winkler, H., Fischer-Colbrie, R. & Weber, A. (1981). Molecular organization of vesicles storing transmitter: chromaffin granules as a model. In: *Chemical Neurotransmission 75 Years*, pp 57–68. Eds: Stjärne, L., Hedqvist, P., Lagercrantz, H. & Wennmalm, Å. Academic Press: London.

Zimmermann, H. (1979). Vesicle recycling and transmitter release. *Neuroscience*, **4**, 1773–1804.

Zimmermann, H. & Whittaker, V.P. (1977). Morphological and biochemical heterogeneity of cholinergic synaptic vesicles. *Nature*, **267**, 633–635.

Zucker, R.S. (1974a). Crayfish neuromuscular junction facilitation activated by constant presynaptic action potentials and depolarizing pulses. *Journal of Physiology*, **241**, 69–89.

Zucker, R.S. (1974b). Excitability changes in crayfish motor neurone terminals *Journal of Physiology*, **241**, 111–129.

4 Signal Transduction Mechanisms

C.D. Benham[1]

*Dept of Pharmacology, SmithKline Beecham Research &
Development, The Frythe, Welwyn, Herts, AL6 9AR. U.K.*

INTRODUCTION

The purpose of this chapter is to lay out the mechanistic basis by which signalling
by autonomic transmitters, after binding to their cell surface receptors, is transduced
into a response in the effector cell. The importance of a rise in the free cytoplasmic
calcium concentration, $[Ca^{2+}]_i$ in triggering smooth muscle and cardiac muscle contraction and in stimulating secretion from gland cells has been recognised for a long
time. However, most of the details of how the electrical excitability of cells and
second messenger systems control influx of Ca^{2+} across the cell membrane and
release of internal stores of Ca^{2+} eluded investigation. Understanding of this aspect
of cell signalling has increased enormously during the 1980s, fuelled by the development of new techniques for examining cell physiology at the single cell level. This has
helped to bridge the gap between biochemistry and physiology allowing some of the
biochemical systems that have been identified and characterised in cell-free systems
to be studied in whole cells.

   Perhaps the most important advance was the development of the patch clamp
technique by Neher, Sakmann and colleagues in the late 1970s (Hamill *et al.* 1981).
For the first time it was possible to study the membrane currents that underlie cell
excitability while precisely controlling both the external and internal cellular environment. Putative internal second messenger systems identified using biochemical
techniques could now be inserted into dialysed cells and the roles of the various
components of the system elucidated.

[1] Present Address
SmithKline Beecham Pharmaceuticals
The Pinnacles
Harlow, Essex CH19 5AD U.K.

The central role of $[Ca^{2+}]_i$ in cell activation has placed a high priority on being able to measure this parameter. The new fluorescent Ca^{2+} indicator dyes developed by Roger Tsien have transformed $[Ca^{2+}]_i$ measuring technology from a preserve of only a few specialised laboratories to a resource that is now available in most University departments of Physiology and Pharmacology. This second major development has greatly increased the pace of work directed to understanding how $[Ca^{2+}]_i$ is regulated in cells. One or other and in some cases a combination of both techniques allied with biochemical and molecular biology techniques for purifying the intermediates of signal transduction pathways outlined in Fig. 4.1 have helped to burn away some of the grey fog that shrouded the events that occurred immediately after receptor activation and before an end response was apparent. The important advances have been in two areas. Firstly, G-proteins have been identified as molecular transducers coupling the majority of receptor molecules to channels or second-messenger-generating enzymes. The second important advance has been the confirmation that inositol phosphates are intracellular messengers with almost ubiquitous distribution in mammalian cells, standing alongside the cyclic nucleotide second messenger systems. This has led to major progress in understanding the mechanisms of Ca^{2+} release in a variety of cell types.

The scope of this chapter is vast, but by concentrating on the important mechanisms found in effector tissues that are autonomically innervated the task has been made a little simpler. Nevertheless it should be pointed out that the chapter

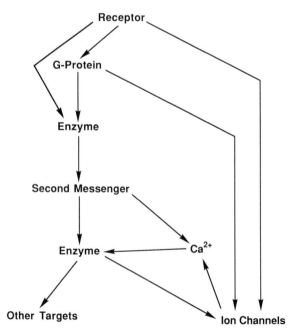

FIGURE 4.1   Generic signal transduction pathways. Receptors are intrinsically coupled to enzymes or ion channels or more often mediate their actions through G-protein activation.

is weighted towards considering the roles of Ca^{2+} in signalling, while cyclic nucleotides and other second messengers are less comprehensively treated. This betrays the author's interests, but can also be justified in that these are the areas where most progress has been made in the last five years.

Needless to say, the general concepts discussed here apply to signalling in other systems. Thus in central neurons, glutamate receptors activate transduction mechanisms analogous to those activated by ATP in the periphery. The transduction pathways activated by catecholamine, muscarinic and purinergic receptors are currently being rapidly unravelled. However, less progress has been made on the mechanisms of actions of peptides that may be co-released such as neuropeptide Y, substance P and vasoactive intestinal polypeptide (VIP). This chapter will consider different stages in the transduction cascades outlined in Fig. 4.1 in turn, in order to provide a different emphasis from the other chapters in this volume.

RECEPTORS

Transmission of signals across the impervious lipid bilayer that surrounds eukaryotic cells is the key process in cellular signal transduction. Molecular biology has identified structural homologies between receptors which have helped to clarify the classification and evolutionary origins of the different classes of molecules that perform this function. Cell surface receptors can on this basis be classified into several groups (Barnard *et al.* 1989).

RECEPTOR FAMILIES

G-protein-coupled receptors

These receptors are coupled to a GTP binding protein that in turn activates an enzyme to release a diffusible second messenger such as cAMP or inositol trisphosphate, or directly regulates ion channel activity. The large number of receptors that are coupled via G proteins are structurally related. Each consists of a single polypeptide chain of 400 to 500 amino acids with a characteristic structural pattern of seven membrane-spanning hydrophobic domains that show considerable sequence homology, indicating that they belong to a receptor superfamily with a common ancestor (Fig. 4.2A) (Dohlman *et al.* 1987). Between the hydrophobic domains that are thought to form alpha helices are hydrophilic loops, of which the one linking transmembrane sequences 5 and 6 is thought to contain the area that interacts with the G-protein. This cytoplasmic loop is the only one with appreciable differences in length between the various receptors in the family. The C-terminal tail in the cytoplasmic domain is also of variable length. Catecholamine, muscarinic, $GABA_B$ and many peptide (vasopressin, oxytocin and tachykinins) receptors belong to this super-family (North, 1989). One of the main goals of receptor molecular biology at the present time is to understand the important features of this highly conserved structure that make it so well suited to signal transduction. There are several approaches to this problem. Firstly, deletion mutants can be produced that

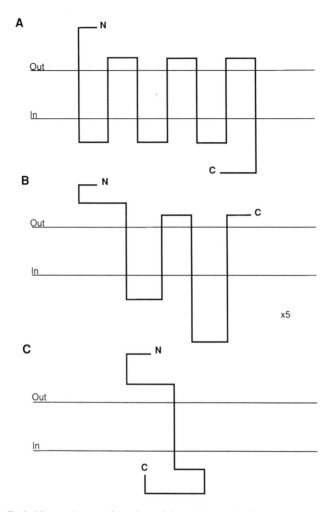

FIGURE 4.2   Probable membrane orientation of three classes of cell membrane receptors. Membrane spanning domains have been deduced from hydrophobic sequences in the primary polypeptide structure. A, The seven membrane spanning sequence structure typical of the G-protein receptor coupled superfamily. B, Structure of one sub-unit of the ACh receptor-channel molecule. All the five sub units have this basic four membrane-spanning sequence structure. C, The much simpler structure of the ANP receptor that is directly coupled to guanylyl cyclase.

lack specific regions of the receptor and the effect on function is then examined in a suitable expression system. Secondly, site-directed mutagenesis can be used to alter single amino acids in the primary sequence. Both approaches have been used to look at the function of voltage gated sodium channels (Stuhmer *et al.* 1989). However, these approaches suffer from the potential drawback that rather than having direct effects on function, they are allosterically modulating activity of another part of the molecule. A further technique is to create chimeras from different receptors. This technique has been used for study of catecholamine receptor structure-function

(Kobilka *et al.* 1988). The great power of this technique is that the experimenter is trying to confer new properties in the chimera rather than just trying to destroy selectively one aspect of function in the deletion and mutation experiments. Thus if for example, α_2-inhibition of adenylate cyclase can be converted to stimulation by incorporation of a part of the β_2-receptor, then this would be strong evidence that that part of the structure was involved in binding to the stimulatory protein G_s. As well as being theoretically elegant, the technique works in practice. Precisely this type of experiment has been used to show that in the β adrenoceptor, the region that includes the fifth and sixth membrane-spanning domain and the second extracellular and third cytoplasmic loop is involved in determining G-protein specifity (Lefkowitz *et al.* 1989).

Receptor–channel complexes

Conceptually and operationally simpler are the agonist-gated ion channels where the receptor is an integral part of an ion channel, and the opening and closing of the channel is directly controlled by agonist binding (Hille, 1989). The receptor-channel molecule has several representatives expressed in neurons (nicotinic acetylcholine (ACh), glutamate, 5-hydroxytryptamine$_3$ (5-HT$_3$), GABA$_A$ and glycine) (Barnard *et al.* 1989) but the P$_{2x}$-purinoceptor is the only clear example in autonomically innervated effector structures.

Although the purinoceptors have not yet been sequenced it is likely to have a similar structure to the acetylcholine, GABA and glycine receptor channels. Perhaps the most important finding from the structural studies is that each receptor class comprises a number of similar, but not identical, molecules that have their own specific (tissue) distribution. Although functional studies of nicotinic receptors had already suggested this, the actual diversity has exceeded expectations particularly for the other channels such as the GABA$_A$ receptor channel complex (Sieghart, 1989).

The archetypal ACh channel is a pentamer of 270 kDa with subunit stoichiometry of α_2, β, γ, δ. These subunits are descendants of a single ancestral gene so that they retain considerable sequence homology. Each subunit has four hydrophobic domains suggesting four membrane-spanning sequences; these are illustrated in Fig. 2B. This allows the subunits to arrange themselves in the membrane like the staves of a barrel, a structure consistent with electron microscope images (Toyoshima & Unwin, 1988). Each subunit is thought to present one membrane spanning sequence towards the centre of the structure creating the channel pore. The creation of ACh receptor/channel chimeras, this time utilising interspecies differences in channel function, has been important in defining the M2 transmembrane region as the part of the subunit that lines the pore (Imoto *et al.* 1986).

Receptor–enzyme complexes

The agonist acts on a receptor which is itself an enzyme catalysing generation of second messenger on the inner surface of the cell membrane. A clear example of this is the activation by atrial natriuretic peptide of the membrane form of guanylyl cyclase. The isolation, sequencing and expression of the complementary DNA clone encoding the membrane form of guanylyl cyclase has revealed that this same protein is the ANP receptor (Chinkers *et al.* 1989). The deduced amino acid sequence suggests that the 1057 amino acids are arranged as single extracellular and intra-cellular domains of about equal size, separated by one membrane-spanning region (Fig. 4.2c). Although such definitive structural evidence does not exist for other members of this class it may be a general transduction system for some peptides. It has been suggested that peptides such as bradykinin can activate phospholipase A_2 directly, leading to arachidonic acid release. This would imply that the brady-kinin receptor binding site is intrinsic to at least one form of PLA_2 (Burch & Axelrod, 1987). A further group of peptide receptors with the same structural motif of a single membrane spanning domain are the growth factor receptors that are coupled to protein tyrosine kinase (Carpenter, 1987). These two families could be considered as a third major superfamily of receptors. This large area of long-term modulation of cell function will not be considered further in this chapter.

As many of the aspects of receptor pharmacology will be discussed in the other chapters in this volume this chapter will immediately move to the processes that follow receptor binding. As laid out in Fig. 4.1 there are now clear pathways mapped out by which the various receptors are coupled to their final effector targets. The various components of these pathways will be considered in turn. The complex interrelationships of $[Ca^{2+}]_i$ with other components of the pathway make sequential analysis of this latter part of the process, outlined in the lower half of the diagram, rather difficult. The ATP gated channel described in the next section is an example of the direct transduction pathway shown on the right side of Figure 4.1. This is not the most widespread form of receptor coupling, but its operational simplicity makes it the obvious place to start a mechanistic survey.

EXTRACELLULAR RECEPTORS DIRECTLY COUPLED TO ION CHANNELS

Conceptually, the simplest mechanism for signal transduction where a rise in $[Ca^{2+}]_i$ is the desired result would be if a cell surface receptor was an intrinsic part of a Ca^{2+}-permeable channel. Thus, ligand binding would directly gate Ca^{2+} influx and cell activation. This type of transduction mechanism which was termed receptor-operated Ca^{2+} entry (Bolton, 1979; Van Breemen et al. 1979) has now been characterised in smooth muscle cells. However, the original hypothesis of a purely Ca^{2+}-permeable conductance has had to be amended slightly because many channels that transport Ca^{2+} into the cell do not share the high selectivity of voltage-gated Ca^{2+} channels. They also admit other cations. These channels can thus have dual excitatory roles, both depolarising the cell due to Na^+ influx which indirectly leads to Ca^{2+} entry and also directly admitting Ca^{2+}.

A body of recent data now supports the idea that P_{2x} purinoreceptors are coupled to a cation permeable channel in vascular and visceral smooth muscle (Benham & Tsien, 1987; Friel, 1988). This conductance has many similarities to the intensively studied nicotinic ACh receptor-channel complex that mediates the endplate current at the neuromuscular junction in skeletal muscle. These channels are permeable to a variety of cations such that the reversal potentials are close to zero mV in normal physiological ion gradients. This means that at the resting membrane potential, an inward current will flow tending to depolarise the cell.

In many vascular smooth muscles, application of ATP causes contraction. Membrane potential recordings show that ATP causes an increase in membrane conductance and depolarisation, which can reach the threshold for action potential generation (Suzuki, 1985). More detailed study of the conductance increase has been performed using isolated single smooth muscle cells and patch clamp techniques. These have shown that ATP opens a cation-permeable channel with a latency of less than 100 ms and can activate unitary currents in isolated outside out membrane patches, suggesting that the receptors are closely coupled to the ion channel (Benham & Tsien, 1987). It is unlikely that a diffusible second messenger is involved but a linkage through a membrane coupled G-protein cannot be ruled out. However, it is tempting to make a direct comparison between this conductance at the smooth muscle neuroeffector junction and ACh at the skeletal neuromuscular junction. The functional analogy is not very precise because the fast signal transmission that the directly coupled receptor channel complex provides is not so important in smooth muscle as it is in skeletal muscle function.

Electrophysiological experiments suggested that the ATP-gated channels of vascular smooth muscle have a Ca^{2+} permeability about 10-fold higher than that of the skeletal muscle ACh-gated channels (Benham & Tsien, 1987). So while the low Ca^{2+} permeability of the ACh channels permits negligible Ca^{2+} entry at the skeletal neuromuscular junction, (the situation may be different for neuronal nicotinic channels), a biologically important Ca^{2+} influx might occur through the ATP-gated channels. A direct demonstration of this has been provided by combining Ca^{2+} measuring dye technology and single cell voltage-clamp. Using this combination of techniques it is possible to look at agonist mediated changes in $[Ca^{2+}]_i$ while excluding any contribution from Ca^{2+} entry through voltage-gated Ca^{2+} channels. These experiments have clearly shown that ATP-evoked currents carry enough Ca^{2+} into the cell to elevate $[Ca^{2+}]_i$ significantly. The rise in $[Ca^{2+}]_i$ is dependent on extracellular Ca^{2+} and is not due to triggering of release from internal stores (Fig. 4.3, Benham, 1989).

This pathway provides a mechanism whereby $[Ca^{2+}]_i$ can be elevated preferentially under the ATP activated channels, so that such a rise will be confined to cells that are exposed to the transmitter. This is in addition to the effects on membrane potential which modulate the activity of voltage-dependent Ca^{2+} channels. In this case, because many effector cell systems are electrically coupled, the depolarisation caused by ATP will spread electrotonically and if the threshold for action potential generation is reached then a propagating wave of excitation will be evoked. This latter case where voltage-gated channels are opened is probably the major excitatory pathway when exogenous ATP is applied to smooth muscle and explains the sensitivity of contractions to voltage-dependent Ca^{2+} channel antagonists such as dihydropyridine derivatives. However, lower level stimulation by ATP which might result from ATP release from sympathetic nerves stimulated at low frequency when action potentials are not evoked, may depend more on the direct route of Ca^{2+} entry providing a steady low-level Ca^{2+} influx (Benham, 1989). Clearly, experimental tests of these suggestions in whole tissues are now required.

a.

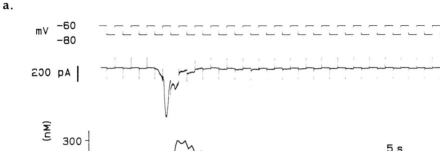

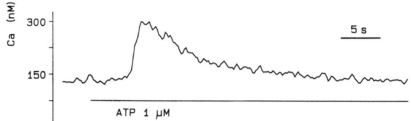

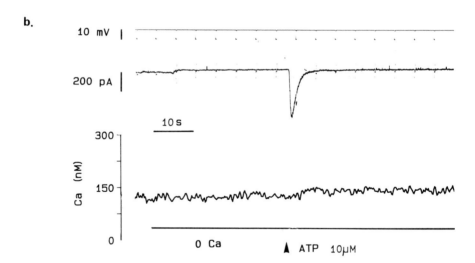

b.

FIGURE 4.3   ATP-gated channels admit sufficient Ca^{2+} to elevate $[Ca^{2+}]_i$ in vascular smooth muscle cells. Simultaneous recording of membrane potential (top), membrane current (middle) and $[Ca^{2+}]_i$ (lower traces) in single smooth muscle cells from rabbit ear artery. a. Cell bathed in saline containing 1.5 mM Ca^{2+}. b, $[Ca^{2+}]_i$ response but not current is abolished in the absence extracellular Ca^{2+}. (From Benham, 1989).

G-PROTEINS

The remainder of the transduction pathways described that are activated by autonomic transmitters mediate their actions through one of a family of closely related proteins that provide a link between the receptor protein and a membrane bound enzyme or ion channel.

G-PROTEIN DIVERSITY

The initial background to this field was provided by studies on the adenylate cyclase system. After the original observation by Rodbell and colleagues (1971) that GTP as well as glucagon was required for adenylate cyclase stimulation, subsequent studies lead to the realisation that the transmembrane signalling complex included at least three proteins; the cell surface receptor, the adenylate cyclase catalyst and a transducer protein. Purification of the transducer revealed a guanine nucleotide binding protein that stimulated guanylyl cyclase which was named G_s. The bidirectional nature of the enzyme regulation was explained by the subsequent purification of G_i, the intermediate necessary for receptor mediated inhibition (Bokoch et al. 1983). The search for other G-protein coupled receptors was then spurred by the discovery of G_o (o for other) Neer et al, 1984), which at that time had no known function. In the last five years molecular biology has defined the diversity of these molecules. A concise recent review is provided by Freissmuth, Casey and Gilman (1989).

Signal-transducing G proteins are heterotrimers made up of α, β and γ subunits. While the α subunits confer substrate specificity, the role of the $\beta\gamma$ subunit appears to be to deactivate α as the oligomer is formed (see below for detailed mechanism). G-protein oligomers are classified by the identity of their α subunit. 12 polypeptides are the products of the nine genes that encode α subunits. The α subunits contain the guanine nucleotide binding domain and have GTPase activity. The G-proteins considered here are all activated when they bind GTP. Consequently in the presence of non-hydrolysable GTP analogues the G-protein is continuously activated as the analogue remains continuously bound (Freissmuth et al. 1989). The α subunits also contain the sites for ADP ribosylation by cholera toxin to give persistent activation of $G_{s\alpha}$ and by pertussis toxin to inhibit the activity of $G_{i\alpha}$ and $G_{o\alpha}$. These toxins have proved to be valuable tools for defining G-protein coupled events and descriminating between different G-proteins through their α subunit specificity (Dolphin, 1987).

$G_{s\alpha}$

Most tissues contain a mixture of two forms of G_s, a long and a short form. These have generally similar properties except that the GDP-dissociation rate for the long form is about 2.5-fold faster than the short form. This means that the long form will have a higher resting GTP occupancy and hence basal activity in the absence of agonist (see below). Thus the relative expression of the subtypes of G_s may determine the resting activity of the pathway and the shape of the agonist-mediated response (Graziano et al. 1989). In addition to the classical coupling with adenylate cyclase, this G-protein also appears to be directly coupled to cardiac L-type Ca^{2+} channels (Mattera et al 1989). This finding may provide an explanation as to how cardiac cells can respond so quickly to β adrenoceptor activation.

$G_{i\alpha}$ (at least 3) and $G_{o\alpha}$

All these subunits are ADP-ribosylated by pertussis toxin so they are candidates for involvement in pathways disrupted by this toxin. In addition to inhibiting adenylate cyclase, G_i also couples muscarinic receptor activation to potassium channel opening in cardiac cells (Yatani *et al*. 1988). In some but not all cells, phospholipase C activation is inhibited by pertussis toxin but conclusive roles for G_i or G_o have not yet been established.

G-PROTEIN REGULATORY CYCLE

A proposed model for the regulatory cycle has been constructed now that individual components have been identified and purified, allowing the system to be reconstituted. In essence the cycle involves the G-protein α subunit cycling between a GDP-bound oligomer that is inactive (1) and an activated, GTP-bound monomer (2) (Fig. 4.4). In the absence of agonist-receptor complex the cycle proceeds slowly around the dotted activation pathway to give low basal activation of the effector molecule such as adenylate cyclase (in the case of G_s). The catalytic rate of conversion of GTP to GDP is about 10-fold faster than the rate of dissociation of GDP which is thus the rate-limiting step and of crucial importance to the regulatory function of G-proteins. If an agonist receptor complex is present then the faster righthand activation path is followed. The binding of agonist to the receptor promotes interaction with the G-protein which dramatically lowers the affinity for GDP causing it to dissociate. This frees the G-protein to bind GTP which has the effect of causing the ternary complex to dissociate (as agonist receptor affinity is

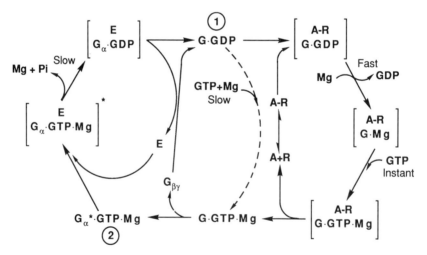

FIGURE 4.4 The regulatory cycle for G-protein-mediated signal transduction. The G-protein α-subunit cycles between the inactive GDP bound oligomer (1) and the activated GTP bound monomer (2). In the absence of agonist the cycle proceeds very slowly on the dashed pathway. G, G protein; E, Effector; A, agonist; R, receptor. Asterisk indicates active species. (modified from Freissmuth *et al*. 1989).

lowered) freeing the receptor to activate more G-protein and the activated α subunit to dissociate from the β complex and interact with its effector target such as the enzyme molecule adenylyl cyclase.

The signal is terminated when GTP is hydrolysed by the α subunit. This process is slow so that the average lifetime of the activated subunit is many seconds similar to the rate of deactivation of membrane bound adenylyl cyclase (Levitski, 1988). This long activated-subunit lifetime allows for considerable signal amplification. Thus there are two points of amplification in the system, the number of α subunits activated by a receptor and the number of effectors activated by each α subunit. Although the above mechanism has much experimental support it should be remembered that the data comes generally from detergent solubilised preparations. Subunit dissociation and exchange have not yet been demonstrated in the plasma membrane.

SECOND MESSENGERS

SIGNALLING THROUGH CYCLIC AMP

Since its discovery by Rall *et al.* (1957), cAMP and the hormone regulated adenylyl cyclase that generates this second messenger have been studied as a model of transmembrane signalling pathways. Cellular levels of cAMP are dependent on the relative rates of its production by adenylyl cyclase and its breakdown by a variety of phosphodiesterases. The target enzyme for cAMP is cAMP-dependent Protein Kinase (cAMP-PrK) (Fig. 4.5). The best characterised target protein is the L-type Ca^{2+} channel in cardiac muscle cells which is upregulated by phosphorylation (see later).

Adenylyl cyclase

Adenylyl cyclase is localised to the inner leaflet of the cell membrane. The same enzyme system seems to be present in a variety of tissues. Tissue specificity lies in the differential expression of receptors that are coupled to the cyclase and the cellular targets of cAMP-PrK. As described in the section on G-proteins, the enzyme is under dual control from stimulatory and inhibitory G proteins G_s and G_i. Binding of $G_{\alpha s}$ to adenylyl cyclase activates the catalytic subunit resulting in increased production of cAMP. Conversely, $G_{\alpha i}$ binding results in inhibition of the catalytic subunit. cAMP-mediated effects have often been identified by examining the actions of cholera toxin and forskolin, agents that increase cAMP levels by activating adenylyl cyclase. Cholera toxin acts at G_s while forskolin activates adenylyl cyclase directly (but also has other actions).

Cyclic nucleotide phosphodiesterases (PDEs)

In contrast to the homogeneous distribution of adenylyl cyclase, it has now been clearly demonstrated that there are multiple isoforms of PDE and that these isoforms are differentially expressed. The isoforms were originally named numerically with Roman numerals according to their order of elution from anion exchange resins.

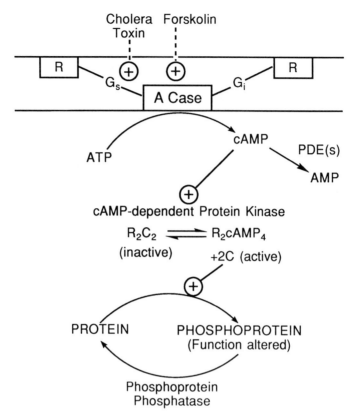

FIGURE 4.5 The cAMP transduction pathway. R, receptor; G_s & G_i, stimulatory and inhibitory GTP regulatory proteins; A Case, adenylyl cyclase; R_2 & C_2, regulatory and catalytic subunits of cAMP-dependent protein kinase.

Unfortunately, these peaks of activity are now known to contain multiple isozymes (for review see Murray, 1990). The growing body of sequence information now suggests that there are five main families of PDE inhibitors. Within these basic groups, sub-families have been distinguished. The twenty isozymes so far identified have specific tissue distribution providing a rationale for selective therapeutic intervention (Beavo & Reifsnyder, 1990). Beavo has proposed that should now be named according to substrate and regulators. One family (V) specifically hydrolyses cGMP. The other four families so far identified that hydrolyse cAMP are:

I. Ca^{2+}/Calmodulin dependent PDE. This family now numbers seven enzymes with similar K_m for both cyclic nucleotides. Isozymes specific to brain, heart and smooth muscle have been identified (Beavo, 1988). Use of relatively selective inhibitors such as vinpocetine, which appears to bind to the catalytic site rather than the calmodulin binding site, suggest that in whole tissue the primary target of this PDE is cGMP.

II. cGMP stimulated PDE. Activity of this enzyme is increased dramatically by micromolar concentrations of cGMP. It is a major form of PDE in many tissues including cardiac muscle cells but is not found in vascular smooth muscle cells (Beavo, 1988).

III. cGMP inhibited PDE. Interest in this enzyme has increased recently as it is the target of a new class of cardiotonic agents (the PDE III inhibitors) such as milrinone (Silver *et al*. 1988). The enzyme has a low (0.1 μM) K_m for both cAMP and cGMP and is inhibited by cGMP with a similar K_i.

IV. cAMP-specific PDE. The substrate properties of this form are similar to III, but it is not inhibited by cGMP or by the milrinone class of PDE inhibitors. The list demonstrates the complexity of the control of cAMP levels and also the close interrelationship with cGMP levels. This provides a partial explanation of the varying consensus as to the roles of the two second messengers in different tissues. The complexity requires that the enzymes are characterised in cell-free, partially purified conditions, but properties deduced in these systems do not take into account important factors such as cellular compartmentation of the second-messenger-generating enzymes and targets. Clearly if a cyclic nucleotide is generated in one compartment by a given receptor then it can only act on targets in the same compartment in intact cells. There is evidence that cAMP is compartmentalised in both cardiac and smooth muscle (Murray *et al*. 1989). Thus, modulation of cAMP levels may not have any end effect if the cAMP level has changed in a different compartment from the site of the target molecule. Monitoring every stage in the transduction pathway is obviously necessary to get a complete picture of the important features of the regulation in any particular cell.

Cyclic AMP-dependent protein kinase (cAMP-PrK)

The cAMP target enzyme, cAMP-PrK, appears to be ubiquitously distributed in mammalian tissues. The inactive holoenzyme is made up of four subunits, two catalytic and two regulatory. When two cAMP molecules bind to each of the regulatory subunits the two catalytic subunits are released and activated by this dissociation step (reviewed by Beebe & Corbin, 1986). Identification of two isoenzymes separated by ion exchange chromatography which differ in their R subunits has now been followed by separation of at least four R subunits and two C subunits using molecular biological techniques (McKnight *et al*. 1988). However, to date the functional significance of this diversity is not known, nor is that of the separate particulate and soluble fractions of the enzyme. The distribution of type I and II isoenzymes show no differences in various vascular tissues (Silver, 1987). In most tissues cAMP-PrK is the predominant form of PrK A, but in vascular smooth muscle cGMP-dependent protein kinase is present in approximately equal amounts.

cAMP-PrK targets

There are a multitude of protein targets whose function is altered when they are phosphorylated by cAMP-Pr-K. The most important ones described in cardiac cells are discussed in the last section of the chapter and are listed in figure 8. Additionally, phosphodiesterase targets and ion channel targets are discussed in more detail in the appropriate sections.

SIGNALLING THROUGH CYCLIC GMP

Unlike cAMP, a clear primary role for this second messenger in autonomic neuro-effector transmission has not been identified. However, as alluded to in the above section, by sharing many of the targets of cyclic AMP and regulating cAMP levels by its actions on PDE, it may counter the actions of cAMP in cardiac cells where ACh has been shown to directly stimulate a rise in cGMP. cGMP levels can be raised by a variety of agonists. In addition to atrial natriuretic peptide (ANP) (Chinkers *et al* 1989) acting on the cell surface, guanylyl cyclase stimulation by nitric oxide is a further important pathway utilised both by endogenous endothelium-dependent relaxing factor (EDRF) and the nitrosovasodilators used therapeutically. Non-adrenergic, non-cholionergic (NANC) nerve stimulation in the bovine retractor penis and in the rat anococcygeus smooth muscles stimulates cGMP production in the smooth muscle and this may be mediated through nitric oxide (Collier & Vallance, 1989).

Guanylyl cyclase

Guanylyl cyclase exists as several isoforms with m.w. of approx 150,000. The most straightforward distinction is between particulate and soluble forms of the enzyme. The compartmentation of the guanylyl cyclases provides an obvious basis for differential activation and selective targetting at effector molecules. In addition to identifying particulate guanylyl cyclase in plasmalemma it has also been seen in endoplasmic reticulum, Golgi and nuclear membranes suggesting there are multiple particulate forms (Waldman & Murad, 1987). ANP stimulation is confined to the particulate guanylyl cyclase and this fraction is relatively insensitive to nitric oxide. Soluble guanylyl cyclase from a variety of tissues has similar molecular properties and has long been known as the substrate for nitrosovasodilators (Murad, 1986). Nitric oxide interacts with the haem group on the enzyme increasing V_{max} by 15-30 fold without affecting the K_m for its substrate, GTP. There is some evidence that free radicals also can directly activate the soluble cyclase.

Phosphodiesterases

Hydrolysis of cGMP can potentially be catalysed by several PDEs with some overlap with those mediating cAMP breakdown. In vascular smooth muscle the most important may be the Ca^{2+}/Calmodulin-dependent PDE. The K_m of this enzyme for cGMP is about 10-fold lower than for cAMP so cGMP may be the physiological substrate. In addition to this group, the type V PDEs are specific for cGMP and at least one form is present in smooth muscle. A relatively specific inhibitor of this group is zaprinast (Beavo & Reifsnyder, 1990).

Cyclic GMP-dependent kinases

Like the analogous cAMP-dependent kinase the protein kinase activated by cGMP requires two nucleotide molecules to bind for activation. However, there are several differences in their properties. Firstly the tissue distribution of the cGMP kinase is localised to smooth muscle, heart, lung, cerebellum and platelets (Walter, 1981) in contrast to the ubiquitous distribution of the cAMP-PrK. Activation of the kinase does not require dissociation of the enzyme into active subunits (Gill *et al.* 1976) and the target specifity of this kinase is, not surprisingly, quite different.

Additional targets for cGMP are at least two PDEs. The cGMP-stimulated cAMP PDE is not found in smooth muscle but this may be an important substrate in cardiac muscle for mediating cGMP effects via control of cAMP. In intestinal smooth muscle a cGMP selective PDE has been identified with as yet unknown function.

Targets for cGMP dependent kinase

Numerous proteins are phosphorylated by cGMP in both broken-cell and whole-tissue preparations, however, to date none of these proteins have been identified (Lincoln, 1989). So, this has not been a successful approach to describing the mechanism of action of cGMP so far. However, the kinase has been shown to be important for mediating the actions of cGMP in smooth muscle as loss of the enzyme during prolonged smooth muscle cell culture results in loss of responsiveness to cGMP and this can be restored by incorporating the kinase into permeabilised cells (Lincoln, 1989). The final effector that mediates the reduction in $[Ca^{2+}]_i$ stimulated by cGMP in smooth muscle seems to be the plasmalemmal Ca^{2+} ATPase (see later) but this protein is not itself phosphorylated. Clearly some as yet unidentified intermediate is required in the transduction pathway.

SIGNALLING THROUGH INOSITOL PHOSPHATES

Within 10 years, the role of inositol phoshates as a second messenger pathway has moved from being a controversial topic in the early 1980s to being today universally accepted as an important and almost ubiquitous pathway within animal cells. The key elements in this pathway are set out in Fig. 4.6. Membrane-bound phosholipase C (PLC) is activated by a G-protein to hydrolyse phosphatidylinositol 4,5-bisphosphate. The products are inositol trisphosphate and diacylglycerol. The targets for these two second messengers are intracellular Ca^{2+} store and protein kinase C, respectively. The major exceptions in mammalian cells where this bifurcating signal pathway is not important are cardiac and skeletal muscle. In these cells Ca^{2+} stores are released by faster, more direct means than the mechanism of $Ins(1,4,5)P_3$ induced Ca^{2+} release. Agonist stimulated phosphoinositide metabolism was first demonstrated by Hokin & Hokin (1953), who showed that ACh induced ^{32}P incorporation into phosholipids in porcine pancreas. Subsequent study of inositide metabolism in a variety of tissues and the correlation with the ability of agonists to mobilise Ca^{2+} alerted Michell (1975) to the possibility that inositide turnover was causally involved in Ca^{2+} mobilisation. Work over the last 10 years (reviewed by Berridge & Irvine 1984, 1989) has provided detailed information on many aspects of this pathway culminating in the very recent elucidation of the primary structure of the $Ins(1,4,5)P_3$ receptor (Furuichi, *et al.*, 1989).

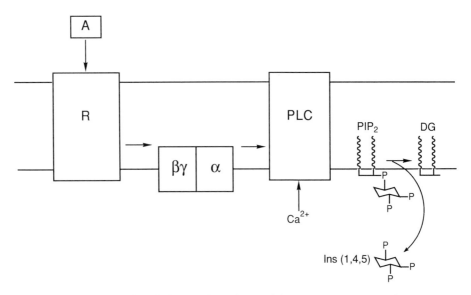

FIGURE 4.6   Generation of inositol 1,4,5 trisphoshate and diacylglycerol (DG). Agonist (A) – receptor (R) binding stimulates α subunit dissociation a G-protein that activates phospholipase C (PLC). PLC cleaves phosphatidyl inositol 4,5-bisphosphate (PIP$_2$) to yield the two messengers. DG remains membrane bound while Ins (1,4,5) P$_3$ is released into the cytoplasm.

The inositol phosphate metabolic pathway

Ins(1,4,5)P$_3$ is formed by cleavage of phosphatidylinositol 4,5-bisphosphate (PtdIns(4,5)P$_2$), which is one of the inositol lipids found in the inner leaflet of the plasma membrane (Fig. 4.6). The phosphodiesterase cleavage also yields 1,2-diacylglycerol which remains localised in the membrane where it can activate protein kinase C. The Ins(1,4,5)P$_3$ formed is released into the cytoplasm, diffuses to the sites of internal Ca^{2+} stores and triggers Ca^{2+} release. Activation of the phosphatidylinositol-specific phosphodiesterase (also called phospholipase C (PLC)) by agonists is achieved by G-protein activation in an analogous manner to the activation of adenylyl cyclase. The specific G-protein has been termed G$_p$ (Cockroft, 1987) although it is not yet clear whether one or more G-proteins can activate PLC in different cells. PLC is also a Ca^{2+}-dependent enzyme.

Some of the key evidence for a causal role of IP$_3$ in Ca^{2+} release was the confirmation of the correct temporal relationship between the appearance of the signals following receptor activation. Important advances were firstly the demonstration that PtdIns(4,5)P$_2$ breakdown was a very early event and was independent of extracellular Ca^{2+} (Michell *et al.*, 1981). Secondly and most importantly, in salivary glands IP$_3$ production followed serotonin stimulation without any detectable lag while the Ca^{2+} dependent physiological response to serotonin was delayed by at least a second (Berridge, 1983). At this time it was not realised that there are more than one naturally occurring IP$_3$. In addition to Ins(1,4,5)P$_3$, Ins(1,3,4)P$_3$ is also found and generally builds up in cells more slowly than Ins(1,4,5)P$_3$. The function of this isomer and the role of Ins(1:2cyc,4,5)P$_3$ is not as yet understood. They may just be biologically inactive intermediates. Ins(1,3,4)P$_3$ may be formed from the hydrolysis of Ins(1,3,4,5)P$_4$ which does seem to have a role in control of Ca^{2+} signalling (see later).

The concentration of Ins(1,4,5)P$_3$ is determined by the balance between its rate of formation from PtdIns(4,5)P$_2$ and its breakdown by hydrolysis to Ins(1,4)P$_2$ or by its phosphorylation to Ins(1,3,4,5)P$_4$. In many tissues the rapid appearance of Ins(1,3,4,)P$_3$ which only appears to be formed by the hydrolysis of Ins(1,3,4,5)P$_4$ suggests that much of the Ins(1,4,5)P$_3$ formed is converted to Ins(1,3,,4,5)P$_4$ (Irvine *et al.* 1986). Eventually all the inositol phoshates formed must be metabolised back to inositol before resynthesis of phosphatidylinositol can occur in the plasma membrane.

Ca^{2+} MOBILISATION AND INOSITOL PHOSPHATE TARGETS

Evidence for the role of Ins(1,4,5)P$_3$ in Ca^{2+} mobilisation has come first from the study of the effect of inositol phosphates in permeabilised cell preparations (Streb *et al.* 1983; Somlyo *et al.* 1985) and more recently by perfusing the compounds into gland cells with patch clamp pipettes in the whole cell mode whilst using Ca^{2+}-activated membrane currents as a monitor of [Ca^{2+}]$_i$ (Morris *et al.* 1987). The concentration of Ins(1,4,5)P$_3$ producing half maximal release is quite consistent in a variety of permeabilised cell preparations ranging from 0.1 to 1 μm (listed in Berridge & Irvine, 1984). The response is not totally specific for this inositol phosphate isomer as other isomers with phosphates in both the 4- and 5- positions (e.g. Ins(2,4,5)P$_3$) are also effective at releasing Ca^{2+}.

Ins(1,4,5)P$_3$ releases Ca^{2+} from a non-mitochondrial internal store which has many characteristics of the endoplasmic reticulum (ER). Another candidate for the Ins(1,4,5)P$_3$ – sensitive Ca^{2+} pool are calciosomes, small membrane vesicles that have calcium pumping properties which have been found in non-muscle peripheral tissues (Volpe *et al.*, 1988). The IP$_3$ receptor also binds heparin and concanavalin A, properties that have aided purification and study of Ins(1,4,5)P$_3$-mediated Ca^{2+} release.

An IP$_3$-binding protein of 260 K has been purified that was immunohisto-chemically localised to particles associated with the ER (Ross *et al.*, 1989). This protein has now been purified and inserted into lipid vesicles where it mediates calcium flux with similar receptor specificity to the endogenous receptor (Ferris *et al.* 1989). Thus it appears that one protein molecule functions both as an Ins(1,4,5)P$_3$ receptor and as a Ca^{2+} channel. The primary structure of this protein has also been elucidated and shares many similarities such as seven putative membrane-spanning domains with the structure of the cell surface G-protein-coupled receptors. In addition to this basic structural homology and quite unexpectedly, the protein is about 50% homologous with the ryanodine receptor-Ca^{2+} channel complex that has been isolated from the sarcoplasmic reticulum of skeletal muscle (Takeshima *et al.* 1989). The ryanodine sensitive channel is the pathway by which IP$_3$ insensitive

Ca^{2+} stores are released (predominant in cardiac muscle) and is described in more detail in the section on Ca^{2+}-induced Ca^{2+} release. Ca^{2+} selective channels that are activated by $Ins(1,4,5)P_3$ have been isolated from vascular smooth muscle membranes and inserted into lipid bilayers. The reconstituted channels have a conductance of about 10 pS and are sensitive to heparin (Erhlich & Watras, 1988), consistent with the properties of the receptor in intact cells. Ca^{2+} stores are discussed in more detail in a later section as is another putative target for inositol phosphates, a plasmalemmal Ca^{2+} channel.

DIACYLGLYCEROL AND PROTEIN KINASE C

Inositol lipid hydrolysis produces diacylglycerol in addition to inositol phosphates. The suggestion by Nishizuka (1984) that this was the endogenous activator of protein kinase C is now universally accepted. Some types of protein kinase C also require phospholipid such as phosphatidylserine and Ca^{2+}. Diacylglycerol increases the affinity of the enzyme for Ca^{2+} so that it can be activated in the absence of any increase in $[Ca^{2+}]_i$. So while enzyme activation may under certain physiological conditions proceed in the absence of any Ca^{2+} signal, it is also likely that synergistic activation by a rise in $[Ca^{2+}]_i$ and diacylglycerol production is an important mechanism of physiological stimulation (Nishizuka, 1986). However, the cleavage of $PtdIns(4,5)P_2$ by PLC is not the only source of diacylglycerol. Observation that the stoichiometry of diacylglycerol and $Ins(1,4,5)P_3$ production was not constant lead to the discovery of synthetic pathways for diacylglycerol leading from phosphatidylcholine. More recent work has suggested other candidates as protein kinase C activators such as arachidonic acid. Clearly these developments allow for a much more flexible role for protein kinase C in signal transduction.

As in other systems, molecular biology techniques have revealed a multiplicity of protein kinase C types. The review by Nishizuka (1988) lists seven subspecies, which can be divided into two groups based on structural similarities. The subspecies show distinct tissue localisation. The γ-subspecies is found only in the brain and spinal cord, while the α and both β forms are found in many tissues but in variable ratios. Providing further support for distinct biological roles for the subspecies is the work characterising the distinct enzymatic properties of the molecules. The β subspecies appear to be the most likely targets for diacylglycerol while the γ-species is the only one that is activated by micromolar levels of arachidonic acid. These differences in properties of the pure subspecies obviously provide a basis for the observed differences in responses of native enzymes from different tissues to activators like phospholipid, Ca^{2+} and diacylglycerol. However, the precise constitution of the protein kinase C activity in any tissue such as heart has not been determined.

Many proteins are phosphorylated by protein kinase C in cell-free systems. *In vivo* it is likely that the physiological targets are more limited. Thus, unravelling the physiological roles of protein kinase C is not easy. As with G-proteins, study of function has been both by examining models of receptor activation where diacyglycerol is known to be produced and by short circuiting the pathway. This can be achieved by certain phorbol esters such as 12-0-tetradecanoylphorbol-13-

acetate (TPA) and phorbol dibutyrate that cause long-term activation of protein kinase C. However, some care is required with interpreting the effects of these compounds as they do have other actions besides stimulation of protein kinase C.

Postulated protein kinase C functions on acute signalling processes can be broken down into actions that are synergistic to the parallel Ca^{2+} mobilisation, such as a role in maintained smooth muscle contraction (Rasmussen $et\,al$, 1987), and a negative feedback role which either involves inhibiting inositol phospholipid breakdown or an increase in the rate of Ca^{2+} efflux from cells. This latter effect could be mediated by an increase in Ca^{2+} ATPase pumping or Na^+/Ca^{2+} exchange (Nishizuka, 1986).

In many cell types, prolonged activation of pathways acting through PLC activation results in a characteristic pattern of $[Ca^{2+}]_i$ signal. α-adrenoceptor activation in vascular smooth muscle is one such example. An initial $[Ca^{2+}]_i$ transient is followed by a return of $[Ca^{2+}]_i$ to close to resting levels while tension is maintained (Morgan & Morgan, 1984). While Ca^{2+} ionophores can mimic the first phase of the contractile response, a slow maintained contraction is induced by the phorbol ester, TPA (Rasmussen $et\,al.$ 1987). The initial $[Ca^{2+}]_i$ transient and tension generation are achieved by Ca^{2+}-stimulated activation of myosin light chain kinase which phosphorylates myosin light chain. It is possible that the subsequent maintained tension in the absence of high $[Ca^{2+}]_i$ is due to protein kinase C phosphorylation of other proteins associated with the contractile proteins. However, the timecourse and magnitude of tension generated by activating this mechanism do not mimic agonist mediated events well. An alternative explanation involving a G-protein-mediated, Ca^{2+}-independent interaction with the contractile proteins may turn out to be the true mechanism (Kitazawa $et\,al.$ 1989).

While a true steady plateau level of $[Ca^{2+}]_i$ is seen during the maintained phase of the response in some cells, in many others the plateau measured in populations of cells masks individual oscillatory activity as described in the next section. Protein kinase C may modulate this type of activity by regulating the frequency of oscillation as TPA has been shown to inhibit oscillations in hepatocytes (Woods $et\,al.$, 1987).

CONTROL OF $[Ca^{2+}]_i$: Ca^{2+} INFLUX

Many of the mechanisms controlling Ca^{2+} homeostasis are end targets of the other second messenger systems. These mechanisms can be divided into those affecting Ca^{2+} influx and Ca^{2+} efflux pathways. Ca^{2+} can enter the cytoplasm either from outside the cell across the plasma membrane or by Ca^{2+} release from intracellular Ca^{2+} stores. Ca^{2+} efflux processes remove Ca^{2+} from the cytoplasm by active transport to the outside and back into internal stores.

Ca^{2+} INFLUX THROUGH VOLTAGE-GATED Ca^{2+} CHANNELS

Voltage-dependent Ca^{2+} channels are found in all cell types that are electrically excitable. They appeared early in eukaryotic evolution being found in simple unicellular organisms such as Paramecium as well as the wide variety of cell types within each mammalian body (Hagiwara, 1975). The importance of this signal transduction mechanism in cell physiology and pathology, particularly in the cardio-vascular system, is demonstrated by the growing importance of drugs such as dihydropyridines that block Ca^{2+} entry through one type of Ca^{2+} channel.

Voltage-gated Ca^{2+} channels are controlled by a gating process that is dependent on the membrane potential. That is, the kinetics of channel opening and closing are a consequence of changes in the membrane potential. This gating process both for voltage gated and receptor gated channels is a complex process that is poorly understood. For the voltage-gated channels it is thought that a charged region of the protein that sits within the membrane acts as a voltage sensor. Any change in the membrane field will result in a reorientation of the sensor within the field which can result in the channel flipping from a closed to open formation and vice versa. Purification and sequencing of Na^+ and Ca^{2+} channels has revealed conserved sequences of the channel structure that could fulfill this role. One experimental test of this mechanism is to look for gating currents. These are the charge movements that must occur when the charged voltage sensor moves through the membrane field, generating a small current as the channel opens. The current is difficult to measure as it is much smaller than that due to ion permeation through the channel itself but in spite of these problems it has been described for Ca^{2+} channels (Fenwick, Marty & Neher, 1982).

The electrochemical driving force for movement of Ca^{2+} ions across the cell membrane is high because the equilibrium potential calculated using the Nernst equation is about $+120\,mV$ in unstimulated cells. The actual reversal potential measured for Ca^{2+} currents is much less positive than this, usually about $+50$ to $+70\,mV$ as the channels do not act as perfect Ca^{2+} electrodes. Even so they are still among the most selective of ion channels with selectivity for divalent cations over monovalent cations of about 10^4. The Na^+ channel is at least an order of magnitude less selective for Na^+ over K^+.

The complex profile of voltage-dependent activation and inactivation of both Ca^{2+} and Na^+ channels confers upon them the ability to support regenerative action potential firing allowing both frequency-modulated and amplitude-modulated Ca^{2+} signalling. Transmitter substances can control Ca^{2+} influx through these channels in two ways. Firstly, by direct modulation of the behaviour of the Ca^{2+} channel itself and secondly by regulating membrane potential through the opening and closing of other ion channels which will indirectly regulate Ca^{2+} channel currents. In the following sections the basic properties of Ca^{2+} channels themselves and their modulation will be described followed by consideration of the other conductances that are modulated by transmitters to regulate indirectly Ca^{2+} influx.

Channel types

The belief for many years that there was only one type of voltage-gated Ca^{2+} channel has proven to be too simplistic. There is now considerable evidence that more than one type of voltage gated Ca^{2+} channel exists. The greatest diversity will probably be found in neurones where at least three and possibly four types have been clearly identified by characterisation of unitary current properties. In peripheral

TABLE 4.1
Properties of T-type and L-type Ca^{2+} channels in cardiac cells. (from Tsien, Hess & Nilius, 1987).

Property	T	L
Activation Range	+ve to -60 mV	+ve to -30 mV
Inactivation range	$-$ve to -50 mV	$-$ve to -20 mV
Rate	Fast	Slow (without Ca^{2+})
Mechanism	Voltage dependent	Voltage dependent $[Ca^{2+}]_i$ dependent
g (110 Ca/110 Ba)	8 pS/8 pS	8 pS/25 pS
Cd block	less sensitive	sensitive
Ni block	sensitive	resistant
DHP modulation	only at high conc.	Yes

tissues there appear to be only two clear classes as yet: L-type or high-threshold channels and T-type or low-threshold channels. The properties of these two classes of channels in cardiac cells are summarised in Table 4.1.

L-type channels are sensitive to Ca^{2+} antagonists such as dihydropyridines which modulate the pattern of gating of the channels rather than occluding the channels like the inorganic blocking ions cadmium and cobalt. This up-modulation by agonists and down-modulation by dihydropyridine antagonists has provided a tool to investigate the role of these channels in signal transduction. Experiments on cardiac smooth muscle and gland cells have shown that in many tissues Ca^{2+} entry through L-type Ca^{2+} channels can provide sufficient stimulus to evoke contraction or secretion. It is this type of channel in peripheral tissues that is the main target for modulation of voltage-gated Ca^{2+} influx by transmitters. The basic conductance properties of the pore seem to be highly conserved amongst different cell types, however, there are some differences in the modulating mechanisms that interact with the channels in different cell types. Thus for example the β-adrenergic up-regulation of these channels in cardiac cells is not seen in vascular smooth muscle cells.

The role of T-type or low-threshold channels is more difficult to identify, partly due to the lack of availability of specific blockers of this conductance pathway, although some inorganic ions like nickel have some specificity. The most likely role inferred from the low-threshold activation and rapid inactivation is that the conductance has a pacemaker function. This is achieved because it provides a depolarising conductance pathway that is activated in the right voltage range to bring the membrane potential to threshold for firing the next action potential. In the heart, T-type channels are expressed as a large part of the total Ca^{2+} conductance of the cells only in the sino-atrial node (Bean, 1989). As inactivation is rapid, the amount of Ca^{2+} transported across the membrane may be less important for cell function.

Ca²⁺ channel modulation in cardiac cells

L-type Ca^{2+} channels in cardiac membranes were the first voltage gated channels that were shown to be modulated by a neurotransmitter and consequently much more is known about channel modulation in these cells than in other autonomically innervated organs. β-adrenoceptor activation results in an increase in the amplitude of voltage-gated Ca^{2+} current in intact cardiac preparations (Reuter, 1983). The effect is mediated by a chain of events finally resulting in cyclic-AMP dependent protein kinase phosphorylating the Ca^{2+} channels. The details of this cascade of events are described in the section on cAMP.

This channel phosphorylation could increase current flow through the channels by a variety of mechanisms. The whole cell macroscopic Ca^{2+} current, I_{Ca}, can be described as:

$$I_{Ca} = N \times p_o \times i$$

where N is the total number of functional channels in the cell, i is the unitary current flowing through the individual channels and p_o is the probability that each of the channels will be open at any instant. An increase in any of these parameters will obviously lead to an increase in whole cell current.

Current-voltage relationships for the unitary currents in the absence and presence of catecholamines are identical, showing that there is no change in i, the unitary conductance remains constant at about 25 pS. The probability that any channels will open during a depolarising step is much less than unity and it is p_o that increases on β-adrenoceptor activation so that any channel is more likely to open during a depolarising pulse. At the single channel level, this is seen as a 3-fold reduction in the number of blank sweeps (when no channel openings are seen). A further effect is an increase in the mean open time of single channels so that not only are channels more likely to open, but they are also less likely to close once open. The net result is that whole-cell current increases by about 4-fold following phosphorylation.

Once increased by adenylyl cyclase activation, the current can be decreased by hormones or transmitters that directly inhibit adenylyl cyclase such as ACh acting on ventricular cells (Hescheler, Kameyama & Trautwein, 1986). An alternative mechanism of down regulation is seen in amphibian ventricular myocytes where cAMP concentrations are reduced by cGMP activation of a cAMP-phosphodiesterase (Hartzell & Fischmeister, 1986). However, these mechanisms have little effect on resting activity of the Ca^{2+} current nor do other manouevres designed to reduce channel phosphorylation such as adding specific protein phosphatases or kinase inhibitor (Kameyama *et al.* 1986). These results suggest that cAMP-dependent phosphorylation is not obligatory for the channel to open.

A further mechanism of β-receptor induced up-regulation of Ca^{2+} channels has now been described. This is direct modulation by the G-protein, G_s (Yatani *et al.*, 1987). The effect is much less dramatic than the enhancement produced by cAMP-dependent protein kinase but it does illustrate the diverse targets that exist at the G-protein level of the transduction cascade.

Ca^{2+} channel modulation in smooth muscle cells

Up-regulation of Ca^{2+} channels in vascular smooth muscle cells has been advanced as a theoretical mechanism for increasing Ca^{2+} influx without any membrane depolarisation. An increase in the opening probability at a constant potential could increase Ca^{2+} influx if this potential was in the activation range of the channels (Cohen *et al.*, 1986). A hint that this might be one mechanism of action of noradrenaline in vascular smooth muscle was provided by the observation that vascular tissues exposed to noradrenaline fired Ca^{2+} action potentials much more readily than during electrical stimulation.

Direct studies of voltage gated Ca^{2+} currents in single cells from artery (Benham & Tsien 1988) and portal vein (Pacaud *et al.*, 1989) have confirmed this effect. Ca^{2+} current is increased by a constant factor at all potentials with no shift in activation. The details of the receptor channel coupling are not clear at this stage. In ear artery, not surprisingly responses were larger when GTP was included in the internal solution (Benham & Tsien, 1988) suggesting the involvement of a G-protein. Single-channel studies using cell attached patches on mesenteric artery cells have confirmed an effect on L-type dihydropyridine sensitive channels (Nelson, Standen, Brayden & Worley, 1988). In these experiments, noradrenaline applied in the bathing medium was able to act on channels in the patch of membrane, implying an indirect mechanism of action involving a diffusible second messenger. In portal vein this increase in Ca^{2+} current is mediated by α receptor activation as it is blocked by prazosin but not propranolol. Inhibition of Ca^{2+} currents in portal vein cells has also been reported. This has been shown to be due to noradrenaline mediated Ca^{2+} store release, raising $[Ca^{2+}]_i$ and hence leading to Ca^{2+}-induced inactivation of the Ca^{2+} current as the effect is not seen if $[Ca^{2+}]_i$ is heavily buffered.

MEMBRANE POTENTIAL MODULATION

Cells that use Ca^{2+} entry through voltage-gated Ca^{2+} channels as a transduction mechanism are subject to regulation of Ca^{2+} influx by changes in membrane potential achieved by variations in membrane conductance. This can either affect the rate of discharge and duration of action potentials or the steady influx of Ca^{2+} in cells that are not normally electrically active. For cells where Ca^{2+} influx is through voltage-gated Ca^{2+} channels, changes in membrane conductance that cause depolarisation will increase influx, while those causing hyperpolarisation will decrease influx. However, if a route of Ca^{2+} entry is through pathways that are not voltage dependent, such as the ATP-gated cation channel, then the opposite is the case. This has been elegantly demonstrated in mast cells which do not express voltage-gated Ca^{2+} channels (Penner *et al.* 1988).

In many tissues a variety of membrane conductances are modulated as consequences of activation of one receptor type. This complexity provides, by differential expression of the different membrane conductances, for subtle variations of response of different tissues to activation by the same receptor. Thus, different vascular tissues respond in various ways to α-adrenoceptor activation at the membrane level. This is illustrated in Fig. 4.9 which shows responses of single ear artery cells to noradrenaline application. Noradrenaline can cause depolarisation, hyperpolarisation or no change in membrane potential depending on the mix of conductances available in the cell that are described in this section. Similar tissue diversity is possible in cardiac cells in different regions of the heart in response to elevated cAMP levels.

The conductances that are involved may be divided into those channels that are primarily voltage gated like the Ca^{2+} channels just considered and secondarily modulated by ligands, and those that require an internal or external ligand to bind for activation. This second group may of course display voltage dependence once activated by the ligand.

MODULATION OF VOLTAGE DEPENDENT CURRENTS

Subtle modulation of the voltage gated ion channels that determine the shape of the action potential is particularly important in cardiac muscle cells. Both the length of the cardiac cycle and the amount of Ca^{2+} entering each cell during each cardiac cycle are determined by the overall behaviour of these conductances. The first three examples in this section have all been intensively studied in cardiac cells.

Sodium current

Neuromodulation of the voltage-dependent sodium current has received much less attention than modulation of Ca^{2+} current. The all-or-none nature of the action potential in neurons makes it an unlikely target for modulation in these cells, but in cardiac cells modulation of Na^+ currents could have a variety of subtle functional effects due to modifications in rate of rise of the action potential. Whatever the functional consequences, this conductance is modulated by isoprenaline in cardiac cells and in neurons. Schubert *et al.* (1989) have investigated the mechanism in some detail in cardiac ventricular myocytes. They found that the G-protein G_s coupled the β-adrenergic receptor to sodium channels by two pathways, one membrane-delimited (direct) and the other cytoplasmic (indirect). The indirect pathway was through generation of cAMP by adenylyl cyclase activation. The direct pathway, which could be demonstrated by applying GTP to the inside of isolated membrane patches in the continuous presence of isoprenaline on the outside, is directly analogous to muscarinic activation of the K^+ channel and to G_s modulation of Ca^{2+} current in the same cells.

The modulatory action of both pathways is to shift the inactivation of the Na^+ current to more hyperpolarised potentials. This has the effect of reducing the magnitude of the Na^+ current at any given potential as the cell depolarises during the upstroke of the action potential. This will result in a slower rate of rise of the cardiac action potential. However, the effect is much more pronounced in depolarised tissues, suggesting that it may be more important in depolarised ischaemic tissue.

The pacemaker current, I_f

The pacemaker current, I_f, is a voltage-gated current activated around the diastolic membrane potential that has been identified in cells from cardiac sino-atrial node. It is thought to be responsible for the characteristic pacemaker depolarisation that precedes action potential generation in these cells. The conductance is permeable to both Na^+ and K^+ so that the reversal potential of the current is positive to the diastolic potential. Thus current flow through the channels will depolarise the cell. The conductance is activated at negative potentials and switches off as the cell membrane depolarises with half activation at about $-75\,mV$. Modification of this voltage sensitivity will thus regulate the current flow at any potential and hence the rate of depolarisation in cells where this is an important contributor to membrane conductance as it is in sino-atrial node cells (DiFrancesco, 1985).

The channel is another target for modulation by cAMP. The sympathetic chronotropic effect appears to be mediated by shifting activation to more positive potentials speeding the rate of depolarisation (Brown et al. 1979). In contrast, ACh shifts the activation curve to more hyperpolarised potentials resulting in less I_f being activated so that depolarisation is slowed. The action of ACh is pertussis-toxin sensitive. Catecholamine modulation of I_f is reversed by ACh but direct activation of adenylyl cyclase by forskolin is not reversed. This indicates that ACh acts by inhibiting adenylyl cyclase activity (DiFrancesco & Tromba, 1987).

The delayed rectifier I_K

The delayed rectifier potassium current has been identified in a variety of cells, including cardiac muscle cells. This conductance is important for repolarising the membrane potential during the downstroke of the action potential. This conductance is regulated by adenylyl cyclase in a parallel manner to the regulation of Ca^{2+} channels in the heart. β adrenoceptor activation leads to an increase in the current without any shift in voltage dependence of activation and this increase can be antagonised by ACh application. Experiments with forskolin and non hydrolysable GTP analogues give results as expected if the agonists modulated adenylyl cyclase levels (Harvey & Hume, 1989).

M-current I_M

Another channel that acts somewhat like a slow delayed rectifier is a potassium channel that activates on depolarisation showing no inactivation. Thus, it exerts a repolarising influence. In the presence of acetylcholine acting through muscarinic receptors, the conductance is switched off, resulting in excitation of the cell as this brake, termed the M current, is let off. The current was first described in bullfrog sympathetic neurons (Adams et al. 1982). The mechanism also seems to be important in amphibian gut smooth muscle cells where the current is reduced by ACh and is enhanced by isoprenaline (Sims et al. 1987). However, attempts to identify the same conductance in mammalian gut smooth muscle have so far been unsuccessful. Modulation of voltage-gated channels in the toad stomach cells also seems to be distinct from mammalian cells, so particular care is required when considering these cells as a model system for the mammalian gut.

LIGAND GATED CHANNELS

In addition to the conductances described above, a further class of channels that absolutely require a ligand to bind for activation have important roles, particularly in secretory cells and smooth muscle cells.

Conductances activated by a rise in $[Ca^{2+}]_i$

Although control of ionic conductances by $[Ca^{2+}]_i$ was known to occur (Meech, 1978), the widespread occurrence of these channels in cell membranes has only been revealed by the advent of the patch-clamp technique. Channels activated by a rise in $[Ca^{2+}]_i$ can be classified into three classes, those selective for potassium, chloride and for monovalent cations. In addition to shaping action potentials and Ca^{2+} influx as mentioned above, they also play important roles in electrolyte secretion in exocrine glands (Marty, 1989).

1. Ca^{2+}-activated K^+ channels. As the equilibrium potential for K^+ is between -75 and -100 mV under physiological conditions, activation of K^+ currents will hyperpolarise cells and inhibit Ca^{2+} influx through voltage-gated Ca^{2+} channels. A heterogeneous collection of channels are included in this group of which the most studied are the almost ubiquitous BK channels. These high conductance channels are identified specifically by their sensitivity to charybdotoxin. BK is simply 'big potassium'. Channels of this type are found in an enormous variety of cells, although a notable exception is cardiac muscle. The channels in different tissues are activated over a range of $[Ca^{2+}]_i$, from 10–100 nM in exocrine cells to 10–100 μM in skeletal muscle. These values are for a membrane potential of -50 mV. The channels are also potential dependent, their opening probability increases with depolarisation. The rapid opening rate of the channels following a rise in $[Ca^{2+}]_i$ suggests that Ca^{2+} binds directly to the channel molecule. The combination of voltage and Ca^{2+} dependence makes the channels well suited to exert a rapid hyperpolarising action on the membrane potential.

A distinct second group of K^+-selective channels can be defined by a lower unit conductance (about 10–20 pS) and by being sensitive to block by the bee venom, apamin. Activation of these channels is in the calcium concentration range of 100–1000 nM. This conductance and its activation by α_1 adrenoceptor has been extensively studied in hepatocytes (Capiod & Ogden, 1989). A similar mechanism seems to occur in some smooth muscles. α-adrenoceptor mediated relaxation in parts of the gut such as guinea pig taenia coli muscle is blocked by apamin, suggesting that relaxation is mediated by the hyperpolarisation causing a cessation of the normal spontaneous activity (see Bülbring & Tomita, 1987). One incongruous feature of this response is that the mechanism apparently involves the release of internal stores of Ca^{2+} which might be expected to cause contraction. In the absence of any hard evidence we can only speculate that the $[Ca^{2+}]_i$ signal is localised to a site close to the target channels and does not directly activate the contractile proteins.

2. Ca^{2+}-dependent chloride channels also have a wide distribution including exocrine cells and smooth muscle cells. The channels are activated over a $[Ca^{2+}]_i$ range of 100 to 2000 nM in lacrimal glands. The effect of activating these channels will depend on the chloride equilibrium potential which differs in various cell types. Thus, in vascular smooth muscle cells the chloride equilibrium potential lies at about -40 mV so that channel opening will result in inward current flow (as Cl^- moves out) and membrane depolarisation. It is likely that this conductance underlies the depolarising action of α-receptor activation by noradrenaline in many vascular

smooth muscles. In contrast in many neurons E_{Cl} is much more negative so that chloride currents exert a hyperpolarising action.

3. In contrast to the potassium and chloride channels the non-selective cation channels activated by Ca^{2+} are independent of membrane potential. They also have rather slow open–close kinetics and generally require $[Ca^{2+}]_i$ to rise above 1000 nM before showing any activation. Divalent cation permeability is variable but some conductances are clearly Ca^{2+} permeable. Such channels would seem an obvious candidate to mediate the sustained Ca^{2+} influx following inositol-trisphosphate-induced Ca^{2+} release. However such a straightforward explanation does not seem to hold in most cells.

Many of the properties of these channels have been revealed by studies on isolated membrane patches. Studies on the same conductances in whole cells have shown that in some cases the channel properties are modified relative to behaviour in isolated patches indicating that other cytoplasmic agents in addition to $[Ca^{2+}]_i$ are modulating activity. In some cells BK channel phosphorylation by cAMP-dependent protein kinase has been reported to increase channel opening. In tracheal smooth muscle this seems to be the mechanism by which β-receptor activation hyperpolarises airway smooth muscle leading to muscle relaxation (Kume *et al.*, 1989).

Cation channels gated by other 2nd messengers

In addition to the cation channels that are dependent on a rise in cytosolic Ca^{2+}, other conductances apparently permeant to both divalent and monovalent cations are activated by agonists in smooth muscle. In mammalian intestinal smooth muscle, muscarinic receptors activate a cation conductance that is permeable to Ca^{2+} and in vascular smooth muscle noradrenaline in addition to activating a Ca^{2+}-dependent chloride current, also stimulates a cation current (Amedee *et al* 1990). In contrast to the cation current directly gated by ATP that has a very rapid onset, these conductances have a latency of about 0.5 s at 23°C, suggesting an intermediate step in the pathway. Both these conductances are not abolished by adding Ca^{2+} chelators to the intracellular solution, indicating that $[Ca^{2+}]_i$ is probably not the messenger. The mechanism may well be similar to that seen in exocrine gland cells on muscarinic stimulation (see section on inositol phosphates).

cAMP-dependent chloride channel

In isolated cardiac cells, isoprenaline activates a background current that was suppressed by simultaneous application of ACh. Alteration of the chloride equilibrium potential shifted the reversal potential of this current as expected for a chloride conductance (Harvey & Hume, 1989). Application of the catalytic sub-unit of cAMP Pr-K by internal dialysis was also effective indicating that this is another target of the cAMP messenger pathway (Bahinski *et al.* 1989). Interestingly, this conductance is negligible in the absence of kinase activation and shows little voltage dependence in contrast to the Na^+, Ca^{2+} and K^+ channels that are targets of the same kinase in these cells. The relationship of these channels to those in epithelial cells that are defective in cystic fibrosis will probably be clarified when single channel data is available for these cardiac channels.

The role of these channels is probably to counter the effect of the enhanced Ca^{2+} currents on action potential length. The chloride current will exert a repolarising force on the cell tending to shorten action potentials and thus prevent the possible arrhythmogenic effects of strong sympathetic stimulation.

Muscarinic receptor coupled K^+ channel

ACh applied to cardiac tissue and stimulation of the vagus nerve causes slowing of the heart. This may be attributed to a variety of mechanisms one of which is the activation of a potassium conductance g_K causing hyperpolarisation of pacemaker cells (Noble 1975). In contrast to voltage gated channels whose activity is modulated (see below) this channel shows no resting activity in the absence of ACh. The channels show inward rectification, meaning that they pass current more readily at hyperpolarised potentials making their action more marked at hyperpolarised membrane potentials. However this current can still shorten action potentials at membrane potentials positive to zero mV (Iijima, Irisawa & Kameyama, 1985). ACh-activated K^+ channels have been identified at the single-channel level (Sakmamn, Noma & Trautwein, 1983), and this system has proved useful for studying the mechanism of modulation.

ACh has been shown to have a quite direct coupling to the K^+ channels in that ACh had to be added to the pipette solution in the cell attached patch conformation to see an effect. If ACh is added to the bathing medium in contact with the rest of the cell then no channel activity is seen. Thus a cytoplasmic second messenger such as cAMP can be ruled out. However, if cell-attached patches containing ACh-activated channels are excised then channel activity declined, clearly some diffusible molecule was required for channel activation. The involvement of a pertussis-sensitive G protein first reported by Pfaffinger et al. (1985) provided a strong clue and GTP was soon identified as the soluble factor (Kurachi et al. 1986). Activation of K^+ channels in isolated inside-out membrane patches by addition of exogenous G protein sub-units to the cytoplasmic surface of the membrane was used to identify the active subunit. It is now clear that muscarinic activation is mediated by the α subunit of G_i as reported by Codina et al. (1987). The conflicting report that the $\beta\gamma$ subunit was effective (Logothetis et al. 1987) has now been resolved. It now appears that $\beta\gamma$ stimulates the channels by activating phospholipase A_2 (Kim et al. 1989). This pathway is not involved in muscarinic activation of K^+ channels but the modulation of these channels by arachidonic acid may well be coupled to another as yet unidentified transmitter through a different G-protein.

INTRACELLULAR Ca^{2+} STORES

In addition to the mechanisms regulating Ca^{2+} influx and extrusion across the cell membrane, sequestration of Ca^{2+} also occurs within cells. More than 99% of Ca^{2+} entering cells is immediately buffered by cytoplasmic Ca^{2+} buffers. Little is known about possible regulation of this buffering capacity by transmitters. Storage of Ca^{2+} in specific cytoplasmic organelles has been much more intensively studied.

Although Ca^{2+} is sequestered in mitochondria, current evidence suggests this store only has pathological importance. Signal transduction pathways are coupled to stores that can either be released by inositol 1,4,5 trisphosphate ($Ins(1,4,5)P_3$) or by a rise in $[Ca^{2+}]_i$ itself. Discharge of these stores is regulated by Ca^{2+} selective channels in the store membrane that are gated by $Ins(1,4,5)P_3$ and Ca^{2+} respectively. There is now growing evidence that these distinct release processes are also coupled to structurally distinct stores, although the identity of the stores defined biochemically has not been unequivocally correlated with defined morphological structures.

Ca^{2+}-INDUCED Ca^{2+} RELEASE

Studies using intact and skinned muscle fibres (skeletal, smooth and cardiac) revealed the existence of Ca^{2+} stores in these tissues that could be released by a rise in $[Ca^{2+}]_i$ (reviewed by Endo, 1977). In cardiac tissue, detailed studies using microdissected cardiac fibres suggested a physiological role for Ca^{2+} induced Ca^{2+} release in triggering contraction (Fabiato, 1983). The role of this process has recently been further clarified with the introduction of a new pharmacological tool, ryanodine, which is an alkaloid that binds to the Ca^{2+} release channel and modulates its activity. The channel is located in the sarcoplasmic reticulum membrane and is the pathway by which Ca^{2+} is released into the cytoplasm when a triggering level of Ca^{2+} is reached on the cytoplasmic surface of the sarcoplasmic reticulum membrane. Application of this compound to cardiac cells inhibits contractions without suppressing voltage-gated Ca^{2+} currents (Mitchell et al. 1984). More recent studies measuring Ca^{2+} currents and $[Ca^{2+}]_i$ transients in cardiac myocytes have shown that the rise in $[Ca^{2+}]_i$ does not exactly parallel the Ca^{2+} influx and it is inhibited by ryanodine (Barcenas-Ruiz & Weir, 1987), this is interpreted as further evidence that the Ca^{2+} influx is a trigger for Ca^{2+} release and does not supply all the Ca^{2+} for contraction. However, the primary modulatory site in this pathway where autonomic transmitters act, is the Ca^{2+} influx through the L-type Ca^{2+} channels. This is effective because the amount of Ca^{2+} entry determines the amount of Ca^{2+} release. The functional importance of this mechanism is that it provides a way of raising $[Ca^{2+}]_i$ very rapidly to the micromolar level from a resting level of about 100 nM in cardiac cells. Without cytoplasmic stores, diffusion through the cytoplasm from the sub-sarcolemmal space would presumably seriously slow activation of the contractile proteins. As cardiac muscle contractions are rapid there is no role for a slower and perhaps more variable mechanism provided by inositol phosphate-mediated release.

Ins $(1,4,5)P_3$-SENSITIVE Ca^{2+} STORES AND MULTIPLE POOLS

In smooth muscle and many other cell types both Ca^{2+} release processes operate and appear to be intimately linked in physiological responses that may be initiated by agonist-induced formation of Ins $(1,4,5)P_3$. In these cells, only part of the stored Ca^{2+} seems to be releasable by $Ins(1,4,5)P_3$, typically between 30 and 50% of Ca^{2+} taken up by non mitochondrial stores is directly releasable by $Ins(1,4,5)P_3$, the

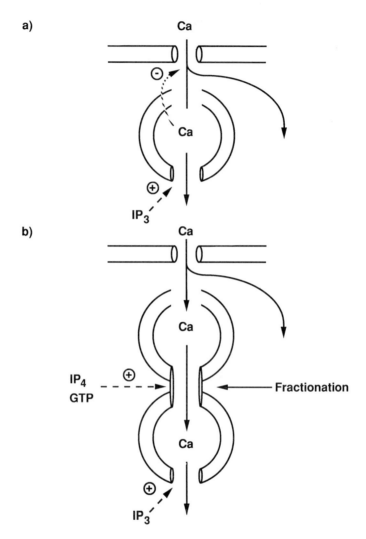

FIGURE 4.7   Capacitative models of Ca^{2+} entry and multiple Ca^{2+} stores. a, Simple model in which Ins (1,4,5) P_3 controls a channel in an intracellular organelle which indirectly regulates Ca^{2+} entry across the plasma membrane by a negative feedback loop. b, A more complex model that includes two inter-connected stores which could explain the results obtained in cell fractionation experiments (see text). From Berridge & Irvine (1989).

remainder can be released on addition of Ca^{2+} ionophores. This evidence has lead to the idea that there are multiple Ca^{2+} pools within cells which may be coupled in such a way that the Ins(1,4,5)P_3-insensitive pool can be used to amplify the Ins(1,4,5)P_3-sensitive one. The Ins(1,4,5)P_3-insensitive pool can be released by caffeine in many cells and physiologically it may be released by a Ca^{2+}-induced Ca^{2+} release mechanism. Studies on permeabilised cells have shown that GTP can release Ca^{2+} from both types of store.

Recent work on membrane vesicles derived from DDT_1MF-2 cells (a smooth muscle cell line), provides evidence to support the idea of two interconnected Ca^{2+} stores (Ghosh *et al.*, 1989) (Fig. 4.7B). The $InsP_3$-releasable store is found in the membrane fraction that also contains the rough endoplasmic reticulum, an observation consistent with the IP_3 receptor localisation studies. The location of the IP_3-insensitive store is not so clear. However, the two stores may be connected anatomically as the actions of GTP differ in whole cells and cell fractions. The authors propose that this behaviour can be explained by a pathway regulated by GTP between the two stores. When this is broken by fractionation, GTP will cause release from the IP_3-insensitive store rather than a net uptake as occurs in intact permeabilised cells (Ghosh *et al.*, 1989). Another proposal is that this is the site of action of IP_4 (Berridge & Irvine, 1989) (Fig. 4.7B).

Inositol phosphate mediated Ca^{2+} entry

Many agonists have dual effects on $[Ca^{2+}]_i$. Firstly, they stimulate $Ins(1,4,5)P_3$-mediated Ca^{2+} release, resulting in a transient rise in $[Ca^{2+}]_i$. Secondly, they generate a sustained plateau of raised $[Ca^{2+}]_i$ that is dependent on extracellular Ca^{2+} and hence due to a stimulated Ca^{2+} influx. Because it is possible to refill Ca^{2+} stores in some tissues without elevating cytoplasmic Ca^{2+}, models such as those outlined in Fig. 4.7 have been proposed with some sort of restricted corridor directing Ca^{2+} entry into the Ca^{2+} store. On maintained stimulation, Ca^{2+} passes straight through the store to maintain an elevated $[Ca^{2+}]_i$.

Evidence to support the idea that both $Ins(1,4,5)P_3$ and $Ins(1,3,4,5)P_4$ are involved in this process has been reported. Some of the most elegant evidence suggests that there is a synergistic role of the two inositol phosphates in mediating Ca^{2+} entry in exocrine gland cells where muscarinic receptors are coupled to this transduction mechanism. Using internally dialysed single lacrimal acinar cells, Morris *et al.* (1987) showed that $Ins(1,4,5)P_3$ and $Ins(1,3,4,5)P_4$ perfused inside the cell together produced a much larger rise in $[Ca^{2+}]_i$ than when either was applied in isolation. The rise in $[Ca^{2+}]_i$ was external Ca^{2+} dependent and not seen in Ca^{2+}-free external solution. This supports the original evidence for a role for IP_4 in regulating Ca^{2+} influx suggested by the work of Irvine and Moor on sea urchin eggs (1986).

Detailed study of this mechanism has been hampered by not knowing the precise location of the channel. The simplest idea, that the channel allows Ca^{2+} to flow directly into the cytosol would suggest that the channels are present in the plasma membrane. Evidence for this has been provided in lymphocytes (Kuno & Gardner, 1987). An alternative scheme requires Ca^{2+} to flow directly into the endoplasmic reticulum before being released into the cytosol. This would explain how smooth muscle Ca^{2+} stores can be refilled without activating the contractile proteins. As the stores must be released before Ca^{2+} can enter from the outside (at least in endothelial cells (Jacob *et al.* 1988), this could explain the requirement for $Ins(1,4,5)P_3$ as well as $Ins(1,3,4,5)P_4$. In this scheme in the model shown in Fig. 4.7A, the Ca^{2+} channel in the cell membrane at the top of the figure would be controlled by IP_4.

[Ca^{2+}]$_i$ oscillations

The ability to measure [Ca^{2+}]$_i$ in single cells rather than cell populations led to the discovery that in many cells sustained stimulation by an agonist leading to Ins(1,4,5)P$_3$ elevation, generates oscillations in [Ca^{2+}]$_i$ rather than a sustained plateau of [Ca^{2+}]$_i$ as population studies had suggested. The oscillations are not due to membrane potential oscillations leading to periodic voltage-dependent entry of Ca^{2+}, but instead are produced by Ca^{2+} discharge from internal stores. In many but not all cell types the oscillations which have a constant amplitude have a frequency that is dependent on the concentration of agonist.

Various oscillator models have been put forward to explain this process (reviewed by Berridge & Irvine, 1989). These may be divided into two basic models. One involves oscillations in Ins(1,4,5)P$_3$ levels being generated by feedback loops regulating its production or degradation. This oscillating Ins(1,4,5)P$_3$ level then directly regulates Ca^{2+} release from the store. Direct measurement of inositol phosphate concentrations in single cells is not yet possible, so we cannot directly test this idea. However in pancreatic acinar cells direct perfusion of the stable phosphorothioate analogue of Ins(1,4,5)P$_3$ did not abolish ACh-induced oscillations (Wakui, Potter & Petersen, 1989), making it less likely, in these cells anyway, that fluctuations in Ins(1,4,5)P$_3$ have an important role. The alternative theories are focussed on Ca^{2+}-induced Ca^{2+} release as the pivotal mechanism. These propose that Ca^{2+} is released from the oscillator store when the store is full and when the cytoplasmic Ca^{2+} has reached a trigger level. The inositol phosphate sensitive store then plays a role as a pathway through which the sustained influx of Ca^{2+} is maintained which feeds the oscillator. As experimental data on a variety of cells favours more than one hypothesis it is likely that there is some diversity in the mechanisms utilised.

Whatever the precise mechanism(s) controlling oscillations, their widespread utilisation in signal transduction prompts the question as to their functional advantages. One of the major roles of the rise in [Ca^{2+}]$_i$ is to activate phosphorylases. If these enzymes have a slow deactivation time relative to the frequency of spiking, then the Ca^{2+} signal will be integrated to a constant level of enzyme activity which is frequency dependent. Several advantages of this approach have been widely discussed (e.g. Jacob, 1990). 1. Sustained levels of high [Ca^{2+}]$_i$ are toxic to cells due to activation of proteases and lipases. Oscillators avoid this danger and provide a mechanism for limiting Ca^{2+} entry which might otherwise get out of control. 2. An oscillator system that can produce a maximal response conserves energy by reducing the total fluxes of Ca^{2+} across the cell membrane. 3. It may be a more accurate form of signal transduction. Many enzymes are activated over a narrow range of [Ca^{2+}]$_i$ which means that either the steady [Ca^{2+}]$_i$ has to be very finely controlled or an oscillator is used.

Ca^{2+} EFFLUX PROCESSES

The rate at which Ca^{2+} is sequestered from the cytoplasm will obviously affect the temporal nature of the Ca^{2+} signal. Ca^{2+} can either be extruded from the cell or taken up into intracellular organelles such as the sarcoplasmic reticulum and stored for re-release. Extrusion pathways from the cell are passive Na^+/Ca^{2+} exchange which depends on the inward Na^+ gradient to drive Ca^{2+} efflux, and the plasmalemmal Ca^{2+}-ATPase, pump that actively extrudes Ca^{2+}. Uptake into the S R is through another structurally distinct Ca^{2+}-ATPase and a further Ca^{2+} ATPase pumps Ca^{2+} into mitochondria. There is considerable evidence that Na^+/Ca^{2+} exchange does have a significant role in Ca^{2+} homeostasis in both vascular and cardiac muscle (reviewed in Brading & Lategan, 1985; Noble, 1984) but there is very little evidence that it is modulated by second messengers. The cytoplasmic concentrations of Ca^{2+} and Na^+ regulate the turnover rate of the exchanger so that indirect modulation through changes in concentrations of these ions will occur. Up-regulation by protein kinase C has been suggested on the basis of phorbol ester stimulation of the exchanger (Vigne et al., 1988).

   In contrast, there is clearer evidence for complex modulation of Ca^{2+}-ATPase activity in vascular smooth muscle, although it is still unclear as to the relative importance of the plasmalemmal and sarcoplasmic reticulum Ca^{2+}-ATPases in Ca^{2+} extrusion. There is considerable variation in the relative amounts of the two pumps in different tissues which indicates tissue-specific importance of the mechanisms. cAMP-and cGMP-dependent protein kinases both regulate the endoplasmic reticulum Ca^{2+}-ATPase by phosphorylation of phosholamban. Thus stimulated uptake of Ca^{2+} into the ER could be one mechanism of β-receptor mediated relaxation in smooth muscle. The plasmalemmal ATPase is stimulated by Ca^{2+}-calmodulin that provides a direct negative feedback regulation of a rise in $[Ca^{2+}]_i$. Additionally, cGMP dependent protein kinase stimulates this ATPase but the pump itself is not phosphorylated. Clearly, an as yet unidentified intermediate is involved in this pathway (Eggermont et al., 1988; Lincoln, 1989). The inability to demonstrate an effect on Ca^{2+} influx mechanisms by cGMP suggests by default that modulation of these efflux mechanisms by cGMP may be a major mechanism by which it reduces $[Ca^{2+}]_i$ and causes smooth muscle relaxation (Lincoln, 1989). The plasmalemmal pump can also be down-regulated, by muscarinic agonists but the transduction mechanism has not been investigated as yet.

INTEGRATION OF PATHWAYS AND RELATION TO PHYSIOLOGY

The development of ever more powerful techniques has generated a wealth of data on putative mechanisms for signal transduction. Many pathways seem to involve a number of branches leading to a host of targets. With this wealth of data it is important but often difficult to maintain a perspective as to which mechanisms and pathways are of most physiological relevance. Study of neuroeffector signal transduction at the single cell level has several drawbacks that should be remembered. Firstly, bath application of transmitter is a quite different stimulus from local release of transmitter possibly directed at specialised post synaptic structures or clusters of receptor. The many experiments described using patch-clamp techniques all suffer from the problem of dialysis of unknown internal cofactors which might alter the balance of mechanisms expressed.

Sometimes the approach of returning to the whole tissue with the knowledge that we have of the underlying mechanisms and re-examining carefully what happens in an intact preparation can be illuminating. In cardiac tissue we have seen that muscarinic receptor activation can have a variety of effects of which the most studied is the G-protein-mediated activation of the potassium conductance. It is widely assumed that it is this mechanism that slows depolarisation in sino-atrial cells resulting in the negative chronotropic effect.

Hirst and colleagues (Campbell *et al.* 1989) have examined this question by comparing closely the effects of vagal stimulation and applied ACh on the electrical properties of pacemaker cells in the guinea-pig sino-atrial node. Applied ACh produced effects expected of an agent increasing a K conductance, in particular cessation of beating was accompanied by a hyperpolarisation of the membrane beyond the normal diastolic potential. In contrast vagal stimulation resulted in the heart stopping with the membrane potential quiescent at a more depolarised level. This is interpreted as evidence that decreasing inward current flow through the inward rectifier (DiFrancesco, 1985) is the more important mechanism in these cells.

CARDIAC MUSCLE AND NORADRENALINE

Activation of β-adrenoceptors on cardiac muscle cells produces a variety of intracellular effects illustrated in Fig 4.8. The positive inotropic action of noradrenaline is undoubtedly primarily mediated by the central adenylyl cyclase pathway leading to phosphorylation of the voltage-gated Ca^{2+} channels. The resulting increase in Ca^{2+} influx during each action potential triggers a larger release from internal stores leading to a stronger contraction. However, the same G-protein that stimulates adenylyl cyclase also appears capable of modulating Ca^{2+} channels and inhibiting current flow through Na^+ channels. We are unsure of the importance of these pathways in physiological responses.

The other point of signalling divergence is at the level of targets for cAMP-dependent protein kinase. Some of these targets are involved in the chronotropic response and its consequences. Pacemaking is speeded by changes in I_f and to accommodate this action potentials are shortened by changes in chloride and potassium currents. Faster relaxation of the myocyte is needed and this is achieved in part by enhanced Ca^{2+} pumping from the cytoplasm and by more rapid dissociation of Ca^{2+} from troponin. Yet other targets within the cell change the metabolic status of the cell. Thus, the ramifications of any particular stimulus to a cell are all addressed as consequences of stimulating the same receptor. This is perhaps a partial explanation of the at times apparently baroque splendour of the signalling pathways that we are unravelling.

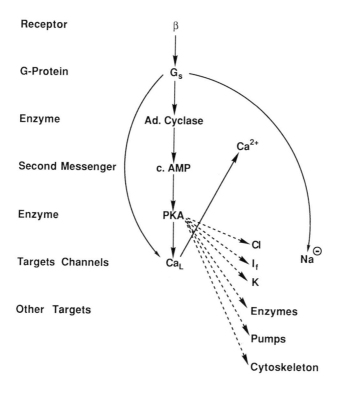

FIGURE 4.8   Multiple transduction pathways emanating from G_s activation by the β-adrenoceptor in cardiac myocytes.

Principle actions of G-protein, G_s are indicated. The most important target of cAMP-dependent protein kinase (PKA) for the inotropic action is the L-type Ca^{2+} channel (Ca_L). Other targets are linked with dotted lines and discussed in the text. The implications of a rise in $[Ca^{2+}]_i$ acting as a second messenger itself are not indicated in this figure.

VASCULAR SMOOTH MUSCLE AND NORADRENALINE

The diversity of cellular actions of noradrenaline on vascular smooth muscle cells exemplifies the problem of trying to relate these mechanisms to physiology but also now provides clues as to the diversity of responses seen within the vascular system. The major pathway activated by α receptors is the IP_3-mediated release of internal Ca^{2+} stores. This rise in $[Ca^{2+}]_i$ could directly stimulate contraction but it also activates a variety of Ca^{2+}-dependent channels in the cell membrane (Fig 4.9). These can modulate Ca^{2+} influx through voltage gated channels by changing membrane potential. In addition, noradrenaline also has a direct modulatory action on voltage gated Ca^{2+} channels in vascular smooth muscle cells.

Some of the diversity in vascular tissue responses to sympathetic nerve stimulation (Bolton & Large, 1987) may thus be due to subtle weighting of the activation of hyperpolarising and depolarising conductances activated by the same central pathway. Other differences could reflect the relative importance of internal and

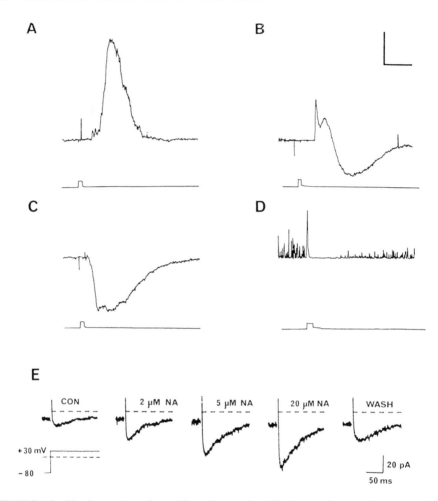

FIGURE 4.9   Membrane channels modulated by noradrenaline in vascular smooth muscle cells. A-D, Internal store release can result in activation of outward, inward or biphasic current responses dependent on the relative activation of Ca^{2+}-activated chloride, potassium and cation conductances (From Amedee *et al.* 1990). E. Noradrenaline upregulates voltage-gated Ca^{2+} currents in the same ear artery cells. Effect of cumulative doses of noradrenaline (NA) in the same cell. (From Benham & Tsien, 1988).

external sources of Ca^{2+}. The strongly hypotensive actions of voltage-gated Ca^{2+} channel antagonists *in vivo* point to an important role for Ca^{2+} influx through these channels in resistance vessels. The role of Ca^{2+} store release is not so clear, particularly as in some tissues store release is associated with relaxation (e.g. ATP-mediated, endothelium-independent relaxation of rabbit portal vein). The development of techniques to localise Ca^{2+} within cells may throw some light in this area as spatial localisation would seem the most likely explanation of this particular paradox.

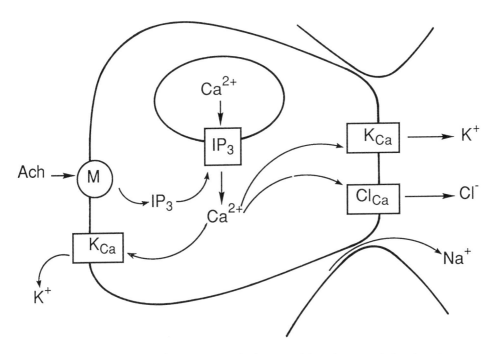

FIGURE 4.10 Model of the major components in the muscarinic pathway regulating electrolyte secretion in lacrimal glands. (modified from Hille, 1989).

CONTROL OF SALT AND FLUID SECRETION

Stimulation of salt and fluid secretion in exocrine glands by muscarinic receptor activation involves a rather different set of effectors but once again the central orchestrator is a rise in $[Ca^{2+}]_i$. The principles of regulation in this system have been outlined in reviews (Marty, 1987; Petersen & Gallacher, 1988). Directional salt transport is achieved in these cells by the heterogeneous channel distribution in the acinar cell membranes. While the $Na^+ - K^+$ pump and Cl^- transporters are essential for maintaining ion gradients, the process appears to be controlled by the Ca^{2+}-dependent Cl^- and K^+ channels described earlier. Their location in the baso-lateral and luminal membranes is shown in Fig. 4.10.

In unstimulated cells the channels are closed and because there is no pathway for Cl^- ions to move across the cell, there can be no salt movement. Muscarinic (M2) receptor activation leads to a rise in $[Ca^{2+}]_i$ by IP_3-mediated release of Ca^{2+} stores and the opening of the luminal Ca^{2+}-dependent channels. This results in an efflux of KCl into the lumen with water following passively. The KCl gradient is then maintained by the basolateral $Na^+ - K^+$ pump and by a Cl^- transporter or channel. The rise in $[Ca^{2+}]_i$ has two other consequences. Firstly, it stimulates protein secretion and secondly, closes gap junctions electrically uncoupling the cells. The relevance of the latter effect is not clear as yet, but it may be somehow linked to an increase in permeability of the tight junctions that normally limit paracellular movement of ions (Marty, 1987).

CONCLUSIONS

Ten years ago our understanding of the mechanisms of signal transduction that follow agonist binding to cell surface receptors was extremely limited. The interest in this area and the investigative tools that have become available has resulted in exciting and exhausting progress. The aim of this chapter has been to review some of that progress.

In the near future, more detailed descriptions of the various signalling pathways will be produced, and with this will come a greater understanding of the significance of some of the diversity of mechanistic options available in cell signalling. This will be coupled with a better idea of the physiological relevance of many of the processes that are now being described at the cellular level. Clearly, by the end of this millenium we will have made further advances towards the goal of an explanation of cell function at the molecular level.

ACKNOWLEDGEMENT

I would like to thank Drs J.E. Merrit and A.J. Hunter for their careful reading and constructive criticism of an earlier version of the manuscript.

REFERENCES

Amedee, T., Benham, C.D., Bolton, T.B., Byrne, N.G. & Large, W.A. (1990). Potassium, chloride and non-selective cation conductances opened by noradrenaline in rabbit ear artery cells. *Journal of Physiology*, **423**, 551–568.

Adams, P.R., Brown, D.A. & Constanti, A. (1982). M-currents and other potassium currents in bullfrog sympathetic neurons. *Journal of Physiology*, **330**, 537–572.

Bahinski, A., Nairn, A.C., Greengard, P. & Gadsby, D.C. (1989). Chloride conductance regulated by cyclic AMP-dependent protein kinase in cardiac myocytes. *Nature*, **340**, 718–721.

Barcenas-Ruiz, L. & Weir, W.G. (1987). Voltage dependence of intracellular $[Ca^{2+}]_i$ transients in guinea pig ventricular myocytes. *Circulation Research* **61**, 148–154.

Barnard, E.A., Darlison, M.G., Marshall, J. & Sattelle, D.B. (1989). Structural characteristics of cation and anion channels directly gated by agonists. In '*Ion Transport*'. Eds. Keeling, D.J. & Benham, C.D. Academic Press. London.

Bean, B.P. (1989). Classes of calcium channels in vertebrate cells. *Ann Review of Physiology*, **51**, 367–384.

Beavo, J.A. (1988). Multiple isoenzymes of cyclic nucleotide phosphodiesterases. *Adv. Sec. Mess. Phosphoprot. Res.* **22**, 1–38.

Beavo, J.A. & Reifsnyder, D.H. (1990). Primary sequence of cyclic nucleotide phosphodiesterase isozymes and the design of selective inhibitors. *TIPS*, **11**, 150–155.

Beebe, S.J. & Corbin, J.D. (1986). Cyclic nucleotide dependent protein kinases. *The Enzymes*, **17**, 43–111.

Benham, C.D. (1989). ATP-activated channels gate calcium entry in single smooth muscle cells dissociated from rabbit ear artery. *Journal of Physiology*, **419**, 689–701.

Benham, C.D. & Tsien, R.W. (1987). Receptor-operated, Ca-permeable channels activated by ATP in arterial smooth muscle. *Nature*, **328**, 275–278.

Benham, C.D. & Tsien, R.W. (1988). Noradrenaline modulation of calcium channels in single smooth muscle cells from rabbit ear artery. *Journal of Physiology*, **404**, 767–784.

Berridge, M.J. (1983). Rapid accumulation of inositol trisphosphate reveals that agonists hydrolyse polyphosphoinositides instead of phosphatidylinositol. *Biochemical Journal*, **212**, 849–858.

Berridge, M.J. & Irvine, R.F. (1984). Inositol trisphosphate, a novel second messenger in cellular signal transduction. *Nature*, **312**, 315–321.

Berridge, M.J. & Irvine, R.F. (1989). Inositol phosphates and cell signalling. *Nature*, **341**, 197–205.

Bokoch, G.M., Katada, T., Northup, J.K., Hewlett, E.L., & Gilman, A.G. (1983). Identification of the predominant substrate for ADP-ribosylation by islet-activating factor. *Journal of Biological Chemistry*, **258**, 2072–2075.

Bolton, T.B. (1979). Mechanisms of action of transmitters and other substances on smooth muscle. *Physiological Reviews*, **59**, 606–718.

Bolton, T.B. & Large, W.A. (1987). Are junction potentials essential? Dual mechanism of smooth muscle cell activation by transmitter released from autonomic nerves. *Quarterly Journal of Experimental Physiology*, **71**, 1–28.

Brading, A.F. & Lategan, T.W. (1985). Na-Ca Exchange in vascular smooth muscle. *Journal of Hypertension*, **3**, 109–116.

Brown, H.F., DiFrancesco, D. & Noble, S.J. (1979). How does adrenaline accelerate the heart? *Nature*, **280**, 235–236.

Bulbring, E. & Tomita, T. (1987). Catecholamine action on smooth muscle. *Pharmacological Reviews*. **39**, 50–96.

Burch, R.M. & Axelrod, J. (1987). Dissociation of bradykinin induced prostaglandin formation from phosphatidylinositol turnover in Swiss 3T3 fibroblasts: Evidence for G protein regulation of phosholipase A2. *Proc. Natl. Acad. Sci. USA*, **84**, 6374–6378.

Campbell, G.D., Edwards, F.R., Hirst, G.D.S. & O'Shea, J.E. (1989). Effects of vagal stimulation and applied acetylcholine on pacemaker potentials in the guinea-pig heart. *Journal of Physiology*, **415**, 57–68.

Capiod, T. & Ogden, D.C. (1989). The properties of calcium-activated potassium ion channels in guinea-pig isolated hepatocytes. *Journal of Physiology*, **409**, 285–295.

Carpenter, G. (1987). Receptors for epidermal growth factor and other polypeptide mitogens. *Annual Review of Biochemistry*, **56**, 881–914.

Chinkers, M., Garbers, D.L., Chang, M-S., Lowe, D.G., Chin, H., Goeddel, D.V. & Schulz, S. (1989). A membrane form of guanylate cyclase is an atrial natriuretic peptide receptor. *Nature*, **338**, 78–83.

Cockroft, S. (1987). Polyphosphoinositide phosphodiesterase: regulation by a novel guanine nucleotide binding protein, G_p. *TIBS*, **12**, 75–78.

Codina, J., Yatani, A., Grenet, D., Brown, A.M. & Birnbaumer, L. (1987). *Science*, **236**, 442–445.

Cohen, C.J., Janis, R.A., Taylor, D.G. & Scriabine, A. (1986). Where do Ca^{2+} antagonists act? In *'Calcium antagonists in cardiovascular disease'* Ed. Opie, L.H. pp 151–163. New York. Raven Press.

Collier, J.G. & Vallance, P. (1989). Second messenger role for NO widens to nervous and immune systems. *TIPS*, **10**, 427–430.

DiFrancesco, D. (1985). The cardiac hyperpolarising-activated current, I_f. Origins and developments. *Progress in Biophysics and Molecular Biology*. **46**, 163–183.

DiFrancesco, D. & Tromba, C. (1987). Acetylcholine inhibits activation of the cardiac hyperpolarising-activated current, I_f. *Pflugers Arch*. **410**, 139–142.

Dohlman, H.G., Caron, M.G. & Lefkowitz, R.J. (1987). A family of receptors coupled to guanine nucleotide regulatory proteins. *Biochemistry*, **26**, 2657–2664.

Dolphin, A.C. (1987). Nucleotide binding proteins in signal transduction and disease. *TINS*, **10**, 53–57.

Eggermont, J.A., Vrolix, M., Wuytack, F. Raeymaekers, L., & Casteels, R. (1988). The $(Ca^{2+} lMg^{2+})$-ATPases of the plasma membrane and of the endoplasmic reticulum in smooth muscle cells and their regulation. *Journal of Cardiovascular Pharmacology*, **12**, supp 5, S51–S55.

Ehrlich, B.E. & Watras, J. (1988). Inositol 1,4,5-trisphosphate activates a channel from smooth muscle sarcoplasmic reticulum. *Nature*, **336**, 583–585.

Endo, M. (1977). Calcium release from the sarcoplasmic reticulum. *Physiological Reviews*, **57**, 71–108.

Fabiato, A. (1983) Calcium induced Ca release from the sarcoplasmic reticulum. *American Journal of Physiology*, **245**, C1–C14.

Fenwick, E.M., Marty, A. & Neher, E. (1982). Sodium and calcium channels in bovine chromaffin cells. *Journal of Physiology*, **331**, 599–635.

Ferris, C.D., Huganir, R.L., Supattatone, S. & Snyder, S.H. (1989). Purified Inositol 1,4,5-trisphosphate receptor mediates calcium flux in reconstituted lipid vesicles. *Nature*, **342**, 87–89.

Freissmuth, M., Casey, P.J. & Gilman, A.G. (1989). G proteins control diverse pathways of trans-membrane signalling. *FASEB Journal*, **3**, 2125–2131.

Friel, D.D. (1988). An ATP sensitive conductance in single smooth muscle cells from the rate vas deferens. *Journal of Physiology*, **401**, 361–380.

Furuichi, T., Yoshikawa, S., Miyawaki, A., Wada, K. Maeda, N. & Mikoshiba, K. (1989). Primary structure and functional expression of the inositol 1,4,5-trisphoshate-binding protein P_{400}. *Nature*, **342**, 32–38.

Ghosh, T.K., Mullaney, J.M., Tarazi, F.I. & Gill, D.L. (1989). GTP-activated communication between distinct inositol 1,4,5-trisphoshate-sensitive and insensitive calcium pools. *Nature*, **340**, 236–239.

Gill, G.N., Holdy, K.E., Walton, G.M. & Kanstein, C.B. (1976). Purification and characterisation of 3':5'-cyclic-GMP-dependent protein kinase. *Proc. Nat. Acad. Sci. U.S.A.* **73**, 3918–3922.

Graziano, M.P., Freissmutm, M. & Gilman, A.G. (1989). Expression of G_{sa} in Escherichia coli: purification and properties of two forms of the protein. *Journal of Biological Chemistry*, **264**, 409–418.

Grynkiewicsz, G., Poenie, M. & Tsien, R.Y. (1985) A new generation of Ca indicators with greatly improved fluorescence properties. *Journal of Biological Chemistry*, **260**, 3440–3450.

Hagiwara, S. (1975). Ca-Dependent action potential. In '*Membranes*' Vol 3. Ed. Eisenman, G. Marcel Dekker, New York.

Hamill, O.P., Marty, A., Neher, E., Sakmann, B. & Sigworth, F.J. (1981). Improved patch-clamp techniques for high resolution current recording from cells and cell free membrane patches. *Pflugers Arch. ges. Physiol.* **391**, 85–100.

Hartzell, H.C. & Fischmeister, R. (1986). Opposite effects of cyclic GMP and cyclic AMP on Ca^{2+} current in single heart cells. *Nature*, **323**, 273–275.

Harvey, R.D. & Hume, J.R. (1989). Autonomic regulation of a chloride current in heart. *Science*, **244**, 983–985.

Hescheler, J., Kameyama, M. & Trautwein, W. (1986). On the mechanism of muscarinic inhibition of the cardiac Ca current. *Pflugers Arch.* **407**, 183–189.

Hille, B. (1989). Ionic channels. Evolutionary origins and modern roles. *Quarterly Journal of Experimental Physiology*, **74**, 785–804.

Hokin, M.R. & Hokin, L.E. (1953). *Journal of Biological Chemistry*, **203**, 967–977.

Iijima, T., Irisawa, H. & Kameyama, M. (1985). Membrane currents and their modification by acetylcholine in isolated atrial cells of the guinea-pig. *Journal of Physiology*, **359**, 485–501.

Imoto, K., Methfessel, C., Sakmann, B., Mishina, M., Mori, Y., Konno, T., Fukkuda, K., Kurasaki, M., Bujo, H., Fujita, Y. & Numa, S. (1986). Location of a delta-subunit region determining ion transport through the acetylcholine receptor channel. *Nature*, **324**, 670–64.

Irvine, R.F. & Moor, R.M. (1986). Microinjection of inositol 1,3,4,5, tetrakisphosphate activates sea urchin eggs by a mechanism dependent on external Ca^{2+}. *Biochemical Journal*, **240**, 917–920.

Irvine, R.F., Letcher, A.J., Heslop, J.P., & Berridge, M.J. (1986). The inositol tris/terakisphosphate pathway- demonstration of Ins(1,4,5)P3 3-kinase activity in animal tissues. *Nature*, **320**, 631–634.

Jacob, R. (1990). Calcium oscillations in electrically non-excitable cells. *Biochimica et Biophysica Acta*, **1052**, 427–438.

Jacob, R., Merritt, J.E., Hallam, T.J. & Rink, T.J. (1988). Repetitive spikes in cytoplasmic calcium evoked by histamine in human endothelial cells. *Nature*, **335**, 40–45.

Kameyama, M., Hescheler, J., Mieskes, G. & Trautwein, W. (1986). The protein specific phosphatase 1 antagonises the β-adrenergic increase of the cardiac calcium current. *Pflugers Archiv.* **405**, 285–293.

Kim, D., Lewis, D.L., Graziadei, L., Neer, E.J., Bar-Sagi, D. & Clapham, D.E. (1989). G-protein $\beta\gamma$-subunits activate the cardiac muscarinic K^+-channel via phosholipase A_2. *Nature*, **337**, 557–560.

Kitazawa, T., Kobayashi, S., Horiuti, K., Somlyo, A.V., Somlyo, A.P. (1989). Receptor coupled, Permeabilised smooth muscle. Role of the phosphatidylinositol cascade, G-proteins, and modulation of the contractile response to Ca^{2+}. *Journal of Biological Chemistry*, **264**, 5339–5342.

Kobilka, B.K., Kobilka, T.S., Daniel, K., Regan, J.W., Caron, M.G. & Lefkowitz, R.J. (1988). Chimeric α_2-, β_2-adrenergic receptors: delineation of domains involved in effector coupling and ligand binding specifity. *Science*, **240**, 1310–1316.

Kuno, M. & Gardner, P. (1987). Ion channels activated by inositol 1,4,5-trisphosphate in plasma membrane of human T-lymphocytes. *Nature*, **326**, 301–304.

Kurachi, Y., Nakajima, T. & Sugimoto, T. (1986). On the mechanism of activation of the muscarinic K^+ channels by adeenosine in isolated atrial cells: involvement of GTP-binding proteins. *Pflugers Arch.* **407**, 264–274.

Kume, H., Takai, A., Tokuno, H. & Tomita, T. (1989). Regulation of Ca^{2+} dependent K^+ channel activity in tracheal myocytes by phosphorylation. *Nature*, **341**, 152–154.

Lefkowitz, R.J., Kobilka, B.K, & Caron, M.G. (1989). The new biology of drug receptors. *Biochemical Pharmacology*, **38**, 2941–2948.

Levitski, A. (1988). From epinephrine to cyclic AMP. *Science*, **241**, 800–806.

Lincoln, T.M. (1989). Cyclic GMP and mechanisms of vasodilation. *Pharmacol. Therapeutics.* **41**, 479–502.

Logothetis, D.E., Kurachi, Y., Galper, J., Neer, E.J. & Clapham, D.E. (1987). The $\beta\gamma$ subunits of GTP binding proteins activate the muscarinic K^+ channel in heart. *Nature*, **325**, 321–326.

Marty, A. (1987). Control of ionic currents and fluid secretion by muscarinic agonists in exocrine glands. *TINS*, **10**, 373-377.

Marty, A. (1989). The physiological role of calcium dependent channels. *TINS*, **12**, 420-424.

Mattera, R., Graziano, J.P., Yatani, A. Graf, R., Codina, J., Gilman, A.G., Birnbaumer, L. & Brown, A.M. (1989). Bacterially synthesized splice variants of the α sub unit of the G protein G_s activate both adenylate cyclase and dihydropyridine sensitive calcium channels. *Science*, **243**, 804-807.

McKnight, G.S., Clegg, C.H. Uhler, M.D. Chrivia, J.C., Cadd, G.G., Correll, L. & Otten, A.D. (1988). Analysis of the cAMP-dependent protein kinase system using molecular geentic techniques. In: *Recent Progress in Hormone Research* **44**, 307-355. Ed. by Clark, J.H. Academic Press, London.

Meech, R.W. (1978). Calcium dependent K-channel activation in nervous tissues. *Annual Reviews of Biophysics and Bioengineering*, **7**, 1-18.

Michell, R.H. (1975). Inositol phospholipids and cell surface receptor function. *Biochem. Biophysica. Acta.* **415**, 81-147.

Michell, R.H., Kirk, C.J., Jones, L.M., Downes, C.P. & Creba, J.A. (1981). The stimulation of inositol lipid metabolism that accompanies calcium mobilisation in stimulated cells: defined characteristics and unanswered questions. *Phil. Trans. R. Soc. B.* **296**, 123-137.

Mitchell, M.R., Powell, T, Terrar, D.A. & Twist, V.W. (1984). Ryanodine prolongs Ca currents while suppressing contraction in rat ventricular muscle cells. *British Journal of Pharmacology*, **81**, 13-15.

Morgan, J.P. & Morgan, K.G. (1984). Stimulus specific patterns of intracellular calcium levels in smooth muscle of ferret portal vein. *Journal of Physiology*, **351**, 155-167.

Morris, A.P., Gallagher, D.V., Irvine, R.F. & Petersen, O.H. (1987). Synergism of inositol trisphosphate and tetrakisphosphate in activating Ca^{2+}-dependent K^+ channels. *Nature*, **330**, 653-655.

Murad, F. (1986). Cyclic guanosine monophosphate as a mediator of vasodilation. *Journal of Clinical Investigation*, **78**, 1-5.

Murray, K.J. (1990). Cyclic AMP and mechanisms of vasodilation. *Pharmac. Ther.* **47**, 329-345.

Murray, K.J., Reeves, M.L. & England, P.J. (1989). Protein phosphorylation and compartments of cyclic AMP in the control of cardiac contraction. *Molec. Cell Biochem.* **89**, 175-179.

Neer, E.J., Lok, J.M. & Wolf, L.G. (1984). Purification and properties of the inhibitory guanine nucleotide regulatory unit of brain adenylate cyclase. *Journal of Biological Chemistry*, **259**, 14222-14229.

Nelson, M.T. Standen, N.B. Brayden, J.E. & Worley, J.F. (1988). Noradrenaline contracts arteries by activating voltage dependent calcium channels. *Nature*, **336**, 382-385.

Nishizuka, Y. (1984). the role of protein kinase C in cell surface signal transduction and tumour promotion. *Nature*, **308**, 693-698.

Nishizuka, Y. (1986). Studies and perpectives of protein kinase C. *Science*, **233**, 305-312.

Nishizuka, Y. (1988). The molecular heterogeneity of protein kinase C and its implications for cellular regulation. *Nature*, **334**, 661-665.

Noble, D. (1975). *The initiation of the heartbeat*. Clarendon press. Oxford.

Noble, D. (1984). The surprising heart; a review of recent progress in cardiac electrophysiology. *Journal of Physiology*, **353**, 1-50.

North, R.A. (1989). Neurotransmitters and their receptors: from the clone to the clinic. *Seminars in the Neurosciences*, **1**, 81-90.

Pacaud, P., Loirand, G., Mironneau, C. & Mironneau, J. (1989). Noradrenaline activates a calcium-activated chloride conductance and increases the voltage-dependent calcium current in cultured single cells of rat portal vein. *British Journal of Pharmacology*, **97**, (1) 139-46.

Penner, R., Mathews, G. & Neher, E. (1988). Regulation of calcium influx by second messengers in rat mast cells. *Nature*, **334**, 499-504.

Petersen, O.H. & Gallacher, D.V. (1988). Electrophysiology of pancreatic and salivary acinar cells. *Annual Reviews of Physiology*, **50**, 65-80.

Pfaffinger, P.J., Martin, J.M., Hunter, D.D., Nathanson, N.M. & Hille, B. (1985). GTP-binding proteins couple cardiac muscarinic receptors to a K channel. *Nature*, **317**, 536-540.

Rall, T.W., Sutherland, E.W. & Berthet, J. (1957). The relationship of epinephrine and glucagon to liver phosphorylase IV. Effect of epinephrine and glucagon on the reactivation of phosphorylase in liver homogenates. *Journal of Biological Chemistry*, **224**, 463-475.

Rassmussen, H., Takuwa, Y. & Park, S. (1987). Protein kinase C in the regulation of smooth muscle contraction. *FASEB Journal.* **1**, 177-185.

Reuter, H. (1983). Calcium channel modulation by neurotransmitters, enzymes and drugs. *Nature*, **301**, 569-574.

Rodbell, M., Birnbaumer, L., Pohl, S.L. & Krans, H.M.J. (1971). The glucagon sensitive adenyl cyclase system in plasma membranes of the rat liver. An obligatory role of guanyl nucleotides in glucagon action. *Journal of Biological Chemistry* **246**, 1877–1882.

Ross, C.A., Meldolesi, J., Milner, T.A., Satoh, T., Supattapone, S. & Snyder, S.H. (1989). Inositol 1,4,5-trisphosphate receptor localised to endoplasmic reticulum in cerebellar Purkinje neurons. *Nature*, **339**, 468–470.

Sakmann, B. Noma, A. & Trautwein, W. (1983). Acetylcholine activation of single muscarinic K^+ channels in isiolated pacemaker cells of the mammalian heart. *Nature*, **303**, 250–253.

Schubert, B., VanDongen, A.M.J., Kirsch, G.E. & Brown, A.M. (1989). β-adrenergic inhibition of cardiac sodium channels by dual G-protein pathways. *Science*, **245**, 516–519.

Sieghart, W. (1989). Multiplicity of $GABA_A$-benzodiazepine receptors. *TIPS*, **10**, 407–411.

Silver, P.J., Lepore, R.E., O'Connor, B., Lemp, B., Hamel, L.T., Bentley, R.G. & Harris, A.L. (1988). Inhibition of the low K_m cyclic AMP phosphodiesterase and activation of the cyclic AMP system in vascular smooth muscle by milrinone. *J. Pharmac. exp. Ther.* **247**, 34–42.

Sims, S.M., Singer, J.J. & Walsh, J.V. (1987). Antagonistic adrenergic-muscarinic regulation of M current in smooth muscle cells. *Science*, **239**, 190–193.

Somlyo, A.V. Bond, M., Somlyo, A.P. & Scarpa, A. (1985). Inositol trisphoshate induced calcium release and contraction in vascular smooth muscle. *Proceedings of the National Academy of Sciences*, **82**, 5231–5235.

Stuhmer, W., Conti, F., Suzuki, H., Wang, X., Noda, M., Yahagi, N., Kubo, H. & Noma, S. (1989). Structural parts involved in activation and inactivation of the sodium channel. *Nature*, **339**, 597–603.

Streb, H., Irvine, R.F., Berridge, M.J. & Schulz, I. (1983). Release of Ca^{2+} from a non-mitochondrial intracellular store in pancreatic acinar cells by inositol-1,4,5-trisphosphate. *Nature*, **306**, 67–69.

Suzuki, H. (1985). Electrical responses of smooth muscle cells of the rabbit ear artery to adenosine trisphosphate. *Journal of Physiology*, **359**, 401–415.

Takeshima, T., Nishimura, S., Matsumoto, T., Ishida, H., Kangawa, K., Minamino, N., Matsuo, H., Ueda, M., Hanaoka, M., Hirose, T. & Numa, S. (1989). Primary structure and expression from complementary DNA of skeletal muscle ryanodine receptor. *Nature*, **339**, 439–445.

Toyoshima, C. & Unwin, N. (1988). Ion channel of acetylcholine receptor reconstructed from images of postsynaptic membrane. *Nature*, **336**, 247–250.

Tsien, R.W., Hess, P. & Nilius, B. (1987). Cardiac calcium currents at the level of single channels. *Experientia*, **43**, 1169–1172.

Van Breemen, C., Aaronson, P., & Loutzenhiser, R. (1979). Na-Ca interactions in mammalian smooth muscle. *Pharmacological Reviews*, **30**, 167–208.

Vigne, P., Breittmayer, J-P., Duval, D., Frelin, C. & Lazdunski, M. (1988). The Na^+/Ca^{2+} antiporter in aortic smooth muscle cells. *Journal Biological Chemistry*, **263**, 8078–8083.

Volpe, P., Krause, K-H., Hashimoto, S., Zorzato, F., Pozzan, T., Meldolesi, J. & Lew, D.P. (1988). Calciosome, a cytoplasmic organelle: The inositol 1,4,5-trisphosphate sensitive Ca^{2+} store of nonmuscle cells? *Proceedings of the National Academy of Sciences. USA.* **85**, 1091–1095.

Wakui, M., Potter, B.V. & Petersen, O.H. (1989). Pulsatile intracellular calcium release does not depend on fluctuations in inositol trisphoshate concentration. *Nature*, **339**, 317–320.

Waldman, S.A. & Murad, F. (1987). Cyclic GMP synthesis and function. *Pharmacological Reviews*, **39**, 163–196.

Walter, U. (1981). Distribution of cyclic GMP-dependent protein kinase in various rat tissues and cell lines determined by a sensitive and specific radioimmunoassay. *Eur. J. Biochem.* **118**, 339–346.

Woods, N.M., Cuthbertson, K.S.R. & Cobbold, P.H. (1987). Phorbol-ester-induced alterations of free calcium ion transients in single rat hepatocytes. *Biochemical Journal*, **246**, 619–623.

Yatani, A., Codina, J., Imoto, Y., Reeves, J.P., Birnbaumer, L. & Brown, A.M. (1987). A G-protein directly regulates mammalian cardiac calcium channels. *Science*, **238**, 1288–1292.

Yatani, A., Mattera, R., Codina, J., Graf, R., Okabe, K., Padrell, E., Iyengar, R., Brown, A.M. & Birnbaumer, L. (1988). The G protein gated atrial K^+ is stimulated by three distinct G_i α-subunits. *Nature*, **336**, 680–682.

5 Transmission: Acetylcholine

Noel J. Buckley and Malcolm Caulfield[1]

Division of Molecular Cell Biology, National Institute for Medical Research, The Ridgeway, Mill Hill, London NW7 1AA

Acetylcholine is the principal excitatory neurotransmitter in the autonomic nervous system. Descriptions of acetylcholine actions and their division into 'muscarine-like' and 'nicotine-like' date back to the turn of the century. Eighty years later, we have gained a more detailed and confused picture of the metabolism, distribution and actions of acetylcholine.

Undoubtedly, one of the most exciting recent developments has been the application of molecular biological procedures to the cloning of the gene families encoding the muscarinic and nicotinic gene families. This revolution has pervaded the whole of pharmacology since it allows the control and study of the expression of single gene products and comparison of their properties with those of endogenous receptors. Much of this chapter is structured around this dualism: the analysis of function in autonomic tissues and the study of individual receptor subtypes expressed in foreign host cells. Diversity is a key issue whether considering the size of nicotinic and muscarinic receptor gene families or the range of cellular functions modulated by acetylcholine. Bringing together these perspectives is one of the most important goals in pharmacology, but at the moment, this is still a distant goal and will remain so until we learn to manipulate the endogenous gene products present in autonomic tissues.

Considerable advances have also been made in understanding the mechanisms of muscarinic and nicotinic receptor mediated actions. Muscarinic actions are mediated by activation of intermediary G-proteins which then act upon various second messengers including cAMP and phosphoinositides, and ion channels permeable to Ca^{2+}, K^+ and Cl^-. In contrast, nicotinic actions mediate short latency, rapid excitatory effects by acting as ligand dependent cation channels in an analogous manner to the nicotinic receptors of the neuromuscular junction of skeletal muscle.

Understanding this diversity of action at the molecular and cellular levels also provides a foundation for developing and screening selective therapeutic agents aimed at regulating many aspects of autonomic dysfunction including cardiac and gastrointestinal disorders. Ready availability of cell lines expressing individual gene products and advances in receptor binding/function will greatly accelerate this process.

---

[1] Department of Pharmacology
University College London
Gower Street
London WC1E 6BT

INTRODUCTION

The autonomic nervous system (ANS) consists of those neural pathways having ganglionic synapses located outside the central nervous system. Organs and tissues innervated by the ANS include smooth muscle (visceral, vascular and others), some types of secretory cell and cardiac muscle. The parasympathetic division of the ANS is formed of neurones arising in the brain stem (eg. Edinger-Westphal nucleus and the dorsomedial nucleus of the vagus) and sacral spinal cord, while the sympathetic division has neurones originating in the intermediolateral columns of the spinal cord. Through the work of Langley, acetylcholine (ACh) has been classically regarded as the neurotransmitter at synapses of the ganglia of the ANS and of the post-ganglionic neuro-effector junctions of the parasympathetic ANS. This picture is an over-simplification and major exceptions to the scheme include the cholinergic innervation of the anatomically sympathetic sweat glands, inhibitory cholinergic components of the sympathetic input to the spleen and cholinergic vasodilator fibres in skin, muscles and the coronary vasculature (reviewed by Campbell, 1970).

Jacobowitz (1974) has constructed a scheme which encompasses most of the present knowledge of cholinergic innervation in the ANS (Figure 5.1).

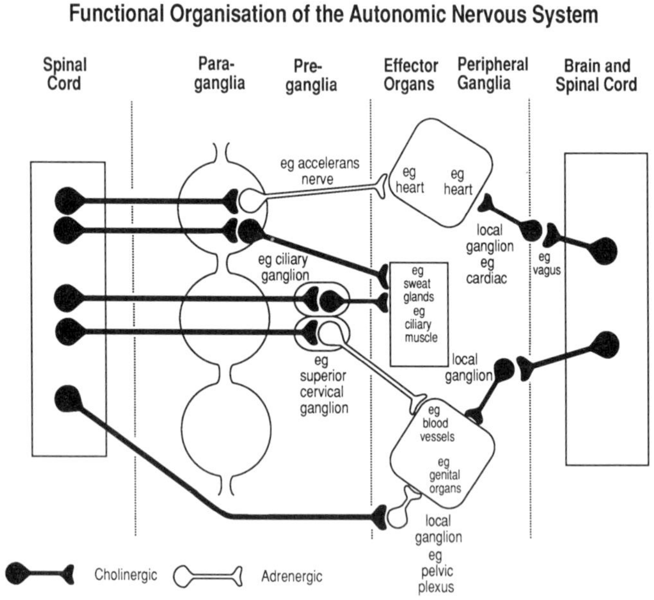

Functional Organisation of the Autonomic Nervous System

FIGURE 5.1 The paravertebral, or paraganglia, are those of the sympathetic chain; fibres arise from the lumbar and thoracic spinal cord. Pre-ganglia are those 'sympathetic' ganglia located remotely from the sympathetic chain, for example the superior cervical ganglion. Peripheral ganglia are remote from the CNS, for example the cholinergic coeliaco-mesenteric ganglia, or the adrenergic hypogastric ganglion. Local peripheral ganglia are those located very close to the target organ, such as intramural cardiac ganglia. The adrenal medulla receives its cholinergic innervation direct from the thoracic spinal cord without intervening synapses and is regarded functionally as resembling a ganglionic neurone.

The evidence for ACh as a neurotransmitter in the autonomic nervous system has been reviewed concisely by Gershon (1970). Early work relied heavily on the use of specific antagonists such as atropine to demonstrate that a given nerve-mediated response was the result of an action of released acetylcholine. Alternatively, potentiation of a response by an anti-cholinesterase (eg. physostigmine, neostigmine) was indicative of a role for ACh. More recently, tools such as radiolabelled choline, histochemical stains for acetylcholinesterase (AChE) and specific antibodies for cholineacetyltransferase (ChAT) have greatly accelerated the identification and characterization of cholinergic innervation in the peripheral nervous system. Acetyl-cholinesterase staining has, however, been demonstrated in non-cholinergic cells and high affinity uptake of (^{3}H)-choline has been shown in glial cells (Massarelli *et al.*, 1974) and in the cardiac ganglion of the horseshoe crab (Ivy *et al.* 1985). This latter tissue is non-cholinergic, but nevertheless has high affinity sodium-dependent uptake of choline which is inhibited by hemicholinium HC-3 and stimulated by potassium depolarisation. It is currently thought that the strongest evidence for cholinergic innervation is the presence of reactivity to specific anti-ChAT antibodies.

ACETYLCHOLINE METABOLISM—SYNTHESIS, RELEASE AND INACTIVATION

Much of the knowledge of ACh synthesis, release and degradation has come from studies of easily accessible and manipulable tissues which contain large numbers of cholinergic neurones; for example, work on skeletal neuromuscular junction and electric eel and *Torpedo* electroplax has provided much information on cholinergic synaptic vesicles, while experiments on brain preparations (particularly synaptosomes) have generated the bulk of data on ACh synthesis and release. In contrast, the cholinergic neurones of the peripheral autonomic nervous system generally form a diffuse innervation of their target organs and are not as amenable to detailed study of biochemical processes. Autonomic ganglia and adrenal chromaffin cells are an exception to this generalization.

A simplified general scheme to describe the dynamics of ACh synthesis, release and inactivation is shown in figure 2. Reference will be made to more detailed data on particular organs and systems in the appropriate section.

It is accepted that ACh is synthesized by the action of choline acetyltransferase, using choline and acetate as precursors (Tucek, 1988). Choline is transported into the cell by a metabolically driven mechanism which has high affinity for choline ($K_T < 10\,\mu M$) and which is dependent on the presence of extracellular sodium ions. There is also a low-affinity transport system ($K_T > 20\,\mu M$) which is probably not associated with ACh synthesis under normal conditions. The source of acetate is from normal metabolic processes, (Tucek, 1988). Uptake of choline is closely coupled to neuronal activity; depolarization of presynaptic terminals and increase in transmitter release greatly accelerates high-affinity choline uptake. This increased uptake results in turn in an increase in the synthesis of ACh and it has been suggested that this arises from a physical coupling of ChAT with the high-affinity choline

Acetylcholine Synthesis, Release and Metabolism

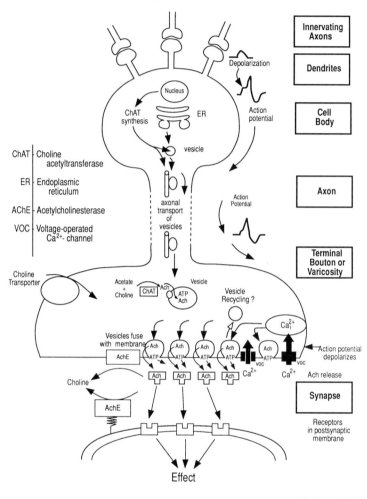

After Agoston (1988)

FIGURE 5.2   The diagram represents schematically the events involved in the synthesis, release and breakdown of ACh in a neurone. Incoming excitatory inputs depolarize dendrites, resulting in action potential firing in the cholinergic cell body. The action potential propagates along the axon and depolarizes the terminal boutons to elicit ACh release. Refer to the text for a resume of the component pathways and mechanisms.

transporter. However, the weight of evidence is that the coupling is kinetic, such that transport of choline into the presynaptic terminal is the rate-limiting factor for ACh synthesis. (Ducis, 1988). Synthesized ACh is stored in specific synaptic vesicles, which have a specific ACh transporter in their membranes. ACh is stored in synaptic vesicles together with ATP, which may function as a counter-ion (Zimmerman, 1988).

Arrival of an action potential at the nerve terminal causes influx of calcium through voltage-sensitive calcium channels, resulting in ACh release. Release probably results from fusion of ACh-containing vesicles with the presynaptic membrane and empty vesicles may then be recycled locally. Released ACh diffuses to receptors on the innervated organs and induces a response. ACh diffusing from the receptor is hydrolysed by acetylcholinesterase (located both pre- and post-synaptically) to choline and acetate. Some of the choline generated is taken up by the presynaptic choline transporter to be re-used in ACh synthesis. Synapses at autonomic neuro-effector junctions are different from the specialized synapses of the skeletal neuro-muscular junction where ACh has only a short distance (20–50 nm) to diffuse from the presynaptic terminal to the receptors and where there is a high concentration of receptors at the point of innervation by a synaptic bouton. Junctions between autonomic cholinergic nerves and heart or smooth muscle cells, for example, are wide (micrometres) and lack anatomical postjunctional special-ization, while the receptors are often fairly uniformly distributed over the surface of the innervated cell (see reviews by Gabella, 1976; Loffelholz & Pappano, 1985; Kilbinger, 1988). The terminal areas of cholinergic fibres innervating smooth and cardiac muscles have many variocosities containing ACh-filled small electron-luscent vesicles. These varicosities represent sites of ACh release *en passage*, rather than synapses in the strict sense (Kilbinger, 1988). The same is not true of cholinergic synapses in autonomic ganglia, where there is a much tighter apposition of pre- and post-synaptic elements and a concentration of ACh receptors around the points of presynaptic innervation (cf Harris *et al.*, 1971; Nishi 1974; Hartzell, 1980).

ACETYLCHOLINE RECEPTORS

The division of acetylcholine receptors into muscarinic and nicotinic subtypes (Dale, 1914) is almost as old as the original concept of 'receptive substances' (Langley, 1905; Ehrlich, 1907). Dale defined muscarinic receptors as those activated by muscarine and blocked by atropine, while nicotinic receptors are activated by nicotine and blocked by curare. Ever since, the classification of receptors into subtypes has been a predilection of pharmacologists, not least because it offers a rationale for under-standing the diversity of agonist induced responses and complex antagonist binding curves. ACh receptors are no exception. Dale's original classification still stands, nearly 80 years later, although there are now several further strata of complexity involving further subdivisions based upon, and reflected in, the different subtypes' ligand binding profiles, pharmacological specificities, gene structure and distribu-tion. All of these aspects will be discussed below.

   Muscarinic receptors are members of a large family of G-protein coupled receptors that work via activation of one or more intermediary GTP-binding proteins (G-proteins). The overall molecular mechanism of activation involves four steps; (1) Agonist binds to and activates the receptor (2) Activated receptor binds to GDP-bound G-protein and catalyses exchange of GTP for GDP (3) Activated G-protein dissociates from the receptor and into GTP-α and $\beta\gamma$ subunits (4) Dissociated

G-protein activates effector mechanisms until GTP is hydrolysed and inactive $\alpha\beta\gamma$ trimer is reformed (for reviews of G-protein action see Gilman, 1987; Neer and Clapham, 1988; Bourne et al., 1990).

Within the autonomic nervous system and its effector organs, muscarinic receptors are widespread and their stimulation can induce a wide array of responses including contraction of smooth muscle, relaxation of cardiac muscle, stimulation of glandular secretion, modulation of ganglionic transmission and pre-synaptic inhibition of neurotransmitter release (discussed in later sections). This diversity is reflected in the range of biochemical and ionic effector mechanisms acted upon by activation of muscarinic receptors. These include (1) Stimulation of phosphoinositide (PI) hydrolysis (2) Inhibition of adenylyl cyclase (AC) (3) Activation of K^+ conductances (4) Inhibition of M-current (5) Stimulation of cGMP production (6) Stimulation of arichidonic acid release (7) Inhibition of Ca^{2+} conductances.

In contrast, cholinergic nicotinic receptors have a more discrete distribution and a consequently more limited role in autonomic function; they are found only within autonomic ganglia and are not present in any effector tissues. Unlike their muscarinic counterparts, nicotinic receptors require no intermediary molecule for their action and the receptor subunits constitute a multimeric ion channel. Their activation leads to only one response–a rise in Na^+ and Ca^{2+} conductance, resulting in a depolarisation of the plasma membrane. However, this one event is responsible for the fast excitatory postsynaptic potential (e.p.s.p.) that mediates ganglionic excitatory transmission.

Hence, the contrast between muscarinic and nicotinic receptors could hardly be greater. On the one hand, muscarinic receptors belong to a supergene family encoding monomeric G-protein coupled receptors that are responsible for a wide array of intracellular responses while on the other hand, nicotinic receptors belong to an entirely different gene family that encodes multi-subunit ion-channel receptors and transduce a singular effector mechanism. It is remarkable that two such divergent families of molecules have evolved that are each capable of being activated by the same amine ligand, ACh.

Our knowledge and understanding of cholinergic receptors has been revolutionized in recent years by molecular cloning studies. These studies have radically altered our perception of receptors in terms of their molecular structure, diversity, pharmacological properties and patterns of expression. For these reasons we shall discuss (1) The cloning of the muscarinic and nicotinic receptor gene families (2) The pharmacological properties of their gene products (3) The patterns of expression of muscarinic and nicotinic receptor genes and gene products.

MOLECULAR CLONING STUDIES

Cloning of the muscarinic receptor gene family

In 1986 Numa and his colleagues (Kubo et al., 1986a) reported the nucleotide sequence of a cDNA clone encoding an m1 receptor. Their cloning strategy was direct; based upon the peptide sequence of tryptic fragments of purified receptor from porcine cerebrum, a series of degenerate oligodeoxynucleotides were used to screen a brain cDNA library and isolate a cDNA clone encoding an m1 receptor. Later that year, the porcine atrial receptor (m2) was cloned using sequence derived from another tryptic fragment to screen an atrial library (Kubo *et al.*, 1986b); the same sequence was subsequently reported by Peralta *et al.* (1987a). Hydrophobicity plots of the predicted amino acid sequence of these receptors yielded a model of the receptor that had seven transmembrane domains, an extracellular N-terminal and an intracellular C-terminal. This structure had previously been predicted for the β-adrenergic receptor (Dixon *et al.*, 1986) and the opsin family (Nathans and Hogness, 1983; Nathans *et al.*, 1986) and has subsequently been seen to be the hallmark of all G-protein coupled receptors that have been cloned to date (over 80 receptor sequences at the time of writing). It was this sequence similarity between the cerebral m1 muscarinic receptor and the β1-adrenergic receptor that provided the impetus for the cloning of a further three muscarinic receptor genes (m3,m4,m5). Sequence comparison of the m1 and β1-adrenergic receptor identified a region of maximal homology in the second transmembrane domain. Using a probe corresponding to this domain to screen cDNA and genomic libraries at low stringency led to the cloning of the rat and human m3, m4 and m5 receptor genes (Bonner *et al.*, 1987; 1988; Peralta *et al.*, 1987b). The fact that all of these genes are intronless in their coding regions greatly facilitated their cloning from genomic libraries. All members of this family are highly related (sequence identity ranges from 85% homology between m2 and m4 to 65% between m1 and m4 within the transmembrane domains – there is much less sequence identity in the N-terminal, C-terminal and large cytoplasmic loop). Since these original cloning studies were undertaken, a number of cognate sequences from other species have been reported (Shapiro *et al.* 1988; Onai *et al.*, 1989; Tietje *et al.*, 1990).

Deletional analysis and site directed mutagenesis studies indicate that the binding site is buried within the hydrophobic transmembrane domains and involves two aspartate residues located in transmembrane helices II and III; mutation of the more distal aspartic acid of the m1 receptor to asparagine resulted in a greatly reduced antagonist affinity compared with the wild type receptor (Fraser *et al.*, 1990). Earlier protein mapping studies had pointed to an acidic residue in transmembrane III being the site of alkylation by the irreversible antagonist, propylbenzylylcholine mustard (PrBCM) (Curtis *et al.*, 1989; Kurtenbach *et al.* 1990). Similar studies have shown that the G-protein coupling domain involves the N- and C-terminal most portions of the large cytoplasmic loop. The original observations came from chimaeras of m1/m2 receptors that demonstrated the involvement of the large intracellular loop (i3) in coupling of the m1 receptor to ACh-induced currents in frog oocytes (Kubo *et al.*, 1988). More subtle deletional analysis showed that only the distal 17 amino acids and proximal 11 amino acids of the i3 loop were necessary (Wess *et al.*, 1989; 1990; Shapiro and Nathanson, 1989). For a more detailed discussion of the molecular structure of the muscarinic receptor see Hulme *et al.*, 1990.

Cloning of the nicotinic receptor gene family

Nicotinic receptors were first isolated from the electric organ of *Torpedo* and were shown to be composed of four different polypeptide chains arranged as a $\alpha_2\beta\gamma\delta$ pentamer (for reviews, see Conti-Tronconi and Raftery, 1982; Popot and Changeux, 1984; Hucho, 1986 for more details). The subsequent cloning of the individual subunits represented a landmark in receptor biology since these were the first neuro-transmitter receptors to be cloned (their cloning predated that of their muscarinic counterparts by half a decade). Primary sequence data derived from peptide mapping was rapidly followed by isolation of cDNAs encoding the α-subunit (Noda *et al.*, 1982; Devillers-Theiry *et al.*, 1983), β-subunit (Noda *et al.*, 1983a), γ-subunit (Noda *et al.*, 1983b) and δ-subunit (Claudio *et al.*, 1983; Noda *et al.*, 1983b). Mammalian cognate sequences isolated from skeletal muscle soon followed (Noda *et al.*, 1983c; La Polla *et al.*, 1984; Takai *et al.*, 1984; Tanabe *et al.*, 1984; Boulter *et al.*, 1985). Hydropathicity plots based upon sequence data predicted four hydrophobic domains of sufficient length to form transmembrane helices (see Guy and Hucho, 1987). Although at one point there were several models proposed to describe the exact arrangement of the transmembrane helices (see Guy and Hucho, 1987), the present consensus is that both the N-terminus and C-terminus are extracellular and the protein spans the membrane four times. In the case of the α- and γ-subunits, immunochemical evidence lends direct support to this model (Ratnam *et al.*, 1986; La Rochelle *et al.*, 1987).

However, it was clear from pharmacological studies and from RNA analyses that the α subunit in neuronal nicotinic receptors was not the same as the α1 subunit of muscle. In 1986 Boulter *et al.* screened a PC12 cDNA library at low stringency using an α1 probe and succeeded in cloning a novel α subunit (α3). Similar strategies rapidly led to the identification of an α2 (Wada *et al.*, 1988b), an α4 subunit (Goldman *et al.*, 1987) and an α5 subunit (Boulter *et al.*, 1990). Identification of these subunits as α-subunits relied upon the presence of conserved cysteine residues at cys192 and cys193. Three further β-subunits have also been cloned; β2 (Deneris *et al.*, 1988), β3 (Deneris *et al.*, 1989) and β4 (Duvoisin *et al.*, 1989; Isenberg and Meyer, 1989; Boulter *et al.*, 1990). Classification of these subunits as β is simply that, obversely to α-subunits, they do not possess cys192 and cys193. All of these subunits share a great degree of sequence identity and, in the case of the α3, α5 and β4 subunits, are found in a gene cluster (Boulter *et al.*, 1990). All deletional and mutagenesis studies conducted to date have been conducted upon the muscular nicotinic receptor subunits (see Connolly, 1989 for review). Unlike their muscarinic counterparts, the binding site of the α-subunit of the nicotinic receptors is associated with the extracellular N-terminal and is critically dependent upon two cysteine residues that are conserved among all nicotinic acid receptors (Kao *et al.*, 1984; Mishina *et al.*, 1985). Further discussion of the molecular structure of muscular nicotinic receptor subunits is beyond the scope of this chapter but reviews can be found in Luyten, 1986; Hucho, 1986; Guy and Hucho, 1987; Connolly, 1989.

This unexpected degree of cholinergic receptor heterogeneity had several immediate implications for understanding the role of nicotinic and muscarinic receptors in autonomic function:

1) The idea evolved that receptor diversity existed to allow discrete coupling of one receptor subtype to one signal transduction mechanism (selective coupling hypothesis), at least for muscarinic receptors. This hypothesis was flawed since there were now more muscarinic receptor subtypes than known signal transduction pathways. This idea clearly was not applicable to nicotinic receptors since only one effector exists.

2) The ligand binding and pharmacological properties of individual receptors could now be assessed since each gene could be introduced into a suitable expression system and its gene product analysed. In other words, pharmacology could turn from examining endogenous gene products to controlling exogenous gene expression. This was particularly useful for autonomic tissues, most of which express a mixture of subtypes.

3) The distribution of the mRNAs encoding each receptor could be mapped throughout the autonomic nervous system.

What are the pharmacological properties of these receptors and how do they relate to those mediating the actions of ACh in the autonomic nervous system?

PHARMACOLOGY

Whereas the great majority of studies of the pharmacological properties of recombinant muscarinic receptors have been carried out using transfected fibroblasts, in the case of neuronal nicotinic receptors, all gene expression studies have been conducted on injected oocytes. This is a consequence of two fundamental differences in the inherent properties of the two classes of receptor. On the one hand, construction of cell lines expressing muscarinic receptors requires the transfection of only one gene and subsequent analysis has largely relied upon biochemical assays (although see Fukuda et al., 1987; 1988; Jones et al., 1988a; 1988b). On the other hand, nicotinic receptors require the co-expression of at least two subunits and analysis of receptor function relies upon electrophysiological recordings of membrane currents – functions most easily carried out by injecting mRNA into frog oocytes.

Muscarinic receptors – gene expression studies

The establishment of stable cell lines expressing a single transfected muscarinic receptor gene has allowed a precise analysis of the second messengers and effector mechanisms employed by each subtype. A number of expression systems have been used to address this question, including mammalian fibroblasts, neuroblastomas and frog oocytes.

Broadly speaking, these experiments have provided a consensus that m1, m3 and m5 receptors tend to couple to stimulation of phosphoinositide hydrolysis via a pertussis toxin (PTX) insensitive G-protein whilst m2 and m4 receptors prefer to couple to inhibition of adenylyl cyclase via a PTX-sensitive G-protein (Bonner et al., 1988; Conklin et al., 1988; Fukuda et al., 1988; Lai et al., 1988; Peralta et al., 1988b; Ashkenazi et al., 1989; Shapiro et al., 1989). However there are several nuances which complicate this rather simplistic picture.

1) It has not yet been possible to assess which receptors couple directly to opening of K^+ channels, as in atria (and some brain stem nuclei) since no cell line has been reported to express this channel.

2) Expression of high levels of receptor (5,000–2,000,000 receptors/cell) in transfected fibroblasts can result in coupling to more than one second messenger. Expression of high receptor densities in CHO fibroblasts transfected with m2 or m4 genes can result in a weak stimulation of phosphatidyl (PI) hydrolysis in addition to the usual inhibition of adenylyl cyclase (Peralta *et al.*, 1988; Ashkenazi *et al.*, 1989). Since the PI response is less sensitive to PTX inhibition than the cyclase inhibition, then it seems likely that different G-proteins are involved in mediating the two responses (Ashkenazi *et al.*, 1989).

3) There are several reports demonstrating a dependence of the nature of the receptor activated response upon the host cell phenotype. One study reports that when m1 receptors are expressed in RAT-1 fibroblasts then activation leads to both a stimulation of PI hydrolysis and an inhibition of adenylyl cyclase; the former via a PTX-insensitive G-protein and the latter via a PTX-sensitive G-protein (Stein *et al.*, 1988). Expression of the chick m4 receptor in CHO cells leads to both stimulation of phosphoinositide metabolism and inhibition of adenylyl cyclase whilst expression in Y1 cells (at equivalent receptor density) leads only to inhibition of adenylyl cyclase (Tietje *et al.*, 1989). Interestingly, a similar dependence of response upon host cell type has also been observed with the D2 dopamine receptor (Vallar *et al.*, 1990).

In order to gain insight into the coupling specificities of muscarinic receptors in neural cells, Fukuda *et al* (1988) transfected muscarinic receptor genes into NG108-15 neuroblastoma × glioma cells. This cell line expresses an endogenous m4 receptor (Peralta *et al.*, 1987b; Lazareno *et al.*, 1990) which couples to inhibition of adenylyl cyclase. NG108-15 cells also express a voltage-activated K^+ current (M-current) that is believed to play a role in damping neuronal firing in autonomic ganglia and CNS neurons (Brown and Adams, 1980). In these cells the native m4 receptor is not coupled to closure of the M-current. Activation of muscarinic receptors on native and transfected NG108-15 cells demonstrated that m1 and m3 receptors couple to inhibition of the M-current but that m2 and m4 receptors do not. Since several autonomic ganglia are known to express m1 and m3 receptors (NJB, unpublished observations), the implication is that these receptors have the potential to mediate the observed muscarinic-receptor inhibition of the M-current which in turn leads to a facilitation of neuronal firing.

A number of caveats apply to interpretation of results obtained with transfected cells – not least is the consideration of the effect of host cell phenotype on the observed response. Since receptors interact with other cellular components, then such variation might be expected as a function of host cell type. One of the most obvious interactions is the coupling between the receptor and the G-protein(s). This will not only be an important determinant of the effector mechanisms to which the activated receptor may couple, but can also affect agonist binding characteristics. The shape of an agonist binding curve is dependent upon the nature of the receptor/G-protein interaction: in the extreme case of a receptor expressed in a host cell where it cannot couple to a G-protein, then the agonist binding curve will appear to indicate binding to a homogeneous population of low-affinity sites. Phenomenonologically, the same result may be produced in gene expression systems that permit high levels of expression (such as SV40-driven genes expressed in COS-7 cells or methotrexate-induced amplification of genes linked to the dihydrofolate reductase gene), such that the great majority of sites may be uncoupled. These examples serve to underline that, as with all such gene expression studies, they can provide invaluable data for assessing the pharmacological potential of any given receptor, but do not directly address the coupling specificity of endogenous receptors. To answer this question, specific probes for each receptor subtype, each G-protein and each effector will be necessary. Only then will it be possible to dissect completely the receptor/transducer/effector pathway in situations where multiple components of the cascade co-exist – as in most situations in the autonomic nervous system.

What do these studies tell us about signal transduction in the autonomic nervous system? Little is known of the second messengers responsible for mediating the effects of muscarinic receptor activation in the ANS and in most cases, receptor activation leads to activation or inhibition of several second messengers, further complicating attempts to identify the specific components of the receptor/second messenger/effector cascade.

Binding Properties of Muscarinic Receptor Subtypes

Although hints of muscarinic receptor heterogeneity have been evident for many years (see Riker and Wescoe, 1951), these ideas were not formalised until the discovery of two selective muscarinic receptor antagonists, pirenzepine (PZP) and AF-DX116. PZP was first used to identify a supopulation of high (M_1) and low (M_2) affinity binding sites in the CNS, gut and salivary glands (Hammer *et al.*, 1980). Initial binding studies showed that M_1 sites predominated in telencephalic regions of the CNS and autonomic ganglia whilst M_2 sites were more prevalent in autonomic effector tissues, brain stem and cardiac muscle (Hammer *et al.*, 1980; Hammer and Giachetti, 1982). Six years later, another selective antimuscarinic agent, AF-DX116 was shown to distinguish between cardiac and glandular muscarinic receptors (Giachetti *et al.*, 1986). These observations have led to the present pharmacological classification of muscarinic receptors into M_1, M_2 and M_3; this subdivision is still useful, even though its limitations have become evident from molecular cloning studies (see later)*. Since the discovery of AF-DX116, a number

---

*M_1, M_2 and M_3 (and latterly M_4) refer to pharmacological subdivisions of muscarinic receptors. m1, m2, m3, m4 and m5 refer to distinct molecular species revealed by molecular cloning studies. They should not be used interchangeably without qualification: A full discussion of the use and limitations of these nomenclatures is beyond the scope of this chapter but essentially the pharmacological distinction is operational in nature and depends upon assay conditions, ligand and receptor source whereas the molecular distinction defines different molecular subtypes.

of other cardioselective compounds have been used to further discriminate M_2 from M_3 receptors. Of these, perhaps the most useful are methoctramine (Melchiorre et al., 1987) and himbacine (Gilani and Cobbin, 1986) which have greater selectivities than AF-DX116. The poor selectivity of these compounds makes it is probable that the identification of M_2 and M_3 receptors was serendipitous, and relied not so much on the subtype selectivity of these compounds, but rather more upon the fortuituous fact that heart and submaxillary gland express a relatively homogeneous class of receptors – a rare finding. The converse selectivity is shown by 4-diphenylacetoxy-N-methylpiperidine methiodide (4-DAMP; Barlow et al., 1976) and hexahydrosiladifenidol (Mutschler and Lambrecht, 1984) and its p-fluoro derivative (Lambrecht et al., 1988). A full discussion of the binding properties of native muscarinic receptors is beyond the scope of this chapter, but see Birdsall and Hulme (1989) for further details. For these reasons, it seems prudent to assume that low concentrations of PZP label predominantly M_1 receptors and that those receptors that remain unlabelled are best thought of as non-M_1. Considering its poor selectivity, use of (^{3}H)-AF-DX116 undoubtedly leads to labelling of multiple subtypes. This perception is lent more weight when a consideration of the binding properties of the full range of cloned receptors is undertaken (see later). With these caveats in mind, then broadly speaking, M_1 receptors are found in sympathetic ganglia, enteric ganglia, intracardiac ganglia and lower amounts in sublingual gland and ileal smooth muscle. M_2 receptors are present in atrial muscle, smooth muscle of airway and gut and autonomic ganglia, whilst M_3 receptors are found in salivary glands, pancreas and smooth muscle.

In a few cases, attempts have been made to correlate results obtained from binding studies with those obtained from functional studies – this has led to both clarification and confusion. Both can be illustrated by a consideration of the pharmacology of gut smooth muscle. Activation of muscarinic receptors on ileal smooth muscle leads to contraction via M_3 receptors (Michel and Whiting, 1988b), yet binding studies reveal a mixed poulation of M_2, M_3 (Giraldo et al., 1987) and a small amount of M_1 receptors (the presence of both m2 and m3 transcripts has also been confirmed by RNA analysis – see later). Binding studies suffer from the limitation of not being able to determine which cell types express individual receptors, so it is not possible to conclude that the anomalous M_1 and M_2 receptors are only present on smooth muscle. However, autoradiographic studies have verified the presence of M_1 and non-M_1 sites on ileal smooth muscle (Buckley and Bumstock, 1986a). The function of the non-M_3 receptors remains unclear.

With notable exceptions (see above), the characterization of receptors by radioligand binding studies has long been the prerogative of workers studying the CNS. The reasons for this are two-fold: (1) autonomic ganglia and effector tissues are either small or frequently express a low concentration of receptor (this may be partially overcome by using tissue from large domestic animals but then functional correlation is lost) and (2) background binding can be high, largely due to the presence of large amounts of connective tissue (see Hulme, 1990 for a discussion of this problem). For these reasons, autoradiographic approaches have been particularly useful in identifying receptor subtypes expressed by autonomic tissues. These studies have borne out and greatly expanded our knowledge of their distribution demonstrated by binding studies and inferred from functional studies (see later).

The establishment of stable cell lines, each expressing a singular muscarinic receptor subtype opened the way for an examination of the binding properties of individual muscarinic receptors (Buckley *et al.*, 1989), a task that was previously impossible due to the widespread coexpression of several genes. These data, taken together with other studies (Fukuda *et al.*, 1987; Akiba *et al.*, 1988; Mei *et al.*, 1989), clearly demonstrate that, whereas each receptor has a unique binding profile, no antagonist has a selectivity greater than five-fold (Buckley *et al.*, 1989). The most useful antagonists for diagnostic purposes appear to be pirenzepine (m1 selective), methoctramine (m2 selective) and hexahydrosiladifenidol (m3 selective). No individual antagonist is selective for m4 or m5 receptors. A salient lesson of this study was that use of any classical antagonist (with the possible exception of low concentrations of PZP) will lead to labelling of multiple subtypes of muscarinic receptor in situations where extensive co-expression occurs – RNA analysis has shown this latter situation to be the norm rather than the exception.

Nicotinic receptors – oocyte studies

Oocytes injected with different combinations of neuronal nicotinic receptor subunit mRNAs have been used to address essentially three questions: (1) Which combinations produce a functional receptor? (2) What are the membrane characteristics of the activiated channel? (3) Is the activated current sensitive to blockade by α-bungarotoxin (α-btx) and/or neuronal bungarotoxin (n-btx)?

Microelectrode recordings from injected oocytes have revealed that any combination of $\alpha2$, $\alpha3$, or $\alpha4$ with either $\beta2$ or $\beta4$ results in a functional receptor activated channel (Boulter *et al.*, 1987; Deneris *et al.*, 1989; Wada *et al.*, 1988b; Duvoisin *et al.*, 1989). Apparently, no functional receptor is produced using $\beta3$ subunits, and $\alpha5/\beta2$, $\alpha5/\beta3$ and $\alpha5/\beta4$ combinations are all non-functional and do not bind α-btx (Boulter *et al.*, 1990).

Ganglionic nicotinic receptors, unlike their muscular correlates, are insensitive to blockade by the snake venom toxin, α-btx (Ravdin and Berg, 1979; Chiappinelli and Dryer, 1984; Obata, 1974). Likewise the nicotinic receptor of PC12 cells is not blocked by α-btx (Patrick and Stallcup, 1977), although α-btx binding sites are present (Patrick and Stallcup, 1977). However, these receptors are blocked by a minor component of the *Bungarus* venom known as n-btx: this fraction was originally known as Bgt3.1 (Ravdin and Berg, 1979) and is equivalent to κ-bungarotoxin (Chiappinelli, 1983). When applied to injected oocytes, n-btx blocks receptors formed from $\alpha3/\beta2$ or $\alpha4/\beta2$ (Boulter *et al.*, 1987) – although the $\alpha4/\beta2$ pair is less sensitive than $\alpha3/\beta2$ – but has no effect on $\alpha2/\beta2$ (Wada *et al.*, 1988b) and $\alpha3/\beta4$ (Duvoisin *et al.*, 1989) combinations. Interestingly, Northern blot analysis of mRNA extracted from PC12 cells reveals the presence of only $\alpha3$, $\alpha5$, $\beta2$ and $\beta4$ transcripts (Boulter *et al.*, 1990). The easiest interpretation is that $\alpha3/\beta4$ receptors form the nicotinic receptor of PC12 cells that is insensitive to blockade by α-btx or n-btx. The possible function of the other permutations is unclear since $\alpha3/\beta2$ is blocked by n-btx but not by α-btx and all combinations involving $\alpha5$ subunits have no channel or toxin binding activity (see above). There is plenty of latitude to accommodate such diversity since PC12 cells express at least three different classes of nicotinic receptor activated channel, although these could also be different states of the same channel (Bormann and Matthaei, 1983). Multiple channels have also been observed in developing chick sympathetic neurons, the characteristics of which change during development (Moss et al., 1989). It would be interesting to know if a parallel change in subunit gene expression occurred. Another unanswered question concerns the possible existence and nature of the mechanisms responsible for directing assembly of appropriate subunits at the expense of inappropriate combinations.

Although PC12 cells are frequently used as prototypic autonomic ganglia, it is worth bearing in mind that, as yet, the only subunits that have been reported in autonomic ganglia are $\beta4$ in rat superior cervical ganglia (Isenberg and Meyer, 1989), $\alpha3$ in chick ciliary ganglia (Boyd et al., 1988) while $\alpha5$ and $\beta4$ transcripts are found in trigeminal ganglia (Duvoisin et al., 1989; Boulter et al., 1990).

When single channel properties of different subunit combinations are examined then further diversity is realised (Papke et al., 1989). When $\alpha2/\beta2$, $\alpha3/\beta2$ and $\alpha4/\beta2$ pairs are considered then each combination exhibits two open states; primary open states are 33.6pS, 15.4pS and 13.3pS whilst secondary open states are 15.5pS, 1.3pS and 5.1pS respectively. However, the relationship between the various subunits, toxin binding sites and channel types and states in autonomic neurons awaits clarification.

DISTRIBUTION

Much of what we know of cholinergic receptor distribution comes from studies using radiolabelled ligands and, in some cases, antibodies. Nearly all of these studies were performed prior to the realisation of the full extent of receptor heterogeneity as revealed by molecular cloning studies. From the preceding discussions it should be clear that none of the pharmacological tools used to label muscarinic and nicotinic receptors has sufficient discriminatory ability to label a singular specific subtype of receptor. The only procedure capable of identifying tissues and cells expressing individual gene products is RNA analysis. Even this approach has its limitations, namely the frequent low levels of gene expression and the fact that it is the site of synthesis, not the final receptor destination that is identified. Undoubtedly, the availability of primary sequence data will accelerate the production of monospecific antibodies that can be used in immunochemical studies to map the distribution of individual gene products. With these provisos in mind then the following sections discuss the use of receptor autoradiography, receptor immunocytochemistry and RNA analysis in mapping cholinergic receptor distribution.

Muscarinic receptor localisation

Most autoradiographic studies have been performed using the non-selective mus-carinic ligands (^{3}H)-N methylscopolamine (^{3}H-NMS), (^{3}H)-quinuclidinylbenzilate (^{3}H-QNB) or the irreversible alkylating agent (^{3}H)-propylbenzylylcholine mustard

(^{3}H-PrBCM). In addition, the selective muscarinic antagonist, (^{3}H)-pirenzepine (^{3}H-PZP) has been used to label subsets of muscarinic receptor. Although initial studies indicated that PZP was M_1 selective, later binding studies on transfected fibroblasts demonstrated a much lower degree of selectivity (see earlier) and it is likely that use of (^{3}H)-PZP also leads to significant labelling of m4 receptors (Buckley *et al.*, 1989). Several studies have shown muscarinic receptors associated with autonomic nerve cell bodies and nerve tracts including sympathetic ganglia (Buckley and Burnstock, 1986a; Wamsley *et al.*, 1981; Yamamura *et al.*, 1984; Zarbin *et al.*, 1982). In these cases, all cell bodies seem to express muscarinic receptors. Intramural ganglia of the gut and atria also express muscarinic receptors (Buckley and Burnstock, 1984a; 1984b; 1986a; 1986b; Hassall *et al.*, 1987; James and Burnstock, 1989) but in the case of enteric ganglia, studies on cultured neurons indicate that only a subpopulation of neurons are labelled. In a number of cases, cultures of dissociated neurons have been used as models to examine the distribution of muscarinic receptors on the cell surface (Buckley and Burnstock, 1984b; 1986; Hassall *et al.*, 1987; James and Burnstock, 1989). Cultured myenteric neurons have also been used to map the distribution of muscarinic receptors over the neuronal cell surface using a combination of autoradiography and immunocytochemistry; these studies demonstrated receptors were distributed over cell soma, neurites, varicosities and growth cones with no evidence of any clustering (Buckley and Burnstock, 1984b; 1986b; Buckley *et al.*, 1988). This absence of clusters contrasts sharply with the situation found with nicotinic receptors on parasympathetic ganglia (see later). Several groups have also reported the distribution of muscarinic receptors on smooth muscle from gut (Buckley and Burnstock 1984a; 1986a), airways (Barnes, 1984; Barnes *et al.*, 1983; Basbaum *et al.*, 1984; Van-Koppen *et al.*, 1988), iris (Hutchins and Hollyfield, 1985). With one exception (Basbaum *et al.*, 1984), all of these studies demonstrate an even distribution of receptors over the smooth muscle with no evidence of any hot spots as seen with nicotinic receptors on skeletal muscle. Likewise, studies on cardiac muscle both in section (Dashwood and Spyer, 1986) and in culture (Hassall *et al.*, 1987) have revealed an even distribution of muscarinic receptors. All of these studies are useful in determining the tissues and cells that express muscarinic receptors, but tell us nothing about the subtype of receptor present.

There has only been one report of Northern blot analysis of peripheral tissues (Maeda *et al.*, 1988). This showed that RNA extracted from ileum, colon, trachea and bladder expressed m2 and m3 mRNA; lacrimal and parotid gland expressed m1 and m3 transcripts and atria expressed only m2 transcripts. As alluded to earlier, the receptors mediating the contractile response of smooth muscle appears to be M_3, yet binding studies reveal a predominance of M_2 sites (Michel and Whiting, 1988b; Giraldo *et al.*, 1987; Lazareno and Roberts, 1989). Clearly this RNA analysis confirms the presence of both m2 and m3 transcripts and poses the question of what is the function of the M_2 receptors? The presence of m3 transcripts in glandular tissue is consistent with binding studies (Hammer *et al.*, 1986; Waelbroek *et al.*, 1987; Michel and Whiting, 1988a), but again, the function of the M_1 receptors remains unclear. So far, no *in situ* hybridisation studies have been reported – such studies would shed light on which cell types express which transcripts and may be helpful in assessing the function of these 'non-functional' receptors.

Nicotinic receptor localisation

In contrast to the situation with muscarinic receptors, there is a dearth of receptor ligands to map the distribution of nicotinic receptors in autonomic ganglia. Although (^{3}H)-nicotine has been used successfully to localise nicotinic receptors in brain, there are no reports of parallel work conducted on autonomic ganglia. The most common tool has been (^{125}I)-α bungarotoxin or horseradish peroxidase conjugates of α-bungarotoxin, but these earlier studies (see Jacob and Berg, 1983; Marshall, 1981; Greene et al., 1973; Gangitano et al., 1979; Messing and Gonatas, 1983) suffer from the severe drawback that the relationship between toxin binding sites and neuronal nicotinic receptors is unclear (see earlier). Less ambiguity is involved in interpreting data using n-btx or a monoclonal antibody MAb35 (Jacob et al., 1984). Using these agents, binding sites can be seen to be localised to synaptic regions of chick ciliary ganglia (Jacob and Berg, 1983; Loring et al., 1988). As expected from functional studies (Brenner and Martin, 1976), axotomy leads to a decline in MAb35 binding sites. This decline is accompanied by a 74% decrease in α3 transcripts (Boyd et al., 1988); no α2 or α4 mRNA could be detected under these conditions.

From the preceding discussion, several points are clear:

1) There is a great diversity of muscarinic and nicotinic receptor gene expression in the autonomic nervous system.
2) Experiments on transfected cell lines and injected oocytes has provided information on the pharmacological properties of each receptor subtype.
3) Pharmacological probes for dissecting the receptor/transducer/effector cascade are insufficiently specific to allow a complete analysis of the cellular functions mediated by each receptor subtype.
4) Only preliminary data are available on the distribution of muscarinic receptor genes in the ANS and even less is known about the subunits and combinations of subunits of nicotinic receptors expressed in the ANS.

So what can we learn about the specific roles of cholinergic receptor subtypes in mediating the effects of ACh in the ANS? The following discussion will attempt to address this issue – details of the endogenous responses can be found in the next section.

Autonomic Ganglia: The fast e.p.s.p. is clearly mediated by a nicotinic receptor but since no comprehensive in situ hydridization studies have been performed and pharmacological tools are available to distinguish amongst the various permutations of subunits possible, then we are no further forward in identifying what combination of subunit(s) are responsible for this effect. However, from the previous discussion, the most likely candidates are the α3/β4 and α4/β2 combinations for the n-btx insensitive and n-btx sensitive sites, respectively. The slow e.p.s.p. is mediated via a muscarinic receptor that has a high affinity for PZP and is responsible for inhibiting the M-current. This clearly points to the m1 receptor. The PI response could be mediated by the m1 and/or m3 receptor. The slow inhibitory postsynaptic potential (i.p.s.p.) is mediated via a receptor that has a lower affinity for PZP and a high affinity for AF-DX116. The most likely candidate is the m2 receptor, although the m4 receptor could also fit the bill.

Presynaptic inhibition of neurotransmitter release: The low affinity of PZP in functional studies clearly indicates that the receptor mediating inhibition of noradrenaline release from sympathetic nerve terminals in heart and from enteric neurons in gut is not M_1. Little more can be concluded. Any of the known second messengers could mediate such inhibition, including hyperpolarization due to K^+ influx, inhibition of cAMP production or release of intracellular Ca^{2+}.

Atria: This is the clearest case for a specific subtype mediating a specific event. Three observations are important, (1) only m2 transcripts are found in atrial muscle, (2) the direct coupling of m2 to K^+ channels in isolated membrane patches, (3) The low affinity of PZP and high affinity of AF-DX116 for the atrial receptor. All of these facts are consistent with the identification of the m2 receptor as the receptor responsible for the actions of acetylcholine on atrial muscle. However, in chick heart, both molecular cloning and pharmacological data indicate the presence of a receptor looking more like an m4 receptor (Tietje *et al.*, 1990).

Smooth muscle and glands: Both pharmacological characterization of the muscular and glandular binding sites and mRNA analysis point to the m3 receptor. The roles of the m1 and m2 receptor transcripts found in gland and smooth muscle, respectively are unknown.

MEDIATION OF FUNCTIONAL EFFECTS

NEURONS – AUTONOMIC GANGLIA

Autonomic ganglia have been shown to possess the biochemical machinery necessary for cholinergic transmission. Choline for ACh synthesis is transported into the presynaptic terminals innervating ganglion cells by a saturable, high affinity system. Activity of the transporter is accelerated by nerve stimulation (Birks & MacIntosh, 1961; Collier & Ilson, 1971; Higgins & Neal, 1978; O'Regan *et al.*, 1982). Paradoxically, preganglionic denervation of the rat superior cervical ganglion does not reduce high affinity choline transport (Bowery & Neal, 1975). This may reflect uptake of choline into glial cells (cf Massarelli *et al.*, 1974) and the relatively low level of high affinity transport in unstimulated ganglia (Higgins & Neal, 1982). The presence of immuno-reactivity to ChAT antibodies has been shown in various ganglia (cf Furness *et al.*, 1985; Costa *et al.*, 1986; James & Burnstock, 1989). ACh can be released by stimulation of the preganglionic nerve or by depolarization with high concentrations of potassium (eg Birks & MacIntosh, 1961; Koelle, 1961; Brown *et al.*, 1970; Higgins & Neal, 1982).

ACh-containing synaptic vesicles were isolated from bovine superior cervical ganglia by Wilson *et al.* (1973) and were found to contain about 2000 ACh molecules per vesicle (4.4 µmol ACh/mg protein). Cholinergic synaptic vesicles from guinea-pig myenteric plexus contained 31 µmol ACh/mg protein (Dowe *et al.*, 1980). These vesicles probably possess the ACh-uptake system found in other tissues, as vesamicol (AH5183), an antagonist of ACh vesicular transport, blocked ACh uptake in preparations from rat superior cervical ganglia and guinea-pig ileum myenteric plexus (Marshall & Parsons, 1987). Both globular and collagen-tailed forms (A12 and G4) of acetylcholinesterase are found in ganglia and are probably located both pre- and post-synaptically (Davis & Koelle, 1978; Gisiger *et al.*, 1978; Ferrand *et al.*, 1986). Considerable quantities of the collagen-tailed form are located intra-cellularly in cultured rat superior cervical ganglion neurones (Ferrand *et al.*, 1986). Cat sympathetic ganglia have also been shown to contain both acetyl- and butyryl-cholinesterase, but the butyryl cholinesterase has an exclusively post-synaptic localization (Davis & Koelle, 1978). As in other tissues, it is likely that synaptic transmission is accompanied by the release of AChE, as high K^+-depolarization causes release of the A12 form of the enzyme (Verdiere *et al.*, 1984). The function of this released AChE is unclear.

Fast synaptic transmission – ionic mechanisms

Stimulation of the pre-ganglionic nerve generates fast and slow membrane potential changes in the post-synaptic nerve. Intracellular recordings have shown that the initial fast excitatory depolarising post-synaptic potential (epsp), which initiates an action potential, begins to rise after 1–2 ms, and has a duration of 50–100 ms in amphibian (Nishi & Koketsu, 1960; Blackman & Purves, 1969; reviewed by Nishi, 1974) and mammalian neurones (Sacchi & Perri, 1971; Bennett & McLachlan, 1972).

The action potential of the ganglionic neuron is generated when the depolarizing epsp activates sufficient voltage-sensitive Na^+ (and to some extent Ca^{2+}) channels. Repolarization of the membrane after an action potential comes about by a combi-nation of inactivation of the voltage-activated Na^+ and Ca^{2+} channels and by activation of potassium channels. Studies of the epsp have usually employed modifications of external solutions to prevent action potential firing (eg, reduction of ACh release or action). The involvement of vesicular ACh release in generation of the epsp was supported by the observations of Blackman *et al.* (1963) who showed that epsp's in frog ganglia ran down in low Ca^{2+}/high Mg^{2+} medium in a fashion consistent with quantal release of ACh. Sacchi & Perry (1976) described the quantal nature of epsp's in rat superior cervical ganglion cells and estimated that each quantum contained 3×10^4 molecules of ACh. Estimates of the number of quanta released from one synaptic terminal range from 0.4 in frog ganglia to 15 in guinea-pig superior cervical ganglia and 35 in the rat, compared with 150–300 at the skeletal neuromuscular junction (Nishi *et al.*, 1967; Blackman & Purves, 1969; Sacchi & Perri, 1971; Sacchi *et al.*, 1978). The dependence of the epsp on ACh release was shown by Bennett & McLachlan (1972), who found that HC-3 (20 μM) abolished the epsp within 5 minutes in guinea-pig superior cervical ganglia stimulated presynaptically at 10 Hz. In the rat superior cervical ganglion, Sacchi *et al.* (1978) measured extracellularly-recorded synaptic compound action potentials and ACh release in the presence of HC-3 (6 μM) and choline-free solution and estimated that ACh release could be reduced 3-fold before failure of synaptic transmission occurred.

Analysis of the effect of changing membrane potential by current injection on the fast epsp has shown that the epsp changes from a depolarization to a hyper-polarization when the membrane potential is more positive than $\sim -10\,mV$ and is accompanied by a decrease in membrane resistance, suggesting that the postsynaptic membrane of ganglion cells becomes more permeable to Na^+ and K^+ ions during the epsp (reviewed by Nishi, 1974; Skok *et al.*, 1989). It has generally been thought that activation of nicotinic receptors on ganglion neurones increases the conductance of Na^+ and K^+ channels, and that any influx of Ca^{2+} is secondary to the depolar-ization-induced opening of voltage-sensitive calcium channels (cf. Skok *et al.*, 1989). However, Tokimasa & North (1984) showed in voltage-clamped bullfrog B cells (in the presence of scopolamine to block muscarinic receptor effects) that iontophoresed ACh induced a nicotinic inward current and an outward current. It seems likely that the outward current resulted from opening of Ca^{2+}-dependent K^+-channels by Ca^{2+} entering directly via nicotinic receptor channels.

Voltage-clamp studies of ganglion neurones have demonstrated the inward fast excitatory postsynaptic current (e.p.s.c.) responsible for the epsp in amphibian (Kuba and Nishi, 1979; MacDermott *et al.*, 1980; Marshall, 1985; Lipscombe & Rang, 1988), and mammalian (Selyanko *et al.*, 1979; Rang, 1981; Derkach *et al.*, 1983) ganglion cells. Information about the rate of closure of ACh-induced channels can be obtained from the decay of the synaptic current, as the activated presynaptic neuron delivers an almost instantaneous pulse of ACh at a high concentration to the postsynaptic receptors, which opens the nicotinic channels in a virtually synchronous fashion. A curve fitted to the synaptic decay will describe the channel open time distribution and the mean channel open time will be given by the time constant (τ) of e.p.s.c. decay. Single-exponential decays of epsc's gave values ranging from 4.5 ms to 9.5 ms in amphibian and rabbit ganglia (Kuba & Nishi, 1979; MacDermott *et al.*, 1980; Derkach *et al.*, 1983; Marshall, 1985; Lipscombe & Rang, 1988) although Rang (1981) described a bi-exponential decay of epsc in rat parasympathetic ganglion neurons. Analysis of the ACh-induced current fluctuations in voltage-clamped ganglion neurons gave estimates of single channel conductances from 17 pS to 36 pS (Rang, 1981; Derkach *et al.*, 1983; Ogden *et al.* 1984; Marshall, 1985; Cull-Candy & Mathie, 1986).

Single-channel conductances have also been measured in membrane patches of ganglion neurones (Ogden *et al.*, 1984; Schofield *et al.*, 1985; Derkach *et al.*, 1987; Mathie *et al.*, 1987) and values were found ranging from 35 pS to 42 pS. Ogden *et al.* (1984) have suggested that single-channel conductances may have been under-estimated in analyses of ACh-induced current fluctuations because of poor voltage clamping of remote dendrites, especially at high frequencies. Channel openings occur in bursts and the burst lengths can be fitted by a short duration component (0.32 ms – Derkach *et al.*, 1987; 0.42 ms – Mathie *et al.*, 1987) and a long duration component (8.54 ms – Derkach *et al.*, 1987; 11.9 ms – Mathie *et al.*, 1987) in rat sympathetic ganglia. The short burst durations probably correspond to the open time of a single channel, although longer durations of channel opening may result from binding to the receptor of two (rather than one) molecules of ACh (cf. Colquhoun & Sakmann, 1985). Nevertheless the longer burst duration corresponds most clearly with the apparent channel open time estimated from studies of epsc decay, so it is likely that an ACh molecule released from a presynaptic terminal and activating a receptor produces considerably more than one opening of the channel per activation.

The overall conclusion of these studies of ACh mediation of the fast epsp and the underlying mechanisms is that the receptors and channels involved behave in a similar way qualitatively to the ACh receptor-channel units of the skeletal neuromuscular junction (cf. Colquhoun & Sakmann, 1985; reviewed by Skok *et al.*, 1989). The picture emerging from molecular biological studies of neuronal nicotinic ACh receptors complements and extends this view. Thus, the ion channel is likely to be intrinsic to the receptor structure, probably resulting from arrangement of receptor subunits to form a pore (see earlier).

Fast synaptic transmission – pharmacology of the receptors

Fast excitatory transmision in autonomic ganglia is mediated by an action of ACh on nicotinic receptors. It has been known for more than forty years that the ganglionic nicotinic receptors differ from the nicotinic receptors at the skeletal neuromuscular junction, as hexamethonium selectively blocks ganglionic nicotinic effects, while decamethonium blocks neuromuscular nicotinic effects (Barlow & Ing, 1948; Paton & Zaimis, 1949). Use of classical pharmacological techniques to classify ganglionic nicotinic receptors by antagonist potency has been hindered by the realization that many compounds previously thought to act by competing with ACh at the receptor also possess the ability to directly block the ACh-receptor channels (cf. Ascher *et al.*, 1979; Brown, 1980). The upshot is that most nicotinic ganglion blockers cannot be used to derive apparent receptor-binding constants from rightward shifts of agonist dose-response curves using Schild analysis, except in situations where the ACh effect measured is not the result of many intervening steps between receptor occupation and response. This is illustrated by the apparently competitive behaviour of some nicotinic antagonists against agonist-stimulated (parasympathetic ganglion-mediated) contractions of intestinal smooth muscle (see Brown, 1980) while the same compounds (mecamylamine, pempidine, hex-amethonium) appear non-competitive against agonist-depolarizations in rat sympathetic ganglia (Caulfield *et al.*, 1990) and show channel-blocking activity in rat parasympathetic ganglia (Ascher *et al.*, 1979). Comparisons of the effects of nicotinic antagonists on e.p.s.c. decay in frog ganglia have shown that these receptors and their channels bear more pharmacological similarity to the receptor/ionophore of the skeletal neuromuscular junction than to the receptors on mammalian auto-nomic neurones (Lipscombe & Rang, 1988). The isolation of α-bungarotoxin (α-btx) the neuromuscular blocking component of the venom of *Bungarus multicinctus* and its radiolabelling, allowed the direct visualization of nicotinic receptor binding sites at the neuromuscular junction and other sites (Changeux *et al.*, 1970) and ultimately led to the purification and characterization of the receptor subunit structure (see earlier). Binding experiments with radiolabelled α-bungarotoxin also showed labelling of sites in autonomic ganglia (eg. Fumagalli *et al.*, 1976) con-centrated in sub-synaptic areas (reviewed by Skok *et al.*, 1989), leading to the view that the α-btx sites were the same as the nicotinic receptors mediating the fast excitatory action of ACh. However, it soon became apparent that α-btx was unable to antagonize the action of exogenously applied nicotinic agonists (Brown & Fumagalli, 1977) or nicotinic synaptic transmission in mammalian autonomic ganglia (reviewed by Brown, 1979). Furthermore, other snake neurotoxins with neuromuscular-blocking properties, such as venom fractions of *Naja melanoleuca* and *Dendroaspis iridis*, were ineffective against ganglionic nicotinic responses (Brown, 1979). Chick ciliary ganglion neurons were reported to be more susceptible to block of transmission by α-btx (Chiappinelli & Zigmond, 1978). More recently, toxins have been identified which selectively block ganglionic nicotinic receptors. Surugatoxin is a non-polypeptide toxin from shellfish which selectively blocks carbachol depolarizations in rat sympathetic ganglia and ACh responses in para-sympathetic ganglia, probably by a competitive action (Ascher *et al.*, 1979; Brown *et al.*, 1976). Unfortunately, surugatoxin does not have a high enough binding affinity to be a useful radioligand for receptor binding studies. A 15 kD (dimeric) fraction of *Bungarus multicinctus* venom has been identified which selectively blocks agonist responses and nicotinic synaptic transmission in mammalian (Quik & Lamarca, 1982; Chiappinelli & Dyer, 1984; Sah *et al.*, 1987; Mathie *et al.*, 1988) and avian ganglia (Chiappinelli *et al.*, 1987; reviewed by Loring & Zigmund, 1988).

The toxin is potent in chick ciliary ganglion preparations, producing complete block of synaptic transmission at a concentration of 20–50 nM (Quik & Lamarca, 1982; Loring & Zigmond, 1988). Higher concentrations are needed to block transmission in rat sympathetic ganglia (1.5 μM: Chiappinelli & Dyer, 1984), although this may result from limited access of toxin to the receptor in the intact preparation. Lower concentrations of toxin are effective against synaptic transmission between cultured rat sympathetic ganglia and heart cells (400 nM – Sah *et al.*, 1987) and against ACh-induction of channel activity in rat cultured ganglion cell membrane patches (100 nM – Mathie *et al.*, 1988). Lophotoxin (Sorensen *et al.*, 1987) and κ-flavitoxin (Chiappinelli *et al.*, 1987) also selectively block nicotinic effects in ganglia and may be useful as tools for defining neuronal nicotinic receptors. The bungarotoxin fraction now designated neuronal bungarotoxin (n-btx) appears to compete with ACh for the receptor, as it did not alter individual ACh channel characteristics in patch-clamp recordings from rat sympathetic ganglion neurons (Mathie *et al.*, 1988). Accordingly, the effects of n-btx are prevented by pre-incubation with di-hydro-β-erythroidine (a competitive nicotinic receptor antagonist – Sah *et al.*, 1987; Loring & Zigmond, 1988). Radio-iodinated n-btx has been used in autoradiographic studies to identify binding sites in avian (Loring & Zigmond, 1987) and mammalian (Loring *et al.*, 1988) ganglion neurons. There are two [125]I n-btx binding sites, one of which (about 20% of total binding) is occluded by α-btx. The α-btx/n-btx type of site is located on extra-synaptic areas of the neuronal membrane, while the non-αbtx/n-btx site is concentrated at synapses. The concentration of n-btx sites in the synaptic membrane is approximately 3–5 fold lower than the concentration of α-btx sites (15–20,000/μm^2) at skeletal neuromuscular junction synapses (Loring & Zigmond, 1987; Loring *et al.*, 1988). Comparison of single channel currents and currents produced by a single quantum of released ACh shows that each quantum opens about 100 channels in ganglionic neurons (Rang, 1981; Derkach *et al.*, 1983). This is about twenty times lower than the corresponding receptor/channel activation at muscle end-plates (Skok *et al.*, 1989). Given that the quantal content of ACh molecules released from ganglionic nerve terminals (12–30,000 – Nishi *et al.*, 1967; Sacchi & Perry, 1976) is similar to that at the end-plate, it appears that there is a discrepancy between the number of ^{125}I n-btx binding sites and functional nicotinic receptor sites in the post-synaptic membrane of ganglion cell synapses. Perhaps some of the n-btx sites represent non-functional receptors.

Finally, the α-btx binding sites in autonomic ganglia should not be dismissed as insignificant. They are concentrated at synapses and the binding of the toxin is prevented by low concentrations of nicotinic antagonists (Fumagalli *et al.*, 1976). Molecular biological studies have shown considerable sequence homology between neuronal and neuromuscular junction nicotinic receptors (see earlier), so it seems feasible that the α-btx site on ganglionic neurones may represent true nicotinic receptors which have changed configuration (i.e. are not functional and do not recognize n-btx). Given that α-btx-bound receptors are known to recycle at the end-plate membrane, it is possible that the α-btx labelled sites on ganglion neurons represent receptors which are being recycled.

Further characterization of the nicotinic receptors on ganglionic neurons may come from the use of specific antibodies. Jacob *et al.* (1984) demonstrated that monoclonal antibodies to skeletal muscle and *Electrophorus* electric organ ACh receptors also bound to neurones of the chick ciliary ganglia. The sites were concentrated around the synapse and had a different distribution of α-btx binding sites. Sargent & Pang (1989) isolated sixty monoclonal antibodies against ACh receptors from Torpedo electric organ. Of these, 41 antibodies also bound to nicotinic receptors at the myoneural end plate. Incubation with frog cardiac ganglia revealed that 8 of these monoclonal antibodies bound to ganglion cell membranes, concentrated in the synaptic areas. Electron microscopy studies of one of these antibodies showed that binding was highest in postsynaptic membranes.

Slow synaptic transmission – mechanisms and pharmacology

Presynaptic stimulation of ganglionic inputs results not only in fast synaptic events, mediated by the nicotinic actions of ACh, but also generates slow potential changes in the ganglionic neurons which can modulate neuronal excitability over considerable periods. The slow potential changes become more marked with repetitive pre-ganglionic stimulation in the presence of nicotinic receptor blockers (eg. d-tubocurarine, dihydro β-erythroidine) to suppress the e.p.s.p. and action potential spike. Intracellular recording has shown slow hyperpolarizing and slow depolarizing potentials occurring with a considerable delay (100–400 ms) after preganglionic stimulation and these potentials have been called slow inhibitory (s-i.p.s.p.) and slow excitatory (s-e.p.s.p.) postsynaptic potentials, respectively (reviewed by Nishi, 1974). Both the s-i.p.s.p. and the s-e.p.s.p. are blocked by atropine at low concentrations and are mimicked by ACh application, indicating that they are generated by actions of ACh on muscarinic receptors (Koketsu & Nishi, 1967; Libet & Tosaka, 1969; Nishi, 1974). Studies of radiolabelled antagonist binding have demonstrated sites with muscarinic receptor characteristics in rat (Burt,

1978; Sinicropi *et al.*, 1979), rabbit (Koval *et al.*, 1982) human (Watson *et al.*, 1984) and ox (Hammer & Giachetti, 1982) sympathetic neurons and in guinea pig myenteric plexus neurons (Buckley & Burnstock, 1986). Only 20–40% of the sites are localized on the neuronal somata, suggesting that the bulk of muscarinic receptors are in dendrites (Koval *et al.*, 1982; Buckley & Burnstock, 1986).

Slow synaptic transmission – muscarinic depolarization (slow epsp)

Indications of the ionic mechanisms resulting from ACh action which generate the slow epsp came initially from studies with intracellular electrodes. Kobayashi & Libet (1968) showed in frog ganglia that the s-epsp decreased with membrane hyper-polarization and increased on depolarization, with either no conductance change or a conductance decrease. These findings have been confirmed by other workers in both amphibian and mammalian ganglia (reviewed by Nishi, 1974; Skok *et al.*, 1989). Weight & Votava (1970) specifically suggested that the s-epsp resulted from a reduction in K^+ channel conductance, on the basis of their intracellular study of frog ganglia.

Studies of the action of muscarinic agonists in voltage-clamped amphibian and mammalian sympathetic ganglia cells showed that the depolarization (and hence the s-epsp) largely resulted from inhibition of a voltage-dependent K^+-current, dubbed the M-current (I_M -Brown & Adams, 1980; Adams *et al.*, 1982; reviewed by Brown, 1988). The current is activated slowly at potentials more positive than $-70\,mV$ (τ_{min} 100 ms, Adams *et al.*, 1982) and (unlike many other voltage-activated currents) does not show any discernible inactivation. This means that normally the M-current will act as a 'brake' on neuronal excitability, tending to oppose membrane depolar-ization by pushing membrane potential towards the equilibrium potential for K^+ ($\sim -88\,mV$) as M-channels open (Adams *et al.*, 1982). Thus, I_M will effectively 'voltage-clamp' the membrane at around the rest potential ($-70\,mV$). Addition of a muscarinic agonist, or presynaptic stimulation to release ACh, will cause an inward current at rest, as I_M is suppressed, then the depolarization itself will open more M-channels, which can be closed by the agonist, thereby producing more depolar-ization, and so on. (Adams *et al.*, 1982; Brown & Selyanko, 1985b; Brown, 1988). Because of its voltage-dependence, a diagnostic characteristic of depolarizing responses mediated by M-current inhibition is that they will not reverse to a hyper-polarization when the membrane potential is held more negative than the threshold potential for I_M activation (ie. -60 to $-70\,mV$), simply because the M-channels will be shut at these negative potentials. The overall result of M-current suppression is an increase in neuronal excitability and (usually) repetitive neuronal firing of the disinhibited neuron. Analysis of the relative contributions of I_M inhibition and inhibition of other potassium current(s) (particularly I_{AHP}, the current underlying the spike after-hyperpolarization) to the increase in neuronal excitability seen with muscarinic agonists in frog and rat ganglia indicates that I_M inhibition plays a major part, although suppression of I_{AHP} adds to the disinhibitory effect (Brown, 1988). M-currents have now been demonstrated in sympathetic ganglia of frog, rat, rabbit and guinea-pig (reviewed by Brown, 1988).

The situation in myenteric and parasympathetic neurons is interesting. Rat submandibular ganglion cells do not exhibit an M-current (Ascher *et al.*, 1979; Rang, 1981). In intracellular recordings from cells of the mudpuppy cardiac ganglion, Hartzell *et al.* (1977) found that stimulation of the vagal preganglionic input or ACh application in the presence of a nicotinic blocker did not produce a depolarization. This suggests that there is no muscarinic reduction of M-current in these neurones. However, Allen & Burnstock (1990) have shown that muscarine depolarized guinea-pig cardiac ganglion neurons and these neurones possess a current with characteristics of I_M so it seems likely that depolarization results from M-current inhibition (T.G.J. Allen, personal communication).

In intracellular recordings from S-type neurons of the guinea-pig myenteric plexus and neurons of the submucous plexus, presynaptic stimulation can elicit a slow-e.p.s.p. in about 25% of cells which is blocked by muscarinic antagonists. Application of muscarinic agonists (ACh, muscarine, oxotremorine) depolarizes the neurons which exhibit a s-epsp (North & Tokimasa, 1982; North *et al.*, 1985; Surprenant, 1986; Galligan *et al.*, 1989). Galligan *et al.* (1989) found evidence that at least part of the muscarinic depolarization in guinea-pig myenteric plexus neurons was the consequence of M-current inhibition. In contrast, Mihara & Nishi (1989) found that muscarinic depolarization of neurons in the submucous plexus of the guinea-pig caecum reversed direction when the membrane was held at potentials more negative than the equilibrium potential for K^+ ions. This suggests that M-current inhibition did not contribute significantly to muscarinic depolarization of these cells.

There is now evidence that muscarinic inhibition of the M-current in ganglion neurons is mediated by G-protein activation (reviewed by Brown, 1990). In voltage-clamp studies of cultured bullfrog and rat sympathetic neurons, inclusion of GTPγS or Gpp(NH)P in the patch pipettes resulted in slow inhibition of I_M (Brown *et al.*, 1988; Pfaffinger, 1988; Brown *et al.*, 1989); when muscarine was applied to these cells, I_M was inhibited and the inhibition did not reverse when the muscarine was washed out (Brown *et al.*, 1988). Inclusion of GDP-β-S in the patch pipette also inhibited muscarine inhibition of I_M (Brown *et al.*, 1988; Brown *et al.*, 1989). The G-protein mediating inhibition of I_M is not susceptible to ADP-ribosylation by PTX suggesting that G_t, G_o or G_i are not involved (Brown *et al.*, 1988; Pfaffinger, 1988; Brown *et al.*, 1989).

Given the probable involvement of a G-protein-coupled mechanism and the long latency between muscarinic agonist application (or presynaptic stimulation of cholinergic inputs) (eg. Brown & Selyanko, 1985a,b) and I_M inhibition, it seems likely that reduction of I_M by cholinergic agonists is the consequence of changes in the level of an intracellular second messenger which influences the opening of M-channels. Changes in cAMP or cGMP are probably not involved in muscarinic I_M inhibition, because injections of high concentrations of metabolically stable cAMP or cGMP analogues do not influence basal I_M or muscarinic-inhibition of I_M in bullfrog and rat neurones (Adams *et al.*, 1982; Brown *et al.*, 1989). Another possibility is that muscarinic I_M suppression is mediated via G-protein stimulation of phospholipase C activity. Muscarinic agonists have been shown to increase the production of inositol polyphosphates (including IP_3) in rat (Bone *et al.*, 1984;

Horwitz *et al.*, 1985; Cahill & Perlman, 1986) and chick (Bhave *et al.*, 1988) sympathetic neurons and extracts of rat superior cervical ganglion contain phospholipase C (Horwitz *et al.*, 1985). Although the cellular localization of the agonist-stimulated phospholipase C activity is not known, it is reasonable to regard the products of this pathway (ie. IP$_3$ and DAG) as possible intracellular mediators of I$_M$ suppression. IP$_3$ is probably not the mediator, as intracellular injection of IP$_3$ had no effect on M-current or muscarinic M-current inhibition in frog (Brown *et al.*, 1988) or rat (Brown *et al.*, 1989) sympathetic ganglion cells. Production of DAG by muscarinic agonist phospholipase C activation would be expected to activate protein kinase C, which could phosphorylate the M-channel protein(s) and thereby modulate M-channel function. In accord with this, muscarinic agonists have been found to enhance phosphorylation of unidentified proteins in rat sympathetic ganglia (Cahill & Perlman, 1986). Phorbol esters are protein kinase C (PKC) activators and hence would be expected to mimic the action of endogenous DAG, so if phospholipase C activation mediates muscarinic I$_M$-suppression, phorbol esters should themselves suppress I$_M$. Brown *et al.* (1988) described a partial reduction of I$_M$ and a block of muscarine's effects on I$_M$ by 1 μM phorbol dibutyrate (PDBu) applied to frog sympathetic neurones. Tsuji *et al.* (1987) also found inhibition of the M-current in bullfrog sympathetic ganglia cells after application of PDBu (1 μm) or OAG (>5 μM). Brown *et al.* (1989) found in rat cultured sympathetic neurones that PDBu (0.5–2 μM) suppressed I$_M$, but not to the same extent as muscarine. The concentrations of PDBu found to be effective in these studies are about 5–10 times higher than the minimum concentrations required to activate PKC (cf. Grove *et al.*, 1990). A further test of the role of PKC activation in muscarinic M-current reduction is to see whether the muscarinic agonist effect is reduced by inhibitors of protein kinase C, such as staurosporine. Grove *et al.* (1990) found that the extracellularly-recorded depolarizations evoked by PDBu (>0.1 μM) in rat sympathetic ganglia were blocked by staurosporine, suggesting that these depolarizing responses resulted from PKC activation. However, similar depolarizations produced by muscarine were reduced by staurospine to a much smaller extent (Grove *et al.*, 1990). As most of the muscarine-induced depolarization of rat sympathetic neurones is thought to result from M-current inhibition (Constanti & Brown, 1981) the results of Grove *et al.* (1990) argue against an involvement of PKC in muscarinic agonist I$_M$ inhibition. A direct test of the effects of PKC inhibition on the muscarinic M-current response in ganglia cells is awaited with interest.

The receptor subtype mediating muscarinic receptor-induced ganglion depolarization of various neurons is probably of the M$_1$ subtype, having a high affinity for pirenzepine and a low affinity for gallamine and methoctramine (Brown *et al.*, 1980; North *et al.*, 1985; Field & Newberry, 1989; Allen & Burnstock, 1990). M-current inhibition by muscarinic agonists in amphibian and mammalian neurons also has an M$_1$ receptor profile (Brown, 1988; Tsuji & Kuba, 1988; Marrion *et al.*, 1989). Attempts have been made to define the receptor subtype mediating the muscarinic s-epsp and in the main they support the view that s-epsp results from activation of ganglionic M$_1$ receptors (Ashe & Yarosh, 1974; North *et al.*, 1985;

Yavari & Weight, 1987; Yarosh *et al.*, 1988b; Mochida & Kobayashi, 1988a; Mochida & Kobayashi, 1988b). Interpretation of the results of these studies is complicated by the lack of knowledge of the true concentration of agonist (ACh, released presynaptically) at the receptors mediating the s-epsp. If the ACh concentration is near – or supramaximal, or is not uniform at all the receptors, then 'block' of the response by pirenzepine (for example) cannot be taken as proof that M_1 receptors mediate that response. The 'blocking' concentration of antagonist will simply be a reflection of the ACh concentration achieved during the epsp.

Muscarinic receptor activation can inhibit (and possibly induce) other ion currents which could result in neuronal depolarization or enhanced excitability (reviewed by North, 1989):

ACh and muscarinic agonists can reduce the background, or leak potassium conductance, which is a conductance not dependent on membrane potential. This results in a depolarization when the resting membrane potential is more positive than the equilibrium potential for K^+ ions and it will reverse polarity at the K^+ equilibrium potential. This effect of muscarinic agonists has been demonstrated indirectly in frog sympathetic neurons and neurons of guinea-pig myenteric plexus (cf. North, 1989). It is notable that the phorbol ester PDBu can reduce leak potassium current to a greater extent than muscarine in frog (Brown *et al.*, 1988) and rat (Brown *et al.*, 1989) sympathetic neurons, and it is likely, therefore, that this contributes to the depolarizing actions of PDBu (Grove *et al.*, 1990), and may infer that reduction of 'leak' by muscarine results indirectly from DAG effects on PKC. Analysis of agonist-antagonist interactions has shown that the muscarinic inhibition of 'leak' K^+-current in enteric neurons is probably via M_1 receptors (North, 1989).

Another K^+-current in ganglion neurons is a Ca^{2+}-activated conductance associated with generation of the after hyperpolarization (AHP) which follows an action potential (cf Adams *et al.*, 1986; Cassell & McLachlan, 1987; Brown, 1988; North 1989). Pennefather *et al.* (1985) used 2-electrode voltage-clamp techniques in bullfrog ganglion neurons to demonstrate a combination of I_M suppression, which increased the tendency of the neuron to fire, and suppression of I_{AHP}, which enabled the neuron to fire more frequently.

There is also the possibility that muscarinic receptor stimulation could, under some circumstances, increase conductance (eg. to Na^+ ions) to induce an inward current at or near rest potential. This would increase excitability (cf. North, 1989). For example, in bullfrog sympathetic neurons, muscarine decreases I_M and in 11% of cells tested, 12-fold higher muscarine concentrations increase conductance and induce an inward current which is exaggerated on hyperpolarization, which seems consistent with an increase in conductance to Na^+ ions (Tsuji & Kuba, 1988).

Slow synaptic transmission – muscarinic hyperpolarization (slow ipsp)

Slow hyperpolarizations following presynaptic stimulation in the presence of nicotinic blockers have been demonstrated in autonomic ganglia of many species (Nishi, 1974). Exogenous application of muscarinic agonists can also produce hyperpolarizations under certain conditions (eg. Hartzell *et al.*, 1977; Gallagher *et al.*, 1982; North & Tokimasa, 1982; Dodd & Horn, 1983; Cole & Shinnick-Gallagher,

1984; Field & Newberry, 1989; Mihara & Nishi, 1989; Allen & Burnstock, 1990). The cholinergic hyperpolarization probably results from an increase in potassium conductance, at least in frog sympathetic (Dodd & Horn, 1983) and mudpuppy cardiac ganglion cells (Hartzell *et al.*, 1977). Selyanko *et al.* (1990) have shown that muscarine can enhance a slow transient 'A-like' K^+ current in frog sympathetic neurons, which may contribute to muscarinic hyperpolarization. The mechanism(s) underlying the s-ipsp in other ganglionic neurons is less clear, but in rat sympathetic neurons, muscarine produced an outward current and a conductance decrease (when I_M was blocked by external Ba^{2+} or internal Cs^+ ions), which reversed at the equilibrium potential for Cl^- ions. This suggests that muscarine could hyperpolarize the cell by reducing chloride conductance. This current was also seen when synaptic inputs were activated (Brown & Selyanko, 1985a; Brown & Selyanko, 1985b). M_1-muscarinic receptors are not involved in the muscarinic hyperpolarization and are probably not involved in the s-ipsp. The high potency of gallamine, AF-DX116 and methoctramine suggests M_2 receptor subtype mediation (Ashe & Yarosh, 1984; Yarosh *et al.*, 1988b; Mochida & Kobayashi, 1988b; Field & Newberry, 1989; Yavari & Weight, 1987; Allen & Burnstock, 1990).

Previous suggestions that the muscarinic hyperpolarization and s-ipsp were the result of ACh releasing dopamine (or some other catecholamine) from small intensely fluorescent (SIF) cells (see Nishi, 1974) have not been supported by subsequent studies. Responses produced by ACh or preganglionic stimulation are not reduced by adrenergic receptor antagonists and the s-ipsp, but not the ACh response, is rapidly reduced in low Ca^{2+}/high Mg^{2+} medium (eg. Hartzell *et al.*, 1977; Gallagher *et al.*, 1982; Dodd & Horn, 1983; Cole & Shinnick-Gallagher, 1984; Yavari & Weight, 1988). Nevertheless, histochemical work has shown that bullfrog C neurons innervate SIF cells, which depolarize in response to presynaptic stimulation and ACh (Dunn & Marshall, 1985b). Recent studies of submucous plexus neurons of the guinea-pig ileum (Mihara & Nishi, 1989) and rabbit sympathetic ganglia (Mochida *et al.*, 1988) have demonstrated s-ipsp responses sensitive to α_2-adrenergic receptor antagonists (yohimbine, idazoxan), consistent with the presence of hyperpolarizing α_2-adrenergic receptors in ganglionic neurons (Brown & Caulfield, 1979).

Muscarinic agonists have been shown to reduce voltage-activated calcium currents (presumably 'N'-type) in sympathetic (Belluzi *et al.*, 1985; Wanke *et al.*, 1987; Song *et al.*, 1989), and parasympathetic neurons (Tse *et al.*, 1990), although negative results have been reported (Bley & Tsien, 1990). The degree of inhibition of the Ca^{2+}-current is small (rarely more than 40%) and it is possible that the inclusion of high concentrations of tetraethylammonium to block K^+-currents may obscure the muscarinic inhibition of I_{ca}. The effect of ACh on Ca^{2+}-currents may be reflected in changes in the form of the action potential, the AHP and in reduced Ca^{2+}-spikes (cf. Akasu & Koketsu, 1982). The muscarinic suppression of the Ca^{2+}-current (Wanke *et al.*, 1987; Song *et al.*, 1989) and the effects on action potential profile (Mochida & Kobayashi, 1988a) are blocked by pertussis toxin and intracellular GDPβS, while intracellular GTPγS is inhibitory itself, thereby pointing to G-protein mediation of this response. If the response is mediated by changes in levels of an intracellular second messenger, it is probably not cAMP, because muscarinic agonists still inhibit the Ca^{2+} current or change action potential profile when cAMP is elevated by forskolin or inclusion of cAMP in the pipette solution (Wanke *et al.*, 1987; Mochida & Kobayashi, 1988a; Song *et al.*, 1989). Stimulators of protein kinase C (phorbol esters and DAG analogues) did not mimic the inhibitory action of ACh on Ca^{2+} currents in rat sympathetic neurons (Wanke *et al.*, 1987) but did mimic ACh effects on action potential profile in extracellular recordings from rabbit sympathetic ganglia (Mochida & Kobayashi, 1988c). Wanke *et al.* (1987) have suggested that the receptor subtype concerned is pirenzepine-sensitive (ie. M_1), while Mochida & Kobayashi (1986, 1988b) found that M_2-receptor antagonists (but not pirenzepine) antagonized the action potential profile effects of ACh. Further studies are needed to clarify this issue.

The functional significance of the muscarinic inhibition of Ca^{2+}-currents in ganglion neurons is not entirely clear. It seems that the reduction in calcium influx is not solely responsible for muscarinic inhibition of Ca^{2+}-dependent K^+-currents, such as I_{AHP} (see North, 1989). However, ACh did reduce the high-K^+ depolariza-tion evoked rise in intracellular Ca^{2+} by 33%, as measured by Fura-2 fluorescence (Wanke *et al.*, 1987). Perhaps the presence of ACh receptors mediating inhibition of voltage-sensitive Ca^{2+}-currents does not necessarily point to a function for those receptors in synaptic transmission. The neurons of autonomic ganglia innervate organs where pre-synaptic inhibition of transmitter output by muscarinic agonists is readily evident (review by Starke *et al.*, 1989) and one possible explanation of such inhibition is that it results from a reduction in the amount of Ca^{2+} entering nerve terminals. The presence on ganglionic neuronal cell bodies of a muscarinic receptor mechanism for inhibiting voltage-activated calcium channels may just reflect a ubiquitous distribution of this mechanism over the whole neuron. The mechanism then only becomes functionally relevant at the terminal projections of the neuron.

PRESYNAPTIC RECEPTORS – MODULATION OF TRANSMITTER RELEASE

The role of cholinergic 'autoreceptors' in modulating transmitter output is covered in a recent comprehensive review by Starke *et al.* (1989).

Muscarinic receptors: Most of the work on presynaptic receptors in the peripheral nervous system has been done with the guinea-pig ileum preparation, where (^{3}H)-choline is readily taken up into the presynaptic nerve terminals, and is incor-porated into ACh, so that the efflux of (^{3}H)-ACh can be monitored, released (generally) by high K^+-depolarisation or field stimulation of the neurons (eg. Szerb, 1976). The output of (^{3}H)-ACh can then be depressed by application of exogenous muscarinic agonists and this inhibition is reversed by muscarinic receptor antagonists (eg. Kilbinger, 1977; Fosbraey and Johnson, 1980a; Kilbinger and Wessler, 1980). The depression of ACh release by exogenous agonists is reflected in a decrease of the contractile response evoked by field stimulation (Fosbraey and Johnson, 1980b; Kilbinger and Nafziger, 1985). Evidence that sufficient ACh to activate the inhibi-tory muscarinic receptors is released during nerve stimulation comes from observa-tions that muscarinic antagonists increase the output of ACh and enhance the stimulation-induced contractions (eg. Gustaffsson *et al.*, 1980; Kilbinger & Wessler, 1983). ACh-induced inhibition of stimulated efflux is also seen in synaptosomes prepared from guinea-pig myenteric plexus (Briggs & Cooper, 1982) so it is likely that the inhibitory effects on the intact preparation result from an action at the nerve terminals of the neuro-effector junction. Receptor binding sites, possibly destined to become functional presynaptic inhibitory receptors have been shown to be transported from the cell body towards the terminals in the rat vagus nerve (Zarbin *et al.*, 1982). Presynaptic inhibition of cholinergic transmission mediated by muscarinic receptors has now been demonstrated in a wide range of peripheral tissues, including heart (Wetzel & Brown, 1985), bronchi (Aas & Fonnum, 1986), bladder (D'Agostino *et al.*, 1986) and iris (Gustaffsson *et al.*, 1980).

In contrast, it seems that inhibitory muscarinic receptors are not present on the cholinergic terminals of autonomic ganglia (Kato *et al.*, 1975). However, North's group (North *et al.*, 1985; see Surpranant, 1986 for review) has shown in submucous and myenteric plexus neurons of the guinea-pig that ACh can decrease the amplitude of the nicotinic f-epsp, taken to reflect a decrease in ACh output from the presynaptic terminals. Inhibitory presynapic muscarinic receptors are also found on non-cholinergic neurones (so-called hetero-receptors); for example, ACh can inhibit the stimulation-evoked release of (^{3}H)-noradrenaline from many tissues (Vanhoutte, 1977), thereby antagonizing the effects of sympathetic nerve activity, such as β-receptor mediated arterial relaxation (Cohen *et al.*, 1984), α-receptor mediated contraction (O'Rourke & Vanhoutte, 1987), iris contraction (Bognar *et al.*, 1988), and cardiac rate acceleration (Fuder *et al.*, 1982). The electrophysiological consequences of activation of presynaptic inhibitory muscarinic receptors at adrenergic sites have been demonstrated by Komori & Suzuki (1987), who showed in rabbit saphenous artery cells that excitatory junction potentials evoked by perivascular nerve stimulation were reduced by ACh. The s-ipsp in guinea-pig enteric neurons, which is mediated by release of noradrenaline, was also decreased by muscarinic agonists (North *et al.*, 1985). There is some evidence that activation of appropriately localized cholinergic neurons can inhibit noradrenaline release at certain sites (cf. Vanhoutte, 1977).

The inaccessibility of presynaptic nerve terminals has hindered direct studies of the mechanisms involved in muscarinic receptor inhibition of transmitter output. Experiments measuring effects on transmitter release cannot be interpreted unequivocally, as any changes could be secondary to effects on any of the events leading up to release, rather than on the muscarinic receptor mechanism. The most plausible hypothesis is that muscarinic receptor activation affects ion channels in the nerve terminals which either prevents action propagation, for example by opening K^+ channels and hyperpolarizing the membrane, or reduces calcium influx at the terminal membrane, either of which would reduce transmitter release. There are numerous examples of such actions on ion channels in neuronal cell bodies. (See *Slow Synaptic transmission – mechanisms and pharmacology*). Even so, it is conceivable that biochemical alterations in nerve terminals could influence transmitter output in an inhibitory fashion. For example, receptor-mediated elevation of intracellular cAMP could result in kinase activation and phosphorylation of a protein central to the release process. Phosphodiesterase inhibitors, forskolin and stable membrane-permeant cAMP analogues all increase ACh release from peripheral cholinergic neurones (eg. Yau *et al.*, 1987; Alberts & Ogren, 1988; Alberts, 1989) but claims that inhibitory muscarinic receptors work by decreasing cAMP levels in nerve terminals are speculative at best.

Definition of the receptor subtype(s) involved in muscarinic presynaptic inhibition has largely been based on the effects of antagonists on the position of dose-response relationships for exogenous agonists reducing transmitter release. The consensus is that nearly all presynaptic muscarinic receptors, whether they are on cholinergic or non-cholinergic nerves, are not of the M_1 subtype. That is, the pA_2 or log K_B for pirenzepine at these receptors is in the range 6.5–7.0 (Fuder *et al.*, 1982; Kilbinger *et al.*, 1984; North *et al.*, 1985; O'Rourke & Vanhoutte, 1987; Fuder *et al.*, 1989; Eglen & Whiting, 1990). However, in chicken heart, Jeck *et al.* (1988) found that low concentrations of pirenzepine (30 nM) increased evoked ACh release, implying the presence of presynaptic inhibitory M_1 receptors. This result must be viewed with some caution, as the low effective concentrations of antagonist could simply reflect low concentrations of synaptically-released agonist. A similar criticism can be made of the study by Kawashima *et al.* (1988), who suggested that M_1 receptors mediated the enhancement of ACh output from guinea-pig ileum due to 100 nM P2P. Studies using other selective muscarinic antagonists have suggested that M_2 receptors are involved at the adrenergic terminals of the iris, with pA_2 values of 6.7 for pirenzepine, 8.5 for himbacine, 6.9 for hexahydrodifenidol and 7.9 for methroctramine (Fuder *et al.*, 1989). In ileum, the inhibitory effects of cholinergic agonists on noradrenaline release are probably also M_2 receptor-mediated as pirenzepine has low affinity at this site ($pA_2 = 6.6-7$). High affinity for AF-DX116 (pA_2 7.3–7.6) and low affinity for pirenzepine ($pA_2 = 6.6$) suggests that the receptor at adrenergic terminals of rat heart is also of the M_2 subtype (Fuder *et al.*, 1982; Dammann *et al.*, 1989). The cholinergic terminals of the guinea-pig ileum have M_3 subtype muscarinic inhibitory receptors, with pA_2 values of 6.7 for pirenzepine (Halim *et al.*, 1982), 8.7–8.8 for 4-DAMP (Halim *et al.*, 1982; North *et al.*, 1985), 6.7 for AF-DX116 (Dammann *et al.*, 1989) and 8.1 for hexahydrosiladifenidol (Fuder *et al.*, 1985). Indirect evidence, based on enhancement of nerve-evoked effects by the non-competitive M_2-receptor blocker gallamine and by methoctramine suggests that there may be an inhibitory M_2 receptor on cholinergic nerve terminals in bronchial and tracheal smooth muscle (Fryer & Madagan, 1984; Blader *et al.*, 1985; Barnes *et al.*, 1988 Aas & Maclagan, 1990). However, it ought to be borne in mind that both of these antagonists have allosteric effects on muscarinic receptors, which may contribute to the enhanced responses.

A further complication arises from the possibility that many of the effects previously interpreted as involving M_1 receptors, because of high affinity for pirenzepine, may in fact be mediated by M_4 receptors, which also have high affinity for pirenzepine (Lazareno *et al.*, 1990). This may be particularly important for presynaptic muscarinic receptors, as voltage-activated Ca^{2+}-currents in a neuron-like cell line (NG108–15 neuroblastoma × glioma hybrid) which expresses M_4 receptor mRNA (Lazareno *et al.*, 1990) are inhibited by ACh and other muscarinic agonists (Higashida *et al.*, 1990; Caulfield *et al.*, 1991) and this effect is antagonized by pirenzepine with high affinity (Caulfield *et al.*, 1991). It has been suggested that inhibitory muscarinic presynaptic effects in the rabbit vas deferens (which are inhibited by pirenzepine with high affinity) are mediated by M_4, rather than M_1 receptors (Micheletti *et al.*, 1990). Discrimination between M_1 and M_4 receptor effects, both of which have high affinity for pirenzepine, is probably best achieved by himbacine, which has higher affinity for M_4 receptors ($-\log K_B$ 8.2–8.8) than for M_1 receptors ($-\log K_B$ 7.2; Lazareno *et al.*, 1990; Caulfield *et al.*, 1991).

While most autonomic ganglia do not have inhibitory muscarinic receptors on the cholinergic terminals, there is good evidence for nicotinic receptors at these sites in amphibian and mammalian ganglia, as application of ACh or other nicotinic agonists can depolarize the nerve terminals, and this effect is blocked by nicotinic receptor blockers (Koketsu & Nishi, 1967; Ginsborg, 1971; Caulfield *et al.*, 1990). Pharmacologically, the receptors appear similar to those at the postsynaptic membrane, as various antagonists have similar potencies in antagonizing agonist depolarizations at both sites (Caulfield *et al.*, 1990). There have been proposals that these presynaptic nicotinic receptors play a functional role in synaptic transmission by 'positive feedback' enhancement of ACh release (Koelle, 1961). Cholinergic agonists do enhance nerve stimulated-evoked ACh output from ganglia but this enhancing effect of ACh on ACh release is only seen in the presence of an anticholinesterase (Brown *et al.*, 1970; Collier & Katz, 1970; Collier & Katz, 1975), whereby the agonist-evoked ACh release comes from a 'surplus', non-physiological pool of ACh (cf Brown *et al.*, 1970; Brown, 1980).

HEART

Most of the parasympathetic (vagal) innervation of the heart is concentrated in the atria, rather than the ventricles. The cardiac ganglia, which send out short postganglionic processes, are concentrated in the interatrial septum, the points of entry of the venae cavae, and the roots of the aortic, pulmonary and coronary arteries. The ventricles are innervated, as evidenced by the presence of ACh, AChE and ChAT, albeit at lower levels than in atria (41% in chicken, 19% in rabbit, cat and rat). In mammals, there is good evidence that cholinergic innervation is concentrated in atria and around the elements of the pacemaker and conducting tissue (ie. sino-atrial node, Purkinje fibres and atrio-ventricular node). Further details of cholinergic innervation and function in heart can be found in the comprehensive review of Loffelholz & Pappano (1985). The innervation is very diffuse, with long varicosities and seemingly little differentiation between junctional and extrajunctional areas of the postsynaptic membrane (Hartzell, 1980). Although there is a high-affinity choline transport system in heart (Loffelholz & Pappano, 1985), it seems that (at least in chick heart) choline can also arise from breakdown of choline-containing phospholipids, probably by the action of phospholipase A_2. This efflux of choline is enhanced by carbachol and could represent a way of increasing ACh precursor availability during high levels of neuronal activity (Lindmar *et al.*, 1988).

The main functional effects of vagal stimulation (and of exogenous muscarinic agonists) are a reduction in heart rate and a decrease in force of ventricular contractions. The latter effect only occurs in the presence of tonic stimulation, for example by resting noradrenaline release from cardiac sympathetic nerves, or by addition of exogenous β-adrenergic receptor agonists. Heart contraction and the maintenance of the cardiac rhythm results from cyclic changes in the membrane potential of pacemaker tissue, communicated in a co-ordinated fashion throughout the heart. Action potential firing in the pacemaker and conducting tissue in turn generates action potentials in cardiac muscle cells, which contract. Each excitatory cycle is generated initially by the diastolic pacemaker depolarization, which itself results from hyperpolarization-induced activation of an inward cation-selective current, I_f. When threshold is reached, an action potential fires as a consequence of the opening of voltage-sensitive Na^+ and Ca^{2+} channels. The slow maintained part of the action potential is a combination of an increase in I_{Ca} and of Na/Ca exchange, the net result of which is the slow inward current, I_{si}. The resultant influx of Ca^{2+} into the cell in turn causes more Ca^{2+} release from intracellular stores, which achieves intracellular Ca^{2+} levels sufficient to cause activation of the contractile apparatus (cf. Hartzell, 1984; Hartzell, 1988). Many workers have reported that ACh inhibits adenylate cyclase and can lower cAMP levels in heart cells, whereas adrenergic receptor agonists have the converse effect (see Brown, 1982; Loffelholz & Pappano, 1985; Nathanson, 1987; Hartzell, 1988). As the positive inotropic and chronotropic effects of β-agonists could be linked to cAMP influences on phosphorylation of key regulatory proteins (Hartzell, 1984; Ahmad et al., 1989), it seemed reasonable to assume that the negative chronotropic and negative inotropic effects of muscarinic agonists reflected their inhibitory effects on adenylate cyclase (Biegon & Pappano, 1980; Hazeki & Ui, 1981; reviewed by Hartzell, 1988).

ACh on K$^+$ currents

In the sino-atrial node, ACh hyperpolarizes the cells, thereby slowing the rate of diastolic depolarization and slowing heart rate. The suggestion that this hyperpolarization results from an increase in potassium conductance came from observations of increased ^{42}K efflux in the presence of ACh, and reversal of the ACh hyperpolarization at the equilibrium potential for K^+ (reviewed by Brown, 1982). Similar effects of ACh have been demonstrated in Purkinje fibres (Mubagwa & Carmeliet, 1983) and atrial cells (Giles & Noble, 1976). The ACh-induced hyperpolarization occurs with a latency of more than 100 ms following the beginning of ACh application (Hill-Smith & Purves, 1978, cf. Nargeot et al., 1982) and the ACh inhibition of adenylate cyclase and hyperpolarization is blocked by pertussis toxin (Hazeki & Ui, 1981; Sorota et al., 1985; Loffelholz & Pappano, 1985; Hartzell, 1988), so a G-protein activated second messenger mechanism is likely to mediate the ACh-induced potassium conductance increase.

Patch-clamp studies allowed the direct demonstration of an ACh-induced K^+ conductance increase (Noma & Trautwein, 1978; Carmeliet & Mubagwa, 1986a) and it seemed from studies of channels in cell patches (whole cell mode) that the K^+ channels activated by ACh were the same as those responsible for the resting inward-rectifying K^+-conductance, I_{K1}, in nodal cells. Both had conductances of 35–39 pS and similar open times (Sakmann *et al.*, 1983). However, further analysis has shown that I_{K1} can be blocked in a time-dependent fashion by low concentrations of Ba^{2+}, whereas Ba^{2+} block of I_{ACh} is time and voltage-dependent (Carmeliet & Mubagwa, 1986b). Simmons & Hartzell (1988) voltage-clamped single atrial cells and clearly differentiated I_{KACh} from I_{K1}, as I_{KACh} showed slow relaxation (cf Iijima *et al.*, 1985), whereas I_{K1} was virtually instantaneous. The idea that ACh was acting via diffusible intracellular second messengers was not supported by the findings of Soejima & Noma (1984) who patch-clamped isolated atrial cells and found that K^+-channel activity was increased only when ACh was applied inside the patch electrode. Regardless, a number of reports have directly confirmed that an endogenous G protein is an obligatory intermediary for the action of ACh on K^+ channels, as pertussis toxin blocks ACh action and GTP must be present on the inside of the membrane for ACh to be effective (Pfaffinger *et al.*, 1985; Kurachi *et al.*, 1986; Horie & Irisawa, 1989; Clark *et al.*, 1990). ACh-induced channels were also activated by $GTP\gamma S$ and ACh action was blocked by $GDP\beta S$ (Sorota & Hoffman, 1989; Horie & Irisawa, 1989; Clark *et al.*, 1990). Contrarily, Sorota & Hoffmann (1989) found that ACh-induced K^+-channels did not require GTP on the inside membrane surface. A recent exciting development has been the application of G-proteins or purified G-protein subunits to isolated patches of heart cell membrane (reviewed by Hartzell, 1988; Brown & Birnbaumer, 1990 and Szabo & Otero, 1990). In excised patches of atrial cells, Yatani *et al.* (1987a,b) and Codina *et al.* (1987) found that a purified pertussis toxin substrate (ie. a G-protein) from erythrocytes (when pre-incubated with $GTP\gamma S$), activated I_{KACh} channels. Neither G_s or G_t were effective inducers of channels. Similar results were obtained with recombinant $G_{i\alpha}$, $G_{i\alpha-1}$, G_{i-2} and G_{i-3}, expressed in E. coli, thereby eliminating the possibility of activation caused by contaminants of the purified subunits (Yatani *et al.*, 1988a). Controversy arose when Logothetis *et al.* (1987) reported that 20 nM $\beta\gamma$-subunit increased channel activity in isolated atrial cell patches. The effect was not blocked by pertussis toxin treatment and intriguingly, α_{39} and α_{41} subunits were without effect. This was confusing, because $\beta\gamma$-subunits lack GTP-ase activity and the GTP binding site. Also, activation of any receptor-G protein interaction (eg. with β-adrenergic receptors) would be expected to release $\beta\gamma$-subunits, so it was difficult to see how specificity of action could be maintained. Logothetis *et al.* (1988) have now reported that α-subunits (both α_{39} and α_{40}) activate K^+ channels at picomolar concentrations. Cerbai *et al.* (1988) have provided independent confirmation that picomolar levels of α-subunit activates K^+ channels in atrial patches and they also find that much higher concentrations of $\beta\gamma$(30 nM) are effective. Kobayashi *et al.* (1990) extracted and purified GTP-γ-S binding proteins from bovine brain membranes and showed that $\alpha_{i\alpha1-3}$, α_{o1}, α_{o2} (picomolar concentrations) and $\beta\gamma$(nanomolar concentrations) all activated K^+ channels in atrial patches. Confusingly, Okabe *et al.* (1990) have now described *inhibitory* effects on atrial K^+ channels of $\beta\gamma$-subunits extracted from erythrocytes, placenta and bovine brain. This group suggests that *stimulatory* effects reported with $\beta\gamma$ are due to contamination of $\beta\gamma$ with the detergent (CHAPS) used in the extraction, as CHAPS itself increased channel activity (Yatani *et al.*, 1990a). Furthermore, the water-soluble $\beta\gamma$-subunits of transducin are reported to inhibit atrial K^+ channels (Okabe *et al.*, 1990; Yatani *et al.*, 1990a). A monoclonal antibody to the α-subunit of frog transducin was found to block agonist activation of channels, suggesting that G_α mediated ACh effects (Yatani *et al.*, 1988b). A further twist to this story is the finding that the activation of I_{KACh} channels by $\beta\gamma$-subunits (2 nM) is blocked by an antibody to phospholipase A_2. The antibody is without effect against ACh or GTP-induced channel activation (Kim *et al.*, 1989). The effect of $\beta\gamma$-was blocked by the lipoxygenase inhibitor NDGA and mimicked by arachidonic acid and products of the lipoxygenase pathway (Kurachi *et al.*, 1989; Kim *et al.*, 1989). Overall, it appears that the ACh activation of I_{KACh} channels in atrial cell membranes is the consequence of an action of activated α-subunits. The possible role of $\beta\gamma$-subunits in the ACh response remains mysterious. Recently, Yatani *et al.* (1990b) have fitted another piece into the jigsaw puzzle with their finding that recombinant GAP (GTPase activating protein, not to be confused with GAP-43) inhibits I_{KACh} channels previously activated by carbachol. The inhibitory effect is manifest as a decrease in opening frequency. The block by GAP was prevented by a specific GAP antibody. The interaction of G_K with the channel was not directly inhibited by GAP, as channels preactivated with α-subunit + GTPγS were not inhibited by GAP. The action of GAP is not the result of a direct interaction with G_K, but was rather due to interaction between GAP and ras p21 protein, as an antibody to ras p21 prevented the GAP effect. The antibody by itself has no effect, so it is unlikely that endogenous ras p21 is exerting an influence on the channels. Addition of recombinant GTP-activated ras p21 proteins mimicked the inhibitory effect of GAP, suggesting that ras p21 protein was interacting with endogenous GAP. Yatani *et al.* (1990b) conclude that the GAP/ras p21 complex does not interfere with the action of activated G_K on the channel, but decreases channel opening frequency by interfering with the coupling of G_K to the muscarinic receptor.

Clearly the G-protein mediated actions of ACh (via muscarinic receptors) on atrial K^+ channels have many complexities and it will be some time before the role of the different G-protien subunits is finally elucidated.

ACh on Ca^{2+} currents

ACh decreases the force of contraction of atrial and ventricular tissue in the presence of positive isotropes (reviewed by Brown, 1982; Loffelholz & Pappano, 1985; Hartzell, 1988). This effect is blocked by pertussis toxin (Endoh *et al.*, 1985; Sorota *et al.*, 1985). This results from a decrease in action potential duration which is the consequence of a decrease in I_{si} (eg. Giles & Noble, 1976; Iijima *et al.*, 1985; Carmeliet & Mubagwa, 1986a). The decrease in I_{si} is largely due to a reduction in the voltage-activated calcium current, I_{Ca} (Iijima *et al.*, 1985). I_{ACh} activated by isoprenaline in atrial cells becomes irreversibly activated by agonist when β, γ-imidoguanosine triphosphate (GppNHp) is present (Breitweiser & Szabo, 1985) and the ACh effect is blocked by pertussis toxin (Hescheler *et al.*, 1986). Although it is commonly thought that ACh only reduces I_{Ca} (or I_{si}) which has been elevated, for example, by catecholamines, there is some evidence that ACh can suppress basal I_{Ca} (reviewed by Hartzell, 1988). The effect of ACh on basal I_{Ca} occurs primarily in multicellular preparations (cf Giles & Noble, 1976; Biegon & Pappano, 1980), rather than in patch-clamp recordings from isolated cells (eg. Hescheler *et al.*, 1986). It has been suggested that this may be the consequence of tonic release of noradrenaline in the multicellular preparations, although Biegon & Pappano (1980) were still able to see ACh-induced decreases in basal I_{Ca} in preparations depleted of catecholamines (with reserpine or 6-hydroxydopamine) in the presence of a β-adrenergic receptor antagonist. Hartzell (1988) raises the interesting possibility that recordings from cells using the patch-clamp technique may be influenced by 'wash-out' of a soluble cellular component which normally maintains a high level of basal I_{Ca}, thereby decreasing the ability of ACh to depress the new basal I_{Ca}.

The identity of the putative intracellular mediator of the effects of ACh on I_{Ca} has yet to be determined with certainty, but Fischmeister & Hartzell (1986; 1987) have suggested that ACh may elevate cyclic GMP in ventricular cells, thereby activating a phosphodiesterase which in turn lowers cAMP levels. This is an extension of the proposal that ACh works by reducing cyclic AMP levels, which in turn reduces I_{Ca}. Thus isoprenaline-induced cyclic AMP elevation and increases in the size of the ventricular action potential are both blocked by ACh (eg. Biegon & Pappano, 1980). As in guinea pig ventricle cells (Hescheler *et al.*, 1986), I_{Ca} which was elevated by isoprenaline was inhibited by ACh (Fischmeister & Hartzell, 1986). Perfusion of the patch pipette with cyclic GMP also decreased the elevated I_{Ca}, while 8-bromo-cyclic GMP (which does not stimulate phosphodiesterase) was ineffective (Fischmeister & Hartzell, 1987). There are evidently some problems with this idea. ACh-induced cyclic GMP increases are poorly correlated with the negative inotropic effect (Ingebretsen, 1980) and neither 8-bromo-cyclic GMP or sodium nitroprusside mimicked ACh effects on force of contraction of ventricular strips (Endoh & Shimizu, 1979; Endoh & Yamashita, 1981). Crucially, intracellular application of cyclic GMP (released from a photoactivatable analogue) did not decrease I_{Ca} (Nargeot *et al.*, 1983). Nevertheless, there are mitigating arguments in defence of the cyclic GMP hypothesis, such as the observation that the pool of cyclic GMP elevated by sodium nitroprusside is different from the cyclic GMP pool increased by ACh (Lincoln & Keely, 1981).

The proposed role of ACh-induced cyclic AMP reduction in the inhibitory effects on I_{Ca} is not supported by a number of observations. The positive inotropic effect of cholera toxin or isobutylmethylxanthine which is accompanied by a rise in cyclic AMP, is reversed by ACh without any apparent effect on the elevated cyclic AMP levels (Pappano *et al.*, 1982; Lindemann & Watanabe, 1985; Rardon & Pappano, 1986; Schmied & Korth, 1990). Similarly, forskolin increases action potential duration and cyclic AMP levels, but only the action potential effects are reversed by a muscarinic agonist (Rardon & Pappano, 1986). Likewise, elevation of I_{Ca} in guinea-pig ventricular myocytes by isobutylmethylxanthine or forskolin is reversed by ACh (Hescheler *et al.*, 1986). Hartzell (1988) gives an elegant and critical review of the evidence against cyclic AMP as a mediator of the ACh inhibition of I_{Ca} and rightly points out that the true picture is likely to be a complex interplay of multiple processes, possibly including cyclic AMP levels.

Positive inotropic effects of ACh – inward cation current

High concentrations of muscarinic agonists can exert a positive inotropic effect on atrial (Brown & Brown, 1984) and ventricular muscle (Korth & Kuhlkamp, 1987) although such an effect has not been seen with vagal stimulation (see Hartzell, 1988). This effect is associated with a depolarization which reverses direction at a membrane potential of + 25 mV and is abolished by removal of Na^+ ions from the extracellular solution (Matsumato & Pappano, 1989). It is also accompanied by a rise in intracellular Ca^{2+} (Korth et al., 1988). The response is insensitive to pertussis toxin (Tajima et al., 1987) and may be a result of an increase in IP_3 production, which is also seen at high muscarinic agonist concentrations and is pertussis toxin insensitive (Quist, 1982; Brown & Brown, 1984; Brown et al., 1985b Tajima et al., 1987).

Pacemaker current, I_f

This is a current carried by Na^+ and K^+ ions which is activated by hyperpolarization. It plays a role in initiating action potentials in Purkinje fibres, where the peak diastolic membrane potential negativity is in the range for activation of I_f, but it has been reported that ACh does not affect I_f in rabbit Purkinje fibres (Carmeliet & Mubagwa, 1986a). However, there is some question whether this is also the case in nodal cells, where the membrane potential does not go more negative than $- 65$ mV under normal conditions. Catecholamines act on I_f to shift the activation curve to more positive potentials. DiFrancesca & Tromba (1987) found in cells of the sino-atrial node that ACh decreased I_f by shifting the activation curve to more negative potentials and inhibited the shift produced by isoprenaline. From this it can be predicted that the ACh effect on I_f in nodal cells will be most prominent in the face of sympathetic neuronal activity. In contrast to the suggestion that ACh does not influence I_f in Purkinje fibres (Carmeliet & Mubagwa, 1986a; Hartzell, 1988), Chang et al. (1990) have found that the isoprenaline-evoked positive shift of the I_f activation curve in dog Purkinje fibres is antagonized by ACh. Mechanisms involved in the ACh effect on I_f probably involve a G-protein, because ACh is ineffective in preparations pre-incubated with pertussis toxin. Once again, the role of cyclic AMP is controversial, because the ACh response was still seen in the presence of IBMX, but was abolished by forskolin (DiFrancesco & Tromba, 1987).

An interesting footnote perhaps puts all of the deliberation over ACh effects on the various ion currents into context. Intracellular recordings from sino-atrial node cells have shown that vagal stimulation slows the rate of action potential discharge and that this effect is atropine-sensitive. However, the only apparent electrophysiological correlate was a decrease in the rate of diastolic depolarization. Action potential amplitude and duration were unchanged. Exogenous application of low concentrations of ACh did not mimic the effects of cholinergic nerve activity, but instead decreased amplitude and duration of action potentials. Effects on action potential frequency and diastolic depolarization rate were only seen at higher concentrations ($> 1\ \mu M$) (Campbell et al., 1989). This raises the tantalizing possibility that the major *physiological* effect of ACh is that on I_f, rather than on I_{Ca} or

I_{KACh}, and is perhaps a salutary warning that observations in systems that are far removed from the physiological may have pharmacological rather than physiological relevance.

Receptors

The muscarinic receptor mediating the negative inotropic and chronotropic effects of ACh and muscarinic agonists in mammalian preparations is of the M_2 subtype (Hammer & Giachetti, 1982). Pirenzepine has a low affinity ($pA_2 = 6.7$, Barlow *et al.*, 1981) as does 4-DAMP ($pA_2 = 8.1$, Barlow *et al.*, 1976). Competitive ligands shown to have high affinity for the heart muscarinic receptor in functional studies include AF.DX 116 ($pA_2 = 7.3$, Micheletti *et al.*, 1988), methoctramine ($pA_2 = 8$, Melchiorre, 1988) and himbacine ($pA_2 = 8.2$, Gilani & Cobbin, 1986). An interesting aside is that, while the muscarinic receptor in mammalian heart is clearly an M_2 receptor, mRNA found in chicken heart codes for a receptor homologous to the mammalian receptor (Tietje *et al.*, 1990, Lazareno *et al.*, 1990), probably explaining the curiously high pA_2 values previously found for pirenzepine in this tissue (7.8 – negative inotropic effect, Jeck *et al.*, 1988; 7.4, adenylate cyclase inhibition, Brown *et al.*, 1985a).

SMOOTH MUSCLE

Gut

Cholinergic nerves supplying the gut make *en passant* synaptic contacts with smooth muscle cells (Gabella, 1976), releasing ACh to cause contraction. The cholinergic terminals can be visualised with antibodies to ChAT (cf Costa *et al.*, 1986) or AChE staining (Burnstock & Bell, 1974) and uptake of choline by a high affinity transporter is readily demonstrable (reviewed by Kilbinger, 1988). The picture is considerably more complex in the ganglia of the digestive tract, which are localized in the myenteric and submucous plexuses. While the ganglion cells receiving vagal inputs have apparent pathways responsible for initiating smooth muscle contraction, there are also complex interactions with local neuronal networks which modulate gut status during (for example) peristalsis. North (1982) has provided a concise review of the role of cholinergic neurones in the nerve plexuses of the gut. (Figure 5.3).

In vivo studies of intestinal motility have shown that ACh (given close-arterially) can evoke both contraction and relaxation or inhibition of contractions and that all effects of ACh are blocked by atropine. The relaxant effects are blocked by tetrodotoxin (Fox *et al.*, 1985). Micheletti *et al.* (1988) reported muscarinic inhibitory effects in isolated duodenum longitudinal muscle which were also tetrodotoxin-sensitive. These observations confirm the presence of ACh-responsive neurones releasing inhibitory factor(s) in gut. Micheletti *et al.* (1988) suggested that one of the inhibitory transmitters could be GABA, because ACh relaxations were blocked by bicucculine. Komori & Suzuki (1988) give a flavour of the complex nature of cholinergic effects in the gut in their report of ACh effects on stomach excitatory junction potentials and contractions:

Cholinergic Influences on Gut Motility

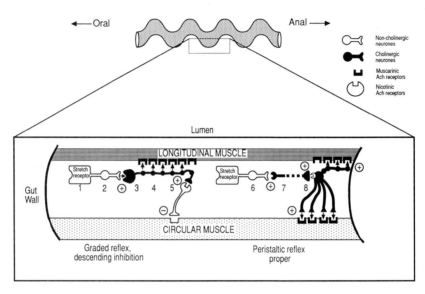

FIGURE 5.3 1. Radial distension of gut activates stretch receptors (1), causing firing in non-cholinergic afferent neurons (2), which stimulates cholinergic neurons (3). These release ACh from varicosities (4), causing contraction of longitudinal muscle. The cholinergic neurons also stimulate other non-cholinergic neurons (5), mediating descending inhibition of circular muscle contractility. 2. Afferent sensory neurons (6-non-cholinergic) excite cholinergic interneurons (7), which project orally and excite cholinergic neurons (8), which cause longitudinal and circular muscle contraction. Cells (5) and (8) are probably S-type cells (see text), receiving excitatory input via ACh stimulation of nicotinic receptors; in this respect, they resemble postganglionic neurons. Cells (2), (3), (6) and (7) do not have a cholinergic input and may be AH-type cells.

'Physostigmine potentiated the enhancement and the depression after the enhancement, while atropine potentiated the initial depression and inhibited the enhancement and the depression after the enhancement of both slow waves and oscillatory contractions.'

It has been known for many years that ACh-induced contraction of gut muscle is accompanied by depolarization and an increase in membrane conductance to Na^+ ions, which results in action potential firing (see Kuriyama, 1970; Bolton, 1976). More recently, patch-clamp studies of single muscle cells have shown that ACh or muscarinic agonists induce an inward current which persists when the membrane potential is held at the equilibrium potential for K^+ ions (Benham et al., 1985). This suggests that the current is the result of changes in Na^+ or possibly Ca^{2+} conductance, but not K^+ conductance. Inoue & Isenberg (1990a) have studied cells of guinea-pig ileum and made similar findings. The situation was not straightforward, as complicated bursts of ACh-induced transient outward currents appeared at membrane potentials more positive than $-30\,mV$. These may be due to ACh elevating intracellular Ca^{2+}, in turn opening Ca^{2+}-activated-K^+ channels, as caffeine (which releases Ca^{2+} from intracellular stores), had similar effects (Benham et al., 1985). Clapp et al. (1987) showed that ACh and muscarine increased a voltage-activated Ca^{2+} current in amphibian stomach cells and this would be expected to contribute to ACh-induced excitability increases. However, while the patch pipette contained Cs^+ to block outward currents, the extracellular medium contained only 5 mM TEA, so it is possible that apparent enhancement of I_{Ca} with ACh could in fact be suppression of an inadequately-blocked outward K^+ current. Sims et al. (1985) have also demonstrated muscarinic agonist suppression of an M-like K^+ current in amphibian stomach cells, although it is not clear what role this effect plays in ACh contractions, or whether it is present in other species or other parts of the digestive tract. Kuriyama (1970) has reviewed the literature showing that ACh can also induce contractions in gut muscle depolarized by high K^+ medium and has suggested that these contractions may result from release of Ca^{2+} from intracellular stores. This has now been directly demonstrated with fura-2 (Himpens & Somlyo, 1988) and Quin-2 (Grider & Makhlouf, 1988) imaging of intracellular Ca^{2+} levels. There is also a component of the elevation of intracellular Ca^{2+} resulting from Ca^{2+} influx through voltage-activated channels, although this is present in longitudinal msucle cells, but not in circular muscle cells (Grider & Makhlouf, 1988). The likely involvement of G-proteins and second messengers in excitatory ACh effects in gut muscle was suggested by the long latency between ACh application and muscle depolarization (Purves, 1974; Bolton, 1976) or induction of inward current (Benham et al., 1985). However, Lim & Bolton (1988) found that the carbachol-evoked inward current was unaffected by inclusion of either GTP γ-S or GppNHp in the patch pipette. NaF/AlCl$_3$ was also without effect, so Lim & Bolton (1988) concluded that a G-protein did not mediate the muscarinic inward current. Also Komori & Bolton (1990) have shown that the ACh inward current in rabbit jejunal cells is affected only minimally by inclusion of GTP γ-S or GDPβ-S in the patch pipette. This conflicts with the findings by Inoue & Isenberg (1990c), who reported that ACh-induced inward current was blocked by GDPβ-S or pertussis toxin in the patch pipette and was mimicked by inclusion of GTP γ-S in the pipette solution. However, I_{ACh} may be modulated by Ca_i^{2+}, as it was abolished by EGTA in the patch pipette (Lim & Bolton, 1988; Inoue & Isenberg, 1990b) and was enhanced when intracellular Ca^{2+} was elevated by Ca^{2+} influx after induction of voltage-sensitive I_{Ca} (Inoue & Isenberg, 1990b). Thus, ACh may open Ca^{2+}-sensitive membrane channels to elicit I_{ACh} secondary to an ACh-induced increase in intracellular Ca^{2+}. The likeliest candidate connecting ACh receptor activation with increases in intracellular Ca^{2+} is an increase in IP$_3$. Accordingly, application of IP$_3$ or carbachol to β-escin permeabilized guinea-pig ileum causes contraction which is blocked by heparin (Kobayashi et al., 1989). In ileum permeabilized with Staphylococcus aureus toxin, carbachol and IP$_3$ caused contraction and release of Ca^{2+} from intracellular stores and these effects were inhibited by GDPβ-S, suggesting a G-protein mediates the action of muscarinic receptor stimulation (Kitazawa et al., 1989). This conflicts with the suggestion that the muscarinic inward current results from a G-protein-independent elevation of intracellular Ca^{2+} (Lim & Bolton, 1988).

The muscarinic receptors of gut smooth muscle have been the subject of many pharmacological investigations. They have low affinity for pirenzepine ($pA_2 = 6.7$), AF-DX116 ($pA_2 = 6.6$), methoctramine ($pA_2 = 6.3$) and himbacine ($pA_2 = 7.4$) and high affinity for 4-DAMP ($pA_2 = 8.9$) and hexahydrodifenidol ($pA_2 = 8.2$), showing that they are of the M_3 subtype (Lazareno & Roberts, 1989 and references therein).

In the main, ACh-induced contraction of non-gastrointestinal smooth muscle involves similar electrophysiological changes, biochemical mechanisms and receptors to those demonstrated in gut muscle. For example, muscarinic agonists stimulate phospholipase C in trachea by a G-protein-linked mechanism in tracheal and uterine muscle (Marc et al., 1988; Meurs et al., 1988) and the M_3 muscarinic receptor subtype is involved (Barnes et al., 1988; Monferini et al., 1988).

Blood vessels

ACh both contracts and relaxes arteries and arterioles. Although many of the vascular beds and vessels which respond to ACh do not have a functional innervation, it is clear that in some areas (eg. coronary arteries, hindlimb muscles, facial skin) stimulation of nerve trunks produces vasodilatation which is blocked by atropine. The cholinergic vasodilator neurones concerned are quite distinct from the sympathetic vasoconstrictor fibres, as the two responses can be elicited at different stimulus intensities. Differential vasodilation or constriction of hindlimb muscle vascular beds or coronary beds can also be separately evoked by stimulation of distinct vasodilator or vasoconstrictor centres in the brainstem (reviewed by Campbell, 1970). Direct proof of the vasodilator action of cholinergic neurones has come from the elegant study of Neild et al. (1990), who visualized single neurons and adjacent arterioles in the submucous plexus of guinea-pig intestine. Vessels constricted with PGF_2 or phenylephrine dilated in response to stimulation of adjacent neurons and the dilator effect was mimicked by muscarine. Muscarine- and nerve-evoked dilatations were blocked by tetrodotoxin and atropine. It is possible that the vasodilator effects of ACh are propagated by intercellular communication (Segal & Duling, 1986). The relaxant effects of ACh on vascular smooth muscle are now known to result from an action of ACh on vascular endothelial cells, which then release a short-lived relaxant factor, endothelium-derived relaxing factor (EDRF), (reviewed by Furchgott, 1984). Removal of endothelium results in loss of the relaxant response to ACh and the relaxant factor can be detected in donor strips from which the endothelium has been removed (Furchgott & Zawadcki, 1980). The requirement for endothelium has been demonstrated for ACh-induced vasodilatation *in vivo* (Angus et al., 1983). Furchgott suggested that EDRF may be nitric oxide (NO) because endothelium-independent vasodilators such as sodium nitroprusside and glyceryltrinitrate are known to generate free NO. Both NO-releasing vasodilators and EDRF-induced vascular relaxation were blocked by haemoglobin and methylene blue, and both increased cyclic GMP levels in blood vessels (see Furchgott, 1984). Palmer et al. (1988a) used a chemiluminescence assay based on the reaction of NO with ozone to directly demonstrate the release of NO (evoked by bradykinin) from aortic endothelial cells. It is likely that the NO is derived from the terminal guanidinium nitrogen of L-arginine (Palmer *et al.*, 1988b). In rabbit aorta and coronary artery, Rees *et al.* (1989) estimated that the ACh-induced reduction in coronary perfusion pressure could be accounted for by the amount of NO released by ACh, although this interpretation was not supported by the finding that N^G-monomethyl-L-arginine, an inhibitor of NO production, totally abolished NO release by ACh but only inhibited the ACh relaxation by 55–70% (Rees *et al.*, 1989).

The effect of ACh on endothelial cells which results in NO release has not been identified, but ACh does increase intracellular Ca^{2+} levels in these cells (Danthuluri *et al.*, 1988; Busse *et al.*, 1988). ACh also hyperpolarizes the cells, probably by opening Ca^{2+}-activated K^+ channels (Busse *et al.*, 1988; Olesen *et al.*, 1988). This effect is unlikely to involve a G-protein, as neither pertussis toxin treatment or inclusion of GTP γ-S in the patch pipette altered ACh responses (Olesen *et al.*, 1988). The ACh-induced relaxation of blood vessels is accompanied by membrane hyperpolarization, which reverses at the equilibrium potential for K^+ ions and is therefore probably the result of an increase in K^+ conductance (Kuriyama & Suzuki, 1978; Kitamura & Kuriyama, 1979; Komori & Suzuki, 1987). The ACh-hyperpolarization disappears on removal of the endothelium (Bolton *et al.*, 1984; Taylor *et al.*, 1988). However, relaxation can occur without hyperpolarization (eg with substance P – Bolton & Clapp, 1986). The results of Brayden & Large (1986) do not concur with this scheme, as they found in catecholamine-depleted rabbit lingual artery that the field-stimulation evoked dilatation was sensitive to atropine, but was not totally eliminated by removal of endothelium. ACh-hyperpolarization and a nerve-induced inhibitory junction potential were seen in preparations either with or without endothelium. Similar results were obtained in cat posterior auricular arteries (Brayden & Bevan, 1985). Thus ACh seems to be able to produce vasodilatation by endothelium-dependent and endothelium-independent routes, depending on the location of the vessel. It is also emerging that there are differences between EDRF and NO. Miller & Vanhoutte (1989) found that NO relaxation of dog femoral artery was enhanced by superoxide dismutase (SOD) and decreased by methylene blue. An equivalent ACh relaxation was unaffected by SOD and was inhibited to a much lesser extent by methylene blue.

The endothelium-dependent effect of ACh is possibly mediated by an EDRF-induced increase in cyclic GMP levels in the vessel (eg. Rapoport & Murad, 1983; Taylor *et al.*, 1988). The receptors mediating the ACh relaxation are of the M_3 subtype, with high affinity for 4-DAMP ($pA_2 = 9$) and intermediate affinity for pirenzepine ($pA_2 = 7$), (Eglen & Whiting, 1985, reviewed by Eglen & Whiting, 1990). Pelc *et al.* (1988) showed that ACh-induced coronary artery dilatation was sensitive to pirenzepine, but not AF.DX 116.

High concentrations of ACh produce contraction of arterial smooth muscle or vasoconstriction, especially in the absence of endothelium (Kuriyama & Suzuki, 1978; Bolton *et al.*, 1984; Komori & Suzuki, 1987; O'Rourke & Vanhoutte, 1987; Duckles, 1988). The ACh contraction is blocked by pertussis toxin preincubation, unlike the relaxation (Hohlfeld *et al.*, 1990), so a G-protein seems to be involved. The mechanism for the contraction is not known and the receptors involved have yet to be clearly defined. The receptors mediating coronary artery contraction may be M_3 (Eglen & Whiting, 1990) although O'Rourke & Vanhoutte (1987) reported a pA_2 of 8.1 in dog saphenous vein, implying that M_1 receptors mediate the response.

GLANDS

Adrenal glands

The catecholamine-secreting cells of the adrenal gland are pharmacologically and ontogenetically related to autonomic ganglia neurones. Release of catecholamines is evoked primarily by the action of ACh released from the splanchnic nerve activating chromaffin cell nicotinic receptors. Both chromaffin cells and ganglionic neurones originate from neural crest cells. Nicotinic receptor-evoked secretion of catecholamines results from depolarization activating voltage-operated Ca^{2+} channels (cf Cena et al., 1983) which elevate intracellular Ca^{2+} levels (Sasakawa et al., 1985; Misbahuddin & Oka, 1988). Catecholamine release is accompanied by release of neuropeptide Y (Hexum et al., 1987), atrial natriuretic peptide (Okazaki et al., 1989), methionine-enkephalin (Cherdchu and Hexum, 1988) and AChE (Mizobe & Livett, 1983). Patch-clamp studies of chromaffin cells have shown that the ACh-evoked increased current and depolarization is similar to that seen in ganglion neurones (eg. Kidokoro et al., 1982; Fenwick et al., 1982; Claphman & Neher, 1984; Inoue & Kuriyama, 1989). Single channel conductance was 44 pS and membrane resistance was so high that a single channel opening could evoke an action potential (Fenwick et al., 1982). Some of the stimulating effects of ACh on catecholamine release through nicotinic receptor activation could be mediated by increases in IP_3 (Eberhard & Holz, 1987; TerBush et al., 1988) which in turn releases Ca^{2+} from intracellular stores (Sasakawa et al., 1985; Yamada et al., 1988). Nicotinic receptor activation has dramatic effects on protein phosphorylation, primarily through Ca^{2+}/calmodulin-activated kinases (Haycock et al., 1988). An intriguing aspect of the nicotinic receptor/channel complex in chromaffin cells is its sensitivity to modulation by other substances likely to be released locally or be present intracellularly. Substance P increases the rate of inactivation of I_{ACh} (Clapham & Neher, 1984), cyclic AMP analogues and a phosphodiesterase inhibitor increase I_{ACh} (Higgins & Berg, 1988) and glucocorticoids decrease I_{ACh} (Inoue & Kuriyama, 1989). As might be expected, the pharmacological characteristics of the nicotinic receptors of adrenal chromaffin cells are more similar to those of autonomic ganglion cells than to skeletal muscle receptors. They are not blocked by α-btx (Douglas et al., 1967) but are blocked by surugatoxin (Bourke et al., 1988).

There are also muscarinic receptors on the adrenal chromaffin cells of most mammals and their stimulation results in catecholamine secretion, although this is not the case with bovine chromaffin cells (reviewed by Marley, 1987). Muscarinic receptor stimulation produces catecholamine release without Ca^{2+} influx, but elevates internal Ca^{2+} by IP_3-induced release from intracellular stores (Misbahuddin & Oka, 1988; Yamada et al., 1988; Nakazato et al., 1988a). The muscarinic receptors on the chromaffin cell may be of the M_1 subtype, as pirenzepine is a potent antagonist (IC_{50} 70 nM) of the ACh response in perfused guinea-pig adrenal glands (Nakazato *et al.*, 1988b).

Other exocrine glands

ACh and cholinergic stimulation produces secretion of sweat, tears, saliva and enzymes (eg from exocrine pancreas). The mechanisms involved are probably similar at the various sites (reviewed by Petersen & Maruyama, 1984; Petersen & Findlay, 1987).

In sweat glands, voltage-clamp studies have shown that cholinergic agonists open K^+ channels in the serosal membrane (Larsen *et al.*, 1990). This results in Na^+ influx via an amiloride-sensitive $Na^+/K^+/Cl^-$ co-transporter, then extrusion of Na^+ by the Na^+/K^+ pump. Cl^- also leaves the cell by passing out through Cl^- channels which are probably also activated by ACh. The changes in K^+ and Cl^- conductance are secondary to IP_3-induced elevation of intracellular Ca^{2+}, caused by muscarinic receptor activation of phospholipase C (Doughney *et al.*, 1989).

A similar situation exists in salivary glands; patch-clamp studies have demonstrated ACh-induced opening of high conductance (BK – 250pS) Ca^{2+}-activated K^+ channels (Maruyama *et al.*, 1982; Gallagher & Morris, 1987). This change in K^+ conductance can be seen as a release of K^+ from glands (Petersen, 1970). Muscarinic stimulation elevates IP_3 and causes not only a release of Ca^{2+} from intracellular stores, but also increases Ca^{2+} influx (Dehaye *et al.*, 1988; Takemura & Putney, 1989). The cholinergic stimulation is accompanied by oscillations of intracellular Ca^{2+} which probably reflect feedback inhibition of released Ca^{2+} on IP_3-induced Ca^{2+} release processes. The oscillations in Ca^{2+} levels are mirrored by oscillations of membrane current as Ca^{2+}-dependent K^+ channel activity oscillates (Gray, 1988). Muscarinic agonists also increase recovery of salivary gland cell vesicles from acid load, possibly by activating Na^+/H^+ transport, and this may serve to compensate the loss of HCO_3^- through the open anion channels (Manganel & Turner, 1989; Melvin *et al.*, 1988). The agonist-induced acidification is blocked by amiloride and by a calcium chelator, suggesting that it is also secondary to changes in Ca^{2+} in the cell (Melvin *et al.*, 1988). Muscarinic stimulation elicits other changes in salivary gland cells which could influence secretion processes, including decreasing coupling (measured by dye) between cells (Sasaki *et al.*, 1988) and increasing the turnover of cyclic AMP (Deeg *et al.*, 1988). Amylase secretion from glands is accelerated by carbachol in parallel with increased ^{86}Rb efflux and IP_3 production (Dehaye *et al.*, 1988). Functional studies of amylase release and IP_3 production have demonstrated that the receptors are of the M_3 subtype, with high affinity for hexahydrosiladifenidol ($pA_2 = 8.1$) and low affinity for pirenzepine ($pA_2 = 7.1$) and AF-DX116 ($pA_2 = 6.2$, Dehaye *et al.*, 1988).

Extensive studies of lacrimal gland cells have demonstrated increased opening of the Ca^{2+}-activated BK K^+ channels in the presence of ACh (Trautmann & Marty, 1984; Marty *et al.*, 1986) and at higher concentrations opening of Ca^{2+}-dependent Cl^- channels (Marty *et al.*, 1986; Evans & Marty, 1986; Horn & Marty, 1988; Marty & Tan, 1989). Field stimulation of glands or ACh produces hyperpolarization which is blocked by atropine and increased by neostigmine (Pearson & Petersen, 1984). These effects are due to elevation of intracellular Ca^{2+} and they are mimicked by inclusion of IP_3 in the patch pipette. There is also a delayed rise in Ca^{2+} originating from an influx through receptor-operated Ca^{2+}-channels (Marty & Tan, 1989). G-protein coupling is suggested by the finding that GTP γ-S in the patch pipette causes irreversible activation in the presence of ACh (Evans & Marty, 1986). Similar changes in Ca^{2+}-dependent K^+ and Cl^- channels, resulting in cell hyperpolarization and K^+ efflux, occur in cells of the exocrine pancreas (Nishiyama & Petersen, 1975; Petersen & Maruyama, 1984; Petersen & Singh, 1985; Petersen & Findlay, 1987). These changes are also accompanied by increased enzyme secretion.

The ion fluxes evoked by ACh and resulting in fluid and electrolyte secretion from glands are summarized in Figure 4.

A well-known effect of ACh is to increase the output of acid from the stomach. Activation of cholinergic nerves increases acid output which is blocked by atropine, tetrodotoxin and by the histamine H_2 receptor antagonist metiamide (Angus & Black, 1982). Atropine-sensitive increases in pepsin output have also been shown

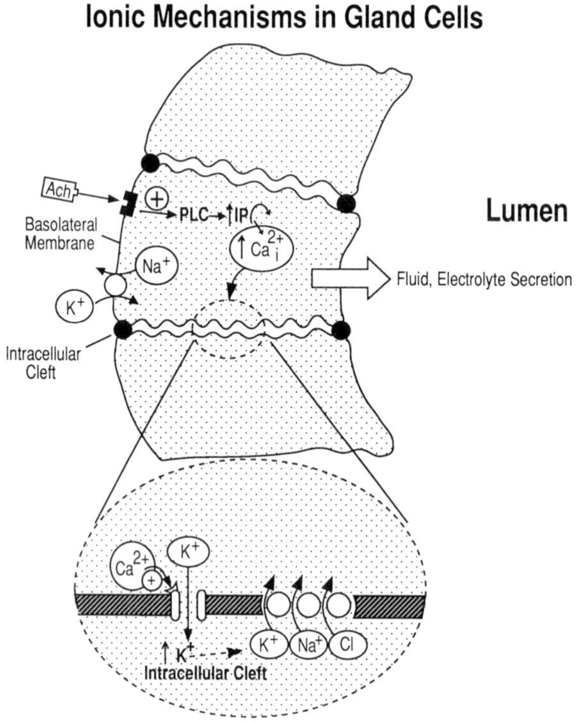

Ionic Mechanisms in Gland Cells

FIGURE 5.4 Muscarinic receptors on the basolateral membrane of a gland cell are stimulated by ACh released from a neurone, causing elevated IP turnover, in turn elevating intracellular Ca^{2+} (via release from intracellular stores). Increased Ca_i^{2+} activates K^+ channels, allowing K^+ efflux and elevating K^+ concentrations in the intracellular cleft. Fluid and electrolyte secretion occur as a consequence of secondary ion fluxes, largely through the $K^+/Na^+/Cl^-$ co-transporter.

following vagal activation (Hirschowitz *et al.*, 1983). The stomach parietal cell has receptors for all three secretagogues, (histamine, gastrin and ACh), but the reason that histamine H_2-receptor antagonists can block gastrin and ACh-induced acid and pepsin output may be *via* interactions of these substances on the release of other active agents from other cells. For example, ACh inhibits the release of somatostatin from non-parietal cells (Soll 1984; Makhlouf, 1984). Decrease of somatostatin release will in turn increase gastrin release (Makhlouf, 1984). Thus, the final outcome at the parietal cell may depend on a balance between contributions from gastrin, ACh and histamine, such that alterations of the contribution of any one of these factors could inhibit responses to one of the other agents.

ACh also increases bombesin output which can increase gastrin secretion (Makhlouf, 1984; figure 5.5).

In vivo studies have shown that pirenzepine is almost as potent as atropine in inhibiting vagal or muscarinic agonist-evoked acid and pepsin secretion (Hirschowitz *et al.*, 1983). *In vitro*, ACh effects on acid secretion from parietal cells, or on somatostatin secretion are antagonized by relatively high concentrations of

Cholinergic Influences on Acid Secretion

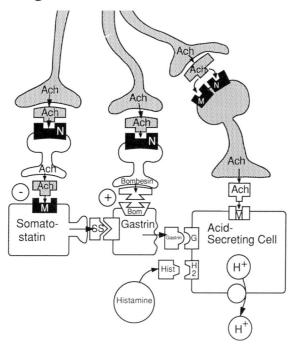

FIGURE 5.5   The action of ACh, released from the vagus, on acid secretion in the stomach is the result of a complex interaction between cells secreting other neuromodulators/hormones, including histamine, gastrin, bombesin and somatostatin. The effects of ACh can be direct on the acid-secreting cell, or indirect. The indirect actions can be via inhibition of release (eg somatostatin) or by increasing release (eg bombesin). Excitatory nicotinic and muscarinic receptors are probably involved at a ganglion-like synapse between the vagal input and the cholinergic innervation of the acid-secreting cell.

pirenzepine ($>$ 30 nM), suggesting that the muscarinic receptors or target cells are not M_1 receptors. The high potency of pirenzepine *in vivo* may reflect antagonism of excitatory M1 receptors located on a ganglion-like neurone interposed between the vagal input and the target cell.

TROPHIC AND ADAPTIVE RESPONSES OF CHOLINERGIC NEURONES AND TARGETS

The classical view is that denervation or removal of a neurotransmitter input to receptors on the target cell results in an increase in the sensitivity of the target cell to transmitter, generally by an increase in the number of receptors. This is not the case with muscarinic receptors in rat sympathetic ganglion neurones (Burt, 1978; Sinicropi *et al.*, 1979). However, axotomy does produce a large decrease in receptor numbers (Sinicropi *et al.*, 1979), and also decreases the nicotinic epsp (Purves, 1975). Denervation does not affect ganglionic sensitivity to muscarinic or nicotinic stimulants (Brown, 1969), although there is a report of an increase in muscarinic receptors after denervation of cat sympathetic ganglia (Taniguchi *et al.*, 1983). Dunn & Marshall (1985b) convincingly showed that denervation of frog ganglia did not increase the sensitivity to ACh, measured by depolarizations in response to iontophoretic agonist application. This stands in contrast to the observations of Kuffler *et al.* (1971) and Dennis & Sargent (1979) on cardiac parasympathetic ganglia of the frog. Changes of responsivity during development of innervation are complex (reviewed by Schuetze & Role, 1987) but ACh sensitivity increases in parallel with the appearance of innervation and this change is mimicked by conditioned medium from spinal cord cells, suggesting that the presynaptic neuron is releasing a trophic factor (Role, 1988). The increase in sensitivity is the consequence of complex changes in the conductance properties and open/closed kinetics of ACh receptor operated ion channels (Moss *et al.*, 1989). Denervation produces other curious effects, for example the development of cross-connections between ganglion cell neurons in parasympathetic ganglia (Taniguchi *et al.*, 1983). This does not occcur in sympathetic ganglia (Voyvodic, 1987).

 For some time it has been known that various cells secrete a factor which induces the expression of cholinergic features in neurons. Rich sources of the factor are usually targets for cholinergic innervation (eg. heart cells) and media conditioned by such cells greatly increases ACh synthesis, ChAT levels, high affinity choline uptake, ACh release and function of cholinergic synapses of ganglion cells in culture (reviewed by Patterson, 1978; Richardson, 1988). Responsiveness of neurons to the muscarinic effects of ACh is seen in four times the number of cells cultured with heart conditioned media (Nishi & Willard, 1988). The factor also suppresses the development of 'adrenergic' phenotypic characteristics in ganglion neurones (eg. Potter *et al.*, 1983). This plasticity is also seen *in vivo*; for example, in rat eccrine sweat glands the innervation after birth is by adrenergic fibres and after 18–21 days the innervation gradually becomes cholinergic in nature. This is not due to ingrowth of cholinergic neurones and loss of the original adrenergic innervation, as 6-hydroxydopamine treatment to destroy the adrenergic input also prevents development of the later cholinergic characteristic. Thus, the neurones change their phenotype during development (Landis, 1983). An even more striking illustration was provided by Schotzinger & Landis (1988) who took footpad skin (where the innervation is normally cholinergic) and transplanted it to hairy skin, where innervation is normally adrenergic. The previously adrenergic innervation then became cholinergic (Schotzinger & Landis, 1988). Purification of the cholinergic differentiation factor showed it was a basic membrane-bond glycoprotein, with a molecular weight of about 29 kD (Fukada, 1985; Wong & Kessler, 1987). Yamamori *et al.* (1989) have identified sufficient of the amino acid sequence of the peptide to clone sequences of complementary DNA which reveal that the cholinergic differentiation factor is identical to leukaemic inhibitory factor (LIF), a haematopoietic factor which had previously been shown to induce macrophage differentiation and inhibit proliferation of a leukaemic myeloid cell line. It can be expected that the availability of pure cholinergic differentiation factor (LIF) will rapidly generate much more information on its sources, secretion and mechanism of action.

REFERENCES

Aas, P. and Fonnum, F. (1986). Presynaptic inhibition of acetylcholine release. *Acta Physiol. Scand.,* **127**, 335–342.

Aas, P. and Maclagan, J. (1990). Evidence for prejunctional M_2 muscarinic receptors in pulmonary cholinergic nerves in the rat. *Br. J. Pharmacol.,* **101**, 73–76.

Adams, P.R., Jones, S.W., Pennefather, P., Brown, D.A., Koch, C. and Lancaster, B. (1986). Slow synaptic transmission in frog sympathetic ganglia. *J. Exp. Biol.,* **124**, 259–285.

Agoston, D.V. (1988a). Cholinergic co-transmitters. In V.P. Whittaker (Ed.), *Handbook of experimental pharmacology, Vol. 86* (pp. 479–533). Berlin and Heidelberg: Springer Verlag.

Ahmad, Z., Green, F.J, Subuhi, H.S and Watanabe, M. (1989). Autonomic regulation of type 1 protein phosphatase in cardiac muscle. *J. Biol. Chem.,* **264**, 3859–3863.

Akasu, T. and Koketsu, K. (1982). Modulation of voltage-dependent currents by muscarinic receptor in sympathetic neurones of bullfrog. *Neurosci. Lett.,* **29**, 41–45.

Akiba, I., Maeda, A., Bujo, H., Nakai, J., Mishina, M. and Numa, S. (1988). Primary structure of porcine muscarinic acetylcholine receptor III and antagonist binding properties. *FEBS Lett.,* **235**, 257–265.

Alberts, P. (1989). Effects of N6,2′-O-dibutyryladenosine 3′, 5′-cyclic monophosphate, adenosine, and of oxotremorine and 3-isobutyl-1-methylxanthine on the electrically evoked [³H]acetylcholine secretion in the guinea-pig ileum myenteric plexus. *Acta Physiol. Scand.,* **137**, 489–496.

Alberts, P. and Ogren, V.R. (1988). Interaction of forskolin with the effect of atropine on [³H]acetylcholine secretion in guinea-pig ileum myenteric plexus. *J. Physiol. Lond.,* **395**, 441–453.

Allen, T.G.J. and Burnstock, G. (1990). M_1 and M_2 muscarinic receptors mediate excitation and inhibition of guinea-pig intracardiac neurones in culture. *J. Physiol.,* **442**, 463–480.

Angus, J.A. and Black, J.W. (1982). The interaction of choline esters, vagal stimulation and H2-receptor blockade on acid secretion *in vitro. Eur. J. Pharmacol.,* **80**, 217–224.

Angus, J.A., Campbell, G.R., Cocks, T.M. and Manderson, J.A. (1983). Vasodilatation by acetylcholine is endothelium-dependent: a study by sonomicrometry in canine femoral artery *in vivo. J. Physiol.,* **344**, 209–222.

Ascher, P., Large, W.A. and Rang, H.P. (1979). Studies on the mechanism of action of acetylcholine antagonists on rat parasympathetic ganglion cells. *J. Physiol.,* **295**, 139–170.

Ashe, J.H. and Yarosh, C.A. (1984). Differential and selective antagonism of the slow-inhibitory postsynaptic potential and slow-excitatory postsynaptic potential by gallamine and pirenzepine in the superior cervical ganglion of the rabbit. *Neuropharmacology.*, **23**, 1321–1329.

Ashkenazi, A., Winslow, J.W., Peralta, E.G., Peterson, G.L., Schimerlik, M.I., Capon, D.J. and Ramachandran, J. (1987). An M2 muscarinic receptor subtype coupled to both adenylyl cyclase and phosphoinositide turnover. *Science,* **238**, 672–675.

Ashkenazi, A., Peralta, E.G., Winslow, J.W., Ramachandran, J. and Capon, D.J. (1988). Functional role of muscarinic acetylcholine receptor subtype diversity. *Cold. Spring. Harb. Symp. Quant. Biol.,* **53 Pt 1**, 263–272.

Ashkenazi, A., Peralta, E.G., Winslow, J.W., Ramachandran, J. and Capon, D.J. (1989). Functionally distinct G proteins selectively couple different receptors to PI hydrolysis in the same cell. *Cell,* **56**, 487–493.

Barlow, R.B. and Ing, H.R. (1948). Curare-like action of polymethylene bis-quaternary ammonium salts. *Br. J. Pharmacol.,* **3**, 298–304.

Barlow, R.B., Berry, K.J., Glenton, P.A.M., Nikolaou, N.M. and Soh, K.S. (1976). A comparison of affinity constants for muscarinic acetylcholine receptors in guinea-pig atrial pacemaker cells at 29°C and in ileum at 29°C and 37°C. *Br. J. Pharmacol.,* **58**, 613–621.

Barlow, R.B., Caulfield, M.P., Kitchen, R., Roberts, P.M. and Stubley, J.K. (1981). The affinities of pirenzepine and atropine for functional muscarinic receptors in guinea-pig atria and ileum. *Br. J. Pharmacol.,* **73**, 183P–184P.

Barnes, P.J. (1984). Localization and function of airway autonomic receptors. *Eur. J. Respir. Dis. Suppl.,* **135**, 187–197.

Barnes, P.J., Basbaum, C.B. and Nadel, J.A. (1983). Autoradiographic localization of autonomic receptors in airway smooth muscle. Marked differences between large and small airways. *Am. Rev. Respir. Dis.,* **127**, 758–762.

Barnes, P.J., Minette, P. and Maclagan, J. (1988). Muscarinic receptor subtypes in airways. *Trends. Pharmacol. Sci.,* **9**, 412–416.

Basbaum, C.B., Grillo, M.A. and Widdicombe, J.H. (1984). Muscarinic receptors: evidence for a nonuniform distribution in tracheal smooth muscle and exocrine glands. *J. Neurosci.,* **4**, 508–520.

Belluzi, O., Sacchi, O. and Wanke, E. (1985). Identification of delayed potassium and calcium currents in the rat sympathetic neurone under voltage clamp. *J. Physiol.,* **358**, 109–129.

Benham, C.D., Bolton, T.B. and Lang, R.J. (1985). Acetylcholine activates an inward current in single mammalian smooth muscle cells. *Nature,* **316**, 345–347.

Bennett, M.R. and McLachlan, E. (1972). An electrophysiological analysis of the storage of acetylcholine in preganglionic nerve terminals. *J. Physiol.,* **221**, 657–668.

Bhave, S.V., Machotra, R.K., Wakade, T.D. and Wakade, A.R. (1988). Formation of inositol trisphosphate by muscarinic agents does not stimulate transmitter release in cultured sympathetic neurons. *Neurosci. Lett.,* **90**, 234–238.

Biegon, R.L. and Pappano, A.J. (1980). Dual mechanism for inhibition of calcium-dependent action potentials by acetylcholine in avian ventricular muscle. *Circ. Res.,* **46**, 353–362.

Birdsall, N.J.M. and Hulme, E.C. (1989). The binding properties of muscarinic receptors. In J.H. Brown (Ed.), *The muscarinic receptors* (pp. 31–92). Clifton, N.J.: Humana Press.

Birks, R. and MacIntoh, F.C. (1961). Acetylcholine metabolism of a sympathetic ganglion. *Can. J. Biochem. Physiol.,* **39**, 787–827.

Blaber, L.C., Fryer, A.D. and Maclagan, J. (1985). Neuronal muscarinic receptors attenuate vagally-induced contraction of feline bronchial smooth muscle. *Br. J. Pharmacol.,* **86**, 723–728.

Blackman, J.G. and Purves, R.D. (1969). Intracellular recordings from ganglia of the thoracic sympathetic chain of the guinea-pig. *J. Physiol.,* **203**, 173–185.

Blackman, J.G., Ginsborg, B.L. and Ray, C. (1963). On the quantal release of the transmitter at a sympathetic synapse. *J. Physiol.,* **167**, 402–415.

Bley, K.R. and Tsien, R.W. (1990). Inhibition of Ca and K channels in sympathetic neurons by neuropeptides and other ganglionic transmitters. *Neuron,* **2**, 379–391.

Bognar, I.T., Pallas, S., Fuder, H. and Muscholl, E. (1988). Muscarinic inhibition of [^{3}H]-noradrenaline release on rabbit iris *in vitro*: effects of stimulation conditions on intrinsic activity of methacholine and pilocarpine. *Br. J. Pharmacol.,* **94**, 890–900.

Bolton, T.B. (1976). On the latency and form of membrane responses of the smooth muscle to the iontophoretic application of acetylcholine or carbachol. *Proc. Roy. Soc. B,* **194**, 99–199.

Bolton, T.B., Lang, R.J., Takewaki, T. and Clapp, L.H. (1984). Autonomic receptors and cell membrane potential. *Bibl. Cardiol.,* 108–114.

Bolton, T.B. and Clapp, L.H. (1986). Endothelial-dependent relaxant actions of carbachol and substance P in arterial smooth muscle. *Br. J. Pharmacol.,* **87**, 713–723.

Bone, E.A., Fretten, P., Palmer, S., Kirk, C.J. and Michell, R.H. (1984). Rapid accumulation of inositol phosphates in isolated rat superior cervical sympathetic ganglia exposed to V_1-vasopressin and muscarinic cholinergic stimuli. *Biochem. J., 221*, 803–811.

Bonner, T.I., Buckley, N.J., Young, A.C. and Brann, M.R. (1987). Identification of a family of muscarinic acetylcholine receptor genes. *Science, 237*, 527–532.

Bonner, T.I., Young, A.C., Brann, M.R. and Buckley, N.J. (1988). Cloning and expression of the human and the rat m5 muscarinic acetylcholine receptor genes. *Neuron, 1*, 403–410.

Bormann, J. and Matthaei, H. (1983). Three types of acetylcholine induced single channel currents in clonal rat phaeochromocytoma cells. *Neurosci. Lett., 40*, 193–197.

Boulter, J., Luyten, W., Evans, K., Mason, P., Ballivet, M., Goldman, D., Stengelin, S., Martin, G., Heinemann, S. and Patrick, J. (1985). Isolation of a clone coding for the α-subunit of a mouse acetylcholine receptor. *J. Neurosci., 5*, 2545–2552.

Boulter, J., Connolly, J.G., Deneris, E., Goldman, D., Heinemann, S. and Patrick, J. (1987). Functional expression of two neuronal nicotinic acetylcholine receptors from cDNA clones identifies a gene family. *Proc. Nat. Acad. Sci. USA, 84*, 7763–7767.

Boulter, J., O'Shea-Greenfield, A., Duvoisin, R.M., Connolly, J.G., Wada, E., Jensen, A., Gardner, P.D., Ballivet, M., Deneris, E.S., McKinnon, D., Heinemann, S. and Patrick, J. (1990). α-3, α-5, β-4: three members of the rat neuronal nicotinic acetylcholine receptor-related gene family form a gene cluster. *J. Biol. Chem., 265*, 4472–4482.

Bourke, J.E., Dunn, S.J., Marley, P.D. and Livett, B.G. (1988). The effects of neosurugatoxin on evoked catecholamine secretion from bovine adrenal chromaffin cells. *Br. J. Pharmacol., 93*, 275–280.

Bourne, H.R., Sanders, D.A. and McCormick, F. (1990). The GTPase superfamily; a conserved system for diverse cell functions. *Nature, 348*, 125–132.

Bowery, N.G. and Neal, M.J. (1975). Failure of denervation of influence the high affinity uptake of choline by sympathetic ganglia. *Br. J. Pharmacol., 55*, 278P.

Boyd, R.T., Jacob, M.H., Couturier, S., Ballivet, M. and Berg, D.K. (1988). Expression and regulation of neuronal acetylcholine receptor mRNA in chick ciliary ganglia. *Neuron, 1*, 495–502.

Brayden, J.E. and Bevan, J.A. (1985). Neurogenic muscarinic vasodilation in the cat. An example of endothelial cell-independent cholinergic relaxation. *Circ. Res., 56*, 205–211.

Brayden, J.E. and Large, W.A. (1986). Electrophysiological analysis of neurogenic vasodilatation in the isolated lngual artery of the rabbit. *Br. J. Pharmacol., 89*, 163–171.

Breitwieser, G.E. and Szabo, G. (1985). Uncoupling of cardiac muscarinic and β-adrenergic receptors from ion channels by a guanine nucleotide analogue. *Nature, 317*, 538–540.

Brenner, H.R. and Martin, A.R. (1976). Reduction in acetylcholine sensitivity of axotomized ciliary ganglion cells. *J. Physiol. Lond., 260*, 159–175.

Briggs, C.A. and Cooper, J.R. (1982). Cholinergic modulation of the release of ^{3}H-acetylcholine from synaptosomes of the myenteric plexus. *J. Neurochem., 38*, 501–508.

Brown, A.M. and Birnbaumer, L. (1990). Ionic channels and their regulation by G protein subunits. *Ann. Rev. Physiol., 52*, 197–213.

Brown, D.A. (1969). Responses of normal and denervated cat superior cervical ganglia to some stimulant compounds. *J. Physiol., 201*, 225–236.

Brown, D.A. (1979). Neurotoxins and the ganglionic (C6) type of receptor. In B. Ceccarelli and F. Clementi (Eds.), *Advances in cytopharmacology, Vol. 3* (pp. 225–230). New York: Raven Press.

Brown, D.A. (1980). Locus and mechanism of action of ganglion-blocking agents. In D.A. Kharkevich (Ed.), *Handbook of experimental pharmacology, Vol. 53* (pp. 185–235). Berlin and Heidelberg: Springer Verlag.

Brown, D.A. (1988). M currents. In T. Narahasi (Ed.), *Ion channels, Vol. 1* (pp. 55–94). New York and London: Plenum Publishing Corp.

Brown, D.A. (1990). G-proteins and potassium currents in neurons. *Ann. Rev. Physiol., 52*, 215–242.

Brown, D.A. and Adams, P.R. (1980). Muscarinic suppression of a novel voltage-sensitive K^+-current in a vertebrate neurone. *Nature, 283*, 673–676.

Brown, D.A. and Caulfield, M.P. (1979). Hyperpolarizing α_2-adrenoceptors in rat sympathetic ganglia. *Br. J. Pharmacol., 65*, 435–445.

Brown, D.A. and Fumagalli, L. (1977). Dissociation of α-bungarotoxin binding and receptor block in the rat superior cervical ganglion. *Brain Res., 129*, 165–168.

Brown, D.A. and Selyanko, A.A. (1985a). Two components of muscarine-sensitive membrane current in rat sympathetic neurones. *J. Physiol. Lond., 358*, 335–363.

Brown, D.A. and Selyanko, A.A. (1985b). Membrane currents underlying the cholinergic slow excitatory post-synaptic potential in the rat sympathetic ganglion. *J. Physiol. Lond.*, **365**, 365–387.

Brown, D.A., Jones, K.B., Halliwell, J.V. and Quilliam, J.P. (1970). Evidence against a presynaptic action of acetylcholine during ganglionic transmission. *Nature*, **226**, 958–959.

Brown, D.A., Garthwaite, J., Hayashi, E. and Yamada, S. (1976). Action of surugatoxin on nicotinic receptors in the superior cervical ganglion of the rat. *Br. J. Pharmacol.*, **58**, 157–159.

Brown, D.A., Forward, A. and Marsh, S. (1980). Antagonist discrimination between ganglionic and ileal muscarinic receptors. *Br. J. Pharmacol.*, **71**, 362–364.

Brown, D.A., Higashida, H., Adams, P.R., Marrion, N.V. and Smart, T.G. (1988). Role of G-protein coupled phophatidylinositide systems in signal transduction in vertebrate neurons: experiments on neuroblastoma hybrid cells and ganglion cells. *Cold Spring Harbor Symp. Quant. Biol.*, **LIII**, 375–383.

Brown, D.A., Marrion, N.V. and Smart, T.G. (1989). On the transduction mechanism for muscarine-induced inhibition of M-current in cultured rat sympathetic neurones. *J. Physiol.*, **413**, 469–488.

Brown, H.F. (1982). Electrophysiology of the sinoatrial node. *Physiol. Rev.*, **62**, 505–530.

Brown, J.H. and Brown, S.L. (1984). Agonists differentiate muscarinic receptors that inhibit cyclic AMP formation from those that stimulate phosphoinositide metabolism. *J. Biol. Chem.*, **259**, 3777–3781.

Brown, J.H., Goldstein, D. and Masters, S.B. (1985a). The putative M_1 muscarinic receptor does not regulate phosphoinositide hydrolysis. Studies with pirenzepine and McN A 343 in chick heart and astrocytoma cells. *Mol. Pharmacol.*, **27**, 525–531.

Brown, J.H., Buxton, I.L. and Brunton, L.L. (1985b). α_1-adrenergic and muscarinic cholinergic stimulation of phosphoinositide hydrolysis in adult rat cardiomyocytes. *Circ. Res.*, **57**, 532–537.

Buckley, N. and Burnstock, G. (1984a). Autoradiographic localisation of muscarinic receptors in guinea-pig intestine: distribution of high and low affinity agonist binding sites. *Brain Res.*, **294**, 15–22.

Buckley, N.J. and Burnstock, G. (1984b). Distribution of muscarinic receptors on cultured myenteric neurons. *Brain Res.*, **310**, 133–137.

Buckley, N.J. and Burnstock, G. (1986a). Autoradiographic localization of peripheral M_1 muscarinic receptors using [^{3}H]pirenzepine. *Brain Res.*, **375**, 83–91.

Buckley, N.J. and Burnstock, G. (1986b). Localization of muscarinic receptors on cultured myenteric neurons: a combined autoradiographic and immunocytochemical approach. *J. Neurosci.*, **6**, 531–540.

Buckley, N.J., Saffrey, M.J., Hassall, C.J. and Burnstock, G. (1988). Localization of muscarinic receptors on peptide-containing neurones of the guinea-pig myenteric plexus in tissue culture. *Brain Res.*, **445**, 152–156.

Buckley, N.J., Bonner, T.I., Buckley, C.M. and Brann, M.R. (1989). Antagonist binding properties of five cloned muscarinic receptors expressed in CHO-K1 cells. *Mol. Pharmacol.*, **35**, 469–476.

Burnstock, G. and Bell, C. (1974). Peripheral autonomic transmission. In J.I. Hubbard (Ed.), *The peripheral nervous system* (pp. 277–327). New York and London: Plenum Press.

Burt, D.R. (1978). Muscarinic receptor binding in rat sympathetic ganglia is unaffected by denervation. *Brain Res.*, **143**, 573–579.

Busse, R., Fichtner, H., Luckhoff, A. and Kohlhardt, M. (1988). Hyperpolarization and increased free calcium in acetylcholine-stimulated endothelial cells. *Am. J. Physiol.*, **255**, H965–H969.

Cahill, A.L. and Perlman, R.L. (1986). Nicotinic and muscarinic agonists, phorbol esters and agents which raise cyclic AMP levels phophorylate distinct groups of proteins in the superior cervical ganglion. *Neurochem. Res.*, **11**, 327–332.

Campbell, G. (1970). Autonomic nervous supply to effector tissues. In E. Bulbring, A.F. Brading, A.W. Jones and T. Tomita (Eds.), *Smooth muscle* (pp. 451–495). London: Edward Arnold.

Campbell, G.D., Edwards, F.R., Hirst, G.D.S. and O'Shea, J.E. (1989). Effects of vagal stimulation and applied acetylcholine on pacemaker potentials in the guinea-pig heart. *J. Physiol. Lond.*, **415**, 57–68.

Carmeliet, E. and Mubagwa, K. (1986a). Changes by acetylcholine of membrane currents in rabbit cardiac Purkinje fibres. *J. Physiol. Lond.*, **371**, 201–217.

Carmeliet, E. and Mubagwa, K. (1986b). Characterization of the acetylcholine-induced potassium current in rabbit cardiac Purkinje fibres. *J. Physiol. Lond.*, **371**, 219–237.

Cassell, J.F. and McLachlan, M. (1987). Muscarinic agonists block five different potassium conductances in guinea-pig sympathetic neurones. *Br. J. Pharmacol.*, **91**, 259–261.

Caulfield, M.P., Barlow, R.B. and Brown, D.A. (1990). Pharmacological studies of pre- and post-synaptic neuronal nicotinic receptors. *Br. J. Pharmacol.*, **99**, 166P.

Caulfield, M.P., Brown, D.A. and Barlow, R.B. (1991). A pharmacological study of the m4 muscarinic receptor mediating inhibition of the voltage-gated Ca-current in neuroblastoma x glioma hybrid (NG 108–15) cells. *Br. J. Pharmacol. (Proc. Suppl.),* **In press.**

Cena, V., Nicolas, G.P., Sanchez-Garcia, P., Kirpekar, S.M. and Garcia, A.G. (1983). Pharmacological dissection of receptor-associated and voltage-sensitive ionic channels involved in catecholamine release. *Neuroscience,* **10,** 1455–1462.

Cerbai, E., Klockner, U. and Isenberg, G. (1988). The α-subunit of the GTP binding protein activates muscarinic potassium channels of the atrium. *Science,* **240,** 1782–1783.

Chang, F., Gao, J., Tromba, C., Cohen, I. and DiFrancesco, D. (1990). Acetylcholine reverses effects of β-agonists on pacemaker current in canine cardiac Purkinje fibers, but has no direct action: a difference between primary and secondary pacemakers. *Circ. Res.,* **66,** 633–636.

Changeux, J.P., Kasai, M. and Lee, C.Y. (1970). The use of a snake venom toxin to characterize the cholinergic receptor protein. *Proc. Natl. Acad. Sci. U.S.A.,* **67,** 1241–1247.

Cherdchu, C. and Hexum, T.D. (1988). Characterization of 1, 1-dimethyl-4-phenylpiperazinium-induced increased proenkephalin processing in bovine chromaffin cells. *Life Sci.,* **43,** 1069–1077.

Chiappinelli, V.A. (1983). Kappa bungarotoxin: a probe for the neuronal nicotinic receptor in the avian ciliary ganglion. *Brain Res.,* **277,** 9–21.

Chiappinelli, V.A. and Dryer, S.E. (1984). Nicotinic transmission in sympathetic ganglia: blockade by the snake venom neurotoxin kappa-bungarotoxin. *Neurosci. Lett.,* **50,** 239–244.

Chiappinelli, V.A. and Zigmond, R.E. (1978). α-bungarotoxin blocks nicotinic transmission in the avian ciliary ganglion. *Proc. Natl. Acad. Sci. U.S.A.,* **75,** 2999–3003.

Chiappinelli, V.A., Wolf, K.M., Debin, J.A. and Holt, I.L. (1987). Kappa-flavitoxin: isolation of a new neuronal nicotinic receptor antagonist that is structurally related to kappa-bungarotoxin. *Brain Res.,* **402,** 21–29.

Clapham, D.E. and Neher, E. (1984). Substance P reduces acetylcholine induced currents in isolated bovine chromaffin cells. *J. Physiol. Lond.,* **347,** 255–277.

Clapp, L.H., Vivaudou, M.B., Walsh, J.V., Jr. and Singer, J.J. (1987). Acetylcholine increases voltage-activated Ca current. *Proc. Natl. Acad. Sci. U.S.A.,* **84,** 2092–2096.

Clark, R.B., Nakajima, T., Giles, W., Kanai, K., Momose, Y. and Szabo, G. (1990). Two distinct types of inwardly rectifying K channels in bullfrog atrial myocytes. *J. Physiol. Lond.,* **424,** 229–251.

Claudio, T., Ballivet, M., Patrick, J. and Heinemann, S. (1983). Nucleotide and deduced amino acid sequences of *Torpedo californica* acetylcholine receptor gamma subunit. *Proc. Nat. Acad. Sci. USA,* **80,** 1111–1115.

Codina, J., Yatani, A., Grenet, D., Brown, A.M. and Birnbaumer, L. (1987). The α-subunit of the GTP binding protein G_K opens atrial potassium channels. *Science,* **236,** 442–445.

Cohen, R.A., Shepherd, J.T. and Vanhoutte, P.M. (1984). Neurogenic cholinergic prejunctional inhibition of sympathetic β-adrenergic relaxation in the canine coronary artery. *J. Pharmacol. Exp. Ther.,* **229,** 417–421.

Cole, A.E. and Shinnick-Gallagher, P. (1984). Muscarinic inhibitory transmission in mammalian sympathetic ganglia mediated by increased potassium conductance. *Nature,* **307,** 270–271.

Collier, B. and Ilson, D. (1977). The effect of preganglionic nerve stimulation on the accumulation of certain analogues of choline by a sympathetic ganglion. *J. Physiol. Lond.,* **264,** 489–509.

Collier, B. and Katz, H.S. (1970). The release of acetylcholine by acetylcholine in the cat's superior cervical ganglion. *Br. J. Pharmacol.,* **39,** 428–438.

Collier, B. and Katz, H.S. (1975). Studies upon the mechanism by which acetylcholine releases surplus acetylcholine in a sympathetic ganglion. *Br. J. Pharmacol.,* **55,** 189–197.

Colquhoun, D. and Sakmann, B. (1985). Fast events in single-channel currents activated by acetylcholine and its analogues at the frog muscle end-plate. *J. Physiol. Lond.,* **369,** 501–557.

Conklin, B.R., Brann, M.R., Buckley, N.J., Ma, A.L., Bonner, T.I. and Axelrod, J. (1988). Stimulation of arachidonic acid release and inhibition of mitogenesis by cloned genes for muscarinic receptor subtypes stably expressed in A9 L cells. *Proc. Natl. Acad. Sci. U.S.A.,* **85,** 8698–8702.

Connolly, J.G. (1989). Structure function relationships in nicotinic acetylcholine receptors. *Comp. Biochem. Physiol. C.,* **93A,** 221–231.

Constanti, A. and Brown, D.A. (1981). M-currents in voltage-clamped mammalian sympathetic neurones. *Neurosci. Lett.,* **24,** 289–294.

Conti Tronconi, B.N. and Raftery, M.A. (1982). The nicotinic cholinergic receptor; correlation of molecular structure with functional properties. *Ann. Rev. Biochem.,* **51,** 491–530.

Costa, M., Furness, J.B. and Gibbins, I.L. (1986). Chemical coding of enteric neurons. *Prog. Brain Res.,* **68,** 217–241.

Cull-Candy, S.G. and Mathie, A. (1986). Ion channels activated by acetylcholine and gamma-amino butyric acid in freshly dissociated sympathetic neurones of the rat. *Neurosci. Lett.*, **66**, 275–280.

Curtis, C.A.M., Wheatley, M., Bansal, S., Birdsall, N.J.M., Eveleigh, P., Pedder, E.K., Poyner, D. and Hulme, E.C. (1989). Propylbenzilylcholine mustard labels an acidic residue in transmembrane helix 3 of the muscarinic receptor. *J. Biol. Chem.*, **264**, 489–495.

D'Agostino, G., Kilbinger, H., Chiari, M.C. and Grana, E. (1986). Presynaptic inhibitory muscarinic receptors modulating [^{3}H] acetylcholine release in the rat urinary bladder. *J. Pharmacol. Exp. Ther.*, **239**, 522–528.

Dale, H.H. (1914). The action of certain esters and ethers of choline and their relation to muscarine. *J. Pharmacol. Exp. Ther.*, **6**, 147–190.

Dammann, F., Fuder, H., Giachetti, A., Giraldo, E., Kilbinger, H. and Micheletti, R. (1989). AF-DX116 differentiates between prejunctional muscarine receptors located on noradrenergic and cholinergic nerves. *Naunyn Schmiedebergs. Arch. Pharmacol.*, **339**, 268–271.

Danthuluri, N.R., Cybulsky, M.I. and Brock, T.A. (1988). ACh-induced calcium transients in primary cultures of rabbit aortic endothelial cells. *Am. J. Physiol.*, **255**, H1549–1553.

Dashwood, M.R. and Spyer, K.M. (1986). Autoradiographic localization of α-adrenoceptors, muscarinic acetylcholine receptors and opiate receptors in the heart. *Eur. J. Pharmacol.*, **127**, 279–282.

Davis, R. and Koelle, G.B. (1978). Electron microscope localization of acetylcholinesterase and butyrylcholinesterase in the superior cervical ganglion of the cat. 1. Normal ganglion. *J. Cell Biol.*, **78**, 785–809.

Deeg, M.A., Graeff, R.M., Walseth, T.F. and Goldberg, N.D. (1988). A Ca-linked increase in coupled cAMP synthesis and hydrolysis is an early event in cholinergic and β-adrenergic stimulation of parotid secretion. *Proc. Natl. Acad. Sci. U.S.A.*, **85**, 7867–7871.

Dehaye, J.P., Marino, A., Soukias, Y., Poloczek, P., Winand, J. and Christophe, J. (1988). Functional characterization of muscarinic receptors in rat parotid acini. *Eur. J. Pharmacol.*, **151**, 427–434.

Deneris, E., Connolly, J., Boulter, J., Wada, K., Wada, E., Swanson, L., Patrick, J. and Heinemann, S. (1988). Identification of a cDNA coding for a subunit common to distinct acetylcholine receptors. *Neuron*, **1**, 45–54.

Deneris, E., Boulter, J., Swanson, L.W., Patrick, J. and Heinemann, S. (1989). β-3: a new member of the nicotinic acetylcholine receptor gene family is expressed in the brain. *J. Biol. Chem.*, **264**, 6268–6272.

Dennis, M.J. and Sargent, P.B. (1979). Loss of extrasynaptic acetylcholine sensitivity upon reinnervation of parasympathetic ganglion. *J. Physiol. Lond.*, **289**, 263–275.

Derkach, V.A., Selyanko, A.A. and Skok, V.I. (1983). Acetylcholine-induced current fluctuations and fast excitatory postsynaptic currents in rabbit sympathetic neurones. *J. Physiol. Lond.*, **336**, 511–526.

Derkach, V.A., North, R.A., Selyanko, A.A. and Skok, V.I. (1987). Single channels activated by acetylcholine in rat superior cervical ganglion. *J. Physiol. Lond.*, **388**, 141–151.

Devilliers-Thiery, A., Giraudat, J., Bentaboulet, M. and Changeux, J.P. (1983). Complete mRNA coding sequence of the acetylcholine binding subunit of *Torpedo marmorata* acetylcholine receptor: a model for the transmembrane organisation of the polypeptide chain. *Proc. Nat. Acad. Sci. USA*, **80**, 2067–2081.

DiFrancesco, D. and Tromba, C. (1987). Acetylcholine inhibits activation of the cardiac hyperpolarizing-activated current, I_f. *Pflugers. Arch.*, **410**, 139–142.

DiFrancesco, D. and Tromba, C. (1988). Muscarinic control of the hyperpolarization-activated current (I_f) in rabbit sino-atrial node myocytes. *J. Physiol. Lond.*, **405**, 493–510.

DiFrancesco, D., Ducouret, P. and Robinson, R.B. (1989). Muscarinic modulation of cardiac rate at low acetylcholine concentrations. *Science*, **243**, 669–671.

Dixon, R.A.F., Kobilka, B.K., Strader, G.J., Benovic, J.L., Dohlman, H.G., Frielle, T., Bolanowski, M.A., Bennett, C.D., Rands, E., Diehl, R.E., Mumford, R.A., Slater, E.E., Sigal, I.S., Caron, M.G., Lefkowitz, R.J. and Strader, C.D. (1986). Cloning of the gene and cDNA for mammalian beta-adrenergic receptor and homology with rhodopsin. *Nature*, **321**, 75–79.

Dodd, J. and Horn, J.P. (1983). Muscarinic inhibition of sympathetic C neurones in the bullfrog. *J. Physiol. Lond.*, **334**, 271–291.

Doughney, C., Pedersen, P.S., McPherson, M.A. and Dormer, R.L. (1989). Formation of inositol polyphosphates in cultured human sweat duct cells in response to cholinergic stimulation. *Biochim. Biophys. Acta*, **1010**, 352–356.

Douglas W.W., Kanno, T. and Sampson, S.R. (1967). Effects of acetylcholine and other medullary secretagogues and antagonists on the membrane potential of adrenal chromaffin cells: an analysis employing technique of tissue culture. *J. Physiol. Lond.*, **188**, 107–120.

Dowe, G.H.C., Kilbinger, H. and Whittaker, V.P. (1980). Isolation of cholinergic synaptic vesicles from the myenteric plexus of guinea-pig small intestine. *J. Neurochem.*, **35**, 993-1003.

Ducis, I. (1988). The high-affinity choline uptake system. In V.P. Whittaker (Ed.), *Handbook of experimental pharmacology* (pp. 410-445). Berlin and Heidelberg: Springer Verlag.

Duckles, S.P. (1988). Vascular muscarinic receptors: pharmacological characterization in the bovine coronary artery. *J. Pharmacol. Exp. Ther.*, **246**, 929-934.

Dunn, P.M. and Marshall, L.M. (1985b). Innervation of small intensely fluorescent cells in frog sympathetic ganglia. *Brain Res.*, **339**, 371-374.

Duvoisin, R.M., Deneris, E.S., Patrick, J. and Heinemann, S. (1989). The functional diversity of the neuronal nicotinic receptors is increased by a novel subunit: β4. *Neuron*, **3**, 487-496.

Eberhard, D.A. and Holz, R.W. (1987). Cholinergic stimulation of inositol phosphate formation in bovine adrenal chromaffin cells: distinct nicotinic and muscarinic mechanisms. *J. Neurochem.*, **49**, 1634-1643.

Eglen, R.M. and Whiting, R.L. (1985). Determination of the muscarinic receptor subtype mediating vasodilatation. *Br. J. Pharmacol.*, **84**, 3-5.

Eglen, R.M. and Whiting, R.L. (1990). Heterogeneity of vascular muscarinic receptors. *Journal of Autonomic Pharmacology*, **10**, 233-245.

Ehrlich, P.H. (1907). Quoted from the collected works of P.H. Ehrlich, In 'Origins of the receptor theory', by J. Pascandola. *Trends. Pharmacol. Sci.*, **1**, 184-192, (1979).

Endoh, M. and Shimizu, T. (1979). Failure of dibutyryl and 8-bromo cyclic GMP to mimic the antagonistic action of carbachol on the positive inotropic effects of sympathomimetic amines in the canine isolated ventricular myocardium. *Jpn. J. Pharmacol.*, **29**, 423-433.

Endoh, M. and Yamashita, S. (1981). Different responses to carbachol, sodium nitroprusside and 8-bromoguanosine 3,5-monophosphate of canine atrial and ventricular muscle. *Br. J. Pharmacol.*, **73**, 393-399.

Endoh, M., Maruyama, M. and Iijima, T. (1985). Attenuation of muscarinic cholinergic inhibition by islet- activating protein in the heart. *Am. J. Physiol.*, **249**, H309-H320.

Evans, M.G. and Marty, A. (1986). Potentiation of muscarinic and α-adrenergic responses by an analogue of guanosine 5'-triphosphate. *Proc. Natl. Acad. Sci. U.S.A.*, **83**, 4099-4103.

Feletou, M. and Vanhoutte, A.M. (1990). Endothelium-dependent hyperpolarization of canine coronary smooth muscle. *Br. J. Pharmacol.*, **93**, 515-524.

Fenwick, E.M., Marty, A. and Neher, E. (1982). A patch clamp study of bovine chromaffin cells and their sensitivity to acetylcholine. *J. Physiol. Lond.*, **331**, 577-597.

Ferrand, C., Clarous, D., Delteil, C. and Weber, M. (1986). Cellular localization of the molecular forms of acetylcholinesterase in primary cultures of rat sympathetic neurons and analysis of the secreted enzyme. *J. Neurochem.*, **46**, 349-358.

Field, J.L. and Newberry, N.R. (1989). Methoctramine and hexahydrodifenidol antagonise two muscarinic responses on the rat superior cervical ganglion with opposite selectivity. *Neurosci. Lett.*, **100**, 254-258.

Fischmeister, R. and Hartzell, H.C. (1986). Mechanism of action of acetylcholine on calcium current in single cells from frog ventricle. *J. Physiol. Lond.*, **376**, 183-202.

Fischmeister, R. and Hartzell, H.C. (1987). Cyclic guanosine 3,5-monophosphate regulates the calcium current in single cells from frog ventricle. *J. Physiol. Lond.*, **387**, 453-472.

Fosbraey, P. and Johnson, E.S. (1980a). Release-modulating acetylcholine receptors on cholinergic neurones of the guinea-pig ileum. *Br. J. Pharmacol.*, **68**, 289-300.

Fosbraey, P. and Johnson, E.S. (1980b). Modulation by acetylcholine of the electrically-evoked release of ^{3}H-acetylcholine from the ileum of the guinea-pig ileum. *Br. J. Pharmacol.*, **69**, 145-149.

Fox, J.E., Daniel, E.E., Jury, J. and Robotham, H. (1985). Muscarinic inhibition of canine small intestinal motility *in vivo. Am. J. Physiol.*, **248**, G526-G531.

Fraser, C.M., Wang, C.D., Robinson, D.A., Gocayne, J.D. and Venter, J.C. (1990). Conserved transmembrane aspartates in muscarinic cholinergic receptors display unique roles in receptor function. *J. Biol. Chem.*, **In Press**.

Fryer, A.D. and Maclagan, J. (1984). Muscarinic inhibitory receptors in pulmonary parasympathetic nerves in the guinea-pig. *Br. J. Pharmacol.*, **83**, 973-978.

Fuder, H., Rink, D. and Muscholl, E. (1982). Sympathetic nerve stimulation in the rat heart. Afinities of N-methylatropine and pirenzepine at pre- and post-synaptic muscarine receptors. *Naunyn Schmiedebergs. Arch. Pharmacol.*, **318**, 210-219.

Fuder, H., Kilbinger, H. and Muller, H. (1985). Organ selectivity of hexahydrosiladifenidol in blocking pre- and postjunctional muscarinic receptors studied in guinea-pig ileum and rat heart. *Eur. J. Pharmacol.,* **113**, 125–127.

Fuder, H., Schoepf, J., Linkel, J., Wesner, M.T., Melchiorre, C., Tacke, R., Mutschler, E. and Lambrecht, G. (1989). Different muscarine receptors mediate the prejunctional inhibition of [^{3}H]-noradrenaline release in rat or guinea-pig iris and the contraction of the rabbit iris sphincter muscle. *Naunyn Schmiedebergs. Arch. Pharmacol.,* **340**, 597–604.

Fukada, K. (1985). Purification and partial characterization of a cholinergic neuronal differentiation factor. *Proc. Natl. Acad. Sci. U.S.A.,* **82**, 8795–8799.

Fukuda, K., Kubo, T., Akiba, I., Maeda, A., Mishina, M. and Numa, S. (1987). Molecular distinction between muscarinic acetylcholine receptor subtypes. *Nature,* **327**, 623–625.

Fukuda, K., Higashida, H., Kubo, T., Maeda, A., Akiba, I., Bujo, H., Mishina, M. and Numa, S. (1988). Selective coupling with K$^+$ currents of muscarinic acetylcholine receptor subtypes in NG108-15 cells. *Nature,* **335**, 355–358.

Fumagalli, L., Derenzis, G. and Miani, N. (1976). Acetylcholine receptors: number and distribution in intact and deafferented superior cervical ganglion of the rat. *J. Neurochem.,* **27**, 47–52.

Furchgott, R.F. (1984). The role of endothelium in the responses of vascular smooth muscle to drugs. *Annu. Rev. Pharmacol. Toxicol.,* **24**, 175–197.

Furchgott, R.F. and Zawadcki, J.V. (1980). The obligatory role of endothelial cells in the relaxation of arterial smooth muscle by acetylcholine. *Nature,* **288**, 373–376.

Furness, J.B., Costa, M. and Keast, J.R. (1984). Choline acetyltransferase and peptide immunoreactivity of submucous neurons in the small intestine of the guinea-pig. *Cell Tissue. Res.,* **237**, 329–333.

Furness, J.B., Costa, M., Gibbins, I.L., Llewellyn-Smith, I.J. and Oliver, J.R. (1985). Neurochemically similar myenteric and submucous neurons directly traced to the mucosa of the small intestine. *Cell Tissue. Res.,* **241**, 155–163.

Gabella, G. (1976). *Structure of the autonomic nervous system.* New York: Wiley.

Gallagher, D.V. and Morris, A.P. (1987). The receptor-regulated calcium influx in mouse submandibular acinar cells is sodium-dependent: a patch-clamp study. *J. Physiol. Lond.,* **384**, 119–130.

Gallagher, J.P., Griffith, W.H. and Shinnick-Gallagher, P. (1982). Cholinergic transmission in cat parasympathetic ganglia. *J. Physiol. Lond.,* **332**, 473–486.

Galligan, J.J., North, R.A. and Tokimasa, T. (1989). Muscarinic agonists and potassium currents in guinea-pig myenteric neurones. *Br. J. Pharmacol.,* **96**, 193–203.

Gangitano, C., Fumagalli, L. and Miani, N. (1979). Appearance of new α-bungarotoxin acetylcholine receptors in cultured sympathetic ganglia chick embryos. *Brain Res.,* **161**, 131–141.

Gershon, M.D. (1970). The identification of neurotransmitters in smooth muscle. In E. Bulbring, A.F. Brading, A.W. Jones and T. Tomita (Eds.), *Smooth muscle* (pp. 496–522). London: Edward Arnold.

Giachetti, A., Micheletti, R. and Montagna, E. (1986). Cardioselective profile of AF-DX116, a muscarine M$_2$ receptor antagonist. *Life. Sci.,* **38**, 1663–1672.

Gilani, S.A.H. and Cobbin, L.B. (1986). The cardio-selectivity of himbacine: a muscarine receptor antagonist. *Naunyn Schmiedebergs. Arch. Pharmacol.,* **332**, 10–20.

Giles, W. and Noble, S.J. (1976). Changes in membrane currents in bullfrog atrium produced by acetylcholine. *J. Physiol. Lond.,* **261**, 103–123.

Gilman, A.G. (1987). G-proteins: transducers of receptor-generated signals. *Ann. Rev. Biochem.,* **56**, 615–649.

Ginsborg, B.L. (1971). On the presynaptic acetylcholine receptors in sympathetic ganglia of the frog. *J. Physiol. Lond.,* **216**, 237–246.

Giraldo, E., Monferini, E., Ladinsky, H. and Hammer, R. (1987). Muscarinic receptor heterogeneity in guinea-pig intestinal smooth muscle: binding studies with AF-DX116. *Eur. J. Pharmacol.,* **141**, 475–477.

Gisiger, V., Vigny, M., Gautron, J. and Rieger, F. (1978). Acetylcholinesterase of rat sympathetic ganglia: molecular forms, localization and effects of denervation. *J. Neurochem.,* **30**, 501–516.

Goldman, D., Deneris, E., Kochhar, A., Patrick, J. and Heinemann, S. (1987). Members of a nicotinic acetylcholine receptor gene family are expressed in different regions of the mammalian central nervous system. *Cell,* **48**, 965–973.

Gray, P.T.A. (1988). Oscillation of free cytosolic calcium evoked by cholinergic and catecholaminergic agonists in rat parotid acinar cells. *J. Physiol. Lond.,* **406**, 35–53.

Greene, L.A., Sytokowski, A.J., Vogel, Z. and Nirenberg, M. (1973). Bungarotoxin used as a probe for acetylcholine receptors in cultured neurons. *Nature,* **243**, 163–166.

Grider, J.R. and Makhlouf, G.M. (1988). Contraction mediated by Ca^{2+} release in circular and Ca^{2+} influx in longitudinal intestinal muscle cells. *J. Pharmacol. Exp. Ther.,* **244**, 432–437.

Grove, E.A., Caulfield, M.P. and Evans, F.J. (1990). Inhibition of protein kinase C prevents phorbol ester- but not muscarine-induced depolarizations in the rat superior cervical ganglion. *Neurosci. Lett.,* **110,** 162–166.

Gustaffsson, L., Hedqvist, P. and Lundgren, G. (1980). Pre- and post-junctional effects of prostaglandin E_2, prostaglandin synthetase inhibitors and atropine on cholinergic neurotransmission in guinea-pig ileum and canine iris. *Acta Physiol. Scand.,* **110,** 401–411.

Guy, H.R. and Hucho, F. (1987). The ion channel of the nicotinic acetylcholine receptor. *Trends. Neurosci.,* **10,** 318–321.

Halim, S., Kilbinger, H. and Wessler, I. (1982). The muscarinic antagonist potency of pirenzepine in smooth muscle of the guinea-pig ileum. *Scand. J. Gastroenterol.,* **17 (Suppl. 72),** 87–93.

Hammer, R. and Giachetti, A. (1982). Muscarinic receptor subtypes: M_1 and M_2 biochemical and functional characterisation. *Life. Sci.,* **31,** 2991–2998.

Hammer, R., Berrie, C.P., Birdsall, N.J.M., Burgen, A.S.V. and Hulme, E.C. (1980). Pirenzepine distinguishes between subclasses of muscarinic receptor. *Nature,* **283,** 90–92.

Hammer, R., Giraldo, E., Schiavi, G.B., Monferini, E. and Ladinsky, H. (1986). Binding profile of a novel cardioselective muscarine receptor antagonist, AF-DX116, to membranes of peripheral tissues and brain in the rat. *Life, Sci.,* **38,** 1653–1662.

Harris, A.J., Kuffler, S.W. and Dennis, M.J. (1971). Differential chemosensitivity of synaptic and extrasynaptic areas on the neuronal surface membrane in parasympathetic neurons of the frog, tested by microapplication of acetylcholine. *Proc. R. Soc. Lond. Biol.,* **177,** 541–553.

Hartzell, H.C. (1980). Distribution of muscarinic acetylcholine receptors and presynaptic nerve terminals in amphibian heart. *J. Cell Biol.,* **86,** 6–20.

Hartzell, H.C., Kuffler, S.W., Stickgold, R. and Yoshikami, D. (1977). Synaptic excitation and inhibition resulting from direct action of acetylcholine on two types of chemoreceptors on individual amphibian parasympathetic neurones. *Journal of Physiology,* **271,** 817–846.

Hassall, C.J., Buckley, N.J. and Burnstock, G. (1987). Autoradiographic localisation of muscarinic receptors on guinea-pig intracardiac neurones and atrial myocytes in culture. *Neurosci. Lett.,* **74,** 145–150.

Haycock, J.W., Greengard, P. and Browning, M.D. (1988). Cholinergic regulation of protein III phosphorylation in bovine adrenal cells. *J. Neurosci.,* **8,** 3233–3239.

Hazeki, O. and Ui, M. (1981). Modification by islet-activating protein of receptor-mediated regulation of cyclic AMP accumulation in isolated rat heart cells. *J. Biol. Chem.,* **256,** 2856–2862.

Hescheler, J., Kameyama, M. and Trautwein, W. (1986). On the mechanism of muscarinic inhibition of the cardiac Ca current. *Pflugers. Arch.,* **407,** 182–189.

Hexum, T.D., Majane, E.A., Russett, L.R. and Yang, H.Y. (1987). Neuropeptide Y release from the adrenal medulla after cholinergic receptor stimulation. *J. Pharmacol. Exp. Ther.,* **243,** 927–930.

Higashida, H., Hashii, M., Fukuda, K., Caulfield, M.P., Numa, S. and Brown, D.A. (1990). Selective coupling of different muscarinic acetylcholine receptors to neuronal calcium currents in DNA-transfected cells. *Proc. R. Soc. Lond. Biol.,* **242,** 68–74.

Higgins, A.J. and Neal, M.J. (1982). Potassium activation of (^{3}H) choline accumulation by isolated sympathetic ganglion of the rat. *Br. J. Pharmacol.,* **77,** 573–580.

Higgins, L.S. and Berg, D.K. (1988). Cyclic AMP-dependent mechanism regulates acetylcholine receptor function on bovine adrenal chromaffin cells and discriminates between new and old receptors. *J. Cell Biol.,* **107,** 1157–1165.

Hill-Smith, I. and Purves, R. (1978). Synaptic delay in the heart; an iontophoretic study. *Journal of Physiology,* **279,** 31–54.

Himpens, B. and Somlyo, A.P. (1988). Free calcium and force transients during depolarization and pharmacomechanical coupling in guinea-pig smooth muscle. *Journal of Physiology,* **395,** 507–530.

Hirschowitz, B.I., Fong, J. and Molina, E. (1983). Effects of pirenzepine and atropine on vagal and cholinergic gastric secretion and gastrin release and on heart rate in the dog. *J. Pharmacol. Exp. Ther.,* **225,** 263–268.

Hohlfeld, J., Liebau, S. and Forstermann, U. (1990). Pertussis toxin inhibits contractions but not endothelium-dependent relaxations of rabbit pulmonary artery in response to acetylcholine and other agonists. *J. Pharmacol. Exp. Ther.,* **252,** 260–264.

Horie, M. and Irisawa, H. (1989). Dual effects of intracellular magnesium on muscarinic potassium channel current in single guinea-pig atrial cells. *J. Physiol. Lond.,* **408,** 313–332.

Horn, R. and Marty, A. (1988). Muscarinic activation of ionic currents measured by a new whole-cell recording method. *J. Gen. Physiol.,* **92,** 145–159.

Horwitz, J., Tsymbalov, S. and Perlman, R.L. (1985). Muscarine stimulates the hydrolysis of inositol-containing phospholipids. *J. Pharmacol. Exp. Ther.,* **233,** 235–241.

Hulme, E.C. (1990). Receptor biochemistry: a practical approach. *Oxford: IRL Press.*

Hulme, E.C., Birdsall, N.J.M. and Buckley, N.J. (1990). Muscarinic receptor subtypes. *Ann. Rev. Pharmacol. Toxicol.,* **30**, 633–673.

Hucho, F. (1986). The nicotinic acetylcholine receptor and its ion channel. *Eur. J. Biochem.,* **158**, 211–226.

Hutchins, J.B. and Hollyfield, J.G. (1985). Acetylcholine receptors in the human retina. *Invest, Ophthalmol. Vis. Sci.,* **26**, 1550–1557.

Iijima, T., Irisawa, H. and Kameyama, M. (1985). Membrane currents and their modification by acetylcholine in isolated single atrial cells of the guinea-pig. *Journal of Physiology,* **359**, 485–501.

Ingebretsen, C.G. (1980). Interaction between α and β adrenergic receptors and cholinergic receptors in isolated perfused rat heart: effects of cAMP protein kinase and phosphorylase. *J. Cyclic. Nucleotide. Protein. Phosphor. Res.,* **6**, 121–132.

Inoue, R. and Isenberg, G. (1990a). Intracellular calcium ions modulate acetylcholine-induce inward current in guinea-pig ileum. *Journal of Physiology,* **424**, 73–92.

Inoue, R. and Isenberg, G. (1990b). Effect of membrane potential on acetylcholine-induced inward current in guinea-pig ileum. *Journal of Physiology,* **424**, 57–72.

Inoue, R. and Isenberg, G. (1990c). Acetylcholine activates nonselective cation channels in guinea-pig ileum through a G-protein. *Am. J. Physiol.,* **258**, C1173–C1178.

Inoue, M. and Kuriyama, H. (1989). Glucocorticoids inhibit acetylcholine-induced current in chromaffin cells. *Am. J. Physiol.,* **257**, C906–C912.

Isenberg, K.E. and Meyer, G.E. (1989). Cloning of a putative neuronal nicotinic acetylcholine receptor subunit. *J. Neurochem.,* **52**, 988–991.

Ivy, M.T., Sukumar, R. and Townsel, J.G. (1985). The characterization of a sodium-dependent high affinity choline uptake system unassociated with acetylcholine biosynthesis. *Comp. Biochem. Physiol.,* **81C**, 351–357.

Jacob, M.H. and Berg, D.K. (1983). The ultrastructural localisation of α-bungarotoxin binding sites in relation to synapses on chick ciliary ganglion neurons. *J. Neurosci.,* **3**, 260–271.

Jacobowitz, D.M. (1974). The peripheral autonomic nervous system. In J.I. Hubbard (Ed.), *The peripheral nervous system* (pp. 87–110). New York and London: Plenum Press.

James, S. and Burnstock, G. (1989). Autoradiographic localization of muscarinic receptors on cultured, peptide-containing neurones from newborn rat superior cervical ganglion. *Brain Res.,* **498**, 205–214.

Jeck, D., Lindmar, R., Loffelholz, K. and Wanke, M. (1988). Subtypes of muscarinic receptor on cholinergic nerves and atrial cells of chicken and guinea-pig hearts. *Br. J. Pharmacol.,* **93**, 357–366.

Jones, S.V., Barker, J.L., Bonner, T.I., Buckley, N.J. and Brann, M.R. (1988a). Electrophysiological characterization of cloned M1 muscarinic receptors expressed in A9 L cells. *Proc. Natl. Acad. Sci. U.S.A.,* **85**, 4056–4060.

Jones, S.V., Barker, J.L., Buckley, N.J., Bonner, T.I., Collins, R.M. and Brann, M.R. (1988b). Cloned muscarinic receptor subtypes expressed in A9 L cells differ in their coupling to electrical responses. *Mol. Pharmacol.,* **34**, 421–426.

Kao, P.N., Dwork, A.J., Kaldany, R.J., Silver, M.L., Wideman, J., Stein, S. and Karlin, A. (1984). Identification of two α-subunit half cysteines specifically labelled by an affinity reagent for the acetylcholine binding site. *J. Biol. Chem.,* **259**, 1162–1165.

Kawashima, K., Fujimoto, K., Suzuki, T. and Oohata, H. (1988). Direct determination of acetylcholine release by radioimmunoassay and presence of presynaptic M_1 muscarinic receptors in guinea-pig ileum. *J. Pharmacol. Exp. Ther.,* **244**, 1036–1039.

Kilbinger, H. (1988). The autonomic cholinergic neuroeffector junction. In V.P. Whittaker (Ed.), *Handbook of experimental pharmacology, Vol. 86* (pp. 581–595). Berlin and Heidelberg: Springer Verlag.

Kilbinger, H. and Nafziger, M. (1985). Two types of neuronal muscarinic receptors modulating acetylcholine release from guinea-pig myenteric plexus. *Naunyn Schmiedebergs. Arch. Pharmacol.,* **328**, 304–309.

Kilbinger, H., Halim, S., Lambrecht, G., Weiler, W. and Wessler, I. (1984). Comparison of affinities of muscarinic antagonists to pre- and post-junctional receptors in the guinea-pig ileum. *Eur. J. Pharmacol.,* **103**, 313–320.

Kilbinger, H. and Wessler, I. (1983). The variation of acetylcholine release from myenteric neurones with stimulation frequency and train length. Role of presynaptic muscarine receptors. *Naunyn Schmiedebergs. Arch. Pharmacol.,* **324**, 130–133.

Kim, D., Lewis, D.L., Graziadei, L., Neer, E.J., Bar-Sagi, D. and Clapham, D.E. (1989). G-protein β-γ subunits activate the cardiac muscarinic K^+-channel via phospholipase A_2. *Nature,* **337**, 557–560.

Kitamura, K. and Kuriyama, H. (1979). Effects of acetylcholine on the smooth muscle cells of isolated main coronary artery of the guinea-pig. *J. Physiol. Lond.*, **293**, 119–133.

Kitazawa, T., Kobayashi, S., Horiuti, K., Somlyo, A.V. and Somlyo, A.P. (1989). Receptor-coupled, permeabilized smooth muscle. Role of the phosphatidylinositol cascade, G-proteins, and modulation of the contractile response to Ca^{2+}. *J. Biol. Chem.*, **264**, 5339–5342.

Kobayashi, S., Kitazawa, T., Somlyo, A.V. and Somlyo, A.P. (1989). Cytosolic heparin inhibits muscarinic and α-adrenergic Ca^{2+} release in smooth muscle. Physiological role of inositol 1,4,5-trisphosphate in pharmacomechanical coupling. *J. Biol. Chem.*, **264**, 17997–18004.

Kobayashi, I., Shibasaki, H., Takahashi, K., Tohyama, K., Kurachi, Y., Ito, H., Ui, M. and Katada, T. (1990). Purification and characterization of five different α-subunits of guanine-nucleotide binding proteins in bovine brain membranes – their physiological properties concerning the activities of adenylate cyclase and atrial muscarinic K channels. *Eur. J. Biochem.*, **191**, 499–506.

Koketsu, K. and Nishi, S. (1967). Characteristics of the slow inhibitory postsynaptic potential of bullfrog sympathetic ganglion cells. *Life. Sci.*, **6**, 1827–1836.

Komori, K. and Suzuki, H. (1987). Heterogeneous distribution of muscarinic receptors in the rabbit saphenous artery. *Br. J. Pharmacol.*, **92**, 657–664.

Komori, K. and Suzuki, H. (1988). Modulation of smooth muscle activity by excitatory and inhibitory nerves in the guinea-pig stomach. *Comp. Biochem. Physiol. C.*, **91**, 311–319.

Komori, S. and Bolton, T.B. (1990). Role of G-proteins in muscarinic receptor inward and outward currents in rabbit jejunal smooth muscle. *J. Physiol.* **427**, 359–419.

Korth, M. and Kuhlkamp, V. (1987). Muscarinic receptors mediate negative and positive inotropic effects in mammalian ventricular myocardium: differentiation by agonists. *Br. J. Pharmacol.*, **90**, 81–90.

Korth, M., Sharma, V.K. and Sheu, S.S. (1988). Stimulation of muscarinic receptors raises free intracellular Ca concentration in rat ventricular myocytes. *Circ. Res.*, **62**, 1080–1087.

Koval, L.M., Derkach, V.A., Selyanko, A.A., Skok, V.I. and Ivanov, A.Y. (1982). Distribution of muscarinic receptors in mammalian sympathetic ganglion: autoradiographic and electrophysiological studies. *J. Auton. Nerv. Syst.*, **6**, 37–46.

Kuba, K. and Nishi, S. (1979). Characteristics of fast excitatory postsynaptic current in bullfrog sympathetic ganglion cells. Effects of membrane potential temperature and Ca ions. *Pflugers. Arch.*, **378**, 205–212.

Kubo, T., Maeda, A., Sugimoto, K., Akiba, I., Mikami, A., Takahashi, H., Haga, T., Haga, K., Ichiyama, A., Kangawa, K., Matsuo, H., Hirose, T. and Numa, S. (1986a). Primary structure of porcine cardiac muscarinic acetylcholine receptor deduced from the cDNA sequence. *FEBS Lett.*, **209**, 367–372.

Kubo, T., Fukuda, K., Mikami, A., Maeda, A., Takahashi, H., Mishina, M., Haga, T., Haga, K., Ichiyama, A., Kangawa, K., Kojima, M., Matsuo, H., Hirose, T. and Numa, S. (1986b). Cloning, sequencing and expression of complementary DNA encoding the muscarinic acetylcholine receptor. *Nature*, **323**, 411–416.

Kubo, T., Bujo, H., Akiba, I., Nakai, J., Mishina, M. and Numa, S. (1988). Location of a region of the muscarinic acetylcholine receptor involved in selective effector coupling. *FEBS Lett.*, **241**, 119–125.

Kuffler, S.W., Dennis, M.J. and Harris, A.J. (1971). The development of chemosensitivity in extra-synaptic areas of the neuronal surface after denervation of parasympathetic ganglion cells. *Proc. R. Soc. Lond. Biol.*, **177**, 555–563.

Kurachi, Y., Nakajima, T. and Sugimoto, T. (1986). On the mechanism of activation of muscarinic K^+ channels by adenosine in isolated atrial cells: involvement of GTP-binding proteins. *Pflugers. Arch.*, **407**, 264–274.

Kurachi, Y., Ito, H., Sugimoto, T., Shimuzu, T., Miki, I. and Ui, M. (1989). Arachidonic acid metabolites as intracellular modulators of the G-protein gated cardiac K^+ channel. *Nature*, **337**, 555–557.

Kuriyama, H. (1970). Effects of ions and drugs on the electrical activity of smooth muscle. In A.F. Brading, A.W. Jones and T. Tomita (Eds.), *smooth muscle* (pp. 366–395). London: Edward Arnold.

Kuriyama, H. and Suzuki, H. (1978). The effects of acetylcholine on the membrane and contractile properties of smooth muscle cells of the rabbit superior mesenteric artery. *Br. J. Pharmacol.*, **64**, 493–501.

Kurtenbach, E., Curtis, C.A.M., Pedder, E.K., Aitken, A., Harris, A.C.M. and Hulme, E.C. (1990). Muscarinic acetylcholine receptors: peptide sequencing identifies residues involved in antagonist binding and disulphide formation. *J. Biol. Chem.*, **265**, 13702–13708.

La Rochelle, W., Wray, B., Sealock, R. and Froehner, S. (1987). Immunochemical demonstration that the amino acids 360–377 of the acetylcholine receptor γ-subunit are cytoplasmic. *J. Cell Biol.*, **100**, 684–691.

Lai, J., Mei, L., Roeske, W.R., Chung, F.Z., Yamamura, H.I. and Venter, J.C. (1988). The cloned murine M1 muscarinic receptor is associated with the hydrolysis of phosphatidylinositols in transfected murine B82 cells. *Life. Sci.*, **42**, 2489–2502.

Lambrecht, G., Feifel, R., Forth, B., Stohmann, C., Tacke, R. and Mutschler, E. (1988). Para-fluoro-hexahydrosiladifenidol: the first M2 β-selective antagonist. *Eur. J. Pharmacol.*, **152**, 193–194.

Landis, S.C. (1983). Development of cholinergic sympathetic neurons: evidence for transmitter plasticity *in vivo*. *Fed. Proc.*, **42**, 1633–1638.

Langley, J.N. (1905). On the reaction of cells and of nerve endings to certain poisons, chiefly as regards the action of striated muscle to nicotine and to curare. *J. Physiol. Lond.*, **33**, 374–413.

LaPolla, R.J., Mixter Mayne, K. and Davidson, N. (1984). Isolation and characterisation of a cDNA clone for the complete protein coding region of the δ-subunit of the mouse acetylcholine receptor. *Proc. Nat. Acad. Sci. USA*, **81**, 7970–7974.

Larsen, E.H., Novak, I. and Pedersen, P.S. (1990). Cation transport by sweat ducts in primary culture. Ionic mechanism of cholinergically evoked current oscillations. *J. Physiol. Lond.*, **424**, 109–131.

Lazareno, S. and Roberts, F.F. (1989). Functional and binding studies with muscarinic M$_2$-subtype selective antagonists. *Br. J. Pharmacol.*, **98**, 309–317.

Lazareno, S., Buckley, N.J. and Roberts, F.F. (1990). Characterisation of muscarinic M4 binding sites in rabbit lung, chicken heart and NG 108-15 cells. *Mol. Pharmacol.*, **38**, 805–815.

Libet, B. and Tosaka, T. (1969). Slow inhibitory and excitatory postsynaptic responses in single cells of mammalian sympathetic ganglion. *J. Neurophsyiol.*, **32**, 43–50.

Lim, S.P. and Bolton, T.B. (1988). A calcium-dependent rather than a G-protein mechanism is involved in the inward current evoked by muscarinic receptor stimulation in dialysed single smooth muscle cells of small intestine. *Br. J. Pharmacol.*, **95**, 325–327.

Lincoln, T.M. and Keely, S.L. (1981). Regulation of cardiac cyclic GMP-dependent protein kinase. *Biochem. Biophys. Acta,* **676**, 230–244.

Lindemann, J.P. and Watanabe, A.M. (1985). Muscarinic cholinergic inhibition of β-adrenergic stimulation of phospholamban phosphorylation and Ca transport in guinea-pig ventricles. *J. Biol. Chem.*, **260**, 13122–13129.

Lindmar, R., Loffelholz, K. and Sandmann, J. (1988). On the mechanism of muscarinic hydrolysis of choline phospholipids in the heart. *Biochem. Pharmacol.*, **37**, 4689–4695.

Lipscombe, D. and Rang, H.P. (1988). Nicotinic receptors of frog sympathetic ganglia resemble pharmacologically those of skeletal muscle. *J. Neurosci.*, **8**, 3258–3265.

Loffelholz, K. and Muscholl, E. (1970). Inhibition by parasympathetic nerve stimulation of the release of the adrenergic transmitter. *Naunyn Schmiedebergs. Arch. Pharmacol.*, **267**, 181–184.

Loffelholz, K. and Pappano, A.J. (1985). The parasympathetic neuroeffector junction of the heart. *Pharmacol. Rev.*, **37**, 1–24.

Lagothetis, D.E., Kurachi, Y., Galper, J., Neer, E.J. and Clapham, D.E. (1987). The $\beta\gamma$-subunits of GTP-binding proteins activate the muscarinic K$^+$ channel in heart. *Nature*, **325**, 321–326.

Logothetis, D.E., Kim, D.H., Northup, J.K., Neer, E.J. and Clapham, D.E. (1988). Specificity of action of guanine nucleotide-binding regulatory protein subunits on the cardiac muscarinic K$^+$ channel. *Proc. Natl. Acad. Sci. U.S.A.*, **85**, 5814–5818.

Loring, R.H. and Zigmond, R.E. (1987). Ultrastructural distribution of ^{125}I-toxin F binding sites on chick ciliary neurons: synaptic localization of a toxin that blocks ganglionic nicotinic receptors. *J. Neurosci.*, **7**, 2153–2162.

Loring, R.H. and Zigmond, R.E. (1988). Characterization of neuronal nicotinic receptors by snake venom neurotoxins. *Trends. Neurosci.*, **11**, 73–78.

Loring, R.H., Sah, D.W.Y., Landis, S.C. and Zigmond, R.E. (1988). The ultrastructural distribution of putative nicotinic receptors on cultured neurons from the rat superior cervical ganglion. *Neuroscience,* **24**, 1071–1080.

Luyten, W.H.M.L. (1986). A model for the acetylcholine binding site of the nicotinic acetylcholine receptor. *J. Neurosci. Res.*, **16**, 51–73.

MacDermott, A.B., Connor, E.A., Dionne, V.E. and Parsons, R.L. (1980). Voltage-clamp study of fast excitatory synaptic currents in bullfrog sympathetic ganglion cells. *J. Gen. Physiol.*, **75**, 39–60.

Maeda, A., Kubo, T., Mishina, M. and Numa, S. (1988). Tissue distribution of mRNAs encoding muscarinic acetylcholine receptor subtypes. *FEBS Lett.*, **239**, 339–342.

Makhlouf, G.M. (1984). Regulation of gastrin and somatostatin secretion by gastric intramural neurones. *Trends Pharmacol. Sci. Suppl.*, 63–68.

Manganel, M. and Turner, R.J. (1989). Agonist-induced activation of Na$^+$/H$^+$ exchange in rat parotid acinar cells. *J. Memb. Biol.*, **111**, 191–198.

Marc, S., Leiber, D. and Harbon, S. (1988). Fluoroaluminates mimic muscarinic- and oxytocin-receptor-mediated generation of inositol phosphates and contraction in the intact guinea-pig myometrium. Role for a pertussis/cholera-toxin-insensitive G protein. *Biochem. J.*, **255**, 705–713.

Marley, P.D. (1987). New insights into the non-nicotinic regulation of adrenal medullary function. *Trends. Pharmacol. Sci.*, 411–413.

Marrion, N.V., Smart, T.G. and Brown, D.A. (1987). Membrane currents in adult rat superior cervical ganglia in dissociated tissue culture. *Neurosci. Lett.*, **77**, 55–60.

Marrion, N.V., Smart, T.G., Marsh, S.J. and Brown, D.A. (1989). Muscarinic suppression of the M-current in the rat sympathetic ganglion is mediated by receptors of the M_1-subtype. *Br. J. Pharmacol.*, **98**, 557–573.

Marshall, I.G. and Parsons, S.M. (1987). The vesicular acetylcholine transport system. *Trends. Neurosci.*, **10**, 174–177.

Marshall, L.M. (1981). Synaptic localisation of α-bungarotoxin binding which blocks nicotinic transmission at frog sympathetic neurons. *Proc. Nat. Acad. SCi. USA*, **78**, 1948–1952.

Marshall, L.M. (1985). Presynaptic control of synaptic channel kinetics in sympathetic neurones. *Nature*, **317**, 621–623.

Marty, A., Evans, M.G., Tan, Y.P. and Trautmann, A. (1986). Muscarinic response in rat lacrimal glands. *J. Exp. Biol.*, **124**, 15–32.

Marty, A. (1987). Control of ionic currents and fluid secretion by muscarinic agonists in exocrine glands. *Trends. Neurosci.*, **10**, 373–377.

Marty, A. and Tan, Y.P. (1989). The initiation of calcium release following muscarinic stimulation in rat lacrimal glands. *J. Physiol. (Lond)*, **419**, 665–687.

Maruyama, Y., Gallacher, D.V. and Petersen, O.H. (1982). Voltage- and Ca-activated K channel in basolateral membranes of mammalian salivary glands. *Nature*, **302**, 827–829.

Massarelli, R., Ciesielski-Treska, J., Ebel, A. and Mandel, P. (1974). Choline uptake in glial cell cultures. *Brain Res.*, **81**, 361–363.

Mathie, A., Cull-Candy, S.G. and Colquhoun, D. (1987). Single channel and whole-cell currents evoked by acetylcholine in dissociated sympathetic neurons of the rat. *Proc. R. Soc. Lond. Biol.*, **232**, 239–248.

Mathie, A., Cull-Candy, S.G. and Colquhoun, D. (1988). The mammalian neuronal nicotinic receptor and its block by drugs. In G.G. Lunt (Ed.), *Neurotox 88: molecular basis of drug and pesticide action* (pp. 393–403). Amsterdam: Elsevier.

Matsumoto, K. and Pappano, A.J. (1989). Sodium-dependent membrane current induced by carbachol in single guinea-pig ventricular myocytes. *J. Physiol. Lond.*, **415**, 487–502.

Mei, L., Lai, J., Roeske, W.R., Fraser, C.M., Venter, J.C. and Yamamura, H.I. (1989). Pharmacological characterisation of the M_1 muscarinic receptors expressed in murine fibroblast B82 cells. *J. Pharmacol. Exp. Ther.*, **248**, 661–670.

Melchiorre, C., Cassinelli, A. and Quaglia, W. (1987). Differential blockade of muscarinic receptor subtypes by polymethylene tetramines. Novel class of selective antagonists of cardiac M_2 muscarinic receptors. *J. Med. Chem.*, **30**, 201–204.

Melchiorre, C. (1988). Polymethylene tetraamines: a new generation of selective muscarinic antagonists. *Trends. Pharmacol. Sci.*, **9**, 216–220.

Melvin, J.E., Moran, A. and Turner, R.J. (1988). The role of HCO_3^- and Na^+/H^+ exchange in the response of rat parotid acinar cells to muscarinic stimulation. *J. Biol. Chem.*, **263**, 19564–19569.

Messing, A. and Gonates, N.K. (1983). Extrasynaptic localisation of α-bungarotoxin receptors in cultured chick ciliary ganglion neurons. *Brain Res.*, **269**, 172–176.

Meurs, H., Roffel, A.F., Postema, J.B., Timmermans, A., Elzinga, C.R., Kauffman, H.F. and Zaagsma, J. (1988). Evidence for a direct relationship between phosphoinositide metabolism and airway smooth muscle contraction induced by muscarinic agonists. *Eur. J. Pharmacol.*, **156**, 271–274.

Michel, A.D. and Whiting, R.L. (1988a). Methoctramine, a polymethylene tetraamine, differentiates three subtypes of muscarinic receptor in direct binding studies. *Eur. J. Pharmacol.*, **145**, 61–66.

Michel, A.D. and Whiting, R.L. (1988b). Methoctramine reveals heterogeneity of M_2 muscarinic receptors in longitudinal ileal smooth muscle membranes. *Eur. J. Pharmacol.*, **145**, 305–311.

Micheletti, R., Schiavone, A. and Giachetti, A. (1988). Muscarinic M_1 receptors stimulate a nonadrenergic noncholinergic inhibitory pathway in the isolated rat duodenum. *J. Pharmacol. Exp. Ther.*, **244**, 680–684.

Micheletti, R., Giudici, L., Turconi, M. and Donetti, A. (1990). 4-DAMP analogues reveal heterogeneity of M_1 muscarinic receptors. *Br. J. Pharmacol.*, **100**, 395–397.

Mihara, S. and Nishi, S. (1989). Muscarinic excitation and inhibition of neurons in the submucous plexus of the guinea-pig caecum. *Neuroscience., 31*, 247–257.

Miller, V.M. and Vanhoutte, P.M. (1989). Is nitric oxide the only endothelium-derived relaxing factor in canine femoral veins. *Am. J. Physiol.,* H1910–H1916.

Misbahuddin, M. and Oka, M. (1988). Muscarinic stimulation of guinea-pig adrenal chromaffin cells stimulates catecholamine secretion without significant increase in Ca^{2+} uptake. *Neurosci. Lett., 87*, 266–270.

Mishina, M., Tobimatsu, T., Imoto, K., Tanaka, K., Fujita, Y., Fukuda, K., Kurasaki, M., Takahashi, H., Morimoto, Y., Hirose, T., Inayama, S., Takahashi, T., Kuno, M. and Numa, S. (1985). Location of functional regions of acetylcholine receptor α-subunit by site directed mutagenesis. *Nature, 313*, 364–369.

Mizobe, F. and Livett, B.G. (1983). Nicotine stimulates secretion of both catecholamines and acetylcholinesterase from cultured adrenal chromaffin cells. *J. Neurosci., 3*, 871–876.

Mochida, S. and Kobayashi, H. (1986). Activation of M_2 muscarinic receptors causes an alteration of action potentials by modulation of Ca entry in isolated sympathetic neurons of rabbits. *Neurosci. Lett., 72*, 199–204.

Mochida, S. and Kobayashi, H. (1988a). GTP-binding proteins mediate the M_2-muscarinic effect on the action potential in isolated sympathetic neurons of rabbits. *Neurosci. Lett., 93*, 247–252.

Mochida, S. and Kobayashi, H. (1988b). A novel muscarinic receptor antagonist AF-DX116 differentially blocks slow inhibitory and slow excitatory postsynaptic potentials in the rabbit sympathetic ganglia. *Life. Sci., 42*, 2195–2201.

Mochida, S. and Kobayashi, H. (1988c). Protein kinase C activators mimic the M_2-muscarinic receptor-mediated effects on the action potential in isolated sympathetic neurons of rabbits. *Neurosci. Lett., 86*, 201–206.

Mochida, S., Mizobe, F., Fisher, A., Kawanishi, G. and Kobayashi, H. (1988). Dual synaptic effects of activating M_1-muscarinic receptors, in superior cervical ganglia of rabbits. *Brain Res., 455*, 9–17.

Monferini, E., Giraldo, E. and Ladinsky, H. (1988). Characterization of the muscarinic receptor subtypes in the rat urinary bladder. *Eur. J. Pharmacol., 147*, 453–458.

Moss, B.L., Schuetze, S.M. and Role, L.W. (1989). Functional properties and developmental regulation of nicotinic acetylcholine receptors on embryonic chicken sympathetic neurons. *Neuron, 3*, 597–607.

Mubagwa, K. and Carmeliet, E. (1983). Effects of acetylcholine on electrophysiological properties of rabbit cardiac Purkinje fibers. *Circ. Res., 53*, 740–751.

Mutschler, E. and Lambrecht, G. (1984). Selective muscarinic agonists and antagonists in functional tests. *Trends Pharmacol. Sci. Suppl.,* 39–44.

Nakazato, Y., Oleshansky, M.A. and Chiang, P.K. (1988a). Effects of muscarinic pharmacophores on the cholinergic regulation of catecholamine secretion from perfused adrenal glands. *Arch. Int. Pharmacodyn. Ther., 293*, 209–218.

Nakazato, Y., Ohga, A., Oleshansky, M., Tomita, U. and Yamada, Y. (1988b). Voltage-independent catecholamine release mediated by the activation of muscarinic receptors in guinea-pig adrenal glands. *Br. J. Pharmacol., 93*, 101–109.

Nargeot, J., Lester, H.A., Birdsall, N.J.M., Stockton, J., Wassermann, N.H. and Erlanger, B.F. (1982). A photoisomerizable muscarinic antagonist. Studies of binding and of conductance relaxations of frog heart. *J. Gen. Physiol., 79*, 657–678.

Nargeot, J., Nerbonne, J.M., Engels, J. and Lester, H.A. (1983). Timecourse of the increase in the myocardial slow inward current after a photochemically generated concentration jump of intracellular cAMP. *Proc. Natl. Acad. Sci. U.S.A., 80*, 2395–2399.

Nathans, J. and Hogness, D.S. (1983). Isolation, sequence analysis and intron-exon arrangement of the gene encoding bovine rhodopsin. *Cell, 34*, 807–814.

Nathans, J., Thomas, D. and Hogness, D.S. (1986). Molecular genetics of human color vision: the genes encoding blue, green and red pigments. *Science, 232*, 193–202.

Nathanson, N.M. (1987). Molecular properties of the muscarinic acetylcholine receptor. *Annu. Rev. Neurosci., 10*, 195–236.

Neer, E.J. and Clapham, D.E. (1988). Role of G protein subunits in transmembrane signalling. *Nature, 333*, 129–134.

Neild, T.O., Shen, K.Z. and Surprenant, A. (1990). Vasodilatation of arterioles by acetylcholine released from single neurones in the guinea-pig submucosal plexus. *J. Physiol. Lond., 420*, 247–265.

Nishi, S. (1974). Ganglionic transmission. In J.I. Hubbard (Ed.), *The peripheral nervous system* (pp. 225–276). New York and London: Plenum Press.

Nishi, S. and Koketsu, K. (1960). Electrical properties and activities of single sympathetic neurons in frogs. *J. Cell. Comp. Physiol., 55*, 15–30.

Nishi, R. and Willard, A.C. (1988). Conditioned medium alters electrophysiological and transmitter-related properties expressed by rat enteric neurons in cell culture. *Neuroscience, 25*, 759-769.

Nishi, S., Soeda, H. and Koketsu, K. (1967). Release of acetylcholine from sympathetic preganglionic nerve terminals. *J. Neurophysiol., 30*, 114-118.

Nishiyama, A. and Petersen, O.H. (1975). Pancreatic acinar cells: ionic dependence of acetylcholine-induced membrane potential and resistance change. *J. Physiol. Lond., 244*, 431-465.

Noda, M., Takahashi, H., Tanabe, T., Toyosato, M., Furutani, Y., Hirosa, T., Asai, M., Inayama, S., Miyata, T. and Numa, S. (1982). Primary structure of the α-subunit precursor of the Torpedo californica acetylcholine receptor deduced from cDNA sequence. *Nature, 299*, 793-797.

Noda, M., Takahashi, H., Tanabe, T., Toyosato, M., Kikyotani, S., Hirosa, T., Asai, M., Takashima, H., Inayama, S., Miyata, T. and Numa, S. (1983a). Primary structures of β and delta subunit precursors of Torpedo californica acetylcholine receptor cDNA sequences. *Nature, 301*, 251-255.

Noda, M., Takahashi, H., Tanabe, T., Toyosatao, M., Kikyotani, S., Furitani, Y., Hirose, T., Takashima, H., Inayama, S., Miyata, T. and Numa, S. (1983b). Structural homology of *Torpedo californica acetylcholine receptor subunits. Nature, 302*, 528-532.

Noda, M., Furutani, Y., Takahashi, H., Toyosato, M., Tanabe, T., Shimizu, S., Kikyotani, S., Kayano, T., Hirose, T., Inayama, S. and Numa, S. (1983c) Cloning and sequence analysis of calf cDNA and human genomic DNA encoding a-subunit precursor of molecular acetylcholine receptor. *Nature, 305*, 818-823.

Noma, M. and Trautwein, W. (1978). Relaxation of the ACh-induced potassium current in the rabbit sinoatrial node. *Pflugers. Arch., 377*, 193-200.

North, R.A. (1982). Electrophysiology of the enteric nervous system. *Neuroscience, 7*, 315-325.

North, R.A. (1989). Muscarinic cholinergic receptor regulation of ion channels. In J.H. Brown (Ed.), *Muscarinic receptor subtypes* (pp. 341-373). Clifton, N.J.: Human Press.

North, R.A. and Tokimasa, T. (1982). Muscarinic synaptic potentials in guinea-pig myenteric neurones. *J. Physiol. Lond., 333*, 151-156.

North, R.A., Slack, B.E. and Surprenant, A. (1985). Muscarinic M_1 and M_2 receptors mediate depolarization and presynaptic inhibition in guinea-pig enteric nervous system. *J. Physiol. Lond., 368*, 435-452.

O'Regan, S., Collier, B. and Israel, M. (1982). Studies on presynaptic cholinergic mechanisms using analogues of choline and acetate. *J. Physiol. (Paris), 78*, 454-460.

O'Rourke, S.T. and Vanhoutte, P.M. (1987). Subtypes of muscarinic receptors on adrenergic nerves and vascular smooth muscle of the canine saphenous vein. *J. Pharmacol. Exp. Ther., 241*, 64-67.

Obata, K. (1974). Transmitter sensitivities of some nerve and muscel cells in culture. *Brain Res., 73*, 71-88.

Ogden, D.C., Gray, P.T.A., Colquhoun, D. and Rang, H.P. (1984). Kinetics of acetylcholine activated ion channels in chick ciliary ganglion neurones grown in tissue culture. *Pflugers. Arch., 400*, 44-50.

Okabe, K., Yatani, A., Evans, T., Ho, Y.K., Codina, J., Birnbaumer, L. and Brown, A.M. (1990). $\beta\gamma$-dimers of G proteins inhibit atrial muscarinic K^+ channel. *J. Biol. Chem., 265*, 12854-12858.

Okazaki, M., Yanagihara, N., Izumi, F., Nakashima, Y. and Kuroiwa, A. (1989). Carbachol-induced cosecretion of immunoreactive atrial natriuretic peptides with catecholamines from cultured bovine adrenal medullary cells. *J. Neurochem., 52*, 222-228.

Olesen, S.P., Davies, P.F. and Clapham, D.E. (1988). Muscarinic-activated K^+ current in bovine aortic endothelial cells. *Circ. Res., 62*, 1059-1064.

Onai, T., FitzGerald, M.G., Arakawa, S., Gocayne, J.D., Urquhart, D.A., Hall, L.M., Fraser, C.M., McCombie, W.R. and Venter, J.C. (1989). Cloning, sequence analysis and chromosome localization of a Drosophila muscarinic acetylcholine receptor. *FEBS Lett., 255*, 219-225.

Palmer, R.M.J., Ferrige, A.G. and Moncada, S. (1988a). Nitric oxide release accounts for the biological activity of endothelium-derived relaxing factor. *Nature, 327*, 524-526.

Palmer, R.M.J., Ashton, D.S. and Moncada, S. (1988b). Vascular endothelial cells synthesize nitric oxide from arginine. *Nature, 333*, 664-666.

Papke, R.L., Boulter, J., Patrick, J. and Heinemann, S. (1989). Single channel currents of rat neuronal nicotinic acetylcholine receptors expressed in Xenopus oocytes. *Neuron, 3*, 589-596.

Pappano, A.J., Matsumoto, K., Tajima, T., Agnarsson, U. and Webb, W. (1988). Pertussis toxin-insensitive mechanism for carbachol-induced depolarization and positive inotropic effect in heart muscle. *Trends Pharmacol. Sci. Suppl.*, 35-39.

Pappano, A.J., Hartigan, P.M. and Coutu, M.D. (1982). Acetylcholine inhibits the positive inotropic effect of cholera toxin in ventricular muscle. *Am. J. Physiol., 243*, H434-H441.

Paton, W.D.M. and Zaimis, E.J. (1949). The pharmacological action of polymethylene bis-trimethylammonium salts. *Br. J. Pharmacol., 4*, 381-400.

Patrick, J. and Stallcup, W. (1977). α-bungarotoxin binding and cholinergic receptor function on a rat sympathetic nerve line. *J. Biol. Chem.*, **252**, 8629-8634.

Patterson, P.H. (1978). Environmental determination of autonomic neurotransmitter functions. *Annu. Rev. Neurosci.*, **1**, 1-17.

Patterson, P.H. and Chun, L.L.Y. (1974). The influence of non-neuronal cells on catecholamine synthesis and accumulation in cultures of dissociated sympathetic neurons. *Proc. Natl. Acad. Sci. U.S.A.*, **71**, 3607-3610.

Pearson, G.T. and Petersen, O.H. (1984). Nervous control of membrane conductance in mouse lacrimal acinar cells. *Pflugers. Arch.*, **400**, 51-59.

Pearson, G.T., Flanagan, P.M. and Petersen, O.H. (1984). Neural and hormonal control of membrane conductance in the pig pancreatic acinar cell. *Am. J. Physiol.*, **247**, G520-G526.

Pelc, L.R., Daemmgen, J.W., Gross, G.J. and Warltier, D.C. (1988). Muscarinic receptor subtypes mediating myocardial blood flow redistribution. *J. Cardiovasc. Pharmacol.*, **11**, 424-431.

Pennefather, P., Lancaster, B., Adams, P.R. and Nicoll, R.A. (1985). Two distinct Ca-dependent K currents in bullfrog sympathetic ganglion cells. *Proc. Natl. Acad. Sci. U.S.A.*, **82**, 3040-3044.

Peralta, E.G., Winslow, J.W., Peterson, G.L., Smith, D.H., Ashkenazi, A., Ramachandran, J., Schimerlik, M.I. and Capon, D.J. (1987a). Primary structure and biochemical properties of an M2 muscarinic receptor. *Science*, **236**, 600-605.

Peralta, E.G., Ashkenazi, A., Winslow, J.W., Smith, D.H., Ramachandran, J. and Capon, D.J. (1987b). Distinct primary structures, ligand-binding properties and tissue-specific expression of four human muscarinic acetylcholine receptors. *EMBO J.*, **6**, 3923-3929.

Peralta, E.G., Winslow, J.W., Ashkenazi, A., Smith, D.H., Ramachandran, J. and Capon, D.J. (1988a). Structural basis of muscarinic acetylcholine receptor subtype diversity. *Trends. Pharmacol. Sci.*, **Suppl**, 6-11.

Peralta, E.G., Ashkenazi, A., Winslow, J.W., Ramachandran, J. and Capon, D.J. (1988b). Differential regulation of PI hydrolysis and adenylyl cyclase by muscarinic receptor subtypes. *Nature*, **334**, 434-437.

Petersen, O.H. (1970). Some factors influencing stimulation-induced release of potassium from cat submandibular gland to fluid perfused through the gland. *J. Physiol.*, **208**, 431-447.

Petersen, O.H. and Findlay, I (1987). Electrophysiology of the pancreas. *Physiol. Rev.*, **67**, 1054-1116.

Petersen, O.H. and Maruyama, Y. (1984). Calcium-activated potassium channels and their role in secretion. *Nature*, **307**, 693-696.

Petersen, O.H. and Singh, J. (1985). Acetylcholine-evoked potassium release in the mouse pancreas. *J. Physiol.*, **365**, 319-329.

Pfaffinger, P. (1988). Muscarine and t-LHRH suppress M-current by activating an IAP-insensitive G-protein. *J. Neurosci.*, **8**, 3343-3353.

Pfaffinger, P.J., Martin, J.M., Hunter, D.D., Nathanson, N.M. and Hille, B. (1985). GTP-binding proteins couple cardiac muscarinic receptors to a K channel. *Nature*, **317**, 536-538.

Popot, J.L. and Changeux, J.P. (1984). The nicotinic receptor of acetylcholine: structure of an oligomeric integral membral protein. *Physiol. Rev.*, **64**, 1162-1239.

Potter, D.D., Furshpan, E.J. and Landis, S.C. (1983). Transmitter status in cultured sympathetic neurones: plasticity and multiple function. *Fed. Proc.*, **42**, 1626-1632.

Purves, R.D. (1974). Muscarinic excitation: a micro-electrophoretic study in cultured smooth cells. *Br. J. Pharmacol.*, **52**, 77-86.

Purves, D. (1975). Functional and structural changes in mammalian sympathetic neurones following interruption of their axons. *J. Physiol.*, **252**, 429-463.

Quik, M. and Lamarca, M.V. (1982). Blockade of transmission in rat sympathetic ganglia by a toxin which co-purifies with α-bungarotoxin. *Brain Res.*, **238**, 385-399.

Quist, E.E. (1982). Evidence for a carbachol-stimulated phosphatidylinositol effect in heart. *Biochem. Pharmac.*, **31**, 3130-3133.

Rang, H.P. (1981). The characteristics of synaptic currents and responses to acetylcholine rat submandibular ganglion cells. *J. Physiol.*, **311**, 23-55.

Rapoport, R.M., Draznin, M.B. and Murad, F. (1983). Endothelium-dependent relaxation in rat aorta may be mediated through cyclic GMP-dependent protein phosphorylation. *Nature*, **306**, 174-176.

Rardon, D.F. and Pappano, A.J. (1986). Carbachol inhibits electrophysiological effects of cyclic AMP in ventricular myocytes. *Am. J. Physiol.*, **251**, H601-H611.

Ratnam, M., Sargent, P.B., Sarin V., Fox, J.L., LeNguyen D., Riviere, J., Criado, M. and Lindstrom, J. (1986). Location of antigenic determinants on primary sequences of subunits of nicotinic acetylcholine receptor by peptide mapping. *Biochemistry*, **25**, 2621-2632.

Ravdin, P.M. and Berg, D.K. (1979). Inhibition of neuronal acetylcholine sensitivity by α-toxins from *Bungarus multicinctus* venom. *Proc. Nat. Acad. Sci. USA*, **76**, 2072–2076.

Rees, D.D., Palmer, R.M., Hodson, H.F. and Moncada, S. (1989). A specific inhibitor of nitric oxide formation from L-arginine attenuates endotelium dependent relaxation. *Br. J. Pharmacol.*, **96**, 418–424.

Richardson, G.P. (1988). Development of the cholinergic synapse: role of trophic factors. In V.P. Whittaker (Ed.), *Handbook of experimental pharmacology, Vol. 86* (pp. 81–105). Berlin and Heidelberg: Springer Verlag.

Riker, W.F. and Wescoe, W.C. (1951). The pharmacology of Flaxedil, with observations on certain analogs. *Ann. N.Y. Acad. Sci.*, **54**, 373–394.

Role, L.W. (1988). Neural regulation of acetylcholine sensitivity in embryonic sympathetic neurons. *Proc. Nat. Acad. Sci. USA*, **85**, 2825–2829.

Sacchi, O. and Perri, V. (1971). Quantal release of acetylcholine from the nerve endings of the guinea-pig superior cervical ganglion. *Pflugers Arch.*, **329**, 207–219.

Sacchi, O. and Perri, V. (1976). Some properties of the transmitter release mechanisms at the rat ganglionic synapse during potassium stimulation. *Brain Res.*, **107**, 275–289.

Sacchi, O., Consolo, S., Perri, G., Prigoni, I., Ladinsky, H. and Perri, V. (1978). Storage and release of acetylcholine in the isolated superior cervical ganglion of the rat. *Brain Res.*, **151**, 443–456.

Sah, D.W.Y., Loring R.H. and Zigmond, R.E. (1987). Long term blockade by toxin F of nicotinic synaptic potentials in cultured sympathetic neurons. *Neuroscience*, **20**, 867–874.

Sakmann, B., Noma, A. and Trautwein, W. (1983). Acetylcholine activation of single muscarinic K^+ channels in isolated pacemaker cells of the mammalian heart. *Nature*, **303**, 250–253.

Sargent, P.B. and Pang, D.Z. (1989). Acetylcholine receptor-like molecules are found in both synaptic and extrasynaptic clusters on the surface of neurons in the frog cardiac ganglion. *J. Neurosci.*, **9**, 1062–1072.

Sasakawa, N., Ishii, K. and Kato, R. (1985). Calcium-independent desensitization of tises in intracellular free Ca^{2+} concentration and catecholamine release in cultured adrenal chromaffin cells. *Biochem. Biophys. Res. Comm.*, **133**, 147–153.

Sasaki, Y., Shida, Y. and Kanno, Y. (1988). Suppression of intercellular communication in acinar cells from rat submandibular gland by cholinergic adrenergic agonists. *Jap. J. Physiol.*, **38**, 531–543.

Schmied, R. and Korth, M. (1990). Muscarinic receptor stimulation and cyclic AMP-dependent effects in guinea-pig ventricular myocardium. *Br. J. Pharmacol.*, **99**, 401–407.

Schofield, G.G., Weight, F.F. and Adler, M. (1985). Single acetylcholine channel currents in sympathetic neurons. *Brain Res.*, **342**, 200–203.

Schotzinger, R.J. and Landis, S.C. (1988). Cholinergic phenotype developed by noradrenergic sympathetic neurons after innervation of a novel cholinergic target *in vivo*. *Nature*, **335**, 637–639.

Schuetze, S.M. and Role, L.W. (1987). Developmental regulation of nicotinic acetylcholine receptors. *Ann. Rev. Neurosci.*, **10**, 403–457.

Segal, S.S. and Duling, B.R. (1986). Flow control among microvessels co-ordinated by intercellular communication. *Science*, **234**, 868–870.

Selyanko, A.A., Derkach, V.A. and Skok, V.I. (1979). Fast excitatory postsynaptic currents in voltage clamped mammalian sympathetic neurons. *J. Auton. Nerv. Syst.*, **1**, 127–137.

Selyanko, A.A., Zidichouski, J.A. and Smith, P.A. (1990). A muscarine-sensitive slow transient outward current in frog autonomic neurones. *Brain Res.*, **524**, 236–243.

Shapiro, R.A. and Nathanson, N.M. (1989). Deletion analysis of the mouse m1 muscarinic acetylcholine receptor: effects on phosphoinositide metabolism and down-regulation. *Biochemistry*, **28**, 8946–8950.

Shapiro, R.A., Scherer, N.M., Habecker, B.A., Subers, E.M. and Nathanson, N.M. (1988). Isolation, sequence, and functional expression of the mouse M1 muscarinic acetylcholine receptor gene. *J. Biol. Chem.*, **263**, 18397–18403.

Simmons, M.A. and Hartzell, H.C. (1988). A quantitative analysis of the acetylcholine-activated potassium current in single cells from frog atrium. *Pflugers Arch.*, **409**, 454–461.

Sims, S.M., Singer, J.J. and Walsh, J.V. (1985). Cholinergic agonists suppress a potassium current in freshly dissociated smooth muscle cells of the toad. *J. Physiol.*, **367**, 503–529.

Sinicropi, D.V., Kauffman, F.C. and Burt, D.R. (1979). Axotomy in rat sympathetic ganglia: reciprocal effects on muscarinic receptor binding and 6-phosphogluconate dehydrogenase activity. *Brain Res.*, **161**, 560–565.

Skok, V.I., Selyanko, A.A. and Derkach, V.A. (1989). *Neuronal acetylcholine receptors*. New York and London: Consultants Bureau.

Soejima, M. and Noma, A. (1984). Mode of regulation of the ACh-sensitive K-channel by the muscarinic receptor in rabbit atrial cells. *Pflugers. Arch.*, **400**, 424–431.

Soll, A.H. (1984). Fundic mucosal muscarinic receptors modulating acid secretion. *Trends. Pharmacol. Sci. Suppl.*, 60–62.

Song, S-Y., Saito, S., Noguchi, K. and Konishi, S. (1989). Different GTP-binding proteins mediate regulation of calcium channels by acetylcholine and noradrenaline in rat sympathetic neurons. *Brain Res.*, **494**, 383–386.

Sorenson, E.M., Culver, P. and Chiappinelli, V.A. (1987). Lophotoxin: selective blockade of nicotinic transmission in autonomic ganglia by a coral neurotoxin. *Neuroscience*, **20**, 875–884.

Sorota, S. and Hoffman, B.F., (1989). Role of G-proteins in the acetylcholine-induced potassium current of canine atrial cells. *Am. J. Physiol.*, **257**, H1516–1522.

Sorota, S., Tsuji, Y., Tajima, T. and Pappano, A.J. (1985). Pertussis toxin treatment blocks hyper-polarization by muscarinic agonists in chick atrium. *Circ. Res.*, **57**, 748–758.

Starke, K., Gothert, M. and Kilbinger, H. (1989). Modulation of neurotransmitter release by presynaptic autoreceptors. *Physiol. Rev.*, **69**, 865–989.

Stein, R., Pinkas-Kramarski, R and Sokolovsky, M. (1988). Cloned m1 muscarinic receptors mediate both adenylate cyclase inhibition and phosphoinositide turnover. *EMBO J.*, **7**, 3031–3035.

Surprenant, A. (1986). Muscarinic receptors in the submucous plexus and their roles in mucosal ion transport. *Trends. Pharmacol. Sci. Suppl.*, 23–27.

Szabo, G. and Otero, A.S. (1990). G Protein mediated regulation of K^+ channels in heart. *Ann. Rev. Physiol.*, **52**, 293–305.

Szerb, J.C. (1976). Storage and release of labelled acetylcholine in the myenteric plexus of the guinea-pig ileum. *Can. J. Physiol. Pharmacol.*, **54**, 12–22.

Tajima, T., Tsuji, Y., Brown J.H. and Pappano, A.J. (1987). Pertussis toxin-insensitive phosphoinositide hydrolysis, membrane depolarization, and positive inotropic effect of carbachol in chick atria. *Circ. Res.*, **61**, 436–445.

Takai, T., Noda, M., Furutani, Y., Takahashi, H., Notaki, M., Shimitsu, S., Kayano, T., Tanabe, T., Tanaka, K., Hirosa, T., Inayama, S. and Numa, S. (1984). Primary structure of γ subunit precursor of calf muscle acetylcholine receptor deduced from the cDNA sequence. *Eur. J. Biochem.*, **143**, 109–115.

Takemura, H. and Putney, J.W., Jr. (1989). Capacitative calcium entry in parotid acinar cells. *Biochem. J.*, **258**, 409–412.

Tanabe, T., Noda, M., Furutani, Y., Takai, T., Takahashi, H., Tanaka, K., Hirosa, T., Inayama, S. and Numa, S. (1984). Primary structure of β-subunit precursor of calf muscle acetylcholine receptor deduced from cDNA sequence. *Eur. J. Biochem.*, **144**, 11–17.

Taniguchi, T., Kurahashi, K. and Fujiwara, M. (1983). Alterations in muscarinic cholinergic receptors after preganglionic denervation of the superior cervical ganglion in cats. *J. Pharmacol. Exp. Ther.*, **224**, 674–678.

Taylor, S.G., Southerton, J.S., Weston, A.H. and Baker, J.R.J. (1988). Endothelium-dependent effects of acetylcholine in rat aorta: a comparison with sodium nitroprusside and cromakalim. *Br. J. Pharmacol.*, **94**, 853–863.

TerBush, D.R., Bittner, M.A. and Holz, R.W. (1988). Ca influx causes rapid translocation of protein kinase C to membranes. Studies of the effects of secretagogues in adrenal chromaffin cells. *J. Biol. Chem.*, **263**, 18873–18879.

Tietje, K.M., Goldman, P.S. and Nathanson, N.M. (1990). Cloning and functional analysis of a gene encoding a novel muscarinic receptor expressed in chick heart and brain. *J. Biol. Chem.*, **265**, 2828–2834.

Tokimasa, T. and North, R.A. (1984). Calcium entry through acetylcholine-channels can activate potassium conductance in bullfrog sympathetic neurons. *Brain Res.*, **295**, 364–367.

Trautmann, A. and Marty, A. (1984). Activation of Ca-dependent K channels by carbamylcholine in rat lacrimal glands. *Proc. Nat. Acad. Sci. USA*, **81**, 611–615.

Tse, A., Clark, R.B. and Giles, W.R. (1990). Muscarinic modulation of calcium current in neurones from the interatrial septum of bull-frog heart. *J. Physiol.*, **427**, 127–149.

Tsuji, S. and Kuba, K. (1988). Muscarinic regulation of two ionic currents in the bullfrog sympathetic neurone. *Pflugers. Arch.*, **411**, 361–370.

Tsuji, S., Minota, S. and Kuba, K. (1987). Regulation of two ion channels by a common muscarinic receptor-transduction system in a vertebrate neuron *Neurosci. Lett.*, **81**, 139–145.

Tucek, S. (1988). Choline acetyltransferase and the synthesis of acetylcholine. In V.P. Whittaker (Ed.), *Handbook of experimental pharmacology, vol. 86* (pp. 125–165). Berlin and Heidelberg: Springer Verlag.

Vallar, L., Muca, C., Magni, M., Albert, P., Bunzo J., Meldolesi, J. and Civelli, O. (1990). Differential coupling of dopaminergic D2 receptors expressed in different cell types. *J. Biol. Chem.*, **265**, 10320-10326.

van-Koppen, C.J., Blankesteijn, W.M., Klaassen, A.B., Rodrigues-de-Miranda, J.F., Beld, A.J. and van-Ginneken, C.A. (1988). Autoradiographic visualization of muscarinic receptors in human bronchi. *J. Pharmacol. Exp. Ther.*, **244**, 760-764.

Vanhoutte, P.M. (1977). Cholinergic inhibition of adrenergic transmission. *Fed. Proc.*, **36**, 2444-2449.

Vanhoutte, P.M., Rubanyi, G.M., Miller, V.M. and Houston, D.S. (1986). Modulation of vascular smooth muscle contraction by the endothelium. *Ann. Rev. Physiol.*, **48**, 307-320.

Verdiere, M., Derer, M. and Poullet, M. (1984). Decrease of tailed assymetric 16S acetylcholinesterase in rat superior cervical ganglion neurons *in vitro* after potassium depolarization: partial antagonist action of a calcium channel blocker. *Neurosci. Lett.*, **52**, 135-140.

Voyvodic, J.T. (1987). Development and regulation of dendrites in the rat superior cerical ganglion. *J. Neurosci.*, **7**, 904-912.

Wada, K., Ballivet, M., Boulter, J., Connolly, J., Wada, E., Deneris, E., Swanson, L.W. and Heinemann, S. (1988a). Primary structure and expression of $\beta2$, a novel subunit of neuronal nicotinic acetylcholine receptors. *Science*, **240**, 330-334.

Wada, K., Ballivet, M., Boulter, J., Connolly, J., Wada, E., Deneris, E., Swanson, L. and Heinemann, S. (1988b). Functional expression of a new pharmacological subtype of brain nicotinic acetylcholine receptor. *Science*, **240**, 330-334.

Waelbroeck, M., Camus, J., Winand, J. and Christophe, J. (1987). Different antagonist binding properties of rat pancreatic and cardiac muscarinic receptors. *Life. Sci.*, **41**, 2235-2240.

Wamsley, T.K., Zarbin, M.A. and Kuhar, M.J. (1981). Muscarinic cholinergic receptors flow in the sciatic nerve. *Brain Res.*, **217**, 155-161.

Wanke, E., Ferroni, A., Malgaroli, A., Ambrosini, A., Pozzan, T. and Meldolesi, J. (1987). Activation of a muscarinic receptor selectively inhibits a rapidly inactivated Ca^{2+} current in rat sympathetic neurons. *Proc. Natl. Acad. Sci. U.S.A.*, **84**, 4313-4317.

Watson, M., Roeske, W.R., Johnson, P.C. and Yamamura, H.I. (1984). [^{3}H]-Pirenzepine identifies putative M$_1$ muscarinic receptors in human stellate ganglia. *Brain Res.*, **290**, 179-182.

Weight, F.F. and Votava, J. (1970). Slow synaptic excitation in sympathetic ganglion cells: evidence for synaptic inactivation of potassium conductance. *Science*, **170**, 755-757.

Wess, J., Brann, M.R. and Bonner, T.I. (1989). Identification of a small intracellular region of the muscarinic m3 receptor as a determinant of selective coupling to PI turnover. *FEBS Lett.*, **258**, 133-136.

Wess, J., Bonner, T.I., Dorje, F. and Brann, M.R. (1990). Delineation of muscarinic receptor domains conferring selectivity of coupling to guanine nucleotide binding proteins and second messengers. *Mol. Pharmacol.*, **38**, 517-523.

Wetzel, G.T. and Brown, J.H. (1985). Presynaptic modulation of acetylcholine release from cardiac parasympathetic neurons. *Am. J. Physiol.*, **248**, H33-H39.

Wilson, W.S., Schulz, R.A. and Cooper, J.T. (1973). The isolation of cholinergic synaptic vesicles from bovine superior cervical ganglion and estimation of their acetylcholine content. *J. Neurochem.*, **20**, 659-667.

Wong, V. and Kessler, J.A. (1987). Solubilization of a membrane factor that stimulates levels of Substance P and choline acetyltransferase in sympathetic neurons. *Proc. Nat. Acad. Sci. USA*, **84**, 8726-8729.

Yamada, Y., Teraoka, H., Nakazato, Y. and Ohga, A. (1988). Intracellular Ca antagonist TMB-8 blocks catecholamine secretion evoked by caffeine and acetylcholine from perfused cat adrenal glands in the absence of extracellular Ca^{2+}. *Neurosci. Lett.*, **90**, 338-342.

Yamamori, T., Fukada, K., Aebersold, R., Korsching, S., Fann, M.I. and Patterson, P.H. (1989). The cholinergic neuronal differentiation factor from heart cells is identical to leukemia inhibitory factor. *Science*, **246**, 1412-1416.

Yamamura, H.I., Watson, M., Wamsley, J.K., Johnson, P.C. and Roeske, W.R. (1984). Light microscopic autoradiographic localization of [^{3}H]pirenzepine and [^{3}H](−)quinuclidinyl benzilate binding in human stellate ganglia. *Life. Sci.*, **35**, 753-757.

Yarosh, C.A., Acosta, C.G. and Ashe, J.H. (1988a). Modification of nicotinic ganglionic transmission by muscarinic slow postsynaptic potentials in the *in vitro* rabbit superior cervical ganglion. *Synapse.*, **2**, 174-182.

Yarosh, C.A., Olito, A.C. and Ashe, J.H. (1988b). AF-DX116: a selective antagonist of the slow inhibitory postsynaptic potential and methacholine-induced hyperpolarization in superior cervical ganglion of the rabbit. *J. Pharmacol. Exp. Ther.*, **245**, 419-425.

Yatani, A., Codina, J., Imoto, Y., Reeves, J.P., Birnbaumer, L. and Brown, A.M. (1987a). A G protein directly regulates mammalian cardiac calcium channels. *Science*, **238**, 1298–1292.

Yatani, A., Codina, J., Brown, A.M. and Birnbaumer, L. (1987b). Direct activation of mammalian atrial muscarinic potassium channels by GTP regulatory protein, G. *Science*, **235**, 207–211.

Yatani, A., Hamm, H., Codina, J., Mazzoni, M.R., Birnbaumer, L. and Brown, A.M. (1988a). A monoclonal antibody to the α-subunit of G_K blocks muscarinic activation of atrial K_+ channels. *Science*, **241**, 828–831.

Yatani, A., Mattera, R., Codina, J., Graf, R., Okabe, K., Padrell, E., Iyengar, R., Brown, A.M. and Birnbaumer, L. (1988b). The G-protein gated atrial K channel is stimulated by three distinct G_i-α subunits. *Nature*, **336**, 680–682.

Yatani, A., Okabe, K., Birnbaumer, L. and Brown, A.M. (1990a). Detergents, dimeric $G_{\beta\gamma}$ and eicosanoid pathways to muscarinic atrial K^+ channels. *Am. J. Physiol.*, **258**, H1507–H1514.

Yatani, A., Okabe, K., Polakis, P., Halenbeck, R., McCormick, F. and Brown, A.M. (1990b). ras p21 and GAP inhibit coupling of muscarinic receptors to atrial channels. *Cell*, **61**, 769–776.

Yau, W.M., Dorsett, J.A. and Youther M.L. (1987). Stimulation of acetylcholine release from myenteric neurons of guinea-pig small intestine by forskolin and cyclic AMP. *J. Pharmacol. Exp. Ther.*, **243**, 507–510.

Yavari, P. and Weight, F.F. (1987). Antagonists discriminate muscarinic excitation and inhibition in sympathetic ganglion. *Brain Res.*, **400**, 133–138.

Yavari, P. and Weight, F.F. (1988). Pharmacological studies in frog sympathetic ganglion: support for the cholinergic monosynaptic hypothesis for slow ipsp mediation. *Brain Res.*, **452**, 175–183.

Zarbin, M.A., Wamsley, J.R. and Kuhar, M. (1982). Axonal transport of muscarinic cholinergic receptors in rat vagus nerve: High and low affinity agonist receptors move in opposite directions and differ in nucleotide sensitivity. *J. Neurosci.*, **2**, 934–941.

Zimmerman, H. (1988). Cholinergic synaptic vesicles. In V.P. Whittaker (Ed.), *Handbook of experimental pharmacology, Vol. 86* (pp. 349–382). Berlin and Heidelberg: Springer Verlag.

6 Transmission: Noradrenaline

Marianne Fillenz

University Laboratory of Physiology, Parks Road, Oxford

Noradrenaline in peripheral sympathetic neurones is stored in two populations of vesicles; in small dense cored vesicles it is co-stored with ATP and in large dense cored vesicles it is co-stored with a variety of peptides and soluble proteins. Although release from both populations of vesicles occurs by Ca^{++}-dependent exocytosis there are differences in the frequency dependence, site and functional implications of release from the two kinds of vesicles. The large vesicles originate in the cell body and reach the nerve terminal by axoplasmic transport; the origin of the small vesicles is still unclear, although most evidence suggests their local formation in the nerve terminal. Released noradrenaline interacts with a number of receptor subtypes, whose location is both pre- and post-synaptic. Adrenoceptors have a variety of effector mechanisms which include G-protein-mediated effects on ion channels as well as intracellular messengers. Prolonged exposure to agonist results in desensitization of the receptors which may be homologous or heterologous and in both cases appears to involve phosphorylation.

Noradrenaline is metabolized by the enzymes monoamine oxidase and catechol-o-methyltransferase, which are both intracellular; metabolism is therefore preceded by reuptake into the nerve terminal or uptake into non-neuronal cells. The synthesis of noradrenaline is catalysed by three enzymes; of these tyrosine hydroxylase is the rate limiting enzyme; it requires tyrosine, oxygen and the cofactor tetrahydrobiopterin, the last being present in subsaturating concentration. Tyrosine hydroxylase is a substrate for a variety of protein kinases and phosphorylation results in a change of its kinetic parameters. In addition to this covalent activation, the transcription of the gene for tyrosine hydroxylase can also be regulated, giving rise to trans-synaptic induction.

Receptors on both the somatodendritic membrane and the nerve terminal play a critical role in the various mechanisms, with a wide range of time courses, which directly or indirectly regulate the release of noradrenaline. The receptors on the cell body control the impulse traffic as well as the transcription of genes for the transmitter-synthesizing enzyme proteins and the precursor molecules for the peptides co-stored with noradrenaline. The presynaptic autoreceptors and heteroreceptors regulate the release of noradrenaline by a rapid effect on ion channels and a slower effect on the rate of noradrenaline synthesis.

KEY WORDS: noradrenaline, adrenoceptors, tyrosine hydroxylase, vesicles

In 1905 Elliott demonstrated that the effect of stimulating the sympathetic nervous system could be mimicked by the injection of an extract of the adrenal gland, which was identified as adrenaline. Subsequently von Euler (1946) showed that the catecholamine stored and released by sympathetic nerves was noradrenaline, not adrenaline.

THE NORADRENERGIC NEURONE

The fluorescence histochemical technique of Hillarp and Falck (Falck *et al.*, 1962) gave great impetus to the study of noradrenergic neurones. With this technique it became possible not only to map the distribution of noradrenaline-containing neurones, but also to describe the morphology of these neurones and map the distribution of noradrenaline within them. A series of studies from Scandinavia (Norberg & Hamberger, 1964; Malmfors, 1965; Sjöstrand, 1965) showed that noradrenaline is present throughout the neurone, albeit unevenly distributed with respect to concentration. This was based on the semiquantitative relation between fluorescence intensity and noradrenaline concentration.

The cell bodies and axons show uniform dim fluorescence. The most striking appearance is that of the terminal parts of the axons which are characterized by strings of brightly fluorescent varicosities. Electron microscope studies show that the varicosities are characterized by the aggregations of subcellular vesicles identical in diameter to those seen at the neuromuscular junction but characterized by an electron dense core (Richardson, 1962; Grillo & Palay, 1962). In addition to the small vesicles, occasional large vesicles are also seen (Fig. 6.1). This suggests that dense-cored vesicles may represent noradrenaline storage organelles.

The ultrastructure of sympathetic cell bodies and non-terminal axons resemble that of other neurones; there are occasional clusters of small dense-cored vesicles in dendrites and axons, but more commonly solitary large dense-cored vesicles are seen (Richards & Da Prada, 1980).

METABOLIC CYCLE OF NORADRENALINE

The metabolic cycle of noradrenaline consists of its synthesis, storage, release, interaction with receptors and metabolism. These processes occur in different subcellular compartments and have their own regulatory mechanisms.

Noradrenaline is synthesized from the amino-acid precursor tyrosine by a series of

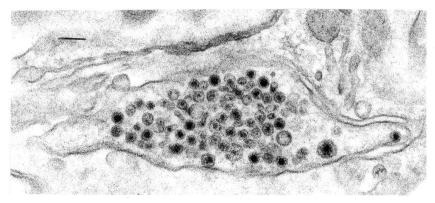

FIGURE 6.1   Electronmicrograph of varicosity of rat vas deferens showing large and small dense-cored vesicles. Calibration 100 nm.

NORADRENALINE   SYNTHESIS

TYROSINE

Tyrosine Hydroxylase
+ O_2 + BH_4

DOPA

L-Aromatic Amino Acid
Decarboxylase

DOPAMINE

Dopamine - β - Hydroxylase

NORADRENALINE

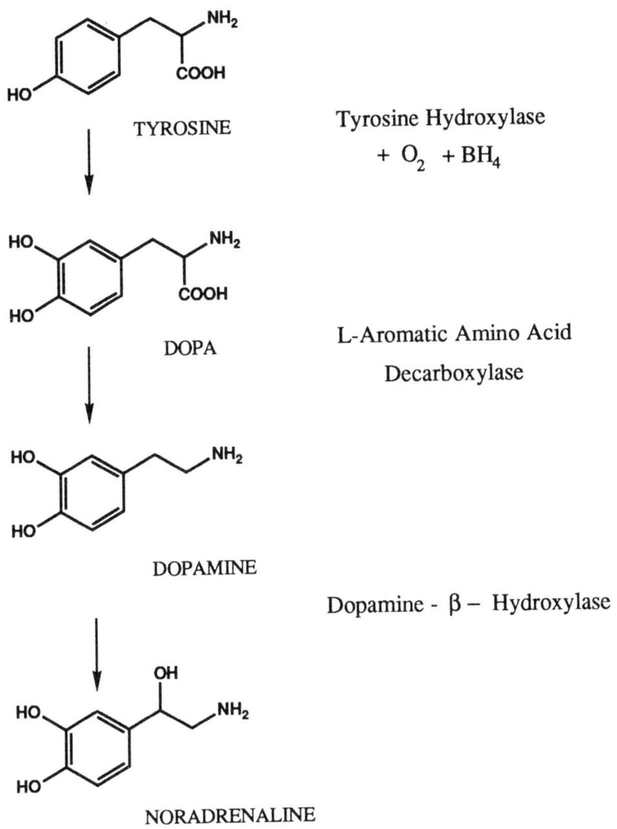

FIGURE 6.2   Pathway for noradrenaline synthesis and the enzymes that catalyze the various steps in the chain of reactions.

enzyme-catalyzed reactions first described by Blaschko (1939). The first step (Figure 6.2) is the hydroxylation of tyrosine to dihydroxyphenylalanine (DOPA), followed by decarboxylation to yield dopamine, which in turn is hydroxylated at the C-2 of the side chain to give noradrenaline. The three enzymes that catalyze these reactions are tyrosine hydroxylase (TH), aromatic amino-acid decarboxylase (DDC) and dopamine-β-hydroxylase (DBH).

There are two pathways for the metabolism of noradrenaline: deamination by the enzyme monoamine oxidase (MAO) and methylation at the C-3 atom on the benzene ring by catechol-O-methyltransferase (COMT) (Figure 6.3). Deamination and methylation can occur in either order. The deamination by MAO leads to the formation of the corresponding aldehydes; these aldehyde intermediates are rapidly

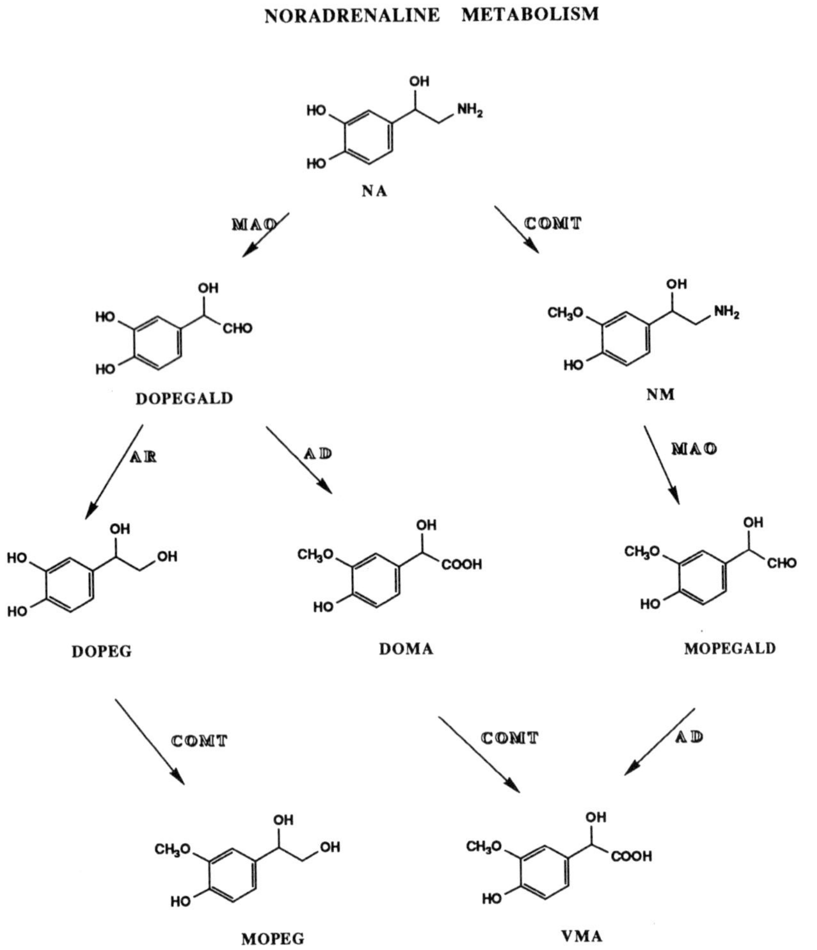

NORADRENALINE METABOLISM

FIGURE 6.3 The various pathways for metabolism of noradrenaline and the enzymes that catalyze the reactions. NA, noradrenaline; DOPEGALD, 3, 4-dihydroxyphenylglycol aldehyde; DOPEG, 3, 4-dihydroxyphenylglycol; MOPEG, 3-methoxy-4-hydroxyphenylglycol; MOPEGALD, 3-methoxy-4-hydroxyphenylglycol aldehyde; NM, normetanephrine; DOMA, 3, 4-dihydroxymandelic acid; VMA, 3-methoxy-4-hydroxymandelic acid; MAO, monoamine oxidase; COMT, catechol-O-methyltransferase; AR, aldehyde reductase; AD, aldehyde dehydrogenase.

metabolized, either by oxidation to the corresponding acid or by reduction to the corresponding alcohol or glycol (Figure 6.3). In some species, such as the rat, these metabolites undergo further sulfate conjunction.

SUBCELLULAR COMPARTMENTALIZATION

A number of techniques, which include differential and density centrifugation as well as immunohistochemistry at the light and electron-microscopic levels, have been used to establish the subcellular localization of the various chemical reactions together with the relevant enzymes.

Of the various enzymes, DDC always appears in the final supernatant and is generally accepted as being an enzyme with an exclusively cytoplasmic localization. Tyrosine hydroxylase is also found largely in the final supernatant after differential centrifugation, although there is always a small percentage of particle-bound enzyme (Laduron & Belpaire, 1968; Oesch, Otten & Thoenen, 1973). Immunohistochemistry at the electron-microscopic level shows tyrosine hydroxylase association with neuro-tubules (Pickel et al., 1975), but it is not clear at present to what extent this is a fixation artefact. Monoamine oxidase has been localized to the outer membranes of mitochondria on the basis of its co-migration in density gradients with a marker enzyme for mitochondria (Hörtnagl et al., 1969). COMT is also largely in solution in the cytoplasm (Brock & Fonnum, 1972) although recently a small proportion of membrane-bound COMT has been described (Rivett, Francis & Roth, 1983).

A key enzyme in noradrenergic neurones is dopamine-β-hydroxylase (DBH), and there has been considerable controversy concerning its subcellular localization (Klein et al., 1979). DBH, like noradrenaline, occurs both in a soluble and a particulate form, and the ratios of these two forms of the enzyme differ in the various parts of the neurone. Brimijoin (1974), using homogenization in both isotonic and hypotonic media, found that in the cell body 70%, and in nonterminal axons 50% of the DBH was particulate; 30–50% appeared in the final supernatant after homogenization in isotonic sucrose. The latter fraction was called sucrose-soluble, and it could represent either DBH in free solution in the cytoplasm or DBH stored in a delicate structure, easily disrupted by homogenization. Administration of the drug reserpine blocks active uptake of dopamine and noradrenaline into vesicles and possibly other membrane-bound structures; under these circumstances, dopamine fails to be converted to noradrenaline (Stjärne & Lishajko, 1966). This argues against a cytoplasmic location for DBH. The presence of DBH immunoreactivity in a reticulum of interconnecting membranes suggests that in axons some of the DBH is localized in the endoplasmic reticulum and may represent the sucrose soluble fraction of DBH (Pickel, Joh & Reis, 1986). Treatment of the particulate fraction with hypotonic media leads to a further separation into membrane-bound DBH and osmotically releasable DBH, which in axons is 20% of the particulate DBH. As will be shown later, the particulate DBH in axons is associated with the large dense-cored axonal vesicles, and 20% of this DBH makes up part of the soluble core of the vesicle. The distribution of DBH in terminal axons is quite different: only 10–15% of the DBH is sucrose-soluble, the rest being particulate, and of that, only 2% is osmotically releasable (Brimijoin, 1974). The identity of the subcellular particles in the nerve terminal that contain DBH will be discussed next.

NORADRENALINE STORAGE

Differential and density centrifugation of tissues with a sympathetic innervation revealed the presence of a low- and high-density population of noradrenaline storage particles. Electron microscopic examination of these fractions confirmed that they contained small and large dense-cored vesicles respectively (Bisby & Fillenz, 1971).

Improvements in technique have resulted in greatly enhanced purification of the large dense-cored vesicles prepared from splenic nerve (Lagercrantz, Klein & Stjärne, 1970) and small dense-cored vesicles prepared from vas deferens of castrated rats (Fried, Lagercrantz & Hökfelt, 1978). These preparations have been used to determine the chemical composition of the cores and of the membranes of these two populations of noradrenaline storage particles.

Although both large axonal vesicles and small terminal vesicles store noradrenaline, there are significant differences in their biochemical composition, which have important implications for their function and life cycle.

VESICLE CORES

Both kinds of vesicles, but particularly the small vesicles, are resistant to mechanical or hypotonic disruption, and separate analyses of core constituents and membrane constituents are therefore difficult. The use of histochemical methods, however, have demonstrated that administration of reserpine, (a drug that leads to depletion of noradrenaline) causes loss of the electron-dense core in the small vesicles, but merely a reduction in the electron density of the core in the large vesicles (Hökfelt, 1966; Van Orden et al., 1966; Farrell, 1968; Tranzer & Thoenen, 1968). This finding implies that whereas the electron density of small vesicles, especially after dichromate fixation, is attributable to noradrenaline, the electron density of the large vesicles is due to components additional to noradrenaline (Thureson-Klein, Klein & Yen, 1973).

Adenosine triphosphate (ATP) is also a core constituent in both small and large vesicles. ATP is found in the fractions of the density gradient prepared from vas deferens of castrated rat that contain the small vesicles (Fried et al., 1978); after pulse labelling with ^{3}H-adenosine, ^{3}H-ATP is found in these same fractions (Aberer et al., 1979). In a histochemical study, both large and small vesicles in sympathetic nerve terminals gave the ATP-specific uranaffin reaction (Richards & Da Prada, 1977). The noradrenaline:ATP ratio is variable, and ranges between 4:1 and 50:1 in both kinds of vesicles (Fried, 1980; Fredholm, Fried & Hedqvist, 1982; Klein, 1982).

Attempts have been made to calculate the noradrenaline content of both small and large vesicles, in terms of both concentration and numbers of molecules. Such calculations have been based on results from a number of different techniques. The estimated figures based on analysis of purified vesicle fractions give noradrenaline concentrations of 0.12–0.22 M and 9000–16000 molecules for large terminal vesicles in the splenic nerve (Klein, 1982); the small vesicles in rat vas deferens are estimated to contain a concentration of 0.036–0.06 M and 900 molecules of noradrenaline (Fried, 1980). An alternative approach is to assay the noradrenaline content of intact tissue and relate it to the number of large and small vesicles in single varicosities measured in electron-micrographs (Hökfelt, 1969; Dahlström, Häggendal & Hökfelt, 1966). According to these calculations a small dense-cored vesicle contains 30000 molecules of noradrenaline (Stjärne, 1989).

Large vesicles show little variation in core size and density, whereas the cores of small vesicles show considerable variation in size: furthermore the mean core size, as well as the core size distribution of small vesicles differs between nerve terminals in different organs (Fillenz & Pollard, 1976). Small vesicles can take up noradrenaline additional to their endogenous content (Fillenz, Howe & West 1976). This uptake is saturable and the maximum noradrenaline content of the vesicles is a measure of their storage capacity; the ratio of the content to storage capacity is a measure of the degree of saturation of this storage capacity. There is a close correlation between the mean core volume derived from electron micrographs and the degree of saturation derived from biochemical measurements (Fillenz & Stanford, 1981). This suggests that in the nerve terminal the small vesicles, with their wide range of core sizes, represent a population that is much more heterogeneous with respect to noradrenaline content than the large vesicles.

In addition to noradrenaline and ATP the cores of the large axonal vesicles also contain soluble proteins, peptides and lipids (Klein, 1982). Thirty eight percent of the total vesicle protein is in the soluble core and of this 53% is the enzyme DBH. Another class of soluble proteins found in the core of the large dense-cored vesicles are the chromogranins (Hagn et al., 1986). Three members of this family are at present recognized, called Chromogranin A, B (secretogranin I) and C (secretogranin II) respectively. These proteins are now known not to be confined to noradrenergic neurones but are found in a wide variety of peptidergic endocrine cells and neurones (Fischer-Colbrie, Hagn & Schober, 1987). The function of these proteins is at present under intense investigation; they may be involved in packaging and organization of the granule matrix, or may act as regulatory proteins after secretion or as precursors of peptide hormones (Huttner, Benedum & Rosa, 1988).

The large vesicles also contain peptides. The presence of these peptides varies in different peripheral noradrenergic neurones (Costa, Furness & Gibbins 1986; Lundberg & Hökfelt, 1986); immunohistochemical studies of sympathetic ganglia have shown that noradrenaline co-exists in the cell bodies of noradrenergic neurones with opioids (Schultzberg et al., 1978), somatostatin (Hökfelt et al., 1977), substance P (Kessler & Black, 1982) and neuropeptide Y (NPY) (Lundberg et al., 1983).

Vesicle fractions of ox splenic nerve and vas deferens contain opioid peptides (Wilson et al., 1980; Klein et al., 1982; De Potter, Coen & De Potter 1987). On density gradients of spleen and vas deferens the opioid peptides are found in regions where the large vesicles equilibrate, but are absent from the less dense regions where the small vesicles are found; opioid activity is also absent from the most purified small vesicle fractions. In purified preparations of large axonal vesicles Leu- and Met-Enkephalin make up half the opioid peptides. The approximate molar ratio of DBH:opioids:noradrenaline is 1:2:300. After isolation of storage vesicles from nerve terminals in rat vas deferens NPY was found in large but not small vesicles (Fried et al., 1985).

If all the soluble constituents are released on exocytosis, release from large and small vesicles will clearly have very different functional implications.

VESICLE MEMBRANES

The chemical composition of the membranes of the two kinds of vesicle is not yet fully determined. In addition to lipids, which are a major constituent of all membranes, the large axonal vesicles contain proteins which include DBH, cytochrome b_{561}, chromomembrin B, adenosine 5'-triphosphatase and Synapsin I.

The presence of DBH in the membrane of large vesicles is based both on distribution pattern in density gradients and on direct enzyme, gel-electrophoretic and radioimmunological assays, carried out on purified vesicle preparations; the evidence for small vesicles rests mainly on analytical density centrifugation. Such experiments have been carried out with homogenates from rat vas deferens, rat heart, rat hypothalamus and dog spleen (Chubb, De Potter & De Schaepdryver 1970; Bisby, Fillenz & Smith, 1973; Belmar, De Potter & De Schaepdryver 1974; Nelson & Molinoff, 1976; De Potter & Chubb, 1977). In all cases there is a DBH peak in the high density region of the gradient, coinciding closely with the high density peak of noradrenaline; there is also a second, smaller peak of DBH in the low-density region whose position coincides much less well with the low-density peak of noradrenaline. There are two possible explanations for this finding; one is that small vesicles have no DBH and the low density DBH peak represents the empty membranes or membrane fragments of large vesicles. This is the view taken by Klein, which he has supported with quantitative arguments (Klein *et al.*, 1979). An alternative explanation is that the disparity in DBH and noradrenaline peak position is due to the wide range of noradrenaline content, and hence density, of the small vesicles. If all the vesicles contain the same amount of DBH, but a significant proportion have a low noradrenaline content, the DBH peak would be expected to occur at a lower density than the noradrenaline peak. This prediction is borne out by the relative saturation of noradrenaline storage capacity and the relative disparities of the DBH and noradrenaline peak positions in density gradients of various tissues. In density gradients of rat vas deferens the DBH and noradrenaline peaks coincide closely (Bisby, Fillenz & Smith, 1973); nerve terminals in this organ have small vesicles whose noradrenaline storage capacity appears more or less fully saturated. In the spleen, however, where the mean core volume of small vesicles suggests a lower percentage saturation (Fillenz & Pollard, 1976), the DBH peak occurs at a significantly lower density than the noradrenaline peak in the gradient (Chubb *et al.*, 1970).

A further study compared ox and rat vas deferens, whose nerve terminals differ in the relative number of small and large vesicles. The fractions containing the large and the small vesicles were assayed both by the usual techniques for DBH activity and also for their ability to synthesize ^{3}H-noradrenaline from ^{3}H-tyrosine. The results provide clear evidence both for the presence of DBH in small vesicles and for the ability of small vesicles to synthesize noradrenaline (Neuman *et al.*, 1984).

Recently a membrane protein has been described called synaptophysin or p38. Immunofluorescence staining with antibodies raised to p38 shows its presence in nerve terminals in the vas deferens as well as most regions of the central nervous system. Immunohistochemistry at the electron microscopic level revealed that p38 was associated with small vesicles but was not present on the membrane of large vesicles (Navone *et al.*, 1986, 1988; De Camilli & Jahn, 1990). These findings have been challenged by Lowe *et al.* (1988) who report that large vesicle membranes also contain p38, although in very much lower concentrations. The composition of membranes of the two vesicle populations has important implications for their respective origins and functions.

NORADRENALINE RELEASE

Release by exocytosis involves the fusion of the vesicle membrane with the plasma membrane in such a way that a channel is formed through which the soluble components of the vesicle core escape into the extracellular compartment. That release of catecholamines occurs by exocytosis was first demonstrated in the adrenal medulla, where soluble proteins from vesicles but not from the cytoplasm appeared in the medium following splanchnic nerve stimulation (Douglas, 1968; Kirshner & Viveros, 1972; Smith, 1972). The finding that electrical stimulation of the splenic nerve caused the release of DBH and chromogranins together with noradrenaline established exocytotic release from large vesicles (De Potter *et al.*, 1969; Geffen, Livett & Rush 1969, 1970).

A powerful argument for exocytotic release of ACh at the skeletal neuromuscular junction was the quantal character of both the spontaneous and stimulus-evoked end plate potential (Katz, 1958). Intracellular recording from smooth muscle cells of the rat vas deferens showed both spontaneous and evoked junction potentials (Burnstock & Holman, 1961, 1962). In contrast to the miniature end plate potentials at the skeletal neuromuscular junction, junction potentials in the vas deferens showed a wide variation in size (Holman, 1970). This was attributed to the multiple innervation by varicosities at varying distances from muscle cells as well as the electrotonic spread of junction potentials from neighbouring muscle cells through low resistance gap junctions. Blakeley and Cunnane (1979) observed transient accelerations during the rising phase of evoked junction potentials recorded intracellularly from guinea pig and mouse vas deferens which they called 'discrete events'. The first time-derivative of the excitatory junction potentials reveals families of such discrete events, whose members have the same latency and peak time but whose distribution of amplitude is multimodal. Brock and Cunnane (1987), using extracellular focal microelectrode recording have monitored spontaneous excitatory junction currents (SEJC) and evoked excitatory junction currents (EJC) in a variety of sympathetically innervated smooth muscle tissues. The SEJC and EJCs are similar in amplitude and time course (Astrand, Brock & Cunnane, 1988) and are thought to represent the effects of single quanta of transmitter.

CO-RELEASE WITH ATP

Whereas initially the spontaneous and evoked excitatory junction potentials were attributed to the release of noradrenaline, it has become clear that they are due to the release of ATP (Sneddon, Westfall and Fedan, 1982; Burnstock, 1990). EJCs are blocked by ATP-receptor antagonists, but unaffected by adrenoceptor antagonists (Sneddon & Burnstock, 1984); furthermore while iontophoresis of ATP results in membrane depolarization, noradrenaline produces no change in membrane potential (Sneddon & Westfall, 1984). SEJCs and EJCs therefore provide evidence for quantal release of ATP; to what extent noradrenaline release is also quantal depends on the degree to which the ratio of NA:ATP in small dense-cored vesicles, which provide the releasable pool, is constant.

A recent study has suggested that ATP and noradrenaline may be derived from different vesicle populations. Although noradrenaline does not produce a change in membrane potential, it gives rise to a slow muscle contraction, which is clearly distinguishable from the fast twitch-like contraction produced by ATP. Nerve stimulation gives rise to a dual response with an initial phasic and a later tonic component. In the rat seminal vesicles the phasic and tonic responses differ in their frequency dependence which led the authors to suggest that the transmitters which mediate these two responses may be derived from different pools (Wali & Greenidge, 1989). A difference in pharmacological sensitivity of the adrenergic and non-adrenergic contractile response also suggests separate vesicular neuronal origins for noradrenaline and ATP in the rabbit vas deferens (Trachte *et al.*, 1989). However in both these studies the evidence for release is based on contractile response and is therefore indirect.

There has been extensive discussion of the intermittency of release from a single varicosity as illustrated by the probability of release which varies between 0.002–0.03 (Cunnane & Stjärne, 1982; Brock & Cunnane, 1987; Åstrand & Stjärne, 1988). This has been contrasted with release at the skeletal neuromuscular junction and at central synapses. However, if probability of release is calculated in terms of the number of vesicles available for release, the release probability at these other synapses is very similar to that found at single varicosities of sympathetic fibres (for a detailed discussion see Stjärne, 1989).

CO-RELEASE WITH PEPTIDES

As mentioned above noradrenaline is co-stored with various peptides in the large dense-cored vesicles. The distribution of the peptides differs in sympathetic neurones; thus some ganglion cells which show tyrosine hydroxylase immunoreactivity (which indicates that they synthesize noradrenaline) also exhibit neuropeptide Y-like immunoreactivity NPY-LI (Lundberg *et al.*, 1983). Electrical stimulation of the splenic nerve at 10 Hz causes release of NPY-LI into the perfusion medium (Lundberg, Hua & Franco-Cereda, 1984b). Similarly co-release of ^{3}H-noradrenaline and met-enkephalin from bovine vas deferens was demonstrated in response to electrical stimulation (De Potter *et al.*, 1987). Physiological activation of the sympathetic nervous system enhances NPY-LI levels in parallel with plasma catecholamines in rats after exposure to stress (Zukowska-Grojec, Konarska & McCarty, 1988), in humans after exercise (Lundberg *et al.*, 1985) and in newborn infants (Lundberg *et al.*, 1986). The NPY-LI in plasma shows a higher correlation with plasma noradrenaline than with plasma adrenaline and is therefore thought to be derived from sympathetic nerve terminals.

The release of noradrenaline and of peptides differs in their frequency dependence. Stimulation of the splenic nerve in the pig at 2 Hz for 2 mins increases the release of noradrenaline, but has little effect on NPY. The same number of stimuli delivered in bursts of 20 Hz for 1 second with 20 second intervals produces no further increase in noradrenaline release but increases NPY release 4–5 fold (Lundberg *et al.*, 1986).

MORPHOLOGICAL STUDIES

Morphological evidence for exocytosis was provided by electron-micrographs which showed both large and small vesicles fused with the plasma membrane and their interiors in communication with the extracellular space (Fillenz, 1971; Thureson-Klein, 1983; Thureson-Klein & Klein, 1990). This confirms that released noradrenaline is derived from both populations of vesicles.

In the central nervous system exocytotic profiles of large dense-cored vesicles are found outside the structurally specialized synaptic areas; this morphological evidence together with the different frequency dependence of noradrenaline and peptide release suggests that there may be differences in the release mechanism of large and small vesicles (Zhu, Thureson-Klein & Klein, 1986).

Quantitative studies of the morphological changes following release yield results which depend on the stimulus parameters. Prolonged electrical stimulation or K^+-evoked release from vas deferens results in a reduction in small dense-cored vesicles, an increase in small agranular vesicles, a decrease in the total number of small vesicles and no change in the number of large dense-cored vesicles (Basbaum & Heuser, 1976; Pollard, Fillenz & Kelly, 1982). Brief (1 min) electrical field stimulation of human omental veins, on the other hand, causes an increase in the number of small vesicles, both dense-cored and agranular. Large dense-cored vesicles also show a small increase in number which is not statistically significant (Thureson-Klein & Stjärne, 1981).

RELATION BETWEEN LARGE AND SMALL VESICLES

There has been extensive speculation concerning the relation between the two populations of vesicles and their respective life cycles. There are two main theories. According to the first proposed by A.D. Smith in 1972, small vesicles are derived from large vesicles by membrane retrieval after the latter have undergone exocytosis. According to the second, large dense-cored vesicles and small dense-cored vesicles have separate origins, the large vesicles being formed in the cell body and carried by rapid axoplasmic transport to the nerve terminal, whereas the small vesicles are formed locally in the nerve terminal, by budding from a special DBH-rich endoplasmic reticulum.

Results obtained with a variety of techniques are relevant to this question.

A comparison of the subcellular distribution of various vesicle markers in cell bodies, axons and varicose terminals of sympathetic noradrenergic neurones is shown in Figure 6.4. The histograms show particle-bound noradrenaline, NPY and DBH as a percentage of the total. The figures are derived by measuring the distribution of each component between the microsomal pellet and the final supernatant after differential centrifugation of tissue homogenates. The appearance in the supernatant implies either that the compound is present in solution in the cytoplasm, or that it has been released during the experimental manipulations from an easily disrupted subcellular compartment. The figures for DBH (Brimijoin, 1974) and for noradrenaline and NPY (Fried et al., 1985) come from different studies; therefore direct comparisons of absolute percentages are difficult. However the changes in different parts

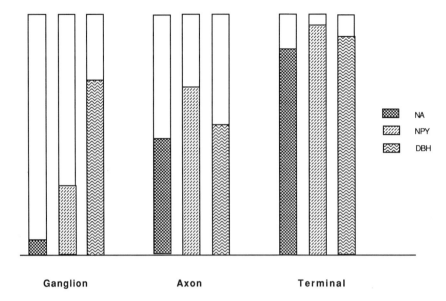

FIGURE 6.4   Schematic diagram to show the relative proportion of particle bound noradrenaline (NA), NPY (neuropeptide Y) and DBH (dopamine-β-hydroxylase) in cell body (ganglion), axon and nerve terminal. In each case the open column is the soluble and the hatched column the particulate proportion of each component.

of the neurone are significant. Both NPY and DBH are synthesized and packaged into vesicles in the cell body; the vesicles at this stage contain little noradrenaline and the molar ratio of particle-bound NA:NPY is 10:1. The 'soluble' NPY is presumably released from endoplasmic reticulum or Golgi regions in the process of homogenization.

In the axon the increase in particle-bound NPY is attributable to the greater number of large dense-cored vesicles compared with the cell body. Although a similar explanation applies to noradrenaline, there is an additional factor since there is an increase in the ratio of NA:NPY. Klein (1982) showed that this is due to synthesis of noradrenaline in the vesicles. Using the purified preparations of large noradrenergic vesicles from proximal, intermediate and intrasplenic portions of the ox splenic nerve, he showed that there was no change in the concentration of ATP, opioids or DBH but a 60% increase in the concentration of noradrenaline as the vesicles travelled down the axon. The decrease in the percentage of particle-bound DBH between cell body and axon is attributable to a new DBH-containing compartment, described by Brimijoin (1974) as sucrose-soluble, which on the basis of electron microscopic immunocytochemistry (Pickel *et al.*, 1976), appears to be DBH-containing endoplasmic reticulum.

The most dramatic change is seen in the nerve terminals where the NA:NPY ratio has increased to 150:1 and the sucrose-soluble DBH has been almost completely replaced by particulate DBH. Furthermore, there is a difference between the particulate DBH in the axon and the terminals: of the axonal particulate DBH, 20% is osmotically releasable, which is characteristic of the large noradrenergic vesicles; in the terminals, only 5% of the particulate DBH is osmotically releasable, which is indicative of a preponderance of small noradrenergic vesicles that have membrane-bound but no soluble DBH.

Changes, similar to those seen in the nerve terminal, occur above a ligature, in that there is an increase in the particulate DBH at the expense of sucrose-soluble DBH and the additional particulate DBH has a low percentage of osmotically releasable DBH. Although early electron-micrographs, fixed with osmium, showed only large dense-cored vesicles above a ligature (Kapeller & Mayor, 1967) a later study in which dichromate fixation, specific for catecholamines, was used, showed both large and small dense-cored vesicles above a ligature (Tomlinson, 1975).

All these results can be readily explained by the separate formation of large vesicles in the cell body and small vesicles by budding from endoplasmic reticulum; appearances in electron-micrographs which suggest such small vesicle formation have been reported by various workers (Fillenz, 1971; Tranzer, 1972; Hökfelt, 1973; Holtzmann, 1977). However none of these findings exclude the formation of small vesicles from large vesicles. A decisive test would be the demonstration of a marker protein present in the membrane of small vesicles but absent from that of the large vesicles. It has been suggested that synaptophysin (p38) is such a marker protein since according to some workers it is present in small but not large vesicles (Navone *et al.*, 1986, 1988; De Camilli & Jahn, 1990). However others have shown this membrane protein to be present in large vesicles, but in much lower concentration (Lowe, Maddedu & Kelly, 1988; Dukler & Fischer-Colbie, 1990). The question therefore is at present unresolved.

ADRENERGIC RECEPTORS

RECEPTOR SUBTYPES

Stimulation of the sympathetic nervous system produces a wide variety of physiological effects, which can be roughly classified into excitatory and inhibitory. This led Cannon and Rosenblueth to postulate the release of two chemical mediators Sympathin E and Sympathin I (1937). However in 1948 Ahlquist showed that the effects of sympathetic stimulation could be mimicked by a series of sympathomimetic amines. The physiological responses fell into two groups according to the order of potency of the amines. This led Ahlquist to the conclusion that there was only a single chemical agent released from the nerve terminal, which interacted with different receptor subtypes, which he called α and β adrenoceptors. Whether the response was excitatory or inhibitory depended on the tissue rather than the identity of the receptor. This introduced pharmacological criteria as the basis of the identification of receptor subtypes, the most reliable criterion being a selective antagonist. There followed the subdivision of β receptors into β_1 and β_2 (Lands *et al.*, 1967) and the subdivision of α receptors into α_1 and α_2 (Berthelson & Pettinger, 1977). Recently each of these α-receptor subtypes has been shown to be heterogeneous. A subdivision of α_1 receptors was first suggested on pharmacological grounds in 1982 (Coates, Jahn & Weetman 1982). Later it was shown that although α_1 receptor stimulation always causes smooth muscle contraction by producing a rise in intracellular Ca^{++}, some α_1-receptor responses did and others did not require extracellular Ca^{++}. Similarly α_2 adrenoceptors, on the basis of differences in antagonist binding, have been subdivided into α_{2A} and α_{2B} (Bylund, 1985; Bylund, Ray-Prenger & Murphy, 1988). Atypical β-adrenoceptor responses in a number of tissues which include adipocytes as well as smooth muscle of stomach fundus, ileum and colon have led to the proposal of a third subtype of β adrenoceptor which has been called the β_3 receptor (Zaagsma & Nahorski, 1989).

The distinction between the various adrenoceptor subtypes was greatly reinforced by the demonstration that they had different effector mechanisms. Thus β_1, β_2 and β_3 receptors stimulate adenyl cyclase, which results in an increase in the intracellular concentration of cyclic AMP (cAMP) and activation of cAMP-dependent protein kinase (PKA). α_2 receptors depress the activity of adenyl cyclase and so reduce the concentration of cAMP. Of the two α_1 receptor subtypes α_{1B} receptors stimulate phospholipase C which leads to the breakdown of phosphatidyl inositol polyphosphates into inositol phosphates, which release Ca^{++} from intracellular stores and diacyl glycerol, which activates protein kinase C. The stimulation of α_{1A} adrenoceptors on the other hand does not trigger inositol phospholipid hydrolysis and appears to control the opening of dihydropyridine-sensitive Ca^{++} channels by an as yet unknown mechanism (Han, Abel & Minneman, 1987). These effector mechanisms regulate the contraction of smooth and cardiac muscle, the opening and closing of ion channels and the activity of a wide variety of enzymes.

A number of different techniques have contributed to the identification and characterization of the various adrenoceptor subtypes. Radiolabelled ligand binding studies have established the characteristics of the binding sites of the receptor irrespective of the functional coupling mechanisms. Both β_1 and β_2 receptors as well as α_2 receptors share certain characteristics which are less well marked in α_1 receptors. Thus antagonists interact with a single binding affinity state of the receptors. Agonist binding curves suggest a two-state binding model, with high and low affinity binding sites. In the presence of guanosine triphosphate (GTP) or its analogues all the binding sites are converted to the low affinity form. The presence of the high affinity binding sites is also influenced by Mg^{++} and Na^+ ions (Hoffman & Lefkowitz, 1980; Bylund & U'Prichard, 1983).

These results are attributable to the role of G proteins which mediate the effect of agonist binding on the activity of enzymes (Figure 6.5). Thus G_s mediates the stimulation of adenyl cyclase by β receptors (both β_1 and β_2), G_i mediates the inhibition of adenyl cyclase by α_2 receptors and G_o mediates the stimulation of phospholipase C by α_1 receptors (Gilman, 1987).

Each of the G proteins that has been implicated in transmembrane signalling reactions is a heterotrimer, with subunits designated α, β and γ. The α subunit of each G protein is unique and serves to distinguish one G protein from another; the $\beta\gamma$ subunit complex of the three G proteins can be interchanged in functional assays (Casey et al., 1988). Each G protein α subunit has a single high affinity binding site

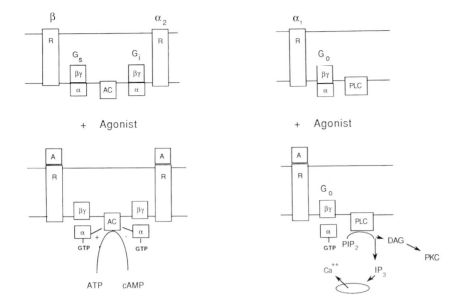

FIGURE 6.5   Schematic representation of the second messenger systems coupled to the various adrenoceptor subtypes. For details see text.

for guanine nucleotides. In the basal state this is occupied by GDP. Agonist binding to the receptor activates the G protein which results in the exchange of GTP for GDP. This leads to the dissociation of the α subunit from the $\beta\gamma$ subunit complex, the activation of adenyl cyclase and phospholipase C in the case of G_s and G_o respectively and the reduction of the affinity of the receptor for the agonist. The α subunit has intrinsic GTPase activity: hydrolysis of GTP to GDP results in deactivation of the G protein.

The inhibition of adenyl cyclase by α_2 receptors is largely dependent on the presence of G_s and may therefore depend on the binding and inactivation of the α_s subunits by the $\beta\gamma$ subunits released when G_i binds GTP. However there is also evidence for some direct inhibition of the enzyme by α_i (Jakobs & Schultz, 1983).

The α subunits of G proteins are substrates for ADP-ribosylation catalyzed by bacterial toxins. This results in activation of G_s by cholera toxins and inactivation of G_i and G_o by pertussis toxin (Casey *et al.*, 1988).

Recent evidence suggests that G proteins in addition to modulating ion channels through intracellular second messengers may also act directly on ion channels. Thus G_s purified from human red cells and preactivated with GTPγS directly facilitates the opening of Ca^{++} channels in skeletal muscle and heart muscle cells (Yatani *et al.*, 1987, 1988). This is a modulatory effect, as G_s will not itself open the channel but increases the frequency of opening when the membrane is depolarized. Since the effect is mediated by G_s it is attributable to β adrenoceptors.

Noradrenaline has also been shown to inhibit Ca^{++} channels in dorsal root ganglion cells and sympathetic ganglia. This effect is mimicked by intracellular application of guanine nucleotides and blocked by pertussis toxin, which suggests the participation of G proteins. The fact that the effect can be obtained in patch clamp experiments using outside-out patches eliminates the participation of intracellular second messengers (Marchetti *et al.*, 1986).

It is clear from these results that not only does noradrenaline produce a variety of effects by virtue of interacting with a number of different adrenoceptor subtypes, but each receptor has access to more than one effector mechanism (Neher & Clapham, 1988). Thus noradrenaline can modulate ion channels both through a direct effect of G proteins associated with the various adrenoceptor subtypes and through ion channel protein phosphorylation by various protein kinases.

MOLECULAR BIOLOGY

The current understanding of the biochemistry of adrenergic receptors has come from the development of several experimental techniques. With the development of photoaffinity labelling of the receptors and of specific ligand-based affinity columns it became possible to purify to homogeneity the various subtypes of adrenergic receptors, to obtain limited protein sequence information, design oligonucleotide probes and clone the genes. This work has revealed a basic similarity in the structure of the various adrenoceptor subtypes (Lefkowitz *et al.*, 1988; Sigal, Dixon & Strader, 1988). They all consist of a single polypeptide chain with molecular weights of 64000–80000. They share the characteristic feature of all G-protein linked receptors of seven membrane-spanning domains which consist of clusters of hydrophobic amino acids. The hydrophilic regions which connect the transmembrane domains alternate as extracellular and cytoplasmic loops, the N terminal being extracellular and the C-terminal cytoplasmic (Figure 6.6).

Techniques such as mutagenesis, limited proteolysis and more recently the construction and expression of chimeric α_2-β_2-adrenergic receptors have thrown light on the functions of the various domains of the receptors. Thus the ligand-binding function resides in the hydrophobic membrane-spanning domains, with different sites for agonist and antagonist binding (Dixon *et al.*, 1987; Kobilka *et al.*, 1988). Deletion of the third cytoplasmic loop leads to a loss of the ability to activate G_s, which suggests that this region of the receptor may constitute part of a binding site that interacts with G proteins (Dixon *et al.*, 1988). Finally the cytoplasmic loop containing the C-terminal has a number of phosphorylation sites (Dohlman *et al.*, 1987).

Gene cloning of adrenoceptors has confirmed and clarified the pharmacological characterization of subtypes. Both β_1 and β_2 adrenoceptors have been cloned from several species (Dixon *et al.*, 1986; Yarden *et al.*, 1986). The human β_1 and β_2 receptor proteins are 54% identical overall and 71% identical in the seven hydrophobic membrane spanning regions (Kobilka *et al.*, 1987a; Frielle *et al.*, 1987). Most recently a human genomic library revealed a gene coding for a polypeptide with 51% homology with β_1 and 46% homology with β_2-adrenoceptors; this has been called the gene for the β_3 receptor (Emorine *et al.*, 1989). Three separate genes for human α_2-receptor proteins have been obtained which map to chromosome 10, 4 and 2. The products of these genes show different pharmacological properties (Kobilka

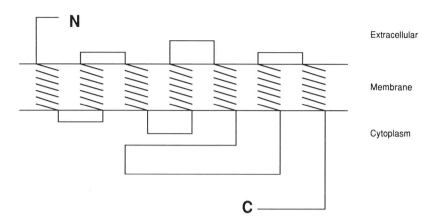

FIGURE 6.6   Schematic representation of the structure of the beta-adrenoceptor showing the seven
transmembrane domains and the extracellular and cytoplasmic loops.

et al., 1987b; Sigal *et al.*, 1988). A gene for the hamster α_1-adrenoceptor has also
been cloned (Cotecchia *et al.*, 1988).

DESENSITIZATION

The response of receptors to agonist stimulation is not constant but is dependent on
the previous history of the receptor. The earliest discovery of the enhanced responsiveness of tissues following denervation was called denervation supersensitivity
(Cannon & Rosenblueth, 1949). This was followed by the discovery of the rapid
reduction in response of cells exposed to adrenergic agonists. It has recently become
clear that such fluctuations in receptor responsiveness are not only drug- or lesion-
induced but occur under physiological conditions – receptors show physiological
sensitization and desensitization.

There has been considerable progress in the understanding of the mechanism of
desensitization (Sibley & Lefkowitz, 1985). This can be further subdivided into
homologous and heterologous desensitization. In the former decreased responsiveness of a receptor is produced by and confined to its own agonist, in the latter
diminished responsiveness of a receptor follows exposure to a variety of agonists and
correspondingly exposure to one agonist affects the responsiveness of a number of
receptors. Both homologous and heterologous desensitization have been most thoroughly investigated in β-adrenergic receptors and have been shown to involve receptor phosphorylation (Sibley *et al.*, 1987).

In homologous desensitization agonist-occupied β_2-adrenoceptors are phosphorylated by a novel cytoplasmic protein kinase called β ARK (β-adrenergic receptor
kinase) (Benovic *et al.*, 1986). This agonist-induced phosphorylation occurs at a
serine- and threonine-rich region of the C-terminal of the receptor (Bouvier *et al.*,
1987). The phosphorylation results in uncoupling of the receptors from the G proteins followed by internalization of the receptor that is detected as a reduction in the
number of binding sites. Removal of the agonist results in a relatively rapid reversal,

brought about by the activity of phosphatases and the re-insertion of the dephos-phorylated receptors into the plasma membrane (Sibley *et al.*, 1986).

Heterologous desensitization is due to phosphorylation of the receptor by cAMP-dependent protein kinase. Thus activation of any receptor coupled to G_s will increase intracellular cAMP concentration, activate PKA and lead to phosphoryla-tion of β-receptors. This phosphorylation, which is confined to serine residues, occurs at a site different from the phosphorylation by βARK. The phosphorylation impairs the receptor-G protein interaction; it therefore reduces the effector response but does not cause receptor internalization and reduction in the number of binding sites. Protein kinase C can also phosphorylate the β-adrenergic receptor *in vitro* (Bouvier *et al.*, 1987). Treatment of intact cells with phorbol esters, which directly activate PKC, results in receptor phosphorylation and desensitization (Sibley *et al.*, 1984).

SENSITIZATION

A chronic decrease in noradrenaline release (produced by drugs such as reserpine), chronic administration of receptor blocking drugs or neurotoxic destruction of noradrenergic neurones by 6-OH-DA administration, all result in a supersensitivity which is most commonly demonstrated in periperal tissues by a shift to the left of the dose-response curve for exogenously administered noradrenaline.

The development of homologous sensitization and its underlying mechanism can-not be studied *in vitro* in the same way as desensitization and may in part represent recovery from a state of partial homologous desensitization.

There are a number of findings which suggest that there is heterologous sensitiza-tion. Thus a chronic increase in intracellular cAMP leads to an increase in the number of β receptors (Porzig, Becker & Reuter, 1982). Stimulation of α_1-receptors in rat pinealocytes, which leads to phosphatidyl inositol breakdown and activation of pro-tein kinase C, potentiates isoprenaline-induced accumulation of cAMP in these cells (Sugden *et al.*, 1985). In frog erythrocytes activation of protein kinase C by phorbol esters, leads to increased production of cAMP in response to isoprenaline, pro-staglandins, and to GTP. This increased responsiveness is paralleled by phosphoryla-tion of the catalytic subunit of adenyl cyclase. Incubation of the purified bovine catalytic subunit in the presence of purified protein kinase C leads to stoichiometric phosphorylation of the enzyme (Yoshimasa *et al.*, 1987).

METABOLISM

Release of noradrenaline and its interaction with receptors is followed by its inactiva-tion. Two enzymes metabolize noradrenaline.

MONOAMINE OXIDASE

The inactivation of adrenaline and other amines by monoamine oxidase (MAO) was first described by Blaschko *et al.* (1937). Differences in substrate and inhibitor specificities of different MAO preparations led to the suggestion that there were two major forms of the enzyme called MAO-A and MAO-B (Johnston, 1968). In the past the differences between MAO-A and MAO-B have been attributed to different lipid environments, or to two different active sites on a single protein. Recent work has shown that phospholipids are not an essential component of the enzyme but increase the affinity of both MAO-A and MAO-B for their substrates (Navarro-Welch & McCanley, 1982).

Progressive purification has yielded two enzymes with different molecular properties and consistent differences in subunit molecular weights, in spite of considerable organ and species variation (Singer, 1987). Monoclonal antibodies to human MAO-B (Denney et al., 1982, 1983) and MAO-A (Kochersperger et al., 1985) have been raised which show no cross reactivity. This confirms that MAO-A and MAO-B are two distinct enzymes.

MAO is present both in neurones and non-neuronal cells. Many cells and tissues contain both forms of the enzyme, although in some sympathetic nerve terminals, e.g. the heart, only MAO-A is present. In addition to the MAO activity localized to the outer mitochondrial membrane (Hörtnagl et al., 1969), MAO-A and B activity has also been found in microsomal fractions (Egashira & Yamamaha, 1981). It has been suggested that this represents newly synthesized enzyme, a precursor of mitochondrial MAO, not yet inserted into the mitochondria.

CATECHOL-O-METHYLTRANSFERASE

A second enzyme, catechol-O-methyltransferase (COMT), inactivates catecholamines by methylation. This effect was first demonstrated by Axelrod in 1957 and the enzyme was partly purified and characterized in 1958 (Axelrod & Tomchick, 1958). COMT in the periphery is extraneuronal and most of it is soluble in the cytoplasm. Recently a small proportion of membrane-bound COMT has been described which has the same V_{max} as the soluble enzyme, but a K_m which is lower by one order of magnitude (Reid et al., 1986).

A comparison of MAO and COMT activity carried out in rat heart, which contains only MAO-A (Grohmann, 1987) showed that neuronal MAO activity is greater than extraneuronal COMT and the latter is more active than extraneuronal MAO. The greater activity of COMT may be attributable to the high affinity membrane bound form of COMT.

NORADRENALINE UPTAKE

Metabolism of the released noradrenaline can only occur after uptake into cells, since in contrast to acetylcholinesterase, the metabolizing enzyme for acetylcholine, neither of the two enzymes responsible for the metabolism of noradrenaline are extracellular.

There are two membrane carriers which are responsible for the rapid termination of action of noradrenaline. The cocaine- and desipramine-sensitive, sodium- and chloride-dependent neuronal uptake, uptake$_1$, and the corticosterone- and

3-O-methyl-isoprenaline-sensitive, sodium- and chloride-independent extraneuronal uptake, uptake$_2$ (Iversen, 1967).

The functional coupling of the neuronal and extraneuronal transport with intracellular MAO can be explored by measuring the effects of selective inhibition of transport systems or enzyme activity. Evidence from such experiments indicates that normally neither of these mechanisms are saturated and that neuronal MAO activity is very high; as a consequence axoplasmic noradrenaline is kept very low (Trendelenburg *et al.*, 1986).

SYNTHESIS

Released and metabolized noradrenaline is replaced by the new synthesis of noradrenaline. The rate-limiting step in noradrenaline synthesis is tyrosine hydroxylation, since of the three enzymes involved in its synthesis tyrosine hydroxylase has the lowest V_{max}. Increase in the rate of noradrenaline synthesis is attributable to an increase in the activity of tyrosine hydroxylase. Tyrosine hydroxylase has been isolated and purified from a number of sources. Vulliet *et al.* (1980), using rat phaeochromocytoma as a source, obtained an enzyme with a specific activity of 360 nmol/min/mg protein, which gave a single band on SDS gels and had a molecular weight of approximately 60000. The native enzyme from rat phaeochromocytoma has an apparent molecular weight of approximately 225000; the enzyme therefore normally exists as a tetramer. cDNAs for tyrosine hydroxylase mRNA prepared from rat phaeochromocytomas have been cloned (Lamoureux *et al.*, 1982) and sequenced (Grima *et al.*, 1985). Based on the sequence of the rat cDNA, the predicted molecular weight of the tyrosine hydroxylase monomer is 55903, in close agreement with the figure derived from the purification of the enzyme.

Tyrosine hydroxylase is a mixed function oxidase, which requires molecular oxygen and a reduced pteridine cofactor as co-substrates (Nagatsu, Levitt & Udenfriend, 1964). The enzyme is inhibited by catechol compounds; this inhibition is competitive with the pterin cofactor. This end-product feedback inhibition of the enzyme provided the earliest theories for the regulation of noradrenaline synthesis (Udenfriend, 1966). It was proposed that the enzyme was normally subject to end-product inhibition and that release of noradrenaline reduced this feedback inhibition. Evidence in support of this mechanism was the demonstration that drug-induced increases in the concentration of cytoplasmic noradrenaline resulted in a depression in the rate of tyrosine hydroxylase activity (Neff & Costa, 1966).

However, physiologically triggered noradrenaline release comes from the vesicular pool and some of this released noradrenaline is taken up again into the cytoplasm and so would increase rather than decrease cytoplasmic noradrenaline concentration. The increased synthesis rate which accompanies acceleration of impulse traffic is unlikely therefore to reduce the end-product inhibition of tyrosine hydroxylase, which is a cytoplasmic enzyme.

There is now extensive evidence that acceleration of tyrosine hydroxylation is brought about by a covalent activation of the enzyme. The earliest evidence for this hypothesis was the report of Morgenroth *et al.* (1974) that after hypogastric nerve stimulation tyrosine hydroxylase in the supernatant of the stimulated vas deferens showed a change in kinetic parameters compared to the unstimulated vas deferens: this change consisted of a decrease in the K_m for reduced cofactor, a reduced K_m for tyrosine, an increase in K_i for noradrenaline and an increase in V_{max}. This led to the theory that the accelerated synthesis resulting from nerve stimulation was due to activation of tyrosine hydroxylase rather than removal of end-product inhibition.

Tyrosine hydroxylase *in vitro* is activated by a variety of unrelated conditions: anions, such as heparin (Kuczenski & Mandell, 1972) phospholipids (Lloyd & Kaufman, 1974) ATP, Mg^{++} and cAMP-dependent protein kinase (Weiner *et al.*, 1977).

The activation by these modulators is remarkably similar and involves a decrease in K_m for cofactor and an increase in K_i for catechols. Measurements of the concentration of tetrahydrobiopterin, the naturally occurring cofactor, in adrenal medullary cells, shows this to be substantially below saturation for the enzyme (Viveros *et al.*, 1981). Changes in K_m for cofactor could therefore regulate the activity of the enzyme.

Limited proteolysis also activates the enzyme. Hoeldke and Kaufmann (1977) used partial trypsinization which led to the isolation of an enzyme with a subunit molecular weight of 34000 rather than 60000. This fragment showed greater activity than the native enzyme (Musacchio, Würzburger & D'Angelo, 1971; Kuczensky, 1973; Vigny & Henry, 1981) and could not be further activated by either phosphatidyl serine or a cAMP-dependent protein phosphorylating system.

These results taken together suggest that the enzyme is normally inhibited and that this inhibition is removed either by cleavage of the molecule or by a charge-dependent conformational change (Vulliet & Weiner, 1981). The characteristics of the enzyme which are affected by the conformational change are the affinity for cofactor and for catechols, and since the interactions between these compounds is competitive, this may in fact represent a single or two closely related sites.

Of all the various manipulations which activate the enzyme, phosphorylation is the most likely to have a physiological significance. Much is now known about the various intracellular protein kinases and the second messengers which activate them. Phosphorylation and activation of tyrosine hydroxylase has been studied both *in situ* and *in vitro*. In the former, intact cells are incubated under conditions which increase the intracellular concentration of the second messengers which activate the various protein kinases; in the latter the purified enzyme is incubated with the purified protein kinases.

These experiments have attempted to discover which protein kinases phosphorylate tyrosine hydroxylase, what the sites of phosphorylation are and what the effects of phosphorylation are on the enzyme.

Both *in situ* and *in vitro* phosphorylation experiments have shown that tyrosine hydroxylase is a substrate for cAMP-dependent protein kinase (PKA), (Vulliet, Langen & Weiner, 1990), Ca^{++}/Calmodulin dependent protein kinase (PKII) (Yamauchi & Fujisawa, 1981; Vulliet, Woodgett & Cohen, 1984), Ca^{++}/

phospholipid-dependent protein kinase (PKC) (Albert *et al.*, 1984) and cGMP-dependent protein kinase (Roskoski, Vulliet & Glass, 1987).

SITES OF PHOSPHORYLATION

The question of the sites of phosphorylation by the various protein kinases is not yet fully resolved. Experiments with the purified enzyme suggested that PKA and PKC phosphorylated the enzyme at the same single site, whereas PKII phosphorylated the enzyme at two sites, one of which was shared with PKA and PKC (Vulliet *et al.*, 1985).

In situ phosphorylation followed by tryptic digestion of the enzyme yielded a number of phosphopeptides (Campbell, Hardie & Vulliet, 1986; Tachikawa *et al.*, 1986; Cahill & Perlman, 1987). Although there are some differences between the results of these studies there are important points of agreement. Thus, PKA and PKC elicit the phosphorylation of only one tryptic peptide. According to some reports PKA and PKC phosphorylate the same site (Albert *et al.*, 1984; McTigue *et al.*, 1985) whereas according to others PKC and PKA phosphorylate different sites (Tachikawa *et al.*, 1987). Activation of PKII by high K^+ or Ca^{++} ionophores always yields a number of phosphopeptides, one of which is the same as that resulting from phosphorylation by PKA. There are also differences in the time course of phosphorylation. Ca^{++}CM-dependent phosphorylation is rapid and transient, whereas cAMP-dependent phosphorylation is slower and maintained (Waymire *et al.*, 1988). The difficulty with the *in situ* phosphorylation experiments is that a rise in intracellular Ca^{++} can, in some cases, result in an increase in intracellular cAMP, which will complicate the results. However the *in vitro* phosphorylation results also suggest that PKA and PKII phosphorylate a common site.

Phosphorylation by PKC produces the same phosphopeptide as PKA, which suggests that phosphorylation occurs at the same sites (Vulliet *et al.*, 1985). Tyrosine hydroxylase purified from rat phaeochromocytoma is phosphorylated and activated by purified cGMP-PK. Phosphorylation by cGMP-PK and PKA is not additive and the phosphopeptides are identical, which suggests that phosphorylation is at the same residue (Roskoski *et al.*, 1987).

In all studies on phosphorylation only serine residues have been found to be phosphorylated. Specific kinases have preferences for certain amino acid sequences in the vicinity of the phosphorylated serine residues. Amino acid sequencing of the phosphopeptides and comparison with the known cDNA for tyrosine hydroxylase suggests that three of the protein kinases (PKA, PKC and cGMP-PK) probably phosphorylate Ser^{40}, whereas PKII phosphorylates Ser^{19} and Ser^{40} (Campbell *et al.*, 1986).

EFFECTS OF PHOSPHORYLATION

In situ phosphorylation by all four protein kinases is accompanied by activation of the enzyme.

Phosphorylation by PKA changes the pH-activity curve of the enzyme by increasing its activity in the high pH range (Okuno *et al.*, 1989) and also changes its kinetic parameters. In control rats supernatant from vas deferens contains two kinetically different forms of tyrosine hydroxylase with a low and high apparent K_m value for cofactor. After phosphorylation, the enzyme has a single K_m corresponding to the high-affinity form. (Weiner, 1979). These results suggest that cAMP-dependent phosphorylation converts the low affinity enzyme to the high-affinity form. Phosphorylation by PKC produces reduction in K_m for cofactor and increase in K_i for catechols, kinetic changes which are very similar to those seen after phosphorylation by PKA (Albert *et al.*, 1984).

It is not known at present what proportion of the enzyme is normally in the high-affinity form and whether the non-phosphorylated enzyme shows any activity *in vivo*.

Although *in situ* phosphorylation by PKII results in activation of tyrosine hydroxylase, purified tyrosine hydroxylase phosphorylated by PKII is not activated, but requires an activator protein (Yamauchi *et al.*, 1981). The main effect of Ca^{++}-dependent activation appears to be an increase in V_{max} with no effect on the shape of the pH activity curve (Okuno *et al.*, 1989); there is little or no effect on affinity of the enzyme for cofactor or catechols. When affinity changes are seen, they could be due to a secondary cAMP-dependent activation.

Activity-dependent acceleration of noradrenaline synthesis is likely to be mediated by Ca^{++}-dependent phosphorylation and activation of tyrosine hydroxylase, since the passage of the action potential is accompanied by Ca^{++} influx through voltage-gated channels. The magnitude and subcellular localization of the increase in intracellular Ca^{++} will depend on the number and distribution of the voltage-gated Ca^{++} channels. They are present in high density in the nerve terminal where they play a key role in noradrenaline release and may be responsible for the parallel increase in noradrenaline synthesis rate. They are also found in the somatodendritic membrane, and antidromic stimulation causes an increase in the rate of tyrosine hydroxylation, although there is no evidence that noradrenaline is normally released from the cell body or dendrites.

Phosphorylation and activation of tyrosine hydroxylase by PKA, PKC and cGMP-dependent protein kinase depends on the stimulation of appropriate receptors and will be dealt with in Section 7.

TRANS-SYNAPTIC INDUCTION

In addition to the relatively rapid covalent activation of tyrosine hydroxylase, the amount of enzyme protein can also change. In adult animals *in vivo* such alterations are the result of changes in preganglionic impulse traffic. Increased preganglionic activity, such as occurs during cold stress or after drugs which lead to reflex activation of the sympathetic nervous system (reserpine, phenoxybenzamine), results in an induction of tyrosine hydroxylase and DBH but not dopa decarboxylase (DDC). The induction is prevented by preganglionic section, by chlorisondamine (a ganglionic nicotinic cholinergic receptor antagonist), by cyclohexamide and by actinomycin D (Thoenen & Otten, 1977). These results suggest that the induction is trans-synaptic and involves gene transcription and translation. Electrical stimulation of the rat superior cervical ganglion through implanted electrodes produces induction only when stimulation is orthodromic but not when it is antidromic (Chalazonitis & Zigmond, 1980). This confirms that it is mediated by an effect of the preganglionic transmitter other than the setting up of action potentials. With the development of techniques for measuring the messenger RNA for tyrosine hydroxylase (tyrosine hydroxylase mRNA) it was possible to demonstrate that reserpine-induced increase in tyrosine hydroxylase was accompanied by an increase in tyrosine hydroxylase mRNA; both effects were prevented by decentralization (Black *et al.*, 1985).

Preganglionic impulse traffic also regulates the rate of synthesis of the peptides co-stored and co-released with noradrenaline. However, preganglionic stimulation has opposite effects on peptides and on tyrosine hydroxylase. Thus decentralization and chlorisonidamine decrease, and phenoxybenzamine increases, the substance P content of adult rat sympathetic ganglia (Kessler & Black, 1982).

There have also been studies on changes in preprosomatostatin mRNA (SS mRNA) and tyrosine hydroxylase mRNA in rat sympathetic neurones both *in vivo* and in culture (Spiegel, Kremer & Kessler, 1989). Preganglionic section results in increased ganglion levels of SS mRNA and decreased levels of tyrosine hydroxylase mRNA. Conversely, K^+-or veratridine-induced depolarization of ganglion cells in culture decrease levels of SS mRNA and increase levels of tyrosine hydroxylase mRNA.

The induction of tyrosine hydroxylase by veratridine or K^+-mediated dopalorization in culture appears to conflict with the failure of antidromic stimulation to produce induction of tyrosine hydroxylase *in vivo*. However the selective induction of tyrosine hydroxylase and DBH, but not DDC, by increased preganglionic cholinergic activity differs from the effect of depolarization *in vitro*, since in the latter case the small increases in tyrosine hydroxylase and DBH are accompanied by increases in DDC and MAO (Otten & Thoenen, 1976a, b).

The opposite effects of preganglionic blockade or section on the synthesis of peptides and of the noradrenaline-synthesizing enzymes indicates that there is separate long-term regulation of noradrenaline and its co-stored peptides *in vivo*.

RECEPTOR-MEDIATED MODULATION OF NORADRENERGIC FUNCTION

According to the classical picture, the receptors on sympathetic noradrenergic neurones are the nicotinic cholinergic receptors on the somato-dendritic membrane, whose activation by acetylcholine released from preganglionic nerve terminals, mediates the rapid transmission of impulses across the preganglionic/postganglionic synapse (Bard, 1968). In this model the amount of noradrenaline released from the nerve terminals and the synthesis rate which replaces the released noradrenaline are determined by the number and frequency of nerve impulses, which in turn are dependent on the amount of acetylcholine interacting with nicotinic cholinergic receptors on the somatodendritic membrane.

A wealth of new discoveries necessitate a fundamental revision of this model. Thus preganglionic fibres release more than one transmitter, each of which interacts with more than one receptor subtype; these receptors on the somato-dendritic membrane have effects other than the setting up of rapid excitatory postsynaptic potentials (e.p.s.p.s.) (which mediate transmission of nerve impulses). In addition to the receptors on the somato-dendritic membrane, there are also receptors on the nerve terminals: these can be subdivided into autoreceptors and heteroreceptors. The autoreceptors respond to the sympathetic neurone's own transmitters, which include purines and peptides in addition to noradrenaline. Heteroreceptors respond to neurotransmitters released by adjacent neurones or to hormones circulating in the blood stream.

The receptors on noradrenergic neurones have many possible functions; I shall confine myself to two:

1. The regulation of noradrenaline release.
2. The regulation of noradrenaline synthesis.

REGULATION OF NORADRENALINE RELEASE

Inhibitory Presynaptic Autoreceptors

The inhibitory presynaptic autoreceptor was discovered in the early 1970s (for a review of the history see Langer, 1977) and identified as a new α-adrenoceptor subtype by Berthelsen and Pettinger (1977). Numerous experiments, in which the release of noradrenaline was estimated from the overflow of endogenous or radiolabelled noradrenaline in response to electrical field stimulation of isolated tissues, have demonstrated the presence of inhibitory presynaptic autoreceptors on nerve terminals innervating a wide variety of tissues in many different species (Starke, 1987). In spite of minor pharmacological differences seen in the various preparations, it is generally accepted that the inhibitory presynaptic adrenoceptors are α_2 receptors.

There are two possible sites for the inhibitory action: stimulation of the α_2 receptors can either depress the propagation of the action potential in terminal varicose fibres or act on the release mechanism. Impairment of impulse-propagation in terminal fibres by α_2 receptor activation was proposed by Stjärne (1978). This hypothesis is supported by the finding that α_2-agonists, such as clonidine or noradrenaline, block antidromic conduction in guinea-pig enteric neurones (Morita & North, 1981) or orthodromic conduction in the guinea-pig myenteric plexus (Kadlec et $al.$, 1986). On the other hand in the guinea pig vas deferens α_2-receptor agonists depress the release of ^{3}H-noradrenaline in response to both electrical field stimulation and to direct depolarization by 160 mM K in the presence of TTX (Alberts, Bartfai & Stjärne, 1981; Stjärne, 1986).

In electrophysiological experiments the effects of α_2-receptor agonists and antagonists on the ATP-mediated excitatory junction potentials (EJP) and excitatory junction currents (EJC) were studied. The α_2-receptor agonist clonidine was found to suppress and the α_2-receptor antagonist yohimbine to enhance the probability of release, measured both as the occurrence of discrete events on the rising phase of the evoked EJP (Blakeley et $al.$, 1986) and as the occurrence of evoked EJCs (Stjärne & Stjärne, 1989). The degree of α_2-receptor mediated autoinhibition of the release of ^{3}H-noradrenaline, measured as overflow from a tissue, is much greater than that of the release of ATP, measured as EJPs or EJCs (Stjärne, 1989). The significance of these differences is at present unclear.

The release of noradrenaline is triggered by the influx of Ca^{++} and the demonstration that the effectiveness of α_2-receptor-mediated inhibition of release is inversely related to stimulation frequency and extracellular Ca^{++} concentration suggests that stimulation of presynaptic α_2 receptors reduces the availability of Ca^{++} for the excitation-secretion coupling (Langer, 1977).

It is not possible at present to determine the localization of α_2 adrenoceptors, i.e. whether they are on varicosities or on the inter-varicosity membrane. Their presence on brain noradrenergic synaptosomes supports the former alternative.

The inhibitory effect on release through a reduced availability of Ca^{++} could result from the opening of a K^+ channel, inhibition of voltage-sensitive Ca^{++} channels or an interference with the action of Ca^{++} beyond its entry into the nerve terminal.

Although there is evidence that α_2 activation of receptors on somato-dendritic membranes opens K^+ channels (Andrade & Aghajanian, 1985) the depression by α_2-adrenoceptor agonists of release by high K^+ concentrations is more easily explained by a depression of voltage-gated Ca^{++} channels. The demonstration that a single receptor can be coupled to more than one effector mechanism and that G proteins can reduce the probability of opening of Ca^{++} channels (see section on Receptors) lends support to the view that it is this mechanism which operates in noradrenergic nerve terminals. Schoffelmeer and Mulder (1983) reported that veratrine releases noradrenaline from brain synaptosome even in Ca^{++}-free media containing EGTA, presumably as a result of mobilization of intracellular Ca^{++} following the influx of Na^+; this veratrine-induced release in the absence of Ca^{++} is reduced by clonidine, which would suggest an effect of α_2-receptor activation beyond the entry of Ca^{++}. However α_2-receptor activation does *not* depress noradrenaline release triggered by the Ca^{++} ionophore A23187, which raises intracellular Ca^{++} but bypasses the voltage-gated Ca^{++} channel (De Langen & Mulder, 1980).

The ATP co-released with noradrenaline is rapidly broken down to adenosine. Adenosine has been found to depress noradrenaline release from noradrenergic nerves in various tissues of rat, rabbit and dog (Hedqvist & Fredholm, 1976; Enero & Saidman, 1977; Su, 1978; Verhaeghe *et al.*, 1977). Although this suggests that adenosine receptors on noradrenergic terminals act as presynaptic inhibitory receptors, to what extent they can be regarded as autoreceptors depends on the contribution made by the ATP co-released from nerve terminals, because there are other sources of ATP and adenosine (Fredholm, 1988).

Another group of compounds co-released with noradrenaline are the opioid peptides which are co-stored with noradrenaline and which include met- and leu-enkephalin. These opioids interact with at least three opioid receptor subtypes – the μ, κ and δ subtypes. Electrically evoked noradrenaline release is depressed by opioid receptor agonists; this effect is much more restricted than that of α_2 presynaptic receptors. So far it has been described in vas deferens of mouse, rat, rabbit, and hamster (Henderson, Hughes & Kosterlitz, 1972; Hughes, Kosterlitz & Leslie, 1975;

Henderson & North, 1976), cat nictating membrane (Dubocovich & Langer, 1980) and cat spleen (Gaddis & Dixon, 1982).

Adenosine A_1 receptors and opioid receptors appear to have effector mechanisms which are similar or identical to those of α_2 receptors in that they cause inhibition of adenyl cyclase by being coupled to G_i proteins. There is recent evidence that at noradrenergic terminals there is mutual interaction between presynaptic α_2-adrenoceptors, opioid receptors and P_1-purinoceptors: when any one of these receptors is activated, the inhibitory effect on activation of either of the remaining receptors is blunted. Such an interaction could occur at the level of the receptors (Ramme et al., 1986; Schoffelmeer, Putters & Mulder, 1986) the G proteins (Allgaier, Hertting & Kugelgen, 1987) or the Ca^{++} channel (Limberger et al., 1988).

In experiments with a neuroblastoma-gliomia hybrid cell line both noradrenaline, acting on α_2 receptors, and the enkephalin analogue [D-Ala^2, D-Leu^5] enkephalin (DADLE), acting on delta-opioid receptors, reversibly depressed the amplitude of the calcium current. Though acting through separate receptors, each coupled to pertussis toxin-sensitive G proteins, the responses to the two agonists occlude each other and are unrelated to the ability of these receptors to inhibit adenyl cyclase (McFadzen & Docherty, 1989). The inhibition of the Ca^{++} current is due to a change in the voltage-dependence of channel activation, such that in the presence of the transmitter, channels are not opened easily by small or moderate depolarizations which results in a slower current activation; in response to large depolarizations the channels open readily (Bean, 1989).

Recently Tsien and his colleagues (Lipscombe, Kongsamut & Tsien 1989) have demonstrated that α-adrenergic inhibition of noradrenaline release from sympathetic neurones is mediated by a clonidine-insensitive α_2-adrenoceptor subtype whose activation alters the kinetics of N-type Ca^{++} channels; this reduces the probability of channel opening and accelerates the rate of channel closing. The action involves a G protein, but appears to be independent of intracellular cAMP or PKC. Such a mechanism for inhibition of transmitter release is rapid and localized.

There is at present no evidence that the opioid receptors on sympathetic noradrenergic terminals act as autoreceptors under physiological conditions, i.e. are activated by endogenous opioids co-released with noradrenaline. In order to establish this, patterns of stimulation would have to be used which are known to release opioids from the large vesicles and the effect of opioid receptor antagonists examined in the presence of adenosine A_1 and α_2-adrenoceptor antagonists. No such experiments have yet been carried out.

In some sympathetic neurones NPY is co-stored and co-released with noradrenaline. There are a number of reports which suggest that NPY may inhibit the release of noradrenaline. The evidence in most of the earlier studies is indirect and based on changes in the response of the effector organ to sympathetic nerve stimulation in the presence and absence of NPY (Lundberg et al., 1982; Lundberg et al., 1984). The interpretation of such experiments is complicated by the fact that NPY has a direct effect on smooth muscle cells. More recently the effect of NPY administration on stimulation-induced overflow of endogenous or ^{3}H-noradrenaline has been measured. NPY causes a reduction in noradrenaline overflow which is not blocked by adrenoceptor antagonists (Lundberg & Stjärne, 1984; Pernow, Saria & Lundberg, 1996). In addition to these *in vitro* experiments, Pernow and Lundberg (1989) have shown that administration of peptide PYY, which has been shown to displace NPY binding, reduces the stimulation-induced overflow of noradrenaline in the pig kidney *in vivo*. A specific NPY receptor antagonist is required in order to discover whether such inhibition occurs under physiological conditions.

Facilitatory Presynaptic Autoreceptors

Most of the presynaptic receptors activated by transmitters co-released with noradrenaline depress the release of noradrenaline; to the extent to which these co-transmitters come from the same source as noradrenaline, their release will be depressed to the same extent. There appears to be one autoreceptor subtype which enhances the release of noradrenaline. A possible role of β-adrenoceptors in a positive feed-back mechanism was suggested in 1975 (Adler-Graschinsky & Langer, 1975). The presence of presynaptic β receptors and their identification as β_2 receptors has been confirmed in pharmacological experiments. A small decrease in transmitter release elicited by nerve stimulation is obtained with propranolol in isolated guinea-pig atria (Adler-Graschinsky & Langer, 1975) and the perfused cat spleen (Celuch, Dubocovich & Langer, 1978).

Recently Kazanitz and Enero (1989) reported that in rat atria, in the presence of yohimbine, propranolol blocks the facilitatory effect of isoprenaline on stimulation-induced release of noradrenaline, but does not modify the effect of stimulation alone. The authors conclude that although β_2 receptors are present, they respond to circulating adrenaline rather than neuronally released noradrenaline.

Presynaptic Heteroreceptors

Pharmacological experiments have revealed the presence of a variety of presynaptic heteroreceptors on peripheral noradrenergic nerve terminals (Langer, 1980). These include inhibitory muscarinic cholinergic, dopaminergic, prostaglandin, serotonergic, GABA and histamine receptors. There are facilitatory angiotensin and nicotinic cholinergic receptors. The physiological role of most of these receptors is poorly understood. In the heart electrical stimulation of the vagus nerve reduces the rate and force of contraction of cardiac muscle, by a direct effect, as well as reducing the release of noradrenaline elicited by simultaneous stimulation of postganglionic sympathetic nerves (Loffelholz & Muscholl, 1970). This suggests that the physiological antagonism between the sympathetic and parasympathetic effects on the heart occur at both the presynaptic and post-synaptic levels.

In addition to stimulation of presynaptic heteroreceptors by neurotransmitters released from adjacent parasympathetic postganglionic fibres in the peripheral autonomic nervous system, these receptors can be activated by circulating hormones. At present the best evidence for such an effect on sympathetic terminals is for an action of angiotensin. Evidence for presynaptic angiotensin receptors which enhance the release of noradrenaline comes from studies which have demonstrated an increase in noradrenaline overflow in vascular beds *in vivo* in the presence of angiotensin (Hughes & Roth, 1971; Zimmerman, 1977). In rabbit heart angiotensin increases electrically evoked release of both noradrenaline and DBH (Starke, 1970). In a recent study Trachte (1988) investigated the effect of angiotensin on the two components of the contractile response to electrical stimulation in the vas deferens: the phasic response attributed to ATP and the slow tonic response attributed to noradrenaline. He found that angiotensin decreases the non-adrenergic and increases the adrenergic component of the response but has no effect on the postsynaptic response to noradrenaline or ATP. The author concludes that the effects are due to the activation of presynaptic receptors and that noradrenaline and ATP are derived from different compartments.

REGULATION OF SYNTHESIS

There is at present no evidence that impulse-evoked release of noradrenaline occurs anywhere other than from the varicosities of terminal axons. The enzymes responsible for noradrenergic synthesis of noradrenaline, like all proteins, are synthesized in the cell body and are therefore present throughout the neurone. Only noradrenaline stored in vesicles is protected from deamination by mitochondrial MAO. Ganglion cells contain few vesicles; for this reason a large proportion of the product of tyrosine hydroxylation proceeds only as far as dopamine and much of this is deaminated to DOPAC. It has been calculated that axoplasmic transport of noradrenaline in vesicles contributes only 1% to the terminal store (Geffen & Rush, 1968). The physiologically significant synthesis of noradrenaline therefore occurs in the nerve terminals.

Noradrenergic terminals make up a very small proportion of the mass of the tissues they innervate. Studies of the regulation of tyrosine hydroxylase, similar to those on nerve terminals of central catecholaminergic neurones, are therefore difficult to carry out. Adrenomedullary cells, phaeochromocytoma cell lines and ganglion cells have been investigated and these studies have demonstrated the phosphorylation of tyrosine hydroxylase by various protein kinases and the kinetic changes resulting from these phosphorylations. The question to be answered is what are the receptors that are coupled to these second messenger systems and how are these receptors stimulated under physiological conditions?

There is quite an extensive literature on the neurotransmitters that regulate tyrosine hydroxylase in adrenomedullary and PC12 cells, but they can tell us little about the regulation of noradrenaline synthesis in sympathetic noradrenergic nerve terminals. There have also been a number of studies on the regulation by neurotransmitters of tyrosine hydroxylase in superior cervical ganglia. Preganglionic stimulation of the superior cervical ganglion causes a marked increase in tyrosine hydroxylase activity as measured by dopa accumulation in the presence of a DDC inhibitor (Ip, Perlman & Zigmond, 1983). In the absence of a dopamine decarboxylase inhibitor preganglionic stimulation leads to an increase in DOPAC concentration (Anden & Grabowska-Anden, 1986), which indicates that although there is accelerated tyrosine hydroxylation, it does not result in increased synthesis of noradrenaline, because in the absence of an adequate number of vesicles, the conversion of dopamine to noradrenaline by DBH becomes the rate limiting step. The contribution of the accelerated tyrosine hydroxylation resulting from preganglionic nerve stimulation to the releasable store of noradrenaline is therefore questionable.

It might be argued that receptors and their intracellular messenger systems found in cell bodies and dendrites of ganglion cells would also be present in the nerve terminals and therefore serve as models for the nerve terminal. The problem is that cell bodies and nerve terminals 'see' quite different neurotransmitters. Thus the cell bodies will receive the neurotransmitters of the preganglionic nerve terminals, which include peptides in addition to acetylcholine and interact with more than one receptor subtype. Although noradrenergic nerve terminals may interact with transmitters from cholinergic fibres, these will be postganglionic parasympathetic not preganglionic sympathetic fibres. This means not only that the co-transmitters of ACh may be quite different, but that they will be activated under different physiological conditions.

These difficulties are illustrated by the effects of muscarinic receptor activation in ganglion cells. The acceleration of tyrosine hydroxylation in the superior cervical ganglion by preganglionic stimulation could be resolved into effects attributable to stimulation by both nicotonic and muscarinic cholinergic receptors. However in sympathetic nerve terminals in the heart, muscarinic receptor activation causes inhibition of transmitter release (see above). Although there is as yet no evidence for the effect muscarinic receptors have on noradrenaline synthesis in peripheral noradrenergic terminals, in central noradrenergic synaptosomes muscarinic receptor activation depresses both release and synthesis of noradrenaline (Birch & Fillenz, 1986).

In addition to the cholinergic element, there is also a non-cholinergic element implicated in the activation of ganglionic tyrosine hydroxylase by preganglionic stimulation. Screening of a large number of peptides revealed a powerful effect of secretin and vasoactive intestinal polypeptide (VIP) on tyrosine hydroxylase activity (Ip, Baldwin & Zigmond, 1984) which is mediated by an increase in cAMP (Ip, Baldwin & Zigmond, 1985).

The effect of these peptides on tyrosine hydroxylase activity in sympathetic nerve terminals has also been studied (by measuring the accumulation of DOPA in the presence of a DDC inhibitor) in iris, submandibular gland and pineal gland (Schwarzschild & Zigmond, 1989). These tissues are innervated by sympathetic nerve terminals whose cell bodies are in the superior cervical ganglion. Both secretin and VIP were found to accelerate tyrosine hydroxylation by a cAMP-dependent mechanism. The effects therefore are the same on the cell body and the terminals. In the ganglion, VIP-like immunoreactivity appears to be localized in the preganglionic fibres and presumably contributes to excitation of the post-ganglionic neurones. VIP in autonomic end-organs, on the other hand, is contained primarily if not exclusively, in varicosities of parasympathetic fibres. The parasympathetic innervation of the iris has effects which are opposite to those of the sympathetic innervation. The stimulation of tyrosine hydroxylase activity in sympathetic nerve terminals in the iris by VIP from parasympathetic fibres, would appear to conflict with this general pattern.

At present, therefore, there is evidence which suggests that noradrenaline synthesis in nerve terminals can be regulated by presynaptic heteroreceptors; the physiological role of neurotransmitter-mediated activation however is not clear. It is possible that some presynaptic receptors could be activated by circulating hormones, where similar effects on cell bodies and terminals would make much more sense.

There is no evidence at present that presynaptic autoreceptors modulate noradrenaline synthesis as well as release in peripheral noradrenergic neurones. In noradrenergic synaptosomes from cortex and hippocampus, β-adrenoceptor stimulation accelerates synthesis, whereas α_2-adrenoceptor stimulation depresses K^+-enhanced noradrenaline synthesis. Although the effects on synthesis are very similar to the effects on noradrenaline release, there is some evidence that the underlying mechanisms are different. Thus, the α_2-adrenoceptor agonist does not inhibit the Ca^{++} ionophore A23187-induced release of noradrenaline (De Langen & Mulder, 1980) but blocks the Ca^{++} ionophore-induced acceleration of synthesis (Birch & Fillenz, 1985). This fits in with the multiple effector mechanisms of receptors; it is likely that the effect on release is mediated by the direct action of the G_i protein on the Ca^{++} channel (which is by-passed by the ionophore) whereas the effect on tyrosine hydroxylase is through an intracellular second messenger.

ROLE OF RECEPTORS IN INDUCTION

As discussed in Section 6 the control of the synthesis of both the noradrenaline-specific enzymes and the peptides co-stored with noradrenaline is mediated by the activation of the somatodendritic nicotinic cholinergic receptors. There has been controversy about the mechanism of induction of tyrosine hydroxylase.

Guidotti and Costa (1977) proposed that in adrenal medullary cells cholinergic stimulation resulted in the translocation of catalytic subunits of protein kinase A to the nucleus where it stimulated gene expression. This does not appear to be the mechanism in sympathetic ganglia; although incubation with membrane-permeating analogues of cAMP or with cholera toxin (which produces prolonged stimulation of adenyl cyclase) causes induction of tyrosine hydroxylase in adrenal medullary cells (Guidotti *et al.*, 1981), it does not cause induction in sympathetic ganglion cells (Thoenen & Otten, 1975; Hefti *et al.*, 1982). The finding that induction in culture mediated by elevated K^+ is blocked by trifluoperazine, a Ca^{++}/CM inhibitor, suggests that Ca^{++} may play a role. However, the fact that NGF-mediated induction is not inhibited by trifluoperazine and that K^+-induced induction in culture differs from *in vivo* induction, leaves the question of the mechanism of the *in vivo* receptor-mediated induction undecided.

CONCLUSION

Noradrenaline released from sympathetic neurones has, what may be, a uniquely wide range of physiological effects. This is attributable to the characteristics of its storage, release, interaction with receptors and synthesis.

Noradrenaline is found in at least two separate populations of vesicles where it is co-stored with different neuroactive chemical compounds. The release from the two kinds of vesicles is regulated by different control mechanisms.

Noradrenaline interacts with a variety of adrenoceptor subtypes, none of which are directly coupled to an ion channel, but between them command most, if not all, the effector mechanisms known at present.

The synthesis of noradrenaline involves a number of catalytic enzymes; in contrast to all the other non-peptide transmitters (except adrenaline) the vesicles in noradrenergic neurones are involved not only in storage and release but also synthesis, because of the vesicular localization of DBH, which catalyzes the last step in synthesis. This means that where there are few vesicles, dopamine-β-hydroxylation may become rate limiting. The activity of tyrosine hydroxylase which normally determines the synthesis rate in the nerve terminal is subject to multiple regulatory mechanisms, mediated by phosphorylation by a variety of protein kinases.

In addition to the relatively rapid covalent activation, the gene expression for tyrosine hydroxylase is also modulated, leading to regulation in the synthesis of enzyme protein.

Release, tyrosine hydroxylase activity and tyrosine hydroxylase induction are regulated by a variety of receptors, situated both on the somato-dendritic and the nerve terminal membrane. These receptors regulate the activity of noradrenergic neurones over time periods which range from milliseconds to several weeks.

ACKNOWLEDGEMENT

I wish to thank Christine Lake for helping to prepare the manuscript for this chapter.

REFERENCES

Aberer, W., Stitzel, R., Winkler, H. and Huber, E. (1979). Accumulation of ^{3}H-ATP in small dense core vesicles of superfused vasa deferentia. *Journal of Neurochemistry*, **33**, 797–801.

Adler-Grachinsky, E. and Langer, S.Z. (1975). Possible role of a β-adrenoceptor in the regulation of noradrenaline release by nerve stimulation through a positive feedback mechanism. *British Journal of Pharmacology*, **53**, 43–50.

Åhlquist, R.P. (1984). A study of adrenotropic receptors. *American Journal of Physiology*, **153**, 586–600.

Albert, K.A., Helmer-Matyjek, E., Nairn, A.C., Muller, T.H., Haycock, J.W., Greene, L.A., Goldstein, M. and Greengard, P. (1984). Calcium/phospholipid-dependent protein kinase (protein kinase C) phosphorylates and activates tyrosine hydroxylase. *Proceedings of the National Academy of Sciences of the United States of America*, **81**, 7713–7717.

Alberts, P., Bartfai, T. and Stjärne, L. (1981). Site(s) and ionic basis of alpha-autoinhibition and facilitation of ^{3}H-NA secretion in guinea-pig vas deferens. *Journal of Physiology*, **312**, 297–334.

Allgaier, C., Hertting, G. and Kugelgen, O.V. (1987). The adenosine receptor mediated inhibition of noradrenaline release possibly involves a N-protein and is increased by α_2-autoreceptor blockade. *British Journal of Pharmacology*, **90**, 403–412.

Anden, N.-E. and Grabowska-Anden, M. (1985). Synthesis and utilization of catecholamines in the rat superior cervical ganglion following changes in the nerve impulse flow. *Journal of Neural Transmission* **64**, 81–92.

Andrade, R. and Aghajanian, G.K. (1985). Opiate- and α_2-adrenoceptor-induced hyperpolarizations in brain slices: reversal by cyclic adenosine 3'5' monophosphate analogues. *Journal of Neuroscience*, **5**, 2359–2364.

Åstrand, P. and Stjärne, L. (1988). On the secretory activity of single varicosities in the sympathetic nerves innervating the rat tail artery. *Journal of Physiology*, **409**, 207–220.

Åstrand, P., Brock, J.A. and Cunnane, T.C. (1988). Time course of transmitter action at the sympathetic neuroeffector junction in rodent vascular and non-vascular smooth muscle. *Journal of Physiology*, **401**, 657–670.

Axelrod, J. and Tomchick, R. (1958). Enzymatic O-methylation of epinephrine and other catecholamines. *Journal of Biological Chemistry*, **233**, 702–705.

Bard, P. (1968). Autonomic nervous system or efferent pathway to visceral effectors. In *Medical Physiology 12th Edition* by Mountcastle, V.B. C.V. Mosby Company, St. Louis.

Basbaum, C.B. and Heuser, J.E. (1979). Morphological studies of stimulated adrenergic axon varicosities in the mouse vas deferens. *Journal of Cell Biology*, **80**, 310–325.

Bean, B. (1989). Neurotransmitter inhibition of neuronal calcium current by changes in channel voltage dependence. *Nature*, **340**, 153–156.

Belmar, J., De Potter, W.P. and De Schaepdryver, A.F. (1974). Subcellular distribution of noradrenaline and dopamine-β-hydroxylase in the hypothalamus of the rat. Evidence for the presence of two populations of noradrenaline storage particles. *Journal of Neurochemistry*, **23**, 607–609.

Benovic, J.L., Strasser, R.H., Caron, M.G. and Lefkowitz, R.J. (1986). β-adrenergic receptor kinase; identification of a novel protein kinase that phosphorylates the agonist-occupied form of the receptor. *Proceedings of the National Academy of Sciences of the United States of America*, **83**, 2797–2801.

Berthelsen, S. and Pettinger, W.P. (1977). A functional basis for classification of alpha-adrenergic receptors. *Life Sciences*, **21**, 595–606.

Birch, P.J. and Fillenz, M. (1985). Stimulation of noradrenaline synthesis by the calcium ionophore A23187 and its modulation by presynaptic receptors. *Neuroscience Letters*, **62**, 187–192.

Birch, P.J. and Fillenz, M. (1986). Muscarinic receptor activation inhibits both release and synthesis of noradrenaline in rat hippocampal synaptosomes. *Neurochemistry International*, **8**, 171–177.

Bisby, M.A. and Fillenz, M. (1971). The storage of endogenous noradrenaline in sympathetic nerve terminals. *Journal of Physiology*, **215**, 163–179.

Bisby, M.A., Fillenz, M. and Smith, A.D. (1973). Evidence for the presence of dopamine-β-hydroxylase in both populations of noradrenaline storage vesicles in sympathetic nerve terminals of the rat vas deferens. *Journal of Neurochemistry*, **20**, 245–248.

Black, I.B., Chikaraishi, D.M. and Lewis, E.J. (1985). Transsynaptic increase in RNA coding for tyrosine hydroxylase in a rat sympathetic ganglion. *Brain Research*, **339**, 151–153.

Blakeley, A.G.H. and Cunnane, T.C. (1979). The packeted release of transmitter from the sympathetic nerves of the guinea-pig vas deferens: an electrophysiological study. *Journal of Physiology*, **296**, 85–96.

Blakeley, A.G.H., Mathie, A. and Petersen, S.A. (1986). Interactions between the effects of yohimbine, clonidine and (Ca)$_0$ on the electrical response of the mouse vas deferens. *British Journal of Pharmacology*, **88**, 807–814.

Blaschko, H. (1939). The specific action of L-Dopa-decarboxylase. *Journal of Physiology*, **96**, 50–51P.

Blaschko, H., Richter, D. and Schlossmann, H. (1937). The inactivation of adrenaline. *Journal of Physiology*, **90**, 1–17.

Bouvier, M., Leeb-Lundberg, L.M.F., Benovic, J.L., Caron, M.G. and Lefkowitz, R.J. (1987). Regulation of adrenergic receptor function by phosphorylation. II. Effects of agonist occupancy on phosphorylation of α_1 and β_2-adrenergic receptors by protein kinase C and the cyclic AMP-dependent protein kinase. *Journal of Biological Chemistry*, **262**, 106–113.

Brimijoin, S. (1974). Local changes in subcellular distribution of Dopamine-β-hydroxylase after blockade of axonal transport. *Journal of Neurochemistry*, **22**, 347–353.

Brock, J.A. and Cunnane, T.C. (1987). Relationship between the nerve action potential and transmitter release from sympathetic postganglionic nerve terminals. *Nature*, **326**, 605–607.

Brock, O.J. and Fonnum, F. (1972). The regional and subcellular distribution of catechol-O-methyltransferase in the rat brain. *Journal of Neurochemistry*, **19**, 2049–2055.

Burnstock, G. (1990). Noradrenaline and ATP as cotransmitters in sympathetic nerves. *Neurochemistry International*, **17**, 357–368.

Burnstock, G. and Holman, M.E. (1961). The transmission of excitation from autonomic nerve to smooth muscle. *Journal of Physiology*, **155**, 115–133.

Burnstock, G. and Holman, M.E. (1962). Spontaneous potentials at sympathetic nerve endings in smooth muscle. *Journal of Physiology*, **160**, 446–460.

Bylund, D.B. (1985). Heterogeneity of α_2 adrenergic receptors. *Pharmacology, Biochemistry and Behavior*, **22**, 835–843.

Bylund, D.B. and U'Prichard, D.C. (1983). Characterization of α_1 and α_2 adrenergic receptors. *International Review of Neurobiology*, **24**, 343–422.

Bylund, D.B., Ray-Prenger, C. and Murphy, T.J. (1988). α_{2A} and α_{2B} adrenergic receptor subtypes: antagonist binding in tissues and cell lines containing only one subtype. *Journal of Pharmacology and Experimental Therapeutics*, **245**, 600–607.

Cahill, A.L. and Perlman, R.L. (1987). Preganglionic stimulation increases the phosphorylation of tyrosine hydroxylase in the superior cervical ganglion by both cAMP-dependent and calcium-dependent protein kinase. *Biochimica et Biophysica Acta*, **930**, 454–462.

Campbell, D.G., Hardie, D.G. and Vulliet, P.R. (1986). Identification of four phosphorylation sites in the N-terminal region of tyrosine hydroxylase. *Journal of Biological Chemistry*, **261**, 10489–10492.

Cannon, W.B. and Rosenblueth, A. (1937). *Autonomic Neuro-Effector Systems*. New York, The Macmillan Company.

Cannon, W.B. and Rosenblueth, A. (1949). The supersensitivity of denervated structures. *Experimental Biology Monographs*, New York: Macmillan.

Casey, P.J., Graziano, M.P., Freissmuth, M. and Gilman, A.G. (1988). Role of G proteins in transmembrane signaling. *Cold Spring Harbor Symposia on Quantitative Biology*, **53**, 203–208.

Celuch, S.M., Dubocovich, M.L. and Langer, S.Z. (1978). Stimulation of presynaptic β-adrenoceptors enhances ^{3}H-noradrenaline release during nerve stimulation in the perfused cat spleen. *British Journal of Pharmacology*, **63**, 97–109.

Chalazonitis, A. and Zigmond, R.E. (1980). Effects of synaptic and antidromic stimulation on tyrosine hydroxylase activity in the rat superior cervical ganglion. *Journal of Physiology*, **300**, 525–538.

Chubb, I.W., De Potter, W.P. and De Schaepdryver, A.F. (1970). Evidence for two types of noradrenergic storage particles in dog spleen. *Nature* (London), **228**, 1203–1204.

Coates, J., Jahn, U. and Weetman, D.F. (1982). The existence of a new subtype of α-adrenoceptor on the rat anococcygeus is revealed by SGD 101/75 and phenoxybenzamine. *British Journal of Pharmacology*, **75**, 549–552.

Costa, M., Furness, J.B. and Gibbins, I.L. (1986). Chemical coding of enteric neurons. *Progress in Brain Research*, **68**, 217–240.

Cotecchia, S., Schwinn, D.A., Randall, R.R., Lefkowitz, R.J., Caron, M.G. and Kobilka, B.K. (1988). Molecular cloning and expression of the cDNA for the hamster α_1-adrenergic receptor. *Proceedings of the National Academy of Sciences of the United States of America*, **85**, 7159–7163.

Cunnane, T.C. and Stjärne, L. (1982). Secretion of transmitter from individual varicosities of guinea-pig and mouse vas deferens: all-or-none and extremely intermittent. *Neuroscience*, **7**, 2565–2576.

Dahlström, A., Häggendal, J. and Hökfelt, J. (1966). The noradrenaline content of the varicosities of sympathetic adrenergic nerve terminals in the rat. *Acta Physiologica Scandinavica*, **67**, 289–294.

De Camilli, P. & Jahn, R. (1990). Pathways to regulated exocytosis in neurons. *Annual Reviews of Physiology*, **52**, 625–647.

Denney, R.M., Fritz, R.R., Patel, N.T. and Abell, C.W. (1982). Human liver MAO-A and MAO-B separated by immunoaffinity chromatography with MAO-B specific monoclonal antibody. *Science*, **215**, 1400–1403.

Denney, R.M., Fritz, R.R., Patel, N.J., Widen, S.G. and Abell, C.W. (1983). Use of a monoclonal antibody for comparative studies of MAO-B in mitochondrial extracts of human brain and peripheral tissues. *Molecular Pharmacology*, **24**, 60–68.

De Langen, C.D.J. and Mulder, A. (1980). On the role of calcium ions in the presynaptic α-receptor mediated inhibition of ^{3}H-noradrenaline release from rat brain cortex synaptosomes. *Brain Research*, **185**, 399–408.

De Potter, W.P. and Chubb, I.W. (1977). Biochemical observations on the formation of small noradrenergic vesicles in the splenic nerve of the dog. *Neuroscience*, **2**, 167–174.

De Potter, W.P., Coen, E.P. and De Potter, R.W. (1987). Evidence for the coexistence of co-release of (met) enkephalin and noradrenaline from sympathetic nerves of the bovine vas deferens. *Neuroscience*, **20**, 855–866.

De Potter, W.P., De Schaepdryver, A.F., Moerman, E.J. and Smith, A.D. (1969). Evidence for the release of vesicle proteins together with noradrenaline upon stimulation of the splenic nerve. *Journal of Physiology*, **204**, 102–104P.

Dixon, R.A.F., Kobilka, B.K., Strader, C.D., Benovic, J.L., Dohlman, H.G., Frielle, T. and Bolanowski, M.A. (1986). Cloning of the gene and cDNA for mammalian β-adrenergic receptor and homology with rhodopsin. *Nature*, **321**, 75–79.

Dixon, R.A.F., Sigal, I.S. and Strader, C.D. (1988). Structure-function analysis of the β-adrenergic receptor. *Cold Spring Harbor Symposia of Quantitative Biology*, **53**, 487–498.

Dixon, R.A.F., Sigal, I.S., Rands, E., Register, R.B., Candelore, M.R., Blake, A.D. and Strader, C.D.

(1987). Ligand binding to the β-adrenergic receptor involves its rhodopsin-like core. *Nature*, **326**, 73.

Dohlman, H.G., Bouvier, M., Benovic, J.L., Caron, M.G. and Lefkowitz, R.J. (1987). The multiple membrane spanning topography of the β_2-adrenergic receptor: Localization of the sites of binding, glycosylation and regulatory phosphorylation by limited proteolysis. *Journal of Biological Chemistry*, **262**, 14282.

Douglas, W.W. (1968). Stimulus-secretion coupling: the concept and clues from chromaffin and other cells. *British Journal of Pharmacology*, **34**, 451–474.

Dubocovich, M.L. and Langer, S.Z. (1980). Pharmacological differentation of presynaptic inhibitory α-adrenoceptors and opiate receptors in the cat nictitating membrane. *British Journal of Pharmacology*, **70**, 383–393.

Egashira, T. and Yamamaha, Y. (1981). Further studies on the synthesis of A-form of monoamine oxidase. *Japanese Journal of Pharmacology*, **31**, 763–770.

Elliott, T.R. (1905). The action of adrenalin. *Journal of Physiology*, **32**, 401–467.

Emorine, L.J., Marullo, S., Briend-Sutren, M.M., Patey, G., Tate, K., Dalavier-Klutchko, C. and Strosberg, A.D. (1989). Molecular characterization of the human β_3-adrenergic receptor. *Science*, **245**, 1118–1121.

Enero, M.A. and Saidman, B.Q. (1977). Possible feed-back inhibition of noradrenaline release by purine compounds. *Naunyn Schmiedebergs Archives of Pharmacology*, **297**, 39–46.

Euler von, U.S. (1946). A specific sympathomimetic ergone in adrenergic nerve fibres (sympathin) and its relation to adrenaline and nor-adrenaline. *Acta Physiologica Scandinavica*, **12**, 73–97.

Falck, B., Hillarp, N.A., Thieme, G. and Torp, A. (1962). Fluorescence of catecholamines and related compounds condensed with formaldehyde . *Journal of Histochemistry and Cytochemistry*, **10**, 348–354.

Farrell, K.E. (1968). Fine structure of nerve fibres in smooth muscle of the vas deferens in normal and reserpinized rats. *Nature*, **217**, 279–281.

Fillenz, M. (1971). Fine structure of noradrenaline storage vesicles in nerve terminals of the rat vas deferens. *Philosophical Transactions of the Royal Society*, **261**, 319–323.

Fillenz, M. and Pollard, R.M. (1976). Quantitative differences between sympathetic nerve terminals. *Brain Research*, **109**, 443–454.

Fillenz, M. and Stanford, S.C. (1981). Vesicular noradrenaline stores in peripheral nerves of the rat and their modification by tranylcypromine. *British Journal of Pharmacology*, **73**, 401–404.

Fillenz, M., Howe, P.R.C. and West, D.P. (1976). Vesicular noradrenaline in nerve terminals of rat heart following inhibition of monoamine oxidase and administration of noradrenaline. *Neuroscience*, **1**, 113–116.

Fischer-Colbrie, R., Hagn, C. and Schober, M. (1987). Chromogranins A, B, C,: widespread constituents of secretory vesicles. *Annals of the New York Academy of Sciences*, **493**, 120–134.

Fredholm, B.B. (1988). Presynaptic adenosine receptors. *ISI Atlas of Science: Pharmacology*, **2**, 257–260.

Fredholm, B.B., Fired, G. and Hedqvist, P. (1982). Origin of adenosine released from rat vas deferens by nerve stimulation. *European Journal of Pharmacology*, **79**, 233–243.

Fried, G. (1980). Small noradrenergic storage vesicles isolated from rat vas deferens – biochemical and morphological characterization. *Acta Physiologica Scandinavica*, Suppl., **493**, Vol. 111, 1–28.

Fried, G., Lagercrantz, H. and Hökfelt, T. (1978). Improved isolation of small noradrenergic vesicles from rat seminal ducts following castration. A density gradient and morphological study. *Neuroscience*, **3**, 1271–1291.

Fried, G., Terenius, L., Hökfelt, T. and Goldstein, M. (1985). Evidence for differential localization of noradrenaline and neuropeptide Y in neuronal storage vesicles isolated from rat vas deferens. *Journal of Neuroscience*, **5**, 450–458.

Frielle, T.M., Collins, S., Daniel, K.W., Caron, M.G., Lefkowitz, R.J. (1987). Cloning of the cDNA for the β_1-adrenergic receptor. *Proceedings of the National Academy of Sciences of the United States of America*, **84**, 7920–7924.

Gaddis, R.R. and Dixon, W.R. (1982). Modulation of peripheral adrenergic neurotransmission by methionine-enkephalin. *Journal of Pharmacology and Experimental Therapeutics*, **221**, 282–288.

Geffen, L.B. and Rush, R.A. (1968). Transport of noradrenaline in sympathetic nerves and the effect of nerve impulses on its contribution to transmitter stores. *Journal of Neurochemistry*, **15**, 925–930.

Geffen, L.B., Livett, B.G. and Rush, R.A. (1969). Immunological localization of chromogranins and their release by nerve impulses. *Journal of Physiology*, **204**, 58–59P.

Geffen, L.B., Livett, B.G. and Rush, R.A. (1970). Immuno-histochemical localization of chromogranins in sheep sympathetic neurons and their release by nerve impulses. In: *New Aspects of Storage and Release Mechanism of Catecholamines*, eds. Schümann, H.G. and Kroneberg, G. (Bayer Symposium II) Springer, Berlin, pp. 58–72.

Gilman, A.G. (1987). Transducers of receptor generated signals. *Annual Review of Biochemistry*, **56**, 615–632.

Grillo, M. and Palay, S.L. (1962). Granule containing vesicles in the autonomic nervous system. In *Electron Microscopy*, ed. S.S. Breese Jr., Vol. 2, p. V-1. Academic Press, New York.

Grima, B., Lamoureux, A., Blanot, F., Biguet, N.F. and Mallet, J. (1985). Complete coding sequence of rat tyrosine hydroxylase mRNA. *Proceedings of the National Academy of Sciences of the United States of America*, **82**, 617–621.

Grohmann, M. (1987). The activity of the neuronal and extraneuronal catecholamine-metabolizing enzymes in the perfused rat heart. *Naunyn Schmiedebergs Archives of Pharmacology*, **336**, 139–147.

Guidotti, A. and Costa, E. (1977). Trans-synaptic regulation of tyrosine 3-mono-oxygenase biosynthesis in rat adrenal medulla. *Biochemical Pharmacology*, **26**, 817–823.

Guidotti, A., Chuang, D.M., Kumakura, K. and Costa, E. (1981). Molecular biology of tyrosine hydroxylase induction. In: *Function and Regulation of Monoamine Enzymes*, eds. Usdin, E., Weiner, N. and Youdim, M.B.H. Macmillan Publishers, London and Basingstoke, UK, pp. 141–148.

Hagn, C., Klein, R.L., Fischer-Colbrie, R., Douglas II, B.H. and Winkler, H. (1986). An immunological characterization of five common antigens of chromaffin granules and of large dense-cored vesicles of sympathetic nerve. *Neuroscience Letters*, **67**, 295–300.

Han, C., Abel, P.W. and Minneman, K.P. (1987). α_1-Adrenoceptor subtypes linked to different mechanisms for increasing intracellular Ca^{2+} in smooth muscle. *Nature*, **329**, 333–335.

Hedqvist, P. and Fredholm, B.B. (1976). Effects of adenosine on adrenergic neurotransmission; prejunctional inhibition and postjunctional enhancement. *Naunyn Schmiedebergs Archives of Pharmacology*, **293**, 217–223.

Hefti, F., Gnahn, H., Schwab, E. and Thoenen, H. (1982). Induction of tyrosine hydroxylase by nerve growth factor and by elevated K concentrations in cultures of dissociated sympathetic neurons. *Journal of Neuroscience*, **2**, 1554–1566.

Henderson, G. and North, R.A. (1976). Depression by morphine of excitatory junction potentials in the vas deferens of the mouse. *British Journal of Pharmacology*, **57**, 341–346.

Henderson, G., Hughes, J. and Kosterlitz, H.W. (1972). A new example of a morphine-sensitive neuroeffector junction: adrenergic transmission in the mouse vas deferens. *British Journal of Pharmacology*, **46**, 764–766.

Hoeldke, R. and Kaufman, S. (1977). Bovine adrenal tyrosine hydroxylase: Purification and properties. *Journal of Biological Chemistry*, **252**, 3160–3169.

Hoffman, B.B. and Lefkowitz, R.J. (1980). Radioligand binding studies of adrenergic receptors: new insights into molecular and physiological regulation. *Annual Review of Pharmacology and Toxicology*, **20**, 581–608.

Hökfelt, T. (1966). The effect of reserpine on the intraneuronal vesicles of the rat vas deferens. *Experientia*, **22**, 56.

Hökfelt, T. (1969). Distribution of noradrenaline storing particles in peripheral adrenergic neurons as revealed by electron microscopy. *Acta Physiologica Scandinavica*, **76**, 427–440.

Hökfelt, T. (1973). On the origin of small adrenergic storage vesicles. Evidence for local formation in nerve endings after chronic reserpine treatment. *Experientia*, **29**, 580–582.

Hökfelt, T., Elfvin, L-G, Elde, R., Schultzberg, M., Goldstein, M. and Luft, R. (1977). Occurrence of somatostatin-like immunoreactivity in some sympathetic noradrenergic neurons. *Proceedings of the National Academy of Sciences of the United States of America*, **74**, 3587–3591.

Holman, M. (1970). Junction potentials in smooth muscle. In: *Smooth Muscle*, ed. E. Bülbring, A. Brading, A. Jones and T. Tomita, pp. 244–288. London: Edward Arnold Ltd.

Holtzmann, E. (1977). The origin of secretory packages, especially synaptic vesicles. *Neuroscience*, **2**, 327–355.

Hörtnagl, H., Hörtnagl, H. and Winkler, H. (1969). Bovine splenic nerve: characterization of noradrenaline-containing vesicles and other cell organelles by density gradient centrifugation. *Journal of Physiology*, **205**, 103–114.

Hughes, J., Kosterlitz, H.W. and Leslie, F. (1975). Effects of morphine on adrenergic transmission in the mouse vas deferens. Assessment of agonist and antagonist potencies of narcotic analgesics. *British Journal of Pharmacology*, **53**, 371–381.

Hughes, J. & Roth, R.H. (1971). Evidence that angiotensin enhances transmitter release during sympathetic nerve stimulation. *British Journal of Pharmacology*, **41**, 239–255.

Huttner, W.B., Benedum, V.M. and Rosa, P. (1988). Biosynthesis, Structure and Function of the Secretogranins/Chromogranins. In *Molecular Mechanisms in Secretion*, eds. N.A. Thorn, M. Treiman and O.H. Petersen, Alfred Benzon Symposium 25, pp. 380–389. Munksgaard, Copenhagen.

Ip, N.Y., Baldwin, C. and Zigmond, R.E. (1984). Acute stimulation of ganglionic tyrosine hydroxylase activity by secretin, VIP and PHI. *Peptides*, **5**, 309–312.

Ip, N.Y., Baldwin, C. and Zigmond, R.E. (1985). Regulation of the concentration of adenosine 3'5'-cyclic monophosphate and the activity of tyrosine hydroxylase in the rat superior cervical ganglion by three neuropeptides of the secretin family. *Journal of Neuroscience*, **5**, 1947–1954.

Ip, N.Y., Perlman, R.L. and Zigmond, R.E. (1983). Acute transsynaptic regulation of tyrosine 3-monooxygenase in the rat superior cervical ganglion: Evidence for both cholinergic and non-cholinergic mechanisms. *Proceedings of the National Academy of Sciences of the United States of America*, **80**, 2081–2085.

Iversen, L.L. (1967). *The Uptake and Storage of Noradrenaline in Sympathetic Nerves*. Cambridge University Press.

Jakobs, K.H. and Schultz, G. (1983). Occurrence of a hormone-sensitive inhibitory coupling component of the adenylate cyclase in S49 lymphoma cyc^{-} variants. *Proceedings of the National Academy of Sciences of the United States of America*, **80**, 3899–3902.

Johnston, J.P. (1968). Some observations upon a new inhibitor of monoamine oxidase in brain tissue. *Biochemical Pharmacology*, **17**, 1285–1297.

Kadlec, O., Somogyi, G.T., Seferna, I., Masek, K. and Vizi, E.S. (1986). Interactions between the duration of stimulation and noradrenaline on cholinergic transmission in the myenteric plexus-smooth muscle preparation. *Brain Research Bulletin*, **16**, 171–178.

Kapeller, K. and Mayor, D. (1967). The accumulation of noradrenaline in constricted sympathetic nerves as studied by fluorescence and electron microscopy. *Proceedings of the Royal Society of London*, **167**, 282–292.

Katz, B. (1958). Microphysiology of the neuromuscular junction. *Johns Hopkins Hospital Bulletin*, **102**, 275–312.

Kazanietz, M.G. and Enero, M.A. (1989). Modulation of noradrenaline release by presynaptic α_2 and β adrenoceptors in rat atria. *Naunyn Schmiedebergs Archives of Pharmacology*, **340**, 274–278.

Kessler, J.A. and Black, I.B. (1982). Regulation of substance P in adult rat sympathetic ganglia. *Brain Research*, **234**, 182–187.

Kirshner, N. and Viveros, O.H. (1972). The secretory cycle in the adrenal medulla. *Pharmacological Review*, **24**, 385–398.

Klein, R.L. (1982). Chemical composition of the large noradrenergic vesicles. In: *Neurotransmitter Vesicles*, ed. Klein, R.L., Lagercrantz, H. and Zimmerman, H. pp. 133–174. New York: Academic Press.

Klein, R.L., Wilson, S.P., Dzielak, D.J., Yang, W-H. and Viveros, O.H. (1982). Opioid peptides and noradrenaline co-exist in large dense cored vesicles from sympathetic nerve. *Neuroscience*, **7**, 2255–2261.

Klein, R.L., Thureson-Klein, A., Chen-Yen, S.-H., Baggett, J.McC., Gasparis, M.S. and Kirksey, D.F. (1979). Dopamine-β-hydroxylase distribution in density gradients: physiological and artefactual implications. *Journal of Neurobiology*, **10**, 291–307.

Kobilka, B.K., Dixon, R.A.F., Frielle, T., Dohlman, H.G., Bolanowski, M.A., Sigal, I.S., Yang-Fent, T.L., Francke, U., Caron, M.G. and Lefkowitz, R.J. (1987a). cDNA for the human β_2-adrenergic receptor: A protein with multiple membrane spanning domains and a chromosomal location shared with the PDGF receptor gene. *Proceedings of the National Academy of Sciences of the United States of America*, **84**, 46.

Kobilka, B.K., Kobilka, T.S., Daniel, K., Regan, J.B., Caron, M.G. and Lefkowitz, R.J. (1988). Chimeric α_2, β_2 adrenergic receptors: delineation of domains involved in effector coupling and ligand binding specificity. *Science*, **240**, 1310–1316.

Kobilka, B.K., Matsui, H., Kobilka, T.L., Yang-Fung, T.L., Francke, V., Caron, M.G., Lefkowitz, R.J. and Regan, J.W. (1987b). Cloning, sequencing and expression of the gene coding for the human platelet α_2-adrenergic receptor. *Science*, **238**, 650–656.

Kochersperger, L.M., Wagnespack, A., Patterson, J.C., Hsieh, C.C.W., Weyler, W., Salach, J.L. and Denney, R.M. (1985). Immunological uniqueness of human monoamine oxidase A and B: new evidence from studies with monoclonal antibodies to human monoamine oxidase. *Journal of Neuroscience*, **5**, 2874–2881.

Kuczenski, R. (1973). Rat brain tyrosine hydroxylase. Activation by limited tryptic proteolysis. *Journal of Biological Chemistry*, **248**, 2261–2265.

Kuczenski, R.T. and Mandell, A.J. (1972). Regulatory properties of soluble and particulate rat brain tyrosine hydroxylase. *Journal of Biological Chemistry*, **247**, 3114–3122.

Laduron, P. and Belpaire, F. (1968). Tissue fractionation and catecholamines II. Intracellular distribution patterns of tyrosine hydroxylase, dopa decarboxylase, dopamine-B-hydroxylase, phenylethanolamine-N-methyl transferase and monoamine oxidase in adrenal medulla. *Biochemical Pharmacology*, **17**, 1127–1140.

Lagercrantz, H., Klein, R.L. and Stjärne, L. (1970). Improvements in the isolation of noradrenaline storage vesicles from bovine splenic nerves. *Life Sciences*, **9**, 639–650.

Lamoureux, A., Faucon Biguet, N., Samolyk, D., Privat, A., Salomon, J.C., Pujol, J.F. and Mallet, J. (1982). Identification of cDNA clones coding for rat tyrosine hydroxylase antigen. *Proceedings of the National Academy of Sciences of the United States of America*, **79**, 3881–3885.

Lands, A.M., Arnold, A., McAuliff, J.R., Luduena, F.P. and Brown, T.G. (1967). Differentiation of receptor systems activated by sympathomimetic amines. *Nature*, **214**, 597–598.

Langer, S.Z. (1977). Presynaptic receptors and their roles in the regulation of transmitter release. *British Journal of Pharmacology*, **60**, 481–497.

Langer, S.Z. (1980). Presynaptic regulation of the release of catecholamines. *Pharmacological Reviews*, **32**, 337–362.

Lefkowitz, R.J., Kobilka, B.K., Benovic, J.L., Bouvier, M., Cotecchia, S., Hausdorff, W., Dohlman, H.G., Regan, J.W. and Caron, M.G. (1988). Molecular biology of adrenergic receptors. *Cold Spring Harbor Symposia of Quantitative Biology*, **53**, 507–514.

Limberger, N., Spath, L. and Starke, K. (1988). Presynaptic α_2-adrenoceptor, opioid k-receptor and adenosine A_1-receptor interactions on noradrenaline release in rabbit brain cortex. *Naunyn Schmiedebergs Archives of Pharmacology*, **338**, 53–61.

Lipscombe, D., Kongsamut, S. and Tsien, R.W. (1989). α-adrenergic inhibition of sympathetic neurotransmitter release mediated by modulation of N-type calcium channel gating. *Nature*, **340**, 639–642.

Lloyd, T. and Kaufman, S. (1974). The stimulation of partially purified bovine caudate tyrosine hydroxylase by phosphatidyl-L-serine. *Biochemical and Biophysical Research Communications*, **59**, 1262–1269.

Löffelholz, K. and Muscholl, E. (1970). Inhibition of parasympathetic nerve stimulation of the release of the adrenergic transmitter. *Naunyn Schmiedebergs Archives of Pharmacology*, **267**, 181–184.

Lowe, A.W., Maddedu, L. and Kelly, R.B. (1988). Endocrine secretory granules and neuronal vesicles have three integral membrane proteins in common. *Journal of Cell Biology*, **106**, 51–59.

Lundberg, J.M. and Hökfelt, T. (1986). Multiple co-existence of peptides and classical transmitters in peripheral autonomic and sensory neurons: functional and pharmacological implications. *Progress in Brain Research*, **68**, 241–262.

Lundberg, J.M. and Stjärne, L. (1984). Neuropeptide Y (NPY) depresses the secretion of ^{3}H-ndoradrenaline and the contractile response evoked by field stimulation in the vas deferens. *Acta Physiologica Scandinavica*, **120**, 477–479.

Lundberg, J.M., Hua, X.Y. and Franco-Cereceda, A. (1984b). Effects of neuropeptide Y (NPY) on mechanical activity and neurotransmission in the heart, vas deferens and urinary bladder of the guinea-pig. *Acta Physiologica Scandinavica*, **121**, 325–332.

Lundberg, J.M., Anggard, A., Theodorsson-Norheim, E. and Pernow, J. (1984a). Guanethidine-sensitive release of neuropeptide-like immunoreactivity in the cat spleen by sympathetic nerve stimulation. *Neuroscience Letters*, **52**, 175–180.

Lundberg, J.M., Terenius, L., Hökfelt, T. and Goldstein, M. (1983). High levels of neuropeptide Y in peripheral noradrenergic neurons in various mammals including man. *Neuroscience Letters*, **42**, 167–172.

Lundberg, J.M., Hemsen, A., Fried, G., Theodorsson-Norheim, E. and Lagercrantz, H. (1986a). Co-release of neuropeptide Y (NPY)-like immunoreactivity and catecholamines in newborn infants. *Acta Physiologica Scandinavica*, **126**, 471–473.

Lundberg, J.M., Rudehill, A., Sollevi, A., Theodorsson-Norheim, E. and Hamberger, B. (1986b). Frequency- and reserpine-dependent chemical coding of sympathetic transmission: differential release of noradrenaline and neuropeptide Y from pig spleen. *Neuroscience Letters*, **63**, 96–100.

Lundberg, J.M., Martinsson, A., Hemsen, A., Theodorssen-Norheim, E., Svedenhag, J., Ekblom, B. and Hjemdhal, P. (1985). Co-release of neuropeptide Y and catecholamines during physical exercise in man. *Biochemical and Biophysical Research Communications*, **133**, 30–36.

Lundberg, J.M., Terenius, L., Hökfelt, T., Martling, C.R., Tatemoto, K., Mutt, V., Polak, J. and Goldstein, M. (1982). Neuropeptide Y (NPY)-like immunoreactivity in peripheral noradrenergic neurons and effects of NPY on sympathetic function. *Acta Physiologica Scandinavica*, **116**, 477–480.

McFadzen, I. and Docherty, R.J. (1989). Noradrenaline- and enkephalin-induced inhibition of voltage sensitive calcium currents in NG 108–15 hybrid cells. *European Journal of Neuroscience*, **1**, 141–147.

McTigue, M., Cremins, J. and Halegoua, S. (1985). Nerve growth factor and other agents mediate phosphorylation and activation of tyrosine hydroxylase. A convergence of multiple kinase activities. *Journal of Biological Chemistry* **260**, 9047-9056.

Malmfors, T. (1965). Studies on adrenergic nerves. *Acta Physiologica Scandinavica*, **64**, Suppl. 248, 1-93.

Marchetti, C., Carbone, E. and Lux, H.D. (1986). Effects of dopamine and noradrenaline on Ca channels of cultured sensory and sympathetic neurons of chick. *Pflügers Archiv. European Journal of Physiology*, **406**, 104-111.

Morgenroth III, V.H., Boadle-Biber, M.C. and Roth, R.H. (1974). Tyrosine hydroxylase: Activation by nerve stimulation. *Proceedings of the National Academy of Sciences of the United States of America*, **71**, 4283-4287.

Morita, K. and North, R.A. (1981). Clonidine activates membrane potassium conductance in myenteric neurones. *British Journal of Pharmacology*, **74**, 419-428.

Musacchio, J.M., Würzburger, R.J. and D'Angelo, G.L. (1971). Different molecular forms of bovine adrenal tyrosine hydroxylase. *Molecular Pharmacology*, **7**, 136-146.

Nagatsu, T., Levitt, M. and Udenfriend, S. (1964). Tyrosine hydroxylase: The initial step in norepinephrine biosynthesis. *Journal of Biological Chemistry* **239**, 2910-2917.

Navarro-Welch, C. and McCauley, R.B. (1982). An evaluation of phospholipids as regulators of monoamine oxidase A and monoamine oxidase B activities. *Journal of Biological Chemistry* **257**, 13645-13649.

Navone, F., Di Gioia, G., Matteoli, M. and De Camilli, P. (1988). Small synaptic vesicles and large dense-cored vesicles of neurons are related to two distinct types of vesicles of endocrine cells. In *Molecular Mechanisms in Secretion*. Alfred Benzon Symposium 25, eds. N.A. Thorn, M. Freiman and O.H. Petersen pp 455-450. Copenhagen: Munksgaard.

Navone, F., Jahn, R., Di Gioia, G., Stukenbrok, H., Greengard, P. and De Camilli, P. (1986). Protein p38: an integral membrane protein specific for small vesicles of neurons and neuroendocrine cells. *Journal of Cell Biology*, **103**, 2511-2527.

Neff, N.H. and Costa, E. (1966). The influence of monoamine oxidase inhibition on catecholamine synthesis. *Life Sciences*, **5**, 951-959.

Neher, E.J. and Clapham, D.E. (1988). Role of G protein subunits in transmembrane signalling. *Nature*, **333**, 129.

Nelson, D.L. and Molinoff, P.B. (1976). Distribution and properties of adrenergic storage vesicles in nerve terminals. *Journal of Pharmacology and Experimental Therapeutics*, **196**, 346-359.

Neuman, B., Wiedermann, C.J., Fischer-Colbrie, R., Schober, M., Sperk, G. and Winkler, H. (1984). Biochemical and functional properties of large and small dense-core vesicles in sympathetic nerves of rat and ox vas deferens. *Neuroscience*, **13**, 921-931.

Norberg, K.A. and Hamberger, B. (1964). The sympathetic adrenergic neuron. Some characteristics revealed by histochemical studies on the intraneuronal distribution of the transmitter. *Acta Physiologica Scandinavica*, **63**, Suppl. 238, 1-42.

Oesch, F., Otten, V. and Thoenen, H. (1973). The relationship between the rate of axoplasmic transport and subcellular distribution of enzymes involved in the synthesis of norepinephrine. *Journal of Neurochemistry*, **20**, 1691-1706.

Okuno, S., Kanayama, Y. and Fujisawa, H. (1989). Regulation of human tyrosine hydroxylase activity. Effects of cAMP-dependent protein kinase, calmodulin-dependent protein kinase II and polyanion. *FEBS Letters*, **253**, 52-54.

Otten, U. and Thoenen, H. (1976a). Mechanisms of tyrosine hydroxylase and dopamine-β-hydroxylase induction in organ cultures of rat sympathetic ganglia by potassium depolarization and cholinomimetics. *Naunyn Schmiedebergs Archives of Pharmacology*, **292**, 153-159.

Otten, U. and Thoenen, H. (1976b). Role of membrane depolarization in transsynaptic induction of tyrosine hydroxylase in organ cultures of sympathetic ganglia. *Neuroscience Letters*, **2**, 93-96.

Pernow, J. and Lundberg, J.M. (1989). Modulation of noradrenaline and neuropeptide Y (NPY) release in the pig kidney *in vivo*: involvement of α_2, NPY and angiotensin receptors. *Naunyn-Schmiedebergs Archives of Pharmacology*, **340**, 379-385.

Pernow, J., Saria, A. and Lundberg, J.M. (1986). Mechanisms underlying pre- and post-junctional effects of neuropeptide Y in sympathetic vascular control. *Acta Physiologica Scandinavica*, **126**, 239-249.

Pickel, V.M., Joh, T.H. and Reis, D.J. (1976). Ultrastructural localization by immunocytochemistry of dopamine-β-hydroxylase within noradrenergic neurons of the rat brain. *Anatomical Record*, **184**, 503.

Pickel, V.M., Joh, T.H., Field, P.M., Becker, C.J. and Reis, D.J. (1975). Cellular localization of tyrosine hydroxylase by immunohistochemistry. *Journal of Histochemistry and Cytochemistry*, **23**, 1-12.

Pollard, R.M., Fillenz, M. and Kelly, P. (1982). Parallel changes in ultrastructure and noradrenaline content of nerve terminals in rat vas deferens following transmitter release. *Neuroscience*, **7**, 1623-1629.

Porzig, H., Becker, C. and Reuter, H. (1982). Effects of specific ligands and of cAMP on β-adrenoceptor properties in intact cardiac cells. *Naunyn Schmiedebergs Archives of Pharmacology*, **319**, R68.

Ramme, D., Illes, P., Späth, L. and Starke, K. (1986). Blockade of α-adrenoceptors permits the operation of otherwise silent opioid k-receptors at the sympathetic axons of rabbit jejunal arteries. *Naunyn Schmiedebergs Archives of Pharmacology*, **334**, 48–55.

Reid, J.J., Stitzel, R.E. and Head, R.J. (1986). Characterization of the O-methylation of catechol oestrogens by intact rabbit thoracic aorta and subcellular fractions thereof. *Naunyn Schmiedebergs Archives of Pharmacology*, **334**, 17–28.

Richards, J.G. and da Prada, M. (1977). Uranaffin reaction: a new cytochemical technique for the localization of adenine nucleotides in organelles storing biogenic amines. *Journal of Histochemistry and Cytochemistry*, **25**, 1322–1336.

Richards, J.G. and Da Prada, M. (1980). Cytochemical Investigations of Subcellular Organelles Storing Biogenic Amines in Peripheral Adrenergic Neurons. In: *Histochemistry and Cell Biology of Autonomic Neurons, SIF Cells and Paraneurons*, eds. Eränkö, O., Soinila, S. and Päivärinta, H. pp 269–278. Baltimore: Raven Press.

Richardson, K.C. (1962). The fine structure of autonomic nerve endings of rat vas deferens. *Journal of Anatomy*, **96**, 427–442.

Rivett, J.A., Francis, A. and Roth, J.A. (1983). Localization of membrane-bound catechol-O-methyltransferase. *Journal of Neurochemistry*, **40**, 1494–1496.

Roskoski Jr., R., Vulliet, P.R. and Glass, D.B. (1987). Phosphorylation of tyrosine hydroxylase by cyclic GMP-dependent protein kinases. *Journal of Neurochemistry*, **48**, 840–845.

Ross, E.M., Wong, S.K.F., Rubenstein, R.C. and Higashijima, T. (1988). Functional domains in the β-adrenergic receptor. *Cold Spring Harbor Symposia of Quantitative Biology*, **53**, 499–506.

Schoffelmeer, A.N.M. and Mulder, A.H. (1983). ³H-noradrenaline release from rat neocortical slices in the absence of Ca^{++} and its presynaptic α_2-adrenergic modulation: a study on the possible role of cyclic AMP. *Naunyn Schmiedebergs Archives of Pharmacology*, **323**, 188–192.

Schoffelmeer, A.N.M., Putters, J. and Mulder, A.H. (1986). Activation of presynaptic α_2-adrenoceptors attenuates the inhibitory effect of μ-opioid receptor agonists on noradrenaline release from brain slices. *Naunyn Schmiedebergs Archives of Pharmacology*, **333**, 377–380.

Schultzberg, M., Lundberg, J.M., Hökfelt, T., Terenius, L., Brandt, J., Elde, R. and Goldstein, M. (1978). Enkephalin-like immunoreactivity in gland cells and nerve terminals of the adrenal medulla. *Neuroscience*, **3**, 1169–1186.

Schwarzschild, M.A. and Zigmond, R.E. (1989). Secretin and vasoactive intestinal peptide activate tyrosine hydroxylase in sympathetic nerve endings. *Journal of Neuroscience*, **9**, 160–166.

Sibley, D.R. and Lefkowitz, R.J. (1985). Molecular mechanisms of receptor desensitization using the β-adrenergic receptor-coupled adenylyl cyclase system as a model. *Nature*, **317**, 124.

Sibley, D.R., Benovic, J.L., Caron, M.G. and Lefkowitz, R.J. (1987). Regulation of transmembrane signaling by receptor phosphorylation. *Cell*, **48**, 913–922.

Sibley, D.R., Nambi, P., Peters, J.R. and Lefkowitz, R.J. (1984). Phorbol diesters promote β-adrenergic receptor phosphorylation and adenylate cyclase desensitization in duck erythrocytes. *Biochemical and Biophysical Research Communications*, **121**, 9730–9397.

Sibley, D.R., Strasser, R.H., Benovic, J.L., Daniel, K. and Lefkowitz, J.R. (1986). Phosphorylation/dephosphorylation of the β adrenergic receptor regulates its functional coupling to adenylate cyclase and subcellular distribution. *Proceedings of the National Academy of Sciences of the United States of America*, **83**, 9408–9412.

Sigal, I.S., Dixon, R.A.F. and Strader, C.D. (1988). Molecular Biology of Adrenergic Receptors. *ISI Atlas of Science: Pharmacology*, **2**, 387–391.

Singer, T.P. (1987). Perspectives in MAO: past, present and future. A Review. *Journal of Neural Transmission*, **Suppl. 23**, 1–23.

Sjöstrand, N.O. (1965). The adrenergic innervation of the vas deferens and the accessory male genital glands. *Acta Physiologica Scandinavica*, **65**, Suppl. 257, 1–82.

Smith, A.D. (1972). Mechanisms involved in the release of noradrenaline from sympathetic nerves. *British Medical Bulletin*, **29**, 123–129.

Sneddon, P. and Burnstock, G. (1984). Inhibition of excitatory junction potentials in guinea-pig vas deferens by A, B-methylene ATP: further evidence for ATP and noradrenaline as co-transmitters. *European Journal of Pharmacology*, **100**, 85–95.

Sneddon, P. and Westfall, D.P. (1984). Pharmacological evidence that ATP and noradrenaline are co-transmitters in the guinea-pig vas deferens. *Journal of Physiology*, **347**, 561–580.

Sneddon, P., Westfall, D.P. and Fedan, J.S. (1982). Co-transmitters in the motor nerves of the guinea-pig vas deferens: electrophysiological evidence. *Science*, **218**, 693–695.

Spiegel, K., Kremer, N.E. and Kessler, J.A. (1989). Differences in the effect of membrane depolarization on levels of preprosomatostatin mRNA and tyrosine hydroxylase mRNA in rat sympathetic neurons *in vivo* and in culture. *Molecular Brain Research*, **5**, 23–29.

Starke, K. (1970). Interactions of angiotensin and cocaine on the output of noradrenaline from isolated rabbit hearts. *Naunyn Schmiedebergs Archives of Pharmacology*, **265**, 383–386.

Starke, K. (1987). Presynaptic α-autoreceptors. *Reviews of Physiology Biochemistry and Pharmacology*, **107**, 73–146.

Stjärne, L. (1978). Facilitation and receptor-mediated regulation of noradrenaline secretion by control of recruitment of varicosities as well as by control of electro-secretory coupling. *Neuroscience*, **3**, 1147–1155.

Stjärne, L. (1986). On the mechanism and scope of adrenoceptor-mediated control of sympathetic neuro-effector transmission. In: *New Aspects of the Role of Adrenoceptors in the Cardiovascular System*, eds. H. Grobecker, A. Phillippu and K. Starke. Berlin Heidelberg New York, Springer, pp. 14–23.

Stjärne, L. (1989). Basic mechanisms and local modulation of nerve impulse-induced secretion of neurotransmitter from individual sympathetic nerve varicosities. *Reviews of Physiology Biochemistry and Pharmacology*, **112**, 1–137.

Stjärne, L. and Lischajko, F. (1966). Drug-induced inhibition of noradrenaline synthesis *in vitro* in bovine splenic nerve tissue. *British Journal of Pharmacology*, **27**, 398–404.

Stjärne, L. and Stjärne, E. (1989). Some pharmacological applications of an extracellular method to study secretion of a sympathetic cotransmitter, presumably ATP. *Acta Physiologica Scandinavica*, **135**, 227–239.

Su, C. (1978). Purinergic inhibition of adrenergic transmission in rabbit blood vessels. *Journal of Pharmacology and Experimental Therapeutics*, **205**, 351–361.

Sugden, D., Vaneeck, J., Klein, D.C., Thomas, T.P. and Anderson, W.B. (1985). Activation of protein kinase C potentiates isoprenalin-induced cyclic AMP accumulation in rat pinealocytes . *Nature*, **314**, 359–361.

Tachikawa, E., Tank, A.W., Yanagihara, N., Mosimann, W. and Weiner, N. (1986). Phosphorylation of tyrosine hydroxylase on at least three sites in rat pheochromocytoma PC12 cells treated with 56 mM K$^+$: Determination of the sites on tyrosine hydroxylase phosphorylated by cyclic AMP-dependent and calcium-calmodulin-dependent protein kinases. *Molecular Pharmacology*, **30**, 476–485.

Tachikawa, E., Tank, A.W., Weiner, D.H., Mosimann, W.F., Yanagihara, N. and Weiner, N. (1987). Tyrosine hydroxylase is activated and phosphorylated on different sites in rat pheochromocytoma PC12 cells treated with phorbol ester and forskolin. *Journal of Neurochemistry*, **48**, 1366–1376.

Thoenen, H. and Otten, V. (1975). Cyclic nucleotides and trans-synaptic enzyme induction: lack of correlation between initial cAMP increase, changes in cAMP/cGMP ratio and subsequent induction. In: *Chemical Tools in Catecholamine Research II, Regulation of CA Turnover*, ed. O. Almgren, A. Carlsson and J. Engel, pp. 275–282. Amsterdam: North Holland Publishing Company.

Thoenen, H. and Otten, V. (1977). Trans-synaptic enzyme induction: ionic requirements and modulatory role of glucocorticoids. In: *Structure and Function of Monoamine Enzymes, ed. E. Usdin, N. Weiner and M.B.H. Youdim, pp. 439–464*. New York: Marcel Dekker, Inc.

Thureson-Klein, Å. (1983). Exocytosis from large and small dense cored vesicles in noradrenergic nerve terminals. *Neuroscience*, **10**, 245–252.

Thureson-Klein, Å. and Klein, R.L. (1990). Exocytosis from neuronal large dense-cored vesicles. *International Review of Cytology*, **121**, 67–126.

Thureson-Klein, Å. and Stjärne, L. (1981). Dense-cored vesicles in actively secreting noradrenergic neurons. In: *Chemical Neurotransmission 75 Years*, eds. Stjärne, L., Hedqvist, P., Lagercrantz, H. & Wennmalm, A. Academic Press, London, pp. 153–164.

Thureson-Klein, Å., Klein, R.L. and Yen, S.S. (1973). Ultrastructure of highly purified nerve vesicles: correlation between matrix density and norepinephrine content. *Journal of Ultrastructure and Molecular Structure Research*, **43**, 18–55.

Tomlinson, D.R. (1975). Two populations of granular vesicles in constricted post-ganglionic sympathetic nerves. *Journal of Physiology*, **245**, 727–735.

Trachte, G.J. (1988). Angiotensin effects on vas deferens and purinergic transmission. *European Journal of Pharmacology*, **146**, 261–269.

Trachte, G.J., Binder, S.B. and Peach, M.J. (1989). Indirect evidence for separate vesicular neuronal origins of norepinephrine and ATP in the rabbit vas deferems. *European Journal of Pharmacology*, **164**, 425–433.

Tranzer, J.P. (1972). A new amine storing compartment in adrenergic axons. *Nature*, **237**, 57–58.

Tranzer, J.P. and Thoenen, H. (1968). Various types of amine storing vesicles in peripheral adrenergic nerve terminals. *Experientia*, **24**, 484–486.

Trendelenburg, U., Cassis, L., Grohmann, M. and Langeloh, A. (1986). The functional coupling of neuronal and extraneuronal transport with intracellular monoamine oxidase. *Journal of Neural Transmission*, Suppl. 23, 91–101.

Udenfriend, S. (1966). Tyrosine hydroxylase. *Pharmacological Reviews*, **18**, 43–51.

Van Orden III, L.S., Bloom, F.E., Barnett, J.R. and Giarman, N.J. (1966). Histochemical and functional relationships of catecholamines in adrenergic nerve endings. I. Participation of granular vesicles. *Journal of Pharmacology and Experimental Therapeutics*, **154**, 185–199.

Verhaeghe, R.H., Vanhoutte, P.M. and Shepherd, J.T. (1977). Inhibition of sympathetic neurotransmission in canine blood vessels by adenosine and adenine nucleotides. *Circulation Research*, **40**, 208–215.

Vigny, A. and Henry, J.-P. (1981). Bovine adrenal tyrosine hydroxylase: Comparative study of native and proteolyzed enzyme, and their interaction with anions. *Journal of Neurochemistry*, **36**, 484–489.

Viveros, O.H., Abou-Donia, M.M., Lee, C.-L., Wilson, S.P. and Nichol, C.A. (1981). Control of tissue tetrahydrobiopterin levels through GTP-cyclohydrolase: a factor in the regulation of monoamine synthesis. In: *Function and Regulation of Monoamine Enzymes*, eds. E. Usdin, N. Weiner and M.B.H. Youdim, London: Macmillan.

Vulliet, P.R. and Weiner, N. (1981). A schematic model for the allosteric activation of tyrosine hydroxylase. In: *Function and Regulation of Monoamine Enzymes*, ed. E. Usdin, N. Weiner and M.B.H. Youdim, pp. 15–24. London: Macmillan.

Vulliet, P.R., Langan, T.A. and Weiner, N. (1980). Tyrosine hydroxylase: A substrate of cyclic AMP-dependent protein kinase. *Proceedings of the National Academy of Sciences of the United States of America*, **77**, 92–96.

Vulliet, P.R., Woodgett, J.R. and Cohen, P. (1984). Phosphorylation of tyrosine hydroxylase by calmodulin-dependent multiprotein kinase. *Journal of Biological Chemistyr*, **259**, 13680–13683.

Vulliet, P.R., Woodgett, J.R., Ferrari, S. and Hardie, D.G. (1985). Characterization of the sites phosphorylated on tyrosine hydroxylase by Ca^{2+}-and phospholipid-dependent protein kinase, calmodulin-dependent multiprotein kinase and cyclic AMP-dependent protein kinase. *FEBS Letters*, **182**, 335–339.

Wali, F.A. and Greenidge, E. (1989). Evidence that ATP and noradrenaline are released during electrical field stimulation of the rat isolated seminal vesicle. *Pharmacological Research*, **21**, 397–404.

Waymire, J.C., Johnston, J.P., Hummer-Lickteig, K., Lloyd, A., Vigny, A. and Craviso, G.L. (1988). Phosphorylation of bovine adrenal chromaffin cell tyrosine hydroxylase. Temporal correlation of acetylcholine's effect on site phosphorylation, enzyme activation and catecholamine synthesis. *Journal of Biological Chemistry* **263**, 12439–12447.

Weiner, N. (1979). Tyrosine-3-monooxygenase (Tyrosine Hydroxylase). In: *Aromatic Amino Acid Hydroxylases and Mental Disease*, ed. M.B.H. Youdim, Chichester, UK: John Wiley and Sons Ltd. *pp. 141–190*.

Weiner, N., Lee, F-L., Barnes, E. and Dreyer, E. (1977). Enzymology of tyrosine hydroxylase and the role of cyclic nucleotides in its regulation. In: *Structure and Function of Monoamine Enzymes*, eds. Usdin, E., Weiner, N. and Youdim, M.B.H. Marcel Dekker Inc., New York, p. 109–148.

Wilson, S.P., Klein, R.L., Chang, K.-J., Gasparis, M.S., Viveros, O.H. and Yang, W.-H. (1980). Are opioid peptides co-transmitters in noradrenergic vesicles of sympathetic nerves? *Nature*, **288**, 707–709.

Yamauchi, T. and Fujisawa, H. (1981). Tyrosine-3-monooxygenase is phosphorylated by Ca^{2+} calmodulin-dependent protein kinase, followed by activation by activator protein. *Biochemical and Biophysical Research Communications*, **100**, 807–813.

Yamauchi, T., Nakata, H. and Fujisawa, H. (1981). A new activator protein that activates tryptophan 5-monooxygenase and tyrosine 3-monooxygenase in the presence of Ca^{2+}-calmodulin-dependent protein kinase. Purification and characterization. *Journal of Biological Chemistry* **256**, 5404–5409.

Yarden, Y., Rodriques, H., Wong, S.K-F *et al.* (1986). The avian β-adrenergic receptor: primary structure and membrane topology. *Proceedings of the National Academy of Sciences of the United States of America*, **83**, 6795–6799.

Yatani, A., Codina, J., Imoto, Y., Reeves, J.P., Birnbaumer, L. and Brown, A.M. (1987). A G protein directly regulates mammalian cardiac calcium channels. *Science*, **238**, 1288.

Yatani, A., Imoto, Y., Codina, J., Hamilton, S., Brown, A.M. and Birnbaumer, L. (1988). The stimulatory G protein of adenylyl cyclase, G_s, also stimulates dihydropyridine-sensitive Ca^{2+} channels: Evidence for direct regulation independent of phosphorylation by cAMP-dependent protein kinase or stimulation by a dihydropyridine agonist. *Journal of Biological Chemistry*, **263**, 9887.

Yoshimasa, T., Sibley, D.R., Bouvier, M., Lefkowitz, R.J. and Caron, M.G. (1987). Cross-talk between cellular signalling pathways suggested by phorbol ester-induced adenylate cyclase phosphorylation. *Nature*, **327**, 67–70.

Zaagsma, J. and Nahorski, S.R. (1990). Is the adipocyte β-adrenoceptor a prototype for the recently cloned atypical 'β_2-adrenoceptor'? *Trends in Pharmacological Sciences*, **11**, 3–7.

Zhu, P.C., Thureson-Klein, J. and Klein, R.L. (1986). Exocytosis from large dense cored vesicles outside the active synaptic zone of terminal subnucleus caudalis: a possible mechanism for neuropeptide release. *Neuroscience*, **19**, 43–54.

Zukowska-Grojec, Z., Konarska, M. and McCarty, R. (1988). Differential plasma catecholamine and neuropeptide Y responses to acute stress in rats. *Life Sciences*, **42**, 1615–1624.

Zimmerman, B.G. (1977). Actions of angiotensin on adrenergic nerve endings. *Federal Proceedings*, **37**, 199–202.

7 Transmission: Purines

Charles H.V. Hoyle

Department of Anatomy and Developmental Biology, University College London, Gower Street, London WC1E 6BT

Following the development of Burnstock's Purinergic Hypothesis, which put forward the proposal that ATP or a related nucleotide is a neurotransmitter in non-adrenergic, non-cholinergic neurones, there has been such an interest in non-adrenergic, non-cholinergic neurobiology that there is now extensive evidence that ATP is utilised as a neurotransmitter in the autonomic nervous system. Evidence is presented that ATP is localised in and released from sensory, enteric, sympathetic and parasympathetic neurones, and acts as a neuromuscular transmitter in many autonomically innervated organs. Although several different subtypes of P_2-purinoceptor have been described, the subtypes of receptor by which purinergic transmission is effected are predominantly the P_{2X}- and P_{2Y}-purinoceptors. In a few areas adenosine rather than ATP may be the purinergic transmitter, and in these cases a P_1-purinoceptor mediates the transmission. In many cases, especially within the sympathetic division, ATP appears to be a co-transmitter, often with noradrenaline, and it has also been implicated as a co-transmitter with acetylcholine and neuropeptides. ATP can act as a neuromodulator, at prejunctional or postjunctional sites, and can either act as ATP *per se* or be partially degraded to adenosine and modulate via P_1-purinoceptors. Despite the increase in understanding of physiological roles of ATP, the therapeutic potential in manipulating purinergic transmission remains an area which presents itself for future research.

INTRODUCTION

The aim of this chapter is to give an account of the principles and variations of purinergic transmission in the autonomic nervous system, i.e. to review the evidence of sympathetic and parasympathetic nerves that utilise adenosine 5′-triphosphate (ATP) as a transmitter. Sensory purinergic transmission is covered, both peripherally and within the spinal cord, and models of purinergic neurobiology from somatic nerves are also called upon. The major emphasis is on ATP, its storage, release, inactivation and pharmacology, and the types of nerve that utilise ATP, rather than the types of organ that are innervated. There is less emphasis on adenosine, adenosine 5′-monophosphate (AMP) and adenosine 5′-diphosphate (ADP), largely because these compounds are not used as neurotransmitters in the periphery to the same extent as ATP, and partly because of the restriction of space within this article.

367

A feature of many purinergic nerves is that ATP is a cotransmitter with noradrenaline, acetylcholine or neuropeptides (Burnstock, 1990): the principles of cotransmission are covered in greater detail in Chapter 2 by Morris and Gibbins. Mechanisms of signal transduction are also covered in detail by Benham in Chapter 4. A further feature of the utilisation of ATP as a neurotransmitter, which is possibly unique to purinergic transmission, is that when ATP is degraded by extracellular enzymes, substances which can have a potent local activity (principally adenosine) are formed. This is in contrast to aminergic and peptidergic systems where, as far as we know, transmitter substances are broken down into inactive products.

In recent years, coupled with the volume of research into the neurobiology and pharmacology of purine compounds, many books and review articles have been written (e.g. see White, 1988), and amongst these there are large areas of overlap. After the original review, in which Burnstock put forward the purinergic hypothesis (Burnstock, 1972), there are three review articles in particular which, taken together encompass the most important literature on purinergic neurobiology, without excessive duplication, and which cover different ground, or the same ground from a different perspective than this article. Stone (1981a) covers details of purine biosynthesis, and of transmission within the central nervous system; White (1988) deals with purine neurobiology on an organ-by-organ basis; Silinsky (1989) covers cellular and molecular aspects of purinoceptors and their activation.

In this chapter, receptor subclassification is discussed before neurotransmission, because an understanding of the main subclasses of the ATP-receptor (i.e. P_{2X}- and P_{2Y}-purinoceptors) is necessary in order to be able to evaluate the pharmacological evidence for or against ATP being a transmitter. Localisation and storage of ATP are discussed in their logical place, before release of ATP. Purinergic transmission from sensory nerves is treated with a historical perspective, before sympathetic and parasympathetic transmission, and finally, neuromodulation by ATP is considered.

Within the context of this article, intrinsic neurones, i.e. those whose cell bodies lie in visceral intramural ganglia (exemplified by the myenteric plexus), are treated as being parasympathetic, although it is appreciated that the vast majority of intrinsic neurones do not lie directly postganglionically with respect to anatomically preganglionic parasympathetic nerves.

RECEPTORS FOR ADENOSINE AND ADENINE NUCLEOTIDES

The pharmacological activity of purine compounds has been studied since at least the late 1920's when Drury and Szent-Györgi (1929) observed cardioinhibitory and hypotensive actions of AMP. During the early 1930's several groups studied the pharmacological action of purines in various tissues, including many cardiovascular structures and the uterus (Lidner and Rigler, 1931; Wedd, 1931; Deuticke, 1932; Ostern and Parnas, 1932; Gaddum and Holtz, 1933). The first indications, contained within a single study, that adenine nucleosides and nucleotides may have different pharmacological actions (and therefore different receptors) were presented by Gillespie

(1933), who described ATP as being more potent than its non-phosphorylated derivatives in causing relaxation of the guinea-pig ileum, and adenosine as being more potent than its phosphorylated derivatives in causing coronary vasodilatation, or inducing hypotension in cats and rabbits.

The theoretical consideration that there are distinct classes of purinoceptors was put forward by Burnstock (1978). A division into P_1- and P_2-purinoceptors was proposed on the basis of differential rank orders of agonist potency, differential antagonism and different transduction mechanisms. This terminology has gained general acceptance, and now subclassifications of P_1- and P_2-purinoceptors are recognised.

Receptors for adenosine have been subclassified into subtypes A_1, A_2 and A_3, these are all types of P_1-purinoceptor, and selective agonists and antagonists have been developed, especially for the A_1 and A_2 subtypes, (Burnstock and Buckley, 1985; Cooper and Londos, 1988; Silinsky, 1989; Kennedy, 1990). The P_2-purinoceptor has been subdivided into two major classes (P_{2X} and P_{2Y}) (Burnstock and Kennedy, 1985a) and two further classes (P_{2T} and P_{2Z}) (Gordon, 1986). There have been some claims that ATP acts on receptors that do not fit into any of these four classes, and additional P_{2S} and P_3 subclasses have been proposed, but as yet these have received little support. The possibility that there are classes of receptors for adenine dinucleotides is also discussed below, and a report of a receptor for an adenine trinucleotide is covered under the heading 'Receptors for Adenine Oligonucleotides'. The distribution of subtypes of ATP receptors in many tissues and organs is summarised in Table 7.1, and a schematic diagram of purinoceptor classes is shown in Figure 7.1.

TABLE 7.1
Distribution of subtypes of P_2-purinoceptors, and P_3-purinoceptors

P_{2X}-purinoceptors

Smooth muscle
vas deferens

guinea-pig	Fedan et al., 1981, 1982
	Meldrum and Burnstock, 1983
	Sneddon and Westfall, 1984
	Sneddon and Burnstock, 1984
rat	Allcorn et al., 1985
	Bulloch and McGrath, 1988
mouse	Stjärne and Åstrand, 1984, 1985
	Allcorn et al., 1985
	Åstrand et al., 1988
urinary bladder	
guinea-pig	Kasakov and Burnstock, 1983
	Westfall et al., 1983
	Hourani et al., 1985
	Moss and Burnstock, 1985
	Fujii, 1988
rabbit	Hoyle and Burnstock, 1985
	Fujii, 1988
rat	Moss et al., 1987
cat	Theobald, 1982, 1986
	Theobald and Hoffman, 1986
mouse	Holt et al., 1985

TABLE 7.1 *Continued*

ferret	Moss and Burnstock, 1985
marmoset	Moss and Burnstock, 1985
human	Hoyle *et al.*, 1989
blood vessels	
rabbit	
mesenteric artery	Ishikawa, 1985
	Kügelgen and Starke, 1985
	Mathieson and Burnstock, 1985
	Burnstock and Warland, 1987b
	Ramme *et al.*, 1987
central ear artery	Kennedy and Burnstock, 1985b
saphenous artery	Burnstock and Warland, 1987a
portal vein	Kennedy and Burnstock, 1985a
rat	
femoral artery	Kennedy *et al.*, 1985, 1986
tail artery	Sneddon and Burnstock, 1985
	Neild and Kotecha, 1986
coronary vasculature	Fleetwood and Gordon, 1987
	Hopwood and Burnstock, 1987
pancreas vasculature	Bertrand *et al.*, 1987
peripheral vasculature	Flavahan *et al.*, 1985
portal vein	Reilly and Burnstock, 1987
dog	
mesenteric artery	Muramatsu, 1987
	Machaly *et al.*, 1988
	Muramatsu *et al.*, 1989
saphenous artery	Houston *et al.*, 1987
basilar artery	Muramatsu and Kigoshi, 1987
guinea-pig	
saphenous artery	Cheung and Fujioka, 1986
mesenteric artery	Nagao and Suzuki, 1988
seminal vesicles	
guinea-pig	Meldrum and Burnstock, 1985a
nictitating membrane	
cat	Duval *et al.*, 1985
colon	
cat	Hedlund *et al.*, 1983
rectum	
chick	Meldrum and Burnstock, 1985b
Cardiac muscle	
frog	Hoyle and Burnstock, 1986
toad	Bramich and Campbell, 1989
Fibroblasts	
human skin	Okada *et al.*, 1984
mouse L cells	Okada *et al.*, 1984

P$_{2Y}$-purinoceptors

Smooth muscle	
blood vessels	
rabbit	
mesenteric artery	Mathieson and Burnstock, 1985
	Burnstock and Warland, 1987b
portal vein	Kennedy and Burnstock, 1985a
	Reilly *et al.*, 1987
dog	
basilar artery	Muramatsu and Kigoshi, 1987
middle cerebral artery	Muramatsu and Kigoshi, 1987

TABLE 7.1 *Continued*

rat	
coronary vasculature	Fleetwood and Gordon, 1987
	Hopwood and Burnstock, 1987
stomach	
rabbit	Beck *et al.*, 1988
colon	
guinea-pig	Kerr and Krantis, 1979
taenia coli	
guinea-pig	Gough *et al.*, 1973
	Satchell and Maguire, 1975
internal anal sphincter	
guinea-pig	Crema *et al.*, 1983
Cardiac muscle	
rat	Björnsson *et al.*, 1989
Endothelial cells	
rat femoral artery	Kennedy *et al.*, 1985
dog coronary artery	Houston *et al.*, 1987
Sensory neurones	
cat and rat	Krishtal *et al.*, 1983
	Krishtal and Marchenko, 1986
Pancreas	
rat B-cells	Chapal and Loubatières-Mariani, 1983
	Bertrand *et al.*, 1987
	Loubatières-Mariani and Chapal, 1988
Erythrocytes	
turkey	Boyer *et al.*, 1989
	Cooper *et al.*, 1989
Hepatocytes	
rat	Keppens and De Wulf, 1986
Lung	
rat alveolar type II cells	Rice and Singleton, 1987, 1989

P$_{2T}$-purinoceptors

Platelets	
	Haslam and Rosson, 1975
	Cusack *et al.*, 1979, 1985

P$_{2Z}$-purinoceptors

Mast cells	Cockroft and Gomperts, 1979a,b, 1980
	Bennett *et al.*, 1981
	Tatham *et al.*, 1988
Lymphocytes	
	Schmidt *et al.*, 1984
Macrophages	
	Cameron, 1984
	Steinberg and Silverstein, 1987
Erythrocytes	
	Parker and Snow, 1972
Neutrophils	
	Kuhns *et al.*, 1988
Ascites tumour cells	
	Hempling *et al.*, 1969
Contractile vacuole	
amoeba	Pothier *et al.*, 1987

TABLE 7.1 *Continued*

P$_{2S}$-purinoceptors

Guinea-pig ileum

Wiklund and Gustafsson, 1988a,b

P$_3$-purinoceptors

Sympathetic nerve terminals
  tail artery
    rat   Shinozuka *et al.*, 1988
  sympathetic chain
    frog   Silinsky and Ginsborg, 1983
Parasympathetic nerve terminals
  ileum
    guinea-pig   Moody and Burnstock, 1982
                             Moody *et al.*, 1984
                             Wiklund and Gustafsson, 1984, 1986, 1988a
    bladder
      cat   Theobald and De Groat, 1988
Smooth muscle
  stomach
    guinea-pig   Okwuasaba *et al.*, 1977
                             Frew and Lundy, 1982b
Cardiac muscle
  rat   Burnstock and Meghji, 1983
                             Moody *et al.*, 1984

The receptor classifications shown were not necessarily made by the original authors. The P$_{2X}$-purinoceptor classification was made according to antagonism by ANAPP$_3$, α,β-methylene ATP or agonist potency order. The P$_{2Y}$-purinoceptor classification was made according to antagonism by reactive blue 2, or agonist potency order. The P$_{2Z}$-purinoceptor classification was made according to activation by ATP^{4-}, antagonism by divalent cations, or lack of activity of analogues of ATP with altered phosphate chains. The P$_3$-purinoceptor classification was made according to activation by ATP or its analogues being blocked by methylxanthines.

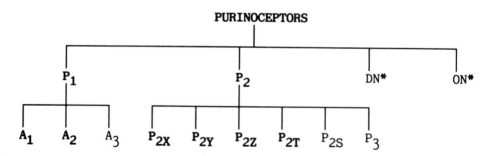

FIGURE 7.1 Schematic diagram of purinoceptor classification. The receptor subclasses in bold have received general support, but as yet those not in bold (P$_{2S}$ and P$_3$), have not. The P$_3$-purinoceptor also has some characteristics of P$_1$-purinoceptors in being antagonised by derivatives of methylxanthine, however, it is activated by ATP and some analogues of ATP (see text). *Receptors for adenine dinucleotides (DN) and oligonucleotides (ON) are variants of purinoceptors, but proposals for their nomenclature have not been made.

P_1-PURINOCEPTORS

At the P_1-purinoceptor, the rank order of agonist potency is: adenosine $\geq$ AMP $\gg$ ADP $\geq$ ATP. This receptor is blocked by methylxanthine derivatives such as caffeine and theophylline. Activation of the P_1-purinoceptor usually leads to changes in intracellular activity of adenylate cyclase. None of these attributes is shared by the P_2-purinoceptor, which has a potency sequence of ATP $\geq$ ADP $\gg$ AMP $\geq$ adenosine. The P_2-purinoceptor is not antagonised by methylxanthines, and its activation may invoke synthesis of prostaglandins.

Extracellular adenosine-receptors, i.e. P_1-purinoceptors, were divided into A_1 and A_2 subclasses on the basis of inactivation and activation, respectively, of adenylate cyclase (Van Calker, Muller and Hampbrecht, 1979). The R_i and R_a subclasses defined by Londos, Cooper and Wolff (1980) correspond directly to the A_1 and A_2 subclasses. The A_1-purinoceptor has a greater affinity for adenosine than the A_2-purinoceptor, and at the A_1-purinoceptor N^6-substituted analogues of adenosine are orders of magnitude more potent than $5'$-substituted analogues. For A_2-purinoceptors the converse rank order of agonist potency occurs. An A_3 subclass has also been described (Ribeiro and Sebastião, 1986; Sebastião and Ribeiro, 1988, 1989). This receptor is not linked to an adenylate cyclase transduction mechanism, and affects mobilisation of calcium instead, it has an agonist profile with 2-chloroadenosine being less potent than N^6- or $5'$-substituted analogues of adenosine. Typically these receptors mediate prejunctional inhibition of release of transmitter and negative tropisms in the heart by inhibiting transmembrane calcium ion fluxes (Ribeiro and Sebastião, 1986).

P_{2X}- AND P_{2Y}-PURINOCEPTORS

At the P_{2X}-purinoceptor, analogues of ATP in which the phosphate chain has been modified, such as α,β-methylene ATP or β,γ-methylene ATP, are orders of magnitude more potent than analogues in which the purine nucleus has been modified, such as 2-methylthio ATP, or than ATP *per se*. On this basis, in the vas deferens, urinary bladder and several types of blood vessels, the contractile responses are typically mediated via P_{2X}-purinoceptors (Fedan *et al.*, 1982; Kasakov and Burnstock, 1983; Meldrum and Burnstock, 1983; Burnstock, Cusack and Meldrum, 1985; Kennedy and Burnstock, 1985a,b; Kennedy, Delbro and Burnstock, 1985; Mathieson and Burnstock, 1985; Burnstock and Warland, 1987b; Fleetwood and Gordon, 1987; Hopwood and Burnstock, 1987; Houston, Burnstock and Vanhoutte, 1987; Reilly and Burnstock, 1987; Kennedy, 1988, 1990).

At the P_{2Y}-purinoceptor the rank order of ATP-analogue potency is reversed: 2-methylthio ATP is orders of magnitude more potent than α,β-methylene ATP or β,γ-methylene ATP, and more potent than ATP *per se*. The relaxant responses to ATP in the guinea-pig taenia coli (Gough, Maguire and Satchell, 1973; Satchell and Maguire, 1975; Maguire and Satchell, 1979; Burnstock *et al.*, 1983) and in many blood vessels (Kennedy and Burnstock, 1985a; Mathieson and Burnstock, 1985; Burnstock and Warland, 1987b; Fleetwood and Gordon, 1987; Hopwood and Burnstock, 1987; Reilly, Saville and Burnstock, 1987) are typically mediated via

P_{2Y}-purinoceptors. In some blood vessels the P_{2Y}-purinoceptor is located on the endothelium rather than on the vascular smooth muscle, and activation of the receptor stimulates release of a relaxant factor (Kennedy, Delbro and Burnstock, 1985; Martin *et al.*, 1985; Houston, Burnstock and Vanhoutte, 1987; Hopwood *et al.*, 1989; Taylor *et al.*, 1989). In the rat nodose ganglion there is an excitatory action of ATP, that is likely to be mediated via P_{2Y}-purinoceptors, since ATP is approximately 10 or 100 times more potent than α,β-methylene ATP or β,γ-methylene ATP, respectively (Krishtal, Marchenko and Pidoplichko, 1983; Krishtal and Marchenko, 1986).

P_{2X}- and P_{2Y}-purinoceptors display differential stereoselectivity (Burnstock *et al.*, 1983). The enantiomer of β,γ-methylene ATP, β,γ-methylene L-ATP, appears to be a selective agonist of the P_{2X}-purinoceptor in the urinary bladder and vas deferens because it is potent in these preparations but inert in the taenia coli (Cusack and Hourani, 1984; Hourani, 1984; Hourani, Welford and Cusack, 1985; Hourani, Loizou and Cusack, 1986; Reilly and Burnstock, 1987; Cusack *et al.*, 1987). In the formation of β,γ-methylene ATP, an electronegative anhydride oxygen atom has been replaced by an electropositive methylenic carbon atom; although the resultant molecule is isosteric it is no longer isopolar. Halogenation of the methylenic carbon atom results in analogues which are isosteric and isopolar (Blackburn, Kent and Kolkmann, 1981). Both difluoro-β,γ-methylene ATP and dichloro-β,γ-methylene ATP are more potent than β,γ-methylene ATP at the P_{2X}-purinoceptor in the bladder and the P_{2Y}-purinoceptor in the taenia coli: L-enantiomers of these compounds are inert in the taenia coli, but not in the bladder (Cusack *et al.*, 1987). Thus a characteristic of the P_{2Y}-purinoceptor is its marked stereospecificity between pairs of enantiomers.

Amongst a series of phosphorothioate analogues of ATP, in which an oxygen atom linked to either the α, β or γ terminal phosphorus atom has been replaced with a sulphur atom, the P_{2Y}-purinoceptor in the taenia coli shows a marked stereoselectivity, while the P_{2X}-purinoceptor in the urinary bladder does not (Burnstock, Cusack and Meldrum, 1984). However, the P_{2X}-purinoceptor in the urinary bladder differs from that in the vas deferens, because although there is no selectivity amongst diastereoisomers of phosporothioate analogues in either preparation, there is a marked stereoselectivity in the vas deferens between the enantiomers of ATP and its 2-substituted analogues, 2-methylthio ATP and 2-chloro ATP: such is not the case in the bladder (Stone, 1985; Burnstock, Cusack and Meldrum, 1985).

An analogue of ADP, adenosine 5'-(2-fluorodiphosphate) (ADPβF) has been claimed to be a specific agonist of P_{2Y}-purinoceptors (Hourani *et al.*, 1988). However, in the rabbit jugular vein, ADPβF may interact with P_1-purinoceptors (Wood *et al.*, 1989).

Another important distinguishing feature of the P_{2X}-purinoceptor is its readiness to be desensitised by α,β-methylene ATP. In many preparations, particularly the urinary bladder, vas deferens and blood vessels, contractile responses to ATP are blocked following desensitisation to α,β-methylene ATP (Kasakov and Burnstock, 1983; Meldrum and Burnstock, 1983; Kennedy and Burnstock, 1985a; Kennedy,

Delbro and Burnstock, 1985; Houston, Burnstock and Vanhoutte, 1987; Reilly and Burnstock, 1987). The P_{2Y}-purinoceptor does not readily desensitise to α,β-methylene ATP.

As yet no competitive antagonist has been developed that is specific for either P_{2X}- or P_{2Y}-purinoceptors. Arylazidoaminopropionyl ATP (ANAPP$_3$) is a photo-affinity analogue of ATP, and when activated by ultraviolet light it covalently binds within the domain of its attachment. ANAPP$_3$ has been shown to block responses to ATP or its analogues, mediated via P_{2X}-purinoceptors (Hogaboom, O'Donnell and Fedan, 1980; Fedan et al., 1982; Frew and Lundy, 1982a; Westfall et al., 1983), but has not been shown to antagonise ATP to a useful extent in P_{2Y} systems (Westfall et al., 1982; Frew and Lundy, 1982a,b). Further, ANAPP$_3$ may interact with P_1-purinoceptors (Frew and Lundy, 1986).

The anthraquinone sulphonic acid derivative, reactive blue 2, has been used as an antagonist of responses mediated via P_2-purinoceptors (Kerr and Krantis, 1979; Choo, 1981; Crema et al., 1983; Manzini, Maggi and Meli, 1985; Manzini, Hoyle and Burnstock, 1986b). However, it is only useful within a narrow range of concentrations, above which it has non-specific effects. Although reactive blue 2 can selectively inhibit P_{2Y}-mediated responses in the gut and some blood vessels (Manzini, Maggi and Meli, 1985; Manzini, Maggi and Meli, 1986b; Burnstock and Warland, 1987b, Reilly, Saville and Burnstock, 1987; Houston, Burnstock and Vanhoutte, 1987; Hopwood and Burnstock, 1987; Hopwood et al., 1989; Taylor et al., 1989), it also antagonises P_{2X}-purinoceptors in the bladder (Choo, 1981). In combination with agonist potency studies, reactive blue 2 has been used to identify P_{2Y}-purinoceptors in other systems, such as rat alveolar type II cells and rabbit stomach (Rice and Singleton, 1987, 1989; Beck et al., 1988).

The trypanocidal drug, suramin, has recently been discovered as a P_2-purinoceptor antagonist, but it is effective at both P_{2X}- and P_{2Y}-purinoceptors in the vas deferens, urinary bladder and taenia coli (Dunn and Blakeley, 1988; Den Hertog, Nelemans and Van Den Akker, 1989; Hourani and Chown, 1989), and it does not have any selectivity for either subtype (Hoyle, Knight and Burnstock, 1990). Potentiation of P_{2X}-mediated events but not P_{2Y}-mediated events has also been reported (Hourani and Chown, 1989; Hoyle, Knight and Burnstock, 1990).

P_{2T}-PURINOCEPTORS

As defined by Gordon (1986), the P_{2T}-purinoceptor exists on platelets (Thrombocytes) and is activated specifically by ADP rather than ATP. Adenosine does not act on this receptor (Haslam and Rosson, 1975), and ATP is a competitive antagonist (Macfarlane and Mills, 1975). Diadenosine polyphosphates also inhibit responses to ADP, but the type of antagonism has not been defined (Harrison, Brossmer and Goody, 1975). In some respects the P_{2T}-purinoceptor is similar to the P_{2Y}-purinoceptor, in that analogues with modified phosphate chains and L-enantiomers of analogues are less potent or inert, while 2-substituted analogues of ADP are very potent (Cusack, Hickman and Born, 1979; Cusack, Hourani and Welford, 1985).

P_{2Z}-PURINOCEPTORS

Again as defined by Gordon (1986), the P_{2Z}-purinoceptor is activated by ATP^{4-}, i.e. ATP dissociated from divalent cations. Acting at the P_{2Z}-purinoceptor ATP mediates secretion from mast cells (Cockroft and Gomperts, 1979a,b, 1980; Bennett, Cockroft and Gomperts, 1981), and causes inhibition of activity of natural killer lymphocytes and monocyte-derived macrophages (Cameron, 1984; Schmidt, Ortaldo and Herberman, 1984). Responses at this receptor are attenuated by raising the extracellular concentration of magnesium ions (thereby increasing the concentration of complexed ATP). The dinucleotide P^1,P^4-diadenosine tetraphosphate can activate this receptor because it possesses four negative charges, but analogues of ATP that do not bear four negative charges on the phosphate chain are inactive (Tatham, Cusack and Gomperts, 1988). The P_{2Z}-purinoceptor also mediates the increase in permeability to monovalent cations induced by ATP in TA3 ascites tumour cells, canine erythrocytes and murine macrophages (Hempling, Stewart and Gasic, 1969; Parker and Snow, 1972; Steinberg and Silverstein, 1987). Also, the enhanced generation of superoxide anions by human polymorphonuclear neutrophils is effected by ATP acting, probably, on a P_{2Z}-purinoceptor (Kuhns *et al.*, 1988).

P_{2S}- AND P_3-PURINOCEPTORS

Both these receptor subclasses are recent proposals, and to date neither has been shown to be selectively or specifically antagonised by any agent. Although they have apparently distinctive pharmacological profiles, as yet neither has gained general acceptance.

In the guinea-pig ileum, 2-methylthio ATP and α,β-methylene ATP are roughly equipotent at causing contraction of the longitudinal muscle, and are an order of magnitude more potent than ATP itself (Wiklund and Gustafsson, 1988a,b). The contractions are not affected by either reactive blue 2 or desensitisation. Therefore it was proposed that this receptor represents a non-P_{2X}, non-P_{2Y} subclass, named P_{2S} (Wiklund and Gustafsson, 1988b).

In the rat tail artery, adenosine and ATP, and their analogues 2-chloroadenosine and β,γ-methylene ATP, all inhibit release of noradrenaline from sympathetic nerves (Shinozuka, Bjur and Westfall, 1988). The effects of all these compounds are antagonised by the P_1-purinoceptor antagonist 8-(p-sulpho) phenyltheophylline. This methylxanthine-sensitive adenine nucleotide receptor was termed P_3 (Shinozuka, Bjur and Westfall, 1988). Actions of ATP or its analogues, which are blocked by P_1-purinoceptor antagonists but which are not mediated via catabolism to adenosine, have been shown in the dog basilar artery (Muramatsu *et al.*, 1981), and on prejunctional parasympathetic nerve terminals in the guinea-pig ileum (Wiklund and Gustafsson, 1984, 1986, 1988a; Wiklund, Gustafsson and Lundin, 1985). Similarly in the rat heart, the negative inotropic action of ATP is blocked by P_1-purinoceptor antagonism but ATP is acting, at least in part, *per se* (Burnstock and Meghji, 1983; Moody, Meghji and Burnstock, 1984). Inhibition of ganglionic transmission and neuromuscular transmission in the cat urinary bladder, by adenosine and adenine nucleotides, may also be mediated via a P_3-purinoceptor since this activity is blocked by caffeine or theophylline (Theobald and De Groat, 1988).

RECEPTORS FOR ADENINE DINUCLEOTIDES

The suggestion that the adenine dinucleotide, β-nicotinamide adenine dinucleotide phosphate (NADP), acts on a population of purinoceptors, distinct from P_1- or P_2-purinoceptors (Hoyle, 1990), is based on the following observations. In the urinary bladder, vas deferens and renal vascular bed, in all of which activation of P_{2X}-purinoceptors causes contractile responses, NADP does not (Hashimoto, Kumakura and Tanemura, 1964; Stone, 1981b; Willemot and Paton, 1981). In the coronary and femoral vascular beds, in which vasodilatation is mediated via activation of P_{2Y}-purinoceptors, NADP is inert. Thus, NADP is not a P_2-purinoceptor agonist. Therefore, in preparations in which it is active, such as the guinea-pig heart and taenia coli or in the bladder and vas deferens, where it causes depression of autonomic neuromuscular transmission (Schrader, Rubio and Berne, 1975; Romanenko, 1980; Romanenko et al., 1980; Stone, 1981b; Willemot and Paton, 1981; Burnstock and Hoyle, 1985), NADP must be acting via a different class of receptor.

The α,ω-adenine dinucleotide polyphosphate, P^1,P^3-diadenosine triphosphate (AP3A), and its homologue, P^1,P^4-diadenosine tetraphosphate, (AP4A), both produce relaxations in the isolated rabbit mesenteric artery. Relaxations to AP3A are independent of endothelial cells, but relaxations to AP4A are not (Busse, Ogilvie and Pohl, 1988). In this vessel there are P_{2Y}-purinoceptors on the smooth muscle, that mediate relaxation, but not on endothelial cells, and there are P_{2X}-purinoceptors, also on the smooth muscle, that mediate contractile responses (Mathieson and Burnstock, 1985; Burnstock and Warland, 1987b). Thus, AP4A does not act on P_{2Y}-purinoceptors. Further, AP3A does not cause contractions in this vessel (Busse, Ogilvie and Pohl, 1988), and therefore it does not act on P_{2X}-purinoceptors. Hence, it is a reasonable proposal that these compounds also act via a novel class of purinoceptor.

RECEPTORS FOR ADENINE OLIGONUCLEOTIDES

The trinucleotide, pppA2'p5'A2'p5'A (2'-5'P_3A_3), in which one moiety of ATP is concatenated with two moieties of adenosine via 2'-5' phosphodiester linkages, has antiviral actions and mimics many of the effects of interferon (see Li and Liu, 1985). It enhances macrophage phagocytosis and activates natural killer lymphocytes. A radioligand binding study has shown that rat peritoneal mast cells possess a distinct receptor-like site for 2'-5'P_3A_3 that can be blocked by ATP (Li and Liu, 1985).

LOCALISATION AND STORAGE OF ATP

One of the problems of identifying the storage location of ATP is that because it is so ubiquitous within cells, great care is needed to ensure that, for example, in the subfractionation of cells to produce purified preparations of selected organelles, the ATP being measured is not contaminated by ATP from another source (Fried et al., 1984). Because ATP is so common, the occurrence of ATP and its synthesising enzymes within neurones is not necessarily a useful guide as to whether or not ATP is serving as a neurotransmitter. Quinacrine has been used to identify purinergic nerves, but because it can also bind to other chemicals, the evidence supplied by quinacrine should not be regarded as definitive of ATP location: however, it can be corroborative with other lines of evidence. For identifying ATP within transmitter vesicles within neurones, the uranaffin reaction may be more reliable (see below).

VESICULAR AND GRANULAR STORAGE OF ATP

Adrenal medullary cells and platelets have both been used as models of cellular storage of ATP. In adrenal medullary cells, ATP is co-stored and co-released from chromaffin granules along with noradrenaline (Douglas and Poisner, 1966; Douglas, 1968; Stevens *et al.*, 1972; Smith, 1972). Further, adrenal medullary cells arise from embryonic neural crest and are postsynaptic to preganglionic sympathetic nerves. These paraneuronal characteristics have led to their use as models for neurosecretory mechanisms (Coupland, 1965; Nagasawa, 1977; Winkler and Westhead, 1980; Viveros and Wilson, 1983). Platelets, although not paraneuronal, have been used as models of subcellular, granular distribution of ATP and 5-HT (Richards and Da Prada, 1977, 1980; Da Prada, Obrist and Pletscher, 1978).

In preparations of secretory granules, or vesicles, of sympathetic nerves, the ratio of noradrenaline to ATP has been reported to range from 4:1 to 60:1 in large granular vesicles (75–90 nm diameter), and from 20:1–60:1 in small granular vesicles (45–55 nm diameter) (Lagercrantz, Fried and Dahlin, 1975; Lagercrantz, 1976; Fried *et al.*, 1984). The ATP that is stored in these granules is probably the ATP that is employed during purinergic transmission. However, the occurrence of ATP within neuronal vesicles or storage granules helps the storage of noradrenaline by providing osmotic stability within the granule, due to a physico-chemical reaction between noradrenaline and ATP (Da Prada, Obrist and Pletscher, 1975). From studies of sympathetic ganglia in culture, it has been determined that less than 15% of ATP within a neurone is involved with uptake and storage of noradrenaline (Wakade and Wakade, 1985).

In cholinergic motor nerves that supply the electric organs of the elasmobranch *Torpedo marmorata* and the teleosts *Narcine brasiliensis* and *Electrophorus electricus*, co-storage with ATP has been studied. These tissues provide a relatively large volume of pre- and post-junctional structures. In vesicles from motoneurones, ATP represents over 80% of the nucleotide content, and the molar ratio of acetylcholine to ATP varies from 4:1 to 10:1 (Dowdall, Boyne and Whittaker, 1974; Boyne, 1976; Zimmermann, 1978, 1982; Israël *et al.*, 1979; Zimmermann and Bokor, 1979).

LOCALISATION OF ATP BY THE URANAFFIN REACTION AND QUINACRINE

The uranaffin reaction has been used to localise the storage of adenine nucleotides within cells (Richards and Da Prada, 1977; Richards, 1978; Böck, 1980). In this reaction the divalent cation of uranium dioxide (UO_2^{2+}) complexes with phosphate groups of nucleotides and forms an electron-dense precipitate. In normal platelets and megakaryocytes there is a strong reaction within storage granules, but in platelets and megakaryocytes taken from animals with purine-storage deficiency there is no reaction. Depletion of 5-HT from platelets with reserpine does not cause depletion of ATP nor of uranaffin reactivity. Also, vesicles containing uranaffin products, and therefore purine nucleotides, are seen in sympathetic nerve terminals innervating the vas deferens, in nerve cell bodies of the superior cervical ganglion, and in secretory granules of adrenal medullary cells (Richards and Da Prada, 1977).

In the guinea-pig ileum, the uranaffin reaction products can be seen in small vesicles with diameters of 40–60 nm in nerve terminals in the myenteric plexus, submucous plexus, deep muscular plexus and circular muscle layer. Large granular vesicles, 80 nm diameter, tend to be unstained. Reactivity is unaffected by reserpine pretreatment or surgical denervation of extrinsic nerves. Thus the reaction product is predominantly within intrinsic neurones supplying the gut (Wilson, Furness and Costa, 1979). In sympathetic nerve terminals innervating the iris and vas deferens, both small and large granular vesicles are uranaffin positive (Richards and Da Prada, 1980).

Quinacrine, which is a fluorescent acridine dye, physically binds adenine nucleotides, particularly ATP, and has been used to label tissues in which ATP is stored (Irvin and Irvin, 1954; Da Prada, Richards and Lorez, 1978; Ålund and Olson, 1979b; Böck, 1980). Just how specific quinacrine is when it is used for labelling neurones is unknown (Olson, Ålund and Norberg, 1976). For instance, quinacrine will bind to DNA, due the high concentrations of purine bases. However, quinacrine does bind compounds with transmitter-like properties, and when complexed it is released from neurones in a transmitter-like fashion (Ålund and Olson, 1979a, 1980). In several studies, quinacrine has been found to stain subpopulations of neurones in tissues with purinergic or non-adrenergic, non-cholinergic innervation (Olson, Ålund and Norberg, 1976; Burnstock et al., 1978a, b; Burnstock, Crowe and Wong, 1979; Crowe and Burnstock, 1981).

TRANSMISSION FROM SENSORY NERVES

Although the hypothesis of purinergic transmission was not put forward until the 1970's (Burnstock et al., 1970; Burnstock, 1972; Burnstock, 1975), the first evidence that ATP might be a neurotransmitter came from studies of sensory innervation in the 1950's (Holton and Perry, 1951; Holton and Holton, 1953; Holton and Holton, 1954; Andrews and Holton, 1958; Holton, 1959).

DETERMINATION OF RELEASE OF ATP FROM SENSORY NERVES

The original model used for these studies was the capillary bed in the rabbit ear, supplied by the central ear artery. Following sympathectomy by surgical removal of the superior cervical ganglion, sensory nerves supplying this vascular bed can be stimulated antidromically by applying electrical pulses to the great auricular nerve (Holton and Perry, 1951). Using short periods of stimulation, under the prevalent condition of elevated vascular tone due to the sympathectomy, there is only a dilatation of the capillary bed and not of the larger arterial elements (Hilton and Holton, 1954). The dilatation evoked by stimulation of the sensory nerves is mimicked by arterial infusion of ATP or ADP, and like the responses to ATP it is antagonised by cholinesterase inhibitors (eserine or physostigmine). Other agents that cause dilatation are not blocked by these cholinesterase inhibitors, and further, purine compounds can be detected in the venous effluent (Holton and Holton, 1953). Extracts of dorsal root ganglia, which contain the cell bodies of the sensory nerves, contain ATP at high concentrations (Holton and Holton, 1954), and using a microassay, the concentration of ATP in single cells isolated from rat dorsal root ganglia has been evaluated at approximately 1.7 mM (Fukuda, Fujita and Ohsawa, 1983). ATP is found in the venous effluent following stimulation of the sensory nerve, along with metabolites of ATP (Holton, 1959), and further, mechanical stimulation of the sensory nerve or stimulation of cutaneous afferents in the ear itself results in capillary dilatation and release of ATP (Holton, 1959).

More recently, advanced pharmacological and electrophysiological techniques have added to the evidence that ATP and adenosine may be neurotransmitters from sensory nerves in the spinal cord. By use of microelectrodes inserted into nerve cell bodies in the cat spinal cord at laminae I and II of the substantia gelatinosa, which receive inputs from sensory afferent nerve fibres, it has been determined that those cell bodies that are postsynaptically activated in response to mechanical stimulation of the skin (due to C-fibre afferent input) are also excited by ATP (Fyffe and Perl, 1984). In the same study, dorsal horn cells that selectively receive a cutaneous nociceptive input displayed little or no response to ATP. Similarly, ATP has been found to be mostly excitatory on non-nociceptive-receiving nerve cell bodies, rather than nociceptive or wide-dynamic range units in the cat spinal cord (Salter and Henry, 1985).

In synaptosomal preparations of nerve terminals from the dorsal horn of the spinal cord, from guinea-pigs and rats, ATP can be released in response to depolarising stimuli or the sensory neurotoxin capsaicin (White, Downie and Leslie, 1985; Sweeney, White and Sawynok, 1989), but it was also observed that the amount of ATP released is low relative to the population of sensory nerves. This implies that ATP is only contained in a proportion of sensory nerve terminals in the spinal cord. In dissociated cell cultures of dorsal horn neurones, ATP excites only a subpopulation of 25–30% of cells, which would be consistent with only a subpopulation of neurones secreting ATP within the spinal cord (Dodd, Jahr and Jessell, 1984; Jahr and Jessell, 1984). From these synaptosomal preparations larger quantities of adenosine are released in response to depolarising stimuli or morphine (Sweeney, White and Sawynok, 1989), and it has been suggested that the theophylline-sensitive anaesthesia induced by morphine is due to release of adenosine, from neurones, within the spinal cord (De Lander and Hopkins, 1986). These observations are supported by the findings that AMP, which like adenosine is also a P_1-purinoceptor agonist, depresses wide dynamic range units that receive nociceptive input in the cat spinal cord (Salter and Henry, 1985), and that P_1-purinoceptor antagonists, such as caffeine or theophylline derivatives, block the depression of nociception induced by mechano-stimulation, at the level of the spinal cord (Salter and Henry, 1987).

FLUORIDE-RESISTANT ACID PHOSPHATASE AND ADENOSINE DEAMINASE IN SENSORY NERVES

Two purine-handling enzymes, fluoride-resistant acid phosphatase (FRAP) and adenosine deaminase (ADA) have been found in non-overlapping populations of small type B cells in dorsal root ganglia (Knyihar, 1971; Nagy and Hunt, 1982; Nagy *et al.*, 1984; Nagy and Daddona, 1985). FRAP is transported from the cell body to the dorsal horn (Knyihar, 1971) and is found in nerve terminals in the deeper layers of the substantia gelatinosa (Nagy and Hunt, 1982). Similarly, ADA-containing sensory fibres project only to layers I and II in the substantia gelatinosa. Although the presence of FRAP or ADA is not evidence that these neurones utilise ATP or adenosine as transmitters, there is a correlation between the projection of these enzyme-containing neurones and the layers of the substantia gelatinosa in which postsynaptic responses to ATP and adenosine have been observed.

In summary, ATP appears to be released from the peripheral terminals of sensory nerves, and from the central terminals of non-nociceptive afferents in the spinal cord. Adenosine may be released from mechanoreceptive or interneurones in the spinal cord, and modulate nociceptive input.

TRANSMISSION FROM SYMPATHETIC AND PARASYMPATHETIC NERVES

Release of ATP from nerves has been measured directly, by using the sensitive firefly luciferin-luciferase luminometric assay, or by high performance liquid chromatogaphy or by measuring ^{3}H-overflow after preincubation with ^{3}H-adenine or ^{3}H-adenosine. Release of ATP can be shown indirectly by pharmacological means, whereby neurogenic responses are selectively blocked in parallel with those due to exogenous ATP. Within the sympathetic division, tissues in which purinergic transmission has been identified by direct or indirect means are given in Table 7.2. Within the parasympathetic division, tissues in which purinergic transmission has been identified are given in Table 7.3.

DIRECT DETERMINATION OF RELEASE OF ATP FROM SYMPATHETIC NERVES

In the guinea-pig taenia coli, following pre-incubation with ^{3}H-adenosine, electrical stimulation of the perivascular sympathetic nerves causes relaxation and overflow of ^{3}H-ATP. The relaxation and release are blocked by the neurotoxin, tetrodotoxin, and the sympatholytic, guanethidine (Su, Bevan and Burnstock, 1971; Kuchii, Miyahara and Shibata, 1973). In varicosities isolated from the guinea-pig myenteric plexus, although ATP and noradrenaline are not always released in parallel, it is

TABLE 7.2

Sympathetic effector tissues in which ATP has been suggested to be a neurotransmitter, from direct determination or from indirect determination.

Guinea-pig		
ileum	direct	White and Al-Humayyd, 1983
		Hammond et al., 1988
taenia coli	direct	Su et al., 1971
		Kuchii et al., 1973
vas deferens	direct, indirect	Fedan et al., 1981
		Sneddon et al., 1982
		Meldrum and Burnstock, 1983
		Sneddon and Burnstock, 1984
		Sneddon and Westfall, 1984
		Stjärne and Åstrand, 1984
		Allcorn et al., 1985
		Stjärne and Åstrand, 1985
		Kirkpatrick and Burnstock, 1987
		Åstrand et al., 1988
		Cunnane and Manchanda, 1988
seminal vesicles	indirect	Meldrum and Burnstock, 1985a
mesenteric artery	indirect	Ishikawa, 1985
		Nagao and Suzuki, 1988
saphenous artery	indirect	Cheung and Fujioka, 1986
Rabbit		
ear capillaries	direct	Holton, 1959
ear artery	indirect	Kennedy and Burnstock, 1985b
		Kennedy et al., 1986
mesenteric artery	indirect	Ishikawa, 1985
		Kügelgen and Starke, 1985
pulmonary artery	direct	Katsuragi and Su, 1980, 1982a,b
saphenous artery	indirect	Burnstock and Warland, 1987a
		Warland and Burnstock, 1987a
hind limb	indirect	Shimada and Stitt, 1984
Rat		
vas deferens	indirect	Allcorn et al., 1985
		Bulloch and McGrath, 1988
tail artery	direct, indirect	Sneddon and Burnstock, 1985
		Neild and Kotecha, 1986
		Westfall et al., 1987
		Åstrand et al., 1988
SCG*	direct, indirect	Potter et al., 1983
		Wolinsky and Patterson, 1985
		Furshpan et al., 1986
		McCaman and McAfee, 1986
		Tolkovsky and Suidan, 1987
cardiovascular system	indirect	Bulloch and McGrath, 1988
		Schlicker et al., 1989
		Flavahnan et al., 1985
Cat		
bladder	indirect	Theobald, 1986
		Theobald and Hoffman, 1986
nictitating membrane	indirect	Duval et al., 1985
Dog		
basilar artery	indirect	Muramatsu and Kigoshi, 1987
mesenteric artery	indirect	Muramatsu, 1987
		Machaly et al., 1988
		Muramatsu et al., 1989
cutaneous veins		Flavahan and Vanhoutte, 1986

TABLE 7.2 Continued

Mouse		
vas deferens	indirect	Stjärne and Åstrand, 1984
		Allcorn et al., 1985
		Stjärne and Åstrand, 1985
		Åstrand et al., 1988
Fish		
chromatophores	direct	Kumazawa and Fujii, 1984
		Kumazawa et al., 1984
		Fujii, 1986
		Fujii and Oshima, 1986
		Kumazawa and Fujii, 1986
Toad		
heart	indirect	Bramich and Campbell, 1989
		Bramich et al., 1989
Frog		
heart	indirect	Hoyle and Burnstock, 1986

* Superior cervical ganglion

evident that a proportion of ATP is co-stored and co-released with noradrenaline (White and Al-Humayyd, 1983; Al-Humayyd and White, 1985a; Hammond, Mac-Donald and White, 1988). Hence in sympathetic nerves supplying the intestine, ATP and noradrenaline co-exist and may co-transmit.

When tissues are exposed to ^{3}H-adenine or ^{3}H-adenosine, ATP is usually the predominant product into which the ^{3}H is incorporated (Su, Bevan and Burnstock, 1971; Su, 1975; Maire, Medilanski and Straub, 1982; Kumazawa and Fujii, 1984; Wolinsky and Patterson, 1985; McCaman, 1986; Tolkovsky and Suidan, 1987), although AMP may be a predominant product (McCaman and McAfee, 1986). If pre-loaded tissues are stimulated to evoke release of purines from neuronal stores, the amount of ATP recovered may be substantially less than that released, due to its rapid degradation by ectoenzymes such as 5'-nucleotidase (Holton, 1959; Silinsky and Hubbard, 1973; Su, 1975; Maire, Medilanski and Straub, 1982, 1984; Wolinsky and Patterson, 1985; MacDonald and White, 1985; Kumazawa and Fujii, 1986; Richardson and Brown, 1987). Ectonucleotidases are bound to smooth muscle membranes in the stomach, small intestine, blood vessels, vas deferens and urinary bladder, membranes of vascular endothelial cells, and membranes of satellite cells and nerve cells in sympathetic ganglia (Burger and Löwenstein, 1970; Nair and Wiechert, 1980; Pearson, Carleton and Gordon, 1980; Cusack, Pearson and Gordon, 1983; Forsman and Elfvin, 1984; Kwan and Kostka, 1984; Kwan and Ramlai, 1985; Kwan and Sipos, 1985; Pearson, Coade and Cusack, 1985; Pearson and Cusack, 1985; Hourani and Chown, 1989). The distribution of ectonucleotidases within purinergically innervated tissues suggests that these enzymes may be important for the degradation of endogenously released ATP. High-affinity uptake systems for adenosine, that are also well-represented in autonomic effector tissues and neurones, are a salvage system. Adenosine is taken up and is subsequently re-incorporated into ATP, in a manner analogous to the uptake and recycling of choline by cholinergic nerves.

TABLE 7.3

Parasympathetic effector tissues in which ATP has been suggested to be a neurotransmitter from direct or indirect determination.

Guinea-pig		
ileum	direct	White, 1982
		White and Leslie, 1982
		Al-Humayyd and White, 1985a,b
taenia coli	direct, indirect	Su et al., 1971
		Kuchii et al., 1973
		Burnstock et al., 1978b
		Satchell, 1981, 1982
gall bladder	direct	Takahashi et al., 1987
urinary bladder	direct, indirect	Burnstock et al., 1978a,b
		Kasakov and Burnstock, 1983
		Westfall et al., 1983
		Hourani, 1984
		Moss and Burnstock, 1985
		Fujii, 1988
Rabbit		
stomach	indirect	Beck et al., 1988
portal vein	direct	Burnstock et al., 1979
urinary bladder	indirect	Dean and Downie, 1978
		Hoyle and Burnstock, 1985
		Fujii, 1988
Cat		
colon	indirect	Hedlund et al., 1983
urinary bladder	indirect	Theobald, 1982, 1986
		Theobald and Hoffman, 1986
vesical ganglia*	indirect	Akasu et al., 1984
Rat		
duodenum	indirect	Manzini et al., 1985 1986a
caecum	indirect	Manzini et al., 1986b
bladder	indirect	Burnstock et al., 1972
		Choo and Mitchelson, 1980
		Choo, 1981
Human		
bladder	indirect	Hoyle et al., 1990
Mouse		
bladder	indirect	Holt et al., 1985
Ferret		
bladder	indirect	Moss and Burnstock, 1985
Marmoset		
bladder	indirect	Moss and Burnstock, 1985
Chick		
ileum	indirect	Ahmad et al., 1978
rectum	indirect	Meldrum and Burnstock, 1985b
Turkey		
gizzard	direct	Burnstock et al., 1970
Toad		
stomach	direct	Burnstock et al., 1970

* adenosine rather than ATP

Similarly, release of ATP has been measured following stimulation of sympathetic nerves supplying many blood vessels, the vas deferens and fish chromatophores, as well as from sympathetic neurones in culture (Su, 1975; Westfall, Stitzel and Rowe, 1978; Katsuragi and Su, 1980, 1982a,b; Levitt and Westfall, 1982; Kumazawa and Fujii, 1984, 1986; Kumazawa et al., 1984; Wolinsky and Patterson, 1985; McCaman and McAfee, 1986; Kirkpatrick and Burnstock, 1987; Tolkovsky and Suidan, 1987; Westfall, Sedaa and Bjur, 1987; Kasakov et al., 1988).

Under some conditions, e.g. hypoxic stress, or during development, sympathetic neurones may release adenosine rather than ATP. This has been shown directly (Wakade and Wakade, 1985; McCaman and McAfee, 1986; Tolkovsky and Suidan, 1987) and indirectly (Potter, Furshpan and Landis, 1983). In this latter case, sympathetic ganglia were co-cultured with rat atrial myocytes, the activity of which was used as an *in situ* bioassay of substances released from the sympathetic nerves (Potter, Furshpan and Landis, 1983; Furshpan, Potter and Matsumoto, 1986). However, it should be pointed out that in the rat atrium, the actions of ATP can be blocked by P_1-purinoceptor antagonists (Burnstock and Meghji, 1983), so in the co-culture system the bioassay could have been detecting ATP in addition to adenosine.

INDIRECT DETERMINATION OF RELEASE OF ATP FROM SYMPATHETIC NERVES

In sympathetically innervated tissues, ATP-release has been examined indirectly by studying the mechanical activity or membrane potential of the effector tissue, and by blocking responses to nerve stimulation in parallel with those to exogenous ATP, in a selective manner.

The majority of such indirect studies have been carried out on the vas deferens and blood vessels. In both these types of tissue, ATP causes contractions mediated via the P_{2X}-purinoceptor subtype. Noradrenaline, also released from sympathetic nerves, also causes a contraction. Electrophysiologically, ATP evokes rapid depolarisation and generation of action potentials that underlie the mechanical contractions, while noradrenaline causes little or no depolarisation, i.e. responses to ATP are electromechanico coupled but responses to noradrenaline are pharmaco-mechanico coupled. In response to electrical stimulation of sympathetic nerve terminals there are usually two elements, firstly a transient depolarisation, or excitatory junction potential (EJP), and secondly a slow depolarisation. In response to a train of pulses consecutive EJPs facilitate and action potentials are generated when their threshold is breached, slow depolarisations only become pronounced following trains of pulses at frequencies above 2 Hz. The slow depolarizations are sensitive to adrenoceptor blockade by α_1-adrenoceptor antagonists, such as prazosin. EJPs are resistant to adrenoceptor blockade, and are blocked by $ANAPP_3$ or by desensitising P_{2X}-purinoceptors with α,β-methylene ATP, and are therefore purinergic (Figure 7.2).

In the guinea-pig vas deferens, photoactivated $ANAPP_3$ blocks the mechanical and electrophysiological responses to exogenous ATP and field stimulation of non-adrenergic sympathetic nerves (Fedan et al., 1981; Sneddon, Westfall and Fedan,

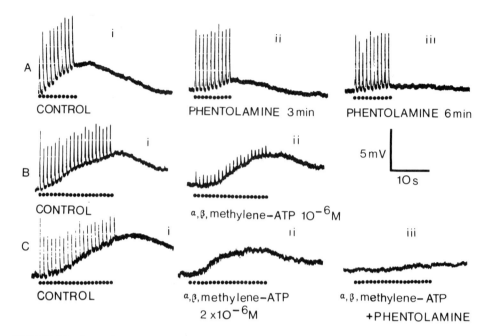

FIGURE 7.2 Intracellular recordings of the electrical responses of single smooth muscle cells of the rat tail artery to field stimulation of the sympathetic motor nerves (the pulse width was 0.1 ms at 0.5 Hz, indicated by a black spot). (Ai) Control response of the muscle. Note that for each individual stimulus there was a rapid depolarisation, and as the train of pulses progressed a slow depolarisation developed. Similar responses were obtained in (Bi) and (Ci), which are also control responses in Krebs solution. (Aii) and (Aiii) show the effect of phentolamine (2 μM) added to the Krebs solution. The fast depolarisations were not reduced, but there was a progressive reduction in the amplitude of the slow depolarisation, which was almost abolished by 6 min. In (Bii) the tissue was incubated in α, β-methylene ATP (1 μM) for over 15 min. The fast depolarisations were reduced but the slow depolarisation was unaffected. (Cii) shows the effect of a higher concentration of α, β-methylene ATP (2 μM). In this case the fast depolarisations are abolished, but the slow depolarisation persists, although it is slightly reduced in amplitude. Subsequent addition of phentolamine (1 μM), in the presence of α, β-methylene ATP (2 μM), abolished both types of neurogenic responses (Ciii). From Sneddon and Burnstock (1984), reproduced with permission of the publishers.

1982; Frew and Lundy, 1982a; Sneddon and Westfall, 1984). Similarly, in the guinea-pig saphenous artery, ANAPP$_3$ blocks EJPs (Cheung and Fujioka, 1986). Using α,β-methylene ATP to desensitise P$_{2X}$-purinoceptors the same results (i.e. selective blockade of purinergic transmission) have been obtained in the vas deferens of several species and in several types of blood vessel (Meldrum and Burnstock, 1983; Sneddon and Burnstock, 1984, 1985; Stjärne and Åstrand, 1984, 1985; Kügelgen and Starke, 1985; Ishikawa, 1985; Allcorn, Cunnane and Kirkpatrick, 1985; Kennedy, Saville and Burnstock, 1986; Neild and Kotecha, 1986; Burnstock and Warland, 1987a; Muramatsu, 1987; Muramatsu and Kigoshi, 1987; Ramme *et al.*, 1987; Warland and Burnstock, 1987; Åstrand, Brock and Cunnane, 1988; Cunnane and Manchanda, 1988; Saville and Burnstock, 1988; Machaly, Dalziel and Sneddon, 1988; MacKenzie, Kirkpatrick and Burnstock, 1988; Nagao and Suzuki, 1988). EJPs in the sympathetically innervated toad sinus venosus are also blocked by α,β-methylene ATP (Bramich, Edwards and Hirst, 1989). Exceptions to non-adrenergic EJPs being sensitive to desensitisation with α,β-methylene ATP are those in the chick rectum (Komori and Ohashi, 1988a,b) and the guinea-pig coronary artery (Keef and Kreulen, 1988). The coronary vasculature represents an exception in any case, since in most animals both ATP and noradrenaline cause vasodilatation rather than vasoconstriction. Other examples of α,β-methylene ATP being used to show purinergic sympathetic transmission include the seminal vesicles (Meldrum and Burnstock, 1985a), the cat nictitating membrane (Duval, Hicks and Langer, 1985), frog heart and toad heart (Hoyle and Burnstock, 1986; Bramich and Campbell, 1989).

One of the criticisms of the use of α,β-methylene ATP has been that although it does not interfere with postjunctional adrenoceptors, it might prevent release of noradrenaline. However, several studies have shown that α,β-methylene ATP does not affect stimulated overflow of ^{3}H-noradrenaline (Kasakov et al., 1988; Yamamoto, Cune and Takasaki, 1988; Muramatsu, Ohmura and Oshita, 1989; Kügelgen, Schöffel and Starke, 1989).

DIRECT DETERMINATION OF RELEASE OF ATP FROM PARASYMPATHETIC NERVES

Increased levels of ATP and related nucleotides were detected chromatographically in the superfusate from turkey gizzard and perfusate of toad stomach, following stimulation of intrinsic nerves (Burnstock et al., 1970). This was the original work that led to the purinergic hypothesis (Burnstock, 1972). Since then, release of ATP has been detected directly, in response to stimulation of parasympathetic nerves supplying the guinea-pig urinary bladder, taenia coli and gall bladder, and rabbit portal vein (Su, Bevan and Burnstock, 1971; Burnstock et al., 1978a,b; Burnstock, Crowe and Wong, 1979; Rutherford and Burnstock, 1978; Levitt and Westfall, 1982; Takahashi et al., 1987). ATP does not appear to be released from isolated parasympathetic nerve trunks in response to electrical field stimulation (Maire, Medilanski and Straub, 1982, 1984), so almost certainly it comes from nerve terminals. Also, in synaptosomal preparations of nerve varicosities from guinea-pig myenteric plexus, ATP-release can be elicited by depolarisation, in a calcium-dependent manner, suggesting neurotransmitter function (White and Leslie, 1982; White and Al-Humayyd, 1983; Al-Humayyd and White, 1985a,b).

INDIRECT DETERMINATION OF RELEASE OF ATP FROM PARASYMPATHETIC NERVES

As in the sympathetically innervated vas deferens, and many blood vessels, excitatory responses to ATP in the urinary bladder are mediated via P_{2X}-purinoceptors. Electrophysiologically, stimulation of the postganglionic parasympathetic nerves evokes a non-cholinergic EJP (Creed, Ishikawa and Ito, 1983; Hoyle and Burnstock, 1985;

Fujii, 1988). Cholinergic transmission is represented by a slower depolarisation (Creed, Ishikawa and Ito, 1983; Fujii, 1988). These non-cholinergic EJPs are abolished following desensitisation with α,β-methylene ATP, and are therefore purinergic (Hoyle and Burnstock, 1985; Fujii, 1988). In studies of mechanical activity, cholinergic responses are blocked by atropine, and the resultant non-cholinergic element of neurally evoked responses can be blocked by either photoactivated ANAPP$_3$ or following desensitisation to α,β-methylene ATP. This is true for many species, including guinea-pig, mouse, rabbit, ferret, marmoset, cat and man (Theobald, 1982; Kasakov and Burnstock, 1983; Westfall et al., 1983; Holt, Cooper and Wyllie, 1985; Hoyle and Burnstock, 1985; Moss and Burnstock, 1985; Katsuragi, Kuratomi and Furukawa, 1986; Theobald, 1986; Theobald and Hoffman, 1986; Fujii, 1988; Hoyle, Chapple and Burnstock, 1989; Brading and Williams, 1990). Desensitisation with the structurally related analogue, β,γ-methylene ATP, and its enantiomer β,γ-methylene L-ATP produce similar results (Hourani, 1984). In the rat distal colon in vivo, in the presence of atropine, stimulation of the pelvic nerves causes contractions that are mimicked by high doses of ATP and α,β-methylene ATP. Following tachyphylaxis to α,β-methylene ATP, these contractile responses are lost, and inhibitions are unmasked. This implies that either at the neuromuscular junction or perhaps within a parasympathetic ganglion, ATP acts as a transmitter onto P$_{2X}$-purinoceptors (Hedlund et al., 1983).

Tachyphylaxis to ATP, rather than desensitisation to a stable analogue, has been used with moderate success. In the rabbit urinary bladder, duodenum and colon, guinea-pig fundus and urinary bladder, and rat duodenum, tachyphylaxis to ATP is accompanied by an attenuated or abolished response to stimulation of intramural inhibitory nerves (Burnstock et al., 1970; Burnstock, Dumsday and Smythe, 1972; Okwuasaba, Hamilton and Cook, 1977; Dean and Downie, 1978; Maggi, Manzini and Meli, 1984; Levin, Rugieri and Wein, 1986; Manzini, Maggi and Meli, 1986a). In the guinea-pig and rat urinary bladder, tachyphylaxis to ATP is accompanied by inhibition of purinergic nerve stimulation when prostaglandin synthesis has been prevented by indomethacin (Burnstock et al., 1978a; Choo and Mitchelson, 1980).

In most regions of the gastrointestinal tract ATP causes relaxation, typically mediated via P$_{2Y}$-purinoceptors. Unfortunately, good competitive antagonists are lacking, but reactive blue 2 and suramin (Figure 7.3) have been used to provide supporting evidence for parasympathetic purinergic neuromuscular transmission. Reactive blue 2 selectively antagonises responses to ATP, or α,β-methylene ATP, and non-adrenergic, non-cholinergic neurogenic responses in the rabbit stomach, rat urinary bladder, duodenum and caecum, and in guinea-pig internal anal sphincter and colon (Kerr and Krantis, 1979; Choo, 1981; Crema et al., 1983; Manzini, Maggi and Meli, 1985, 1986b; Beck et al., 1988). Suramin antagonises relaxant responses to ATP and nerve stimulation in the guinea-pig taenia coli, and excitatory responses to ATP and nerve stimulation in the urinary bladder (Hoyle, Knight and Burnstock, 1990).

Incubation of tissues with the enzyme nucleotide pyrophosphatase selectively antagonises responses to exogenous ATP and electrical stimulation of intramural nerves in the guinea-pig taenia coli and rat duodenum (Satchell, 1981, 1982; Manzini,

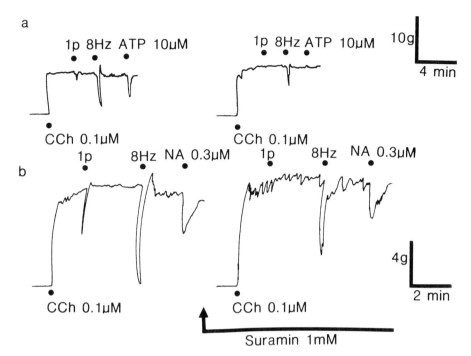

FIGURE 7.3 Antagonism of non-adrenergic, non-cholinergic inhibitory neuromuscular transmission and responses to ATP in the isolated guinea-pig taenia coli. Standard tone was induced by application of carbachol (CCh, 0.1 μM). (a) Responses to electrical field stimulation (single pulse, 1p), 8 Hz for 10 s, and ATP (EC_{50} concentration) in the absence (left-hand panel) and presence of suramin (1 mM) (right-hand panel). (b) Responses to electrical field stimulation, as in (a) and noradrenaline (EC_{50} concentration) in the absence (left-hand panel) and presence (right-hand panel) of suramin (1 mM). Note the attenuation of responses to ATP in parallel with those of electrical field stimulation, while responses to noradrenaline are unaffected. From Hoyle, Knight and Burnstock (1990), reproduced with permission of the publishers.

Maggi and Meli, 1985). Again this suggests that ATP is a neuromuscular transmitter in these preparations.

High-affinity binding sites for ATP have been determined in the bladder using radioactive ligands (Levin, Jacoby and Wein, 1983; Bo and Burnstock, 1989), that are thought to represent P_2-purinoceptors.

Finally, in parasympathetic ganglia on the surface of the cat bladder it has been suggested that adenosine may be a transmitter. This is because adenosine mimics the slow inhibitory postsynaptic potential (s.i.p.s.p): caffeine and adenosine deaminase block responses to adenosine in parallel with blockade of the s.i.p.s.p, and dipyridamole potentiates both responses (Akasu, Shinnick-Gallagher and Gallagher, 1984).

ATP AS A CO-TRANSMITTER

Arising from the volume of evidence that ATP is utilised as a neurotransmitter by sympathetic nerves, is the fact that ATP is co-stored and co-released with noradrenaline (Burnstock, 1976, 1986, 1990). Two drugs have been used as tools, namely 6-hydroxydopamine (6-OHDA) and guanethidine. Both these agents are sympatholytics, causing degeneration of nerve terminals. Although reserpine disrupts storage and release of catecholamines and 5-HT, it does not have such an effect on ATP (Figure 7.4). Evidence that ATP and noradrenaline are co-transmitters comes from the observations that they are released together (determined either by direct or indirect methods), and that the release of both is blocked by guanethidine or 6-OHDA (Figure 7.4). Such results have been found especially in the vas deferens and blood vessels (Fedan *et al.*, 1981; Cheung, 1982; Allcorn, Cunnane and Kirkpatrick, 1985; Kügelgen and Starke, 1985; Kirkpatrick and Burnstock, 1987; Muramatsu and Kigoshi, 1987; Warland and Burnstock, 1987; Nagao and Suzuki, 1988; Saville and Burnstock, 1988).

In the sympathetic ganglion/cardiac myocyte co-culture preparation, adenosine is released from neurites that also release acetylcholine and/or noradrenaline (Potter, Furshpan and Landis, 1983; Furshpan, Potter and Matsumoto, 1986). It has yet to be established whether this co-transmission is a phenotypic expression during development, or whether it is an artefact of the experimental conditions. Another example of ATP being a co-transmitter with acetylcholine in sympathetic nerves occurs in the silurid catfish, in the innervation of chromatophores (Fujii, 1986).

In the parasympathetic division, evidence for co-transmission is less forthcoming, due mostly to the lack of investigative tools. However, ultrastructural evidence and the use of a botulinum toxin as a parasympathetolytic agent, suggests that in the urinary bladder ATP cotransmits with acetylcholine (Hoyes, Barber and Martin, 1975; MacKenzie, Burnstock and Dolly, 1982).

NEUROMODULATION BY ATP

ATP released from autonomic nerves may act as a neuromodulator at prejunctional or postjunctional sites. As a prejunctional modulator, ATP may act on receptors located on nerve terminals, and enhance or inhibit further release of transmitter

---

FIGURE 7.4 Co-transmission in guinea-pig vas deferens, determined by direct measurement of release of ATP using the firefly lantern luciferin-luciferase assay. A. Release of endogenous ATP from control (n = 32), 6-hydroxydopamine-pretreated (6-OHDA, n = 7), tetrodotoxin-treated (TTX, 1.6 μM, n = 7), and guanethidine-treated (GUAN, 5 μg.ml^{-1}, n = 7) vasa deferentia, during electrical field stimulation of the sympathetic motonerves at 8 Hz (0.5 ms, 50 V). Upper panel: mean $\pm$ s.e. nmol ATP released.min^{-1}.g^{-1} tissue. Lower panel: Contractile responses to field stimulation. The preparations were stimulated for 1 min as denoted by the upward brackets. ***P < 0.001. Note that the two sympatholytics, guanethidine and 6-OHDA, almost abolished the release of ATP, as did preventing neuronal action potential discharge with TTX, which shows that the ATP is being released from sympathetic neurones. B. Release of endogenous ATP from control (n = 32) and reserpine-treated (n = 12) vasa deferentia, during electrical field stimulation at 8 Hz (0.5 ms, 20 V). Upper panel: mean $\pm$ s.e nmol ATP released.min^{-1}.g^{-1} tissue. Lower panel: Contractile responses to electrical field stimulation. The preparations were stimulated for 1 min as indicated by the upward brackets. Note that in the reserpine-treated tissue, in which noradrenaline is depleted, the second slow phase of the mechanical response has gone, but that the initial fast phase and the release of ATP have been maintained. From Kirkpatrick and Burnstock (1987), reproduced with permission of the publishers.

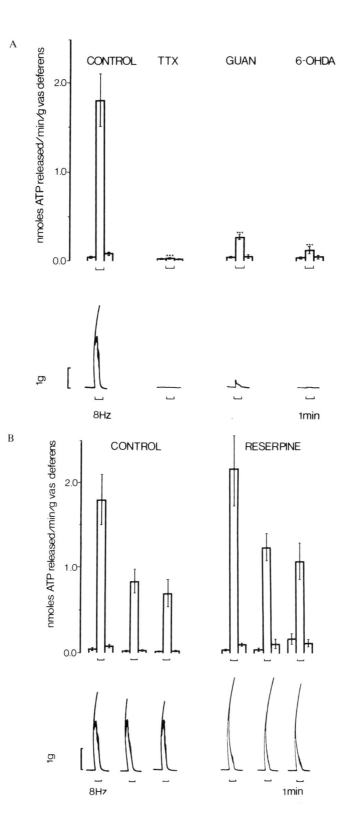

substances from these nerves. In many situations ATP is degraded by ectoenzymes, and the nascent adenosine or AMP may be responsible for the apparent action of ATP. As a postjunctional modulator, ATP might affect the responses evoked by another transmitter, such as noradrenaline or acetylcholine, by causing a change in the affinity of the receptor for the transmitter.

PREJUNCTIONAL NEUROMODULATION BY ATP

In the guinea-pig ileum, cholinergic excitatory neuromuscular transmission can be inhibited by adenosine or adenine nucleotides acting on nerve terminals to prevent release of acetylcholine (Takagi and Takayanagi, 1972; Dowdle and Maske, 1980). Cholinergic ganglionic transmission and neuromuscular transmission in the cat urinary bladder are also inhibited by ATP (Theobald and De Groat, 1988). The receptor involved appears to be a P_1-purinoceptor (or perhaps P_3-purinoceptor) because the actions of adenosine, ATP and some analogues of ATP are blocked by methylxanthines (Dowdle and Maske, 1980; Moody and Burnstock, 1982; Wiklund and Gustàfsson, 1984; Katsuragi et al., 1985; Wiklund , Gustafsson and Lundin, 1985; Frew and Lundy, 1986). Although the action of ATP involves breakdown to adenosine or AMP, that can subsequently activate the prejunctional P_1-purinoceptor, ATP can also act per se, shown by the observations that it is potentiated following inhibition of endogenous 5'-nucleotidase activity (Moody, Meghji and Burnstock, 1984; Wiklund and Gustafsson, 1986). There may also be a prejunctional P_2-purinoceptor, activation of which enhances release of acetylcholine (Moody and Burnstock, 1982). These observations are supported by results of experiments on ileal synaptosomal preparations, from which ^{3}H-acetylcholine-release is inhibited by adenosine and low concentrations of ATP, in a theophylline-sensitive manner, and from which acetylcholine-release can be stimulated by higher concentrations of ATP (Reese and Cooper, 1982).

ATP acts on a prejunctional receptor with a very similar pharmacology in the rat tail artery. Here release of noradrenaline from sympathetic nerves is inhibited by adenosine, ATP and some analogues of ATP, and all these agonists are antagonised by blocking P_1-purinoceptors (Shinozuka, Bjur and Westfall, 1988). Theophylline-sensitive prejunctional inhibition by ATP has been reported in the rabbit pulmonary artery. In this instance the facts that adenosine and ATP inhibit neurogenic release of ^{3}H-purines, and that theophylline enhances this release, have been taken as evidence of prejunctional autoinhibition, i.e. ATP released from nerve terminals can feed back onto receptors on these terminals to inhibit further release (Katsuragi and Su, 1982b). Similarly, in the sympathetically innervated rat and guinea-pig vas deferens, the prejunctional inhibition by ATP, analogues of ATP and adenine dinucleotides is sensitive to P_1-purinoceptor antagonism (Stone, 1981b, 1985).

In the examples quoted above, the type of purinoceptor at which ATP is acting appears to be a P_1-purinoceptor rather than a P_2-purinoceptor. The analogue of ATP, α,β-methylene ATP, is either weakly active or inactive, and responses to purine nucleotides are blocked by antagonists of the P_1-purinoceptor rather than antagonists of the P_2-purinoceptor such as reactive blue 2, suramin or receptor-desensitisation (Moody and Burnstock, 1982; Silinsky and Ginsborg, 1983; Wiklund and Gustafsson, 1988a,b). Shinozuka, Bjur and Westfall (1988) proposed a P_3-purinoceptor for this pharmacological profile, but until the actions of adenosine can be separated pharmacologically from those of ATP, or its analogues, such a nomenclature will not receive strong support.

However, in the mouse vas deferens, the prejunctional inhibition of adrenergic transmission by ATP is not affected by the blocking of P_1-purinoceptors (Kügelgen, Schöffel and Starke, 1989). Likewise in frog sympathetic ganglia ATP acts distinctly, inhibiting release of acetylcholine, while adenosine is inert (Silinsky and Ginsborg, 1983).

POSTJUNCTIONAL MODULATION BY ATP

The earliest demonstration of postjunctional modulation by ATP was probably that of Buchtal and Kahlson (1944), who showed that in addition to adenine nucleotides being able to cause tetanic contractions in the feline tibialis anterior muscle, ATP could also potentiate the responses to acetylcholine. ATP also potentiates cholinergic activity in the rat diaphragm (Ewald, 1976) and a possible physiological source for this ATP is from the terminals of the phrenic nerve (Silinsky and Hubbard, 1973; Silinsky, 1975). Likewise, in the frog sartorius muscle, ATP potentiates acetylcholine (Saji, Escalon de Motta and Del Castillo, 1975; Akasu, Hirai and Koketsu, 1981).

In the autonomic nervous system and its effectors, in frog sympathetic ganglia ATP depresses release of acetylcholine (Akasu, Hirai and Koketsu, 1983a; Silinsky and Ginsborg, 1983) but enhances the affinity of the postjunctional nicotinic receptor for acetylcholine (Akasu, Hirai and Koketsu, 1983b; Akasu and Koketsu, 1985). In the frog spinal ganglia ATP potentiates responses to γ-aminobutyric acid (GABA) (Morita *et al.*, 1984).

In the vas deferens, subthreshold concentrations of ATP potentiate responses to noradrenaline and sympathetic nerve stimulation (Kazic and Milosavljevic, 1980; Huidboro-Toro and Parada, 1988). Similar results were not obtained by Holck and Marks (1978), who found that adenosine and AMP, but not ATP, could potentiate the actions of noradrenaline. In the mesenteric arteries of rabbits and rats, ATP and noradrenaline show postjunctional synergistic interactions (Krishnamurty and Kadowitz, 1982; Lukackso and Blumberg, 1982), and in the rat this seems to involve the P_{2X}-purinoceptor and α_1-adrenoceptor (Ralevic and Burnstock, 1990). In the guinea-pig and rat portal vein, ATP potentiates responses to noradrenaline (Kennedy and Burnstock, 1986).

Neuromodulatory actions of ATP are not restricted to vertebrates. For example, ATP can interact with cholinoceptors in the heart of the fresh-water mussel *Anodonta cygnea*; ATP at very low concentrations inhibits the cardioinhibitory action of acetylcholine by antagonising the cholinoceptor, and it has been suggested that ATP could be released from the cardiac innervation (Turpaev, Nistratova and Putintseva, 1967; Nistratova, 1981). Purine nucleosides and nucleotides can enhance neuromuscular transmission in the intestine of the spiny lobster, *Panulirus argus*, and depress transmission in the body wall of the sea-squirt *Ciona intestinalis* (Hoyle and

Greenberg, 1988), although in neither of these preparations is the actual transmitter substance known.

SUMMARY

ATP is known to be synthesised within many types of cells, including neurones, where its intracellular concentration can be of the order of millimolar, and in which it can be localised within secretory granules. Using techniques for the direct measurement of ATP, and indirect pharmacological methods, ATP has been shown to be released from the peripheral and central ends of sensory neurones, as well as from enteric, sympathetic and parasympathetic neurones. As a neurotransmitter ATP is possibly unique, in that it is degraded into substances (principally adenosine) which can have a potent local activity. In many cases, especially within the sympathetic division, ATP appears to be a co-transmitter, often with noradrenaline, and it has also been implicated as a co-transmitter with acetylcholine and neuropeptides. Coupled with demonstrations of neuronal release of ATP there have been significant developments in the pharmacology of purine compounds, such that for ATP there are possibly 6 recognisable subclasses of the P_2-purinoceptor, while for its related nucleoside, adenosine, there are also several subclasses of the P_1-purinoceptor. Unclassified receptors for purine dinucleotides and oligonucleotides have been described, and their pharmacological profiles are being developed too. Within autonomically innervated tissues ATP has neuromodulatory actions at prejunctional and postjunctional sites, which cannot necessarily be ascribed to adenosine formed after extracellular degradation of ATP. Despite the increase in understanding of physiological roles of ATP, the therapeutic potential in manipulating purinergic transmission remains an area which presents itself for future research.

REFERENCES

Ahmad, A., Singh, R.C.P. and Garg, B.D. (1978) Evidence of non-cholinergic nervous transmission in chick ileum. *Life Sciences*, **22**, 1049–1058.
Akasu, T. and Koketsu, K. (1985) Effect of adenosine triphosphate on the sensitivity of the nicotinic acetylcholine receptor in the bullfrog sympathetic ganglion cell. *British Journal of Pharmacology*, **84**, 525–531.
Akasu, T., Hirai, K. and Koketsu, K. (1981) Increase of acetylcholine-receptor sensitivity by adenosine triphosphate: a novel action of ATP on ACh-sensitivity. *British Journal of Pharmacology*, **74**, 505–507.
Akasu, T., Hirai, K. and Koketsu, K. (1983a) Modulatory actions of ATP on bullfrog sympathetic ganglion cells. *Brain Research*, **258**, 313–317.
Akasu, T., Hirai, K. and Koketsu, K. (1983b) Modulatory action of ATP on nicotinic transmission in bullfrog sympathetic ganglia. In *Physiology and Pharmacology of Adenosine Derivatives*, edited by J.W. Daly, Y. Kuroda, J.W. Phillis, H. Shimizu and M. Ui, pp. 165–171. New York: Raven Press.
Akasu, T., Shinnick-Gallagher, P. and Gallagher, J.P. (1984) Adenosine mediates a slow hyperpolarizing synaptic potential in autonomic neurones. *Nature*, **311**, 62–65.
Al-Humayyd, M. and White, T.D. (1985a) Adrenergic and possible nonadrenergic sources of adenosine 5′-triphosphate release from nerve varicosities isolated from ileal myenteric plexus. *Journal of Pharmacology and Experimental Therapeutics*, **233**, 796–800.

Al-Humayyd, M. and White, T.D. (1985b) 5-Hydroxytryptamine releases adenosine 5'-triphosphate from nerve varicosities isolated from the myenteric plexus of the guinea-pig ileum. *British Journal of Pharmacology*, **84**, 27–34.

Allcorn, R.J., Cunnane, T.C. and Kirkpatrick, K. (1985) Actions of α,β-methylene ATP and 6-hydroxydopamine on the sympathetic neurotransmission in the vas deferens of the guinea pig, rat and mouse: support for co-transmission. *British Journal of Pharmacology*, **89**, 647–659.

Ålund, M. and Olson, L. (1979a) Depolarization-induced decreases in fluorescence intensity of gastrointestinal quinacrine-binding nerves. *Brain Research*, **166**, 121–137.

Ålund, M. and Olson, L. (1979b) Quinacrine affinity of endocrine cell systems containing dense core vesicles as visualised by fluorescence microscopy. *Cell and Tissue Research*, **204**, 171–186.

Ålund, M. and Olson, L. (1980) Release of [^{14}C]-quinacrine from peripheral and central nerves. *Journal of the Autonomic Nervous System*, **2**, 281–294.

Andrews, T.M. and Holton, P. (1958) The substance P and adenosinetriphosphate (ATP) contents of sensory nerve on degeneration. *Journal of Physiology*, **143**, 45–46P.

Åstrand, P., Brock, J.A. and Cunnane, T.C. (1988) Time course of transmitter action at the sympathetic neuroeffector junction in rodent vascular and non-vascular smooth muscle. *Journal of Physiology*, **401**, 657–670.

Beck, K., Calamai, F., Staderini, G. and Susini, T. (1988) Gastric motor responses elicited by vagal stimulation and purine compounds in the atropine-treated rabbit. *British Journal of Pharmacology*, **94**, 1157–1166.

Bennett, J.P., Cockroft, S. and Gomperts, B.D. (1981) Rat mast cells permeabilized with ATP secrete histamine in response to calcium ions buffered in the micromolar range. *Journal of Physiology*, **317**, 335–345.

Bertrand, G., Chapal, J., Loubatières-Mariani, M.M. and Roye, M. (1987) Evidence for two different P_2-purinceptors on β cell and pancreatic vascular bed. *British Journal of Pharmacology*, **91**, 783–787.

Björnsson, O.G., Monck, J.R. and Williamson, J.R. (1989) Identification of P_{2Y}-purinoceptors associated with voltage-activated cation channels in cardiac ventricular myocytes of the rat. *European Journal of Biochemistry*, **186**, 395–404.

Blackburn, G.M., Kent, D.E. and Kolkmann, F. (1981) Three new β, γ-methylene analogues of adenosine triphosphate. *Journal of the Biochemical Society: Chemical Communications*, 1189–1190.

Bo, X. and Burnstock, G. (1989) [^{3}H]-α,β-methylene ATP, a radioligand labelling the P_2-purinoceptor. *Journal of the Autonomic Nervous System*, **28**, 185–188.

Böck, P. (1980) Adenine nucleotides in the carotid body. *Cell and Tissue Research*, **206**, 279–290.

Boyer, J.L., Downes, C.P. and Harden, T.K. (1989) Kinetics of activation of phospholipase C by P_{2Y} purinergic receptor agonists and guanine nucleotides. *Journal of Biological Chemistry*, **264**, 884–890.

Boyne, A.F. (1976) Isolation of synaptic vesicles from *Narcine brasiliensis* electric organ – some influences on release of vesicular acetylcholine and ATP. *Brain Research*, **114**, 481–491.

Brading, A.F. and Williams, J.H. (1990) Contractile responses of smooth muscle strips from rat and guinea-pig urinary bladder to transmural stimulation: effects of atropine and α, β-methylene ATP. *British Journal of Pharmacology*, **99**, 493–498.

Bramich, N. and Campbell, G. (1989) Vagal and sympathetic control of sino-atrial conduction in the heart of the toad, *Bufo marinus*. *Proceedings of the Australian Physiological and Pharmacological Society*, **20**, 37P.

Bramich, N. Edwards, F.R. and Hirst, G.D.S. (1989) Effect of sympathetic nerve stimulation on pacemaker cells of toad sinus venosus. *Proceedings of the Australian Physiological and Pharmacological Society*, **20**, 114P.

Buchtal, F. and Kahlson, G. (1944) The action of adenosine triphosphate and related compounds on mammalian skeletal muscle. *Acta Physiologica Scandinavica*, **8**, 317–324.

Bulloch, J.M. and McGrath, J.C. (1988) Blockade of vasopressor and vas deferens responses by α, β-methylene ATP in the pithed rat. *British Journal of Pharmacology*, **94**, 103–108.

Burger, R.M. and Löwenstein, J.M. (1970) Preparation and properties of 5'-nucleotidase from smooth muscle of small intestine. *Journal of Biological Chemistry*, **245**, 6274–6280.

Burnstock, G. (1972) Purinergic nerves. *Pharmacological Reviews*, **24**, 509–560.

Burnstock, G. (1975) Purinergic transmission. In *Handbook of Psychopharmacology*, edited by L.L. Iversen and S.H. Snyder, Vol. 5, pp. 131–194. New York: Plenum Press.

Burnstock, G. (1976) Do some nerve cells release more than one transmitter? *Neuroscience*, **1**, 239–248.

Burnstock, G. (1978) A basis for distinguishing two types of purinergic receptor. In *Cell Membrane Receptors for Drugs and Hormones: A Multidisciplinary Approach*, edited by R.W. Straub and L. Bolis, pp. 107–118. New York: Raven Press.

Burnstock, G. (1986) Purines as cotransmitters in adrenergic and cholinergic neurones. *Progress in Brain Research*, **68**, 193–203.

Burnstock, G. (1990) Cotransmission. The Fifth Heymans Lecture – Ghent, February 17, 1990. *Archives of International Pharmacodynamics and Therapeutics*, **304**, 7–33.

Burnstock, G. and Buckley, N.J. (1985) The classification of receptors for adenine nucleotides. In *Methods Used in Adenosine Research (Methods in Pharmacology Series)*, edited by D.M. Paton, pp. 193–212. New York: Plenum Press.

Burnstock, G. and Hoyle, C.H.V. (1985) Actions of adenine dinucleotides on the guinea-pig taenia coli: NAD acts indirectly on P_1-purinoceptors; NADP acts like a P_2-purinoceptor agonist. *British Journal of Pharmacology*, **84**, 825–831.

Burnstock, G. and Kennedy, C. (1985) Is there a basis for distinguishing more than one type of P_2-purinoceptor. *General Pharmacology*, **16**, 433–440.

Burnstock, G. and Meghji, P. (1983) The effect of adenyl compounds on the rat heart. *British Journal of Pharmacology*, **79**, 211–218.

Burnstock, G. and Warland, J.J.I. (1987a) A pharmacological study of the rabbit saphenous artery *in vitro*: a vessel with a large purinergic contractile response to sympathetic nerve stimulation. *British Journal of Pharmacology*, **90**, 111–120.

Burnstock, G. and Warland, J.J.I. (1987b) P_2-purinoceptors of two subtypes in the rabbit mesenteric artery: reactive blue 2 selectively inhibits responses mediated via the P_{2Y}- but not the P_{2X}-purinoceptor. *British Journal of Pharmacology*, **90**, 383–391.

Burnstock, G., Campbell, G., Satchell, D. and Smythe, A. (1970) Evidence that ATP or a related nucleotide is the transmitter substance released by non-adrenergic inhibitory nerves in the gut. *British Journal of Pharmacology*, **40**, 668–688.

Burnstock, G., Dumsday, B. and Smythe, A. (1972) Atropine-resistant excitation of the urinary bladder: the possibility of transmission via nerves releasing a purine nucleotide. *British Journal of Pharmacology*, **44**, 451–461.

Burnstock, G., Cocks, T., Crowe, R. and Kasakov, L. (1978a) Purinergic innervation of the guinea-pig urinary bladder. *British Journal of Pharmacology*, **63**, 125–138.

Burnstock, G., Cocks, T., Kasakov, L. and Wong, H. (1978b) Direct evidence for ATP release from non-adrenergic, non-cholinergic ('purinergic') nerves in the guinea-pig taenia coli and bladder. *European Journal of Pharmacology*, **49**, 145–149.

Burnstock, G., Crowe, R. and Wong, H. (1979) Comparative and histochemical evidence for purinergic inhibitory innervation of the portal vein of the rabbit, but not guinea-pig. *British Journal of Pharmacology*, **65**, 377–388.

Burnstock, G., Cusack, N.J., Hills, J.M., MacKenzie, I. and Meghji, P. (1983) Studies on the stereoselectivity of the P_2-purinoceptor. *British Journal of Pharmacology*, **79**, 907–913.

Burnstock, G., Cusack, N.J. and Meldrum, L.A. (1984) Effects of phosphorothioate analogues of ATP, ADP and AMP on guinea-pig taenia coli and urinary bladder. *British Journal of Pharmacology*, **82**, 369–374.

Burnstock, G., Cusack, N.J. and Meldrum, L.A. (1985) Studies on the stereoselectivity of the P_2-purinoceptor on the guinea-pig vas deferens. *British Journal of Pharmacology*, **84**, 431–434.

Busse, R., Ogilvie, A. and Pohl, U. (1988) Vasomotor activity of diadenosine triphosphate and diadenosine tetraphosphate in isolated arteries. *American Journal of Physiology*, **254**, H828–H832.

Cameron, D.J. (1984) Inhibition of macrophage mediated cytotoxicity by exogenous adenosine 5′-triphosphate. *Journal of Clinical Laboratory Immunology*, **15**, 215–218.

Chapal, J. and Loubatières-Mariani, M-M. (1983) Purinergic receptors and insulin release. *Journal of Pharmacology (Paris)*, **14**, 385–394.

Cheung, D.W. (1982) Two components in the cellular response of rat tail arteries to nerve stimulation. *Journal of Physiology*, **328**, 461–468.

Cheung, D.W. and Fujioka, M. (1986) Inhibition of the excitatory junction potential in the guinea-pig saphenous artery by ANAPP$_3$. *British Journal of Pharmacology*, **89**, 3–5.

Choo, L.K. (1981) The effect of reactive blue 2, an antagonist of ATP, on the isolated urinary bladders of the guinea-pig and rat. *Journal of Pharmacy and Pharmacology*, **33**, 248–250.

Choo, L.K. and Mitchelson, F. (1980) The effect of indomethacin and adenosine 5′-triphosphate on the excitatory innervation of the rat urinary bladder. *Canadian Journal of Physiology and Pharmacology*, **58**, 1042–1048.

Cockroft, S. and Gomperts, B.D. (1979a) Activation and inhibition of calcium-dependent histamine secretion by ATP ions applied to rat mast cells. *Journal of Physiology*, **296**, 299–243.

Cockroft, S. and Gomperts, B.D. (1979b) ATP induces nucleotide permeability in rat mast cells. *Nature*, **279**, 541–542.

Cockroft, S. and Gomperts, B.D. (1980) The ATP^{4-} receptor of rat mast cells. *Biochemical Journal*, **188**, 789–798.

Cooper, D.M.F. and Londos, C. (1988) *Receptor Biochemistry and Methodology, Vol. II: Adenosine Receptors*, New York: Alan R. Liss Inc.

Cooper, C.L., Morris, A.J. and Harden, T.K. (1989) Guanine nucleotide-sensitive interaction of a radiolabelled agonist with a phospholipase C-linked P_{2Y}-purinergic receptor. *Journal of Biological Chemistry*, **264**, 6202–6206.

Coupland, R.E. (1965) *The Natural History of the Chromaffin Cell*, London: Longmans, Green and Co. Ltd.

Creed, K.E., Ishikawa, S. and Ito, Y. (1983) Electrical and mechanical activity recorded from rabbit urinary bladder in response to nerve stimulation. *Journal of Physiology*, **338**, 149–164.

Crema, A., Frigo, G.M., Lecchini, S., Manzo, L., Onori, L. and Tonini, M. (1983) Purine receptors in the guinea-pig internal anal sphincter. *British Journal of Pharmacology*, **78**, 599–603.

Crowe, R. and Burnstock, G. (1981) Comparative studies of quinacrine-positive neurones in the myenteric plexus of stomach and intestine of guinea-pig, rabbit and rat. *Cell and Tissue Research*, **221**, 93–107.

Cunnane, T.C. and Manchanda, R. (1988) Electrophysiological analysis of the inactivation of sympathetic transmitters in the guinea-pig vas deferens. *Journal of Physiology*, **404**, 349–364.

Cusack, N.J. and Hourani, S.M.O. (1984) Some pharmacological and biochemical interactions of the enantiomers of adenylyl $5'$-(β,γ-methylene)-diphosphate with the guinea-pig urinary bladder. *British Journal of Pharmacology*, **82**, 155–159.

Cusack, N.J., Hickman, M.E. and Born, G.V.R. (1979) Effects of D- and L-enantiomers of adenosine, AMP and ADP and their 2-chloro and 2-azido analogues on human platelets. *Proceedings of the Royal Society of London, Series B*, **206**, 139–144.

Cusack, N.J. Pearson, J.D. and Gordon, J.L. (1983) Stereoselectivity of ectonucleotidases on vascular endothelial cells. *Biochemical Journal*, **214**, 975–981.

Cusack, N.J., Hourani, S.M.O. and Welford, L.A. (1985) Characterisation of ADP receptors. *Advances in Experimental Medical Biology*, **192**, 29–39.

Cusack, N.J., Hourani, S.M.O., Loizou, G.D. and Welford, L.A. (1987) Pharmacological effects of isopolar phosphonate analogues of ATP on P_2-purinoceptors in guinea-pig taenia coli and urinary bladder. *British Journal of Pharmacology*, **90**, 791–795.

Da Prada, M., Obrist, R. and Pletscher, A. (1975) Accumulation of acetylcholine and aromatic monoamines by interaction with adenosine-$5'$-triphosphate. *Journal of Pharmacy and Pharmacology*, **27**, 63–66.

Da Prada, M., Richards, J.G. and Lorez, H.P. (1978) Blood platelets and biogenic monoamines: biochemical, pharmacological, and morphological approaches. In *Platelets: A Multidisciplinary Approach*, edited by G. de Gaetano and S. Garattini, pp. 331–353. New York: Raven Press.

Dean, D.M. and Downie, J.W. (1978) Contribution of adrenergic and 'purinergic' neurotransmission to contraction in rabbit detrusor. *Journal of Pharmacology and Experimental Therapeutics*, **207**, 431–437.

De Lander, G.E. and Hopkins, C.J. (1986) Spinal adenosine modulates descending antinociceptive pathways stimulated by morphine. *Journal of Pharmacology and Experimental Therapeutics*, **239**, 88–93.

Den Hertog, A., Nelemans, A. and Van Den Akker, J. (1989) The inhibitory action of suramin on the P_2-purinoceptor response in smooth muscle cells of guinea-pig taenia caeci. *European Journal of Pharmacology*, **166**, 531–534.

Deuticke, H.J. (1931) Über der Einfluss von Adenosin und Adenosinphosphosäuren auf den isolierten Meerschweinchenuterus. *Pflügers Archiv*, **230**, 537–555.

Dodd, J., Jahr, C.E. and Jessell, T.M. (1984) Neurotransmitters and neuronal markers at sensory synapses in the dorsal horn. In *Advances in Pain Research Therapy, Vol. 6*, edited by L. Kruger and J.C. Liebeskind, pp. 105–121. New York: Raven Press.

Douglas, W.W. (1968) Stimulus-secretion coupling: the concept and clues from chromaffin and other cells. *British Journal of Pharmacology*, **34**, 451–474.

Douglas, W.W. and Poisner, A.M. (1966) Evidence that the secretory adrenal chromaffin cell releases catecholamines directly from ATP-rich granules. *Journal of Physiology*, **183**, 236–248.

Dowdall, M., Boyne, A.F. and Whittaker, V.P. (1974) Adenosine triphoshate: a constituent of cholinergic synaptic vesicles. *Biochemical Journal*, **140**, 1–12.

Dowdle, E.B. and Maske, R. (1980) The effects of calcium concentration on the inhibition of cholinergic neurotransmission in the myenteric plexus of guinea-pig ileum by adenine nucleotides. *British Journal of Pharmacology*, **71**, 245–252.

Drury, A.N. and Szent-Györgi, A. (1929) The physiological activity of adenine compounds with special reference to their action upon mammalian heart. *Journal of Physiology*, **68**, 213–237.

Dunn, P.M and Blakeley, A.G.H. (1988) Suramin: a reversible P_2-purinoceptor antagonist in the mouse vas deferens. *British Journal of Pharmacology*, **93**, 243–245.

Duval, N., Hicks, P.E. and Langer, S.Z. (1985) Inhibitory effects of α,β-methylene ATP on nerve-mediated contractions of the nictitating membrane in reserpinised cats. *European Journal of Pharmacology*, **110**, 373–377.

Ewald, D.A. (1976) Potentiation of postjunctional cholinergic sensitivity of rat diaphragm muscle by high-energy phosphate adenine nucleotides. *Journal of Membrane Biology*, **29**, 47–65.

Fedan, J.S., Hogaboom, G.K., O'Donnell, J.P., Colby, J. and Westfall, D.P. (1981) Contribution by purines to the neurogenic response of the vas deferens of the guinea pig. *European Journal of Pharmacology*, **69**, 41–53.

Fedan, J.S., Hogaboom, G.K., Westfall, D.P. and O'Donnell, J.P. (1982) Comparison of the effects of arylazidoaminopropionyl ATP (ANAPP3), an ATP antagonist, on response of the smooth muscle of the guinea-pig vas deferens to ATP and related nucleotides. *European Journal of Pharmacology*, **85**, 277–290.

Flavahan, N.A. and Vanhoutte, P.M. (1986). Sympathetic purinergic vasoconstriction and thermosensitivity in a canine cutaneous vein. *Journal of Pharmacology and Experimental Therapeutics*, **239**, 784–789.

Flavahan, N.A., Grant, T.L., Greig, J. and McGrath, J.C. (1985) Analysis of the α-adrenoceptor-mediated, and other, components in the sympathetic vasopressor responses of the pithed rat. *British Journal of Pharmacology*, **86**, 265–274.

Fleetwood, G. and Gordon, J.L. (1987) Purinoceptors in the rat heart. *British Journal of Pharmacology*, **90**, 219–227.

Forsman, C.A. and Elfvin, L-G. (1984) Cytochemical localization of 5'-nucleotidase in guinea-pig sympathetic ganglia. *Journal of Neurocytology*, **13**, 339–349.

Frew, R. and Lundy, P.M. (1982a) Effect of arylazido aminopropionyl ATP (ANAPP3), a putative ATP antagonist, on ATP responses of isolated guinea pig smooth muscle. *Life Sciences*, **30**, 259–267.

Frew, R. and Lundy, P.M. (1982b) Evidence against ATP being the non-adrenergic, non-cholinergic transmitter in the guinea-pig stomach. *European Journal of Pharmacology*, **81**, 333–336.

Frew, R. and Lundy, P.M. (1986) Arylazido aminopropionyl ATP (ANAPP3): interactions with adenosine receptors in longitudinal smooth muscle of the guinea-pig ileum. *European Journal of Pharmacology*, **123**, 395–400.

Fried, G., Lagercrantz, H., Klein, R. and Thureson-Klein, A. (1984) Large and small noradrenergic vesicles – origin, contents, and functional significance. In *Catecholamines : Basic and Peripheral Mechanisms.*, edited by E. Usdin, pp. 45–53. New York: Alan R. Liss, Inc.

Fujii, R. (1986) Mechanisms controlling chromatophore movements in fishes. *Zoological Science*, **3**, 961.

Fujii, K. (1988) Evidence for adenosine triphosphate as an excitatory transmitter in guinea-pig, rabbit and pig urinary bladder. *Journal of Physiology*, **404**, 39–52.

Fujii, R. and Oshima, N. (1986) Control of chromatophore movements in teleost fishes. *Zoological Science*, **3**, 13–47.

Fukuda, J., Fujita, Y. Ohsawa, K. (1983) ATP content in isolated mammalian nerve cells assayed by a modified luciferin-luciferase method. *Journal of Neuroscience Methods*, **8**, 295–302.

Furshpan, E.J., Potter, D.D. and Matsumoto, S.G. (1986) Synaptic functions in rat sympathetic neurons in microcultures: III. A purinergic effect on cardiac myocytes. *Journal of Neuroscience*, **6**, 1099–1107.

Fyffe, R.E.W. and Perl, E.R. (1984) Is ATP a central synaptic mediator for certain primary afferent fibres from mammalian skin? *Proceedings of the National Academy of Sciences, USA*, **81**, 6890–6893.

Gaddum, J.H. and Holtz, P. (1933) The localization of the action of drugs on the pulmonary vessels of dogs and cats. *Journal of Physiology*, **77**, 139–158.

Gillespie, J.H. (1933) The biological significance of the linkages in adenosine triphosphate acid. *Journal of Physiology*, **80**, 345–359.

Gordon, J.L. (1986) Extracellular ATP: effects, sources and fate. *Biochemical Journal*, **233**, 309–319.

Gough, G.R., Maguire, M.H. and Satchell, D.G. (1973) Three new adenosine triphosphate analogues. Synthesis and effects on isolated gut. *Journal of Medicinal Chemistry*, **16**, 1188–1190.

Hammond, J.R., MacDonald, W.F. and White, T.D. (1988) Evoked secretion of [^{3}H]noradrenaline and ATP from nerve varicosities isolated from the myenteric plexus of the guinea-pig ileum. *Canadian Journal of Physiology and Pharmacology*, **66**, 369–375.

Harrison, M.J., Brossmer, R. and Goody, R.S. (1975) Inhibition of platelet aggregation and the platelet release reaction by α,ω diadenosine polyphosphates. *FEBS Letters*, **54**, 57–60.

Hashimoto, K., Kumakura, S. and Tanemura. I. (1964) Mode of action of adenine, uridine and cytidine nucleotides and 2, 6-bis(diethanolamino)-4,8-dipiperidino-pyrimidino(5,4-*d*) pyrimidine on the coronary, renal and femoral arteries. *Arzneimittel Forschung* **14**, 1252–1254.

Haslam, R.J. and Rosson, G.M. (1975) Effects of adenosine on levels of adenosine cyclic 3′,5′-monophosphate in human blood platelets in relation to adenosine incorporation and platelet aggregation. *Molecular Pharmacology*, **11**, 528–544.

Hedlund, H., Fändriks, L., Delbro, D. and Fasth, S. (1983) Blockade of non-cholinergic non-adrenergic colonic contraction in response to pelvic nerve stimulation by large doses of α,β-methylene ATP. *Acta Physiologica Scandinavica*, **119**, 451–454.

Hempling, H.G., Stewart, C.C. and Gasic, G. (1969) The effect of exogenous ATP on the electrolyte content of TA3 ascites tumour cells. *Journal of Cell Physiology*, **73**, 133–140.

Hilton, S.M. and Holton, P. (1954) Antidromic vasodilatation and blood flow in the rabbit's ear. *Journal of Physiology*, **125**, 138–147.

Hogaboom, G.K., O'Donnell, J.P. and Fedan, J.S. (1980) Purinergic receptors: photoaffinity analog of adenosine triphosphate is a specific adenosine triphosphate antagonist. *Science*, **208**, 1273–1276.

Holck, M.I. and Marks, B.H. (1978) Purine nucleoside and nucleotide interactions on normal subsensitive α adrenoceptor responsiveness in guinea-pig vas deferens. *Journal of Pharmacology and Experimental Therapeutics*, **205**, 104–117.

Holt, S.G., Cooper, M. and Wyllie, J.H. (1985) Evidence for purinergic transmission in mouse bladder and for modulation of responses to electrical stimulation by 5-hydroxytryptamine. *European Journal of Pharmacology*, **116**, 105–111.

Holton, P. (1959) The liberation of adenosine triphosphate on antidromic stimulation of sensory nerves. *Journal of Physiology*, **145**, 494–504.

Holton, F.A. and Holton, P. (1953) The possibility that ATP is a transmitter at sensory nerve endings. *Journal of Physiology*, **119**, 50–51P.

Holton, F.A. and Holton, P. (1954) The capillary dilator substance in dry powders of spinal roots: a possible role of adenosine triphosphate in chemical transmission from nerve endings. *Journal of Physiology*, **126**, 124–140.

Holton, P. and Perry, W.L.M. (1951) On the transmitter responsible for antidromic vasodilatation in the rabbits ear. *Journal of Physiology*, **114**, 240–251.

Hopwood, A.M. and Burnstock, G. (1987) ATP mediates coronary vasoconstriction via P_{2X}-purinoceptors and coronary vasodilatation via P_{2Y}-purinoceptors in the isolated perfused rat heart. *European Journal of Pharmacology*, **136**, 49–54.

Hopwood, A.M., Lincoln, J., Kirkpatrick, K.A. and Burnstock, G. (1989) Adenosine 5′-triphosphate, adenosine and endothelium-derived relaxing factor in hypoxic vasodilatation of the heart. *European Journal of Pharmacology*, **165**, 323–326.

Hourani, S.M.O. (1984) Desensitization of the guinea-pig urinary bladder by the enantiomers of adenylyl 5′-(β,γ-methylene)-diphosphonate and by substance P. *British Journal of Pharmacology*, **82**, 161–164.

Hourani, S.M.O. and Chown, J.A. (1989) The effects of some possible inhibitors of ectonucleotidases on the breakdown and pharmacological effects of ATP in the guinea-pig urinary bladder. *General Pharmacology*, **20**, 413–416.

Hourani, S.M.O., Welford, L.A. and Cusack, N.J. (1985) L-AMP-PCP, an ATP receptor agonist in guinea-pig bladder, is inactive on taenia coli. *European Journal of Pharmacology*, **108**, 197–200.

Hourani, S.M.O., Loizou, G.D. and Cusack, N.J. (1986) Pharmacological effects of L-AMP-PCP on ATP receptors in smooth muscle. *European Journal of Pharmacology*, **131**, 99–103.

Hourani, S.M.O., Welford, L.A., Loizou, G.D. and Cusack, N.J. (1988) Adenosine 5-(2-fluorodiphosphate) is a selective agonist at P_2-purinoceptors mediating relaxation of smooth muscle. *European Journal of Pharmacology*, **147**, 131–136.

Houston, D.A., Burnstock, G. and Vanhoutte, P.M. (1987) Different P_2-purinergic receptor subtypes of endothelium and smooth muscle in canine blood vessels. *Journal of Pharmacology and Experimental Therapeutics*, **241**, 501–506.

Hoyes, A.D., Barber, P. and Martin, B.H. (1975) Comparative ultrastructure of nerves innervating muscle of body of bladder. *Cell and Tissue Research*, **164**, 133–144.

Hoyle, C.H.V. (1990) A review of the pharmacological activity of adenine dinucleotides in the periphery: possible receptor classes and transmitter function. *General Pharmacology*, **21**, 827–831.

Hoyle, C.H.V. and Burnstock. G. (1985) Atropine-resistant excitatory junction potentials in rabbit bladder are blocked by α,β-methylene ATP. *European Journal of Pharmacology*, **114**, 239–240.

Hoyle, C.H.V. and Burnstock, G. (1986) Evidence that ATP is a neurotransmitter in the frog heart. *European Journal of Pharmacology*, **124**, 285–289.

Hoyle, C.H.V. and Greenberg, M.J. (1988) Actions of adenylyl compounds in invertebrates from several phyla: evidence for internal purinoceptors. *Comparative Biochemistry and Physiology*, **90C**, 113–122.

Hoyle, C.H.V., Chapple, C. and Burnstock, G. (1989) Isolated human bladder: evidence for an adenine dinucleotide acting on P_{2X}-purinoceptors and for purinergic transmission. *European Journal of Pharmacology*, **174**, 115–118.

Hoyle, C.H.V., Knight, G.E. and Burnstock, G. (1990) Suramin antagonizes responses to P_2-purinoceptor agonists and purinergic nerve stimulation in the guinea-pig urinary bladder and taenia coli. *British Journal of Pharmacology*, **99**, 617–621.

Huidboro-Toro, J.P and Parada, S. (1988) Co-transmission in the rat vas deferens: postjunctional synergism of noradrenaline and adenosine 5'-triphosphate. *Neuroscience Letters*, **85**, 339–344.

Irvin, J.L. and Irvin, E.M. (1954) The interaction of quinacrine with adenine nucleotides. *Journal of Biological Chemistry*, **210**, 45–56.

Ishikawa, S. (1985) Actions of ATP and α,β-methylene ATP on the neuromuscular transmission and smooth muscle membrane of the rabbit and guinea-pig mesenteric arteries. *British Journal of Pharmacology*, **86**, 777–787.

Israël, M., Dunant, Y., Lesbats, B., Manaranche, R., Marsal, J. and Meunier, F. (1979) Rapid acetylcholine and adenosine triphosphate oscillations triggered by stimulation of the *Torpedo* electric organ. *Journal of Experimental Biology*, **81**, 63–73.

Jahr, C.E. and Jessell, T.M. (1984) ATP excites a subpopulation of rat dorsal horn neurones. *Progress in Neuro-Psychopharmacology and Biological Psychiatry*, **8**, 503–505.

Kasakov, L. and Burnstock, G. (1983) The use of the slowly degradable analog, α,β-methylene ATP, to produce desensitization of the P_2-purinoceptor: effect on non-adrenergic, non-cholinergic responses of the guinea-pig urinary bladder. *European Journal of Pharmacology*, **86**, 291–294.

Kasakov, L., Ellis, J., Kirkpatrick, K., Milner, P. and Burnstock, G. (1988) Direct evidence for concomitant release of noradrenaline, adenosine 5'-triphosphate and neuropeptide Y from sympathetic nerve supplying the guinea-pig vas deferens. *Journal of the Autonomic Nervous System*, **22**, 75–82.

Katsuragi, T. and Su, C. (1980) Purine release from vascular adrenergic nerves by high potassium and a calcium ionophore, A23187. *Journal of Pharmacology and Experimental Therapeutics*, **215**, 685–690.

Katsuragi, T. and Su, C. (1982a) Release of purines and noradrenaline by ouabain and potassium chloride from vascular adrenergic nerves. *British Journal of Pharmacology*, **77**, 625–629.

Katsuragi, T. and Su, C. (1982b) Augmentation by theophylline of [^{3}H]purine release from vascular adrenergic nerves: evidence for presynaptic autoinhibition. *Journal of Pharmacology and Experimental Therapeutics*, **220**, 152–156.

Katsuragi, T., Furuta K., Harada, T. and Furukawa, T. (1985) Cholinergic neuromodulation by ATP, adenosine and its N^6-substituted analogues in guinea-pig ileum. *Clinical and Experimental Pharmacology and Physiology*, **12**, 73–78.

Katsuragi, T., Kuratomi, L. and Furukawa, T. (1986) Clonidine-evoked selective P_1-purinoceptor antagonism of contraction of guinea-pig urinary bladder. *European Journal of Pharmacology*, **121**, 119–122.

Kazic, T. and Milosavljevic, D. (1980) Interaction between adenosine triphosphate and noradrenaline in the isolated vas deferens of the guinea-pig. *British Journal of Pharmacology*, **71**, 93–98.

Keef, K.D. and Kreulen, D.L. (1988) Electrical responses of guinea pig coronary artery to transmural stimulation. *Circulation Research*, **62**, 585–595.

Kennedy, C. (1988) Possible roles for purine nucleotides in perivascular neurotransmission. In *Nonadrenergic Innervation of Blood Vessels, Vol. I. Putative Neurotransmitters*, edited by G. Burnstock and S.G. Griffith, pp. 65–76. Boca Raton, Florida: CRC Press, Inc.

Kennedy, C. (1990) P_1- and P_2-purinoceptor subtypes – an update. *Archives Internationales Pharmacodynamie et de Therapie*, **303**, 30–50.

Kennedy, C. and Burnstock, G. (1985a) Evidence for two types of P_2-purinoceptor in longitudinal muscle of the rabbit portal vein. *European Journal of Pharmacology*, **111**, 49–56.

Kennedy, C. and Burnstock, G. (1985b) ATP produces vasodilation via P_1 purinoceptors and vasoconstriction via P_2 purinoceptors in the isolated rabbit central ear artery. *Blood Vessels*, **22**, 145–155.

Kennedy, C. and Burnstock, G. (1986) ATP causes postjunctional potentiation of noradrenergic contractions in the portal vein of guinea-pig and rat. *Journal of Pharmacy and Pharmacology*, **38**, 307–309.

Kennedy, C., Delbro, D. and Burnstock, G. (1985) P_2-purinoceptors mediate both vasodilation (via the endothelium) and vasoconstriction of the isolated rat femoral artery. *European Journal of Pharmacology*, **107**, 161–168.

Kennedy, C., Saville, V.L. and Burnstock, G. (1986) The contributions of noradrenaline and ATP to the responses of the rabbit central ear artery to sympathetic nerve stimulation depend on the parameters of stimulation. *European Journal of Pharmacology*, **122**, 291–300.

Keppens, S. and Wulf, D.E. (1986) Characterization of the liver P_2-purinoceptor involved in the activation of glycogen phosphorylase. *Biochemical Journal*, **240**, 367–371.

Kerr, D.I.B. and Krantis, A. (1979) A new class of ATP antagonist. *Proceedings of the Australian Physiological and Pharmacological Society*, **10**, 156P.

Kirkpatrick, K. and Burnstock, G. (1987) Sympathetic nerve-mediated release of ATP from the guinea-pig vas deferens is unaffected by reserpine. *European Journal of Pharmacology*, **138**, 207–214.

Knyihar, E. (1971) Fluoride-resistant acid phosphatase system of nociceptive dorsal root afferents. *Experientia*, **27**, 1205–1207.

Komori, S. and Ohashi, H. (1988a) Membrane potential responses to ATP applied by pressure ejection in the longitudinal muscle of chicken rectum. *British Journal of Pharmacology*, **95**, 1157–1164.

Komori, S. and Ohashi, H. (1988b) Some membrane properties of the circular muscle of chicken rectum and its non-adrenergic innervation. *Journal of Physiology*, **401**, 417–435.

Krishnamurty, V.S.R. and Kadowitz, P.J. (1982) Influence of adenosine triphosphate on the isolated perfused mesenteric artery of the rabbit. *Canadian Journal of Physiology and Pharmacology*, **61**, 1409–1417.

Krishtal, O.A. and Marchenko, S.M. (1986) Pharmacological properties of the receptor for ATP in the sensory neurons of rats. *Biomedical Research*, **7**, supplement, 79–81.

Krishtal, O.A., Marchenko, S.M. and Pidoplichko, V.I. (1983) Receptor for ATP in the membrane of mammalian sensory neurones. *Neuroscience Letters*, **35**, 41–45.

Kuchii, M., Miyahara, J.T. and Shibata, S. (1973) [³H]-Adenine nucleotide and [³H]-noradrenaline release evoked by electrical field stimulation, perivascular nerve stimulation and nicotine from the taenia of the guinea-pig caecum. *British Journal of Pharmacology*, **49**, 258–266.

Kügelgen, I.V. and Starke, K. (1985) Noradrenaline and adenosine triphosphate as co-transmitters of neurogenic vasoconstriction in rabbit mesenteric artery. *Journal of Physiology*, **367**, 435–455.

Kügelgen, I.V., Schöffel, E. and Starke, K. (1989) Inhibition by nucleotides acting at presynaptic P_2-receptors of sympathetic neuro-effector transmission in the mouse isolated vas deferens. *Naunyn Schmiedeberg's Archives of Pharmacology*, **340**, 522–532.

Kuhns, D.B., Wright, D.G., Nath, J., Kaplan, S.S. and Basford, R.E. (1988) ATP induces transient elevations of $[Ca^{2+}]$: in human neutrophils and primes these cells for enhanced O_2-generation. *Laboratory Investigations*, **58**, 448–453.

Kumazawa, T. and Fujii, R. (1984) Concurrent release of norepinephrine and purines by potassium from adrenergic melanosome-aggregating nerve in *Tilapia*. *Comparative Biochemistry and Physiology*, **78C**, 263–266.

Kumazawa, T. and Fujii, R. (1986) Fate of adenylic co-transmitter released from adrenergic pigment-aggregating nerve to *Tilapia* melanophore. *Zoological Science*, **3**, 599–603.

Kumazawa, T., Oshima, N., Fujii, R. and Miyashita, Y. (1984) Release of ATP from adrenergic nerves controlling pigment in tilapia melanophores. *Comparative Biochemistry and Physiology*, **78C**, 1–4.

Kwan, C.Y. and Kostka, P. (1984) A Mg^{2+}-independent high affinity Ca^{2+}-stimulated adenosine triphosphatase in the plasma membrane of rat stomach smooth muscle. *Biochimica et Biophysica Acta*, **776**, 209–216.

Kwan, C.Y. and Ramlal, T. (1985) 5′-Nucleotidase activity in smooth muscle membranes of rat vas deferens. *Molecular Physiology*, **8**, 277–292.

Kwan, C.Y. and Sipos, S.N. (1985) Inhibition of smooth muscle 5′-nucleotidase by imidazole and its reversal by magnesium. *Biochimica et Biophysica Acta*, **831**, 167–171.

Lagercrantz, H. (1976) On the composition and function of large dense cored vesicles in sympathetic nerves. *Neuroscience*, **1**, 81–92.

Lagercrantz, H., Fried, G. and Dahlin, I. (1975) An attempt to estimate the *in vivo* concentrations of noradrenaline and ATP in sympathetic large dense core nerve vesicles. *Acta Physiologica Scandinavica*, **94**, 136–138.

Levin, R.M., Jacoby, R. and Wein, A.J. (1983) High-affinity, divalent ion-specific binding of ³H-ATP to homogenate derived from rabbit urinary bladder. *Molecular Pharmacology*, **23**, 1–7.

Levin, R.M., Rugieri, M.R. and Wein, A.J. (1986) Functional effects of the purinergic innervation of the rabbit urinary bladder. *Journal of Pharmacology and Experimental Therapeutics*, **263**, 452–457.

Levitt, B. and Westfall, D.P. (1982) Factors influencing the release of purines and norepinephrine in the rabbit portal vein. *Blood Vessels*, **19**, 30–40.

Li, B. and Liu, X. (1985) Mechanism of interferon action: the discovery of pppA2′p5′A2′p5′A receptor. *Scientia Sinica B.*, **28**, 697–706.

Lidner, F. and Rigler, R. (1930) Über die Beeinflussung der Weite der Herzkranzgefässe durch Produkte des Zelkernstoffwechsels. *Pflügers Archiv*, **226**, 697–708.

Londos, C., Cooper, D.M.F. and Wolff, J. (1980) Subclasses of external adenosine receptors. *Proceedings of the National Academy of Sciences, USA*, **77**, 2551–2554.

Loubatières-Mariani, M.M. and Chapal, J. (1988) Purinergic receptors involved in the stimulation of insulin and glucagon secretion. *Diabete et Metabolisme (Paris)*, **14**, 119–126.

Lukackso, P. and Blumberg, A. (1982) Modulation of the vasoconstrictor response to adrenergic stimulation by nucleosides and nucleotides. *Journal of Pharmacology and Experimental Therapeutics*, **222**, 344–351.

MacDonald, W.F. and White, T.D. (1985) Nature of extrasynaptosomal accumulation of endogenous adenosine by K^+ and veratridine. *Journal of Neurochemistry*, **45**, 791–797.

Macfarlane, D.E. and Mills, D.C.B. (1975) The effects of ATP on platelets: evidence against the central role of released ADP in primary aggregation. *Blood*, **46**, 309–320.

Machaly, M., Dalziel, H.H. and Sneddon, P. (1988) Evidence for ATP as a cotransmitter in dog mesenteric artery. *European Journal of Pharmacology*, **147**, 83–91.

MacKenzie, I., Burnstock, G. and Dolly, J.O. (1982) The effects of purified botulinum toxin Type A on cholinergic, adrenergic and non-adrenergic atropine-resistant autonomic neuromuscular transmission. *Neuroscience*, **7**, 997–1006.

MacKenzie, I., Kirkpatrick, K.A. and Burnstock, G. (1988) Comparative study of the actions of AP_5A and α,β-methylene ATP on nonadrenergic, noncholinergic neurogenic excitation in the guinea-pig vas deferens. *British Journal of Pharmacology*, **94**, 699–706.

Maggi, C.A., Manzini, S. and Meli, A. (1984) Evidence that $GABA_A$ receptors mediate relaxation of rat duodenum by activating intramural nonadrenergic-noncholinergic nuerones. *Journal of Autonomic Pharmacology*, **4**, 77–85.

Maguire, M.H. and Satchell, D.G. (1979) The contribution of adenosine to the inhibitory actions of adenine nucleotides on the guinea-pig taenia coli: studies with phosphate-modified adenine nucleotide analogs and dipyridamole. *Journal of Pharmacology and Experimental Therapeutics*, **211**, 626–631.

Maire, J.C., Medilanski, J. and Straub, R.W. (1982) Uptake of adenosine and release of adenine derivatives in mammalian non-myelinated nerve fibres at rest and during activity. *Journal of Physiology*, **323**, 589–602.

Maire, J.C., Medilanski, J. and Straub R.W. (1984) Release of adenosine, inosine and hypoxanthine from rabbit non-myelinated nerve fibres at rest and during activity. *Journal of Physiology*, **357**, 67–77.

Manzini, S., Maggi, C.A. and Meli, A. (1985) Further evidence for involvement of adenosine 5′-triphosphate in non-adrenergic non-cholinergic relaxation of the isolated rat duodenum. *European Journal of Pharmacology*, **113**, 399–408.

Manzini, S., Maggi, C.A. and Meli, A. (1986a) Pharmacological evidence that at least two different non-adrenergic non-cholinergic inhibitory systems are present in the rat small intestine. *Euroepan Journal of Pharmacology*, **123**, 229–236.

Manzini, S., Hoyle, C.H.V. and Burnstock, G. (1986b) An electrophysiological analysis of the effect of reactive blue 2, a putative P_2-purinoceptor antagonist, on inhibitory junction potentials of rat caecum. *European Journal of Pharmacology*, **127**, 197–204.

Martin, W., Cusack, N.J. Carleton, S. and Gordon, J.L. (1985) Specificity of P_2-purinoceptor that mediates endothelium-dependent relaxation of the pig aorta. *European Journal of Pharmacology*, **108**, 295–299.

Mathieson, J.J.I. and Burnstock, G. (1985) Purine-mediated relaxation and constriction of isolated rabbit mesenteric artery are not endothelium-dependent. *European Journal of Pharmacology*, **118**, 221–229.

McCaman, M.W. (1986) Uptake and metabolism of [^{3}H]adenosine by *Aplysia* ganglia and by individual neurons. *Journal of Neurochemistry*, **47**, 1026–1031.

McCaman, M.W. and McAfee, D.A. (1986) Effects of synaptic activity on the metabolism and release of purines in the rat superior cervical ganglion. *Cellular and Molecular Neurobiology*, **6**, 349–362.

Meldrum, L.A. and Burnstock, G. (1983) Further evidence that ATP acts as a cotransmitter with noradrenaline in sympathetic nerves supplying the guinea-pig vas deferens. *European Journal of Pharmacology*, **92**, 161–164.

Meldrum, L.A. and Burnstock, G. (1985a) Evidence that ATP is involved as a co-transmitter in the hypogastric nerve supplying the seminal vesicle of the guinea-pig. *European Journal of Pharmacology*, **110**, 363–366.

Meldrum, L.A. and Burnstock, G. (1985b) Investigations into the identity of the non-adrenergic, non-cholinergic excitatory transmitter in the smooth muscle of chicken rectum. *Comparative Biochemistry and Physiology*, **81C**, 307–309.

Moody, C.J. and Burnstock, G. (1982) Evidence for the presence of P_1-purinoceptors on cholinergic nerve terminals in the guinea-pig ileum. *European Journal of Pharmacology*, **77**, 1–9.

Moody, C.J., Meghji, P. and Burnstock, G. (1984) Stimulation of P_1-purinoceptors by ATP depends partly on its conversion to AMP and adenosine and partly on direct action. *European Journal of Pharmacology*, **97**, 47–54.

Morita, K., Katayama, Y. Koketsu, K. and Akasu, T. (1984) Actions of ATP on the soma of bullfrog of primary afferent neurons and its modulatory action on the GABA-induced response. *Brain Research*, **293**, 360–363.

Moss, H.E. and Burnstock, G. (1985) A comparative study of the electrical field stimulation of the guinea-pig, ferret and marmoset urinary bladder. *European Journal of Pharmacology*, **114**, 311–316.

Moss, H.E., Lincoln, J. and Burnstock, G. (1987) A study of bladder dysfunction during streptozotocin-induced diabetes in the rat using an *in vitro* whole-bladder preparation. *Journal of Urology*, **138**, 1279–1284.

Muramatsu, I. (1987) The effects of reserpine on sympathetic, purinergic neurotransmission in the isolated mesenteric artery of the dog: a pharmacological study. *British Journal of Pharmacology*, **91**, 467–474.

Muramatsu, I. and Kigoshi, S. (1987) Purinergic and non-purinergic innervation in the cerebral arteries of the dog. *British Journal of Pharmacology*, **92**, 901–908.

Muramatsu, I., Fujiwara, M., Miura, A. and Sakakibara, Y. (1981) Possible involvement of adenine nucleotides in sympathetic neuroeffector mechanisms of dog basilar artery. *Journal of Pharmacology and Experimental Therapeutics*, **216**, 401–409.

Muramatsu, I., Ohmura, T. and Oshita, M. (1989) Comparison between sympathetic adrenergic and purinergic transmission in the dog mesenteric artery. *Journal of Physiology*, **411**, 227–243.

Nagao, T. and Suzuki, H. (1988) Effects of α,β-methylene ATP on electrical responses produced by ATP and nerve stimulation in smooth muscle cells of the guinea-pig mesenteric artery. *General Pharmacology*, **19**, 799–805.

Nagasawa, J. (1977) Exocytosis: the common release mechanism of secretory granules in glandular cells, neurosecretory cells, neurons and paraneurons. *Archivum Histologia Japonica*, **40**, supplement, 31–47.

Nagy, J.I. and Daddona, P.E. (1985) Anatomical and cytochemical relationships of adenosine deaminase-containing primary afferent neurons in the rat. *Neuroscience*, **15**, 799–813.

Nagy, J.I. and Hunt, S.P. (1982) Fluoride-resistant acidphosphatase-containing neurons in dorsal root ganglia are separate from those containing substance P or somatostatin. *Neuroscience*, **7**, 89–97.

Nagy, J.I., Buss, M., Labella, L.A. and Daddona, P.E. (1984) Immunohistochemical localization of adenosine deaminase in primary afferent neurons of the rat. *Neuroscience Letters*, **48**, 133–138.

Nair, V. and Wiechert, R.J. (1980) Substrate specificity of adenosine deaminase – function of the 5′-hydroxyl group of adenosine. *Bioorganic Chemistry*, **9**, 423–433.

Neild, T.O. and Kotecha, N. (1986) Effects of α,β-methylene ATP on membrane potential, neuromuscular transmission and smooth muscle contraction in the rat tail artery. *General Pharmacology*, **17**, 461–464.

Nistratova, S.N. (1981) Mechanisms of adaptation of the cardiac muscle of molluscs to nervous influences. In *Physiology and Biochemistry of Adaptations in Marine Animals, Proceedings XIV Pacific Sciences Congress*, pp. 161–169. Vladivostok.

Okada, Y., Yada T., Ohno-Shosaku, T., Oiki, S., Ueda, S. and Machida, K. (1984) Exogenous ATP induces electrical membrane responses in fibroblasts. *Experimental Cell Research*, **152**, 552–557.

Okwuasaba, K.K., Hamilton, J.T. and Cook, M.A. (1977) Relaxations of the guinea-pig fundic strip by adenosine, adenine nucleotides and electrical stimulation: antagonism by theophylline and desensitization to adenosine and derivatives. *European Journal of Pharmacology*, **46**, 181–194.

Olson, L., Ålund, M. and Norberg, K-A. (1976) Fluorescence-microscopical demonstration of a population of gastrointestinal nerve fibres with a selective affinity for quinacrine. *Cell and Tissue Research*, **171**, 407–423.

Ostern, P. and Parnas, J.K. (1932) Über die Auswertung von Adenosinderivaten am überlebenden Froschherz. *Biochem Zeitschrift*, **248**, 389–397.

Parker, J.C. and Snow, R.L. (1972) Influence of external ATP on permeability of dog red blood cells. *American Journal of Physiology*, **223**, 888–893.

Pearson, J.D. and Cusack, N.J. (1985) Investigation of the preferred Mg(II)-adenine-nucleotide complex at the active site of ectonucleotidases in intact vascular cells using phosphorothioate analogues of ADP and ATP. *European Journal of Biochemistry*, **151**, 373-375.

Pearson, J.D., Carleton, J.S. and Gordon, J.L. (1980) Metabolism of adenine nucleotides by ectoenzymes of vascular endothelial and smooth-muscle cells in culture. *Biochemical Journal*, **190**, 421-429.

Pearson, J.D., Coade, S.B. and Cusack, N.J. (1985) Characterization of ectonucleotidases on vascular smooth-muscle cells. *Biochemical Journal*, **230**, 503-507.

Pothier, F., Forget, J., Sullivan, R. and Couillard, P. (1987) ATP and the contractile vacuole in *Amoeba proteus*: mechanism of action of exogenous ATP and related nucleotides. *Journal of Experimental Zoology*, **243**, 379-387.

Potter, D.D., Furshpan, E.J. and Landis, S.C. (1983) Transmitter status in cultured rat sympathetic neurons: plasticity and multiple transmitter functions. *Federation Proceedings*, **42**, 1626-1632.

Ralevic, V. and Burnstock, G. (1990) Postjunctional synergism of noradrenaline and adenosine 5′-triphosphate in the mesenteric arterial bed of the rat. *European Journal of Pharmacology*, **175**, 291-299.

Ramme, D., Regenold, J.T., Starke, K., Russe, R. and Illes, P. (1987) Identification of the neuroeffector transmitter in jejunal branches of the rabbit mesenteric artery. *Naunyn Schmiedeberg's Archives of Pharmacology*, **336**, 267-273.

Reese, J.H. and Cooper, J.R. (1982) Modulation of the release of acetylcholine from ileal synaptosomes by adenosine 5′-triphosphate. *Journal of Pharmacology and Experimental Therapeutics*, **223**, 612-616.

Reilly, W.M. and Burnstock, G. (1987) The effect of ATP analogues on the spontaneous electrical and mechanical activity of rat portal vein longitudinal muscle. *European Journal of Pharmacology*, **138**, 319-325.

Reilly, W.M., Saville, V.L. and Burnstock, G. (1987) An assessment of the antagonistic activity of reactive blue 2 at P_1- and P_2-purinoceptors: supporting evidence for purinergic innervation of the rabbit portal vein. *European Journal of Pharmacology*, **140**, 47-53.

Ribeiro, J.A. and Sebastião, A.M. (1986) Adenosine receptors and calcium: basis for proposing a third (A_3) adenosine receptor. *Progress in Neurobiology*, **26**, 179-209.

Rice, W.R. and Singleton, F.M. (1987) P_{2Y}-purinoceptor regulation of surfactant secretion from rat isolated alveolar type II cells is associated with mobilization of intracellular calcium. *British Journal of Pharmacology*, **91**, 833-838.

Rice, W.R. and Singleton, F.M. (1989) Reactive blue 2 selectively inhibits P_{2Y}-purinoceptor-stimulated surfactant phopholipid secretion from rat isolated alveolar type II cells. *British Journal of Pharmacology*, **97**, 158-162.

Richards, J.G. (1978) Techniques for localizing transmitter substances and secretory granules. In *Peripheral Neurendocrine Interaction*, edited by E.E. Coupland and W.G. Forssman, pp. 1-13. Berlin, Heidelberg, New York: Springer Verlag.

Richards, J.G. and Da Prada, M. (1977) Uranaffin reaction: a new cytochemical technique for the localization of adenine nucleotides in organelles storing biogenic amines. *Journal of Histochemistry and Cytochemistry*, **25**, 1322-1336.

Richards, J.G. and Da Prada, M. (1980) Cytochemical investigations on subcellular organelles storing biogenic amines in peripheral adrenergic neurons. In *Histochemistry and Cell Biology of Autonomic Neurons, SIF Cells and Paraneurons*, edited by O. Eränkö, pp. 269-278. New York: Raven Press.

Richardson, P.J. and Brown, S.J. (1987) ATP release from affinity-purified rat cholinergic nerve terminals. *Journal of Neurochemistry*, **48**, 622-630.

Romanenko, A.V. (1980) Vitamin PP, nicotinamide nucleotides and neuromuscular transmission in guinea-pig caecum longitudinal muscle (taenia coli) *Ukrainskii Biochimicheskii Zhurnal (Kiev)*, **52**, 624-627.

Romanenko, A.V., Baidan, L.V., Khalmuradov, A.G. and Shuba, M.F. (1980) The mechanism of action of vitamin PP and nicotinamide nucleotides on neuromuscular transmission in taenia coli of guinea-pig. *Doklady Akademii Nauk, USSR*, **243**, 493-496.

Rutherford, A. and Burnstock, G. (1978) Neuronal and non-neuronal components in the overflow of labelled adenine compounds from guinea-pig taenia coli. *European Journal of Pharmacology*, **48**, 195-206.

Saji, Y., Escalon de Motta, G. and Del Castillo, J. (1975) Depolarization and potentiation of responses to acetylcholine elicited by ATP on frog muscle. *Life Science*, **16**, 945-954.

Salter, M.W. and Henry, J.L. (1985) Effects of adenosine 5′-monophosphate and adenosine 5′-triphosphate on functionally identified units in the cat spinal dorsal horn. Evidence for a differential effect of adenosine 5′-triphosphate on nociceptive vs non-nociceptive units. *Neuroscience*, **15**, 815-825.

Salter, M.W. and Henry, J.L. (1987) Evidence that adenosine mediates the depression of spinal dorsal horn neurons induced by peripheral vibration in the cat. *Neuroscience*, **22**, 631-650.

Satchell, D.G. (1981) Nucleotide pyrophosphatase antagonises responses to adenosine 5'-triphosphate and non-adrenergic, non-cholinergic inhibitory nerve stimulation in the guinea-pig isolated taenia coli. *British Journal of Pharmacology*, **74**, 319-321.

Satchell, D. (1982) Selective antagonism of inhibitory responses to ATP and non-adrenergic nerve stimulation in the guinea-pig taenia coli by nucleotide pyrophosphatase. *Clinical and Experimental Pharmacology and Physiology*, **8**, 642-643.

Satchell, D.G. and Maguire, M.H. (1975) Inhibitory effects of adenine nucleotide analogs on the isolated guinea-pig taenia coli. *Journal of Pharmacology and Experimental Therapeutics*, **195**, 540-548.

Saville, V.L. and Burnstock, G. (1988) Use of reserpine and 6-hydroxydopamine supports evidence for purinergic cotransmission in the rabbit ear artery. *European Journal of Pharmacology*, **155**, 271-277.

Schlicker, E., Urbanek, E. and Göthert, M. (1989) ATP, α,β-methylene ATP and suramin as tools for characterization of vascular P_{2X} receptors in the pithed rat. *Journal of Autonomic Pharmacology*, **9**, 371-380.

Schmidt, A., Ortaldo, J.R. and Herberman, R.B. (1984) Inhibition of human natural killer cell reactivity by exogenous adenosine 5'-triphosphate. *Journal of Immunology*, **132**, 146-150.

Schrader, J., Rubio, R. and Berne, R.M. (1975) Inhibition of slow action potentials of guinea-pig atrial muscle by adenosine: a possible effect on Ca^{2+} influx. *Journal of Molecular and Cellular Cardiology*, **7**, 427-433.

Sebastião, A.M. and Ribiero, J.A. (1988) On the adenosine receptor and adenosine inactivation at the rat diaphragm neuromuscular junction. *British Journal of Pharmacology*, **94**, 109-120.

Sebastião, A.M. and Ribiero, J.A. (1989) 1,3,8- and 1,3,7-Substituted xanthines: relative potency as adenosine receptor antagonists at the frog neuromuscular junction. *British Journal of Pharmacology*, **96**, 211-219.

Shimada, S.G. and Stitt, J.T. (1984) An analysis of the purinergic component of active muscle vasodilatation obtained by electrical stimulation of the hypothalamus in rabbits. *British Journal of Pharmacology*, **83**, 577-589.

Shinozuka, K., Bjur, R. and Westfall, D.P. (1988) Characterization of prejunctional purinoceptors on adrenergic nerves of the rat caudal artery. *Naunyn Schmiedeberg's Archives of Pharmacology*, **338**, 221-227.

Silinsky, E.M. (1975) On the association between transmitter secretion and the release of adenine nucleotides from mammalian motor nerve terminals *Journal of Physiology*, **247**, 145-162.

Silinsky, E.M. (1989) Adenosine derivatives and neuronal function. *The Neurosciences*, **1**, 155-165.

Silinsky, E.M. and Hubbard, J.I. (1973) Release of ATP from rat motor nerve terminals. *Nature*, **243**, 404-405.

Silinsky, E.M. and Ginsborg, B.L. (1983) Inhibition of acetylcholine release from preganglionic frog nerves by ATP but not adenosine. *Nature*, **305**, 327-328.

Smith, A.D. (1972) Subcellular localisation of noradrenaline in sympathetic neurones. *Pharmacological Reviews*, **24**, 435-440.

Sneddon, P. and Burnstock, G. (1984) Inhibition of excitatory junction potentials in guinea-pig vas deferens by α,β-methylene ATP: further evidence for ATP and noradrenaline as cotransmitters. *European Journal of Pharmacology*, **100**, 85-90.

Sneddon, P and Burnstock, G. (1985) ATP as a co-transmitter in rat tail artery. *European Journal of Pharmacology*, **106**, 149-152.

Sneddon, P. and Westfall, D.P. (1984) Pharmacological evidence that adenosine triphosphate and noradrenaline are co-transmitters in the guinea-pig vas deferens. *Journal of Physiology*, **347**, 561-580.

Sneddon, P., Westfall, D.P. and Fedan, J.S. (1982) Cotransmitters in the motor nerves of the guinea-pig vas deferens: electrophysiological evidence. *Science*, **218**, 693-695.

Steinberg, T.H. and Silverstein, S.C. (1987) Extracellular ATP^{4-} promotes cation fluxes in the J774 mouse macrophage cell line. *Journal of Biological Chemistry*, **262**, 3118-3122.

Stevens, P., Robinson, R.L., Van Dyke, K. and Stitzel, R. (1972) Studies on the synthesis and release of adenosine triphosphate-8-^{3}H in the isolated perfused cat adrenal gland. *Journal of Pharmacology and Experimental Therapeutics*, **181**, 463-471.

Stjärne, L. and Åstrand, P. (1984) Discrete events measure single quanta of adenosine 5'-triphosphate secreted from sympathetic nerves of guinea-pig and mouse vas deferens. *Neuroscience*, **13**, 21-28.

Stjärne, L. and Åstrand, P. (1985) Relative pre- and postjunctional roles of noradrenaline and adenosine

5′-triphosphate as neurotransmitters of the sympathetic nerves of guinea-pig and mouse vas deferens. *Neuroscience*, **14**, 929–946.

Stone, T.W. (1981a) Physiological roles for adenosine and adenosine 5′-triphosphate in the nervous system. *Neuroscience*, **6**, 523–555.

Stone, T.W. (1981b) Actions of adenine nucleotides on the vas deferens, guinea-pig taenia caeci and bladder. *European Journal of Pharmacology*, **75**, 93–102.

Stone, T.W. (1985) The activity of phosphorothioate analogues of ATP in various smooth muscle systems. *British Journal of Pharmacology*, **84**, 165–173.

Su, C. (1975) Neurogenic release of purine compounds in blood vessels. *Journal of Pharmacology and Experimental Therapeutics*, **195**, 159–166.

Su, C., Bevan, J.A. and Burnstock, G. (1971) [^{3}H]adenosine triphosphate: release during stimulation of enteric nerves. *Science*, **173**, 337–339.

Sweeney, M.I., White, T.D. and Sawynok, J. (1989) Morphine, capsaicin and K$^+$ release purines from capsaicin-sensitive primary afferent nerve terminals in the spinal cord. *Journal of Pharmacology and Experimental Therapeutics*, **248**, 447–454.

Takagi, K. and Takayanagi, I. (1972) Effect of N^6, 2′-O-3′, 5′-cyclic adenosine monophosphate, 3′, 5′-cyclic adenosine monophosphate and adenosine triphosphate on acetylcholine output from cholinergic nerves in guinea-pig ileum. *Japanese Journal of Pharmacology*, **27**, 33–36.

Takahashi, T., Kusunoki, M., Ishikawa, Y., Kantoh, M., Yamamura, T. and Utsunomiya, J. (1987) Adenosine 5′-triphosphate release by electrical nerve stimulation from the guinea-pig gallbladder. *European Journal of Pharmacology*, **134**, 77–82.

Tatham, P.E.R., Cusack, N.J. and Gomperts, B.D. (1988) Characterization of the ATP^{4-}-receptor that mediates permeabilization of rat mast cells. *European Journal of Pharmacology*, **147**, 13–21.

Taylor, E.M., Parsons, M.E., Wright, P.W., Pipkin, M.A. and Howson, W. (1989) The effects of adenosine triphosphate and related purines on arterial resistance vessels *in vitro* and *in vivo*. *European Journal of Pharmacology*, **161**, 121–133.

Theobald, R.J. (1982) Arylazido aminopropionyl ATP (ANAPP$_3$) antagonism of cat urinary bladder contractions. *Journal of Autonomic Pharmacology*, **2**, 175–179.

Theobald, R.J. (1986) The effect of arylazido aminopropionyl ATP (ANAPP$_3$) on inhibition of pelvic nerve evoked contractions of the cat urinary bladder. *European Journal of Pharmacology*, **120**, 351–354.

Theobald, R.J. and De Groat, W.D. (1988) The effect of purine nucleotides on transmission in vesical parasympathetic ganglia in the cat. *Journal of Autonomic Pharmacology*, **9**, 167–187.

Theobald, R.J. and Hoffman, V. (1986) Long-lasting blockade of P$_2$-receptors of the urinary bladder *in vivo* following photolysis of arylazidoaminopropionyl ATP, a photoaffinity label. *Life Science*, **38**, 1591–1595.

Tolkovsky, A.M. and Suidan, H.S. (1987) Adenosine 5′-triphosphate synthesis and metabolism localised in neurites of cultured sympathetic neurons. *Neuroscience*, **23**, 1133–1142.

Turpaev, T.M., Nistratova, S.N. and Putintseva, T.G. (1967) Characteristics of the escape of the heart of *Anodonta sp.* and *Helix pomatia*. *Journal of Evolutionary Biochemistry and Physiology, USSR*, **3**, 40–46.

Van Calker, D., Müller, M. and Hampbrecht, B. (1979) Adenosine regulates via two different types of receptors, the accumulation of cyclic AMP in cultured brain cells. *Journal of Neurochemistry*, **33**, 999–1005.

Viveros, O.H. and Wilson, S.P. (1983) The adrenal chromaffin cell as a model to study the co-secretion of enkephalins and catecholamines. *Journal of the Autonomic Nervous System*, **7**, 41–58.

Wakade, A.R. and Wakade, T.D. (1985) Sympathetic neurons grown in culture generate ATP by glycolysis: correlation between ATP content and [^{3}H]norepinephrine uptake and storage. *Brain Research*, **359**, 397–401.

Warland, J.J.I. and Burnstock, G. (1987) Effects of reserpine and 6-hydroxydopamine on the adrenergic and purinergic components of the sympathetic nerve responses of the rabbit saphenous artery. *British Journal of Pharmacology*, **92**, 871–880.

Wedd, A.M. (1931) The action of adenosine and certain related compounds on the coronary flow of the perfused heart of the rabbit. *Journal of Pharmacology and Experimental Therapeutics*, **41**, 355–366.

Westfall, D.P., Sitzel, P.E. and Rowe, J.N. (1978) The postjunctional effects and neural release of purine compounds in guinea-pig vas deferens. *European Journal of Pharmacology*, **50**, 27–38.

Westfall, D.P., Hogaboom, G.K., Colby, J., O'Donnell, J.P. and Fedan, J.S. (1982) Direct evidence against a role of ATP as the nonadenergic noncholinergic inhibitory neurotransmitter in guinea-pig taenia coli. *Proceedings of the National Academy of Sciences, USA*, **79**, 7041–7045.

Westfall, D.P., Hogaboom, G.K., Colby, J., O'Donnell, J.P. and Fedan, J.S. (1983) Evidence for a contribution by purines to the neurogenic response of the guinea-pig urinary bladder. *European Journal of Pharmacology*, **87**, 415–422.

Westfall, D.P., Sedaa, K. and Bjur, R. (1987) Release of endogenous ATP from rat caudal artery. *Blood Vessels*, **24**, 125–127.

White, T.D. (1982) Release of ATP from isolated myenteric varicosities by nicotinic agonists. *European Journal of Pharmacology*, **79**, 333–334.

White, T.D. (1988) Role of adenine compounds in autonomic neurotransmission. In *Pharmacology and Therapeutics. Vol. 38.* edited by C. Bell. pp. 129–168. London: Pergamon Press.

White, T.D. and Al-Humayyd, M. (1983) Acetylcholine releases ATP from varicosities isolated from guinea-pig myenteric plexus. *Journal of Neurochemistry*, **40**, 1069–1075.

White, T.D. and Leslie, R.A. (1982) Depolarization-induced relase of adenosine 5′-triphosphate from varicosities derived from the myenteric plexus of the guinea-pig small intestine. *Journal of Neuroscience*, **2**, 206–215.

White, T.D., Downie, J.W. and Leslie, R.A. (1985) Characteristics of K^+ and veratridine-induced release of ATP from synaptosomes prepared from dorsal and ventral spinal cord. *Brain Research*, **334**, 372–374.

Wiklund, N.P. and Gustafsson, L.E. (1984) Prejunctional adenosine-like and postjunctional ADP-like effect of ATP in guinea-pig ileum. *Acta Physiologica Scandinavica*, **120**, 12A.

Wiklund, N.P. and Gustafsson, L.E. (1986) Neuromodulation by adenine nucleotides, as indicated by experiments with inhibitors of nucleotide inactivation. *Acta Physiologica Scandinavica*, **126**, 217–223.

Wiklund, N.P. and Gustafsson, L.E. (1988a) Agonist and antagonist characterization of the P_2-purinoceptors in the guinea-pig ileum. *Acta Physiologica Scandinavica*, **132**, 15–21.

Wiklund, N.P. and Gustafsson, L.E. (1988b) Indications for P_2-purinoceptor subtypes in guinea-pig smooth muscle. *European Journal of Pharmacology*, **148**, 361–370.

Wiklund N.P. Gustafsson, L.E. and Lundin, L. (1985) Pre- and postjunctional modulation of cholinergic neuroeffector transmission by adenine nucleotides. Experiments with agonist and antagonist. *Acta Physiologica Scandinavica*, **125**, 681–691.

Willemot, J. and Paton, D.M. (1981) Metabolism and presynaptic inhibitiory activity of 2′, 3′ and 5′-adenine nucleotides in rat vas deferens. *Naunyn Schmiedeberg's Archives of Pharmacology*, **317**, 110–114.

Wilson, A., Furness, J.B. and Costa, M. (1979) A unique population of uranaffin-positive intrinsic nerve endings in the small intestine. *Neuroscience Letters*, **14**, 303–308.

Winkler, H. and Westhead, E. (1980) The molecular organization of adrenal chromaffin granules. *Neuroscience*, **5**, 1803–1823.

Wolinsky, E.J. and Patterson, P.H. (1985) Potassium-stimulated purine release by cultured sympathetic neurons *Journal of Neuroscience*, **5**, 1680–1687.

Wood, B.E., Squire, A., O'Connor, S.E. and Leff, P. (1989) ADP-β-F is not a selective P_{2Y}-receptor agonist in rabbit jugular vein. *British Journal of Pharmacology*, **98**, supplement, 794P.

Yamamoto, R., Cune, W.H. and Takasaki, K. (1988) Reassessment of the blocking activity of prazosin at low and high concentrations on sympathetic neurotransmission in the isolated mesenteric vasculature of rats. *Journal of Autonomic Pharmacology*, **8**, 303–309.

Zimmermann, H. (1978) Turnover of adenine nucleotides in cholinergic synaptic vesicles of the *Torpedo* electric organ. *Neuroscience*, **3**, 827–836.

Zimmermann, H. (1982) Co-existence of adenosine 5′-triphosphate and acetylcholine in the electromotor synapse. In *Co-transmission*, edited by A.C. Cuello, pp. 243–259. London: MacMillan Press.

8 Transmission: Peptides

G.J. Dockray

Physiological Laboratory, University of Liverpool, Brownlow Hill, PO Box 147, Liverpool L69 3BX

Peptides are a large and diverse group of potential transmitter substances that are well represented in autonomic neurons, particularly those in the gut. In many instances they are produced in neurons making classical transmitters e.g. the combination of noradrenaline/neuropeptide Y, and of acetylcholine/vasoactive intestinal peptide. The mechanisms of release and action of neuropeptides are, in principle, similar to those of classical transmitters, but peptides often act for longer, and over greater distances than classical transmitters, suggesting a modulatory function. Peptides are made from large precursors, but the processing mechanisms are variable and frequently several different active products can be produced from the same precursor. Good progress is now being made with the development of antagonists for several neuropeptides and these will greatly facilitate future progress in the elucidation of neuropeptide functions.

KEY WORDS: Peptides, peptidergic, co-existence, non-adrenergic, non-cholinergic

INTRODUCTION

Peptides are by far the largest and most diverse group of substances that act as mediators of autonomic neuron activity (Table 8.1). The mechanisms of peptidergic transmission differ in a number of important ways from those involving so-called classical autonomic transmitters i.e. acetylcholine and noradrenaline. One consequence of this is that the criteria that are often applied to determine whether a particular substance is a neurotransmitter (Werman, 1966; Orrego, 1979), must be modified for the neuropeptides (Dockray, 1987a). It should not be thought, however, that peptide and classical transmitters function independently. For the most part the neurons of the parasympathetic and sympathetic divisions that release peptides do so together with acetylcholine or noradrenaline (Hökfelt *et al*, 1986; Lundberg and Hökfelt, 1983; Furness *et al*, 1989; Costa, Furness and Gibbins, 1986). This may not extend, however, to what Langley (1921) first called the enteric division of the autonomic nervous system, where there are likely to be neurons that act through the release of peptides but not classical transmitters (Dockray, 1987a; Furness and Costa,

409

TABLE 8.1
Major Autonomic Neuropeptides.

Peptide	Structurally Related Peptides	Distinguishing Features in the Autonomic Nervous System
Substance P	Neurokinin A and B	Potential enteric excitatory transmitter; NK1 and 2 receptors cloned.
Neuropeptide Y (NPY)	PYY and PP	Present in many noradrenergic sympathetic neurons, and potential modulator of sympathetic transmission.
Vasoactive Intestinal Polypeptide (VIP)	Secretin, PHI, GRF, GIP Glucagon, helodermin/helospectins	Localized with acetylcholine in some parasympathetic postganglionic neurons, potential inhibitory transmitter on gut motility, and stimulant of intestinal secretion.
Met and Leu enkephalin	Many opioid variants including dynorphins	Candidate transmitters for presynaptic inhibition in both gut and prevertebral ganglia.
Gastrin Releasing Peptide (GRP)	Neuromedin B and bombesin	Mediator of vagal release of gut hormones, and potential mitogen.
Cholecystokinin (CCK)	Gastrin	Mainly present in CNS neurons and gut endocrine cells. Potential excitatory transmitter in some autonomic systems; excellent orally active antagonists available.
Calcitonin Gene Related Peptide (CGRP)	Pancreatic β amylin	Mainly present in primary afferent neurons and some enteric neurons. May mediate (with substance P) the effects of antidromic afferent stimulation.
Galanin	?	Interesting inhibitory effects on pancreatic hormone release. Physiology, still largely unexplored.
Somatostatin	?	Widespread inhibitory effects.

1989). Indeed, much of the work that has led up to the recognition of autonomic peptidergic transmitter mechanisms has been done on enteric neurons.

There are several reasons why the enteric nervous system has been of particular importance in identifying and clarifying peptide-mediated autonomic effects. For one thing, enteric neurons are a rich source of a wide variety of biologically active peptides (Costa and Furness, 1989) and it has often proved easier to isolate and chemically characterize neuropeptides from the gut compared with other tissues. The idea that enteric neurons might exert their actions by releasing biologically active peptides has come from several initially quite separate lines of evidence. First, the early work on substance P established that this material was peptide or protein in nature and that it was associated with enteric neurons as it was absent from tissue extracts prepared from the aganglionic segment of colon in Hirschsprung's disease (Ehrenpreis and Pernow, 1953). Second, ultrastructural studies of enteric neurons revealed populations of secretory granules that had profiles which were clearly distinct from those of typical cholinergic and adrenergic neurons but quite closely resembled those in the well characterized peptidergic nerve fibres of the posterior pituitary (Baumgarten, Holstein and Owman, 1970); subsequently up to ten different granule types were described in enteric nerve fibres, which also supported the idea that neurochemical transmission in the gut might be mediated by many different substances (Cook and Burnstock, 1976; Gabella, 1979). Third, studies of transmission in guinea pig taenia coli by Burnstock *et al* (1963) indicated the existence of a class of inhibitory transmitter that was neither adrenergic nor cholinergic (NANC); in itself this is not evidence for peptidergic as opposed say to purinergic or serotonergic transmission (or some other NANC-transmitter), the important point is that it encouraged the idea that other active substances might have autonomic transmitter functions.

The peptides became strong candidates for mediators of NANC transmission when, in the 1970's, a wide variety of them were isolated and chemically characterized from brain and gut (Mutt, 1982). In the case of the gut, it was at first thought that many had hormonal functions. This indeed proved to be the case for some new peptides. But many others were soon shown to occur in peripheral neurons, including those of all three divisions of the autonomic nervous system, as well as afferent neurons (Hökfelt *et al*, 1980; Dockray, 1988). The present account will deal first with the distinguishing features of peptide-mediated transmission, and second with the individual peptides (or groups of peptides) that are recognized to occur in autonomic neurons. About a dozen well characterized peptides or groups of related peptides are represented in the autonomic nervous system, although we know enough to be fairly sure that there are likely to be others that have not yet been properly characterized. The synthesis, release, metabolism, receptors and pharmacology of the major peptides will be considered here. Rather different considerations apply to a diverse range of other peptides whose release is influenced by autonomic reflexes e.g. gastrin, atrial natriuretic peptide, the endothelins, or which like nerve growth factor (NGF) are required for the survival and maturation of peripheral neurons, and these peptides are beyond the scope of the present chapter. Similarly, details of the distribution of individual peptides and their electrophysiological properties will be covered elsewhere in this series, as will details of the actions of different peptides in gut, cardiovascular and respiratory systems, urinogenital tract and skin.

PRINCIPLES OF PEPTIDERGIC TRANSMISSION

In common with classical transmitters, peptide transmitters are stored in vesicles or granules, released on depolarization and act at plasma membrane receptors. In the broadest terms, therefore, there are many similarities between peptidergic and classical autonomic transmission. But there are also important differences. In particular, the synthesis of neuropeptides, unlike that of the classical transmitters, does not occur in nerve terminals or endings. This is because the mechanisms for mRNA translation, and for sequestration of peptides into secretory granules, are limited to the cell soma (Fig. 1). Neuropeptides are therefore produced initially in the cell body, packaged there into granules, and subsequently transported intra-axonally to sites of

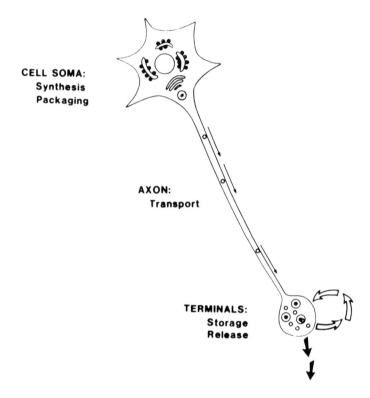

CELL SOMA:
Synthesis
Packaging

AXON:
Transport

TERMINALS:
Storage
Release

FIGURE 8.1   Schematic representation of the intracellular pathway taken by neuropeptides from initial synthesis in the cell body to release at the terminal. Note that the same neuron may produce both neuropeptides and classical transmitters and that the two may be segregated to separate granule populations. There are no known mechanisms for reuptake of peptide transmitter (filled arrows) but classical transmitters or their biosynthetic precursors may be taken up and recycled (open arrow) at the nerve terminal. (Reproduced from Dockray 1989c).

release. There are no known mechanisms for re-uptake and re-use of peptide transmitters. This means that for a unit amount of protein synthesis in the cell body there is a single (or at least strictly limited) opportunity to deliver transmitter to its target. In contrast, in the case of acetylcholine or noradrenaline the synthesis of a unit amount of protein in the form of enzymes such as choline acetyltransferase or dopamine β hydroxylase, offers the potential for production of a great many molecules of transmitter. The obligation imposed by the exclusive synthesis of peptides in the cell body and the inevitable constraints that follow from the necessity for axonal transport, together raise the possibility that peptides might act in a modulatory capacity or in the control of relatively long-term events, rather than as rapidly acting transmitters. Although there may be some exceptions, there are certainly a great many illustrations of the modulatory and long-term effects mediated by neuropeptides.

NEUROPEPTIDE IDENTIFICATION AND DETERMINATION

The discovery of many autonomic neuropeptides e.g. substance P and vasoactive intestinal polypeptide (VIP) (von Euler and Gaddum, 1931; Said and Mutt, 1970) was made by bioassay of tissue extracts, frequently using preparations based on the measurement of smooth muscle tension *in vitro*, or changes in blood pressure; only later was the physiology of these substances worked out, and then often after their distribution had been established. Bioassay preparations are still useful for identifying the release of endogenous neuropeptides in intact systems. Over the last 10–15 years, however, other approaches have been particularly important for the discovery of neuropeptides. Chemical assays for substances with the structural features of regulatory peptides have been exploited by Tatemoto and Mutt (1978; 1980); C-terminal α amides are a common feature of active peptides and assays for α amides first revealed substances such as neuropeptide Y (NPY) and peptide histidine isoleucine amide (PHI). The application of molecular biological approaches from the late 1970's onwards has also led to the discovery of several new peptides. In some cases the new peptide was shown to share a precursor with a known peptide e.g. neurokinin A and substance P; in other cases the discovery of a new mRNA species suggested the existence of a novel peptide, e.g. calcitonin gene related peptide (CGRP). None of these approaches has yet been fully exploited and all might in the future reveal novel transmitter peptides.

The methods referred to above do not lend themselves to routine assays for studies of peptide distribution or release. Immunochemical methods are by far the most convenient and widely used techniques for these purposes. All the major known autonomic neuropeptides have been localized by immunohistochemistry, and can be measured by specific and sensitive radioimmunoassays. The techniques for raising antibodies are well developed and these days it is fairly straightforward to obtain an antibody that reacts with the region of a small peptide that is of particular interest. This is usually the biologically active site, but in biosynthetic and metabolism studies it is often of value to have antibodies to other regions of the active peptide or its precursor.

SYNTHESIS

Neuropeptides, like other biologically active peptides, are characteristically synthesised initially as large precursors. The latter vary in size from about 100 to several hundred residues (Docherty and Steiner, 1982). The coding DNA sequences (exons) for mammalian neuropeptides are frequently interspersed with stretches of noncoding DNA (introns). The process of linking exons into continuous mRNA sequences is under regulatory control. The potential therefore exists for different cells expressing a particular gene to generate alternative mRNA species (and therefore different peptide products) by differential splicing of exons; this is illustrated by one of the tachykinin genes and by the calcitonin/CGRP gene (Table 2). Biologically active peptides may be just a few residues long so that following mRNA translation the biosynthetic process invariably involves extensive cleavage of the precursor. The first

TABLE 8.2
Major Neuropeptide Genes Expressed in Autonomic Neurons.

Gene	Products of Alternative mRNA Splicing	Major Peptide Products*	Other Important Products†
Preprotachykinin A	α-PPT-A β-PPT-A γ-PPT-A	SP-11 Neurokinin-A 10	Neuropeptide K
PreproVIP	–	VIP-28 PHI(M)-27	PHV-42, VIP-GKR
PeproNeuropeptide Y	–	NPY-27	–
Preproenkephalin A	–	Met-enk 5 Leu-enk 5 MERF-7 MERGL-8	N-terminally extended forms (particularly MERF/MERGL).
Preprocalcitonin/GRP	Calcitonin	CGRP-37	–
PreproGRP	3 species varying 3' ends	GRP27	GRP 10, 23
PreproCCK	–	CCK8	CCK 22, 33, 39, 58
PreproSomatostatin	–	SOM 14	SOM 28
PreproGalanin	–	Galanin 27	–

* numbers refer to number of residues in chain. † Other well characterized products generated by alternative pathways of post-translational processing of a common precursor; apart from N-terminally extended opioid peptides, all those listed are biologically active.

cleavage step is virtually obligatory, and involves the removal of an N-terminal signal sequence from the precursor. This sequence ensures sequestration of the precursor into the endoplasmic reticulum, and once it has served this purpose it is completely destroyed. The peptide products of all subsequent processing steps remain within the secretory pathway. For the most part, post-translational processing occurs in the Golgi and immature secretory granules. In addition to cleavage reactions, there are also modifications such as amidation, phosphorylation, glycosylation and sulphation that are introduced into the peptide products (Dockray, 1987b). Most of these processing steps are completed by the time the secretory granule leaves the cell soma. The pathways of post-translational processing are not, however, invariant; as a consequence different cells making the same precursor can produce different products (Table 8.2).

Biologically active peptides can often be grouped into families of substances with related structures and biological activities. There are three different principles that can be applied in recognising such families (Tables 8.1 and 8.2). First, different pathways of post-translational processing may yield products from a particular precursor that vary in chain length or in some other modification, but which are ultimately encoded by the same stretch of DNA, e.g. the forms of cholecystokinin (CCK) of 8, 33, 39, and 58 residues. Second, several similar biologically active peptides may be produced from a single precursor although they are derived from different stretches of DNA within a single gene e.g. VIP and PHI. Third, different genes can encode related peptides, e.g. the different opioid peptides from pro-opiomelanocortin, pro-enkephalin A and prodynorphin. The latter two instances are thought to arise during evolution as a consequence of the duplication of genomic sequences and their subsequent divergence through independent mutation (Dockray, 1989a). In functional terms it is important to recognize that the gene encoding a particular precursor can give rise to a variety of different peptides, i.e. peptide secreting cells seldom produce a unique active product; indeed, since many neurons and endocrine cells express several genes encoding different peptides, they may produce a wide range of peptides. The complete characterization of the transmitter events at junctions involving peptide-producing neurons therefore requires a knowledge both of the genes expressed and of the characteristic patterns of mRNA splicing and post-translational processing of the various precursor molecules in the prejunctional cell.

STORAGE AND RELEASE

Peptide-containing secretory granules can frequently be recognized on the basis of an electron-dense core. This is attributable to the fixation of protein within the granule which might include the active product, inactive peptide from the same precursor, and packaging proteins such as the chromogranins. It is known that in the case of some hormonal peptides there is a differential segregation of peptides within the granule core, e.g. in β-islet cell granules insulin occurs in the granule core and C-peptide in a clear halo region (Orci, 1982; Michael *et al*, 1987). Care is needed in the interpretation of the evidence because there are well documented examples of changes in granule morphology as post-translational processing proceeds. Thus for example, mature granules in G-(gastrin) cells are electron-lucent, but immature granules are electron dense (Rahier, Pauwels and Dockray, 1987). It seems that with progressive proteolytic cleavage of the precursor the small peptide products are either less readily fixed or lose their capacity to aggregate in electron-dense complexes. In some autonomic neurons it is possible to identify electron dense granules that are associated with peptide immunoreactivity, and smaller clear vesicles which are thought to contain classical transmitters e.g. VIP and acetylcholine in postganglionic parasympathetic fibres, and NPY and noradrenaline in sympathetic fibres (Johansson and Lundberg, 1981; Lundberg *et al*, 1981; Fried *et al*, 1985). It is possible that peptidergic granules arrive from the cell soma and are not recharged, whereas classical transmitters are repackaged at the nerve terminal into a distinct population of small clear vesicles.

The calcium dependent release of neurotransmitter (whether classical or peptide) by exocytosis of secretory granules following nerve terminal depolarization is widely recognized as a fundamental property of neurons. *In vitro*, release is often evoked experimentally by electrical field stimulation, or by application of depolarizing concentrations of potassium, e.g. 50 mM (Iversen *et al*, 1980). *In vivo*, electrical stimulation of nerve trunks evokes neuropeptide release which can often be measured in the venous outflow from an organ. Both approaches have been widely used to study the major neuropeptides. The release of classical and peptide transmitter from the same terminal may in some cases be dissociated by varying the frequency of stimulation. Thus in cat salivary gland, high frequency stimulation appears to be required to release VIP, while acetylcholine is released at lower frequency stimulation (Lundberg, 1981; Andersson *et al*, 1982). Given the possibility outlined above, that classical and peptide transmitters may be localized to different vesicles, it seems possible that the pattern of stimulation selects separate populations of vesicles for exocytosis.

METABOLISM

Termination of the action of neuropeptides probably depends on a variety of phenomena including internalization and degradation of receptor-bound peptide, diffusion away from the target cell, and metabolism by proteolytic enzymes. The latter have been studied in some detail. A relatively small number of enzymes is now thought to account for the degradation of most neuropeptides (Turner, Matsas and Kenny, 1985; Bunnett, 1987). Endopeptidase 24.11 (EC 3.4.24.11), also sometimes known as enkephalinase, cleaves to the N-terminus of hydrophobic residues, and acts on many biologically active peptides (Matsas, Kenny and Turner, 1984). It has been cloned and sequenced (Malfroy *et al*, 1987; Erdos and Skidgel, 1989), and although initially studied in kidney it is now recognized to be widely distributed in brain and in peripheral tissues including gut from which it has been isolated (Bunnett *et al*, 1988). The expression of the gene controlling endopeptidase 24.11 is regulated, and in particular appears to be under the inhibitory control of protein kinase C as phorbol ester depresses mRNA levels and the appearance of newly synthesized enzyme in several cell lines (Werb and Clark, 1989). Inhibitors of this enzyme are available, e.g. phosphoramidon and thiorphan, and have been shown to shift the dose-response curve for active peptides to the left, indicating the importance of this enzyme in peptide degradation. A second important enzyme is angiotensin converting enzyme (ACE) or peptidyl dipeptidase (EC 3.4.15.1), which sequentially cleaves C-terminal dipeptides and is inhibited by captopril. Its name reflects the fact that angiotensin I was the first important substrate to be discovered; ACE too is widely distributed and, in addition to angiotensin I, readily cleaves enkephalin, substance P, and other neuropeptides. A third important group of degrading enzyme are the aminopeptidases that are inhibited by bestatin. The use of a cocktail of enzymes inhibitors e.g. thiorphan, bestatin and captopril provides a valuable way to enhance neuropeptide recovery in release studies, and helps characterize metabolic pathways. Because particular peptidases act at well characterized peptide bonds it has been possible to develop degradation-resistant analogues of many neuropeptides.

RECEPTOR MECHANISMS AND PHARMACOLOGICAL MANIPULATION

The mediation of neuropeptide effects by second messenger systems has been intensively studied. There are many well-worked examples of peptides that mobilize cAMP, e.g. VIP, or that act via activation of phospholipase C to cause hydrolysis of phosphatidyl inositol bisphosphate yielding the putative second messengers, diacylglycerol and inositol trisphosphate (IP$_3$) e.g. GRP, CCK. The latter increases intercellular Ca^{++} by modifying release of Ca^{++} from intracellular stores, and may also play a part in controlling Ca^{++} entry (Berridge and Irvine, 1989). One important consequence of neuropeptide actions is the gating of ion channels which may be mediated by intracellular second messengers, or more directly by G-proteins. The channels gated by classical and peptide transmitters can be the same, e.g. the closure of K$^+$ channels mediating the M-current by substance P and acetylcholine in autonomic neurons and smooth muscle (Adams, Brown and Jones, 1983; Sims, Walsh and Singer, 1986).

The gene sequences encoding the receptors for two neuropeptides, namely substance P and neurokinin A, have now been elucidated (Yokota et al, 1989). It has become clear that these belong to a super-family which also incorporates the receptors for classical neurotransmitters, namely muscarinic and β adrenergic receptors. The common structural features of these receptors, and the fact that they activate related G-protein and second messenger systems, provides powerful evidence for the view that the mechanisms of action of neuropeptides are in principle similar to those of classical transmitters. The molecular biological characterization of new receptors is presently a major activity. One approach that deserves mention at the moment is the use of the polymerase chain reaction for amplification of DNA sequences. By using oligonucleotide primers for conserved regions it is likely that the sequences encoding many new receptors belonging to the superfamily mentioned above will soon be identified (Hershey and Krause, 1990).

Considerable attention has been given to the development of antagonists for neuropeptide transmitters. In spite of this, relatively few useful compounds have been produced so far. In part the problem in developing peptide antagonists arises from the size of some neuropeptides and the difficulty in modelling their structures. One commonly used approach is to substitute D-residues or Pro to produce a more rigid structure. The products of this approach are, however, still peptides and so are not really active orally, and even on parenteral administration may be rapidly cleared. The development of non-peptide antagonists that act at peptide receptors has only recently started to make progress; currently, CCK antagonists are the best example to come from this approach (Freidinger, 1989). In this case, useful non-peptide antagonists based on the benzodiazepine ring structure were developed after it was realized that a naturally occurring fungal metabolite with weak CCK antagonist properties was structurally similar to the benzodiazepines.

The difficulty in developing neuropeptide antagonists has encouraged other strategies for blocking neuropeptide actions, one of which is immunoneutralization. Antibodies to the substance of interest may be used either in vivo or in vitro to bind endogenous peptide and so prevent the access of ligand to receptor. It has recently been shown that it is possible to immunize rats with CGRP and VIP, and that the auto-antibodies so produced will reduce the action of the endogenous neuropeptide (Louis et al, 1989b; Forster, 1990). It is important to realize that the antibody must be directed to the biologically active part of the peptide. It is difficult, however, to be sure of the concentrations of antibody at the junctional sites and in the case of polyclonal antibodies it is also difficult to be sure of specificity *in vivo*. For these reasons care is needed in the interpretation of data, particularly if negative results are obtained.

TEMPORAL AND SPATIAL ACTIVITY DOMAINS

It is often thought that peptides act over greater distances and for longer times than classical transmitters. A good example of this is provided by the studies of Jan and Jan (1982; 1983) on ganglionic transmission in the bullfrog. In the 9th sympathetic ganglion of the bullfrog there are two different types of ganglion cell – large B cells and smaller C cells. These cells receive inputs from two different preganglionic fibres. Stimulation of the 3rd, 4th and 5th spinal nerves produces a fast nicotinic excitatory postsynaptic potential (EPSP) in B cells and a slow muscarinic EPSP. Stimulation of 7th and 8th spinal nerves produces a similar fast EPSP but in this case a non-cholinergic long slow EPSP. The latter is seen in both types of ganglion cell. Several lines of evidence indicate that the non-cholinergic slow EPSP is mediated by leutinizing hormone releasing hormone (LHRH). Thus application of LHRH to ganglion cells mimics the slow EPSP. Antagonists to LHRH block the action both of exogenous LHRH and of stimulation of 7th and 8th nerves. Finally it is possible to localize LHRH to nerve fibres in the ganglion. However, the interesting point about the distribution of these fibres is that they occur around the C-cells, but not B-cells. Thus B-cells apparently possess an LHRH input, even though they are some way from the LHRH terminals. This example indicates first that peptides can act as long-term mediators of excitability and second, that they are able to diffuse over relatively large distances to exert their effects.

TACHYKININS

Substance P is one of the most intensively studied neuropeptides of the peripheral nervous system. It was the first biologically active peptide to be localized to enteric neurons, and more recently tachykinin receptors were the first neuropeptide receptors to be cloned and sequenced. The term 'tachykinin' covers the three mammalian members of the group, and also a variety of peptides from lower vertebrates and invertebrates, e.g. physaelemin, eledoisin, kassinin, some of which were chemically characterized before substance P itself. The mammalian tachykinins are substance P, which is an undecapeptide, neurokinin A which comes from the same precursor as substance P, and neurokinin B which comes from a different precursor. The common structural feature that unites these peptides is the C-terminal sequence-Phe-X-Gly-Leu-Met-NH$_2$ (Table 8.3). The relative potency of different tachykinins for different receptor types is determined by the identity of the residue at position 4 from the C-terminus, and to some extent by the N-terminal sequence.

TABLE 8.3
Amino acid sequences for the Tachykinins*.

Substance P	Arg – Pro – Lys – Pro – Gln – Gln – Phe – Phe – Gly – Leu – Met – NH$_2$
Neurokinin A	His – Lys – Thr – Asp – Ser – Phe – Val – Gly – Leu – Met – NH$_2$
Neurokinin B	Asp – Met – His – Asp – Phe – Phe – Val – Gly – Leu – Met – NH$_2$
Physalaemin	Glp – Ala – Asp – Pro – Asn – Lys – Phe – Tyr – Gly – Leu – Met – NH$_2$
Kassinin	Asp – Val – Pro – Lys – Ser – Asp – Gln – Phe – Val – Gly – Leu – Met – NH$_2$
Eledoisin	Glp – Pro – Ser – Lys – Asp – Ala – Phe – Ile – Gly – Leu – Met – NH$_2$

* The mammalian tachykinins, and selected sub-mammalian peptides are shown.

SYNTHESIS

There are two genes that encode the precursors for mammalian preprotachykinins: preprotachykinin A and B (PPT-A and PPT-B). The PPT-A gene encodes both substance P and neurokinin A (Nawa *et al*, 1983; Nawa, Kotani and Nakanishi, 1984; Krause *et al*, 1987 Nakanishi, 1987; Harmar *et al* 1986); the latter was independently isolated from bovine spinal cord and has also been called substance K and neuromedin L (Table 8.4). The bovine, human and rat PPT-A gene sequences are closely similar. Each consists of seven exons; alternative pathways of mRNA splicing lead to the production of three different mRNA species, designated α, β and γ (Fig. 2). In α PPT-A mRNA, an exon that encodes neurokinin A is omitted so that substance P is the sole (known) biologically active product of this precursor. The relevant exon is included in β PPT-A mRNA, so this encodes both substance P and neurokinin A. The third species, γ-PPT A mRNA, resembles β PPT-A, but has a 45 base deletion such that it encodes both substance P and neurokinin A, but a longer N-terminally extended form of neurokinin A, known as neuropeptide K and originally isolated from brain by Tatemoto *et al* (1985), is excluded. All three mRNA species have been identified in enteric neurons (Sternini *et al*, 1989). It seems that in neurons of the guinea pig myenteric plexus producing β-PPT A, neuropeptide K might be transiently generated as an intermediate in the conversion of the precursor to neurokinin A (Deacon *et al*, 1987). Studies following the incorporation of ^{35}S Met into peptides synthesized by rat dorsal root ganglia have identified the major products of immunoprecipitation as substance P and neurokinin A (Harmar and Keen, 1984);

TABLE 8.4
Naturally Occurring Tachykinins, Their Precursors and Receptors.

Peptide	Alternative Name	Precursor	Preferred Receptor
Substance P	–	α, β, γ PPT-A	NK1
Neurokinin A	Neuromedin L Neurokinin α Substance K	β, γ PPT-A	NK2
Neuropeptide K	–	β PPT-A	?
Neurokinin B	Neuromedin K Neurokinin β	PPT-B	NK3

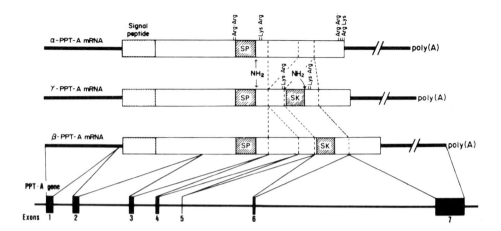

FIGURE 8.2   The relationships between the genomic sequence encoding preprotachykinin A and the three mRNA species produced by differential splicing. There are two main biologically active products (substance P and neurokinin A, also known as substance K). Both are encoded by β and γ PPT-A, but only substance P by α PPT-A. An N-terminally extended form of neurokinin A, neuropeptide K, is encoded by β but not γ PPT-A (Reproduced from Nakanishi, 1987).

indeed the latter was identified by this approach even before it was chemically characterized (Harmar, Schofield and Keen, 1980).

The organization of exons and introns in the PPT-B gene resembles that of the PPT-A gene. Two mRNA species are produced by alternative splicing of the PPT B gene; these differ only in their 5′ untranslated region, so that the peptide precursors are identical (Kotani et al, 1986). Neurokinin B is the only known active product from PPT-B; on present evidence it is produced in central neurons but not in the periphery, even though PPT B mRNA was reported to occur in the gut.

The PPT A gene is normally expressed in a variety of enteric neurons of both myenteric and submucous plexuses, and is expressed in sensory neurons including visceral afferents that form the first arm of autonomic reflexes. The latter include vagal afferents serving the thorax, and spinal afferents from a variety of abdominal organs, as well as skin. Nerve growth factor increases the expression of PPT-A mRNA in cultured dorsal root ganglion cells (Lindsay and Harmar, 1989). There are some postganglionic parasympathetic neurons, e.g. those of the otic ganglion in the rat, that contain substance P, but in general non-enteric autonomic neurons are not normally sites of PPT-A gene expression. In superior cervical ganglia, however, the expression of the gene is promoted following section of the preganglionic nerve (Kessler et al, 1981; Black et al, 1984). The concentrations of substance P in the ganglion also increased after treatment with the ganglion blocker chlorisondamine, and were reduced by treatment with phenoxybenzamine which reflexly stimulates the ganglion. Since cultured superior cervical ganglia also have increased substance P concentrations, it seems that withdrawal of the preganglionic input is a major factor in promoting expression of the gene in these sympathetic neurons.

RELEASE AND METABOLISM

The first information on the release of substance P from autonomic neurons came from studies of smooth muscle responses in atropine-treated preparations of guinea pig ileum following electrical field stimulation (Ambache, Verney and Aboo Zar, 1970; Franco, Costa and Furness, 1979). In this type of experiment it is difficult to quantify precisely the material released, or for that matter to obtain information on its chemical identity. More detailed information has since come from the demonstration of immunoreactive substance P release into the circulation of the vascularly perfused guinea pig ileum in response to an increase in luminal pressure; hexamethonium inhibited this response and the ganglion stimulant dimethylphenylpiperazinium increased it, indicating that enteric substance P neurons have a cholinergic-nicotinic input (Donnerer, Holzer and Lembeck, 1984; Donnerer et al, 1984). Studies of a variety of other tissues in vitro, e.g. the guinea pig ileum longitudinal muscle-myenteric plexus, and dog muscularis mucosa, indicate that immunoreactive substance P is released by high frequency field stimulation by a calcium-dependent mechanism (Baron, Jaffe and Gintzler, 1983; Angel, Go and Szurszewski, 1984; Holzer, 1984).

A population of the small diameter primary afferent nerve fibres serving many organs e.g. skin, cardiovascular, respiratory and genitourinary tract contains substance P which can be released by capsaicin (Hua et al, 1986; Geppetti et al, 1988). There is good evidence from the use of antagonists and antibodies that substance P released from these neurons contributes to the increased local blood flow, and increased vascular permeability seen on application of capsaicin (see below). Although capsaicin is a specific stimulant of small-diameter primary afferents its effects are by no means solely attributable to substance P release, and for example it is also clear that capsaicin releases CGRP from primary afferents.

Neutral endopeptidase 24.11 degrades substance P by cleavage at Gln^6-Phe^7, Phe^7-Phe^8 and Gly^9-Leu^{10}. Substitution of some of these residues yields analogues that are resistant to degradation. Inhibitors of endopeptidase 24.11, e.g. leucine thiorphan, shift to the left the dose-response curve for substance P in the rat small intestine, so that this enzyme is likely to be important for degradation of substance P released from enteric neurons (Djokic et al, 1989). In the guinea pig gall bladder the action of substance P in vitro was increased by thiorphan, but that of neurokinin A or neurokinin B was not, so that different peptidases mediate the degradation of different tachykinins (Maggi et al, 1989).

RECEPTORS AND PHARMACOLOGY

Three types of receptor for the tachykinin group of peptides are suggested by the evidence available at present. Several different nomenclatures have been used in the past to describe these receptors; the presently agreed terms for them are neurokinin (NK) 1, 2 and 3 (Lee et al, 1982; Buck et al, 1984; Watson, 1984; Burcher et al, 1986; Lee et al, 1986; Regoli et al, 1987; Jacoby, 1988). The multiple tachykinin receptors were initially identified on the basis of their characteristic distributions, and relative affinities for different naturally occurring and synthetic agonists. More

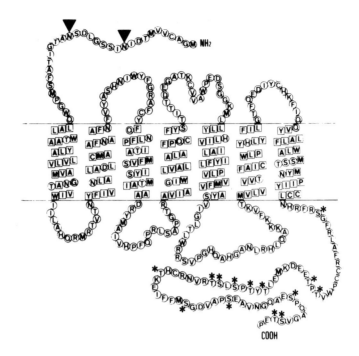

FIGURE 8.3 The predicted aminoacid sequence and organisation of the bovine gastric muscle NK 2 receptor. The predicted structure is based on the cDNA sequence. It is proposed that there are seven transmembrane spanning regions, a glycosylated extracellular domain (▲) and a phosphorylated intracellular domain (*). (Reproduced from Masu et al, 1988).

recently spectacular progress has been made with the molecular characterization of these receptors by cloning and sequencing (Fig. 8.3); in addition some selective antagonists have become available. The bovine gastric muscle NK2 receptor was first identified by the expression in *Xenopus* oocytes of size-fractionated mRNA (Masu *et al*, 1987). Subsequently the rat NK1 receptor was cloned from brain and the rat NK2 receptor from gastric muscle (Yokota *et al*, 1989). The identification of the cloned materials as encoding the NK1 and 2 receptors was made after transfection of Cos-7 cells and binding studies of the expressed receptor. The NK1 and 2 receptor proteins belong to the group that includes rhodopsin, muscarinic and β-adrenergic receptors. They have seven transmembrane-spanning domains, potential C-terminal cytoplasmic phosphorylation sites and N-terminal extracellular domains with potential glycosylation sites. The V and VI transmembrane domains are well conserved and these regions may be important for G-protein coupling. The second messenger systems activated by tachkinin receptors involve phosphatidyl inositol hydrolysis and calcium mobilization (Watson and Downes, 1983; Mantyh *et al*, 1984; Hunter, Goedert and Pinnock, 1985; Holzer and Lippe, 1984).

The rank order of potency of agonists at the three main mammalian receptors are as follows: at NK1 sites substance P > neurokinin A > neurokinin B; at NK2 sites, neurokinin A > neurokinin B > substance P, and at NK3 sites, neurokinin

B > neurokinin A > substance P. Thus the natural ligand for the NK1 receptor is substance P, for NK2 it is neurokinin A, and for NK3 it is neurokinin B (Table 4). Many active tachykinin analogues have been described, including molecules with improved agonist selectivity (Wormser *et al*, 1986; Drapeau *et al*, 1987 a,b); for example, [β − A1a^8] neurokinin A (4-10) is a specific NK2 agonist, and Me-Phe^7 NKB 4-10 is a NK3 agonist (Maggi *et al*, 1990). Tachykinin analogues with antagonist properties, of which D-Pro^2, D-Trp^{7,9} SP was the first example (Holmdahl *et al*,1981; Leander *et al*, 1981), have been shown to be useful in reversing the effects of substance P on inflammatory reactions and smooth muscle motility. More recently conformationally constrained cyclic analogues have been developed that are selective NK2 antagonists (Williams *et al*, 1988).

MAIN AUTONOMIC FUNCTIONS

There is good evidence to suggest a major function for tachykinins as important transmitters of (a) enteric neurons, (b) certain post-ganglionic parasympathetic neurons, and (c) primary afferent neurons. The evidence that tachykinins play a role in mediating non-cholinergic excitatory effects in intestinal peristalsis is impressive (Bartho and Holzer, 1985; Furness and Costa, 1989). Distension of the intestine produces an ascending excitation that has both cholinergic nicotinic and muscarinic elements. When these are blocked there remains an excitatory pathway which is probably mediated by the tachykinins; the latter has been demonstrated in guinea pig and human ileum, and rat colon, and seems to be particularly important in mediating the effects of distension at higher pressures (Grider 1989 a, b; Holzer, 1989). The tachykinin effects in the peristaltic reflex appear to be exerted both within the myenteric plexus and directly on smooth muscle. The smooth muscle receptors are blocked by antagonists like D Arg^{1}D-Pro^{2}D-Trp^{7,9} SP, but neuronal receptors are resistant to blockade by this antagonist (Taylor and Bywater, 1986; Laufer *et al*, 1985). Substance P is also a potent secretagogue, notably on salivary acinar cells, and this property was in fact used as the basis for bioassays in monitoring the isolation of the peptide. There is a physiological basis to the action, because at least in the rat, substance P occurs in parotid post-ganglionic cholinergic neurons and may mediate non-cholinergic nerve-stimulated secretion (Gallacher, 1983).

The tachykinins are excellent candidates for mediating some of the peripheral responses seen after afferent nerve stimulation. These effects are simulated by capsaicin, and their tachykinin component can be inhibited by tachykinin antagonists, and by antibodies (Holzer, 1988; Louis *et al*, 1989 a). In particular substance P is likely to play a prominant role in mediating the local increases in blood flow and capillary permeability in the skin that occur on stimulation of small diameter afferents. It is also likely to be a mediator of events at the peripheral terminals of afferent fibres in airways, genito-urinary tract, and gut (Lembeck and Holzer, 1979; Lundberg and Saria, 1983; Rosza, Varro and Jancso, 1985; Barnes, 1987; Rozsa, Mattila and Jacobson, 1988). Substance P released from primary afferents may also be important in modulating sympathetic neuron discharge in certain prevertebral ganglia. Tachykinins are present in the collaterals of sensory fibres that pass through these ganglia (Matthews and Cuello, 1982). In the guinea pig inferior mesenteric ganglia, substance P has been shown to mimic the slow EPSP seen after stimulation of either pre- or post-ganglionic nerve trunks following cholinergic blockade; this effect is capsaicin-sensitive, and is inhibited by substance P antagonists (Jiang, Dun and Karczmar, 1982; Tsunoo, Konishi and Otsuka, 1982; Otsuka *et al*, 1982).

VASOACTIVE INTESTINAL POLYPEPTIDE AND PHI

Vasoactive intestinal polypeptide (VIP) was discovered by Said and Mutt in extracts of porcine intestine. They characterized the peptide as a 28-residue molecule related in structure to secretin and glucagon, and they noted its potent vasodilator activity and actions on a variety of systems, including respiratory tract and gut (Said and Mutt, 1970; 1972). VIP is found in many cholinergic post-ganglionic parasympathetic neurons, but it also occurs in many instrinsic enteric neurons independently of acetylcholine. The peptide belongs to a large family that includes not only secretin and glucagon, but also gastric inhibitory polypeptide (GIP), growth hormone releasing factor (GRF), and the reptilian peptides – helospectins and helodermins. Another member of the group is PHI, which comes from the same precursor as VIP and has VIP-like actions. These two peptides are the only members of the family to occur widely in peripheral neurons. Porcine, bovine, human and rat VIP are identical in sequence, and guinea pig, chicken, dogfish and cod VIP differ in only five (or fewer) substitutions (Dimaline, 1989). The substitutions are predominantly conservative and do not greatly influence biological activity; VIP is therefore highly conserved which is consistent with important biological functions.

SYNTHESIS

The cDNA sequence for human VIP from a neuroblastoma was elucidated by Itoh *et al* (1983) and shown to indicate a precursor for both VIP and a peptide similar to PHI that had previously been isolated and characterized from hog intestine by Tatemoto and Mutt (1980). The human peptide differed from PHI in its C-terminal residue and has been called PHM (peptide with N-terminal His, and C-terminal Metamide). Porcine PHI shares 13 and 27 residues with VIP, but is less closely related to the other members of the family. The genomic sequence encoding prepro VIP/PHI consists of 9 Kb with seven exons; separate exons encode the signal peptide, the VIP- and the PHI-encoding regions (Bodner, Fridkin and Gozes, 1985; Linder *et al*, 1987). A cAMP-regulatory element occurs -86 to -70 nucleotides upstream of the transcribed region (Tsukada *et al*, 1987), and the abundance of VIP mRNA in human neuroblastoma cell lines has been shown to be increased by cAMP, and also by phorbol esters (Hayakawa *et al*, 1984; Ohawa *et al*, 1985).

Post-translational processing of the VIP/PHI precursor gives rise to VIP and PHI and three other peptides: C- and N-terminal flanking peptides, and a bridging peptide. An alternative post-translational cleavage pathway produces a 42-residue peptide consisting of PHI with a C-terminal extension that includes some of the bridging peptide sequence; the C-terminal residue of the peptide so produced is valine, hence the name PHV-42 (Yiangou *et al*, 1987). It appears that PHV-42 is more potent than PHI in several bioassays, including inhibition of force and frequency of uterine contractions, relaxation of bronchus and gastric smooth muscle. Several VIP-variants have been identified in gut extracts by RIA and ion exchange chromatography (Dimaline and Dockray, 1978). One of the variants in pig duodenum has been characterized as VIP extended at the C-terminus by Gly-Lys-Arg; this represents a biosynthetic intermediate which is generated by pro VIP cleavage but failure to complete the sequence of processing events leading to C-terminal amidation. The variant has been shown to have approximately 50% of the potency of VIP for binding to liver membranes (Gafvelin *et al*, 1988).

RELEASE AND METABOLISM

The plasma concentrations of VIP are generally low and rather constant (< 20 pmol.1^{-1}). The circulating material in mammals is thought to represent spillover of peptide released from nerve endings. Greatly elevated concentrations do, however, occur in patients with the VIP-secreting tumours of the pancreatic cholera or watery diarrhoea-hypokalaemia-achlorhydria (WDHA) syndrome. The assay of VIP in the venous outflow of various tissues has been used to study the mechanisms regulating neuronal VIP release, for example, electrical stimulation of preganglionic nerves produces release of VIP from the pancreas, gastrointestinal tract and salivary glands. Hexamethonium blocks these responses, indicating that VIP is released from post ganglionic neurons (Lundberg *et al*, 1980; Andersson *et al*, 1982; 1983; Fahrenkrug *et al*, 1978). As expected from their common biosynthetic origin, VIP- and PHI-immunoreactivities are released in parallel (Lundberg *et al*, 1984; Yasui *et al*, 1987; Holst *et al*, 1987). The release of VIP into the gastric venous outflow can be evoked by stimulation of the central cut-end of the vagus which suggests that vago-vagal reflexes may be mediated by VIP (Ito, Ohga and Ohta, 1988a, b). There is direct evidence for tetrodotoxin-sensitive, calcium dependent, release of immunoreactive VIP following field stimulation in microdissected rabbit ileum layers, and dog muscularis mucosa *in vitro* (Angel, Go and Szurszewski, 1984; Gaginella, O'Dorisio and Hubel, 1981); in rat ileum, however, nerve-stimulated release of VIP was reported to be independent of extracellular calcium (Belai, Ralevic, and Burnstock, 1987).

There are several reports indicating that in hepatocytes, colon or pancreas cell lines, and enterocytes VIP is internalized after binding to receptors and is then degraded (Misbin *et al*, 1982; Svoboda *et al*, 1988). One of the first steps in the enzymatic degradation of VIP, at least in pig enterocytes, is the removal of the N-terminal residue of VIP by an aminopeptidase (Nau, Ballmann, and Conlon, 1987); it has been suggested that this step inactivates the peptide, and that it occurs prior to internalization.

RECEPTORS AND PHARMACOLOGY

The VIP receptors of the pancreas, intestinal mucosa, gut smooth muscle, fat and liver cells have been intensively studied. The binding of VIP is specific, saturable, reversible, temperature-dependent and associated with increases in adenylate cyclase activity and intracellular cAMP (Amiranoff *et al*, 1978; Binder, Lemp and Gardner, 1980; Bitar and Makhlouf, 1982; Broyart *et al*, 1981; Christophe, Conlon, and Gardner, 1976; Dupont *et al*, 1980; Robberecht, Conlon and Gardner, 1976; Robberecht *et al*, 1982; Schwartz *et al*, 1974). The VIP receptor typically shows higher affinity for VIP than secretin. In guinea pig pancreatic acinar cells there are at least two classes of binding site: one with relatively high affinity for secretin and lower affinity for VIP, the other with high affinity for VIP and lower affinity for secretin (Christophe *et al*, 1976; Zhou, Gardner and Jensen, 1987). It seems probable that PHI interacts with VIP receptors and the evidence from studies of rat enterocytes and guinea pig pancreatic acinar cells suggests it has 5-25 fold lower affinity than VIP (Bataille *et al*, 1980; Jensen *et al*, 1981).

The co-localization of VIP and acetylcholine in many postganglionic parasympathetic neurons has encouraged the search for functional interactions between the two substances. It is clear that there is potentiation between VIP (and other cAMP-activating substances) and acetylcholine (or other Ca^{++}-mobilizing substances) in the stimulation of pancreatic and salivary acinar cells (see below). In cat superior cervical ganglia, VIP selectively facilitated muscarinic transmitter mechanisms (Kawatani, Rutigliano and de Groat, 1985). In part these effects probably reflect interactions at the second messenger level; it has, however, been suggested that VIP increases the binding of muscarinic ligands to cat submandibular gland cells (Lundberg, Hedlund and Bartfai, 1982), although the mechanisms involved remain unknown. Prolonged blockade of muscarinic receptors is said to increase sensitivity of salivary acinar cells both to acetylcholine and VIP, but again by unknown mechanisms (Hedlund, Abens, and Bartfai, 1983).

The chemical properties of VIP receptors have been fairly extensively studied in cross-linking experiments. The binding sites on rat enterocyte membranes correspond to two proteins of 73,000 and 33,000 molecular weight; the high molecular weight form appears to correspond to a high affinity receptor and the lower weight form to a low affinity receptor. Studies in human colonic epithelium, rat liver and guinea pig pancreas suggest species and tissue differences in the molecular weights of different receptor populations which at least in part may be attributable to differential glycosylation (Luis *et al*, 1988; El Battari *et al*, 1987; McArthur *et al*, 1987).

Conformational studies of VIP indicate residues 1-4 and 7-10 might form β-turns, and the 12-28 sequence a helix. Both C- and N-terminal regions are known to be required for full activity (Bodanszky, Klausner, and Said, 1973). Some progress has been made in the development of antagonists, and of selective agonists, for different VIP/PHI/secretin receptors. Thus D-Phe[4] PHI is said to be a highly selective agonist for the VIP-preferring receptors in rat pancreas, as compared with secretin receptors (Robberecht *et al*, 1987). Two analogues have been reported to act as VIP antagonists: GRF (N-Ac-Tyr[1], D-Phe[2]) -GRF (1-29)-amide, and 4-Cl-

DPhe⁶Leu¹⁷VIP. Both antagonized pancreatic VIP receptors but their affinities are low and they are only of limited use (Pandol *et al*, 1986; Waelbroeck *et al*, 1985). There is a pressing need for antagonists that can be used *in vivo*, although it is worth noting that in the absence of useful antagonists there has been widespread and successful use of immunoblockade as a tool to examine the physiological functions of VIP.

MAJOR AUTONOMIC FUNCTIONS

The possible transmitter functions of VIP have been examined in a wide variety of systems including gut, reproductive tract and airways; in a number of instances it is clear that VIP is a co-mediator with acetylcholine of parasympathetic post-ganglionic effects (Fahrenkrug, 1989; Fahrenkrug, Ottesen and Palle, 1988; Barnes, 1987; Lundberg *et al*, 1987). Although PHI or PHV42 are co-released with VIP, they generally have lower activity and for the most part a distinct role has not been attributed to them. Two systems where the physiology of VIP has been well studied deserve particular attention. First, VIP and acetylcholine are together able to account for the effects of parasympathetic stimulation on salivary secretion and blood flow (Lundberg, 1981). It has long been recognized that atropine blocks the salivary secretory response to parasympathetic stimulation, but not the increase in blood flow. It is now thought that at lower frequencies of stimulation acetylcholine is released and acts as the main mediator of the secretory response; VIP appears to potentiate the action of acetylcholine, but more importantly plays a primary role in mediating the atropine-resistant vasodilator response and is released by higher frequency stimulation (Andersson *et al*, 1982).

Second, in the gut, VIP is a major candidate for the NANC transmitter mediating secreto-motor responses of the intestinal epithelium (Krejs *et al*, 1979; Keast, 1987), and relaxation of some sphincters (Biacini, Walsh and Behar, 1984, 1985) and smooth muscle. Early in digestion there is a reflex vago-vagal relaxation of the stomach, which appears to be mediated by VIP acting directly on gastric smooth muscle (Bitar and Makhlouf, 1982; Grider *et al*, 1985b; De Beurme and Lefebvre, 1988; Ito, Ohga and Ohta, 1988 a,b). VIP is also a candidate for the mediator of the descending inhibitory component in the intestinal peristaltic reflex. The presently available evidence suggests that more than one substance might mediate the inhibitory effect on intestinal circular smooth muscle produced by nerve stimulation (Niel, Bywater and Taylor, 1983). One of the substances involved could be a purine and VIP may be one of the others (Hills, Collis and Burnstock, 1983; MacKenzie and Burnstock, 1980; Grider and Makhlouf, 1986; Grider *et al*, 1985a; Costa, Furness and Humphreys, 1986). In part the relevant evidence involves the use of the potassium channel blocker apamin; this inhibits the action of purines, but not VIP. However, apamin also reduces the release of VIP so care is needed in interpreting findings based on the use of this compound to block the response to endogenously released substances (Sjöqvist *et al*, 1983; Grider and Makhlouf, 1987).

NEUROPEPTIDE Y (NPY)

Tatemoto and Mutt discovered NPY using their assay for C-terminally amidated peptides. It is structurally related to pancreatic polypeptide (PP) and to PYY from intestine (Tatemoto, Carlquist and Mutt, 1982). In retrospect, it seems likely that NPY is the neuropeptide previously localized by antibodies to avian PP (Lundberg *et al*, 1982). Neuropeptide Y is particularly important in the autonomic nervous system because it occurs in many of the sympathetic adrenergic neurons that supply blood vessels, and there is evidence for NPY modulation of sympathetic transmission at several levels.

SYNTHESIS

Neuropeptide Y is 36 residues in length, has a tyrosine (designated Y in the single letter notation) at the N-terminus and a tyrosine amide at the C-terminus (hence neuropeptide Y). The primary aminoacid sequence is identical in man, guinea pig, rabbit and rat (O'Hare *et al*, 1988; Corder, Gaillard, and Böhlen, 1988), and there is a single substitution in pig NPY (Tatemoto, 1982). In anglerfish pancreas there is a related peptide that has about 65% sequence homology with mammalian NPY, but possesses a C-terminal dipeptide Tyr-Gly, suggesting a failure of conversion to Tyr-amide during biosynthesis (Andrews *et al*, 1985).

The nucleotide base sequence encoding NPY was deduced first from cDNA cloned from a human phaeochromocytoma (Minth *et al*, 1984) and later from rat brain (Higuchi, Yang and Sabol, 1988). The precursors consist of 97 and 98 residues respectively, and give rise to a single copy of NPY (Fig. 8.4). The genomic sequence for human NPY consists of four exons; the first is non-coding, the second encodes the signal sequence and most of the NPY region, except the C-terminal tyrosine, the third, most of the rest of the precursor, and fourth the extreme C-terminal hexapeptide (Minth, Andrews and Dixon, 1986). DNA sequences that lie within 530 bases upstream of the coding region exert transcriptional control. In PC12 cells and mouse neuroblastoma N18TG-2 cells, the expression of the NPY gene was increased by

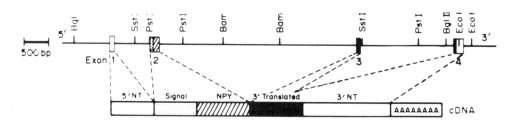

FIGURE 8.4 The organisation of the gene encoding neuropeptide Y showing the relationship between exons, mRNA, and the predicted peptide precursor. The latter contains a single copy of NPY (hatched) coming immediately after the signal sequence (stippled); a separate exon codes for an N-terminal sequence (filled). (Reproduced from Minth et al 1986).

glucocorticoids (dexamethasone) and by cAMP (Higuchi, Yang and Sabol, 1988); in addition, activation of protein kinase C by phorbol ester produced a pronounced potentiation of the cAMP-induced expression of the NPY gene.

RELEASE AND METABOLISM

Circulating concentrations of NPY vary markedly between species. In rat plasma, concentrations are reported as 1 pmol.ml^{-1} compared with 10–80 fmol.ml^{-1} in man. This material may be derived from adrenal chromaffin cells, or from spill-over of sympathetic post-ganglionic secretions. Stress and exercise increase plasma concentrations (Lundberg et al, 1985), and it remains a possibility that the high concentrations sometimes reported are due to stress in handling experimental animals (Castagne et al, 1987). NPY is released from large dense vesicles (Fried et al, 1985) by electrical stimulation of sympathetic nerves, but it is worth pointing out that there are cases e.g. the pig pancreas, where parasympathetic (vagal) stimulation causes NPY release – in this case probably from intrinsic neurons in the pancreas (Sheik et al, 1988).

RECEPTORS AND PHARMACOLOGY

There are only 8 of 36 residues that are conserved in the known avian and mammalian forms of NPY, PP and PYY, but conformational studies indicate that this group of substances has a well-conserved tertiary structure (Glover et al, 1985). This consists of a helical N-terminal region (1–8) rich in proline, linked through a β turn to an α helix that runs from residues 14 to 32. Some analogues in which the central region of 7–17 are replaced with 8 amino-octanoic acid stabilized with Cys e.g. D-Cys^7 Aoc$^{8-17}$ Cys^{20} NPY, retain activity indicating that residues 7–17 do not bind the receptor (Krstenansky et al, 1989).

  The NPY binding sites on rat CNS membranes have been shown in cross-linking experiments to correspond to proteins of 58,000 and 35,000 molecular weight, but the relationship between the two forms remains unknown (Mannon et al, 1989). Recent receptor binding and pharmacological studies with CNS membranes and several different cell lines suggest that there are two types of receptor for NPY, which can be distinguished on the basis of their affinity for C-terminal fragments of NPY. Sheikh, Håkanson, and Schwartz (1989) and Sheikh et al (1989) have characterized the Y_1 receptor as having affinities in the nanomolar range, and showing little or no affinity for C-terminal fragments such as 13–36 NPY. The Y_2 receptor has an affinity in the sub-nanomolar range, and binds C-terminal fragments (Fig. 8.5). Recently Fuhlendorff et al (1990) synthesized an NPY analogue that consisted of the N-terminal sequence of NPY and the C-terminal hexapeptide of PP i.e. Leu^{31} Pro^{34} NPY; this molecule was a selective high affinity agonist at Y_1 receptors. It is known that there are both pre and post-junctional sites of action of NPY and that the two receptors are likely to be different. It seems that the Y_1 receptors identified in binding studies may correspond to the postjunctional sites and Y_2 receptors are

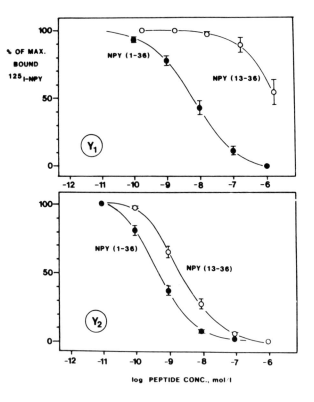

FIGURE 8.5   The relative affinities of NPY and its C-terminal fragment (13–36) for the Y1 and Y2 NPY receptors. The Y1 receptor has nanomolar affinity for NPY but low affinity for C-terminal fragments, the Y2 receptor has sub-nanomolar affinities for NPY and about 10-fold lower affinity for the C-terminal fragment. (Reproduced from Schwartz et al 1989).

likely to be prejunctional (Wahlestedt, Yanaihara and Håkanson, 1986). In addition, however, there may be other relevant receptors, since on rat jejunal membranes there are binding sites for ^{125}I PYY that show about 4-fold lower affinity for NPY (Laburthe *et al*, 1986). At these sites NPY appears to act by inhibition of adenyl cyclase, via a pertussis toxin-sensitive mechanism, indicating the involvement of a G protein. The high affinity binding sites on CNS membranes are also known to be sensitive to guanine nucleotides (Unden and Bartfai, 1984).

MAIN AUTONOMIC FUNCTIONS

There is a rapidly developing body of evidence to indicate that NPY is an important modulator of sympathetic transmission, for example in vasomotor control (Edvinsson, 1988), and in the vas deferens (Stjarne, Lundberg and Åstrand, 1986). The vasomotor effects are readily related to the observation that the sympathetic nerves containing NPY preferentially supply blood vessels (Lindh *et al*, 1986;

Macrae, Furness and Costa, 1986; Furness and Costa, 1987); there are in addition, however, non-adrenergic (intrinsic) NPY-containing neurons in the gut which presumably function independently of the sympathetic nervous system. Three separate sites of action of NPY are distinguishable. First, NPY on its own has variable and often quite weak excitatory post-junctional effects. Second there is a strong post-junctional potentiating interaction between NPY and noradrenaline (and also other substances, e.g. ATP and histamine), that is seen for example in rabbit gastroepiploic artery (Ekblad *et al*, 1984; Edvinsson *et al*, 1984) and mouse vas deferens (Stjarne, Lundberg and Åstrand, 1986). Third, NPY acts presynaptically to inhibit noradrenaline release, for example in the rat vas deferens and the sympathetic innervation of blood vessels e.g. femoral and portal veins (Dahlöf *et al*, 1985; Lundberg *et al*, 1985). At low frequencies of stimulation NPY and noradrenaline are thought to co-operate in producing vasoconstriction, but at higher frequencies of stimulation NPY would act to depress transmitter release. In addition to prejunctional effects on sympathetic ganglionic fibres, NPY may also act prejunctionally on vagal post-ganglionic fibres e.g. in heart, or intrinsic neurons in the intestine (Holzer *et al*, 1987) and colon (Wiley and Owyang, 1987).

OPIOID PEPTIDES

The first of the opioid peptides to be chemically characterized were the pentapeptides Leu and Metenkephalin (Leu-enk and Met-enk) which Hughes *et al* (1975) isolated and sequenced from bovine brain. They used the inhibition of electrically evoked twitches of the guinea pig ileum longitudinal muscle-myenteric plexus preparation to monitor isolation. Endogenous opioid systems are now recognized to be complex, because there are many different naturally occurring opioids and at least three types of opioid receptor. Studies of the physiology of endogenous opioids have been generally facilitated by the availability of antagonists of which naloxone is a well known example.

SYNTHESIS

Three separate genes encode opioid peptides. The organization of the genomic sequences is rather similar and suggests a common evolutionary origin (Fig. 8.6). One gene encodes the precursor, preproenkephalin (preproenkephalin A) that is processed to both Met- and Leu-enkephalin. A second (preproopiomelanocortin) encodes the precursor containing the sequence of Met-enk that gives rise to β endorphin and ACTH (Nakanishi *et al*, 1979). This precursor is not processed to Met-enk; its major opioid product is β endorphin or related peptides. Finally the preprodynorphin (preproenkephalin B) gene encodes a precursor containing the sequence of Leu-enk and gives rise to the dynorphins (Kakidani *et al*, 1982).

Proenkephalin A contains four copies of Met-enk, one of Leu-enk and one each of the C-terminally extended variants Met-enk Arg^6Phe^7 (MERF) and Met-enk $Arg^6Gly^7Leu^8$ or MERGL (Noda *et al*, 1982a; Gubler *et al*, 1982; Comb *et al*, 1982);

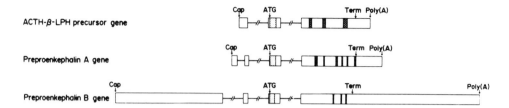

FIGURE 8.6   Schematic representation of the structures of the three genes encoding opioid peptides. Note that although there are minimal amino acid sequence similarities apart from the shared sequence of Met- and Leu-enkephalin, the three genes are organised in similar ways. The stippled box represents the signal sequence; enkephalin sequence are shown in hatched hoxes, and in the case of proopiomelanocortin (ACTH-β-LPH precursor) cross-hatched boxes show the sequence of other active peptides notably melanocyte stimulating hormone (MSH). (Reproduced from Numa, 1984).

these peptides would be released by cleavage at pairs of basic residues, followed by carboxypeptidase B-like removal of C-terminal basic residues. Antibodies specific for MERF and MERGL provide useful specific probes for the localization of pro-enkephalin A, and for the characterization of processing pathways. There are marked differences between tissues and species in the patterns of proenkephalin A post-translational processing. The major immunoreactive forms of MERF and MERGL in many peripheral e.g. enteric, neurons often correspond to the hepta- and octapeptides (Giraud, Williams and Dockray, 1984). But in some gut neurons, and in bovine adrenal medulla chromaffin cells, the authentic hepta- and octapeptides are minority species and N-terminally extended forms of these peptides predominate (Kojima *et al*, 1982; Udenfriend and Kilpartrick, 1983; Lewis *et al*, 1980). Many of these peptides lack opioid activity and their function is unknown. The larger forms of MERF and MERGL in bovine adrenal medulla and in pig stomach and duodenum are also of interest because they occur in phosphorylated and glycosylated forms (Watkinson, Dockray and Young, 1988; Watkinson *et al*, 1988a, b, 1989).

The high concentrations of peptides derived from proenkephalin A in bovine adrenal chromaffin cells (Lewis *et al*, 1980) have attracted considerable attention, and this has become an important system for studying the control of proenkephalin A gene expression. The gene consists of three exons, of which one (exon 3) encodes virtually all of the precursor containing the enkephalin sequences (Noda *et al*, 1982b). In bovine adrenal chromaffin cells, proenkephalin A mRNA concentrations increase in response to depolarization by K^+, and nicotinic stimulation; this response depends on raised intracellular Ca^{++} and is blocked by Ca^{++} channel blockers like dihydropyridine (Pruss and Stauderman, 1988; Kley *et al*, 1987).

The dynorphin precursor contains three copies of the sequence of Leu-enk in each case extended at the C-terminus by Arg-Arg (Kakidani *et al*, 1982). C-terminally extended forms of Leu-enk, namely dynorphin (1–8) and (1–17), and β neo-endorphin, are however the major products. Dynorphin was originally isolated from hog pituitary, but the complete sequence of dynorphin 1–17 was first published for material isolated from hog duodenum (Tachibana *et al*, 1982).

RELEASE AND METABOLISM

Prolonged electrical field stimulation of guinea pig ileum longitudinal muscle-myenteric plexus preparation produces gradual fade in the magnitude of the twitch response. This fade is reversed by naloxone, indicating that field stimulation releases both acetylcholine and opioid peptides, and that the latter in turn act to depress acetylcholine release (Puig *et al*, 1977). The effect is seen clearest with higher frequencies of stimulation. Direct estimation of enkephalin by RIA of the perfusates of the guinea pig ileum longitudinal muscle-myenteric plexus preparation supports the observations of stimulus-induced release (Schultz *et al*, 1977). Since naloxone enhances the peristaltic reflex response of the guinea pig ileum it would appear that endogenous opioids might normally act to regulate acetylcholine (or substance P) release (Bartho *et al*, 1989; Donnerer, Holzer and Lembeck, 1984). It should, however, be noted that while distension released dynorphin it was reported to inhibit Met-enk release from guinea pig ileum so that the precise identity of the opioid peptides that modulate peristaltic reflexes remains to be determined (Donnerer, Holzer and Lembeck, 1984; Clark and Smith, 1981).

The enkephalins are degraded rapidly in blood with a half life of a minute or less. Most released enkephalin is unlikely therefore to survive in the circulation for very long. The first step in degradation in plasma is removal of the N-terminal Tyr residue which causes complete loss of bioactivity (Hambrook *et al*, 1976). Neutral endopeptidase 24.11 (enkephalinase) inactivates enkephalins by cleavage of the Gly-Phe peptide bond. The same bond is also cleaved by angiotensin converting enzyme. Inhibitors of endopeptidase 24.11 e.g. thiorphan and acetorphan appear to enhance the activity of endogenous opioids. Interpretation of the data with these compounds is complicated by the fact that they can act both centrally and peripherally. Moreover, since they may decrease the degradation of many different peptides it is useful to use naloxone to establish a primary action agonist opioids.

Many enkephalin analogues with improved stability have been developed; in particular substitution of a D-residue (e.g. D-Ala) at position 2 decreases cleavage of the N-terminal Tyr. A C-terminal blocking group, e.g. amide, stabilizes the C-terminus to carboxypeptidases. Many of these analogues are useful for pharmacological experiments, but their relative affinities for different receptor types may differ from the parent compound.

RECEPTORS AND PHARMACOLOGY

Pharmacological studies in the chronic spinal dog and on the mouse vas deferens first suggested the occurrence of multiple opioid peptides. Radioligand binding work has since allowed these receptors to be defined in detail (Kosterlitz and Patterson, 1985). Three different types are now usually distinguished (μ, δ and κ). All endogenous opioid peptides act to some extent at the three sites, but they also show selectivity. It seems likely that β endorphin is selective for μ receptors, Met- and Leu-enk are the natural ligands at δ receptors, and to a lesser extent μ sites, and dynorphin is the main natural ligand at κ sites (Corbett *et al*, 1982; Chavkin and Goldstein, 1981). The analogue Tyr-D-Ala-Gly-MePhe-Gly-ol (DAGO) is often used as a selective agonist at μ

receptors; D-Ala², D-Leu⁵ enkephalin (DADLE) and D-Pen², D-Pen⁵ enkephalin (DPDPE) are often used as selective peptide agonists at δ receptors. Naloxone is a relatively selective antagonist for μ sites, but at least *in vivo* it is often used in doses that also block δ and κ sites.

The guinea pig ileum myenteric plexus is rich in μ, but also contains κ receptors; while the submucous plexus is rich in δ-receptors (Chavkin and Goldstein, 1981; Kachur *et al*, 1980; Cherubini and North, 1985). Both μ and δ receptors are coupled to a potassium conductance in submucous plexus neurons, the coupling involves a G-protein and is independent of cAMP and protein kinase C (North *et al*, 1987). In contrast κ receptor stimulation appears to involve inhibition of Ca^{++} channel opening (Cherubini and North, 1985).

MAIN AUTONOMIC FUNCTIONS

The therapeutic actions of opiates on the gut have been known for centuries, and with the characterization of endogenous opioid systems the functional significance of these effects has become clearer (Burks *et al*, 1988; Coupar, 1987). The endogenous opioid peptides probably act at several sites in the gut. The well known bioassay of opioids on field-stimulated guinea pig ileum depends on a presynaptic action to depress acetylcholine release; as already mentioned there may be a corresponding action of endogenous opioids acting in the peristaltic reflex to depress excitatory transmitter release. At least in some species there is a dense enkephalinergic innervation of the pyloric sphincter, and there is evidence that opioids mediate the sphincter contraction in response to duodenal acidification (Reynolds, Ouyang, and Cohen, 1985). In addition to controlling motility, there are direct opioid effects on submucous neurons that probably control epithelial function and this could account for the antisecretory activity of the opioids (Keast, 1987). Finally, opiates depress ganglionic transmission by presynaptic effects. There is opioid peptide immunoreactivity in the preganglionic innervation of the prevertebral ganglia, and in the projection from the gut to these ganglia (Dalsgaard *et al*, 1982; Vincent *et al*, 1984; Lindh, Hökfelt and Elfvin, 1988). It seems possible that the endogenous opioids might have a presynaptic inhibitory function modulating the release of acetylcholine at ganglion cells (Konishi, Tsunoo and Otsuka, 1979, 1981; Shu, Love and Szurszewski, 1987).

BOMBESIN-GASTRIN RELEASING PEPTIDE

The bombesin-GRP group of peptides were discovered first in amphibian skin and later in gut neurons and endocrine cells in higher vertebrates (see Walsh, 1989; Dockray, 1987a). In the mammals these peptides occur in peripheral neurons and endocrine cells in the lung. The terms 'gastrin releasing peptide' (GRP) and 'mammalian bombesin' are both used to describe the mammalian peptides. There are several amphibian representatives of the family and at least two mammalian representatives. In bombesin, the C-terminal tripeptide is His-Leu-Met-NH₂, but in some other amphibian peptides, notably ranatensin and litorin it is -His-Phe-Met-NH₂ or

TABLE 8.5
Amino Acid Sequences of Bombesin-related Peptides and Some Antagonists*.

GRP	Met – Tyr – Pro – Arg – Gly – Asn – His – Trp – Ala – Val – Gly – His – Leu – Met – NH_2
Neuromedin B	Gly – Asn – Leu – Trp – Ala – Thr – Gly – His – Phe – Met – NH_2
Bombesin	Glp – Gln – Arg – Leu – Gly – Asn – Gln – Trp – Ala – Val – Gly – His – Leu – Met – NH_2
Litorin	Glp – Gln – Trp – Ala – Val – Gly – His – Phe – Met – NH_2
Camble et al antagonist	$(CH_3)_2$ CHCH – His – Trp – Ala – Val – DAla – His – Leu – $NHCH_3$
Heimbrook et al antagonist	CH_3 CO – His – Trp – Ala – Val – Gly – His – Leu – OCH_3 CH_3

* The sequences of the two mammalian peptides, representative amphibian peptides and of two recently reported antagonists (Camble et al. 1989; Heimbrook et al, 1989) are shown. Note that only the C-terminal 14 residues of GRP are shown.

(phyllolitorins) Ser-Leu/Phe-Met-NH_2 (Table 5). The mammalian GRPs share a common C-terminal heptapeptide with bombesin, and the second mammalian peptide (neuromedin B) resembles ranatensin (Minamino et al, 1983).

SYNTHESIS

Porcine GRP was first characterized as a 27-residue peptide (GRP 27), but C-terminal fragments of 10 and 23 residues have also been isolated from dog intestine (McDonald et al, 1979; Reeve et al, 1983). A cDNA sequence for human GRP which was elucidated by Spindel et al. (1984) indicated a precursor of 148 aminoacid residues containing a single copy of GRP27 located immediately after the signal sequence. In addition Spindel et al. found a second slightly shorter mRNA species, that was shown to have a deletion of 19 bases yielding a frameshift and a new stop codon so that this mRNA encoded a precursor that was 27 residues longer (Spindel et al, 1986). A third mRNA species has since been identified in small cell carcinoma of the lung cell lines (Sausville et al, 1986). It appears that there is a single copy of the human GRP gene that has three exons: one encodes the N-terminal region of preproGRP including a substantial part of the GRP27 region; a second encodes most of the remaining region of the C-terminal portion of the precursor and the third encodes the extreme C-terminal part of the precursor. There are alternative 5' donor and 3' acceptor sites for the second intron and differential splicing accounts for the three different mRNA species mentioned above (Spindel, Zilberberg and Chin, 1987). The predicted peptides from the three mRNA species have been identified in extracts of small cell carcinoma of the lung, and in human foetal lung extracts (Cuttita et al, 1988; Reeve et al, 1988). It may be that there are species differences in mRNA splicing, since in rat there is a single mRNA species. The expression of the GRP gene is probably under the control of a GC-rich promotor region, and an upstream cAMP regulatory site. Lebacq-Verheyden et al (1988) have suggested that different promotors are active in brain, and in both brain and duodenum.

The cDNA's encoding neuromedin B and ranatensin have been cloned from human hypothalamus and Rana pipiens skin. The predicted precursors are 76 and 82 residues respectively. As in preproGRP, the sequence of neuromedin B immediately follows the signal peptide; the ranatensin precursor is similarly organized. The neuromedin-B mRNA abundance in rat stomach and colon is relatively high but little is known of the expression and functions of this member of the bombesin family (Krane *et al*, 1988).

RELEASE AND METABOLISM

Depending on the species, stimulation of vagal or splanchnic nerves releases GRP-immunoreactivity. Thus electrical stimulation of the vagus produces a prompt release of GRP into the venous outflow of the antrum and fundus of the stomach in anaesthetized pigs, but splanchnic stimulation had no effect (Knuhtsen *et al*, 1984). In the conscious calf, concentrations of immunoreactive bombesin in arterial plasma increased by about $100 \, \mathrm{pmol.l^{-1}}$ with the stimulation of splanchnic, but not vagal nerves; the release was blocked by hexamethonium, but not by atropine or adrenergic antagonists (Bloom and Edwards, 1984). There is little evidence that plasma bombesin concentrations are physiologically regulated. In canine colon muscularis mucosa, immunoreactive bombesin was released by electrical field stimulation by a calcium dependent tetrodotoxin-sensitive mechanism (Angel, Go and Szurszewski, 1984). The clearance rate of GRP varies with the form: GRP-10 is removed more rapidly than the 23 or 27-residue peptides (Bunnett *et al*, 1985).

RECEPTORS AND PHARMACOLOGY

The minimal fragment of bombesin with significant biological activity is the C-terminal heptapeptide. There is now some evidence for the existence of both GRP- and neuromedin B-preferring receptors. In pancreas there is a receptor with ten times higher affinity for bombesin than neuromedin B, but in oesophagus smooth muscle, there is a site with three fold higher affinity for neuromedin B compared with GRP (von Schrenck *et al*, 1989). Pancreatic acinar cells respond to bombesin with Ca^{++} efflux and cGMP accumulation (Deschodt-Lanckman *et al*, 1976). The threshold concentration for increases in intracellular Ca^{++} (0.1 nM) match those for increases in inositol 1,4,5 trisphosphate (IP_3) which suggests that IP_3 mobilizes Ca^{++}. However, amylase release is half maximal at 20 pM, and appears not to be dependent on a rise in Ca_i^{++}. The precise intracellular signals that mediate the effect of low concentrations of bombesin in the pancreas therefore remain uncertain (Bruzzone, 1989). In addition to acute secretory effects, bombesin-related peptides are also mitogens in a number of cell lines. The available evidence indicates that these effects involve G-proteins and mediation by the inositol polyphosphate signalling pathway (Weber *et al*, 1985; Heslop *et al*, 1986; Letterio, Coughlin and Williams, 1986; Lloyd *et al*, 1989). These effects may be of importance because in at least some human small cell lung cancer lines, both bombesin and its receptors are expressed, so that bombesin might be an autocrine growth factor (Cuttita *et al*, 1985).

In recent years excellent progress has been made in the identification of bombesin analogues with antagonist properties (Table 5). Coy *et al* (1988) stabilized the final peptide bond in bombesin to give the analogue Leu^{13}-ψ CH$_2$NH Leu^{14} bombesin, which was an antagonist with an IC_{50} of 35 nM on pancreatic acinar cells, and 18 nM on mitogenic activity in Swiss 3T3 cells. Other analogues in which the immediate C-terminal sequence is modified have also been shown to be useful antagonists (Table 5). Camble et al (1989) found that N-isobutyryl [D-Ala^{24}] GRP 20–26 NHMe had an IC_{50} of 2 nM on 3T3-cell receptor binding and in vivo antagonized bombesin-evoked pancreatic secretion, and Heimbrook et al (1989) showed N-acetyl GRP 20–26 $OCH_2$26-0CH_2-CH_3 to inhibit bombesin-evoked 3T3-cell mitogenesis and receptor binding (IC_{50}, 4 nM) and to reduce the increase in Ca_i^{++} in H345 small cell lung cancer lines; it also inhibited bombesin stimulated gastrin release in rat in vivo. The availability of antagonists such as these that are active in vivo will plainly facilitate future physiological studies.

MAIN AUTONOMIC FUNCTIONS

The best characterized autonomic action of GRP is in mediating neuronal release of the gastric pyloric hormone, gastrin (Walsh, 1989). Electrical stimulation of the vagus, field stimulation of the stomach, or protein in the stomach lumen release gastrin by a nervous mechanism that is not blocked by atropine (Dockray and Gregory, 1989). The evidence that GRP mediates these effects is based on the observation that GRP nerves are present in the antral mucosa, that GRP is a good stimulant of gastrin release and that GRP antibodies inhibit the gastrin response to luminal protein and field stimulation of rat antrum (Schubert, Bitar and Makhlouf, 1982; Schubert et al, 1985). In addition to peripheral actions of bombesin/GRP peptides it should also be noted that these substances have potent effects on autonomic outflow following central administration. It is of interest that while bombesin acts peripherally to stimulate acid secretion (via gastrin) it acts centrally (via the vagus) to inhibit gastric acid secretion (Tache and Gunion, 1985).

CHOLECYSTOKININ

Cholecystokinin (CCK) is well represented in endocrine cells of the intestinal mucosa and in central neurons, but in addition some autonomic neurons contain CCK. The hormonal functions of CCK in the control of the gall bladder and pancreas are well recognized, but its function as a peripheral transmitter is less well understood. There is, however, an increasing body of evidence to indicate that CCK can modulate autonomic functions by reflex actions on visceral afferents.

SYNTHESIS

There are a number of different molecular forms of CCK that vary in chain length by extension at the N-terminus. The major form found in central and peripheral neurons is the sulphated octapeptide, CCK8. The larger forms of 22, 33, 39 and 58 residues all terminate in CCK 8 and are the predominant forms found in gut endocrine cells (Dockray et al, 1989; Tatemoto et al, 1984; Eysselein et al, 1984; Eng *et al*, 1984). The nucleotide base sequence encoding preproCCK indicates precursors of 114 (pig) or 115 (rat, mouse, man) residues (Deschenes *et al*, 1984; Takahashi *et al*, 1985; Gubler *et al*, 1984; Kuwano *et al*, 1984). These contain a single copy of CCK8, and are conserved not only in the CCK8 region but also in sequences to its C- and N-terminals. The available data indicate that the different forms of CCK are produced by alternative patterns of post-translational processing and that the same gene is expressed in both neurons and endocrine cells.

Several of the co-products of CCK8 production have been identified (Eng *et al*, 1983). One that is of interest corresponds to the extreme C-terminus of pro CCK. This contains two tyrosine residues which, like the tyrosine in CCK8, are found in the sulphated state (Eng *et al*, 1986; Varro *et al*, 1986). The consensus sequence for sulphotransferase activity consists of acidic residues N-terminal to the tyrosine, and all the sulphated tyrosines in the CCK precursor conform to this pattern. The sulphation of CCK8 is of particular importance in determining potency at different receptor types. On present evidence all naturally occurring CCK8 is in the sulphated form.

RELEASE AND METABOLISM

There have been many studies of the mechanisms controlling release of CCK from gut endocrine cells and central neurons, but there has been little work on the control of CCK release from peripheral neurons. In the isolated vascularly perfused guinea pig ileum CCK is mainly found in enteric neurons, and distension of the ileum has been shown to evoke CCK release (Donnerer *et al*, 1985).

The degradation of CCK by CNS membranes has been studied by many groups. Endopeptidase 24.11 cleaves the Asp-Phe NH_2, and Gly-Trp bonds in CCK8 (Zuzel *et al*, 1985; Najdovski *et al*, 1985). In addition there are other peptidases that degrade CCK, these include a thiol protease, an amidopeptidase, a serine protease and a third endopeptidase (Rose, Camus and Schwartz, 1988; Matsas, Turner and Kenny, 1984; Steardo *et al*, 1985). For the most part these enzymes have been identified in the CNS and whether or not they function in the periphery remains uncertain.

RECEPTORS AND PHARMACOLOGY

Cholecystokinin and gastrin, share a common C-terminal pentapeptide amide structure, which is known to include the minimal fragment of the two peptides with biological activity. There are three major groups of receptors that interact with these peptides (Jensen *et al*, 1989). Affinity for different receptors is determined by the structure immediately to the N-terminal side of the common tetrapeptide amide sequence. The CCK A-receptor, which is found on pacreatic acinar cells, gall bladder and intestinal smooth muscle, enteric and some central neurons, shows high affinity for CCK but not gastrin; the crucial feature determining high affinity at this receptor is a sulphated tyrosine residue at position 7 from the C-terminus, either desulphation or shifting the sulphated tyrosine to position 6 or 8 from the C-terminus dramatically lowers affinity (Table 8.6). Recent studies suggest that there are high and low affinity CCK A receptors in pancreatic acinar cells, and that some CCK analogues (e.g.

TABLE 8.6
Classification of CCK receptors.

Receptor	Distribution	Relative Affinity for Natural Ligands	Antagonist
CCK-A	Pancreas, gall bladder enteric and vagal neutrons some CNS neurons	CCK >>> gastrin	L364, 718 CR 1409
CCK-B	CNS neurons	CCK > gastrin	L365, 260
G	Parietal cells, some gastric smooth muscle	CCK = gastrin	L365, 260

JM-180) distinguish the two (Matozaki, Martinez and Williams, 1989). It seems that the high affinity receptor is linked to amylase release, but not IP_3 mobilization. The low affinity receptor is associated with mobilization of IP_3 and inhibition of amylase release. The CCK B-receptor is located on CNS neurons and shows only about 10-fold higher affinity for CCK compared with gastrin. The third receptor type is known as the G-receptor and resembles (and may even be identical to) the CCK B-receptor. On parietal cells it mediates the action of gastrin in stimulating acid secretion and shows little or no discrimination between gastrin and CCK.

Excellent progress has been made in recent years in the development of specific antagonists for the gastrin-CCK group of peptides. The earliest antagonists included dibutyryl cyclic guanosine monophosphate and the glutaramic acid derivative, proglumide (Jensen et al, 1989). These have very low affinity and do not distinguish between receptor types, but in recent years several high affinity selective antagonists have been described e.g. CR 1409 (Makovec et al, 1987) and the benzodiazepine derivatives L-364,718 and L-365,260 (Evans et al, 1986; Freidinger, 1989; Freidinger et al, 1989). Thus, L-364,718 is highly selective for CCK A-receptors and is orally active; L-365,260 does not discriminate between G and CCK B-receptors but shows significantly lower affinity for CCK A-receptors (Fig. 8.7).

MAIN AUTONOMIC FUNCTIONS

Peptides of the CCK group have many gastrointestinal actions but insofar as these are of physiological importance they probably reflect effects exerted by the circulating hormone. In the guinea pig ileum it is now clear that CCK occurs in a population of neurons, and that in vitro CCK acts on A-type receptors on myenteric plexus neurons to release acetylcholine and substance P which in turn contract longitudinal smooth muscle. However, there is no convincing evidence that these effects occur during peristalsis and their physiological significance is uncertain (Bartho et al, 1989).

In recent years a substantial body of evidence has been produced to indicate that visceral afferents are likely to be a physiological target for circulating CCK in modulating autonomic outflow. In the rat, circulating CCK has been shown to

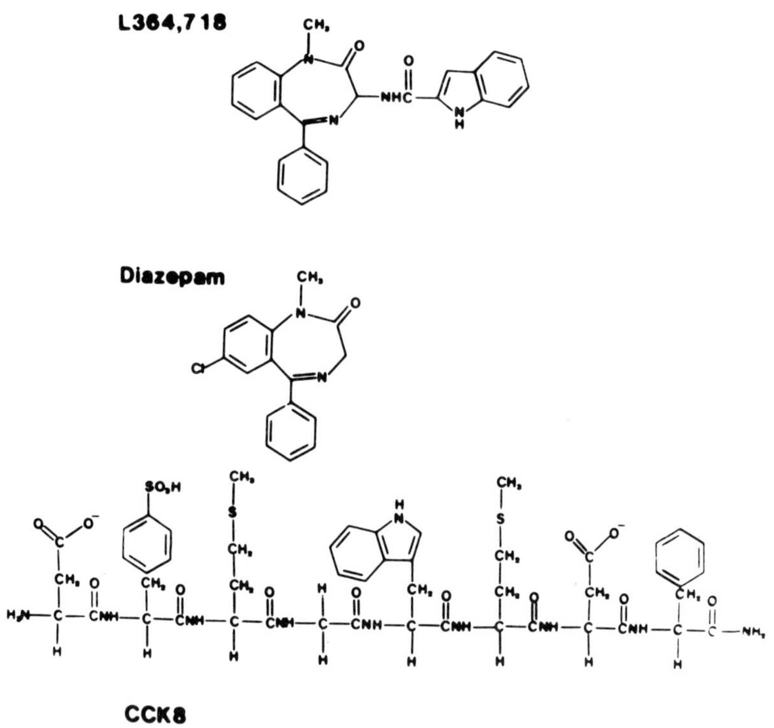

L364,718

Diazepam

CCK8

FIGURE 8.7   The structure of the benzodiazepine CCK-A antagonist, L364,718, showing its relationship to diazepam (which is not a CCK antagonist) and to the primary sequence of CCK8. A related compound, L365,260 is a selective antagonist for CCK-B and gastrin receptors. As a group these antagonists are highly potent, specific and orally active.

stimulate vagal afferents that otherwise function as gastric mechanoreceptors (Raybould, Gayton and Dockray, 1988; Dockray, 1988), presumably by acting at the putative receptor sites on vagal nerve fibres identified by autoradiography (Zarbin *et al*, 1981; Moran *et al*, 1987). Stimulation of this afferent pathway is associated with inhibition of food intake and reflex relaxation of the body of the stomach (Weller, Smith and Gibbs, 1990; Dockray, 1989b; Green *et al*, 1988; Forster *et al*, 1990). These effects of CCK are blocked by nerve section, capsaicin (which lesions small diameter afferents) and L364,718. The efferent side of vago-vagal reflexes triggered by CCK may be mediated in part by stimulation of VIP release from post-ganglionic gastric nerves, and this appears to be important for the inhibition of gastric emptying caused by CCK (Forster, 1990).

SOMATOSTATIN

Somatostatin was discovered and first characterised as a hypothalamic factor that inhibited the release of growth hormone (Brazeau *et al*, 1973). Subsequently, somatostatin was found to occur widely in endocrine cells in the gut and pancreas where it is generally thought to act in a paracrine manner to inhibit adjacent endo or exocrine cells (Yamada and Chiba, 1989). In addition, however, some autonomic neurons express the somatostatin gene – these include enteric neurons and a sub-set of sympathetic post-ganglionic neurons.

SYNTHESIS

The two main forms of somatostatin are peptides of 14 and 28 residues (Pradayrol *et al*, 1978; Schally *et al*, 1980). These have similar biological properties, but the larger form is cleared more slowly than the smaller one (Seal *et al*, 1982). The 28-residue peptide is an N-terminally extended form of somatostatin-14 which is derived from the same precursor by alternative patterns of post-translational processing. The primary amino acid sequence of somatostatin-14 is extremely well conserved across a range of species, including birds, fish and mammals. In some lower vertebrates, however, there are other somatostatin-like peptides (Dockray, 1989a). The cDNA sequences of human and rat preprosomatostatin indicate a peptide precursor of 116 residues with a single copy of somatostatin (Goodman *et al*, 1982; Shen, Pictet and Rutter, 1982). In the genomic sequence there is a single intron (Tavianini *et al*, 1984; Shen and Rutter, 1984). The gene sequence includes a cAMP-consensus sequence in the 5' region at -48 to -41. Recently Zhu *et al* (1989) have purified from rat brain a DNA-binding protein specific for this region and have characterized it as a 43 kDa molecule with a cAMP-dependent phosphorylation site; it is proposed that this protein acts as the transcription factor regulating expression of the somatostatin gene.

RELEASE AND METABOLISM

There have been many studies on the release of somatostatin from gut and pancreatic endocrine cells, but the release from autonomic neurons is less well understood and is complicated by the fact that it is difficult to be sure that released material (at least from the gut) is truly autonomic in origin. Immunoreactive somatostatin is released from guinea pig ileum by distension, but since tetrodotoxin only partly reversed this release it is possible that some of the secreted material was endocrine cell in origin (Donnerer, Holzer and Lembeck, 1984). Immunoreactive somatostatin in the myenteric ganglia of guinea pig ileum can be released by nicotinic stimulation and appears to be inhibited by VIP (Grider, 1989c).

RECEPTORS AND PHARMACOLOGY

Somatostatin inhibits many different cell types, and there is ample evidence to suggest that it inhibits cAMP formation (Yamada and Chiba, 1989). However, this is unlikely to be the whole story. Thus somatostatin hyperpolarizes submucous neurons in guinea pig ileum by increasing an inwardly rectifying potassium conductance by a mechanism that is not thought to be mediated by cAMP, but probably involves a

G-protein (Mihara, North and Suprenant, 1987). Moreover there is some evidence for more than one type of somatostatin receptor. Somatostatin, but not SMS 201-995 (see below), inhibited the proliferation of the Mia PaCa-2 pancreatic cancer cell line; since in other systems SMS 201-995 mimics the action of somatostatin it would appear that the receptor on Mia PaCa-2 cells is not the same as that on cells responding to SMS 201-995. The antiproliferative effect seen in Mia PaCa-2 cells may be mediated by inhibition of tyrosine kinase activity (Liebow *et al*, 1989). The widespread inhibitory actions of somatostatin have encouraged the idea that it might be a therapeutically useful starting point for the development of compounds suitable for use in supressing inappropriate hormone or neurotransmitter release (Moreau and DeFeuds, 1987). One such compound, SMS 201-995, is a long acting agonist (Bauer *et al*, 1982). It is used to treat patients with endocrine tumours, and appears to act in the short term to depress hormone release and over slightly longer periods to cause tumour regression (Gorden, 1989; Maton, Gardner and Jensen, 1989).

MAIN AUTONOMIC FUNCTIONS

There are several potential autonomic transmitter functions for somatostatin. In the toad *Bufo marimus*, somatostatin released from vagal cholinergic neurons accounts for the atropine-resistant inhibition of heart rate (Campbell *et al*, 1982). Somatostatin occurs in a sub-population of noradrenergic sympathetic post-ganglionic neurons that appear to project to the submucous plexus in the gut, and to the pylorus (Hökfelt *et al*, 1977; Costa and Furness, 1984; Lindh *et al*, 1986; Macrae, Furness and Costa, 1986; Furness and Costa, 1987). The functional interactions between somatostatin and noradrenaline in this case are largely unknown. However, submucous neurons in the guinea pig ileum, including many VIP-neurons, show a slow inhibitory junctional potential that may be mediated by somatostatin released from neurons originating in the myenteric plexus (Bornstein, Costa and Furness, 1988; Suprenant, 1989).

CALCITONIN GENE-RELATED PEPTIDE

The sequence of CGRP was first predicted from a transcript generated by alternative mRNA splicing of the calcitonin gene (Amara *et al*, 1982). Immunochemical methods using antibodies to the predicted sequence revealed the material in primary afferent neurons, CNS and intrinsic gut neurons. CGRP is an extremely potent vasodilator and together with substance P probably mediates the vasodilator response to antidromic stimulation of afferent neurons.

SYNTHESIS

Rosenfeld *et al* (1983) postulated that the alternative mRNA transcript of the calcitonin gene would give rise to a peptide of 37 residues with a C-terminal amide (Fig. 8.8). The native peptide was then identified using antibodies to a synthetic

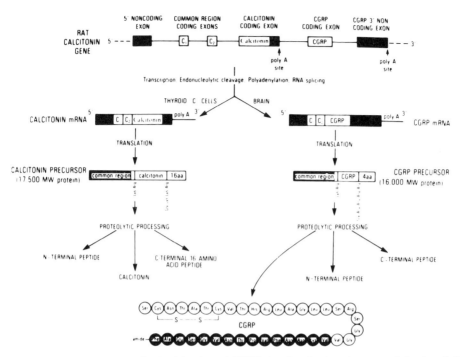

FIGURE 8.8   The biosynthesis of calcitonin and CGRP showing the shared exons and the site of alternative splicing that gives rise to the two mRNA species encoding calcitonin and CGRP. (Reproduced from Rosenfeld et al, 1983).

tetradecapeptide corresponding to the predicted C-terminus. The sequence of human α CGRP was confirmed by isolation of the peptide from a medullary thyroid carcinoma, and characterization using fast-atom bombardment mass spectrometry (Morris *et al*, 1984).

There are six exons in the rat calcitonin/CGRP gene. The first three exons are transcribed in both calcitonin and CGRP mRNAs. A separate (fourth) exon encodes calcitonin, and two (fifth and sixth) are expressed in the CGRP mRNA. Control of the alternative splicing pathways appears to be exerted by a *cis*-acting sequence of the calcitonin-specific 3′ splice junction; in CGRP-producing cells, a sequence in the calcitonin-specific acceptor region is postulated to suppress the production of calcitonin mRNA transcripts (Emeson *et al*, 1989). It is now clear, however, that there is also a second gene which encodes a CGRP-like peptide (Amara *et al*, 1985; Steenbergh *et al*, 1985; Edbrooke *et al*, 1985). The nucleotide sequence in this case predicts a peptide precursor of 134 residues that could yield an homologous 37-residue peptide, but does not contain a calcitonin-encoding region. The forms of CGRP derived from the two genes are known as α and β. In the rat the two forms differ in the substitution of Lys for Glu at the third position from the C-terminus; in man both forms have Glu at this position. Antibodies to rat α CGRP may cross-react poorly with human and rat β CGRP. The calcitonin/αCGRP gene is expressed in sensory neurons, whereas in rat intestine intrinsic neurons β CGRP mRNA predominates (Mulderry et al, 1988). A related peptide, amylin (or islet amyloid protein), is produced in pancreatic β cells and occurs in high concentrations in the amyloid deposits of diabetic patients, but its functional significance is still uncertain (Westermark et al, 1986; Nishi et al, 1990; Leffert et al, 1989).

Many afferent neurons that produce CGRP co-store the material with substance P (Gibson et al, 1984; Lee et al, 1985; Gulbenkian et al, 1986; Ju et al, 1987). The two peptides are packaged in the same granules, and transported together intra-axonally to terminal regions in gastrointestinal tract, biliary tract, skin and cardiovascular systems (Gulbenkian et al, 1986; Gibbins et al, 1985). In rat dorsal root ganglia, virtually all substance P neurons appear to contain CGRP; but further populations of CGRP cells contain other peptides e.g. somatostatin and galanin (Ju et al, 1987). In guinea pig dorsal root ganglia, over 95% of all CGRP immunoreactive neurons contained substance P (Gibbins et al, 1985). In other neurons e.g. intrinsic neurons of the gut, CGRP is found together with combinations of CCK, choline acetyltransferase, somatostatin and NPY, and of enkephalin, VIP, GRP and dynorphin (Furness et al, 1989), and in the ventral horn motoneurons CGRP is produced together with acetylcholine.

RELEASE AND METABOLISM

The plasma concentrations of CGRP in normal subjects are less than 10 pmol.1^{-1}, but patients with medullary thyroid carcinoma often have elevated concentrations (mean 114 pmol.1^{-1}) in the peripheral circulation (Mason et al, 1986). In the anaesthetized rat, basal concentrations of CGRP are 12-20 pmol.1^{-1} and these are stimulated 15-fold by capsaicin indicating that circulating material may be derived from primary afferents (Emson and Zaidi, 1989). In cultured trigeminal neurons, depolarisation by high K$^+$ evoked release of CGRP by a calcium-dependent mechanism (Mason et al, 1984). It has been reported that in rat enteric neurons in vitro CGRP is not released by high K$^+$, although field stimulation produced a Ca^{++} evoked release (Belai and Burnstock, 1988). The precise cellular mechanisms that could account for these observations are still unclear, and further work is needed on this issue.

RECEPTORS AND PHARMACOLOGY

The receptor mechanisms that mediate the action of CGRP are relatively complex and different mechanisms are involved in different cell types. Thus, in coeliac ganglion neurons, CGRP evoked fast, slow and biphasic depolarizations; the mechanism of the fast response is still unknown, but it is of interest since fast responses are relatively unusual for peptide agonists (Dun and Mo, 1988). In the guinea pig pancreas and in gastric smooth muscle, CGRP increases intracellular cAMP (Seifert et al, 1985; Zhou et al, 1986; Maton et al, 1988), and its effects are typical of those of agonists acting via cAMP in these cells. A different type of mechanism may be involved in the cardiovascular effects of CGRP. Thus Nelson *et al* (1990) have recently reported that the vasodilator effects of CGRP are mediated by opening of a K^+ channel of the ATP regulated type.

Calcitonin shows a low affinity in displacing radiolabelled CGRP from CNS binding sites, and conversely CGRP has a low affinity for displacement of calcitonin from its binding sites (Goltzman and Mitchell, 1985). Lynch and Kaiser (1988) have argued that since both calcitonin and CGRP have an amphiphilic alpha helix starting at residue 8 this might account for their common properties. Foord and Craig (1987) have characterized the CGRP receptor from human placenta; immobilized β-CGRP was used to characterize receptor subunits of 62,000–68,000 molecular weight. Sucrose gradient centrifugation studies of the native receptor suggested a molecular weight of 240,000.

There are reports that rat and human α or β CGRP's differ in biological activity (Beglinger *et al*, 1988; Foord and Craig, 1987). In some cases the differences are minor, but in others they are important, for instance β human CGRP, but not α human CGRP inhibited acid secretion in man (Beglinger *et al*, 1988). The receptor basis for these differences remains unknown.

MAIN AUTONOMIC FUNCTIONS

It is clear from the work reviewed above that CGRP influences a wide variety of cells and tissues, including pancreatic acinar cells, gastric mucosa, gut smooth muscle and blood vessels. The vasodilator actions of CGRP are particularly striking, since as little as 15 pmol injected intradermally in man produces a long lasting increase in skin blood flow (Brain *et al*, 1985). CGRP may therefore act together with substance P to mediate the local increases in blood flow seen on antidromic afferent stimulation, and in neurogenic inflammatory responses. On its own CGRP does not increase capillary permeability, but its vasodilator action probably accounts for its ability to potentiate the increase in capillary permeability caused by substance P (Gamse and Saria *et al*, 1985). Immunoneutralization reports provide clear-cut evidence that CGRP contributes to the extravasation seen in response to inflammatory challenges (capsaicin, mustard oil) in the rat (Louis *et al*, 1989a,b). There may be a further interaction between CGRP and substance P, in that substance P stimulation of mast cells releases proteases that readily degrade CGRP and so terminate its effects (Brain and ·Williams, 1988).

The CGRP that occurs in motoneurons has been postulated to have several subtle influences of neurotransmission at the neuromuscular junction. Thus there is evidence that CGRP acts via cAMP to increase synthesis of the β subunit of the acetylcholine receptor channel (New and Mudge, 1986; Fontaine, Klarsfeld, and Changeux, 1987); it also regulates desensitization of the receptor-channel probably by cAMP-modified phosphorylation (Mulle *et al*, 1988; Miles, Greengard and Huganir, 1989). Whether or not these effects are physiological remains to be seen.

GALANIN

Tatemoto *et al* (1983) discovered and named galanin as a peptide from hog intestine with an N-terminal Gly residue and C-terminal alanine amide. It has been found in a wide variety of enteric, autonomic and sensory neurones. Although it has many inhibitor actions its physiological functions remain poorly understood.

SYNTHESIS

Galanin is a 29 residue peptide that has a slight resemblance to the tachykinins, but is probably not related to these or other known regulatory peptides. The cDNA for the galanin precursor has been cloned from pig adrenal medulla and indicates that preprogalanin has 123 residues and contains a single copy of galanin, coming immediately after the signal peptide (Rökaeus and Brownstein, 1986). In rat pituitary tumours, oestrogen stimulated the appearance of an mRNA species that clearly resembles porcine galanin mRNA, but the control of galanin gene expression in neurons is still largely unexplored (Vrontakis *et al*, 1987). The sequence of the precursor suggests that the post-translational processing mechanisms giving rise to galanin closely resemble those encountered in the biosynthesis of many other regulatory peptides, and involve cleavage at pairs of basic residues and the action of peptide α amidating enyzme on a C-terminal Gly-extended intermediate.

RELEASE AND METABOLISM

There have been few studies of galanin release and metabolism. In the anaesthetized dog, Dunning and Taborsky (1989) reported that stimulation of the mixed pancreatic nerve released immunoreactive galanin into the venous outflow.

RECEPTORS AND STRUCTURE-ACTIVITY RELATIONSHIPS

There have been several studies of the inhibitory action of galanin on insulin release (Ahren, Rorsmann and Berggen, 1988). In Rin m5F cells, galanin inhibits the stimulatory effects of glucose, possibly by a mechanism involving a pertussis-toxin sensitive G-protein and activation of ATP-sensitive K channels leading to hyperpolarization (Sharp *et al.*, 1989); this effect can be demonstrated in isolated membrane protein indicating that it is not second-messenger mediated (Dunne *et al*, 1989). It may well be the case, however, that galanin also acts by other mechanisms since it inhibited adenylate cyclase in Rin m5F cells again by a pertussis-sensitive mechanism (Amiranoff *et al* 1988).

Fragments of the N-terminal of galanin, i.e. 1–10 and 1–20, had virtually full biological activity in causing contraction of rat jejunum longitudinal muscle *in vitro* and in inhibiting dog intestinal circular muscle (Ekblad *et al*, 1985; Fox *et al* 1988). There are, however, marked species and tissue variations in the actions of galanin fragments and at the present time the structure-activity relationships, or receptor properties remain incompletely explored.

MAIN AUTONOMIC FUNCTIONS

The presence of galanin in both intrinsic gut neurones and in post-ganglionic sympathetic neurons, together with the available information on its activity in a wide variety of tissues, suggests that it deserves consideration as a peptide mediator of autonomic function (Lindh, Lundberg, and Hökfelt, 1989; Furness *et al*., 1987; Melander *et al*, 1985; Rokaeus, 1987). At present its best studied effects concern the inhibitory control of pancreatic hormone release, and smooth muscle motility, but in both cases many further studies will be needed to define the physiological circumstances in which these effects are important.

PROSPECTS

In addition to the peptides described here, there are a number of other regulatory peptides that are thought to occur in autonomic neurons. These include atrial natriuretic peptide, TRH, vasopressin, and endothelin. For the most part, little is known of the autonomic functions of these peptides although this situation could change quite quickly. In particular it is plain that the methods for molecular characterization of peptidergic mechanisms in the autonomic nervous system are well established. Good progress is also now being made in the development of pharmacologically useful compounds to manipulate peptidergic systems. In the future it should be possible to extend these approaches more widely in seeking to elucidate the diverse roles of peptides as mediators and modulators of autonomic function.

ACKNOWLEDGEMENTS

The help of Christine Carter in the preparation of the manuscript is gratefully acknowledged.

REFERENCES

Adams, P.R., Brown, D.A. and Jones, S.W. (1983). Substance P inhibits the M-current in bullfrog sympathetic neurones. *British Journal of Pharmacology,* **79**, 330–333.

Ahren, B., Rorsman, P., and Berggen, P-O. (1988). Galanin and the endocrine pancreas. *FEBS Letters,* **229**, 233–237.

Amara, S.G., Jonas, V., Rosenfeld, M.G., Ong, E.S., and Evans, R.M. (1982). Alternative RNA processing in calcitonin gene expression generates mRNAs encoding different polypeptide products. *Nature,* **298**, 240–244.

Amara, S.G., Arriza, J.L., Leff, S.E., Swanson, L.W., Evans, R.M. and Rosenfeld, M.G. (1985). Expression in brain of a messenger RNA encoding a novel neuropeptide homologous to calcitonin gene-related peptide. *Science* **229**, 1094–1097.

Ambache, N., Verney, J. and Aboo Zar, M. (1970). Evidence for the release of two atropine-resistant spasmogens from Auerbach's plexus. *Journal of Physiology* **207**, 761–782.

Amiranoff, B., Laburthe, M., Dupont, C. and Rosselin, G. (1978). Characterization of vasoactive intestinal peptide-sensitive adenylate cyclase in rat intestinal epithelial cell membranes. *Biochimica et Biophysica Acta.* **544**, 474–481.

Amiranoff, B., Lorinet, A-M., Lagny-Pourmir, I., and Laburthe, M. (1988). Mechanism of galanin-inhibited insulin release: Occurrence of a pertussis-toxin-sensitive inhibition of adenylate cyclase. *European Journal of Biochemistry.* **177**, 147–152.

Andersson, P-O., Bloom, S.R., Edwards, A.V. and Järhult, J. (1982). Effects of stimulation of the chorda tympani in bursts on submaxillary responses in the cat. Journal of Physiology **322**, 469–483.

Andersson, P-O., Bloom, S.R., Edwards, A.V., Järhult, J., and Mellander, S. (1983). Neural vasodilator control in the rectum of the cat and its possible mediation by vasoactive intestinal polypeptide. *Journal of Physiology* **344**, 49–67.

Andrews, P.C., Hawke, D., Shively, J.E. and Dixon, J.E. (1985). A nonamidated peptide homologous to porcine peptide YY and neuropeptide YY. *Endocrinology* **116**, 2677–2681.

Angel, F., Go, V.L.W., and Szurszewski, J.H. (1984). Innervation of the muscularis mucosae of canine proximal colon. *Journal of Physiology* **357**, 93–108.

Barnes, P.J. (1987). Neuropeptides in human airways: Function and clinical implications. *American Review of Respiratory Disease.* **136**, S77–S83.

Baron, S.A., Jaffe, B.M., and Gintzler, A.R. (1983). ₂lease of substance P from the enteric nervous system: direct quantification and characterization. *Journal of Pharmacology and Experimental Therapeutics* **227**, 365–368.

Bartho, L. and Holzer, P. (1985). Search for a physiological role of substance P in gastrointestinal motility. *Neuroscience* **16**, 1–32.

Bartho, L., Holzer, P., Leander, S., and Lembeck, F. (1989). Evidence for an involvement of substance P, but not cholecystokinin-like peptides, in hexamethonium-resistant intestinal peristalsis. *Neuroscience* **28**, 211–217.

Bataille, D., Gespach, C., Laburthe, M., Amiranoff, B., Tatemoto, K., Vauclin, M., Mutt, V. and Rosselin, G. (1980). Porcine peptide having N-terminal histidine and C-terminal isoleucine amide (PHI): Vasoactive intestinal peptide (VIP) and secretin-like effects in different tissues from the rat. *FEBS Letts* **114**, 240–242.

Bauer, W., Briner, U., Doepfner, Wolfgang, D., Haller, R., Huguenin, R., Marbach, P., Petcher, T.J., and Pless, J. (1982). SMS 201–995: A very potent and selective octapeptide analogue of somatostatin with prolonged action *Life Sciences* **31**, 1133–1140.

Baumgarten, H.G., Holstein, A.F. and Owman, C. (1970). Auerbach's plexus of mammals and man: Electron-microscopic identification of three different types of neuronal processes in myenteric ganglia of the large intestine from Rhesus monkeys, guinea pigs and man. Z. *Zellforsch* **106**, 376–397.

Beglinger, C., Born, W., Hilderbrand, P., Ensinck, J.W., Burkhardt, F., Fischer, J.A. and Gyr, K. (1988). Calcitonin gene-related peptides I and II and calcitonin: Distinct effects on gastric acid secretion in humans. *Gastroenterology* **95**, 958–965.

Belai, A., and Burnstock, G. (1988). Release of calcitonin gene-related peptide from rat enteric nerves is Ca^{2+}-dependent but is not induced by K^+ depolarization. *Regulatory Peptides* **23**, 227–235.

Belai, A., Ralevic, V. and Burnstock, G. (1987). VIP release from enteric nerves is independent of extracellular calcium. *Regulatory Peptides* **19**, 79–89.

Berridge, M.J., and Irvine, R.F. (1989). Inositol phosphates and cell signalling. *Nature 341*, 197–205.

Biancani, P., Walsh, J.H., and Behar, J. (1984). Vasoactive intestinal polypeptide: A neurotransmitter for lower esophageal sphincter relaxation. *Journal of Clinical Investigation* **73**, 963–967.

Biancani, P., Walsh, J., and Behar, J. (1985). Vasoactive intestinal peptide: A neurotransmitter for relaxation of the rabbit internal anal sphincter. *Gastroenterology* **89**, 867–874.

Binder, H.J., Lemp, G.F., and Gardner, J.D. (1980). Receptors for vasoactive intestinal peptide and secretin on small intestinal epithelial cells. *American Journal of Physiology* **238**, G190–G196.

Bitar, K.N., Bradford, P., Putney, J.W. Jr., Makhlouf, G.M. (1986). Cytosolic calcium during contraction of isolated mammalian gastric muscle cells. *Science* **232**, 1143–1145.

Bitar, K.N., and Makhlouf, G.M. (1982). Relaxation of isolated gastric smooth muscle cells by vasoactive intestinal peptide. *Science* **216**, 531–533.

Black, I.B., Adler, J.E., Dreyfus, C.F., Jonakait, G.M., Katz, D.M., LaGamma, E.F., Markey, K.M. (1984). Neurotransmitter plasticity at the molecular level. *Science* **225**. 1266–1270.

Bloom, S.R. and Edwards, A.V. (1984). Characterization of the neuroendocrine responses to stimulation of the splanchnic nerves in bursts in the conscious calf. *Journal of Physiology* **346**, 533–545.

Bodanszky, M., Klausner, Y.S. and Said, S.I. (1973). Biological activities of synthetic peptides corresponding to fragments of and to the entire sequence of the vasoactive intestinal peptides. *Proceedings of the National Academy of Science, U.S.A.* **70**, 382–384.

Bodner, M., Fridkin, M. and Gozes, I. (1985). Coding sequences for vasoactive intestinal peptide and PHM-27 peptide are located on two adjacent exons in the human genome. *Proceedings of the National Academy of Sciences U.S.A.* **82**, 3548–3551.

Bornstein, J.C., Costa, M. and Furness, J.B. (1988). Intrinsic and extrinsic inhibitory synaptic inputs to submucous neurones of the guinea-pig small intestine. *Journal of Physiology* 398, 371–390.

Brain, S.D. and Williams, T.J. (1988). Substance P regulates the vasodilator activity of calcitonin gene-related peptide. *Nature* 335, 73–75.

Brain, S.D., Williams, T.J., Tippins, J.R., Morris, H.R., and MacIntyre, I. (1985). Calcitonin gene-related peptide is a potent vasodilator. *Nature* 313, 54–56.

Brazeau, P., Vale, W., Burgus, R., Ling, N., Butcher, M., Rivier, J., Guillemin, R. (1973). Hypothalamic polypeptide that inhibits the secretion of immunoreactive pituitary growth hormone. *Science* 179, 77–79.

Broyart, J.P., Dupont, C., Laburthe, M. and Rosselin, G. (1981). Characterization of vasoactive intestinal peptide receptors in human colonic epithelial cells. *Journal of Clinical Endocrinology and Metabolism* 52, 715–721.

Bruzzone, R., (1989). Mechanism of action of bombesin on amylase secretion: Evidence for a Ca^{2+}-independent pathway. *European Journal of Biochemistry* 179, 323–331.

Buck, S.H., Burcher, E., Shults, C.W., Lovenberg, W., O'Donohue, T.L. (1984). Novel pharmacology of substance K-binding sites: A third type of tachykinin receptor. *Science* 226, 987–989.

Bunnett, N.W. (1987). Postsecretory metabolism of peptides. *American Review of Respiratory Disease*. 136, S27–S34.

Bunnett, N.W., Clark, B., Debas, H.T., del Milton, R.C., Kovacs, T.O.G., Orloff, M.S., Pappas, T.N., Reeve, J.R. Jr., Rivier, J.E., and Walsh, J.H. (1985). Canine bombesin-like gastrin releasing peptides stimulate gastrin release and acid secretion in the dog. *Journal of Physiology* 365, 121–130.

Bunnett, N.W., Turner, A.J., Hryszko, J., Kobayashi, R., and Walsh, J.H. (1988). Isolation of endopeptidase-24.11 (EC 3.4.24.11, 'enkephalinase') from the pig stomach: Hydrolysis of substance P, gastrin-releasing peptide 10,[Leu^5] enkephalin, and [Met^5] enkephalin. *Gastroenterology* 95, 952–957.

Burcher, E., Buck, S.H., Lovenberg, W., and O'Donohue, T.L. (1986). Characterization and auto-radiographic localization of multiple tachykinin binding sites in gastrointestinal tract and bladder. *Journal of Pharmacology and Experimental Therapeutics*. 236, 819–831.

Burks, T.F., Fox, D.A., Hirning L.D., Shook, J.E. and Porreca, F. (1988). Regulation of gastrointestinal function by multiple opioid receptors. *Life Sciences* 43, 2177–2181.

Burnstock, G., Campbell, G., Bennett, M. and Holman, M.E. (1963). Inhibition of the smooth muscle of the taenia coli. *Nature*, 200, 581–583.

Camble, R., Cotton, R., Dutta, A.S., Garner, A., Hayward, C.F., Moore, V.E., and Scholes, P.B. (1989). N-isobutyryl-His-Trp-Ala-Val-D-Ala-His-Leu-NHMe (ICI 216140) a potent *in vivo* antagonist analogue of bombesin/gastrin releasing peptide (BN/GRP) derived from the C-terminal sequence lacking the final methionine residue. *Life Sciences* 45, 1521–1529.

Campbell, G., Gibbins, I.L., Morris, J.L., Furness, J.B., Costa, M., Oliver, J.R., Beardsley, A.M. and Murphy, R. (1982). Somatostatin is contained in and released from cholinergic nerves in the heart of the toad *Bufo marinus*. *Neuroscience* 7, 2013–2023.

Castagne, V., Corder, R., Gaillar, R. and Morm de, P. (1987). Stress-induced changes of circulating neuropeptide Y in the rat: Comparison with catecholamines. *Regulatory Peptides* 19, 55–63.

Chavkin, C. and Goldstein, A. (1981). Demonstration of a specific dynorphin receptor in guinea-pig ileum myenteric plexus. *Nature* 291, 59–60.

Cherubini, E. and North, R.A. (1985). μ and K opioids inhibit transmitter release by different mechanisms. *Proceedings of the National Academy of Sciences, U.S.A.* 82, 1860–1863.

Christophe, J.P., Conlon, T.P. and Gardner, J.D. (1976). Interaction of porcine vasoactive intestinal peptide with dispersed pancreatic acinar cells from the guinea pig: Binding of radio-iodinated peptide. *Journal of Biological Chemistry* 251, 4629–4634.

Clark, S.J., and Smith, T.W. (1981). Peristalsis abolishes the release of methionine-enkephalin from guinea-pig ileum *in vitro*. *European Journal of Pharmacology* 70, 421–424.

Comb, M., Seeburg, P.H., Adelman, J., Eiden, L. and Herbert, E. (1982). Primary structure of the human Met- and Leu-enkephalin precursor and its mRNA. *Nature* 295, 663–666.

Cook, R.D., and Burnstock, G. (1976). The ultrastructure of Auerbach's plexus in the guinea-pig. *J. Neurocytol.* 5, 171–194.

Corbett, A.D., Paterson, S.J., McKnight, A.T., Magnan, J., and Kosterlitz, H.W. (1982). $Dynorphin_{1-8}$ and $dynorphin_{1-9}$, are ligands for the K-subtype of opiate receptor. *Nature*, 299, 79–81.

Corder, R., Gaillard, G.C., and Böhlen, P. (1988). Isolation and sequence of rat peptide YY and neuropeptide Y. *Regulatory Peptides* 21, 253–261.

Costa, M. and Furness, J.B. (1984). Somatostatin is present in a subpopulation of noradrenergic nerve fibres supplying the intestine. *Neuroscience* 13, 911–919.

Costa, M. and Furness, J.B. (1989). Structure and neurochemical organization of the enteric nervous system. In *Handbook of Physiology. The Gastrointestinal System, volume II,* edited by G.M. Makhlouf. pp 97–109. New York. American Physiological Society.

Costa, M., Furness, J.B., and Gibbins, I.L. (1986). Chemical coding of enteric neurons. *Progress in Brain Research.* **68**, 217–239.

Costa, M., Furness, J.B. and Humphreys, C.M.S. (1986). Apamin distinguishes two types of relaxation mediated by enteric nerves in the guinea-pig gastrointestinal tract. *Naunyn Schmiedeberg's Archives of Pharmacology.* **332**, 79–88.

Coupar, I.M. (1987). Opioid action on the intestine: The importance of the intestinal mucosa. *Life Sciences* **41**, 917–925.

Coy, D.H., Heinz-Erian, P., Jiang, N-Y., Sasaki, Y., Taylor, J., Moreau, J-P., Wolfrey, W.T., Gardner, J.D. and Jensen, R.T. (1988). Probing peptide backbone function in bombesin: A reduced peptide bond analogue with potent and specific receptor antagonist activity. *Journal of Biological Chemistry* **263**, 5056–5060.

Cuttitta, F., Carney, D.N., Mulshine, J., Moody, T.W., Fedorko, J., Fischler, A., Minna, J.D. (1985). Bombesin-like peptides can function as autocrine growth factors in human small-cell lung cancer. *Nature* **316**, 823–826.

Cuttita, F., Fedorko, J., Gu, J., Lebacq-Verheyden, A-M., Linnoila, R.I. and Battey, J.F. (1988). Gastrin-releasing peptide gene-associated peptides are expressed in normal human fetal lung and small cell lung cancer: A novel peptide family found in man. *Journal of Clinical Endocrinology and Metabolism* **67**, 576–583.

Dahlöf, C., Dahlöf, P., Tatemoto, K., and Lundberg, J.M. (1985). Neuropeptide Y (NPY) reduces field stimulation-evoked release of noradrenaline and enhances force of contraction in the rat portal vein. *Naunyn-Schmiedeberg's Archives of Pharmacology* **328**, 327–330.

Dalsgaard, C-J., Hökfelt, T., Elfvin, L-G., and Terenius, L. (1982). Enkephalin-containing sympathetic preganglionic neurons projecting to the inferior mesenteric ganglion: Evidence from combined retrograde tracing and immunohistochemistry. *Neuroscience* 7, 2039–2050.

Deacon, C.F., Agoston, D.V., Nau, R., and Conlon, J.M. (1987). Conversion of neuropeptide K to neurokinin A and vesicular colocalization of neurokinin A and substance P in neurons of the guinea-pig small intestine. *Journal of Neurochemistry* **48**, 141–146.

De Beurme, F.A., and Lefebvre, R.A. (1988). Vasoactive intestinal polypeptide as possible mediator of relaxation in the rat gastric fundus. *Journal of Pharmacy and Pharmacology.* **40**, 711–715.

Deschenes, R.J., Lorenz, L.J., Haun, R.S., Roos, B.A., Collier, K.J., and Dixon, J.E. (1984). Cloning and sequence analysis of a cDNA encoding rat preprocholecystokinin. *Proceedings of the National Academy of Sciences, U.S.A.* **81**, 726–730.

Deschodt-Lanckman, M., Robberecht, P., De Neef, P., Lammens, M. and Christophe, J. (1976). *In vitro* action of bombesin and bombesin-like peptides on amylase secretion, calcium efflux, and adenylate cyclase activity in the rat pancreas. *Journal of Clinical Investigation* **58**, 891–898.

Dimaline, R. (1989). Vasoactive intestinal peptide. In, *Comparative Physiology of Regulatory Peptides.* edited by S. Holmgren, pp 150–173, London: Champman and Hall.

Dimaline, R., and Dockray, G.J. (1978). Multiple immunoreactive forms of vasoactive intestinal peptide in human colonic mucosa. *Gastroenterology* **75**, 387–392.

Djokic, T.D., Sekizawa, K., Borson, D.B., and Nadel, J.A. (1989). Neutral endopeptidase inhibitors potentiate substance P-induced contraction in gut smooth muscle. *American Journal of Physiology* **256**, G39–G34.

Docherty, K. and Steiner D.F. (1982). Post-translational proteolysis in polypeptide hormone biosynthesis. *Annual Review of Physiology.* **44**, 625–638.

Dockray, G.J. (1987a). Physiology of Enteric Neuropeptides. In *Physiology of the Gastrointestinal Tract, Second Edition.* edited by L.R. Johnson, pp 41–66. New York: Raven Press.

Dockray, G.J. (1987b). The biosynthesis of regulatory peptides. *American Review of Respiratory Disease* **136**, S9–S15.

Dockray, G.J. (1988). Regulatory peptides and the neuroendocrinology of gut-brain relations. *Quarterly Journal of Experimental Physiology* **73**, 703–727.

Dockray, G.J. (1989a). Comparative neuroendocrinology of gut peptides. In *Handbook of Physiology – The Gastrointestinal System II,* edited by G.M. Makhlouf, pp 133–170. New York: American Physiological Society.

Dockray, G.J. (1989b). The integrative functions of CCK in the upper gastrointestinal tract. In *The Neuropeptide Cholecystokinin (CCK),* edited by J. Hughes, G.J. Dockray and G. Woodruff. pp 232–239. Chichester: Ellis Horwood.

Dockray, G.J. (1989c). Peptide neurotransmitters. In *Neuronal Communications, Physiological Society*

Study Guide, edited by W. Winlow. pp 108–129. Manchester: Manchester University Press.

Dockray, G.J. and Gregory, R.A. (1989). Gastrin. In *Handbook of Physiology – The Gastrointestinal System volume II,* edited by G.M. Makhlouf, pp 311–336. New York: American Physiological Society.

Dockray, G.J., Dimaline, R., Pauwels, S. and Varro, A. (1989). Gastrin and CCK-related peptides. In *Peptide Hormones as Prohormones: Processing, Biological Activity, Pharmacology,* edited by J. Martinez, pp 244–284. Chichester: Ellis Horwood.

Donnerer, J., Bartho, L., Holzer, P. and Lembeck, F. (1984). Intestinal peristalsis associated with release of immunoreactive substance P. *Neuroscience* **11**, 913–918.

Donnerer, J., Holzer, P., and Lembeck, F. (1984). Release of dynorphin, somatostatin and substance P from the vascularly perfused small intestine of the guinea-pig during peristalsis. *British Journal of Pharmacology.* **83**, 919–925.

Donnerer, J., Meyer, D.K., Holzer, P. and Lembeck, F. (1985). Release of cholecystokinin – immunoreactivity into the vascular bed of the guinea-pig small intestine during peristalsis. *Naunyn Schmiedeberg's Archives of Pharmacology* **328**, 324–326.

Drapeau, G., d'Orleans-Juste, P., Dion, S., Rhaleb, N-E., and Regoli, D. (1987a). Specific agonists for neurokinin B receptors. *European Journal of Pharmacology.* **136**, 401–403.

Drapeau, G., d'Orleans-Juste, P., Dion, S., Rhaleb, N-E., Rouissi, N-E. and Regoli, D. (1987b). Selective agonists for substance P and neurokinin receptors. *Neuropeptides* **10**, 43–54.

Dun, N.J., and Mo, N. (1988). Calcitonin gene-related peptide evokes fast and slow depolarizing responses in guinea-pig coeliac neurones. *Neuroscience Letters* **87**, 157–162.

Dunne, M.J., Bullett, M.J., Li, G., Wollheim, C.N. and Petersen, O.H. (1989). Galanin activates nucleotide-dependent K^+ channels in insulin-secreting cells via a pertussis toxin-sensitive G-protein. *The EMBO Journal* **8**, 413–420.

Dunning, B.E., and Taborsky, G.J. Jr. (1989), Galanin release during pancreatic nerve stimulation is sufficient to influence islet function. *American Journal of Physiology* **256**, E191–E198.

Dupont, C., Laburthe, M., Broyart, M., Batille, D. and Roselin, G. (1980). Cyclic AMP production in isolated colonic epithelial crypts: A highly sensitive model for the evaluation of vasoactive intestinal peptide action in human intestine. *European Journal of Clinical Investigation* **10**, 67–76.

Edbrooke, M.R., Parker, D., McVey, J.H., Riley, J.H., Sorenson, G.D., Pettengill, O.S., and Craig, R.K. (1985). Expression of the human calcitonin/CGRP gene in lung and thyroid carcinoma. *The EMBO Journal* **4**, 715–724.

Edvinsson, L. (1988). The effects of neuropeptide Y on the circulation. *ISI Atlas of Science: Pharmacology* **2**, 357–361.

Edvinsson, L., Ekblad, E., Håkanson, R., and Wahlestedt, C. (1984). Neuropeptide Y potentiates the effect of various vasoconstrictor agents on rabbit blood vessels. *British Journal of Pharmacology* **83**, 519–525.

Ehrenpreis, T. and Pernow, B. (1953). On the occurrence of substance P in the rectosigmoid in Hirschsprung's disease. *Acta Physiologica Scandinavia.* **27**, 380–388.

Ekblad, E., Edvinsson, L., Wahlestedt, C., Uddman, R., Håkanson, R., and Sundler, F. (1984). Neuropeptide Y co-exists and co-operates with noradrenaline in perivascular nerve fibres. *Regulatory Peptides* **8**, 225–235.

Ekblad, E., Håkanson, R., Sundler, F. and Wahlestedt, C. (1985). Galanin: Neuromodulatory and direct contractile effects on smooth muscle preparations. *British Journal of Pharmacology* **86**, 241–246.

El Battari, A., Luis, J., Martin, J-M., Fantini, J., Muller, J-M., Marvaldi, J. and Pichon, J. (1987). The vasoactive intestinal peptide receptor on intact human colonic adenocarcinoma cells (HT29–D4): Evidence for its glycoprotein nature. *Biochemical Journal* **242**, 185–191.

Emeson, R.B., Hedjran, F., Yeakley, J.M., Guise, J.W. and Rosenfeld, M.G. (1989). Alternative production of calcitonin and CGRP mRNA is regulated at the calcitonin-specific splice acceptor. *Nature* **341**, 76–80.

Emson, P.C., and Zaidi, M. (1989). Further evidence for the origin of circulating calcitonin generelated peptide in the rat. *Journal of Physiology* **412**, 297–308.

Eng, J., Shiina, Y., Pan, Y-C.E., Blacher, R., Chang, M., Stein, S., and Yalow, R.S. (1983). Pig brain contains cholecystokinin octapeptide and several cholecystokinin desoctapeptides. *Proceedings of the National Academy of Sciences, U.S.A.* **80**, 6381–6385.

Eng, J., Du, B-H., Pan, Y-C.E., Chang, M., Hulmes, J.D. and Yalow, R.S. (1984). Purification and sequencing of a rat intestinal 22 amino acid C-terminal CCK fragment. *Peptides* **5**, 1203–1206.

Eng, J., Gubler, U., Raufman, J-P., Chang, M., Hulmes, J.D., Pan, Y-C.E., and Yalow, R.S. (1986). Cholecystokinin-associated COOH-terminal peptides are fully sulfated in pig brain. *Proceedings of the National Academy of Sciences, U.S.A.* **83**, 2832–2835.

Erdos, E.G. and Skidgel, R.A. (1989). Neutral endopeptidase 24.11 (enkephalinase) and related regulators of peptide hormones. *The FASEB Journal* 3, 145–151.

Evans, B.E., Bock, M.G., Rittle, K.E., DiPardo, R.M., Whitter, W.L., Veber, D.F., Anderson, P.S. and Freidinger, R.M. (1986). Design of potent, orally effective, nonpeptidal antagonists of the peptide hormone cholecystokinin. *Proceedings of the National Academy of Sciences, U.S.A.* 83, 4918–4922.

Eysselein, V.E., Reeve, J.R. Jr., Shively, J.E., Miller, C. and Walsh, J.H. (1984). Isolation of a large cholecystokinin precursor from canine brain. *Proceedings of the National Academy of Sciences, U.S.A.* 81, 6565–6568.

Fahrenkrug, J. (1989). Vasoactive intestinal peptide. In *Handbook of Physiology: Gastrointestinal System, volume II,* edited by G.M. Makhlouf, pp 611–629. New York, American Physiological Society.

Fahrenkrug, J., Galbo, H., Holst, J.H. and Schaffalitzky de Muckadell, O.B. (1978). Influence of the autonomic nervous system on the release of vasoactive intestinal polypeptide from the porcine gastrointestinal tract. *Journal of Physiology* 280, 405–422.

Fahrenkrug, J., Ottesen, B. and Palle, C. (1988). Vasoactive intestinal polypeptide and the reproductive system. *Annals of the New York Academy of Sciences* 527, 393–404.

Fontaine, B., Klarsfeld, A. and Changeux, J-P. (1987). Calcitonin gene-related peptide and muscle activity regulate acetylcholine receptor α-subunit mRNA levels by distinct intracellular pathways. *Journal of Cell Biology* 105, 1337–1342.

Foord, S.M. and Craig, R.K. (1987). Isolation and characterization of human calcitonin-gene-related peptide receptor. *European Journal of Biochemistry* 170, 373–379.

Forster, E.R. (1990). The role of vasoactive intestinal polypeptide in the control of gastric emptying in the rat. *Journal of Physiology* 424, 12P.

Forster, E.R., Green, T., Elliot, M., Bremner, A. and Dockray, G.J. (1990). Gastric emptying in rats: role of afferent neurones and cholecystokinin. *American Journal of Physiology* 258, G552–G556.

Fox, J.E.T., Brooks, B., McDonald, T.J., Barnett, W., Kostolanska, F., Yanaihara, C., Yanaihara, N. and Rökaeus, A. (1988). Actions of galanin fragments on rat, guinea-pig, and canine intestinal motility. *Peptides* 9, 1183–1189.

Franco, R., Costa, M., and Furness, J.B. (1979). Evidence for the release of endogenous substance P from intestinal nerves. *Naunyn Schmiedebergs Archives of Pharmacol.* 306, 195–201.

Freidinger, R.M. (1989). Cholecystokinin and gastrin antagonists. *Medicinal Research Reviews* 9, 271–290.

Freidinger, R.M., Bock, M.G., DiPardo, R.M., Evans, B.E., Rittle, K.E., Whitter, W.L., Veber, D.F., Anderson, P.S., Chang, R.S.L., and Lotti, V.J. (1989). Development of selective nonpeptide CCK-A and CCK-B/gastrin receptor antagonists. In *The Neuropeptide Cholecystokinin (CCK),* edited by J. Hughes, G.J. Dockray and G. Woodruff, pp 123–132. Chichester: Ellis Horwood.

Fried, G., Terenius, L., Hökfelt, T., and Goldstein, M. (1985). Evidence for differential localization of noradrenaline and neuropeptide Y in neuronal storage vesicles isolated from rat vas deferens. *Journal of Neuroscience* 51, 450–458.

Fuhlendorff, J., Gether, U., Aakerlund, L., Langeland-Johansen, N., Thogersen, H., Melberg, S.G., Olsen, U.B., Thastrup, O. and Schwartz, T.W. (1990). [Leu[31], Pro[34]] Neuropeptide Y: A specific Y_1 receptor agonist. *Proceedings of the National Academy of Science, U.S.A.* 87, 182–186.

Furness, J.B. and Costa, M. (1987). *The Enteric Nervous System.* pp 1–290, Edinburgh: Churchill Livingstone.

Furness, J.B. and Costa, M. (1989). Identification of transmitters of functionally defined enteric neurons. In *Handbook of Physiology, The Gastrointestinal system volume I,* edited by J.D. Wood, pp 387–402. New York: American Physiological Society.

Furness, J.B., Costa, M., Rökaeus, Å., McDonald, T.J. and Brooks, B. (1987). Galanin-immunoreactive neurons in the guinea-pig small intestine: Their projections and relationships to other enteric neurons. *Cell and Tissue Research* 250, 607–615.

Furness, J.B., Morris, J.L., Gibbins, I.L. and Costa, M. (1989). Chemical coding of neurons and plurichemical transmission. *Annual Review of Pharmacological Toxicology.* 29, 289–306.

Gabella, G. (1979). Innervation of the gastrointestinal tract. International Review of Cytology. 59, 129–193.

Gafvelin, G., Andersson, M., Dimaline, R., Jornvall, H. and Mutt, V. (1989). Isolation and characterization of a variant form of vasoactive intestinal peptides. *Peptides* 9, 469–474.

Gaginella, T.S., O'Dorisio, T.M. and Hubel, K.A. (1981). Release of vasoactive intestinal polypeptide by electrical field stimulation of rabbit ileum. *Regulatory Peptides* 2, 165–174.

Gallacher, D.V. (1983). Substance P is a functional neurotransmitter in the rat parotid gland. *Journal of Physiology* 342, 483–498.

Gamse, R. and Saria, A. (1985). Potentiation of tachykinin-induced plasma protein extravasation by calcitonin gene-related peptide. *European Journal of Pharmacology* **114**, 61–66.

Geppetti, P., Maggi, C.A., Perretti, F., Frilli, S., and Manzini, S. (1988). Simultaneous release by bradykinin of substance P- and calcitonin gene-related peptide immunoreactivities from capsaicin-sensitive structures in guinea-pig heart. *British Journal of Pharmacology* **94**, 288–290.

Gibbins, I.L., Furness, J.B., Costa, M., MacIntyre, I., Hillyard, C.J., and Girgis, S. (1985). Co-localization of calcitonin gene-related peptide-like immunoreactivity with substance P in cutaneous, vascular and visceral sensory neurons of guinea pigs. *Neuroscience Letters* **57**, 125–130.

Gibson, S.J., Polak, J.M., Bloom, S.R., Sabate, I.M., Mulderry P.M., Ghatei, M.A., McGregor, G.P., Morrison, J.F.B., Kelly J.S., Evans, R.M. and Rosenfeld, M.G. (1984). Calcitonin gene-related peptide immunoreactivity in the spinal cord of man and of eight other species. *Journal of Neuroscience* **4**, 3103–3111.

Giraud, A.S., Williams, R.G., and Dockray, G.J. (1984). Evidence for different patterns of post-translational processing of pro-enkephalin in the bovine adrenal, colon and striatum indicated by radioimmunoassay using region-specific antisera to Met-enk-Arg^6-Phe^7 and Met-enk-Arg^6-Gly^7-Leu^8. *Neuroscience Letters* **46**, 223–228.

Glover, I.D., Barlow, D.J., Pitts, J.E., Wood, S.P., Tickle, I.J., Blundell, T.L., Tatemoto, K., Kimmel, J.R., Wollmer, A., Strassburger, W. and Zhang, Y-S. (1985). Conformational studies on the pancreatic polypeptide hormone family. *European Journal of Biochemistry* **142**, 379–385.

Goltzman, D. and Mitchell, J. (1985). Interaction of calcitonin and calcitonin gene-related peptide at receptor sites in target tissues. *Science* **227**, 1343–1345.

Goodman, R.H., Jacobs, J.W., Dee, P.C. and Habener, J.F. (1982). Somatostatin-28 encoded in a cloned cDNA obtained from a rat medullary thyroid carcinoma. *Journal of Biological Chemistry* **257**, 271–288.

Gorden, P. (1989). Somatostatin and somatostatin analogue (SMS 201–995) in treatment of hormone-secreting tumors of the pituitary and gastrointestinal tract and non-neoplastic diseases of the gut. *Annals of Internal Medicine* **110**, 35–50.

Green, T., Dimaline, R., Peikin, S. and Dockray, G.J. (1988). Action of the cholecystokinin antagonist L364,718 on gastric emptying in the rat. *American Journal of Physiology* **255**, G685–G689.

Grider, J.R. (1989a). Tachykinins as transmitters of ascending contractile component of the peristaltic reflex. *American Journal of Physiology* **257**, G709–G714.

Grider, J.R. (1989b). Identification of neurotransmitters regulating intestinal peristaltic reflex in humans. *Gastroenterology* **97**, 1414–1419.

Grider, J.R. (1989c). Somatostatin release from isolated ganglia of the myenteric plexus. *American Journal of Physiology* **257**, G313–G315.

Grider, J.R. and Makhlouf, G.M. (1986). Colonic peristaltic reflex: Identification of vasoactive intestinal peptide as mediator of descending relaxation. *American Journal of Physiology* **251**, G40–G45.

Grider, J.R. and Makhlouf, G.M. (1987). Prejunctional inhibition of vasoactive intestinal peptide release. *American Journal of Physiology* **253**, G7–G12.

Grider, J.R., Cable, M.B., Bitar, K.N., Said, S.I. and Makhlouf, G.M. (1985a). Vasoactive intestinal peptide: Relaxant neurotransmitter in tenia coli of the guinea pig. *Gastroenterology* **89**, 36–42.

Grider, J.R., Cable, M.B., Said, S.I., and Makhlouf, G.M. (1985b). Vasoactive intestinal peptide as a neural mediator of gastric relaxation. *American Journal of Physiology* **248**, G73–G78.

Gubler, U., Seeburg, P., Hoffman, B.J., Gage, L.P., and Udenfriend, S. (1982). Molecular cloning establishes proenkephalin as precursor of enkephalin-containing peptides. *Nature* **295**, 206–208.

Gubler, U., Chua, A.O., Hoffman, B.J., Collier, K.J., and Eng, J (1984). Cloned cDNA to cholecystokinin mRNA predicts an identical preprocholecystokinin in pig brain and gut. *Proceedings of the National Academy of Sciences U.S.A.* **81**, 4307–4310.

Gulbenkian, S., Merighi, A., Wharton, J., Varndell, I.M. and Polak, J.M. (1986). Ultrastructural evidence for the coexistence of calcitonin gene-related peptide and substance P in secretory vesicles of peripheral nerves in the guinea pig. *Journal of Neurocytology,* **15**, 535–542.

Hambrook, J.M., Morgan, B.A., Rance, M.J. and Smith, C.F.C. (1976). Mode of deactivation of the enkephalins by rat and human plasma and rat brain homogenates. *Nature* **262**, 782–783.

Harmar, A.J. and Keen, P. (1984). Rat sensory ganglia incorporate radiolabelled amino acids into substance K (neurokinin α) *in vitro*. *Neuroscience Letters* **51**, 387–391.

Harmar, A., Schofield, J.G. and Keen, P. (1980). Cyclohexmimide-sensitive synthesis of substance P by isolated dorsal root ganglia. *Nature* **284**, 267–269.

Harmar, A.J., Armstrong, A., Pascall, J.C., Chapman, K., Rosie, R., Curtis, A., Going, J., Edwards, C.R.W. and Fink, G. (1986). cDNA sequence of human β-preprotachykinin, the common precursor to substance P and neurokinin A. *FEBS Letters* **208**, 67–72.

Hayakawa, Y., Obata, K-I., Itoh, N., Yanaihara, N. and Okamoto, H. (1984). Cyclic AMP regulation of pro-vasoactive intestinal polypeptide/PHM-27 synthesis in human neuroblastoma cells. *Journal of Biological Chemistry* **25**, 9207-9211.

Hedlund, B., Abens, J., Bartfai, T. (1983) Vasoactive intestinal polypeptide and muscarinic receptors: Supersensitivity induced by long-term atropine treatment. *Science* **220**, 519-521.

Heimbrook, D.C., Saari, W.S., Balishin, N.L., Friedman, A., Moore, K.S., Riemen, M.W., Kiefer, D.M., Rotberg, N.S., Wallen, J.W., and Oliff, A. (1989). Carboxyl-terminal modification of a gastrin releasing peptide derivative generates potent antagonists. *Journal of Biological Chemistry* **264**, 11258-11262.

Hershey, A.D., and Krause, J.E. (1990). Molecular characterization of a functional cDNA encoding the rat substance P receptor. *Science* **247**, 958-962.

Heslop, J.P., Blakeley, D.M., Brown, K.D., Irvine R.F. and Berridge, M.J. (1986). Effects of bombesin and insulin on inositol (1,4,5) trisphosphate and inositol (1,3,4) trisphosphate formation in Swiss 3T3 cells. *Cell* **47**, 703-709.

Higuchi, H., Yang H-Y.T., and Sabol, S.L. (1988). Rat neuropeptide Y precursor gene expression: mRNA structure, tissue distribution, and regulation by glucocorticoids, cyclic AMP, and phorbol ester. *Journal of Biological Chemistry* **263**, 6288-6295.

Hills, J.M., Collis, C.F., and Burnstock, G. (1983). The effects of vasoactive intestinal polypeptide on the electrical activity of guinea-pig intestinal smooth muscle. *European Journal of Pharmacology* **88**, 371-376.

Hökfelt, T., Elfvin, L.G., Elde, R., Schultzberg, M., Goldstein, M. and Luft, R (1977). Occurrence of somatostatin-like immunoreactivity in some peripheral sympathetic noradrenergic neurons. *Proceedings of the National Academy of Sciences, U.S.A.* **74**, 3587-3591.

Hökfelt, T., Johansson, O., Ljungdahl, Å., Lundberg, J.M. and Schultzberg, M. (1980). Peptidergic neurons. *Nature* **284**, 515-521.

Hökfelt, T., Everitt, B., Meister, B., Melander T., Schalling, M., Johansson, O., Lundberg, J.M., Hulting, A-L., Werner, S., Cuello, C., Hemmings, H., Ouimet, C., Walaas, I., Greengard, P., and Goldstein, M. (1986). Neurons with multiple messengers with special reference to neuroendocrine systems. *Recent Progress in Hormone Research* **42**, 1-70.

Holmdahl, G., Håkanson, R., Leander, S., Rosell, S., Folkers, K., Sundler, F. (1981). A substance P antagonist, [D-Pro^2, D-Trp^{7,9}]SP, inhibits inflammatory responses in the rabbit eye. *Science* **214**, 1029-1031.

Holst, J.J., Fahrenkrug, J., Knuhtsen, S., Jensen, S.L., Nielson, O.V., Lundberg, J.M. and Hokfelt, T.(1987). VIP and PHI in the pig pancreas: coexistence, corelease, and cooperative effects. *American Journal of Physiology* **252**, G182-G189.

Holzer, P. (1984). Characterization of the stimulus-induced release of immunoreactive substance P from the myenteric plexus of the guinea pig small intestine. *Brain Research* **297**, 127-136.

Holzer, P. (1988). Local effector functions of capsaicin-sensitive sensory nerve endings: involvement of tachykinins, calcitonin gene-related peptide and other neuropeptides. *Neuroscience* **24**, 739-768.

Holzer, P. (1989). Ascending enteric reflex: Multiple neurotransmitter systems and interactions. *American Journal of Physiology* **256**, G540-G545.

Holzer, P. and Lippe, I.Th. (1984). Substance P can contract the longitudinal muscle of the guinea-pig small intestine by releasing intracellular calcium. *British Journal Pharmacology* **82**, 259-267.

Holzer, P., Lippe, I.Th., Bartho, L. and Saria, A. (1987). Neuropeptide Y inhibits excitatory enteric neurons supplying the circular muscle of the guinea pig small intestine. *Gastroenterology* **92**, 1944-1950.

Hua, X-Y., Saria, A., Gamse, A., Gamse, R., Theodorsson-Norheim, E., Brodin, E. and Lundberg, J.M. (1986). Capsaicin induced release of multiple tachykinins (substance P, neurokinin A, and eledoisin-like material) from guinea-pig spinal cord and ureter. *Neuroscience* **19**, 313-319.

Hughes, J., Smith, T.W., Kosterliz, H.W., Fothergill, L.A., Morgan, B.A. and Morris, H.R. (1975). Identification of two related pentapeptides from the brain with potent opiate agonist activity. *Nature* **258**, 577-579.

Hunter, J.C., Goedert, M. and Pinnock, R.D. (1985). Mammalian tachykinin-induced hydrolysis of inositol phospholipids in rat brain slices. *Biochemical and Biophysical Research Communications* **127**, 616-622.

Ito, S., Ohga, A. and Ohta, T. (1988a). Gastric vasodilatation and vasoactive intestinal peptide output in response to vagal stimulation in the dog. *Journal of Physiology* **404**, 669-682.

Ito, S., Ohga, A. and Ohta, T. (1988b). Gastric relaxation and vasoactive intestinal peptide output in response to reflex vagal stimulation in the dog. *Journal of Physiology* **404**, 683-693.

Itoh, N., Obata, K-I., Yanaihara, N., Okamoto, H. (1983). Human preprovasoactive intestinal poly-peptide contains a novel PHI-27-like peptide, PHM-27. *Nature* **304**, 547–549.

Iversen, L.L., Lee, C.M., Gilbert, R.F., Hunt, S. and Emson, P.C. (1980). Regulation of neuropeptide release. *In, Neuroactive Peptides*, edited by A. Burgen, H.W. Kosterlitz, L.L. Iversen, pp 91–111, London: The Royal Society.

Jacoby, H.I. (1988). Gastrointestinal tachykinin receptors. *Life Sciences* **43**, 2203–2208.

Jan, L.Y. and Jan, Y.N. (1982). Peptidergic transmission in sympathetic ganglia of the frog. *Journal of Physiology* **327**, 219–246.

Jan, Y.N., and Jan, L.Y. (1983). A LHRH-like peptidergic neurotransmitter capable of 'action at a distance' in autonomic ganglia. *Trends in Neurosciences* **6**, 320–325.

Jensen, R.T., Tatemoto, K., Mutt, V., Lemp, G.F. and Gardner, J.D. (1981). Actions of a newly isolated intestinal peptide PHI on pancreatic acini. *American Journal of Physiology* **241**, G498–G502.

Jensen, R.T., von Schrenck, T., Yu, D-H., Wank, S.A., and Gardner, J.D. (1989). Pancreatic chole-cystokinin (CCK) receptors: Comparison with other classes of CCK receptors. In *The Neuro-peptide Cholecystokinin (CCK)*, edited by J. Hughes, G.J. Dockray, and G. Woodruff, pp 150–162. Chichester: Ellis Horwood.

Jiang, Z-G., Dun, N.J., Karczmar, A.G. (1982). Substance P: A putative sensory transmitter in mammalian autonomic ganglia. *Science* **217**, 739–741.

Johansson, O. and Lundberg, J.M. (1981). Ultrastructural localization of VIP-like immunoreactivity in large dense-core vesicles of 'cholinergic-type' nerve terminals in cat exocrine glands. *Neuroscience* **6**, 847–862.

Ju, G., Hökfelt, T., Brodin, E., Fahrenkrug, J., Fischer, J.A., Frey, P., Elde, R.P. and Brown, J.C. (1987). Primary sensory neurons of the rat showing calcitonin gene-related peptide (CGRP) immunoreactivity and their reaction to substance P-, somatostatin-, galanin-, vasoactive intestinal polypeptide-, and cholecystokinin-immunoreactive ganglion cells. *Cell Tissue Research* **247**, 417–431.

Kachur, J.F., Miller, R.J. and Field, M. (1980). Control of guinea pig intestinal electrolyte secretion by a δ-opiate receptor. *Proceedings of the National Academy of Sciences U.S.A.* **77**, 2753–2756.

Kakidani, H., Furutani, Y., Takahashi, H., Noda, M., Morimoto, Y., Hirose, T., Asai, M., Inayama, S., Nakanishi, S., Numa, S. (1982). Cloning and sequence analysis of cDNA for porcine β-neoendorphin/dynorphin precursor. *Nature* **298**, 245–249.

Kawatani, M., Rutigliano, M., de Groat, W.C. (1985). Depolarization and muscarinic excitation induced in a sympathetic ganglion by vasoactive intestinal polypeptide. *Science* **229**, 879–881.

Keast, J.R. (1987). Mucosal innervation and control of water and ion transport in the intestine. *Review of Physiological and Biochemical Pharmacology*. **109**, 2–59.

Kessler, J.A., Adler, J.E., Bohn, M.C., Black, I.B. (1981). Substance P in principal sympathetic neurons: Regulation by impulse activity. *Science* **214**, 335–336.

Kley, N., Loeffler, J-P., Pittius, C.W., and Höllt, V. (1987). Involvement of ion channels in the induction of proenkephalin A gene expression by nicotine and cAMP in bovine chromaffin cells. *Journal of Biological Chemistry* **262**, 4083–4089.

Knuhtsen, S., Holst, J.J., Knigge, U., Olesen, M. and Nielsen, O.V. (1984). Radioimmunoassay, pharmacokinetics and neuronal release of gastrin-releasing peptide in anesthetized pigs. *Gastroenterology* **87**, 372–378.

Kojima, K., Kilpatrick, D.L., Stern, A.S., Jones, B.N., and Udenfriiend, S. (1982). Proenkephalin: A general pathway for enkephalin biosynthesis in animal tissues. *Archives of Biochemistry and Biophysics* **215**, 638–643.

Konishi, S., Tsunoo, A. and Otsuka, M. (1981). Enkephalin as a transmitter for presynaptic inhibition in sympathetic ganglia. *Nature* **294**, 80–82.

Konishi, S., Tsunoo, A. and Otsuka, M. (1979). Enkephalins presynaptically inhibit cholinergic trans-mission in sympathetic ganglia. *Nature* **282**, 515–516.

Kosterlitz, H.W., and Peterson, S.J. (1985). Types of opioid receptors: Relation to antinociception. *Philosophical Transactions of the Royal Society London*. **B308**, 291–297.

Kotani, H., Hoshimaru, M., Nawa, H. and Nakanishi, S. (1986). Structure and gene organization of bovine neuromedin K precursor. *Proceedings of the National Academy of Sciences, USA*. **83**, 7074–7078.

Krane, I.M., Naylor, S.L., Helin-Davis, D., Chin, W.W. and Spindel, E.R. (1988). Molecular cloning of cDNAs encoding the human bombesin-like peptide neuromedin B. *Journal of Biological Chemistry* **263**, 13317–13323.

Krause, J.E., Chirgwin, J.M., Carter, M.S., Xu, Z.S., and Hershey, A.D. (1987). Three rat preprota-chykinin mRNAs encode neuropeptides substance P and neurokinin A. *Proceedings of the National Academy of Science, U.S.A.* **84**, 881–885.

Krejs, G.J., Fordtran, J.S., Bloom, S.R., Fahrenkrug, J., Schaffalitzky de Muckadell, O.B., Fischer, J.E., Humphrey, C.S., O'Dorisio, T.M., Said, S.I., Walsh, J.H. and Shulkes, A.A. (1979). Effect of VIP infusion on water and ion transport in the human jejunum. *Gastroenterology* **78**, 722-727.

Krstenansky, J.L., Owen, T.J., Buck, S.H., Hagaman, K.A. and McLean, L.R. (1989). Centrally truncated and stabilized porcine neuropeptide Y analogs: Design, synthesis, and mouse brain receptor binding. *Proceedings of the National Academy of Science, U.S.A.* **86**, 4377-4381.

Kuwano, R., Araki, K., Usui, H., Fukui, T., Ohtsuka, E., Ikehara, M., Takahashi, Y. (1984). Molecular cloning and nucleotide sequence of cDNA coding for rat brain cholecystokinin precursor. *Journal of Biochemistry* **96**, 923-926.

Laburthe, M., Chenut, B., Rouyer-Fessard, C., Tatemoto, K., Couvineau, A., Servin, A. and Amiranoff, B. (1986). Interaction of peptide YY with rat intestinal epithelial plasma membranes: Binding of the radioiodinated peptide. *Endocrinology* **118**, 1910-1917.

Langley, J.N. (1921). *The Autonomic Nervous System. Part 1.* Cambridge, England, W. Heffer and Sons.

Laufer, R., Wormser, U., Friedman, Z.Y., Gilon, C., Chorev, M., and Selinger, Z. (1985). Neurokinin B is a preferred agonist for a neuronal substance P receptor and its action is antagonized by enkephalin. *Proceedings of the National Academy of Science, U.S.A.* **82**, 7444-7448.

Leander, S., Håkanson, R., Rosell, S., Folkers, K., Sundler, F. and Tornqvist, K. (1981). A specific substance P antagonist blocks smooth muscle contractions induced by non-cholinergic non-adrenergic nerve stimulation. *Nature* **294**, 467-469.

Lebacq-Verheyden, A-M., Krystal, G., Sartor, O., Way, J. and Battey, J.F. (1988). The rat prepro gastrin releasing peptide gene is transcribed from two initiation sites in brain. *Molecular Endocrinology* **2**, 556-563.

Lee, C-M., Iversen, L.L., Hanley M.R., and Sandberg, B.E.B. (1982). The possible existence of multiple receptors for substance P. *Naunyn-Schmiedebergs Archives of Pharmacology* **318**, 281-287.

Lee, C-M., Campbell, N.J., Williams, B.J. and Iversen, L.L. (1986). Multiple tachykinin binding sites in peripheral tissues and in brain. *European Journal of Pharmacology* **130**, 209-217.

Lee, Y., Kawai, Y., Shiosaka, S., Takami, K., Kiyama, H., Hillyard, C.J., Girgis, S., MacIntyre, I., Emson, P.C., and Tohyama, M. (1985). Coexistence of calcitonin gene-related peptide and substance P-like peptide in single cells of the trigeminal ganglion of the rat: immunohistochemical analysis. *Brain Research* **330**, 194-196.

Leffert, J.D., Newgard, C.G., Okamoto, H., Milburn J.L., Luskey, K.L. (1989). Rat amylin: Cloning and tissue-specific expression in pancreatic islets. *Proceedings of the National Academy of Sciences U.S.A.* **86**, 3127-3130.

Lembeck, F. and Holzer, P. (1979). Substance P as a neurogenic mediator of antidromic vasodilatation and neurogenic plasma extravasation. *Naunyn-Schmiedeberg's Archives of Pharmacology* **310**, 175-183.

Letterio, J.J., Coughlin, S.R., Williams, L.T. (1986). Pertussis toxin-sensitive pathway in the stimulation of c-myc expression and DNA synthesis by bombesin. *Science* **234**, 1117-1119.

Lewis, R.V., Stern, A.S., Kimura, S., Rossier, J., Stein, S., Udenfriend, S. (1980). An about 50,000-dalton protein in adrenal medulla: A common precursor of [Met]- and [Leu]-enkephalin. *Science* **208**, 1459-1461.

Liebow, C., Reilly, C., Serrano, M., and Schally, A.V. (1989). Somatostatin analogues inhibit growth of pancreatic cancer by stimulating tyrosine phosphatase. *Proceedings of the National Academy of Sciences, U.S.A.* **86**, 2003-2007.

Linder, S., Barkhem, T., Norberg, A., Persson, H., Schalling M., Hökfelt, T. and Magnusson, G. (1987). Structure and expression of the gene encoding the vasoactive intestinal peptide precursor. *Proceedings of the National Academy of Sciences, U.S.A.* **84**, 605-609.

Lindh, B., Hökfelt, T., Elfvin, L-G., Terenius, L., Fahrenkrug, J., Elde, R. and Goldstein, M. (1986). Topography of NPY-, somatostatin-, and VIP-immunoreactive, neuronal subpopulations in the guinea pig celiac-superior mesenteric ganglion and their projection to the pylorus. *Journal of Neuroscience* **6**, 2371-2383.

Lindh, B., Hökfelt, T. and Elfvin, L-G. (1988). Distribution and origin of peptide-containing nerve fibers in the celiac superior mesenteric ganglion of the guinea pig. *Neuroscience* **26**, 1037-1071.

Lindh, B., Lundberg, J.M. and Hökfelt, T. (1989). NPY-, galanin-, VIP/PHI-, CGRP- and substance P-immunoreactive neuronal subpopulations in cat autonomic and sensory ganglia and their projections. *Cell and Tissue Research* **256**, 259-273.

Lindsay, R.M., and Harmar, A.J. (1989). Nerve growth factor regulates expression of neuropeptide genes in adult sensory neurons. *Nature* **337**, 362-364.

Lloyd, A.C., Davies, S.A., Crossley, I., Whitaker, M., Houslay, M.D., Hall, A., Marshall, C.J. and Wakelam, J.O. (1989). Bombesin stimulation of inositol 1,4,5-trisphosphate generation and intracellular calcium release is amplified in a cell line overexpressing the N-ras proto-ocogene. *Biochemical Journal* **260**, 813–819.

Louis, S.M., Jamieson, A., Russell, N.J.W., and Dockray, G.J. (1989a). The role of substance P and calcitonin gene-related peptide in neurogenic plasma extravasation and vasodilatation in the rat. *Neuroscience* **32**, 581–586.

Louis, S.M., Johnstone, D., Russell, N.J.W., Jamieson, A. Dockray, G.J. (1989b). Antibodies to calcitonin-gene related peptide reduce inflammation induced by topical mustard oil, but not that due to carrageenin in the rat. *Neuroscience Letters* **102**, 257–260.

Lundberg, J.M. (1981). Evidence for coexistence of vasoactive intestinal polypeptide (VIP) and acetylcholine in neurons of cat exocrine glands. *Acta Physiologica Scandinavica.* **496**, 1–57.

Lundberg, J.M., and Hökfelt, T. (1983). Coexistence of peptides and classical neurotransmitters. *Trends in Neuroscience* **6**, 325–333.

Lundberg, J.M. and Saria, A. (1983). Capsaicin-induced desensitization of airway mucosa to cigarette smoke, mechanical and chemical irritants. *Nature* **302**, 251–253.

Lundberg, J.M., Anggard, A., Fahrenkrug, J., Hökfelt, T. and Mutt, V (1980). Vasoactive intestinal polypeptide in cholinergic neurons of exocrine glands: Functional significance of coexisting transmitters for vasodilatation and secretion. *Proceedings of the National Academy of Sciences U.S.A.* **77**, 1651–1655.

Lundberg, J.M., Fried, G., Fahrenkrug, J., Holmstedt, B., Hökfelt, T., Lagercrantz, H., Lundgren, G. and Ånggård, A. (1981). Subcellular fractionation of cat submandibular gland: Comparative studies on the distribution of acetylcholine and vasoactive intestinal polypeptide (VIP). *Neuroscience* **6**, 1000–1010.

Lundberg, J.M., Hedlund, B., and Bartfai, T. (1982). Vasoactive intestinal polypeptide enhances muscarinic ligand binding in cat submandibular salivary gland. *Nature* **295**, 147–149.

Lundberg, J.M., Terenius, L., Hökfelt, T., Martling, C.R., Tatemoto, K., Mutt, V., Polak, J., Bloom, S., and Goldstein, M. (1982). Neuropeptide Y (NPY)-like immunoreactivity in peripheral noradrenergic neurons and effects of NPY on sympathetic function. *Acta Physiol. Scand.* **116**, 477–480.

Lundberg, J.M., Fahrenkrug, J., Larsson, O. and Anggard, A. (1984). Corelease of vasoactive intestinal polypeptide and peptide histidine isoleucine in relation to atropine-resistant vasodilatation in cat submandibular salivary gland. *Neuroscience Letters,* **52**, 37–42.

Lundberg, J.M., Torssell, L., Sollevi, A., Pernow, J., Theodorsson Norheim, E., Ånggård, A., and Hamberger, B. (1985). Neuropeptide Y and sympathetic vascular control in man. *Regulatory Peptides* **13**, 41–52.

Lundberg, J.M., Lundblad, L., Martling, C-R., Saria, A. and Stjärne, P., and Ånggård, A. (1987). Coexistence of multiple peptides and classic transmitters in airway neurons: Functional and pathophysiologic aspects. *American Review of Respiratory Disease* **136**, S16–S22.

Luis, J., Martin, J-M., El Battari, A., Marvaldi, J. and Pichon, J. (1988). The vasoactive intestinal peptide (VIP) receptor: recent data and hypothesis. *Biochimie* **70**, 1311–1322.

Lynch, B., and Kaiser, E.T. (1988). Biological properties of two models of calcitonin gene related peptide with idealized amphiphilic α-helices of different lengths *Biochemistry* **27**, 7600–7607.

McArthur, K.E., Wood, C.L., O'Dorisio, M.S., Zhou, Z-C., Gardner, J.D., Jensen, J.T. (1987). Characterization of receptors for VIP on pancreatic acinar cell plasma membranes using covalent cross-linking. *American Journal of Physiology* **252**, G404–G412.

Mackenzie, I., and Burnstock, G. (1980). Evidence against vasoactive intestinal polypeptide being the non-adrenergic, non-cholinergic inhibitory transmitter released from nerves supplying the smooth muscle of the guinea-pig taenia coli. *European Journal of Pharmacology* **67**, 255–264.

Macrae, I.M., Furness, J.B. and Costa, M. (1986). Distribution of subgroups of noradrenaline neurons in the coeliac ganglion of the guinea pig. *Cell and Tissue Research* **244**, 173–180.

McDonald, T.J., Jörnvall, H., Nilsson, G., Vagne, M., Ghatei, M., Bloom, S.R. and Mutt, V. (1979). Characterization of a gastrin releasing peptide from porcine non-antral gastric tissue. *Biochemical and Biophysical Research Communications* **90**, 227–233.

Maggi, C.A., Patacchini, R., Renzi, D., Santicioli, P., Regoli, D., Rovero, P., Drapeau, G., Surrenti, C., and Meli, A. (1989). Effect of thiorphan on response of the guinea-pig gallbladder to tachykinins. *European Journal of Pharmacology* **165**, 51–61.

Maggi, C.A., Giuliani, S., Ballati, L., Rovero, P., Abelli, L., Manzini, S., Giachetti, A. and Meli, A. (1990). *In vivo* pharmacology of [βAla^8] neurokinin A-(4–10), a selective NK-2 tachykinin receptor agonist. *European Journal of Pharmacology* **177**, 81–86.

Makovec, F., Bani, M., Cereda, R., Chiste, R., Pacini, M.A., Revel, L., Rovati, L.A., Rovati, L.C. and Setnikar, I. (1987). Pharmacological properties of lorglumide as a member of a new class of cholecystokinin antagonists. *Arzneimittel-Forschung Drug Research* **37**, 1265–1268.

Malfroy, B., Schofield, P.R., Kuang, W-J., Seeburg, P.H., Mason, A.J., Henzel, W.J. (1987). Molecular cloning and amino acid sequence of rat enkephalinase. *Biochemical Biophysical Research Communications.* **144**, 59–56.

Mannon, P.J., Taylor, I.L., Kaiser, L.M. and Nguyen, T.D. (1989). Cross-linking of neuropeptide Y to its receptor on rat brain membranes. *American Journal of Physiology* **256**, G637–G643.

Mantyh, P.W., Pinnock, R.D., Downes, C.P., Goedert, M. and Hunt, S.P. (1984). Correlation between inositol phospholipid hydrolysis and substance P receptors in rat CNS. *Nature* **309**, 795–797.

Mason, R.T., Peterfreund, R.A., Sawchenko, P.E., Corrigan, A.Z., Rivier, J.E., Vale, W.W. (1984). Release of the predicted calcitonin gene-related peptide from cultured rat trigeminal ganglion cells. *Nature* **308**, 653–655.

Mason, R.T., Shulkes, A., Zajac, J.D., Fletcher, A.E., Hardy, K.J., and Martin, T.J. (1986). Basal and stimulated release of calcitonin gene-related peptide (CGRP) in patients with medullary thyroid carcinoma. *Clinical Endocrinology* **25**, 675–685.

Masu, Y., Nakayama, K., Tamaki, H., Harada, Y., Kuno, M. and Nakanishi, S. (1987). cDNA cloning of bovine substance-K receptor through oocyte expression system. *Nature* **329**, 836–838.

Masu, Y., Tamaki, H., Yokota, Y., and Nakanishi, S. (1988). Tachykinin precursors and receptors: Molecular genetic studies. *Regulatory Peptides* **22**, 9–12.

Maton, P.N., Sutliff, V.E., Zhou, Z-C., Collins, S.M., Gardner, J.D. and Jensen, R.T. (1988). Characterization of receptors for calcitonin gene-related peptide on gastric smooth muscle cells. *American Journal of Physiology* **254**, G789–G794.

Maton, P.N., Gardner, J.D., Jensen, R.T. (1989). Use of long-acting somatostatin analog SMS 201–995 in patients with pancreatic islet cell tumors. *Digestive Diseases and Sciences* **34**, 28S–39S.

Matozaki, T., Martinez, J. and Williams, J.A. (1989). A new CCK analogue differentiates two functionally distinct CCK receptors in rat and mouse pancreatic acini. *American Journal of Physiology* **257**, G594–G600.

Matsas, R., Kenny, A.J., and Turner, A.J. (1984). The metabolism of neuropeptides: The hydrolysis of peptides, including enkephalins, tachykinins and their analogues, by endopeptidase 24.11. *Biochemical Journal* **223**, 433–440.

Matsas, R., Turner, A.J. and Kenny, A.J. (1984). Endopeptidase 24.11 and aminopeptidase activity in brain synaptic membranes are jointly responsible for the hydrolysis of cholecystokinin octapeptide (CCK-8). *FEBS Letters* **175**, 124–128.

Matthews, M.R., and Cuello, A.C. (1982). Substance P-immunoreactive peripheral branches of sensory neurons innervate guinea pig sympathetic neurons. *Proceedings of the National Academy of Sciences, U.S.A.* **79**, 1668–1672.

Melander, T., Hökfelt, T., Rökaeus, Å., Fahrenkrug, J., Tatemoto, K. and Mutt, V. (1985). Distribution of galanin-like immunoreactivity in the gastro-intestinal tract of several mammalian species. *Cell and Tissue Research* **239**, 253–270.

Michael, J., Carroll, R., Swift, H.H., and Steiner, D.F. (1987). Studies on the molecular organization of rat insulin secretory granules. *Journal of Biological Chemistry.* **262**, 16531–16535.

Mihara, S., North, R.A., and Surprenant, A. (1987). Somatostatin increases an inwardly rectifying potassium conductance in guinea-pig submucous plexus neurons. *Journal of Physiology* **390**, 335–355.

Miles, K., Greengard, P., and Huganir, R.L. (1989). Calcitonin gene-related peptide regulates phosphorylation of the nicotinic acetylcholine receptor in rat myotubes. *Neuron* **2**, 1517–1524.

Minamino, N., Kangawa, K., Matsuo, H. (1983). Neuromedin B: A novel bombesin-like peptide identified in procine spinal cord. *Biochemical and Biophysical Research Communications* **114**, 541–548.

Minth, C.D., Bloom, S.R., Polak, J.M. and Dixon, J.E. (1984). Cloning, characterization, and DNA sequence of a human cDNA encoding neuropeptide tyrosine. *Proceedings of the National Academy of Science, U.S.A.* **81**, 4577–4581.

Minth, C.D., Andrews, P.C., and Dixon, J.E. (1986). Characterization, sequence and expression of the cloned human neuropeptide Y gene. *Journal of Biological Chemistry* **26**, 11974–11979.

Misbin, R.I., Wolfe, M.M., Morris, P., Buynitzky, S.J. and McGuigan, J.E. (1982). Uptake of vasoactive intestinal peptide by rat liver. *American Journal of Physiology* **243**, G103–G111.

Moran, T.H., Smith, G.P., Hostetler, A.M. and McHugh, P.R. (1987). Transport of cholecystokinin (CCK) binding sites in subdiaphragmatic vagal branches. *Brain Research* **415**, 149–152.

Moreau, J.P., and DeFeudis, F.V. (1987). Pharmacological studies of somatostatin and somatostatin-analogues: Therapeutic advances and perspectives. *Life Sciences* **40**, 419–437.

Morris, H.R., Panico, M., Etienne, T., Tippins, J., Girgis, S.I. and MacIntyre, I. (1984). Isolation and characterization of human calcitonin gene-related peptide. *Nature* **308**, 746–748.

Mulderry, P.K., Ghatei, M.A., Spokes, R.A., Jones, P.M., Pierson, A.M., Hamid, Q.A., Kanse, S., Amara, S.G., Burrin, J.M., Legon, S., Polak, J.M. and Bloom, S.R. (1988). Differential expression of αCGRP and β-CGRP by primary sensory neurons and enteric autonomic neurons of the rat. *Neuroscience* 25, 195–205.

Mulle, C., Benoit, P., Pinset, C., Roa, M. and Changeux, J-P. (1988). Calcitonin gene-related peptide enhances the rate of desensitization of the nicotinic acetylcholine receptor in cultured mouse muscle cells. *Proceedings of the National Academy of Sciences, U.S.A.* 85, 5728–5732.

Mutt, V. (1982). Gastrointestinal Hormones: A field of increasing complexity. *Scandinavian Journal of Gastroenterology* 17, suppl. 77. 132–152.

Najdovski, T., Collette, N. and Deschodt-Lanckman, M. (1985). Hydrolysis of the C-terminal octapeptide of cholecystokinin by rat kidney membranes: Characterization of the cleavage by solubilized endopeptidase-24.11. *Life Sciences* 37, 827–834.

Nakanishi, S. (1987). Substance P precursor and kininogen: Their structures, gene organizations and regulation. *Physiological Reviews* 67, 1117–1142.

Nakanishi, S., Inoue, A., Kita, T., Nakamura, M., Chang, A.C.Y., Cohen, S.N. and Numa, S. (1979). Nucleotide sequence of cloned cDNA for bovine corticotropin-β-lipotropin precursor. *Nature* 278, 423–427.

Nau, R., Ballmann, M., and Conlon, J.M. (1987). Binding of vasoactive intestinal polypeptide to dispersed enterocytes results in rapid removal of the NH_2-terminal histidyl residue. *Molecular Endocrinology* 52, 97–103.

Nawa, H., Hirose, T., Takashima, H., Inayama, S., and Nakanishi, S. (1983). Nucleotide sequences of cloned cDNAs for two types of bovine brain substance P precursor. *Nature* 306, 32–36.

Nawa, H., Kotani, H., and Nakanishi, S. (1984). Tissue-specific generation of two preprotachykinin mRNAs from one gene by alternative RNA splicing. *Nature* 312, 729–734.

Niel, J.P., Bywater, R.A.R., and Taylor, G.S. (1983). Apamin-resistant post-stimulus hyperpolarization in the circular muscle of the guinea-pig ileum. *Journal of the Autonomic Nervous System.* 9, 565–569.

Nelson, M.T., Huang, Y., Brayden, J.E., Hescheler, J., Standen, N.B. (1990). Arterial dilations in response to calcitonin gene-related peptide involve activation of K^+ channels. *Nature* 344, 770–773.

New, H.V., and Mudge, A.W. (1986). Calcitonin gene-related peptide regulates muscle acetylcholine receptor synthesis. *Nature.* 323, 809–811.

Nishi, M., Sanke, T., Nagamatsu, S., Bell, G.I., Steiner, D.F. (1990). Islet amyloid polypeptide: A new β-cell secretory product related to islet amyloid deposits. *Journal of Biological Chemistry* 265, 4173–4176.

Noda, M., Furutani, Y., Takahashi, H., Toyosato, M., Hirose, T., Inayama, S., Nakanishi, S., and Numa, S. (1982a). Cloning and sequence analysis of cDNA for bovine adrenal preproenkephalin. *Nature* 295, 202–206.

Noda, M., Teranishi, Y., Takahashi, H., Toyosato, M., Notake, M., Nakanishi, S., and Numa, S. (1982b). Isolation and structural organization of the human preproenkephalin gene. *Nature* 297, 431–434.

North, R.A., Williams, J.T., Surprenant, A., Christie, M.J. (1987). μ and δ receptors belong to a family of receptors that are coupled to potassium channels. *Proceedings of the National Academy of Sciences, U.S.A.* 84, 5487–5491.

Numa, S. (1984). Neuropeptide precursors and acetylcholine receptor: Molecular structures and gene expression. In *Proceedings of the IUPHAR 9th International Congress of Pharmacology, edited by W. Paton, J. Mitchell, and P. Turner; MacMillan, London*, 2, 273–288.

O'Hare, M.M.T., Tenmoku, S., Aakerlund, L., Hilsted, L., Johnsen, A. and Schwartz, T.W. (1988). Neuropeptide Y in guinea pig, rabbit, rat and man. Identical amino acid sequence and oxidation of methionine-17. *Regulatory Peptides* 20, 293–304.

Ohawa, K., Hayakawa, Y., Nishizawa, M., Yamagami, T., Yamamoto, H., Yanaihara, N. and Okamoto, H. (1985). Synergistic stimulation of VIP/PHM-27 gene expression by cyclic AMP and phorbol esters in human neuroblastoma cells. *Biochemical Biophysical Research Communications.* 32, 885–891.

Orci, L. (1982). Macro- and Micro-domains in endocrine pancreas. *Diabetes* 31, 538–565.

Orrego, F. (1979). Criteria for the identification of central neurotransmitters and their application to studies with some nerve tissue preparations *in vitro*. *Neuroscience* 4, 1037–1057.

Otsuka, M., Konishi, S., Yanagisawa, M., Tsunoo, A., and Akagi, H. (1982). Role of substance P as a sensory transmitter in spinal cord and sympathetic ganglia. In *Substance P in the Nervous System.* CIBA Foundation Symposium 91. pp 13–34. London: Pitman.

Pandol, S.J., Dharmsathaphorn, K., Schoeffield, M.S., Vale, W. and Rivier, J. (1986). Vasoactive intestinal peptide receptor antagonist [4Cl-D-Phe^6, Leu^{17}] VIP. *American Journal of Physiology* **250**, G553–557.

Pradayrol, L., Chayvialle, J.A., Carlquist, M. and Mutt, V. (1978). Isolation of a porcine intestinal peptide with C-terminal somatostatin. *Biochemical and Biophysical Research Communications* **85**, 701–708.

Pruss, R.M., and Stauderman, K.A. (1988). Voltage-regulated calcium channels involved in the regulation of enkephalin synthesis are blocked by phorbol ester treatment. *Journal of Biological chemistry* **263**, 13173–13178.

Puig, M.M., Gascon, P., Craviso, G.L. and Musacchio, J.M. (1977). Endogenous opiate receptor ligand: Electrically induced release in the guinea pig ileum. *Science* **195**, 419–420.

Rahier, J., Pauwels, S. and Dockray, G.J. (1987). Biosynthesis of gastrin: Localization of the precursor and peptide products using electron microscopic–immunogold methods. *Gastroenterology* **92**, 1146–1152.

Raybould, H.E., Gayton, R.J. and Dockray, G.J. (1988). Mechanisms of action of peripherally administered cholecystokinin octapeptide on brain stem neurons in the rat. *Journal of Neuroscience* **8**, 3018–3024.

Reeve, J.R., Walsh, J.H. Jr., Chew, P., Clark, B., Hawke, D. and Shively, J.E. (1983). Amino acid sequences of three bombesin-like peptides from canine intestine extracts. *Journal of Biological Chemistry* **258**, 5582–5588.

Reeve, J.R. Jr., Cuttitta, F., Vigna, S.R., Heubner, V., Lee, T.D., Shively, J.E., Ho, F-J., Fedorko, J., Minna, J.D. and Walsh, J.H. (1988). Multiple gastrin-releasing peptide gene-associated peptides are produced by a human small cell lung cancer line. *Journal of Biological Chemistry* **263**, 1928–1932.

Regoli, D., Drapeau, G., Dion, S., and d'Orleans-Juste, P. (1987). Pharmacological receptors for substance P and neurokinins. *Life Sciences* **40**, 109–117.

Reynolds, J.C., Ouyang, A., Cohen, S. (1985). Opiate nerves mediate feline pyloric response to intraduodenal amino acids. *American Journal of Physiology* **248**, G307–G312.

Robberecht, P., Conlon, T.P. and Gardner, J.D. (1976). Interaction of porcine vasoactive intestinal peptide with dispersed pancreatic acinar cells from the guinea pig: structural requirements for effects of vasoactive intestinal peptide and secretin on cellular adenosine 3'5' monophosphate. *Journal of Biological Chemistry.* **251**, 4635–4639.

Robberecht, P., Chatelain, P., Waelbroeck, M. and Christophe, J. (1982). Heterogeneity of VIP-recognizing binding sites in rat tissues. In *Vasoactive Intestinal Peptide*, edited by S. Said, pp 323–332. New York, Raven Press.

Robberecht, P., Coy, D.H., De Neef, P., Camus, J-C., Cauvin, A., Waelbroeck, M. and Christophe, J. (1987). [D-Phe^4] peptide histidine-isoleucinamide ([D-Phe^4] PHI), a highly selective vasoactive-intestinal peptide (VIP) agonist, discriminates VIP-preferring from secretin-preferring receptors in rat pancreatic membranes. *European Journal of Biochemistry* **165**, 243–249.

Rökaeus, Å (1987). Galanin: a newly isolated biologically active neuropeptide. *Trends in Neuroscience* **10**, 158–164.

Rökaeus, Å., and Brownstein, M.J. (1986). Construction of a porcine adrenal medullary cDNA library and nucleotide sequence analysis of two clones encoding a galanin precursor. *Proceedings of the National Academy of Science, U.S.A.,* **83**, 6287–6291.

Rose, C., Camus, A., and Schwartz, J.C. (1988). A serine peptidase responsible for the inactivation of endogenous cholecystokinin in brain. *Proceedings of the National Academy of Sciences, U.S.A.* **85**, 8326–8330.

Rosenfeld, M.G., Mermod, J-J., Amara, S.G., Swanson, L.w., Sawchenko, P.E., Rivier, J., Vale, W.W. and Evans, R.M. (1983). Production of a novel neuropeptide encoded by the calcitonin gene via tissue-specific RNA processing. *Nature* **304**, 129–135.

Rozsa, Z., Varro, A. and Jancso, G. (1985). Use of immunoblockade to study the involvement of peptidergic afferent nerves in the intestinal vasodilatory response to capsaicin in the dog. *European Journal of Pharmacology* **115**, 59–64.

Rozsa, Z., Mattila, J. and Jacobson, E.D. (1988). Substance P mediates a gastrointestinal thermoreflex in rats. *Gastroenterology* **95**, 265–276.

Said, S.I., and Mutt, V. (1970). Polypeptide with broad biological activity: Isolation from small intestine. Science 169: 1217–1218.

Said, S.I. and Mutt, V. (1972). Isolation from porcine-intestinal wall of a vasoactive octacosapeptide related to secretin and to glucagon. *European Journal of Biochemistry* **28**, 199–204.

Sausville, E.A., Lebacq-Verheyden, A-M., Spindel, E.R., Cuttitta, F., Gazdar, A.F., and Battey, J.F. (1986). Expression of the gastrin-releasing peptide gene in human small cell lung cancer: Evidence for alternative processing resulting in three distinct mRNAS. *Journal of Biological Chemistry* **261**, 2451-2457.

Schally, A.V., Huang, W-Y., Chang, R.C.C., Arimura, A., Redding, T.W., Millar, R.P., Hunkapiller, M.W. and Hood, L.E. (1980). Isolation and structure of pro-somatostatin: A putative somatostatin precursor from pig hypothalamus. *Proceedings of the National Academy of Sciences U.S.A.* **77**, 4489-4493.

Schubert, M.L., Bitar, K.N., and Makhlouf, G.M. (1982). Regulation of gastrin and somatostatin secretion by cholinergic and noncholinergic intramural neurons. *American Journal of Physiology* **243**, G442-G447.

Schubert, M.L., Saffouri, B., Walsh, J.H. and Makhlouf, G.M. (1985). Inhibition of neurally mediated gastrin secretion by bombesin antiserum. *American Journal of Physiology* **248**, G456-G462.

Schultz, R., Wuster, M., Simantov, R., Snyder, S. and Hertz A. (1977). Electrically stimulated release of opiate-like material from the myenteric plexus of the guinea pig ileum. *European Journal of Pharmacology* **41**, 347-348.

Schwartz, T.W., Fuhlendorff, J., Langeland, N., Thogersen, H., Jorgensen, J. Ch., and Sheikh, S.P. (1989). Y₁ and Y₂ receptor for NPY – the evolution of PP-fold peptide and their receptors. In *Neuropeptide Y-XIV Nobel Symposium* edited by V. Mutt, T. Hökfelt, K. Fuxe and J.M. Lundberg, pp 143, New York: Raven Press.

Schwartz, C.J., Kimberg, D.V., Sheerin, H.E., Field, M. and Said, S.I. (1974). Vasoactive intestinal peptide stimulation of adenylate cyclase and active electrolyte secretion in intestinal mucosa. *Journal of Clinical Investigation.* **54**, 536-544.

Seal, A., Yamada, T., Debas, H., Hollinshead, J., Osadchey, B., Aponte, G. and Walsh, J. (1982). Somatostatin-14 and −28: clearance and potency on gastric function in dogs. *American Journal of Physiology* **243**, G97-G102.

Seifert, H., Sawchenko, P., Chesnut, J., Rivier, J., Vale, W., and Pandol, S.J. (1985). Receptor for calcitonin gene-related peptide: binding to exocrine pancreas mediates biological actions. *American Journal of Physiology* **249**, G147-G151.

Sharp, G.W.G., Marchand-Brustel, Y.L., Yada, T., Russo, L.L., Bliss, C.R., Cormon, M., Monge, L. and Van Obberghen, E. (1989). Galanin can inhibit insulin release by a mechanism other than membrane hyperpolarization or inhibition of adenylate cyclase. *Journal Biological Chemistry* **264**, 7302-7309.

Sheikh, S.P., Holst, J.J., Skak-Nielsen, T., Knigge, U., Warberg, J., Theordorsson-Norheim, E., Hökfelt, T., Lundberg, J.M. and Schwartz, T.W. (1988). Release of NPY in pig pancreas: Dual parasympathetic and sympathetic regulation. *American Journal of Physiology* **255**, G46-G54.

Sheikh, S.P., Håkanson, R., and Schwartz, T.W. (1989). Y₁ and Y₂ receptors for neuropeptide Y. *FEBS Letters* **245**, 209-214.

Sheikh, S.P., O'Hare, M.M.T., Tortora, O., and Schwartz, T.W. (1989). Binding of monoiodinated neuropeptide Y to hippocampal membranes and human neuroblastoma cell lines. *Journal of Biological Chemistry* **264**, 6648-6654.

Shen, L-P., and Rutter, W.J. (1984). Sequence of the human somatostatin I gene. *Science* **224**, 168-171.

Shen, L-P., Pictet, R.L. and Rutter, W.J. (1982). Human somatostatin I: sequence of the cDNA. *Proceedings of the National Academy of Sciences U.S.A.* **79**, 4575-4579.

Shu, H-D., Love, J.A., and Szurszewski, J.H. (1987). Effect of enkephalins on colonic mechanoreceptor synaptic input to inferior mesenteric ganglion. *American Journal of Physiology* **252**, G128-G135.

Sims, S.M., Walsh, J.V. Jr., and Singer, J.J. (1986). Substance P and acetylcholine both suppress the same K^+ current in dissociated smooth muscle cells. *American Journal of Physiology* **251**, G580-G587.

Sjöqvist, A., Fahrenkrug, J., Jodal, M. and Lundgren, O. (1983). Effect of apamin on release of vasoactive intestinal polypeptide (VIP) from cat intestines. *Acta Physiologica Scandinavia.* **119**, 67-76.

Spindel, E.R., Chin, W.W., Price, J., Rees, L.H., Besser, G.M. and Habener, J.F. (1984). Cloning and characterization of cDNAs encoding human gastrin-releasing peptide. *Proceedings of the National Academy of Sciences, U.S.A.* **81**, 5699-5703.

Spindel, E.R., Zilberberg, M.D., Habener, J.F. and Chin, W.W. (1986). Two prohormones for gastrin-releasing peptide are encoded by two mRNAs differing by 19 nucleotides. *Proceedings of the National Academy of Sciences, U.S.A.* **83**, 19-23.

Spindel, E.R., Zilberberg, M.D., and Chin, W.W. (1987). Analysis of the gene and multiple messenger ribonucleic acids (mRNAs) encoding human gastrin-releasing peptide: Alternate RNA splicing occurs in neural and endocrine tissue. *Molecular Endocrinology* **1**, 224-232.

Stjärne, L., Lundberg, J.M. and Åstrand P. (1986). Neuropeptide Y – a cotransmitter with noradrenaline and adenosine 5'-triphosphate in the sympathetic nerves of the mouse vas deferens? A biochemical, physiological and electropharmacological study. *Neuroscience* **18**, 151-166.

Steardo, L., Knight, M., Tamminga, C.A., Barone, P., Kask, A.M. and Chase, T.N. (1985). CCK_{26-33} degrading activity in brain and nonneural tissue: A metalloendopeptidase. *Journal of Neurochemistry* **45**, 784-790.

Steenbergh, P.H., Höppener, J.W.M., Zandberg, J., Lips, C.J.M. and Jansz, H.S. (1985). A second human calcitonin/CGRP gene. *FEBS Letters* **183**, 403-407.

Sternini, C., Anderson, K., Frantz, G., Krause, J.E., and Brecha, N. (1989). Expression of substance P/neurokinin A-encoding preprotachykinin messenger ribonucleic acids in the rat enteric nervous system. *Gastroenterology* **97**, 348-356.

Suprenant, A. (1989). Synaptic transmission in neurons of the submucous plexus. In *Nerves and the Gastrointestinal Tract*, edited by M.V. Singer and H. Goebell, pp 253-263. Lancaster: MPT Press.

Svoboda, M., De Neef, P., Tastenoy, M., and Christophe, J. (1988). Molecular characteristics and evidence for internalization of vasoactive-intestinal-peptide (VIP) receptors in the tumoral rat-pancreatic acinar cell line AR 4-2 J. *European Journal of Biochemistry* **176**, 707-713.

Tache, Y., and Gunion, M. (1985). Central nervous system action of bombesin to inhibit gastric acid secretion. *Life Sciences* **37**, 115-123.

Tachibana, S., Araki, K., Ohya, S., and Yoshida, S. (1982). Isolation and structure of dynorphin, an opioid peptide, from porcine duodenum. *Nature* **295**, 339-340.

Takahasi, Y., Kato, K., Hayashizaki, Y., Wakabayashi, T., Ohtsuka, E., Matsuki, S., Ikehara, M. and Matsubara, K. (1985). Molecular cloning of the human cholecystokinin gene by use of a synthetic probe containing deoxyinosine. *Proceedings of the National Academy of Sciences, U.S.A.* **82**, 1931-1935.

Tatemoto, K. (1982). Neuropeptide Y: Complete amino acid sequence of the brain peptide. *Proceedings of the National Academy of Science, U.S.A.* **79**, 5485-5489.

Tatemoto, K. and Mutt, V. (1978). Chemical determination of polypeptide hormones. *Proceedings of the National Academy of Sciences U.S.A.,* **75**, 4115-4119.

Tatemoto, K. and Mutt, V. (1980). Isolation of two novel candidate hormones using a chemical method for finding naturally occurring polypeptides. *Nature.* **285**, 417-418.

Tatemoto, K., Carlquist, M. and Mutt, V. (1982). Neuropeptide Y – a novel brain peptide with structural similarities to peptide YY and pancreatic polypeptide. *Nature* **296**, 659-660.

Tatemoto, K., Rökaeus, Å., Jörnvall, H., McDonald, T.J., and Mutt, V. (1983). Galanin – a novel biologically active peptide from porcine intestine. *FEBS Letters* **164**, 124-128.

Tatemoto, K., Jörnvall, H., Shiimesmaa, S., Hallden, G. and Mutt, V. (1984). Isolation and characterization of cholecystokinin-58 (CCK-58) from porcine brain. *FEBS Letters* **174**, 289-293.

Tatemoto, K., Lundberg, J.M., Jörnvall, H., and Mutt, V. (1985). Neuropeptide K: Isolation, structure and biological activities of a novel brain tachykinin. *Biochemical Biophysical Research Communications.* **128**, 947-953.

Tavianini, M.A., Hayes, T.E., Magazin, M.D., Minth, C.D. and Dixon, J.E. (1984). Isolation, characterization, and DNA sequence of the rat somatostatin gene. *Journal of Biological Chemistry* **259**, 11798-11803.

Taylor, G. and Bywater, R.A.R. (1986). Antagonism of non-cholinergic excitatory junction potentials in the guinea-pig ileum by a substance P analogue antagonist. *Neuroscience Letters* **63**, 23-26.

Tsukada, T., Fink, J.S., Mandel, G. and Goodman, R.H. (1987). Identification of a region in the human vasoactive intestinal polypeptide gene responsible for regulation by cyclic AMP. *Journal of Biological Chemistry.* **262**, 8743-8747.

Tsunoo, A., Konishi, S. and Otsuka, M. (1982). Substance P as an excitatory transmitter of primary afferent neurons in guinea pig sympathetic ganglia. *Neuroscience* **7**, 2025-2037.

Turner, A.J., Matsas, R., and Kenny, A.J. (1985). Are there neuropeptide-specific peptidases? *Biochemical Pharmacology* **34**, 1347-1356.

Udenfriend, S. and Kilpatrick, D.L. (1983). Biochemistry of the enkephalins and enkephalin-containin peptides. *Archives of Biochemistry and Biophysics* **221**, 309-323.

Unden, A., Bartfai, T. (1984). Regulation of neuropeptide Y(NPY) binding by guanine nucleotides in the rat cerebral cortex. *FEBS Letters* **177**, 125-128.

Varro, A., Young, J., Gregory, H., Cseh, J., and Dockray, G.J. (1986). Isolation, structure and properties of the C-terminal flanking peptide of preprocholecystokinin from rat brain. *FEBS Letters* **204**, 386-390.

Vincent, S.R., Dalsgaard, C-J., Schultzberg, M., Hökfelt, T., Christensson, I. and Terenius, L. (1984). Dynorphin-immunoreactive neurons in the autonomic nervous system. *Neuroscience* **11**, 973-987.

von Euler, U.S. and Gaddum, J.H. (1931). An unidentified depressor substance in certain tissue extracts. *J. Physiol.* **72**, 74–87.

von Schrenck, T., Heinz-Erian, P., Moran, T., Mantey, S.A., Gardner, J.D. and Jensen, R.T. (1989). Neuromedin B receptor in esophagus: Evidence for subtypes of bombesin receptors. *American Journal of Physiology* **256**, G747–G758.

Vrontakis, M.E., Peden, L.M., Duckworth, M.L. and Friesen, H.G. (1987) Isolation and characterization of a complementary DNA (galanin) clone from estrogen-induced pituitary tumor messenger RNA. *Journal of Biological Chemistry* **262**, 16755–16758.

Waelbroeck, M., Robberecht, P., Coy, D.H., Camus, J-C., De Neef, P. and Christophe, J. (1985). Interaction of growth hormone-releasing factor (GRF) and 14 GRF analogs with vasoactive intestinal peptide (VIP) receptors of rat pancreas. Discovery of (N-Ac-Tyr^1,D-Phe^2)-GRF(1-29)-NH$_2$ as a VIP antagonist. *Endocrinology* **116**, 2643–2649.

Wahlestedt, C., Yanaihara, N. and Hakanson, R. (1986). Evidence for different pre- and post-junctional receptors for neuropeptide y and related peptides *Regulatory Peptides* **13**, 307–318.

Walsh, J.H. (1989). Bombesin-like peptides. In *Handbook of Physiology: The Gastrointestinal System, volume II*, edited by G.M. Makhlouf, pp 587–610. New York: American Physiological Society.

Watkinson, A., Dockray, G.J., Young, J. and Gregory, H. (1988a). Characterisation of N-terminally extended met-enkephalin Arg^{6}Gly^{7}Leu^8 variants in the porcine upper digestive tract. *Biochemica et Biophysica Acta.* **955**, 231–235.

Watkinson, A., Dockray, G.J., Young, J. and Gregory, H. (1988b). Proenkephalin A processing in the upper digestive tract: Isolation and characterisation of phosphorylated N-terminally extended Met-enkephalin Arg^{6}Phe^7 variants. *Journal of Neurochemistry* **51**, 1252–1257.

Watkinson, A., Dockray, G.J. and Young, J. (1988c). N-linked glycosylation of a proenkephalin A-derived peptide: Evidence for the glycosylation of an NH$_2$-terminally extended Met-enkephalin. *Journal of Biological Chemistry* **263**, 7147–7152.

Watkinson, A., Young, J., Varro, A., and Dockray, G.J. (1989). The isolation and chemical characterization of phosphorylated enkephalin-containing peptides from bovine adrenal medulla. *Journal of Biological Chemistry* **264**, 3061–3065.

Watson, S.P. (1984). Are the proposed substance P receptor sub-types, substance P receptors? *Life Sciences.* **25**, 797–808.

Watson, S.P., and Downes, C.P. (1983). Substance P induced hydrolysis of inositol phopsholipids in guinea-pig ileum and rat hypothalamus. *European Journal of Pharmacology* **93**, 245–253.

Weber, S., Zuckerman, J.E., Bostwick, D.G., Bensch, K.G., Sikic, B.I. and Raffin, T.A. (1985). Gastrin releasing peptide is a selective mitogen for small cell lung carcinoma *in vitro. Journal of Clinical Investigation* **75**, 306–309.

Weller, A., Smith, G.P., Gibbs, J. (1990). Endogenous cholecystokinin reduces feeding in young rats. *Science* **247**, 1589–1591.

Werb, Z., and Clark, E.J. (1989). Phorbol diesters regulate expression of the membrane neutral metalloendopeptidase (EC 3.4.24.11) in rabbit synovial fibroblasts and mammary epithelial cells. *Journal of Biological Chemistry* **264**, 9111–9113.

Werman, R. (1966). Criteria for identification of a central nervous system transmitter. *Comparative Biochemistry and Physiology.* **18**, 745–766.

Westermark, P., Wernstedt, C., Wilander, E. and Sletten, K. (1986). A novel peptide in the calcitonin gene related peptide family as an amyloid fibril protein in the endocrine pancreas. *Biochemical Biophysical Research Communications* **140**, 827–831.

Wiley, J. and Owyang, C. (1987). Neuropeptide Y inhibits cholinergic transmission in the isolated guinea pig colon: Mediation through α-adrenergic receptors. *Proceedings of the National Academy of Sciences, U.S.A.* **84**, 2047–2051.

Williams, B.J., Curtis, N.R., McKnight, A.T., Maguire, J., Foster, A. and Tridgett, R. (1988). Development of NK-2 selective antagonists. *Regulatory Peptides* **22**, 189.

Wormser, U., Laufer, R., Hart, Y., Chorev, M., Gilon, C. and Selinger, Z. (1986). Highly selective agonists for substance P receptor subtypes. *The EMBO Journal* **5**, 2805–2808.

Yamada, T., and Chiba, T. (1989). Somatostatin. In *Handbook of Physiology – The Gastrointestinal System, Volume II*, edited by G.M. Makhlouf, pp 431–453. New York: American Physiological Society.

Yasui, A., Naruse, S., Yanaihara, C., Ozaki, T., Hoshino, M., Mochizuki, T., Daniel, E.E. and Yanaihara, N. (1987). Corelease of PHI and VIP by vagal stimulation in the dog. *American Journal of Physiology* **253**, G13–G19.

Yiangou, Y., Di Marzo, V., Spokes, R.A., Panico, M., Morris, M., and Bloom, S.R. (1987). Isolation, characterization, and pharmacological actions of peptide histidine valine 42, a novel preprovasoactive intestinal peptide-derived peptide. *The Journal of Biological Chemistry.* **262**, 14010–14013.

Yokota, Y., Sasai, Y., Tanaka, K., Fujiwara, T., Tsuchida, K., Shigemoto, R., Kakizuka, A., Ohkubo, H., and Nakanishi, S. (1989). Molecular characterization of a functional cDNA for rat substance P receptor. *Journal of Biological Chemistry* **264**, 17649–17652.

Zarbin, M.A., Wamsley, J.K., Innis, R.B. and Kuhar, M.J. (1981). Cholecystokinin receptors: Presence and axonal flow in the rat vagus nerve. *Life Sciences* **29**, 697–705.

Zhou, Z-C., Villanueva, M.L., Noguchi, M., Jones, S.W., Gardner, J.D. and Jensen, R.T. (1986). Mechanism of action of calcitonin gene-related peptide in stimulating pancreatic enzyme secretion. *American Journal of Physiology* **251**, G391–G397.

Zhou, Z-C., Gardner, J.D., and Jensen, R.T. (1987). Receptors for vasoactive intestinal peptide and secretin on guinea pig pancreatic acini. *Peptides* **8**, 633–637.

Zhu, Z., Andrisan, O.M., Pot, D.A. and Dixon, J.E. (1989). Purification and characterization of a 43-kDa transcription factor required for rat somatostatin gene expression. *Journal of Biological Chemistry* **264**, 6550–6556.

Zuzel, K.A., Rose, C. and Schwartz, J-C. (1985). Assessment of the role of 'enkephalinase' in cholecystokinin inactivation. *Neuroscience* **15**, 149–158.

9 Transmission: γ-aminobutyric acid (GABA), 5-hydroxytryptamine (5-HT) and Dopamine

Judith M. Hills and Kristjan R. Jessen*

SmithKline Beecham Pharmaceuticals, The Frythe, Welwyn, Hertfordshire, AL6 9AR, U.K.

There is strong evidence for the existence of GABAergic and serotonergic autonomic neurons, while the evidence for dopamine as an atuonomic transmitter is less complete.

The major location of both GABAergic and serotonergic cells is in the enteric nervous system, where they have been found in several species. The role of these neurons in the regulation of gut function is, however, still unclear. Serotonergic neurons may also take part in the regulation of blood flow in certain vascular beds including the mesenteric circulation. The experimental support for dopaminergic autonomic transmission is strongest in blood vessels of the dog kidney and hind limb but is absent for the enteric nervous system.

The development of pharmacological tools of increasing specificity should soon make it possible to define the autonomic involvement of GABA, 5-HT and dopamine more precisely. The extent to which these transmitters are present in the human autonomic nervous system remains an important question.

1. GABA

Strong evidence now indicates that in vertebrates, GABAergic neurons are found in at least one part of the peripheral nervous system, i.e. the myenteric plexus of the gastrointestinal tract. In spite of the fact that some of the earliest pharmacological studies of GABA in the nervous system led to the suggestion that GABA might be involved in enteric neurotransmission (Hobbiger, 1958a, b) subsequent work on this amino acid focused on the central nervous system where GABA was eventually established as a major inhibitory transmitter (Roberts *et al.* 1976).

---

* Department of Anatomy and Embryology, University College, Gower Street, London WC1E 6BT, U.K.

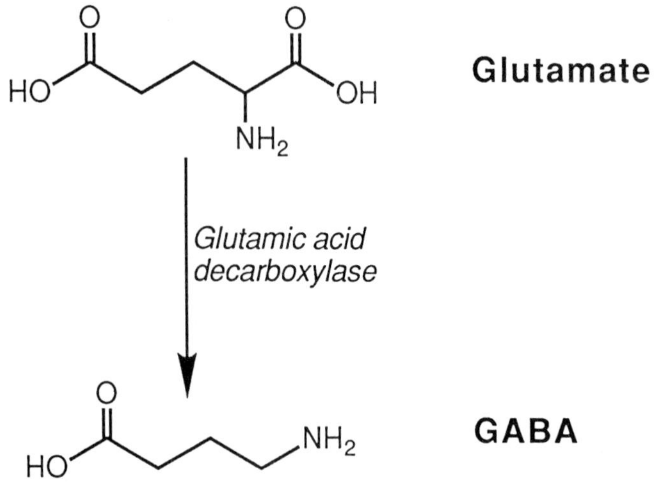

FIGURE 9.1   The synthetic pathway for GABA.

Although the peripheral nervous system was found to possess many GABA-related properties, including GABA receptors (DeGroat, 1970; Bowery and Brown, 1974; Feltz and Rasminsky, 1974; Brown and Marsh, 1978; for review see Hill and Bowery 1986), high affinity glial uptake of GABA (Bowery and Brown, 1972; Schon and Kelly, 1974; for review see Tanaka and Taniyama, 1986), the GABA synthesizing enzyme glutamic acid decarboxylase (GAD; EC 4.1.1.15) (Figure 9.1), the GABA degrading enzyme GABA transaminase (GABA-T; EC 2.6.119), and endogenous GABA (Beart *et al.*, 1974; Osborne *et al.*, 1974; Kanazawa *et al.*, 1976; Bertilson *et al.*, 1976), there was at first no evidence that this related to the presence of GABAergic neurons (for review see Iversen and Kelly, 1975; Roberts *et al.*, 1976; Nistri and Constanti, 1979). In 1979, however, biochemical and autoradiographic experiments on the myenteric plexus of the guinea pig revealed the presence of a distinct population of neurons with some of the key properties of GABAergic cells (Jessen *et al.*, 1979). Since then a substantial amount of data has been obtained, using diverse methods and a large range of animal species, which supports the notion that the myenteric plexus contains intrinsic GABAergic neurons. The possibility that GABAergic neurotransmission takes place in certain other peripheral tissues, including sympathetic ganglia and the bladder, has also been raised, although at present the evidence is less substantial than that pertaining to the myenteric plexus.

GABAERGIC NEURONS IN THE MYENTERIC PLEXUS

Biochemistry

In biochemical measurements of endogenous GABA levels in enteric tissues, the GABA contained in a population of myenteric neurons is inevitably diluted by surrounding tissues. Nevertheless, the amount of GABA in a longitudinal muscle –

myenteric plexus preparation from the guinea pig caecum measured about three times the amount of GABA in a control muscle tissue (Jessen *et al.*, 1979). To circumvent the problem of dilution, endogenous GABA was measured in thin layers of the gut wall. The layer containing the myenteric plexus yielded 4.6 nmole/mg protein which was 15–20 times more than that obtained from surrounding layers not harboring the plexus (Taniyama *et al.*, 1982a). In a developmental study of GABA content of the whole wall of the rat small intestine, a sharp peak in GABA levels was found at postnatal days 7 to 8. The explanation for this is not clear (Gilon *et al.*, 1987). By dissecting the myenteric plexus from the gut wall, it has been shown that the enteric ganglia devoid of smooth muscle are capable of synthesizing ^{3}H-GABA from its precursor ^{3}H-glutamic acid and accumulate the newly synthesized product (Jessen *et al.*, 1979). The amount of ^{3}H-GABA obtained in the myenteric plexus in these experiments reached about 50% of that found in slices of the cerebellum, although this tissue is considered to be rich in GABAergic neurons. In homogenates of the guinea pig myenteric plexus the specific activity of glutamic acid decarboxylase (GAD; EC4.1.1.15) was 21.8 nmole/mg protein per h (Jessen *et al.*, 1979), and in homogenates of a layer of gut colon containing the plexus it was 45.7 nmole/mg protein per h, a value ten times that obtained from surrounding plexus-free layers (Taniyama *et al.*, 1982b; Miki *et al.*, 1983).

Autoradiography

The ability of GABAergic neurons to selectively accumulate ^{3}H-GABA following incubation in low ^{3}H-GABA concentrations has been widely used to identify GABAergic cells in the central nervous system and in the peripheral nervous system of invertebrates. The ^{3}H-GABA is accumulated via high affinity GABA uptake sites and neurons possessing such uptake mechanisms can be visualized by autoradiographic methods. Using this approach, a number of studies have confirmed the presence of a population of ^{3}H-GABA-labelled neuronal cell bodies and fibres in the myenteric plexus of guinea pig, rat and chicken, using cell cultures (Figure 9.2) or whole mounts of the myenteric plexus or sections from the gut wall (Jessen *et al.*, 1979; Krantis and Kerr, 1981a; Jessen *et al.*, 1983; Saffrey *et al.*, 1983; Krantis *et al.*, 1986). The whole-mount preparations showed that the ^{3}H-GABA-labelled fibres were present both inside myenteric ganglia and in the primary, secondary and tertiary nerve plexuses (Krantis and Kerr, 1981a). The destination of some of these labelled fibres can be inferred from observations of ^{3}H-GABA autoradiography in tissue sections (Jessen *et al.*, 1983; Krantis *et al.*, 1986). This showed that heavy autoradiographic labelling was present in the intramuscular nerves that innervate the gut musculature. Electron microscopic autoradiography revealed that the dense autoradiographic labelling was only seen over some axons, other axons, as well as the surrounding smooth muscle being unlabelled. This suggests that, as expected, only a subpopulation of the intramuscular axons have high affinity ^{3}H-GABA uptake sites and thus may be expected to be GABAergic. Since GABA does not have any direct effects on smooth muscle, the role of these fibres in intramuscular nerves is probably the pre-junctional regulation of transmitter release from their neighbours (see below).

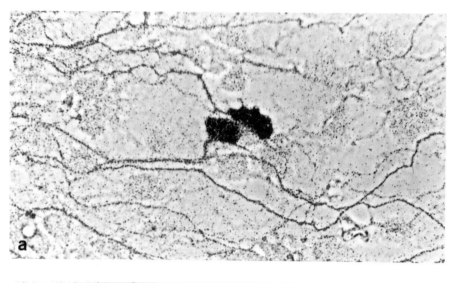

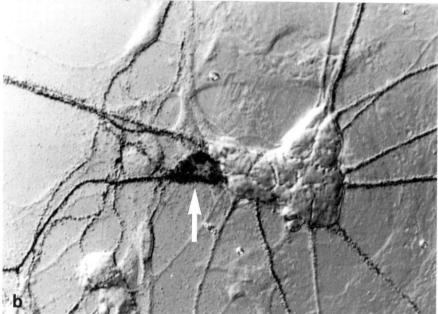

FIGURE 9.2 Autoradiograph showing uptake of ^{3}H-GABA into neurons from the guinea pig myenteric plexus grown in tissue culture. (a) Two heavily labelled neuronal cell bodies and several labelled neurites are visible. Light labelling is seen over glial cells in the background. (b) One neuronal cell body (arrow) out of about seven forming a small aggregate is heavily labelled, illustrating the selectivity of the neuronal ^{3}H-GABA uptake. ($\times 500$). Taken with permission from Jessen *et al*; 1983. Neuroscience Pergamon Press Ltd. Fig. 10. Nature Fig. 1b.

Immunohistochemistry

An alternative method for identification of GABAergic neurons is the use of antibodies to GABA and immunohistochemistry. GABA-immunolabelled enteric neurons were first described in whole-mount preparations of the guinea pig and rat intestine and in the quail gut (Baetge and Gershon, 1986; Jessen *et al.*, 1986) (Figure 9.3). Using either whole mounts or sections many laboratories have confirmed the existence of these cells, and they have now been described in other species, including frogs and cats, and with the help of several different anti-GABA antibodies (Saito and Tanaka, 1986; Davanger *et al.*, 1987; Hills *et al.*, 1987; Furness *et al.*, 1989; Gabriel and Eckert, 1989). The distribution of GABA-immunopositive cells and fibres is in excellent agreement with the results previously obtained with autoradiographic methods and the immunolabelled cells have been taken to be the same population of neurons as those possessing high-affinity GABA uptake sites, although this has not been proved directly. Thus, in the guinea-pig, scattered GABA positive cell bodies are seen in myenteric ganglia, immunolabelled fibres are present to some extent in the ganglia, but chiefly in secondary and tertiary plexuses and, most prominently, in the intramuscular nerves (Figure 9.4). GABA immunopositive neurons are essentially absent from the submucous plexus. At least in the guinea pig and rat, GABA immunolabelled fibres extend throughout the entire length of the gastrointestinal tract. They are essentially confined to the extrinsic musculature, few fibres being found in the submucosa or mucosa. A detailed study has been made on the morphology and projections of GABA-immunolabelled neurons in the guinea pig small intestine following incubation with GABA (Furness *et al.*, 1989). The preincubation enhanced the GABA labelling of cell bodies and dendrites, but did not affect the distribution of immunoreactive structures, an observation which provides further support for the notion that neurons which possess high affinity GABA uptake sites and GABA immunoreactive neurons are the same cell population. The GABA-positive cells had a characteristic morphology, carrying many short dendrites and conforming to the morphology of Dogiel type I neurons. Some of these cells projected to the tertiary plexus while others projected to the circular muscle. Typically the axon of these cells ran in the anal direction for about 0.5 mm before giving rise to branches which joined the circumferentially oriented intramuscular nerves in the circular muscle.

GABA release

Electrical depolarization causes Ca^{++}-dependent and tetrodotoxin-sensitive release of GABA from various gut preparations, as expected if GABA was an enteric neurotransmitter. Release of preloaded ^{3}H-GABA has been obtained from whole segments of cat colon (Taniyama *et al.*, 1982a), from strips of the guinea pig taenia coli containing the myenteric plexus (Jessen *et al.*, 1983) (Figure 9.5), from the guinea pig ileal longitudinal muscle-myenteric plexus preparation (Kerr and Krantis, 1983) and from synaptosomes prepared from guinea pig small intestine (Yau and Verdun, 1983). Direct proof that ^{3}H-GABA can be released from neurons intrinsic to the myenteric plexus came from the observation that elevated potassium releases

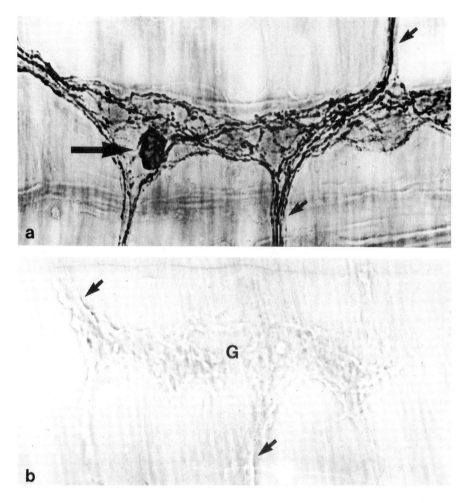

FIGURE 9.3   FIGURE 1(a) Immunoperoxidase micrograph of the rat colon, treated with anti-GABA antibodies. The micrograph shows a single oblong ganglion and three interconnecting nerve strands (two of which are labeled with small arrows). Within the ganglion, an immunopositive neuronal cell body (long arrows) can be seen, as well as numerous immunolabeled nerve fibres, lying among unlabeled neurons and glia. Immunopositive nerve fibres can also be seen in the interconnecting strands. A bundle of anti-GABA-labeled axons appears to traverse the ganglion, running from one of the arrowed, interconnecting strands to the other. (b) Immunoperoxidase micrograph of the rat colon, treated with absorbed anti-GABA antiserum. A single oblong ganglion (G), with associated interconnecting strands (examples labeled with arrows), occupies the centre of the field. No immunopositive structures are present. (× 500.)
Neuroscience 6, 1628–1634, Fig. 1, 1986. Taken with permission from Jessen et al., 1986.

^{3}H-GABA from cultures containing only intrinsic neurons from the guinea pig myenteric plexus (Jessen et al., 1983) (Figure 9.5). It is important that release of endogenous GABA can also be obtained from the gut. Electrical stimulation (Taniyama et al., 1983), pharmacological challenge by certain peptides or hormones (CCK, neurotensin and substance – P) but not others (gastrin and VIP) (Tanaka and Taniyama, 1985; Nakemoto et al, 1987.; Sano et al., 1989) and exposure to ethylene

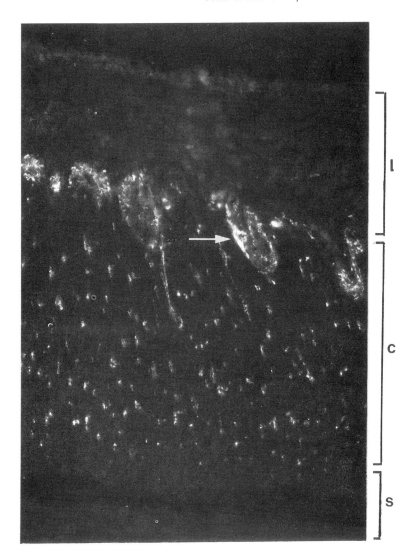

FIGURE 9.4   Immunofluorescence micrograph of guinea-pig colon treated with anti-GABA antiserum. The micrograph shows two immunolabelled myenteric ganglia, one of which contains a clearly immunopositive nerve cell body. Intramuscular nerve bundles in the circular muscle (c) region which has been cut transversely, contain labelled fibres. The submucosa (s) and longitudinal muscle (l) are unlabelled (×320). Hills *et al*; Neuroscience *22*, 301–312, 1987. Fig. 16. Taken with permission from Jessen *et al*, 1987.

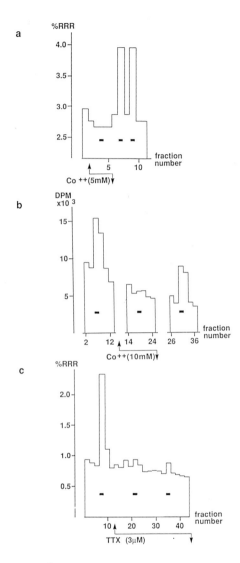

FIGURE 9.5   Evoked release of ^{3}H-GABA from enteric neurones (a) Potassium-induced release of [^{3}H]GABA from cultures of myenteric plexus. Bars indicate 15 min. exposure 62 mMK$^+$ solution. This experiment illustrates the reversible Ca^{2+} dependency of the evoked release; 5mM Co^{2+} (to block Ca^{2+} entry) was present as indicated. The results are expressed as a percentage of radioactivity remaining in the tissue immediately before release. (b) Reversible Ca^{2+} dependence of [^{3}H]GABA release from a strip of taenia coli, evoked by electrical field stimulation (bars, 10Hz, 0.2ms, 45 V). This tissue was incubated in 0.7μM[^{3}H]GABA in Kr-HEPES solution, and 10mM Co^{2+} was present as indicated. The FIGURE shows absolute quantities of radioactivity released (DPM). (c) Blockade of [^{3}H]GABA release by 3μM tetrodotoxin from a strip of taenia coli. Bars indicate three successive periods of electrical field stimulation (10Hz, 0.2ms, 45 V). Tetrodotoxin was present as indicated. The tissue was incubated in 0.7μM [^{3}H]GABA in Krebs solution. The results are expressed as a percentage of radioactivity remaining in the tissue immediately before release. Neuroscience, Figs. 1,6,7,16. Taken with permission from Jessen *et al.*, 1983.

diamine (EDA) (Kerr and Ong, 1984; Stone and Perkins, 1984) which is reported to release GABA from central nerve terminals, all release endogenous GABA from guinea pig gut preparations. Similarly, avermectin B_{1a}, a macrolide anthelmintic that releases GABA from brain synaptosomes (Pong et al., 1980), has GABA-mimetic effects in the ileum, while 3-MPS, a compound which reduces GABA release, also appears to inhibit endogenous GABA release from intestinal preparations (Kerr and Ong, 1986a, b; Ong and Kerr, 1987).

GABA release from enteric neurons is clearly subject to multiple controls. In addition to being stimulated by CCK, neurotensin and substance-P (see above) there is evidence for regulation of GABA release by GABA itself acting via $GABA_A$ receptors (Taniyama et al., 1985). Evoked ^3H-GABA release from the guinea-pig small intestine is also modulated pre-synaptically by α_2-adrenergic and M_1 muscarinic receptors (Hashimoto et al., 1986), both reducing the transmitter release. In addition 5-HT also appears to be able to modulate ^3H-GABA release . 5-HT exerts an excitatory effect on the GABA-ergic neuron via the 5-HT_3 receptor and an inhibitory effect via the 5-HT_{1A} receptor (Shirakawa et al., 1989). One possible mechanism governing GABA release from enteric neurons in the guinea-pig ileum appears to involve protein kinase C since a phorbol ester, TPA, potentiates the high K^+ evoked release of ^3H-GABA (Shuntoh et al., 1989).

The observation that GABA released from enteric neurons in turn affects the release of acetylcholine from cholinergic enteric terminals (Tanaka and Taniyama, 1985; Nakemoto et al., 1987; Sano et al., 1989) has important implications for the role of these cells within the enteric nervous system (see below).

The pharmacology of GABA actions in the gut

Hobbiger (1958a, b) was the first to show that the actions of GABA might not be restricted solely to effects in the central nervous system. From this and other work it became clear that GABA not only possessed excitatory and inhibitory activity in its own right, but also reduced the excitatory effects of other agents applied to isolated gut muscle preparations containing smooth muscle and nerves (Hobbiger et al., 1958a; Inouye et al., 1960). The pharmacological effects of GABA on gastrointestinal preparations were not further investigated until after the demonstration of intrinsic enteric GABA neurons.

GABA receptors in the enteric nervous system as elsewhere may be subdivided into $GABA_A$ receptors which are selectively activated by muscimol and 3-aminopropanesulphonic acid (3-APS) and antagonized by bicuculline, and $GABA_B$ receptors, which are selectively activated by baclofen and antagonized by compounds such as phaclofen (Krantis et al., 1980., Bowery et al., 1981., Ong and Kerr., 1983a).

GABA, and selective $GABA_A$ and $GABA_B$ receptor agonists have diverse effects on the gastrointestinal tract, both in vitro and in vivo. The response to agonists from each receptor subclass fall into two broad categories, inhibitory and excitatory, and are mediated exclusively via the neuronal elements of the gut (see below). In general, the common response to $GABA_B$ receptor activation by for example baclofen, is one of inhibition, mediated usually by a prejunctional reduction of acetlycholine release (Bowery *et al.*, 1981; Kaplita *et al.*, 1982; Bowery *et al.*, 1982; Giotti *et al.*, 1983). Muscle contraction following stimulation of acetylcholine release by $GABA_A$ receptor activation is seen in some gut areas (Krantis *et al.*, 1980; Giotti *et al.*, 1983) while $GABA_A$ receptor agonists can also cause gut muscle relaxation (Ong and Kerr, 1987).

Guinea-Pig Ileum In the guinea-pig ileum, $GABA_A$ and $GABA_B$ receptors are present. Activation of $GABA_A$ receptors causes an initial rapid contraction seen in both electrically stimulated and unstimulated longitudiual muscle preparations (Krantis *et al.*, 1980; Giotti *et al.*, 1983). This excitation which is mimicked by $GABA_A$ receptor agonists such as muscimol and 3-aminopropane sulphonic acid (3-APS), can be blocked by bicuculline, is tetrodotoxin and atropine sensitive and is thus assumed to be cholinergic in nature (Krantis *et al.*, 1980; Giotti *et al.*, 1983). The inhibition of the electrically evoked cholinergically mediated contraction and the relaxant response to GABA is a $GABA_B$ receptor effect since it is mimicked by baclofen and antagonized by both GABA and baclofen tachyphylaxis, while being unaffected by $GABA_A$ receptor antagonists (Bowery *et al.*, 1981; Giotti *et al.*, 1983; Ong and Kerr, 1983a). Electrophysiological analysis of these effects in the guinea-pig ileum myenteric plexus by Cherubini and North (1984a,b) clearly demonstrates the characteristics of GABA-mediated effects in the gut nerve plexuses. GABA selectively and directly depolarized AH type neurones while having no effect on S type neuronal soma. This depolarization is largely blocked by bicuculline and therefore mediated by $GABA_A$ receptors, however there is a small component which may be non-somatic and mediated by interaction at the $GABA_B$ site since baclofen will also depolarize these neurons (Cherubini and North, 1984a). While GABA has no direct effect on S-type neurons, it does reduce their synaptic input by a presynaptic action. This is presumed to be a $GABA_B$ effect, since it is mimicked by baclofen and insensitive to blockade by bicuculline. The ability of baclofen to reduce calcium currents in AH/type2 myenteric neurones (Cherubini and North, 1984a,b) suggests that the mechanism of the presynaptic inhibition of fast and slow excitatory post synaptic potentials (epsps) recorded in S-type neurons may well be via inhibition of terminal calcium influx (Cherubini and North, 1984b).

The interaction between 5-HT and GABA has been investigated in the guinea-pig ileum following initial studies in which Hobbiger (1958a) demonstrated that GABA was able to reduce contractile responses to 5-HT. These results were confirmed by Ong and Kerr (1983b), who further demonstrated that the depression was highly dependent on the time of application of both GABA and 5-HT. When 5-HT was given within 20s of the GABA application, the depression of contractions to 5-HT was said to be mediated via $GABA_A$ receptors, while when 5-HT was applied during the GABA relaxation, the depression of contraction was suggested to be $GABA_B$ receptor mediated (Ong and Kerr, 1983b). The observation that cross desensitization between GABA and 5-HT could be obtained in the guinea-pig ileum led Tonini and colleagues (1983) to suggest that contractile responses to GABA were mediated by action on a 5-HT interneurone. These results could not be substantiated by further experimentation (Ong and Kerr, 1985) which suggests that previous observations probably reflected a physiological antagonism between GABA and 5-HT.

In the central nervous system, benzodiazepines and barbiturates are known to bind to sites on the $GABA_A$ receptor and to modulate the affinity of the receptor complex for GABA and the gating of the chloride channel (Olsen, 1981). The interaction between benzodiazepines and the gut $GABA_A$ receptor has been investigated by several groups. Benzodiazepines bind in a saturable manner to a single class of sites in the ileal longitudinal muscle myenteric plexus preparation ($K_D = 43$ nM, $B_{max} = 229$ fmol/mg protein)(Hullihan et $al.$, 1983). The receptor revealed in these binding studies appears to be more similar to the peripheral benzodiazepine site rather than the central site since clonazepam failed to displace ^{3}H-diazepam binding while Ro5-4864 was a potent displacer (Hullihan et $al.$, 1983). In these studies the effect of diazepam was investigated on the guinea-pig ileal cholinergic contraction in the absence of GABA or $GABA_A$ receptor agonists. Diazepam caused a marked reduction of the contraction, but since diazepam also reduced contractions to carbachol, potassium and calcium, an inhibition of calcium flux rather than a receptor-mediated effect is likely to be the underlying mechanism. Midazolam and diazepam potentiate the bicuculline-sensitive GABA depolarization in guinea-pig ileum myenteric neurones (Cherubini and North, 1985), an effect blocked by the benzodiazepine antagonist Ro 15–1788. The benzodiazepines are more potent at eliciting this functional response than in previously reported binding displacement studies, possibly pointing to the functional receptor belonging to the central and not the peripheral receptor class (Cherubini and North, 1985). In agreement, however, with the previous studies, was the reversible depression by benzodiazepines of the calcium current evoked by direct depolarization of AH-type neurones (Cherubini and North, 1985).

A pre-synaptic $GABA_A$ autoreceptor has been reported on cholinergic terminals in the guinea-pig ileum (Taniyama et $al.$, 1985). Activation of this receptor by muscimol, but not baclofen, results in inhibition of evoked ^{3}H-GABA release from the longitudinal muscle myenteric plexus preparation. The inhibition by muscimol is blocked by bicuculline, may be potentiated by the benzodiazepines, diazepam and clonazepam, and the benzodiazepine potentiation is blocked by Ro-15–1788 (Taniyama et $al.$, 1988). The ileal benzodiazepine binding site in this study appears akin to the central benzodiazepine receptor in that clonazepam is more potent than diazepam, both compounds having IC_{50} values below 1 μM, and the responses are antagonized by the central benzodiazepine receptor antagonist Ro-15–1788. What is of interest, however, is that the central GABA autoreceptor appears to be almost universally of the $GABA_B$ subtype (Pittaluga et $al.$, 1987; Bonanno et $al.$, 1989; Raiteri et $al.$, 1989). In addition, this study confirmed that the stimulation of acetylcholine release mediated by a $GABA_A$ postsynaptic excitation of cholinergic neurons may be modulated by benzodiazepines, this receptor too being of the central type (Taniyama et $al.$, 1988).

Not only do $GABA_A$ receptors in the guinea-pig ileum appear to be modulated by benzodiazepines, but also by barbiturate compounds (Ong and Kerr, 1984; Kerr and Ong, 1986a). The potentiation of $GABA_A$-mediated responses in the intestine thus mirrors that seen in the central nervous system, and suggests that the $GABA_A$

receptor complex is similar if not identical to that reported by Olsen (1981) for the central nervous system.

By virute of its simplicity and accessibility, the guinea-pig ileum has been used extensively to study new compounds for GABA$_A$ and GABA$_B$ receptor agonist and antagonist activity. For example, before the GABA$_B$ antagonist phaclofen was used in studies on the hippocampus to elucidate a physiological role for GABA$_B$ receptors (Dutar and Nicoll, 1988), it was shown to be an antagonist in the ileum (Kerr et al., 1987). Two novel GABA$_B$ receptor agonists which bind to central GABA$_B$ receptors have been recently reported to be pharmacologically active at the GABA$_B$ receptor in the guinea-pig ileum (Hills, et al., 1989; Hills and Howson, 1990).

Colon. The guinea-pig colon has been studied in vitro and in vivo. GABA inhibits the spontaneous contractions in the guinea-pig distal colon in vitro, a response which is insensitive to antagonism by bicuculline (Krantis et al., 1980; Ong and Kerr, 1983a). Ong and Kerr (1983a) reported that the spontaneous contractions are cholinergic in nature and therefore the GABA- and baclofen-mediated relaxation is mediated via inhibition of cholinergic nerve activity. While this probably accounts for a large part of the relaxation seen, ongoing contractions do continue in the presence of atropine (3 μM), and tetrodotoxin (3μM) converts irregular spontaneous activity to regular and presumably myogenic contractions (J.M. Hills, personal observation). It is therefore possible that since Krantis et al. (1980) carried out their experiments in the presence of hyoscine, the GABA inhibition is also due to either stimulation of non-adrenergic non-cholinergic (nanc) inhibitory nerves or inhibition of nanc excitatory nerves. The experiments of Krantis and colleagues (1980) were carried out before the identification of GABA$_A$ and GABA$_B$ receptor subtypes in the intestine. In addition, they report that the GABA relaxation in the distal colon can be partly blocked by bicuculline. We have observed, however, that muscimol (1–100 μM), the GABA$_A$ receptor agonist will relax the guinea-pig colon in a bicuculline (1–10 mM) insensitive manner (J.M. Hills, personal observation). Other reported data also indicate that the GABA relaxation of guinea-pig distal colon is purely a GABA$_B$ receptor mediated effect since it is mimicked by baclofen, blocked by δ-aminovaleric acid (DAVA) and insensitive to bicuculline (Giotti et al., 1985). As in the experiments of Krantis et al., (1980), the relaxation was still observed in the presence of atropine (Giotti et al., 1985). When the same investigators examined GABA actions in vivo given intravenously they found GABA and baclofen to reduce basal tone and spontaneous contractions (Giotti et al., 1985). The relaxation in vivo was bicuculline insensitive, but blocked by DAVA, it could no longer be recorded after atropine treatment and was greater following physostigmine pretreatment (Giotti et al., 1985). The overall conclusions from these studies must be that GABA acts predominantly on GABA$_B$ receptors in the guinea-pig colon to reduce spontaneous activity. Most of this action is via inhibition of cholinergic tone, although other transmitter systems may also be involved.

Peristaltic activity in the guinea-pig colon is modulated by GABA and GABA-like agents. Complete or partial blockade of faecal pellet expulsion from an excised piece of colon occurs with $GABA_A$ receptor antagonists and $GABA_A$ and $GABA_B$ receptor desensitization (Krantis and Kerr, 1981b; Ong and Kerr, 1983a). This result was further investigated by Frigo *et al.*, (1987) who found in contrast that bicuculline at concentrations which did not cause cholinesterase inhibition, increased the acetylcholine (ACh) output and efficiency of the peristaltic reflex. Bicuculline also potentiated cholinergic excitatory and non-adrenergic non-cholinergic inhibitory nerve mediated responses. In agreement with Kerr and colleagues the authors conclude that a GABAergic synapse is involved in the propagation of peristaltic wave (Frigo *et al.*, 1987). A study on peristaltic activity in the guinea-pig ileum confirmed the ability of bicuculline to enhance the efficiency of the peristaltic reflex, however, the effect could not be reproduced using a new highly selective $GABA_A$ receptor antagonist (SR 95531) (Tonini *et al.*, 1989) and such data casts doubt on previous conclusions drawn from studies with bicuculline.

In the rabbit distal colon, GABA stimulates non-adrenergic non-cholinergic inhibitory nerves via $GABA_A$ receptors resulting in relaxation. $GABA_B$ receptor activation in this tissue also causes relaxation but by inhibition of cholinergic activity (Tonini *et al.*, 1989). GABA and baclofen reduce peristaltic efficiency in this tissue, an effect antagonized by DAVA, while SR 95531 is without effect (Tonini *et al.*, 1989). In the cat colon on the other hand, GABA causes an increase in spontaneous contraction amplitude with no after relaxation (Taniyama *et al.*, 1987). The excitatory effect appears to be mediated by a bicuculline-sensitive stimulation of cholinergic neurones. The observation that bicuculline alone reduced contraction amplitude may suggest a modulatory role for endogenous GABA in this tissue (Taniyama *et al.*, 1987). GABA acting on $GABA_B$ receptors reduces cholinergic transmission in the electrical field stimulated guinea-pig taenia coli, although only modest changes in tone occurred in the resting preparation (Ronai *et al.*, 1987). Thus the effects of GABA on colonic tissue are species specific and further work is necessary before the role of GABA in the colon can be fully elucidated.

Duodenum. GABA responses in the duodenum are very different to GABA responses in other areas of the gastrointestinal tract. GABA relaxes the rat duodenum by stimulating intrinsic nanc inhibitory nerves (Maggi *et al.*, 1984; Krantis and Harding, 1987), an effect mimicked by $GABA_A$ receptor agonists and blocked by bicuculline. The nicotinic receptor stimulant DMPP and the muscarinic M_1 receptor agonist McN-A-343 produce a similar effect in this tissue, and this and other evidence has led to the suggestion that GABA neurones under cholinergic control, release GABA which interacts with $GABA_A$ receptors on the nanc inhibitory, possibly purinergic neurones (Maggi *et al.*, 1984; Micheletti *et al.*, 1988). Ethylene diamine (EDA), a compound known to release GABA from nerve terminals, also mimics the effect of GABA. This response is tetrodotoxin resistant and unlike the relaxation seen to exogenous GABA in that it is not blocked by $GABA_A$ antagonists (Maggi *et al.*, 1989).

In the guinea-pig duodenum, GABA and baclofen cause relaxation (Ong and Kerr, 1987). EDA induces ^3H-GABA release from the duodenum and mediates relaxation of the preparation presumed to be via the release of GABA from enteric neurones

(Ong and Kerr, 1987). Although on occasions, a small inhibition of GABA and EDA relaxations was obtained with bicuclline, 3-APS and muscimol exerted inhibition only at concentrations as high as 100μM (Ong and Kerr, 1987). These results are in contrast to those obtained in the guinea-pig duodenum by Barbier et al., (1989), who found that GABA caused contraction via excitation of cholinergic neurones mediated by a $GABA_A$ receptor action. In the same study, baclofen reduced electrically evoked twitch contractions (Barbier et al., 1989).

In a study on the human jejunum in vitro it is reported that GABA and baclofen cause a tetrodotoxin-sensitive inhibition of muscle tone which is mediated by action at the $GABA_B$ receptor since 3-APS and muscimol are ineffective (Gentilini et al., 1988).

A study in conscious animals utilising myoelectrical recording techniques in the jejuno-duodenal region of the rat intestine, concluded that GABA has a dual action on the rat small intestine (Fargeas et al., 1988). GABA and muscimol exert a peripheral action resulting in inhibition of intestinal motility patterns, whereas baclofen acting at a central nervous system locus increased and disrupted duodenal cyclic motility (Fargeas et al., 1988).

Stomach. Although there appear to be no reports on the motility effects of GABA on isolated stomach preparations, a wealth of data now exists which points to both CNS mediated effects of GABA compounds on secretion and motility and an endocrine role for endogenous mucosal GABA. Peripherally administered baclofen causes an atropine sensitive stimulation of gastric acid secretion in the anaesthetized rat which can be blocked by vagotomy (Goto and Debas, 1983; Goto et al., 1985; Andrews and Wood, 1986). Baclofen was shown to stimulate vagal discharge in the rat (Goto et al., 1985), the central locus of action being confirmed by Wood et al., (1987) who mimicked the effect of peripherally administered baclofen by injecting it into the ventromedial hypothalamus. The work of Andrews and colleagues also demonstrated that baclofen acting centrally stimulated gastric motility (Andrews and Wood, 1986; Wood et al., 1987) and that a similar effect also occurred in the ferret stomach (Andrews et al., 1987). Other studies report that centrally administered muscimol also stimulates gastric acid secretion (Levine et al., 1981) and that the effects of intracerebroventricularly administered muscimol and baclofen are blocked by bicuculline administered in the same way (Blandizzi et al., 1988). This data should be considered in the light of earlier investigations showing that bicuculline alone administered into the nucleus ambiguus caused a profound increase in stomach contractility in the cat (Williford et al., 1981).

The effect of GABA and GABA-related compounds on gastrointestinal transit has been studied in the mouse (Sivam and Ho, 1983). Both muscimol and baclofen delayed gastric emptying in the mouse, however, bicuculline itself mimicked the effect of muscimol and in combination produced a greater than additive effect (Sivam and Ho, 1983). Such studies clearly point to the need for further investigation in this area.

Experiments on isolated everted guinea-pig stomach preparations show that GABA will stimulate gastric acid secretion by a $GABA_A$ receptor action (Tsai et al., 1987). The effect of GABA is sensitive to bicuculline, muscarinic blockade and tetrodotoxin, indicating that GABA probably stimulates cholinergic nerves present in the isolated preparation (Tsai et al., 1987). The picture is further complicated by consideration of the effects of GABA and related compounds on gut hormone output. GABA, when added to antral mucosal fragments from the rat stomach stimulates gastrin and inhibits somatostatin release, both effects being sensitive to blockade by bicuculline (Harty and Franklin, 1983; 1986). It would appear that the GABA$_A$ receptor involved in this response is located on cholinergic nerve terminals since the regulation of hormone release is additionally inhibited by muscarinic antagonists and tetrodotoxin and the pattern of the GABA response closely resembles the reported actions of cholinergic agonists on G- and D- cell function (Harty and Franklin, 1986).

The observation that GABA itself was present in gut endocrine cells and may act as a gut hormone was made by Gilon et al., (1987) and Jessen et al., following the demonstration of ^{3}H-GABA uptake and endogenous GABA respectively in some mucosal cells in the rat antrum and duodenum. These observations were extended by the experiments of Davanger et al., (1989) who showed that GABA containing endocrine cells were present in all gut regions even though the highest concentration was present in the antrum pyloricum. The role of these cells is so far unknown. In the light of the above data, it is possible that one of the functions of mucosal GABA may be to modulate the output of gastrin and somatostatin either by a direct hormone action or by a nerve mediated action (Harty and Franklin, 1986).

The role of enteric GABAergic neurons

Much of the pharmacological data has been derived from experiments carried out in guinea-pig ileum longitudinal muscle while the predominant GABAergic innervation in the guinea-pig is within the myenteric plexus and circular muscle. While the pharmacology of myenteric plexus GABA mechanisms in the longitudinal muscle studies provides an indication of overall function, similar experiments are yet to be carried out in circular muscle preparations.

Similarly, very little pharmacological data exist from studies in either rat colon or ileum and yet immunohistochemical studies show an extensive GABAergic innervation in the muscle and nerves of both of these tissues (Jessen et al., 1986., Hills et al., 1987).

Thus the part played by GABAergic neurons in the neural circuits of the gut is still unclear. In the rat, the neurons of the myenteric ganglia are surrounded by many GABA-containing fibres. Such fibres also appear to run for considerable distances from ganglion to ganglion along the long axis of the gut. There is good evidence for GABA receptors on myenteric neurons and their processes, and it seems likely that in the rat, GABAergic neurons might act as conventional interneurons within the myenteric plexus.

In the guinea-pig the distribution of GABAergic fibres is different since they are much more pronounced in the circular muscle than within myenteric ganglia, and a recent study indicates that the ganglia do not contain GABAergic terminals (Pompolo and Furness 1990). Thus, the myenteric GABAergic neurons in the guinea-pig project from the plexus into the circular muscle. These fibres cannot

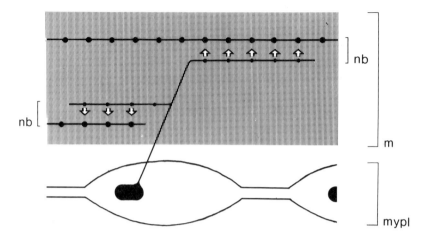

FIGURE 9.6   Possible role of GABAergic neurons in the guinea pig myenteric plexus. The drawing shows ganglia of the myenteric plexus (mypl) and adjoining musculature (m) innervated by bundles of varicose axons (nb). The cell body of a GABAergic neuron sits in a myenteric ganglion, sending its axon into the intramuscular nerve bundles, where GABA, released from varicosities, exerts presynaptic action on neighbouring fibres. Taken with permission from Jessen *et al*; 1983.

directly affect the motility of the muscle since it has no GABA receptors. However, as discussed above, it is well-documented that GABA can modulate release of transmitters including acetylcholine, from enteric nerves. Furthermore, in the circular muscle the GABAergic fibres join the intramuscular axon bundles (the pre-junctional element in the gut neuro-muscular junction (Gabella, 1972)) where they must lie in close proximity to cholinergic and other fibres. We therefore suggest that the sole role of the GABAergic innervation of the circular muscle is the pre-junctional modulation of transmitter release from neighbouring axons (Figure 9.6). In this way GABAergic cells in the myenteric plexus of the guinea-pig would be involved, indirectly, in the modulation of gut motility. The anatomical distribution of GABAergic neurones makes it unlikely that GABA is involved in intestinal epithelial transport, at least in rat and guinea-pig.

GABA AND SYMPATHETIC GANGLIA

The evidence relating to the question of whether sympathetic ganglia contain GABAergic neuronal cell bodies is controversial. In one immunohistochemical study of the rat superior cervical ganglion, clusters of SIF cells (type II) were intensely immunolabelled with GAD antibodies, while the principal cells were GAD negative (Kenny and Arians, 1986). Another study of the same tissue using a different GAD antiserum found a large number of the principal cells to be GAD positive, although clusters of small cells, presumably type II SIF cells, were even more strongly immunolabelled, which is in agreement with the previous report (Häppölä *et al.*, 1987). In a study using two different GABA antibodies however, neither principal cells nor clustered SIF cells were immunolabelled (Kàsa et al., 1988). Instead, scattered small cells were GABA positive. It seems possible that some of them correspond to type I SIF cells, while some of the others have the morphology of satellite glial cells and occurred in close apposition to unlabelled principal neurons. An immunohistochemical study of guinea-pig failed to reveal any GABA antibody binding to neurons in prevertebral sympathetic ganglia (Hills et al., 1988).

High affinity GABA uptake sites are present in sympathetic ganglia as in most other nervous tissue. To date, selective accumulation of ^{3}H-GABA has, however, not been clearly detected in neurons of sympathetic ganglia, in contrast to that seen in the CNS or in myenteric ganglia, and most of the high affinity uptake sites apear to be present on satellite or Schwann cells (Young et al., 1973 Bowery et al., 1979; Wolff et al., 1979).

In conclusion, the existence of GABAergic principal neurons in sympathetic ganglia has not yet been demonstrated unequivocally. Further work is also needed to test the possibility that type I or II SIF cells might be GABAergic.

GABA-immunoreactive nerve fibres have been described in guinea-pig inferior mesenteric and coeliac ganglia (Hills et al., 1988). They were not seen in nerves peripheral to the ganglia, but were present in nerves which provide input to the ganglia from the colon. It was therefore suggested that they originated in the GABAergic neurons of the myenteric plexus. Following proteolytic enzyme treatment, striking GABA-immunolabelling of fibres was also seen around certain neurons in rat superior cervical ganglia (Kàsa et al., 1988; Wolff et al. 1989). Some of these fibres arise outside the ganglia while others may originate from small GABA-immunolabelled cells within the ganglia (see above).

Exogenously applied GABA suppresses ganglionic transmission in both guinea pig prevertebral and rat superior cervical ganglia (Bowery and Hill, 1986). Furthermore, fast depolarizing postsynaptic potentials, which are non-cholinergic and suppressed by the GABA receptor antagonist bicuculline, have been described in rat superior cervical ganglia following stimulation of the preganglionic trunk (Eugene, 1987). There is also evidence that GABA has a trophic role in synaptogenesis in this ganglion (Joò et al., 1987). There is therefore no shortage of possible roles for GABAergic fibres in sympathetic ganglia. The ubiquitous distribution of GABA receptors in nervous tissue means, however, that data on the effects of applied exogenous GABA cannot be regarded as direct evidence for the presence of GABAergic transmission. The presence of GABAergic fibres in sympathetic ganglia needs to be confirmed by other methods, in particular autoradiography of ^{3}H-GABA uptake.

5-HYDROXYTRYPTAMINE (5-HT)

5-Hydroxytryptamine (5-HT) is an ubiquitous indoleamine, the complete function of which in mammalian physiology is still being discovered. Its occurrence is widespread throughout the animal and plant kingdom (Tyce, 1985). 5-HT, first isolated from ox blood in 1948 (Rapoport et al., 1948) was identified as the vasoconstrictor substance in blood (Rapoport et al., 1949) and hence the term 'serotonin' was coined.

'Enteramine' on the other hand, a substance present in the enterochromaffin cells of the gastrointestinal mucosa (Espamer, 1954) was subsequently discovered to be identical to the vasoactive substance in blood (Espamer and Asero, 1952) and was therefore also 5-HT.

Since the early discoveries, 5-HT has been shown to subserve both neurotransmitter and hormonal roles and to be localized widely but discretely. Large amounts of 5-HT are found in the enterochromaffin cells of the gastrointestinal mucosa, the brain, the pineal gland and platelets. In addition 5-HT is thought to have a major peripheral neurotransmitter role in the enteric nervous system, is present in nerves supplying the cerebrovasculature, and may subserve a role in the heart, the lungs, the spleen, the thyroid and in mast cells.

5-HT CONTAINING NEURONES IN THE ENTERIC NERVOUS SYSTEM

The gastrointestinal tract contains large quantities of 5-HT, much of it being located in the enterochromaffin cells of the mucosa (Forsberg and Miller, 1983; Legay et al., 1983). The presence of extraneuronal 5-HT in the gut complicated the task of those proposing an enteric neurotransmitter role for 5-HT. In 1965, Gershon et al., demonstrated the synthesis and release of 5-HT from the myenteric plexus of the mouse intestine (Gershon et al., 1985). The suggestion that the release was from enteric neurones was not immediately accepted. This was confounded by another problem, the inability of the investigators to demonstrate aldehyde-induced fluorescence attributable to 5-HT from untreated animals (Robinson and Gershon, 1971; Dubois and Jacobowitz, 1974). However a fluorescence reaction in the myenteric plexus attributable to 5-HT was demonstrated following treatment with a monoamine-oxidase inhibitor (Robinson and Gershon, 1971). When preparations of myenteric plexus and gut smooth muscle were however obtained free of mucosa, they were shown to contain 5-HT (Feldberg and Toh, 1953) in quite significant quantities (57–110 ng/g) (Juorio and Gabella, 1974; Robinson and Gershon, 1971; Holzer and Skofitsch, 1984; Gershon et al., 1989).

The presence of 5-HT within enteric neurones is now accepted largely because of the application of elegant immunohistochemical techniques. Antibodies directed against 5-HT revealed a dense network of nerve fibres and neuronal cell bodies within the myenteric plexus of guinea-pig (Furness and Costa, 1982; Legay et al., 1984; Saffrey et al., 1984), rats (Nada and Toyohara, 1987), and humans (Kurian et al., 1983). In the guinea-pig small intestine, projection studies have shown neurons containing 5-HT to have their cell bodies located in the myenteric plexus. Axons from these neurons travel in the anal direction for quite long distances and also project through the circular muscle to innervate submucosal ganglia (Furness and Costa, 1982; Furness and Costa, 1987). Such a distribution is consistent with an interneuronal function within the myenteric plexus, and a modulatory function in intestinal secretion in the submucous plexus.

The ultrastructure of 5-HT-containing enteric neurones has been studied both by electronmicrosopic autoradiography following ^{3}H-5HT uptake and immunoelectronmicroscopy. Autoradiographic studies in the guinea-pig and rat myenteric plexus show the nerve fibres that take up ^{3}H-5-HT to contain small clear vesicles and large granular vesicles. The majority of synapses formed are axo-somatic but axodendritic and axo-axonic contacts have also been identified (Gershon and Sherman, 1982a; Feher et al., 1983; Erde et al., 1985; Furness et al., 1988).

Biochemistry

A protein that binds 5-HT, which was first shown to be present in central serotoninergic neurones (Tamir and Huang, 1974) is also present in enteric neurones which utilize 5-HT (Jonakait et al., 1977 and 1979). This serotonin binding protein binds 5-HT very strongly under intracellular conditions (low Na$^+$, CA^{++}), but loses its affinity under extracellular conditions (Gershon and Tamir, 1985; Gershon et al., 1989) and its role is probably that of binding 5-HT within synaptic vesicles.

5-HT is synthesized in all tissues in which it is utilized except platelets (Morrissey et al., 1977; see review by Tyce, 1985; Verbeuren, 1989). It is synthesized from the essential amino acid tryptophan and forms a minor metabolic route for this ingested substance (Figure 9.7). The enzyme responsible for the conversion of tryptophan to 5-hydroxytryptophan is tryptophan hydroxylase, an enzyme found only in tissues utilising 5-HT. L-amino acid decarboxylase, on the other hand, an enzyme which catalyzes the conversion of 5-hydroxytryptophan to 5-HT is non-specific and will decarboxylate not only tryptophan, but also tyrosine, phenylalanine and dihydroxyphenylalanine (see Tyce, 1985; Crillis, 1985). Within the cell both enzymes are associated with the cytosol and to a lesser extent with subcellular organelles (Rodriguez et al., 1964; Green and Sawyer, 1966, Lovenberg et al., 1967).

The regulation of 5-HT synthesis is dependent on the amount of tryptophan available and factors that control the activity of tryptophan hydroxylase. Tryptophan hydroxylase activity is controlled by the calcium-dependent phosphorylation of the enzyme, subsequent to the activation of a specific protein kinase and thus, nerve stimulation which increases intracellular calcium concentration, stimulates the synthesis of 5-HT (see Tyce, 1985).

The main pathway for the catabolism of 5-HT is via oxidative deamination to 5-hydroxyindoleacetaldehyde. The enzyme responsible for this reaction is the A form of monoamine-oxidase (apparent Km in human brain, 70 μM) (Pearce and Roth, 1983), an enzyme localized both intraneuronally on the surface of mitochondria and extraneuronally (Tyce, 1985). 5-Hydroxyindoleacetaldehyde thus formed is rapidly catalyzed by aldehyde dehydrogenase, to 5-hydroxyindoleacetic acid (Km 2.5 μM) (Erwin and Deitrich, 1966; Tyce, 1985). An alternative minor pathway for the conversion of 5-hydroxyindoleacetaldehyde is to the alcohol, 5-hydroxytryptophol, catalyzed by aldehyde reductase (Km 4 μM, (Turner and Tipton, 1972)). Other metabolic pathways which play a lesser role are, sulphation, glucuronidation, N-acetylation and O- and N-methylation (Tyce, 1985; Verbeuren et al., 1988).

The metabolism of 5-HT in enteric neurones has been shown to be similar to that in other tissues. Dreyfus et al., (1977) demonstrated that ^{3}H-5-HT was synthesized from ^{3}H-L-tryptophan in mucosa-free strips of intestinal wall, and that ^{3}H-5-hydroxyindoleacetaldehyde was also produced (Dreyfus et al., 1977; Legay et al., 1983), thus indicating the presence of tryptophan hydroxylase and monoamine

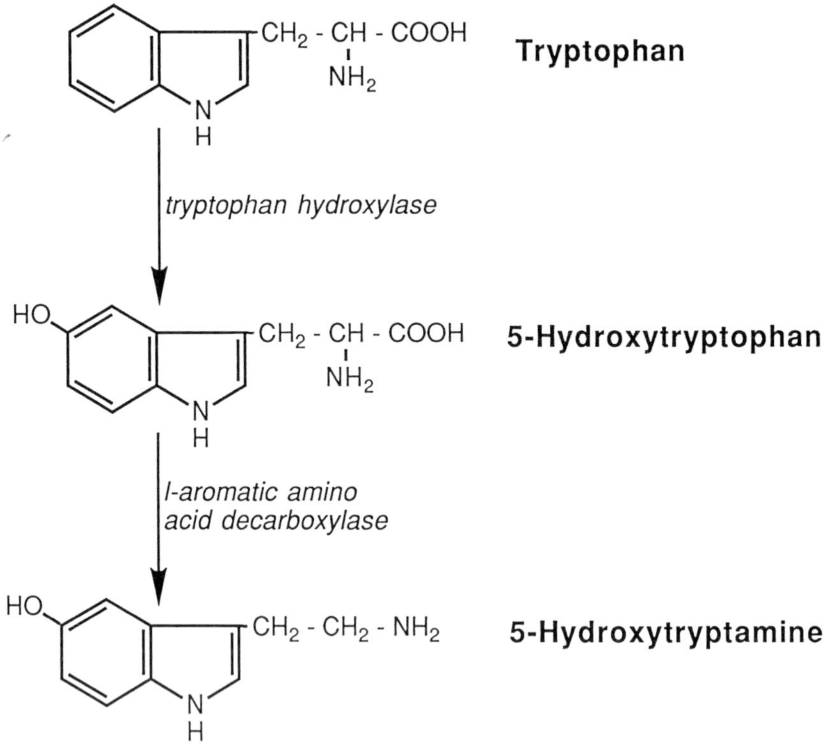

FIGURE 9.7   The synthetic pathway for 5-HT.

oxidase. The same observations were also made in explanted enteric 5-HT containing neurones in organotypic cultures (Dreyfus *et al.*, 1977). Monoamine oxidase does not entirely account for the metabolism of 5-HT since 5-HT O- glucuronide also occurs in the gut (Gershon and Ross, 1966; Gershon *et al.*, 1989).

In addition to metabolism in the liver by monoamine oxidase, 5-HT is removed from the circulation by an active carrier-mediated process into the endothelial cells lining the vasculature (Gillis, 1980; 1985). Functionally, the primary site of this uptake is in the pulmonary vasculature (Gillis, 1985). Coupled with endothelial 5-HT uptake which persists after sympathetic denervation (Iwasawa and Gillis, 1974), 5-HT may be sequestered into sympathetic nerves within the blood vessel wall (Verbeuren *et al.*, 1983). It is suggested that following uptake into sympathetic neurones, 5-HT is rapidly deaminated, and that 5-HT which has previously been incorporated into platelets can subsequently be taken up into sympathetic nerves in the blood vessel wall (Verbeuren *et al.*, 1988).

Extraneuronal uptake and storage of 5-HT without metabolism takes place in platelets (Lingjaerde, 1977), mast cells (Uvnas, 1978), and enteroendocrine cells (Legay *et al.*, 1983).

5-HT uptake into enteric neurones is energy dependent and saturable with a measured Km of about 0.7 μM (Robinson and Gershon, 1971; Gershon and Altman, 1971). The uptake takes place over the entire neuronal surface but is mainly into nerve terminals (Gershon and Sherman, 1982b) and is assumed to be specific for the enteric neurons which utilize 5-HT as their neurotransmitter. The enteric neuronal 5-HT uptake shares common features with that of the central nervous system uptake. It is inhibited by chlorimipramine more potently than imipramine or desmethylimipramine, and is specifically inhibited by fluoxetine and zimelidine (Gershon et al., 1976; Gershon and Jonakait; 1979; Takaki et al., 1985; Gershon et al., 1989). Close analogues of 5-HT are transported by the same uptake process and will inhibit the uptake of 5-HT itself (Gershon et al., 1976; see Gershon et al., 1989). For example, compounds where only the alkyl amino group is substituted, have less affinity for the uptake site, whereas analogues where the ring of the 5-HT molecule is hydroxylated compete with 5-HT for uptake (see review Gershon et al., 1989).

Release

As with the localization of 5-HT in enteric neurones, the demonstration of the release of 5-HT from enteric nerves was hindered by its presence in non-neuronal elements within the gut wall. 5-HT is released from the stomach following vagal stimulation (Paton and Vane, 1963) and 5-HT release increases during peristalsis (Gwee and Yeoh 1968). This release, however, is likely to comprise largely release from the entero-endocrine cells (Paton and Vane, 1963; Tansy et al., 1971). Neuronal release of ^{3}H-5-HT could, however, be corroborated by two types of experimentation. Firstly, 5-HT release occurred on vagal stimulation following asphyxiation of tissue to destroy entero-endocrine cells (Bülbring and Gershon, 1967) and secondly, release could be obtained from isolated preparations containing only myenteric plexus and longitudinal muscle (Schultz and Cartwright, 1974; Jonakait et al., 1979). The release of endogenous 5-HT from the intestine has also been demonstrated and, like that of ^{3}H-5-HT, is calcium dependent (Gershon and Tamir, 1981; Holzer and Skofitsch, 1984).

Pharmacology

When applied to isolated gut smooth muscle preparations, 5-HT causes both direct and indirect excitation and inhibition, depending on regional and species differences (Bulbring and Gershon, 1967; Gershon, 1967; Costa and Furness, 1979). In addition to affecting nerve and muscle, 5-HT also alters mucosal epithelial chloride transport (Hardcastle et al., 1981) by a direct action on crypt cells and by an indirect action mediated by stimulation of submucosal neurones (Cooke and Carey, 1985).

The classification of central and peripheral receptors for 5-HT has come a long way since the early proposal of Gaddum and Picarelli that there were M and D receptor subtypes (Gaddum and Picarelli, 1957). The definition was based on work on the guinea-pig ileum where responses to 5-HT which were blocked by morphine were said to be mediated by 'M' receptors and those blocked by dibenzyline were said to be mediated by 'D' receptors (Gaddum and Picarelli, 1957). The 'M' receptor mediated responses could additionally be blocked by tetrodotoxin and were therefore indirect while those which were 'D' receptor mediated were tetrodotoxin insensitive. Neither morphine nor dibenzyline are specific 5-HT receptor antagonists, and although they have been superseded by more specific agents, the existence of muscle and neuronal 5-HT receptors is widely accepted (Drakontides and Gershon., 1968; Richardson *et al.*, 1985).

At present at least three main classes of 5-HT receptor, (5-HT$_1$, 5-HT$_2$ and 5-HT$_3$) have been identified in the central nervous system by radioligand binding techniques, each utilizing one of three second messenger or ion channel mechanisms as an effector pathway (see Peroutka, 1988, and Watson and Abbott, 1990). The picture in the periphery and more specifically in the enteric nervous system is complex, there not being a direct binding comparison with 5-HT receptor subtypes in the central nervous system.

Although intracellular electrophysiology has helped in the classification of enteric 5-HT receptors to date, much of the identification of individual receptors has been based on isolated tissue experiments. The classical 'D' receptor or muscle receptor is considered to be identical to the 5-HT$_2$ receptor of the central nervous system since it may be blocked by 5-HT$_2$ receptor antagonists (Bradley *et al.*, 1986). The 'M' receptor on the other hand appears not to be a single entity, at present sub-divided into the 5-HT$_3$ (Bucheit *et al.*, 1985) and the 5-HT$_4$ (Bucheit *et al.*, 1985; Craig and Clarke, 1990) receptor.

The neurally mediated effects of 5-HT have been studied using intracellular electrophysiological techniques in the myenteric plexus. 5-HT applied to the surface of myenteric neurones evokes a fast and slow membrane depolarization. The slow depolarizing response is long lasting and associated with a decrease in membrane conductance to potassium (Wood and Mayer, 1979; Johnson *et al.*, 1980). This slow depolarization which can also be evoked by stimulation of interganglionic strands within the plexus (Wood and Mayer, 1979), can be mimicked by hydroxylated indalpines and specifically antagonized by dipeptides of 5-hydroxytryptophan (5-HTP-DP;N-acetyl-5-hydroxytryptophol-5-hydroxytrytophan amide) (Takaki *et al.*, 1985; Mawe *et al.*, 1986; Branchek *et al.*, 1988). The term 5-HT$_{1P}$ has been coined for this peripheral receptor since like central 5-HT$_1$ receptors, it shares a similar high affinity for 5-HT (Mawe *et al.*, 1986). Recent reports also suggest that it may be additionally antagonized by BRL 24924 even though this compound does not displace 5-HT from 5-HT$_1$ receptors (Mawe *et al.*, 1989).

In addition to the slow depolarization, a fast depolarization can be evoked in myenteric neurones by 5-HT and this response appears to be mediated by a different receptor to the slow depolarization. This initial fast 5-HT response can be evoked in both enteric type2/AH and type1/S and submucous neurones and is mediated via a cation channel, which in guinea-pig submucous neurones at least may exist in two conductance states (Johnson *et al.*, 1980; Derkach *et al.*, 1989). The fast response is blocked by ICS 205–930, a compound classically associated with the 5-HT$_3$ receptor (Bradley *et al.*, 1986; Mawe *et al.*, 1986; Nemeth and Gullikson, 1989), and mimicked by 2-methyl-5-HT (Mawe *et al.*, 1986; Gershon *et al.*, 1989). Mawe *et al.*, (1986) suggest that this receptor be classified as a 5-HT$_{2P}$ receptor, even though it is probably identical to the central 5-HT$_3$ receptor (Bradley *et al.*, 1986). The effect of

ICS 205–930 may be consistent with the work of Bucheit *et al.*, (1985), who showed partial blockade of neuronal 5-HT contractions in the guinea-pig ileum with this compound.

In the submucous plexus, the situation is somewhat similar. 5-HT exerts fast depolarising potentials mediated by an increase in sodium conductance via the 5-HT_3 receptor and slow depolarizing potentials mediated by a reduction in potassium conductance via a 5-HT receptor which does not resemble either $5\text{-HT}_{1,2}$or$_3$ subclasses (Surprenant and Crist, 1988). 5-HT_1 receptors do appear to be present on presynaptic elements within the submucous plexus (Galligan *et al.*, 1988).

The existence of two excitatory neuronal receptors for 5-HT in isolated tissue is confirmed in experiments in the guinea-pig ileum (Bucheit *et al.*, 1985; Sanger; 1985; Craig and Clarke, 1990).

Although the recent experiments have served to identify subtypes of neuronal 5-HT receptors, some inconsistencies are still apparent when one considers binding data. Analysis of the binding of ^{3}H-5-HT to membranes derived from myenteric plexus neurones suggests a single receptor site (Branchek *et al.*, 1984; Gershon *et al.*, 1985; Gershon *et al.*, 1989), and not multiple receptor sites as one might except. Further, 5-HTP-DP, 2-methyl-5-HT and the hydroxylated indalpines are the only compounds which displace 5-HT from its binding site (Takaki *et al.*, 1985; Mawe *et al.*, 1986; Gershon *et al.*, 1989), Ketanserin (5-HT_2), ICS 205–930 (5-HT_3) and MDL 72222 (5-HT_3) among others, being ineffective. A 5-HT_3 and 5HT_4 receptor ligand, zacopride, does however bind to specific sites in membranes from both intestinal muscle (Pinkus *et al.*, 1989) and enteric neurones (Gordon *et al.*, 1989).

Conclusion

Considering the wide variety of 5-HT actions, it is not surprising that enteric 5-HT neurones have been claimed to subserve diverse physiological roles in both motor and secretory processes (see Gershon *et al.*, 1989). The role played by the various 5-HT receptors in normal and pathological gut function awaits more data in man. However, while it is now established that 5-HT_3 receptors possess anti-emetic properties, their prokinetic activity is not firmly established (Talley *et al.*, 1989; Nemeth and Gullikson, 1989; Gullikson *et al.*, 1988; Talley *et al.*, 1988; King and Sanger, 1988) and indeed a slowing of colonic transit with GR 38032F has been reported (Talley *et al.*, 1988). Gastrokinetic activity has also been claimed following blockade of the 5-HT_{1p} receptor (Mawe *et al.*, 1989). Such diversity of action to 5-HT indicates that, in the gut plexuses several subtypes of 5-HT receptor are operative, and that each may involved in a different function.

5-HT AND BLOOD VESSELS

5-HT has profound effects in the cardiovascuslar system. In addition to central 5-HT neurones whose efferents control cardiovascular reflexes (Kuhn *et al.*, 1980), there is evidence for a direct innervation of blood vessels by 5-HT (Edvinsson *et al.*, 1983; and Marco *et al.*, 1985). 5-HT also affects the cardiovascular system by virtue of its content in plasma, and the major source of 5-HT which modulates vascular tone is derived from aggregating platelets (see review Houston and Vanhoutte, 1989).

5-HT may be assayed in cerebral blood vessels (Reinhard *et al.*, 1979; Edvinsson *et al.*, 1983), although such measurements are complicated by the presence of 5-HT in platelets. Original studies investigating the innervation of blood vessels, used formaldehyde-induced fluorescence to visualize the fluorophore formed (β-carboline) (Cummings and Felten, 1979). Other workers using this technique attempted to improve its quality by pre-loading with precursors or inhibiting monoamine-oxidase activity (Napoleone *et al.*, 1982; Di Carlo, 1984a), such studies do not however, locate endogenous 5-HT. The visualization of 5-HT by autoradiography, histochemistry and immunohistochemistry has provided more clear cut evidence for serotonergic nerves in cerebral blood vessels (for review see Griffith, 1988). There is also immunohistochemical evidence for the innervation of blood vessels by 5-HT-containing neurones in the mesentery (Griffith and Burnstock, 1983a,b; Dahlstrom and Ahlman, 1983; for review see Dahlstrom *et al*, 1988) and within the arterial supply in the spinal cord (Di Carlo, 1984b). The origin of the 5-HT containing neurones in cerebrovascular nerves appears to be both central from the Raphé nuclei and peripheral from the superior cervical ganglion (for review see Griffith, 1988).

Biochemistry

5-HT uptake and release in nerves of cerebrovascular vessels can be measured (Amenta *et al.*, 1985; Edvinsson *et al.*, 1984; Verbeuren 1983). The uptake is inhibited by specific 5-HT uptake inhibitors and pargylline treatment leads to accumulation of 5-HT, indicating a role for monoamine oxidase in these vessels (Reinhard *et al.*, 1979). There is some debate as to whether the uptake and release occurs in specific 5-HT, or adrenergic neurones.

Pharmacology

The release of 5-HT from a variety of sources as outlined above, is considered important in the control of vascular tone, the overall effect of 5-HT release from aggregating platelets being to limit blood loss following injury (Houston and Vanhoutte, 1989).

5-HT is a powerful vasoconstrictor in its own right and also serves to amplify the effect of other vasoconstrictors (see review, De La Lande, 1989). Most commonly, 5-HT mediates vasoconstriction in the vasculature via interaction with either 5-HT$_1$-like or 5-HT$_2$ receptors, the effect seen being dependent on regional and species differences.

Of the areas where evidence exists for the innervation of blood vessels by serotonergic neurones, namely the cerebrovasculature and mesenteric vessels and the vasa nervorum of sciatic and vagus nerves, only 5-HT receptors in cerebrovascular vessels have been studied in any detail (Griffith, 1988; Fenuik and Humphrey, 1989). The existence of both 5-HT$_2$ (Van Neuten *et al.*, 1981; Muller-Schweinitzer and Engel, 1983 and 'non-5-HT$_2$' (Bradley *et al.*, 1986; Peroutka *et al.*, 1986) receptors mediating contraction has been claimed in the canine basilar artery. On the other hand, Fenuik and Humphrey (1989), claim a heterogeneous population of 5-HT$_2$ and 5-HT$_1$-like receptors in some vessels including the basilar artery, thus making it difficult to draw firm conclusions as to the receptor identity. They also emphasize that the '5-HT$_1$-like' receptor is not synonymous with any of the current 5-HT$_1$ receptor classes (Fenuik and Humphrey, 1989).

It is likely that the genesis of more specific ligands will help to elucidate further the physiological role of the cerebrovascular serotonergic innervation. That there is a function for 5-HT receptors, could hardly be more clearly demonstrated than in the current development of '5-HT$_1$-like' receptor agonists for the treatment of migraine (Fenuik and Humphrey, 1989).

5-HT, SYMPATHETIC, PARASYMPATHETIC AND SENSORY GANGLIA

5-HT receptors on sympathetic, parasympathetic and sensory ganglia may be activated by 5-HT present in plasma or 5-HT released following damage of the blood vessel wall, and there is now evidence that endogenous 5-HT may be present in the ganglia themselves (see Wallis, 1989).

Small intensely fluorescent cells (SIF cells), considered to be either paracrine cells or interneurones, contain 5-HT. Such an observation has been reported in SIF cells of the rat superior cervical ganglion (Verhofstad et al., 1981) and the coeliac superior and inferior ganglia of the guinea-pig (Dun et al., 1985 Ma et al., 1985). Intrinsic 5-HT neurones are now thought to be present in nodose ganglia (Gaudin-Chazal et al., 1978), dorsal root ganglia (Di Carlo, 1983), superior and inferior mesenteric ganglia (Dun et al., 1984; Ma et al., 1985), superior cervical ganglia (Häppölä, 1988) parasympathetic cardiac ganglia (Neel and Parsons, 1986), and ganglia of the gall bladder (Mawe and Gershon, 1989). 5-HT-like immunoreactivity has been identified in nerve fibres, but not cell bodies in trigeminal ganglia (Chouchkov et al., 1988), in spinal preganglionic sympathetic and parasympathetic neurons (Newton and Hamill, 1989), and in cultured sympathetic neurons (Dinah et al., 1987; Soinila et al., 1989).

The origin of 5-HT nerve fibres in sympathetic, parasympathetic and sensory ganglia may be both central and peripheral. It is considered that some fibres may have their cell bodies in the caudal Raphé nuclei (see review, Wallis 1989). Alternatively, projections of intrinsic enteric 5-HT neurones to mesenteric ganglia are well documented (Furness and Costa, 1987) and probably account for the majority of the 5-HT innervation in the prevertebral ganglia.

Little information is currently available on the synthesis of 5-HT in sympathetic, parasympathetic and sensory ganglia. Tryptophan hydroxylase activity has been measured in rat superior cervical ganglia homogenates (Luizzi et al., 1977), and the intensity of 5-HT immunofluorescence is increased in mesenteric ganglia following incubation with L-tryptophan and a monoamine oxidase inhibitor (Dun et al., 1984; Ma et al., 1986).

Of interest is the recent demonstration of the existence of 5-HT in the phrenic nerve of the rat, its release on electrical stimulation and additionally, the presence of its metabolite 5-HIAA in the nerve (Das et al., 1989). Hyperpolarising responses to 5-HT have been recorded in rabbit and rat superior cervical ganglia (Wallis and North

1978; Ireland, 1987) and may well be unrelated to the depolarising responses since the hyperpolarisation is not mediated by the 5-HT$_3$ receptor (Ireland, 1987; Ireland and Jordan, 1987).

Pharmacology

The pharmacology of 5-HT actions on peripheral ganglia has been recently reviewed by Wallis (1989) and will not be dealt with in great detail here.

5-HT receptors are located widely on peripheral ganglia and at a number of sites within any given ganglia (see Wallis, 1989). The predominant effect of 5-HT on the cell bodies of sympathetic pre-ganglionic neurones is a membrane depolarization associated with an increase in membrane resistance probably resulting from closure of potassium channels (Ma and Dun, 1986). The results of experiments studying ganglia *in situ*, point to a 5-HT$_2$ receptor being involved in this response (Kadzielawa, 1983; McCall, 1983; Fozard, 1984) although other studies do not bear this out (see review by Wallis, 1989).

5-HT will also depolarize preganglionic axons via 5-HT$_3$ receptors (Wallis and Dun, 1989), and the same receptor has been implicated in the presynaptic inhibition of ACh release from pre-ganglionic axons (Dun and Karczmar, 1981). Post-ganglionic sympathetic neurones are depolarized by 5-HT acting on a 5-HT$_3$ receptor, thus triggering the release of noradrenaline (see Richardson and Engel, 1986). The response of sympathetic neurones to 5-HT is not always a simple depolarization. For example, guinea-pig coeliac ganglion cells display both fast and slow depolarizations to 5-HT (Wallis & Dun, 1988 Wallis, 1989) and rabbit superior cervical ganglion cells respond to 5-HT with a rapid depolarization often followed by an after hyperpolarization (Wallis and North, 1978; Wallis, 1989). The fast depolarization is thought to be as a result of 5-HT$_3$ receptor-gated conductance increases to Na$^+$ and K$^+$ ions (see Wallis, 1989). The slow depolarization is associated with an increase in membrane resistance and shows different receptor characteristics to the fast response (Kiraly *et al.*, 1983; Wallis, 1989). Hyperpolarising responses to 5-HT have been recorded in rabbit and rat superior cervical ganglia (Wallis and North 1978; Ireland 1987) and may well be unrelated to the depolarising responses since the hyperpolarisation is not mediated by the 5-HT$_3$ receptor (Ireland, 1987; Ireland and Jordan, 1987). Experiments by Gilbert and Newberry (1978) suggest that the hyperpolarisation may due to activation of 5-HT$_1$-like receptors, possibly 5-HT$_{1A}$.

Parasympathetic neurones have been less well studied. 5-HT excites para-sympathetic ganglia in urinary bladder (Saum and DeGroat, 1973; Burnstock *et al.*, 1978), an action attributed at least partly to activation of 5-HT$_3$ receptors (Saxena *et al.*, 1985; Akasu *et al.*, 1987). A presynaptic facilitation of cholinergic transmission mediated via a 5-HT$_1$ receptor has been reported in rabbit vesical pelvic ganglia, an action possibly mediated through activation of adenylate cyclase (Nishimura and Akasu, 1989). An inhibition of cholinergic transmission in cat pelvic vesical ganglia has been attributed to action on a 5-HT$_{1A}$ receptor subtype (Nishimura *et al.*, 1988). The increase in acetylcholine release that occurs in rat bronchi and rabbit heart on exposure to 5-HT is again attributed to 5-HT$_3$ receptor activation

(Aas, 1983), while 5-HT receptors in mouse bladder appear to be 5-HT$_1$-like (Holt *et al.*, 1986).

The soma and axons of sensory neurones are excited by an action on 5-HT$_3$ receptors (Wallis, 1989), the ionic mechanism for which has been well characterized and, as in other tissues which possess 5-HT$_3$ receptors, involves an increase in Na$^+$ and K$^+$ conductance (Higashi and Nishi, 1982; Wallis, 1989). The terminals of sensory neurones are exquisitely sensitive to 5-HT, and activation elicits a number of peripheral reflexes (see Wallis, 1989). The identity of the receptor which mediates stimulation of sensory nerve endings in the periphery has not been classified in every case. The reflex bradycardia caused by 5-HT administration (Bezold-Jarisch), however, is reported to be mediated by stimulation of 5-HT$_3$ receptors (see McQueen and Mir, 1989).

Conclusion

There is a wealth of evidence indicating the presence of 5-HT and multiple 5-HT receptors in sympathetic, parasympathetic and sensory ganglia. As yet, it is impossible to draw firm conclusions as to how important a role such a 5-HT innervation plays in normal or pathological mammalian physiology. However, the presence of 5-HT receptors in sensory pathways may suggest a role for 5-HT in pain conditions and the widespread distribution of 5-HT receptors on sympathetic and parasympathetic neurons points to a modulatory role in peripheral autonomic function.

3. DOPAMINE

The catecholamine dopamine (3-hydroxytyramine (DA)) is established as a neurotransmitter in the CNS (e.g. Hornykiewicz, 1966). It is also the immediate precursor to noradrenaline in adrenergic neurons (Figure 8). Therefore, dopaminergic neurons possess much of the same biochemical machinery as adrenergic ones, the main difference being the higher DA/noradrenaline ratio and lack of the enzyme dopamine-β-hydroxylase (DBH) in the dopaminergic cells. In addition there is evidence that dopaminergic terminals contain high levels of the enzyme DOPA decarboxylase (DDC) in comparison with adrenergic fibres (Bell and McLaughan, 1982). The absence of a simple, positive marker for dopaminergic cells has made unequivocal identification of these neurons difficult in the autonomic nervous system, where they clearly do not constitute more than a small proportion of cells.

A further difficulty in pinpointing dopaminergic nerves within the autonomic nervous system arises from the fact that, both DA and DA receptors are quite widely distributed in peripheral tissues and their locations do not provide a reliable guide to the presence of dopaminergic cells or dopaminergic transmission. A multitude of tissues outside the CNS contain DA in relatively low, but variable amounts and the level of plasma DA is similar to that of adrenaline (Bühler *et al.*, 1978; Van Loon, 1983). The sources of tissue and plasma DA are unclear but could include: (I) SIF cells. These cells are generally present in sympathetic ganglia and may also be found outside ganglia in peripheral tissues (Eränkö and Eränkö, 1971; Chiba and Williams,

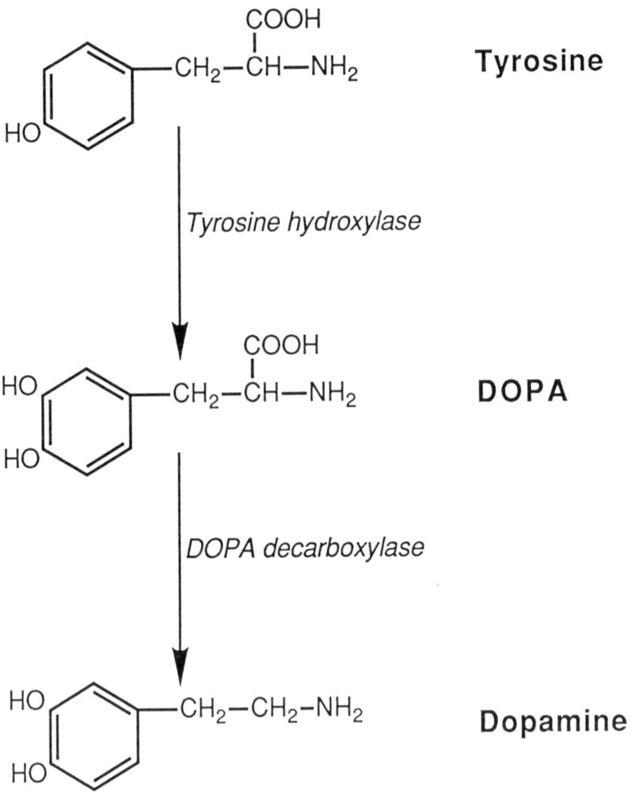

FIGURE 9.8   The sympathetic pathway for Dopamine.

1975; Knight and Bazer, 1979). They store monoamines including, in many cases, DA (Björklund *et al.*, 1970; Neff *et al.*, 1983). (II) Adrenergic neurons (Bell, 1988). DA can be released from adrenergic nerves in certain experimental situations, such as following inhibition of noradrenaline synthesis, or after loading of the nerves with exogenous DOPA (Thoenen *et al.*, 1965; Austin *et al.*, 1967). The evidence for DA release from sympathetic nerves following a more physiological stimulation such as reflex activation is more controversial. Sympathetic discharge following exercise or a change in posture does not consistently induce increase in plasma DA (Van Loon, 1983). Other types of reflex stimulation of the sympathetic system cause more consistent increases in circulating DA. It has been pointed out that many of these stimuli have a "stressfull" association e.g. haemorrhage, thermal stress or enforced restraint (Bell, 1988). (III) Adrenal medulla. The adrenals contain DA as a precursor to noradrenaline and adrenaline, and adrenalectomy results in reduced plasma levels of DA (Bühler *et al.*, 1978; Kvetnansky *et al.*, 1979) (IV) Putative dopaminergic autonomic neurons (see below).

Two types of DA receptors DA_1 and DA_2 receptor, have been described in the periphery (for reviews see e.g. Goldberg *et al.*, 1978; Lokhandwala and Barrett, 1982; Stoof and Kebabian, 1984; Willems, 1985). They are distinct from the DA-1 and DA-2 receptors present in the CNS, although several similarities exist. Peripheral DA receptors are found in many different tissues in two types of location. (I) Postjunctional receptors (predominantly DA_1), found on the smooth muscle muscle in several vascular beds and probably also on juxtaglomerular cells of the kidney, renal tubules and cells of the adrenal cortex. Although DA is well known to have both excitatory and inhibitory effects on various gut preparations, most of them appear to be mediated via adrenergic receptors. The blood vessel DA receptors were first described in kidney and the mesentery but it is now clear that they are present at other sites including the cerebral, coronary and hepatic circulation. The major response associated with activation of these receptors is smooth relaxation and consequent vasodilatation. (II) Prejunctional receptors (predominantly DA_2), found on the nerve terminals of some autonomic nerve endings. Their presence is best established on the sympathetic terminals innervating the cat and dog nictitating membrane and the heart, and the dog femoral artery and the rabbit ear artery. Prejunctional DA receptors have also been described in other vascular beds including those of the kidney and mesentery. Pharmacological activation of prejunctional DA receptors is, in nearly every case, associated with inhibition of noradrenaline release.

It is unlikely that dopaminergic nerves match the presence of peripheral DA receptors at more than at most a few sites, such as the kidney (see below). This raises the question of what function pre- and post-junctional DA receptors might have in other tissues and organs. Much effort has been devoted to this problem, since the strategic localization of DA receptors and the functional consequences of their activation make this a potentially powerful system in the regulation of various body functions. The consensus among the great majority of workers in the field is, however, that the experimental evidence for any of the specific hypotheses advanced in this field is weak (for discussion see Lokhandwala and Barrett, 1982; Willems *et al.*, 1985; Bell, 1988). For instance, DA antagonists do not significantly enhance the adrenergic responses to nerve stimulation in any of the major vascular beds where pre-junctional DA receptors have been described, which fails to support the attractive suggestion that pre-junctional DA-receptors form part of a negative feedback regulation of noradrenaline release (Willems *et al.*, 1985).

The broad issues outlined above, concerning DA metabolism, DA receptors and DA function outside the CNS have been extensively reviewed elsewhere (for refs., see above). The narrower question of whether dopaminergic neurons are present within the autonomic nervous system will now be addressed by reviewing, firstly, the biochemical and histochemical evidence for the existence of dopaminergic neurons in autonomic ganglia and, secondly, the studies of putative dopaminergic nerves and their function in the kidney, since at this site the evidence for dopaminergic innervation is most compelling (Goldberg *et al.*, 1978; Berkowitz, 1982; Lee, 1982; Lokhandwala and Barret, 1982; Neff *et al.*, 1983; Van Loon, 1983; Bell, 1988).

DOPAMINE AND SYMPATHETIC GANGLIA

In the absence of a simple positive marker for dopaminergic neurons (see above), the following indirect methods have been used to look for these cells in sympathetic ganglia: (I) comparison of the DA/noradrenaline ratio in ganglia not thought to contain DA neurons to the DA/noradrenaline ratio in ganglia where the existence of dopaminergic neurons is suspected; (II) comparison of catecholamine fluorescence and DBH immunolabelling of adjacent sections from sympathetic ganglia (catecholamine positive, DBH-negative cells presumed to be dopaminergic); (III) double immunolabelling of the same sections with antibodies to DDC and DBH (DDC-positive, DBH-negative cells presumed to be dopaminergic), and (IV) loading of neurons with DOPA in the presence of monamine oxidase inhibitors following reserpine treatment, since there is evidence from the CNS that this results in the accumulation of DA in dopaminergic but not adrenergic neurons.

The evidence for the existence of dopaminergic autonomic neurotransmission is strongest in vascular beds of the dog kidney and hind limb (Bell and McLaughan, 1982; Bell, 1987). Therefore DA/noradrenaline ratio have been compared in sympathetic ganglia thought to innervate these sites to the ratio in adjacent ganglia projecting to other tissues (Bell and McLaughan, 1982). It was found that the DA/noradrenaline ratio in ganglia projecting to kidney and hind leg vasculature was approx twice (DA being 8–10% of noradrenaline) that of their neighbours (DA 4–5% of noradrenaline). This correlation between high DA/noradrenaline ratio and ganglia, which for other reasons are suspected to give rise to dopaminergic fibres, is worth noting. Other studies, however, have recorded a rather high DA/noradrenaline percentage (7–8%) in the dog superior cervical ganglion, which has not been considered a source of sympathetic dopaminergic neurons (Laverty and Sharman, 1965). Furthermore, DA could reside in cells other than the principal sympathetic neurons, in particular, type I and II SIF cells (Chiba and Williams, 1975) which in many species are thought to contain significant amounts of DA (see above). The question of whether those sympathetic ganglia which have high DA/noradrenaline ratio also harbor dopamine-containing SIF cells has not been examined adequately (see Muller et al., 1984).

When catecholamine fluorescent cells in various dog sympathetic ganglia were assessed for the presence of DBH by immunolabelling of adjacent sections some positive correlation was found between the number of DBH-negative cells, which might be dopaminergic, and DA/noradrenaline ratio of the ganglia (which in turn correlated with ganglia innervating the kidney and hind limb vasculature) (Bell and Muller, 1982; Muller et al., 1984). The significance of the DBH-negative cells is hard to evelute in these experiments because even in ganglia with low DA/noradrenaline ratio a large proportion of neurons was often DBH negative, indicating lack of adequate sensitivity of the immunohistochemical method. In another study, DBH antibodies were used in conjunction with DDC antibodies on the same section of dog sympathetic ganglia in a double-label immunohistochemical assay (Muller et al., 1984). In ganglia providing the renal and hindpaw outflow, 2–5% of the neurons were DDC positive and DBH negative which is the expected phenotype of dopaminergic cells. This study appears, however, to suffer from a methodological problem relating to the use of two rabbit antibodies in a double immunofluorescence test. Furthermore, it seems possible that some of the cells showing the dopaminergic phenotype represent type I chromaffin cells. A related study on the rat superior cervical ganglion using tyrosine hydroxylase (TH) and DBH antibodies revealed a few TH-positive, DBH-negative cells, a phenotype consistent with dopaminergic character (Price, 1984). While this study does not suffer from the methodological problems of the first one, the intensity of the DBH immunostaining varies markedly between cells and it is therefore difficult to judge whether a cell without detectable DBH immunoreactivity lacks the enzyme or merely possesses lower levels.

The rat superior cervical ganglion has also been examined by antibodies against DA (Sakai *et al.*, 1989). Antibody binding was only detected in type I and II SIF cells.

Several neurons in dog sympathetic ganglia which are related to renal and hindpaw outflow react to DOPA loading after reserpine pretreatment as expected from dopaminergic cells (Bell and McLaughan, 1982). Thus a high DA/noradrenaline ratio is restored and catecholamine fluorescence reappears in some cells, the presumed dopaminergic ones. The majority of neurons remained without appreciable fluorescence, as expected for adrenergic neurons under these experimental circumstances.

It seems clear that more work is required to establish whether certain sympathetic ganglia contain dopaminergic principal neurons.

DOPAMINE AND THE KIDNEY

DA stimulates sodium excretion, increases renal filtration rate and causes renal vasodilatation (Goldberg *et al.*, 1978; Lee, 1982). The dog kidney also has a relatively high DA/noradrenaline ratio (Dienerstein *et al.*, 1983). Three lines of histochemical evidence support the proposal that some of the kidney vasculature is innervated by dopaminergic fibres: (I) The catecholamine fluorescence in fibres of the dog kidney reacts to hydrochloric acid treatment as expected if the fluorescence was derived from DA rather than noradrenaline (Dienerstein *et al.*, 1979). (II) DDC positive nerve fibres have been described in dog and human kidney (Dienerstein *et al.*, 1979; Bell *et al.*, 1989). This suggests that these fibres are dopaminergic since studies on the CNS indicate that DDC immunoreactivity is very low or absent in adrenergic nerve terminals but high in dopaminergic ones. (III) Following reserpine treatment, DOPA loading restores catecholamine fluorescence to some nerve fibres in the dog kidney as expected if the fibres are dopaminergic (see above) (Bell *et al.*, 1978).

Furthermore, in some circumstances stimulation of renal nerves produces vasodilation which is blocked by dopamine receptor agonists and results in release of DA in addition to NOR (reviewed in Dienerstein *et al.*, 1983). It seems clear, however, that a substantial amount of the DA efflux is of non-neural origin, since denervation of the kidney suppresses DA release by approx. 50%, while completely abolishing noradrenaline efflux (Stephenson *et al.*, 1982). In accordance with this there is evidence that circulating DOPA, converted to DA by non-neuronal DDC, is the origin of significant amounts of DA secreted from the kidney (Lee, 1982).

In conclusion, there is growing, but not yet conclusive evidence in support of the notion that some of the renal vascular bed receives dopaminergic innervation.

REFERENCES

Aas, P. (1983). Serotonin induced release of acetylcholine from neurones in the bronchial smooth muscle of the rat. *Acta Physiol. Scand.*, **117**, 477–480.

Akasu, T., Hasuo, H. and Tokimasa, T. (1987). Activation of 5-HT$_3$ receptor subtypes causes rapid excitation of rabbit parasympathetic neurones. *Br. J. Pharmacol.*, **91**, 453–455.

Amenta, F., De Rosi, M., Mione, M.C. and Geppetti, P. (1985). Characterisation of ^{3}H 5-hydroxytryptamine uptake within rat cerebrovascular tree. *Eur. J. Pharmacol.*, **112**, 181–186.

Andrews, P.L.R. and Wood, K.L. (1986). Systemic baclofen stimulates gastric motility and secretion via a central action in the rat. *Br. J. Pharmac.*, **89**, 461–467.

Andrews, P.L.R., Bingham, S. and Wood, K.L. (1987). Modulation of the vagal drive to the intramural cholinergic and non-cholinergic neurones in the ferret stomach by baclofen. *J. Physiol.*, **388**, 25–39.

Austin, L., Livett, B.G. and Chubb, I. (1967). Increased synthesis and release of noradrenaline and dopamine during nerve stimulation. *Life Sciences*, **6**, 97–104.

Baetge, G., Gershon, M.D. (1986). GABA in the PNS: Demonstration in enteric neurons. *Brain Res. Bull.*, **16**, 421–424.

Barbier, A.J., Guenaneche, F. and LeFebvre, R.A. (1989). Influence of GABA and ethylene diamine in the guinea-pig duodenum. *J. Auton. Pharmac.*, **9**, 279–291.

Beart, P.M., Kelly, J.S. and Schon, F. (1974). γ-aminobutyric acid in the rat peripheral nervous system, pineal and posterior pituitary. *Biochem. Soc. Trans.*, **2**, 266–268.

Bell, C., Lang, N.J. and Laska, F. (1978) Dopamine-containing vasomotor nerves in the dog kidney. *J. Neurochemistry*, **31**, 77–83.

Bell, C. and McLaughan, E. (1982). Dopaminergic neurons in sympathetic ganglia of the dog. *Proc. R. Soc. Lond.*, B **215**, 175–190.

Bell, C. and Muller, B.D. (1982) Absence of dopamine-β-hydroxylase in some catecholamine-containing sympathetic ganglion cells of the dog: Evidence for dopaminergic autonomic neurons. *Neuroscience Lett.*, **31**, 31–35.

Bell, C. (1987). Dopamine: precursor or neurotransmitter in sympathetically innervated tissues? *Blood Vessels*, **24**, 234–239.

Bell, C. (1988). Dopamine release from sympathetic nerve terminals. *Progress in Neurobiology*, **30**, 193–208.

Bell, C., Bhathal, P.S., Mann, R. and Ryan, G.B. (1989). Evidence that dopaminergic sympathetic axons supply the medullary arterioles of human kidney. *Histochemistry*, **91**, 361–364.

Berkowitz, B.A. (1982) Dopamine and the kidney. *Clin. Sci.*, **62**, 439–448.

Bertilson, L., Suria, A. and Costa, E. (1976). γ-aminobutyric acid in rat superior cervical ganglia. *Nature*, **260**, 540–541.

Blandizzi, C., Bernardini, M.C. and Del Tacca, M. (1988). The effects of GABA agonists and antagonists on rat gastric acid secretion. *Pharmacol. Res. Comm.*, **20**, 419–420.

Bonanno, G., Carazzini, P., Andrioli, G.C., Asaro, D., Pellegrini, G. and Raiteri, M. (1989). Release-regulating autoreceptors of the GABA$_B$-type in human cerebral cortex. *Br. J. Pharmacol.*, **96**, 341–346.

Björklund, A., Cegrell, L., Falck, B., Ritzin, M. and Rosengren, E. (1970). Dopamine containing cells in sympathetic ganglia. *Acta Physiol. Scand.*, **78**, 334–338.

Bowery, N.G. and Brown, D.A. (1972). γ-aminobutyric acid uptake by sympathetic ganglia. *Nature*, **238**, 89–91.

Bowery, N.G. and Brown, D.A. (1974) Depolarizing actions of γ-aminobutyric acid and related compounds on rat superior cervical ganglia *in vitro*. *Br. J. Pharmacol.*, **50**, 205–218.

Bowery, N.G., Brown, D.A., White, R.D. and Yamini, G. (1979) ^{3}H-γ-aminobutyric acid uptake into neuroglial cells of rat superior cervical ganglia. *J. Physiol. Lond.*, **293**, 51–74.

Bowery, N.G., Doble, A., Hill, D.R., Hudson, A.L., Shaw, J.S., Turnbull, M.J. and Warrington, R. (1981). Bicuculline-insensitive GABA receptors on peripheral autonomic nerve terminals. *Eur. J. Pharmacol.*, **71**, 53–70.

Bowery, N.G., Doble, A., Hill, D.R., Hudson, A.L., Shaw, J. and Turnbull, M.J. (1982). Evidence for a second population of GABA receptors which may be located on presynaptic terminals. In: *Presynaptic Receptors: Mechanisms and Functions*, edited by J. de Belleroche. Ellis Harwood Ltd.

Bowery, N.G. and Hill, D.R. (1986). GABA mechanisms in autonomic ganglia. In: *GABAergic Mechanisms in the Mammalian Periphery*, edited by S.L. Erdö and N.G. Bowery. Raven Press: New York.

Bradley, P.B., Humphrey, P.P.A. and Williams, R.H. (1986). Evidence for the existence of 5-hydroxytryptamine receptors, which are not of the 5-HT$_2$ types, mediating contraction of the isolated rabbit basilar artery *Br. J. Pharmacol.*, **87**, 3–4.

Branchek, T.A., Kates, M. and Gershon, M.D. (1984). Enteric receptors for 5-hydroxytryptamine. *Brain Research*, **324**, 107–118.

Branchek, T.A., Mawe, GM., and Gershon, MD. (1988). Characterisation and localisation of a peripheral neural 5-hydroxytrytamine receptor subtype (5-HT$_{IP}$) with a selective agonist, ^{3}H-5-hydroxyindalpine. *J. Neuroscience*. **8**, 2582–2595.

Brown, D.A. and Marsh, S. (1978). Axonal GABA receptors in mammalian peripheral nerve trunks. *Brain Res.*, **156**, 187–191.

Bucheit, K.-H., Engel, G., Mutschler, E. and Richardson, B. (1985). Study of the contractile effect of 5-hydroxytryptamine (5-HT) in the isolated longitudinal muscle strip from the guinea-pig ileum: evidence for two distinct release mechanisms. *Naunyn-Schmied. Arch. Pharmacol.*, **332**, 8–15.

Bühler, H.U., Da Prada, M., Haefely, W. and Piaotti, G.B. (1978). Plasma adrenaline, noradrenaline and dopamine in man and different animal species. *J. Physiol.*, **276**, 311–320.

Bülbring, E. and Gershon, M.D. (1967). 5-Hydroxytryptamine participation in the vagal inhibitory innervation of the stomach. *J. Physiol.*, **192**, 823–846.

Burnstock, G., Cocks, T., Crowe, R. and Kasakov, L. (1978). Purinergic innervation of the guinea-pig urinary bladder. *Br. J. Pharmacol.*, **63**, 125–138.

Cherubini, E. and North, R.A. (1984a). Actions of γ-aminobutyric acid on neurones on guinea-pig myenteric plexus. *Br. J. Pharmacol.*, **82**, 93–100.

Cherubini, E. and North, R.A. (1984b). Inhibition of calcium spikes and transmitter release by γ-aminobutyric acid in the guinea-pig myenteric plexus. *Br. J. Pharmacol.*, **82**, 101–105.

Cherubini, E, and North R.A. (1985). Benzodiazepines both enhance γ-aminobutyrate responses and decrease calcium action potentials in guinea-pig myenteric neurones. *Neuroscience*, **14**, 309–315.

Chiba, T. and Williams, T.H. (1975). Histofluorescence characteristics and quantification of small intensely fluorescent (SIF) cells in sympathetic ganglia of several species. *Cell and Tissue Res.*, **162**, 331–341.

Chouchkov, C., Lazarov, N., and Daviddoff, M. (1988). Serotonin-like immunorectivity in the cat trigeminal ganglion. *Histochemistry*. **88**, 637–639.

Cooke, H.J. and Carey, H.V. (1985). Pharmacological analysis of 5-hydroxytryptamine actions on guinea-pig ileal mucosa. *Eur. J. Pharmacol.*, **111**, 329–327.

Costa, M. and Furness, J.B. (1979). The sites of action of 5-HT in nerve muscle preparations from the guinea-pig small intestine and colon. *Br. J. Pharmacol.*, **65**, 237–248.

Craig, DA., and Clarke, DE., (1990). Pharmacological characterisation of a neuronal receptor for 5-hydroxytryptamine in guinea-pig ileum with properties similar to the 5-hydroxytryptamine$_4$ receptor. *J. Pharmacol. Exp. Ther.* **252**, 1378–1386.

Crillis, C.N. (1985). Peripheral metabolism of serotonin. In: *Serotonin and the Cardiovascular System*, edited by P.M. Vanhoutte, Raven Press: New York. pp. 27–36.

Cummings, J.P. and Felten, D.L. (1979). A raphe dendrite bundle in the rabbit medulla. *J. Comp. Neurol.*, **183**, 1–23.

Dahlstrom, A. and Ahlman, H. (1983). Immunocytochemical evidence for the presence of tryptaminergic nerves of blood vessels, smooth muscle and myenteric plexus in the rat small intestine. *Acta Physiol: Scand.*, **117**, 589–591.

Dahlstrom, A., Nilsson, O., Lundgren, O., and Ahlman, H. (1988) Nonadrenergic, noncholinergic innervation of gastrointestinal vessels: morphological and physiological aspects. In: *Non-adrenergic innervation of blood vessels* Vol II. Ed. G. Burnstock and S. Griffith. pp 143–151.

Das, M., Mohanakumar, KP., Chanhan, SPS., and Ganguly, DK. (1989). 5-hydroxytryptamine in the phrenic nerve diaphragm: evidence for its existence and release. *Neuroscience Letters*. **97**, 345–349.

Davanger, S., Ottersen, O.P., Storm-Mathisen, J. (1987). Immunocytochemical localization of GABA in cat myenteric plexus. *Neurosci. Lett.*, **73**, 27–32.

Davanger, S., Ottersen, O.P. and Storm-Mathisen, J. (1989). GABA-immunoreactive cells in the rat gastrointestinal epithelium. *Anat. Embryol.*, **179**, 221–226.

DeGroat, W.C. (1970). The actions of γ-aminobutyric acid and related amino acids on mammalian autonomic ganglia. *J. Pharmacol. Exp. Ther.*, **172**, 384-396.

De La Lande (1989). Amplification mechanisms in peripheral tissues. In: *The Peripheral Actions of 5-Hydroxytryptamine*, Ed. by J.R. Fozard, Oxford University Press: NY. pp. 123-146.

Derkach, V., Surprenant, A., and North, RA. (1989). 5-HT₃ receptors are membrane ion channels. *Nature.* **339**, 706-709.

Di Carlo, V. (1983). Serotonergic fibres in dorsal roots of the spinal cord. *Neurosci. Lett.*, **43**, 233-244.

Di Carlo, V. (1984a) Perivascular serotonergic neurons: somato dendritic contacts and axonic innervation of blood vessels. *Neurosci. Lett.*, **51**, 295-302.

Di Carlo, V. (1984b). Segmental serotonergic innervation of spinal cord arterial circulation. *Neurosci. Lett.*, **49**, 225-231.

Dienerstein, R.J., Vannice, J., Henderson, R.C., Roth, L.J., Goldberg, L.I. and Hoffman, P.C. (1979). Histofluorescence techniques provide evidence for dopamine containing neuronal elements in the canine kidney. *Science*, **205**, 497-499.

Dienerstein, R.J., Jones, R.T. and Goldberg, L.I. (1983). Evidence for dopamine-containing renal nerves. *Federation Proceedings*, **42**, 3005-3008.

Dinah, WY., Sah, and Matsumoto, SG. (1987). Evidence for serotonin synthesis, uptake and release in dissociated rat sympathetic neurons in culture. *J. Neuroscience.* **7**, 391-399.

Drakontides, A.B. and Gershon, M.D. (1968). 5-HT receptors in the mouse duodenum. *Br. J. Pharmacol.*, **33**, 480-492.

Dreyfus, C.F., Sherman, D., and Gershon, M.D. (1977) Uptake of serotonin by intrinsic neurons of the myenteric plexus grown in organotypic tissue culture. *Brain Research*, **128**, 109-123.

Dubois, A. and Jacobowitz, D.M. (1974). Failure to demonstrate serotonergic neurones in the myenteric plexus of the cat. *Cell Tiss. Res.*, **150**, 493-496.

Dun, N.J. and Karczmar, A.G. (1981). Evidence for a presynaptic inhibitory action of 5-hydroxytryptamine in a mammalian sympathetic ganglion. *J. Pharm. Exptl. Ther.*, **217**, 714-718.

Dun, N.J., Kiraly, M. and Ma, R.C. (1984). Evidence for a serotonin mediated slow excitatory potential in the guinea pig coeliac ganglia. *J. Physiol.*, **351**, 61-76.

Dutar, P. and Nicoll, R.A. (1988). A physiological role for GABA_B receptors in the central nervous system. *Nature*, **332**, 156-158.

Edvinsson, L., Degneurce, A., Duverger, D., Mackenzie, E.T. and Scatton, B. (1983). Central serotonergic nerves project to the pial vessels of the brain. *Nature*, **306**, 55-57.

Edvinsson, L., Birath, E., Uddman, R., Lee, T.J.F., Duverger, D., Mackenzie, E.T. and Scatton, B. (1984). Indole-aminergic mechanisms in brain vessels. Localisation, concentration, uptake and *in vitro* responses of 5-hydroxytryptamine. *Acta Physiol. Scand.*, **121**, 291-299.

Eränkö, O. and Eränkö, L. (1971). Small intensely fluorescent, granule containing cells in the sympathetic ganglion of the rat. *Progress in Brain Research.* **34**, 39-51.

Erde, S.M., Sherman, D.D. and Gershon, M.D. (1985). Morphology and serotonergic innervation of physiologically identified cells of the guinea-pig myenteric plexus. *J. Neurosci.*, **5**, 617-633.

Erwin, V.G. and Deitrich, R.A. (1966). Brain aldehyde dehydrogenase localisation, purification and properties. *J. Biol. Chem.*, **241**, 3533-3539.

Espamer, V. and Asero, B. (1952). Identification of enteramine the specific hormone of the entrochromaffin cell system, as 5-hydroxytryptamine. *Nature*, **169**, 800-801.

Espamer, V. (1954). Pharmacology of indolealkylamines. *Pharmacological Reviews*, **6**, 425-487.

Eugene, D. (1987). Fast non-cholinergic depolarizing postsynaptic potentials in neurons of rat superior cervical ganglia. *Neurosci. Lett.*, **78**, 51-56.

Fargeas, M.J., Fioramonti, J. and Bueno, L. (1988). Central and peripheral action of GABA_A and GABA_B agonists on small intestine motility in rats. *Eur. J. Pharmacol.*, **150**, 163-169.

Feher, E., Leranth, C. and Verhofstad, A.A.J. (1983). Distribution and the fine structure of serotonin-containing fibres in the rat small intestine. *Acta. Biol. Acad. Sci. Hung.*, **34**, 73-80.

Feldberg, W. and Toh, C.C. (1953). Distribution of 5-hydroxytryptamine (serotonin, enteramine) in the wall of the digestive tract. *J. Physiol. London.*, **119**, 352-362.

Feltz, P. and Rasminsky, M. (1974). A model for the mode of action of GABA on primary afferent terminals: depolarizing effects of GABA applied iontophoretically to neurones of mammalian dorsal root ganglia. *Neuropharmacol.*, **13**, 553-563.

Fenuik, W. and Humphrey, P.P.A. (1989). Mechanisms of 5-hydroxytryptamine induced vasoconstriction. In: *The Peripheral Actions of 5-Hydroxytryptamine*, edited by J.N. Foxard, pp. 100-122. Oxford University Press: NY.

Forsberg, E.J. and Miller, R.J. (1983). Regulation of serotonin release from rabbit intestinal entrochromaffine cells. *J. Pharm. Exptl. Therap.*, **227**, 755-766.

Fozard, J.R. (1984). Neuronal 5-HT receptrs in the periphery. *Neuropharmacology*, **23**, 1473–1486.

Frigo, G.M., Galli, A., Lecchini, S. and Marcoli, M. (1987). A facilitatory effect of bicuculline on the enteric neurons in the guinea-pig isolated colon. *Br. J. Pharmac.*, **90**, 31–41.

Furness, J.B. and Costa, M. (1982). Neurons with 5-HT like immunoreactivity in the enteric nervous system: their projections in guinea-pig small intestine. *Neuroscience*, **7**, 341–350.

Furness, J.B. and Costa, M. (1987). Studies of neuronal circuitry of the enteric nervous system. In: *The Enteric Nervous System*, edited by J.B. Furness and M. Costa, Churchill Livingstone. pp. 111–136.

Furness, J.B., Llewellyn-Smith, I.J., Bornstein, J.C. and Costa, M. (1988). Chemical neuroanatomy and the analysis of neuronal circuitry in the enteric nervous system. In: *Handbook of Chemical Neuroanatomy Vol. 6. The Peripheral Nervous System*, edited by A. Bjorklund, T. Hokfelt and C. Owman. Elsevier Science Publishers BV.

Furness, J.B., Trussell, D.C., Pompola, S., Bornstein, J.C., Maley, B.E. and Storm-Mathisen, J. (1989). Shapes and projections of neurons with immunoreactivity for gamma-aminobutyric acid in the guinea-pig small intestine. *Cell Tissue Res.*, **256**, 293–301.

Gabella, G. (1972) Innervation of the intestinal muscular coat. *J. Neurocytol.* **1**, 341–362.

Gabriel, R. and Eckert, M. (1989). Demonstration of GABA-like immunoreactivity in myenteric plexus of frog stomach. *Histochemistry*, **91**, 523–525.

Gaddum, J.H. and Picarelli, Z.P. (1957). Two kinds of tryptamine receptor. *Brit. J. Pharmacol. Chemother.*, **12**, 323–328.

Galligan, JJ., Surprenant, A., Tonini, M, and North, RA. (1988). Differential localisation of 5-HT$_1$ receptors on myenteric and submucosal nervous. *Am. J. Physiol.*, **255** G603–G611.

Gaudin Chazal, G., Dszuta, A., Segu, L., Ternaux, J.P and Puizillout, J.J. (1978). Serotonin containing neurons in the nodose ganglion of the cat. *Waking and Sleeping*, **2**, 149–151.

Gentilini, G., Luzzi, S., Franchi-Micheli, S. Pantalone, D., Cortesini, C. and Zilletti, L. (1988). Effect of γ-aminobutyric acid on human jejunum. *Pharmacol. Res. Comm.*, **20**, 423–424.

Gershon, M.D., Drakontides, A.B. and Ross L.L. (1965). Serotonin: synthesis and release from the myenteric plexus of the mouse intestine. *Science*, **149**, 197–199.

Gershon, M.D., (1967). Effects of tetrodotoxin on innervated smooth muscle preparations. *Br. J. Pharmacol.*, **29**, 259–279.

Gershon, M.D. and Altman, R.F. (1971). An analysis of the uptake of 5-hydroxytryptamine by the myenteric plexus of the small intestine of the guinea-pig. *J. Pharm. Exptl. Ther.*, **179**, 29–41.

Gershon, M.D., Robinson, R. and Ross, L.L. (1976). Serotonin accumulation in the guinea-pig myenteric plexus: ion dependence, structure activity relationship, and the effect of drugs. *J. Pharm. Exptl. Ther.*, **198**, 548–561.

Gershon, M.D. and Jonakait, G.M. (1979). Uptake and release of 5-hydroxytryptamine by enteric serotonergic neurons: effects of fluoxetine (Lilly 110140) and chlorimipramine. *Br. J. Pharmacol.*, **66**, 7–9.

Gershon, M.D. and Tamir, H. (1981). Release of endogenous 5-hydroxytryptamine from resting and stimulated enteric neurons. *Neuroscience*, **6**, 2227–2286.

Gershon, M.D. and Sherman, D. (1982a). Identification of and interactions between nonadrenergic and serotonergic neurites in the myenteric plexus. *J. Comp. Neurol.*, **204**, 407–421.

Gershon, M.D. and Sherman, D. (1982b). Selective demonstration of serotonergic neurones and terminals in electron micrographs: loading with 5,7-dihydroxytryptamine and fixation with NaMnO$_4$. *J. Histochem. Cytochem.*, **30**, 769–773.

Gershon, M.D. and Tamir, H. (1985). Peripheral sources of serotonin binding proteins. In: *Serotonin and the Cardiovascular System*, edited by P.M. Vanhoute, Raven Press: New York. pp. 15–26.

Gershon, M.D., Mawe, G.M. and Branchek, T.A. (1989). 5-HT and enteric neurones. In: *The Peripheral Actions of 5-HT*, Ed. J.R. Fozard, Oxford University Press. pp. 247–273.

Gershon, MD., and Ross, LL. (1966) Radioisotopic studies of the binding, exchange and distribution of 5-hydrotytryptamine synthesised from its radioactive precursor. *J. Physiol.* **186**. 451–476.

Gershon, MD., Takaki, M., Tamir, H., and Branchek, T. (1985) The enteric neural receptor for 5-hydroxytryptamine. *Experientia*. **41**. 863–868.

Gilbert, MJ., and Newberry, NR. (1987) A 5-HT$_1$ like receptor mediates a sympathetic ganglionic hyperpolarisation. *Eur. J. Pharmacol.* **144**, 385–388.

Gillis, C.N. (1980). Metabolism of vasoactive hormones by pulmonary vascular endothelium: possible functional significance. In: *Vascular Neuroeffective Mechanisms*, Ed. J.A. Bevan, T. Godbraind, R.A. Maxwell and P.M. Vanhoute, Raven Press, New York. pp. 304–315.

Gillis, CN. (1985) Peripheral metabolism of serotonin In: *Serotonin and the cardiovascular system*. Ed. P.M. Vanhoutte. Raven Press pp 27–36.

Gilon, G., Reusens-Billen, B., Remacle, C., Janssens de Varebeke, P., Pauwels, G. and Hoet, J.J. (1987).

Localization of high-affinity GABA uptake and GABA content in the rat duodenum during development. *Cell. Tiss. Res.*, **249**, 593–600.

Giotti, A., Luzzi, S., Spagnesi, S. and Zilletti, L. (1983). GABA$_A$ and GABA$_B$ receptor-mediated effects in guinea-pig ileum. *Br. J. Pharmacol.*, **78**, 469–478.

Giotti, A., Luzzi, S., Spagnesi, S., Maggi, C.A. and Zilletti, L. (1985). Modulatory activity of GABA$_B$ receptor, on cholinergic tone in guinea-pig distal colon. *Br. J. Pharmacol.*, **84**, 883–895.

Goldberg, L.I., Volkman, P.H. and Kohli, J.D. (1978). A comparison of vascular dopamine receptors with other dopamine receptors. *Annual Review of Pharmacology and Toxicology*, **18**, 57–79.

Gordon, JC., Barefoot, DS., Sarbin, NS., and Pinkus, LM. (1989). [^{3}H] Zacopride binding to 5-hydroxytryptamine$_3$ sites on partially purified rabbit enteric neuronal membranes. *J. Pharmacol. Exp. Ther*, **251**, 962–968.

Goto, Y. and Debas, H.T. (1983). GABA-mimetic effect on gastric acid secretion. Possible significance in central mechanisms. *Dig. Dis. Sci.*, **28**, 56–60.

Goto, Y., Tache, Y., Debas, H. and Novin, D. (1985). Gastric acid and vagus nerve response to GABA against baclofen. *Life Sci.*, **36**, 2471–2475.

Green, H., Sawyer, J.L. (1966). Demonstration characterisation and assay procedure of tryptophan hydroxylase in rat brain. *Anal. Biochem*, **15**, 53–64.

Griffith, S.G. and Burnstock, G. (1983a). Immunohistochemical demonstration of serotonin in nerves supplying human cerebral and mesenteric blood vessels. Some speculations about their involvement in vascular disorders. *Lancet*, **1**, 561–562.

Griffith, S.G. and Burnstock, G. (1983b). Serotonergic neurons in human fetal intestine: an immunohistochemical study. *Gastroenterology*, **85**, 929–937.

Griffith, S.G. (1988). Serotonin (5-HT) as a neurotransmitter in blood vessels. In: *Nonadrenergic Innervation of Blood Vessels. Vol. 1* Ed. G. Burnstock and S.G. Griffith., CRC. pp. 28–40.

Gullikson, GW., Loeffler, RF., Bianchi, RG., Perkins, WE., Bauer, RF. (1988). Ralationship of 5-HT$_3$ antagonist activity in gastrointestinal prokinetic activity in conscious dogs. *Gastroenterology*. **96**, 869A.

Gwee, M.C.E., Yeoh, T.S. (1968). The release of 5-hydroxytryptamine from rabbit small intestine *in vitro*. *J. Physiol.*, **194**, 817–825.

Häppölä, O. (1988). 5-hydroxytryptamine-immunoreactive neurones and nerve fibers in the superior cervical ganglion of the rat. *Neuroscience*. **27**, 301–307.

Häppölä, O., Päivärinta, H., Soinila, S., Wu, J.Y. and Panula, P. (1987). Localization of L-glutamate decarboxylase and GABA transaminase immunoreactivity in the sympathetic ganglia of the rat. *Neuroscience*, **21**, 271–281.

Hardcastle, J., Hardcastle, P.T.T., and Redfern, J.S. (1981). Action of 5-hydroxytryptamine on intestinal transport in the rat. *J. Physiol.* **320**, 41–55.

Harty, R.F. and Franklin, P.A. (1983). GABA affects the release of gastrin and somatostatin from rat antral mucosa. *Nature*, **303**, 623–624.

Harty, R.F. and Franklin, P.A. (1986). Cholinergic mediation of γ-aminobutyric acid-induced gastrin and somatostatin release from rat antrum. *Gastroenterology*, **91**, 1221–1226.

Hashimoto, S., Tanaka, C. and Taniyama, K. (1986). Presynaptic muscarinic and α-adrenoceptor-mediated regulation of GABA release from myenteric neurones of the guinea-pig small intestine. *Br. J. Pharmacol.*, **89**, 787–792.

Higashi, S. and Nishi, S. (1982). 5-Hydroxytryptamine receptors of visceral primary afferent neurones in rabbit nodose ganglia. *J. Physiol.*, **323**, 543–567.

Hill, D.R. and Bowery, N.G. (1986). GABA-receptor-mediated functional responses in peripheral tissues. In: *GABAergic Mechanisms in the Mammalian Periphery*, edited by S.L. Erdö and N.G. Bowery. Raven Press: New York.

Hills, J.M., Jessen, K.R. and Mirsky, R. (1987). An immunohistochemical study of the distribution of enteric GABA-containing neurons in the rat and guinea-pig intestine. *Neuroscience*, **22**, 301–312.

Hills, J.M., King, B.F., Mirsky, R. and Jessen, K.R. (1988). Immunohistochemical localization and electrophysiological actions of GABA in prevertebral ganglia in guinea-pig. *J. Autonom. Nerv. Syst.*, **22**, 129–140.

Hills, J.M., Dingsdale, R.A., Parsons, M.E., Dolle, R.E. and Howson, W. (1989). 3-Aminopropylphosphinic acid – a potent, selective GABA$_B$ receptor agonist in the guinea-pig ileum and rat anococcygeus muscle. *Br. J. Pharmacol.*, **97**, 1292–1296.

Hills, J.M. and Howson, W. (1990). The GABA$_B$ receptor profile of a series of phosphinic acids – agonist and antagonist activity in a range of peripheral tissues. In: *Peripheral GABAergic Mechanisms* Ed. SL Erdo. Springer-Verlag, Berlin. (In Press).

Hobbiger, F. (1958a). Antagonism by γ-aminobutyric acid to the actions of 5-hydroxytryptamine and nicotine on isolated organs. *J. Physiol.*, **144**, 349–360.

Hobbiger, F. (1958b). Effects of γ-aminobutyric acid on the isolated mammalian ileum. *J. Physiol.*, **142**, 147–164.

Holt, S.E., Cooper, M. and Wyllie, J.H. (1986). On the nature of the receptor mediating the action of 5-hydroxytryptamine in potentiating the responses of the mouse urinary bladder strip to electrical stimulation. *Naunyn Schmiedebergs Arch. Pharmacol.*, **334**, 333–340.

Holzer, P. and Skofitsch, A. (1984). Release of endogenous 5-hydroxytryptamine from the myenteric plexus of the guinea-pig isolated small intestine. *Br. J. Pharmacol.*, **81**, 381–386.

Hornykiewics, O. (1966). Dopamine (3-hydroxytyramine) and brain function. *Pharmacol. Rev.*, **18**, 925–964.

Houston, D.S. and Vanhoute, P.M. (1989). Pathophysiological significance of 5-hydroxytryptamine in the periphery. In: *The Peripheral Actions of 5-Hydroxytryptamine*, edited by J.R. Fozard, OUP N.Y. pp. 377–406.

Hullihan, J.P., Spector, S., Taniguchi, T. and Wang, J.K.T. (1983). The binding of [^{3}H]-diazepam to guinea-pig ileal longitudinal muscle and the *in vitro* inhibition of contraction by benzodiazepines. *Br. J. Pharmac.*, **78**, 321–327.

Inouye, A., Fukuya, M., Tsudiya, K. and Tsujioka, T. (1960). Studies on the effects of γ-aminobutyric acid on the isolated guinea-pig ileum. *Jap. J. Physiol.* **10**, 167–182.

Ireland, SJ. (1987). Origin of 5-hydroxytryptamine-induced hyperpolarisation of the rat superior cervical ganglion and vagus nerve. *Br. J. Pharmac.* **92**, 407–416.

Ireland, SJ., and Jordan, CC. (1987). Pharmacological characterisation of 5-hydroxytryptamine induced hyperpolarisation of the rat superior cervical ganglion. *Br. J. Pharmac.* **92**, 417–427.

Iversen, L.L. and Kelly, J.S. (1975). Uptake and metabolism of γ-aminobutyric acid by neurons and glial cells. *Biochem. Pharmacol.*, **24**, 933–938.

Iwasawa, Y. and Gillis, C.N. (1974). Pharmacological analysis of norepinephrine and 5-hydroxytryptamine removal from the pulmonary circulation: differentiation of uptake sites for each amine. *J. Pharm. Exptl. Ther.*, **188**, 386–393.

Jessen, K.R., Mirsky, R., Dennison, M.E. and Burnstock, G. (1979). GABA may be a neurotransmitter in the vertebrate peripheral nervous system. *Nature.*, **281**, 71–74.

Jessen, K.R., Hills, J.M., Dennison, M.E. and Mirsky, R. (1983) γ-Aminobutyrate as an autonomic neurotransmitter: release and uptake of ^{3}Hγ-aminobutyrate in guinea-pig large intestine and cultured enteric neurons using physiological methods and electron microscopic autoradiography. *Neuroscience*, **10**, 1427–1442.

Jessen, K.R., Hills, J.M. and Saffrey, M.J. (1986): Immunohistochemical demonstration of GABA-ergic neurons in the enteric nervous system. *J. Neurosci.*, **6**, 1628–1634.

Jessen, K.R. Hills, J.M. and Limbrick, A.H. (1988). GABA immunoreactivity and ^{3}H-GABA uptake in mucosal epithelial cells of the rat stomach. *Gut*, **29**, 1549–1556.

Johnson, S.M., Katayama, V. and North, R.A. (1980). Multiple actions of 5-hydroxytryptamine on myenteric neurones of the guinea-pig ileum. *J. Physiol.*, **304**, 459–479.

Jonakait, G.M., Tamir, H., Rapport, M.M. and Gershon, M.D. (1977). Detection of a soluble serotonin binding protein in the mammalian myenteric plexus and other peripheral sites of serotonin storage. *J. Neurochem.*, **28**, 277–284.

Jonakait, G.M., Tamir, H., Gintzler, A.R. and Gershon, M.D. (1979). Release of [^{3}H] serotonin and its binding protein from enteric neurones. *Brain Research*, **174**, 55–69.

Joò, F., Siklos, L., Dames, W. and Wolff, J.R. (1987). Fine structural changes of synapses in the superior cervical ganglion of adult rats after long-term administration of GABA: a morphometric analysis. *Cell Tissue Res.*, **249**, 267–275.

Juorio, A.V. and Gabella, G. (1974). Noradrenaline in the guinea-pig alimentary canal: regional distribution and sensitivity to denervation and reserpine. *J. Neurochem.*, **221**, 851–858.

Kadzielawa, K. (1983). Antagonism of the excitatory effects of 5-hydroxytryptamine on sympathetic preganglionic neurones and neurones activated by visceral afferents. *Neuropharmacology*, **22**, 19–27.

Kanazawa, I., Iversen, L.L. and Kelly, J.S. (1976). Glutamate decarboxylase activity in rat posterior pituitary, pineal gland, dorsal root ganglion and superior cervical ganglion. *J. Neurochem.*, **27**, 1267–1269.

Kaplita, P.V., Walters, D.H. and Triggle, D.J. (1982). γ-aminobutyric acid action on guinea-pig ileal myenteric plexus. *Eur. J. Pharmacol.*, **79**, 43–51.

Kàsa, P., Jo, F., Dobo, E., Wenthold, R.J., Ottersen, D.P., Storm-Mathisen, J. and Wolff, J.R. (1988). Heterogenous distribution of GABA-immunoreactive nerve fibres and axon terminals in the superior cervical ganglion of adult rat. *Neuroscience*, **26**, 635–644.

Kenny, S.L. and Arians, M.A. (1986). The immunofluorescence localization of glutamate decarboxylase in the rat superior cervical ganglion. *J. Auton. Syst.*, **17**, 211–215.

Kerr, D.I.B. and Krantis, A. (1983). Uptake and stimulus evoked release of ^{3}H-γ-aminobutyric acid by myenteric nerves of guinea-pig intestine. *Br. J. Pharmacol.*, **78**, 271–276.

Kerr, D.I.B. and Ong, J. (1984). Evidence that ethylene diamine acts in the isolated ileum of the guinea-pig by releasing endogenous GABA. *Br. J. Pharmacol.*, **83**, 169–178.

Kerr, D.I.B. and Ong, J. (1986a). GABAergic mechanisms in the gut: their role in the regulation of gut motility. In: *GABAergic Mechanisms in the Mammalian Periphery*, edited by S.L. Erdö and N.G. Bowery. Raven Press, New York.

Kerr, D.I.B. and Ong, J. (1986b). γ-Aminobutyric acid-dependent motility induced by avermectin B$_{1a}$ in the isolated intestine of the guinea-pig. *Neurosci. Lett.*, **65**, 7–10.

Kerr, D.I.B., Ong, J., Prager, R.H., Gynther, B.D. and Curtis, D.R. (1987). Phaclofen: a peripheral and central baclofen antagonist. *Brain Research*, **405**, 105–154.

King, FD., and Sanger, GJ. (1988). Gastrointestinal motility enhancing agents. *Ann. Rep. Med. Chem.* **23** 201–210.

Kiraly, M., Ma, RC., and Dun, NJ. (1983). Serotonin mediates a slow excitatory potential in mammalian coeliac ganglion. *Brain research.* **275**, 378–383.

Knight, D.S. and Bazer, G.T. (1979). Visualization of intrarenal catecholamine-containing elements: fluorescence histochemistry and electron microscopy. *J. Auton. Nerv. Syst.*, **1**, 173–181.

Krantis, A., Costa, M., Furness, J.B. and Orbach, J. (1980). γ-aminobutyric acid stimulates intrinsic inhibitory and excitatory nerves in the guinea-pig intestine. *Eur J. Pharmacol.*, **67**, 461–468.

Krantis, A. and Kerr, D.I.B. (1981a). Autoradiographic localisation of ^{3}H-γ aminobutyric acid in the myenteric plexus of the guinea-pig small intestine. *Neurosci. Lett.*, **23**, 263–268.

Krantis, A. and Kerr, D.I.B. (1981b). The effect of GABA antagonism on propulsive activity of the guinea-pig large intestine. *Eur. J. Pharmacol.*, **76**, 111–114.

Krantis, A., Kerr, D.I.B. and Dennis, B.J. (1986). Autoradiographic study of the distribution of (3H) γ-aminobutyrate-accumulating neural elements in the guinea-pig intestine: evidence for transmitter function for γ-amino-butyrate. *Neuroscience*, **17**, 1243–1255.

Krantis, A. and Harding, R.K. (1987). GABA-related actions in isolated *in vitro* preparations of the rat small intestine. *Eur. J. Pharmacol.*, **141**, 291–298.

Kuhn, D.M., Wolf, W.A. and Lovenberg, W. (1980). Review of the role of central serotonergic neuronal system in blood pressure regulation. *Hypertension*, **2**, 243–255.

Kurain, S.S., Feri, G.L., Deweg, J. and Polak, J.M. (1983). Immunocytochemistry of serotonin containing nerves in the human gut. *Histochemistry*, **78**, 523–529.

Kvetnansky, R., Weise, V.K., Thoa, N.B. and Kopin, I.J. (1979). Effects of chronic guanethidine treatment and adrenal medullectomy on plasma levels of catecholamines and corticosterone in forcibly immobilized rats. *J. Pharmacol. Exp. Ther.*, **209**, 287–291.

Laverty, R. and Sharman, D.F. (1965). The estimation of small quantities of 3,4-dihydroxyphenylethylamine in tissues. *Br. J. Pharmacol. Chemother.*, **24**, 538–584.

Lee, H.R. (1982). Dopamine and the kidney. *Clin. Sci.*, **62**, 439–448.

Legay, C., Faudon, M., Hery, F. and Ternaux, J.P. (1983). 5-HT metabolism in the intestinal wall of rat. 1. The mucosa, *Neurochemistry International.*, **5**, 721–7.

Legay, C., Saffrey, M.J. and Burnstock, G. (1984). Coexistence of immunoreactive SP and serotonin in membranes of the gut. *Brain Research*, **302**, 379–382.

Levine, A.S., Morley, J.E., Kneip, J., Grace, M. and Silvis, S.E. (1981). Muscimol induces gastric secretion after central administration. *Brain Research*, **229**, 270–274.

Lingjaerde, O. (1977). Platelet uptake and storage of serotonin. In: *Serotonin in Health and Disease*, Ed. W.B. Essman, Spectrum: New York. pp. 139–199.

Lokhandwala, M.F. and Barrett, R.J. (1982). Cardiovascular dopamine receptors: physiological, pharmacological and therapeutic implications. *J. Auton. Pharmacology.*, **3**, 189–215.

Lovenberg, W., Levine, R.J. and Sjoerdsma, A. (1965). Tryptophan hydroxylase in cell-free extracts of malignant mouse mast cells. *Biochem. Pharmacol.*, **14**, 887–889.

Luizzi, A., Foppan, F.H., Saavedra, J.M., Levi Montalcini, R. and Kopin, I.J. (1977). Gas chromatographic-mass-spectromatic assay of serotonin in rat superior cervical ganglia. Effects of nerve growth factor and 6-hydroxydopamine. *Brain Research*, **133**, 354–357.

Ma, R.C., Horwitz, J., Kiraly, M., Perlman, R.L. and Dun, N.J. (1985). Immunohistochemical and biochemical detection of serotonin in the guinea-pig coeliac-superior mesenteric plexus. *Neurosci. Lett.*, **56**, 107–112.

Ma, R.C. and Dun, N.J. (1986). Excitation of lateral horn neurones of the neonatal rat spinal cord by 5-hydroxytryptamine. *Dev. Brain Research*, **24**, 89–98.

Maggi, C.A. Manzini, S. and Meli, A. (1984). Evidence that GABA$_A$ receptors mediate relaxation of rat duodenum by activating intramural non-adrenergic, non-cholinergic neurones. *J. Auton. Pharmac.*, **4**, 77–85.

Maggi, C.A., Giuliani, S., Santicioli, P., Selleri, S. and Meli, A. (1989). The motor response to ethylenediamine of the rat isolated duodenum: involvement of GABAergic transmission? *Naunyn Schmiedebergs. Arch. Pharmacol.*, **340**, 419–423.

Marco, E.J., Balfagon, G., Salaices, M., Sanchez-Ferrer, C.F. and Marin, J. (1985). Serotonergic innervation of cat cerebral arteries. *Brain Research*, **338**, 137–139.

Mawe, GM., and Gershon, MD. (1989) Structure afferent innervation, and transmitter content of ganglia of the guinea-pig gall bladder. Relationship to the enteric nervous system. *J. Comp. Neurol.* **283**, 374–390.

Mawe, G.M., Branchek, T.A. and Gershon, M.D. (1986). Peripheral neural serotonin receptors: identification and characterisation with specific antagonists and agonists. *Proc. Natl. Acad. SCI.*, **83**, 9799–9803.

Mawe, G.M., Branchek, T.A. and Gershon, M.D. (1989). Blockade of 5HT-mediated slow EPSPS by BRL 24924: gastrokinetic effects. *Am. J. Physiol.*, **257**, 9386–9396.

McCall, R.B. (1983). Serotonergic excitation of sympathetic preganglionic neurones: a micro-inotrophonetic study. *Brain Research*, **289**, 121–127.

McQueen, D.S. and Mir, A.K. (1989). 5-hydroxytryptamine and cardio-pulmonary and carotid body reflex mechanisms. in: *The Peripheral actions of 5-Hydroxytryptamine*, Ed. J.R. Fozard. Oxford Medical Publications: OUP.

Micheletti, R., Schiavene, A. and Giachetti, A. (1988). Muscarinic M$_1$ receptors stimulate a noradrenergic, noncholinergic inhibitory pathway in the isolated rat duodenum. *J. Pharm. Exptl. Ther.*, **244**, 680–684.

Miki, Y., Taniyama, K., Tanaka, C., and Tobe, T. (1983). GABA, glutamic acid decarboxylase and GABA transaminase levels in the myenteric plexus in the intestine of humans and other mammals. *J. Neurochem.*, **40**, 861–865.

Morrissey, J.J., Walker, M.N. and Lovenberg (1977). The absence of tryphophan hydroxylase activity in blood platelets. *Proceedings of the Society for Exptl. Biology and Medicine*, **154**, 496–499.

Muller, B., Harris, T. and Bell, C. (1984). Characterization of chromaffin-like cells in the canine sympathetic chain by enzyme immunohistochemistry and quantitation of their distribution. *Neuroscience*, **13**, 887–899.

Muller, B., Harris, T., Borri Voltattorni, C. and Bell, C. (1984). Distribution of neurons containing DOPA decarboxylase and dopamine-β-hydroxylase in some sympathetic ganglia of the dog: a quantitative study. *Neuroscience*, **11**, 733–740.

Muller-Schweinitzer, E. and Engel, G. (1983). Evidence for mediation by 5-HT$_2$ receptors of 5-hydroxytryptamine-induced contraction of canine basilar artery. *Naunyn-Schmiedebergs. Arch. Pharmacol.*, **324**, 287–292.

Nada, O. and Toyohara, T. (1987). An immunohistochemical study of serotonin containing nerves in the colon of rats. *Histochemistry*, **86**, 229–232.

Nakemoto, M., Tanaka, C. and Taniyama, K. (1987). Release of γ-aminobutyric acid and acetylcholine by neurotensin in the guinea-pig ileum. *Br. J. Pharmacol.*, **90**, 545–551.

Napoleone, P., Sancesaria, G. and Amenta, F. (1982). 5-Hydroxytryptophan uptake into indoleaminergic nerve fibres within rat cerebrovascular tree. *Neurosci. Lett.*, **28**, 57–60.

Neff, N.H., Karoum, F. and Hadjiconstantinou, M. (1983). Dopamine-containing small intensely fluorescent cells and sympathetic ganglion function. *Fed. Proc.* **42**, 3009–3011.

Neel, D.A. and Parsons, R.L. (1986). Catecholamine, serotonin and substance P-like peptide containing intrinsic neurones in the mud puppy parasympathetic cardiac ganglion. *J. Neuroscience*, **6**, 1970–1975.

Nemeth, PR., and Gullikson, GK. (1989). Gastrointestinal motility stimulating drugs and 5-HT receptors on myenteric neurones. *Eur. J. Pharmacol.* **166**, 387–391.

Newton, BW., and Hamill, RW. (1989). Immunohistochemical distribution of serotonin in spinal autonomic nuclei: I.Fiber patterns in the adult rat. *J. Compar. Neuro.* **279**, 68–81.

Nishimura, T., and Akasu, T. (1989). 5-hydroxytryptamine produces presynaptic facilitation of cholinergic transmission in rabbit parasympathetic ganglion. *J. Auton. Nerv. Syst.* **26**, 251–260.

Nishimura, T., Tokimasa, T., and Akasu, T. (1988) 5-hydroxytryptamine inhibits cholinergic transmission through 5-HT$_{1A}$ receptor subtypes in rabbit vesical parasympathetic ganglion. *Brain Research* **442**, 399–402.

Nistri, A. and Constanti, A. (1979). Pharmacological characterization of different types of GABA and glutamate receptors in vertebrates and invertebrates. *Progr. Neurobiol.*, **13**, 117–235.

Olsen, R.W. (1981). GABA-benzodiazepine-barbiturate receptor interactions. *J. Neurochem.,* **37**, 1-13.

Ong, J. and Kerr, D.I.B. (1983a). $GABA_A$ and $GABA_B$-receptor-mediated modification of intestinal motility. *Eur. J. Pharmacol.,* **86**, 9-17.

Ong, J. and Kerr, D.I.B. (1983b). Interactions between GABA and 5-hydroxytryptamine in the guinea-pig ileum. *Eur. J. Pharmacol.,* **94**, 305-312.

Ong, J. and Kerr, D.I.B. (1984). Potentiation of $GABA_A$-receptor-mediated responses by barbiturates in the guinea-pig ileum. *Eur. J. Pharmacol.,* **103**, 327-332.

Ong, J. and Kerr, D.I.B. (1985). Evidence that 5-hydroxytryptamine does not mediate GABA-induced contractile responses in the guinea-pig proximal ileum. *Eur. J. Pharmacol.,* **106**, 665-668.

Ong, J. and Kerr, D.I.B. (1987). Comparison of GABA-induced responses in various segments of the guinea-pig intestine. *Eur. J. Pharmacol.,* **134**, 349-353.

Osborne, N.N., Wu, P.H. and Neuhoff, V. (1974). Free amino acids and related compounds in the dorsal root ganglia and spinal cord of rat as determined by micro dansylation procedure. *Brain Research.,* **74**, 175-181.

Paton, W.D.M. and Vane, J.R. (1963). An analysis of the response of the isolated stomach to electrical stimulation and to drugs. *J. Physiol.,* **165**, 10-46.

Pearce, L.B. and Roth, J.A. (1983). Human brain monoamine oxidase. Solubilization and kinetics of inhibition by octyglucoside. *Arch. Biochem. Biophys.,* **224**, 464-472.

Peroutka, SJ. (1988). 5-Hydroxytryptamine receptor subtypes: Molecular, biochemical and physiological characterisation. *Trends in Neurosci.* **11**, 496-500.

Peroutka, S.J., Huang, S. and Allen, G.S. (1986). Canine basilar artery contractions mediated by 5-hydroxytryptamine$_{1a}$ receptor. *J. Pharm. Exptl. Ther.,* **237**, 901-906.

Pinkus, LM., Sarbin, NS., Barefoot, DS., and Gordon, JC. (1989). Association of [^{3}H] Zacopride with 5-HT$_3$ binding sites. *Eur. J. Pharmacol,* **168**, 255-362.

Pittaluga, A., Asaro, D., Pellegrini, G., and Raiteri, M. (1987). Studies on ^{3}H-GABA and endogenous GABA release in rat cerebral cortex suggest the presence of autoreceptors of the $GABA_B$ type. *Eur. J. Pharmacol.,* **144**, 45-52.

Pompolo, S., and Furness, J.B. (1990) Ultrastructure and synaptology of neurons immunoreactive for γ-aminobutyric acid in the myenteric plexus of the guinea-pig small intestine. *J. Neurocytol.* **19**, 539-549.

Pong, S.S., Wang, C.C. and Fritz, L.C. (1980). Studies on the mechanism of action of avermectin B$_{1a}$: stimulation of release of γ-aminobutyric acid from brain synaptosomes. *J. Neurochem.,* **34**, 351-358.

Price, J. (1984): Immunohistochemical evidence for dopaminergic neurons in the rat superior cervical ganglion. *Proc. R. Soc. Lond.,* **B 222**, 357-362.

Raiteri, M., Bonanno, G. and Fedele, E. (1989). Release of γ-[^{3}H] aminobutyric acid (GABA) from electrically stimulated rat cortical slices and its modulation by $GABA_B$ autoreceptors. *J. Pharmacol. Exptl. Ther.,* **250**, 648-653.

Rapoport, M.M., Green, A.A. and Page, I.H. (1948). Serum vasoconstrictor (Serotonin) IV Isolation and characterisation. *J. Biol. Chem.,* **176**, 1243-1251.

Rapoport, M.M. (1949). Serum vasoconstrictor (Serotonin) The presence of creatinine in the complex. A proposed structure for the vasoconstrictor principle. *J. Biol. Chem.,* **180**, 961-969.

Reinhard, J.F., Liebermann, J.E., Schlosberg, A.J. and Moskowitz, M.A. (1979). Serotonin neurons project to small blood vessels in the brain. *Science,* **206**, 85-87.

Richardson, BP., and Engel, G. (1986). The pharmacology and function of 5-HT$_3$ receptors. *Trends in Neurosciences.* **7**. 424-428.

Richardson, B.P., Engel, G., Donatsch, P. and Stadler, P.A. (1985). Identification serotonin M-receptor subtypes and their specific blockade by a new class of drugs. *Nature* **316**, 216-231.

Roberts, E., Chase, T.N., and Tower, D.B., editors (1976). *GABA in Nervous System Function.* Raven Press: New York.

Robinson, R. and Gershon, M.D. (1971). Synthesis and uptake of 5-HT by the myenteric plexus of the small intestine of the guinea-pig. *J. Pharm. Exptl. Ther.,* **179**, 29-41.

Rodriguez, D.L., Arnaiz, G. and De Robertis E. (1964). 5-Hydroxytryphan decarboxylase activity in nerve endings of the rat brain. *J. Neurochem.,* **11**, 213-219.

Ronai, A.Z., Kardos, J. and Simonyi, M. (1987). Potent inhibitory $GABA_B$ receptors in stimulated guinea-pig taenia coli. *Neuropharmacology,* **26**, 1623-1627.

Saffrey, M.J., Marcus, N., Jessen, K.R. and Burnstock, G. (1983). Distribution of neurons with high affinity uptake sites for GABA in the myenteric plexus of the guinea-pig, rat and chicken. *Cell Tissue Res.,* **234**, 231-235.

Saffrey, M.J., Legay, C. and Burnstock, G. (1984). Development of 5-HT like immunoreactive neurones in cultures of the myenteric plexus from the guinea-pig caecum. *Brain Research,* **304**, 105-116.

Saito, N. and Tanaka, C. (1986). Immunohistochemical demonstration of GABA containing neurons in the guinea-pig ileum using purified GABA antiserum. *Brain Research,* **376**, 78–84.

Sakai, M., Kani, K., Yoshida, M. and Nagatsu, I. (1989). Dopaminergic cells in the superior cervical ganglion of the rat: light and electron microscopic study using an antibody against dopamine. *Neurosci. Lett.,* **96**, 157–162.

Sanger, GJ. (1985) Three different ways in which 5-hydroxytryptamine can affect cholinergic activity in guinea-pig isolated ileum. *J. Pharm. Pharmacol.* **37**, 584–586.

Sano, I., Taniyama, K. and Tanaka, C. (1986) Cholecystokinin, but not-gastrin, induces γ-aminobutyric acid release from myenteric neurons of the guinea-pig ileum. *J. Pharmacol. Exptl. Ther.,* **248**, 378–383.

Saum, WR., and De Groat, WC. (1973). The actions of 5-hydroxtryptamine on the urinary bladder and vesical autonomic ganglion in the cat. *J. Pharmacol. Exp. Ther.* **185**, 70–83.

Saxena, P.R., Heiligers, J., Mylecharone, E.J. and Tio, R. (1985). Excitatory 5-hydroxytryptamine receptors in the cat urinary bladder are of the M and $5-HT_2$ type. *J. Auton. Pharmacol.,* **5**, 101–107.

Schon, F. and Kelly, J.S. (1974). The characterization of ^{3}H-GABA uptake into the satellite glial cells of rat sensory ganglia. *Brain Research,* **66**, 289–300.

Schultz, R. and Cartwright, C. (1974). Effect of morphine on serotonin release from the myenteric plexus of the guinea-pig. *J. Pharm. Exptl. Ther.,* **190**, 420–430.

Shirakawa, J., Takeda, K., Taniyama, K. and Tanaka, C. (1989). Dual effects of 5-hydroxytryptamine on the release of γ-aminobutyric acid from myenteric neurones of the guinea-pig ileum. *Br. J. Pharmac.,* **98**, 339–341.

Shuntoh, H., Taniyama, K. and Tanaka, C. (1989). Involvement of protein kinase C in the Ca^{2+}-dependent vesicular release of GABA from central and enteric neurons of the guinea-pig. *Brain Research,* **483**, 384–388.

Sivam, S.P., and Ho, IK. (1983). GABAergic drugs, morphine and morphine tolerance: a study in relation to nociception and gastrointestinal transit in mice. *Neuropharmacology,* **22**, 767–774.

Soinila, S, Ahonen, M., Lahtinen, T., and Happola, O. (1989). Developmental changes in 5-hydroxytryptamine immunoreactivity of sympathetic cells. *Int. J. Devl. Neuroscience.* **7**, 553–563.

Stephenson, R.K., Sole, M.J. and Baines, A.D. (1982). Neural and extraneural catecholamine production by rat kidneys. *Am. J. Physiol.,* **242**, F261–F266.

Stoof, J.C. and Kebabian, J.W. (1984). Two dopamine receptors: biochemistry, physiology and pharmacology. *Life Sciences,* **35**, 2281–2296.

Stone, T.W. and Perkins, M.N. (1984). Ethylenediamine as a GABA-mimetic. *Trends in Pharmacol. Sci.,* **5**, 241–243.

Surprenant, A., and Christ, J. (1988). Electrophysiological characterisation of functionally distinct 5-hydroxytryptamine receptors on guinea-pig submucousplexus. *Neuroscience.* **24**, 283–285.

Takaki, M., Branchek, T., Tamir, H. and Gershon, M.D. (1985). Specific antagonism of enteric neural serotonin receptors by dipeptides of 5-hydroxytryptophan: evidence that serotonin is a mediator of slow synaptic excitation in the myenteric plexus. *J. Neuroscience,* **5**, 1769–80.

Talley, NJ., Phillips, SF., Haddad, A., Miller, LJ., Twomey, C., Zinsmeister, AR., and Ciociola, A. (1989). Effect of selective $5-HT_3$ antagonist (GR 38032F) on small intestinal transit and release of gastrointestinal peptides. *Digestive Dis and Sci.* **34**, 1511–1515.

Talley, NJ., Phillips, SF., Miller, LJ., Haddad, A. (1988). A specific $5-HT_3$ receptor antagonist (GR-C507/75) delays colonic transit and inhibits postprandial neurotensin (NT) release. *Gastroenterology.* **95**, 891A.

Tamir, H. and Huang, I.L. (1974). Bindings of serotonin to soluble protein from synaptosomes. *Life Science.,* **14**, 83–89.

Tanaka, C. and Taniyama, K. (1985). Substance P provoked γ-aminobutyric acid release from the myenteric plexus of the guinea-pig small intestine. *J. Physiol. (Lond.),* **362**, 319–329.

Tanaka, C. and Taniyama, K. (1986). GABA transport in peripheral tissues: uptake are efflux. In: *GABAergic Mechanisms in the Mammalian Periphery,* edited by S.L. Erdö and N.G. Bowery. Raven Press: New York.

Taniyama, K., Hanada, S. and Tanaka, C. Autoreceptors regulate γ-[^{3}H] aminobutyric acid release from the guinea-pig small intestine. (1985) *Neurosci. Lett.,* **55**, 245–248.

Taniyama, K., Kusonoki, M., Saito, N. and Tanaka, C. (1982a). Release of γ-aminobutyric acid from cat colon. *Science,* **217**, 1038–1040.

Taniyama, K., Miki, Y. and Tanaka, C. (1982b): Presence of γ-aminobutyric acid and glutamic acid decarboxylase in Auerbach's plexus of cat colon. *Neurosci. Lett.,* **29**, 53–56.

Taniyama, K., Miki, Y., Kusonoki, M., Saito, N. and Tanaka, C. (1983). Release of endogenous and labelled GABA from isolated guinea-pig ileum. *Am. J. Physiol.,* **245**, G717–G721.

Taniyama, K., Saito, N., Miki, Y. and Tanaka, C. (1987). Enteric γ-aminobutyric acid-containing neurons and the relevance to motility of the cat colon. *Gastroenterology*, **93**, 519–525.

Taniyama, K., Hashimoto, S., Hanada, S. and Tanaka, C. (1988). Benzodiazepines and barbiturate potentiate the pre- and postsynaptic γ-aminobutyric acid (GABA)$_A$-receptor mediated response in the enteric nervous system. *J. Pharm. Exptl. Ther.*, **245**, 250–256.

Tansy, M.F., Rothman, G., Bartlett, G., Farbert, P. and Hohenleitner, F.J. (1971). Vagal adrenergic degranulation of the enterochromaffin cell system in guinea-pig duodenum. *J. Pharmacol. Science*, **60**, 81–85.

Thoenen, H., Haefely, W., Gey, K.F. and Hürhimann, A. (1965). Diminished effect of sympathetic nerve stimulation in cats pretreated with disulfiram; liberation of dopamine as sympathetic transmitter. *Life Sci.*, **4**, 2033–2038.

Tonini, M., Onari, L., Lechini, S., Frigo, G.M., Perucca, E., Saltarelli, P. and Crema, A. (1983). 5-HT mediated GABA excitatory responses in the guinea-pig proximal ileum. *Naunyn-Schmiedebergs. Arch. Pharm.*, **324**, 180–184.

Tonini, M, De Petris, G., Onari, L., Manzo, L., Rizzi, C. and Crema, A. (1989). The role of GABA$_A$ receptor function in peristaltic activity of the guinea-pig ileum: a comparative study with bicuculline, SR 95531 and picrotoxin. *Br. J. Pharmacol.*, **97**, 556–562.

Tsai, L.H., Taniyama, K. and Tanaka, C. (1987). γ-Aminobutyric acid stimulates acid secretion from the isolated guinea-pig stomach. *Am. J. Physiol.*, **253**, 9601–9606.

Turner, A.J., Tipton, K.F. (1972). The characterisation of two nicotinamide-adenine di-nucleotide phosphate-linked aldehyde reductases from pig brain. *Biochem. J.*, **130**, 765–772.

Tyce, G.M. (1985). Biochemistry of serotonin. In: *Serotonin and the Cardiovascular System*, edited by P.M. Vanhoute. Raven Press: New York.

Uvnas, B. (1978). Chemistry and storage function of mast cell granules. *Dermatology*, **71**, 76–80.

Van Loon, G.R. (1983). Plasma dopamine regulation and significance. *Federation Proceedings*, **42**, 3012–3018.

Van Neuten, J.M., Janssen, P.A.J., Van Beek, J., Xhonneux, R., Verbeuren, T.J. and Vanhoute, P.M. (1981). Vascular effects of ketanserin R(41468) a novel antagonist of 5-HT$_2$ serotonergic receptors. *J. Pharmacol. Exptl. Ther.*, **218**, 217–230.

Verbeuren, T.J., Jordaens, F.H. and Herman, A.G. (1983). Accumulation and release of [3]H 5-hydroxytryptamine in saphenous veins and cerebal arteries of the dog. *J. Pharm. Exptl. Ther.*, **226**, 579–588.

Verbeuren, T.H., Jordaens, F.H., Butt, H. and Herman, A.G. (1988). The endothelium inhibits the penetration of serotonin and norepinerphrine in the isolated canine saphenous vein. *J. Pharm. Exptl. Ther.*, **244**, 276–282.

Verbeuren, T.J. (1989). Synthesis, storage, release and metabolism of 5-hydroxytryptamine in peripheral tissues. In: *The Peripheral Actions of 5-Hydroxytryptamine*, edited by J.R. Fozard, Oxford University Press. pp. 1–25.

Verhofstad, A.A.J., Steinbusch, H.W.M., Penke, B., Varga, J. and Joosten, H.W.J. (1981). Serotonin immunoreactive cells in the superior cervical ganglionic of the rat. Evidence of the existence of separate serotonin and catecholamine containing small ganglion cells. *Brain Research*, **212**, 39–49.

Wallis, DI., and Dun, NJ. (1988) A comparison of fast and slow depolarisations evoked by 5-HT in guinea-pig coeliac ganglion cells *in vitro*. *Br. J. Pharmac.* **93**, 110–120.

Wallis, D. (1989). Interaction of 5-hydroxytryptamine with autonomic and sensory neurones. In: *The Peripheral Actions of 5-Hydroxytryptamine*, Edited J.R. Fozard. Oxford Medical Publications: NY.

Wallis, D.I. and Dun, N.J. (1989). Presynaptic action of 5-hydroxytryptamine on autonomic ganglia. In: *Presynaptic Regulation of Neurotransmitter Release*, edited by J.J. Feigenbaum and M. Hanani. Freund Publishing Co: London. Tel. Aviv.

Wallis, DI., and North, RA. (1978). Intracellular recording of responses of rabbit superior cervical ganglion cells to 5-hydroxytryptamine applied by iontophoresis. *Neuropharmacology* **17**. 1023–1028.

Watson, S. and Abbot, A. (1990). TIPS Receptor Nomenclature Supplement *Trends in Pharmacol. Sci.* **11**. pp 16.

Willems, J.L., Buylaert, W.A., Lefebvre, R.A. and Bogaert, M.G. (1985). Neuronal dopamine receptors on autonomic ganglia and sympathetic nerves and dopamine receptors in the gastrointestinal system. *Pharmacological Reviews*, **37**, 165–216.

Williford, D.J., Ormsbee, H.S., Norman, N., Harmon, J.W., Garney, T.Q., DiMicco, J. and Gillis, R.A. (1981). Hindbrain GABA receptors influence parasympathetic outflow to the stomach. *Science*, **214**, 193–194.

Wolff, J.R., Joò, F., Dames, W. and Fehèr, O. (1979). Induction and maintenance of free postsynaptic membrane thickenings in the adult superior cervical ganglion. *J. Neurocytol.*, **8**, 549–563.

Wolff, J.R., Kàsa, P., Dobo, E., Wenthold, R.J. and Joò, F. (1989). Quantitative analysis of the number and distribution of neurons richly innervated by GABA-immunoreactive axons in the rat superior cervical ganglion. *J. Comp. Neurol*, **282**, 264–273.

Wood, J.D. and Mayer, C.J. (1979). Serotonergic activation of tonic type enteric neurons in the guinea pig small bowel. *J. Neurophysiology*, **422**, 582–593.

Wood, K.L., Addae, J.I., Andrews, P.L.R. and Stone, T.W. (1987). Injection of baclofen into the ventromedial hypothalamus stimulates gastric motility in the rat. *Neuropharmacology*, **26**, 1191–1194.

Yau, W.M. and Verdun, P.R. (1983). Release of γ-aminobutyric acid from myenteric plexus synaptosomes. *Brain Research*, **278**, 271–273.

Young, J.A.C., Brown, D.A., Kelly, J.S. and Schon, F. (1973). Autoradiographic localization of sites of ^{3}H-γ-aminobutyric acid accumulation in peripheral autonomic ganglia. *Brain Research*, **63**, 479–486.

Index